T0406015

Organizational Structure of Science of Synthesis*

* A complete description of the full classification principles can be found in the **Science of Synthesis Guidebook.**

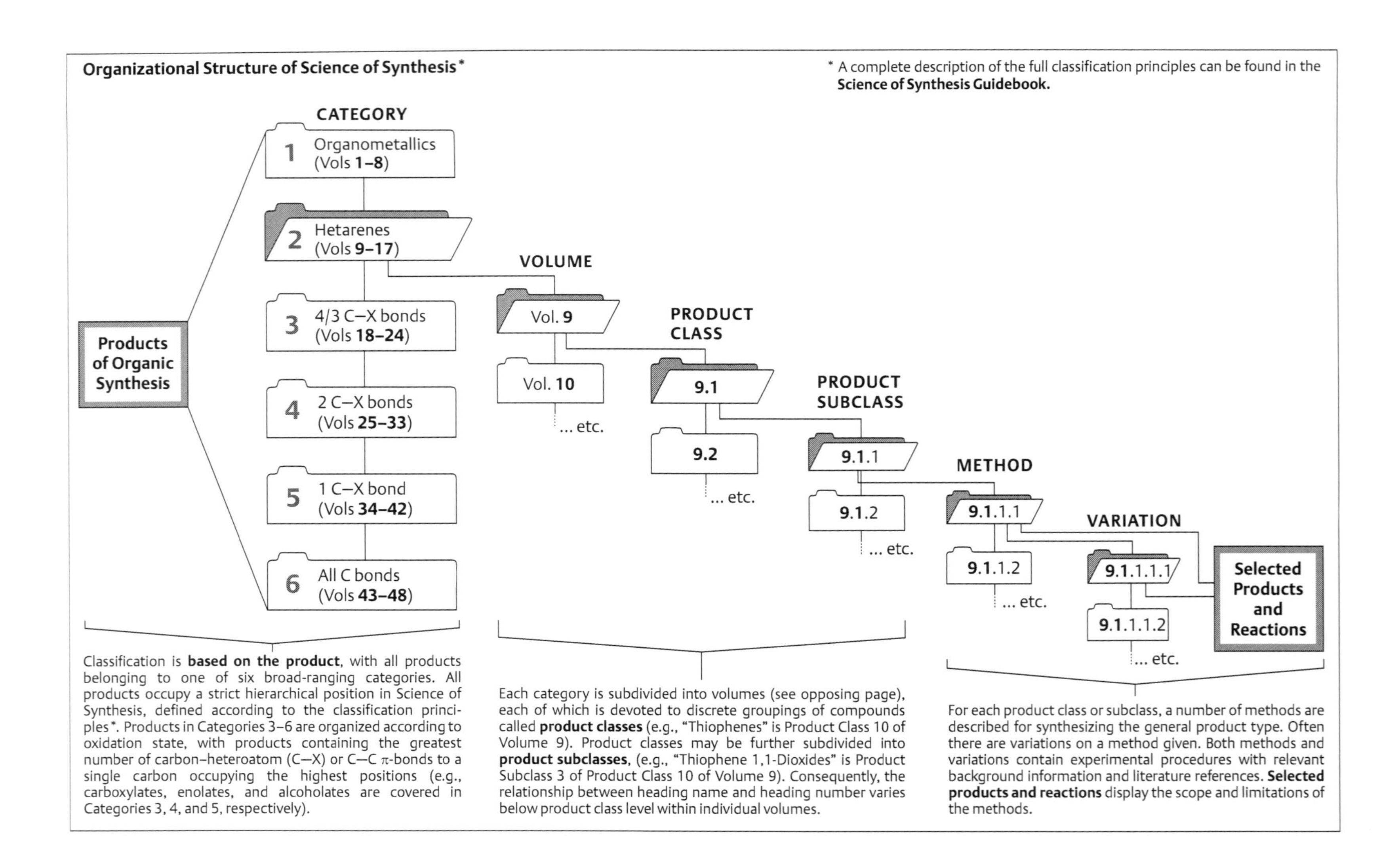

Classification is **based on the product**, with all products belonging to one of six broad-ranging categories. All products occupy a strict hierarchical position in Science of Synthesis, defined according to the classification principles*. Products in Categories 3–6 are organized according to oxidation state, with products containing the greatest number of carbon–heteroatom (C–X) or C–C π-bonds to a single carbon occupying the highest positions (e.g., carboxylates, enolates, and alcoholates are covered in Categories 3, 4, and 5, respectively).

Each category is subdivided into volumes (see opposing page), each of which is devoted to discrete groupings of compounds called **product classes** (e.g., "Thiophenes" is Product Class 10 of Volume 9). Product classes may be further subdivided into **product subclasses**, (e.g., "Thiophene 1,1-Dioxides" is Product Subclass 3 of Product Class 10 of Volume 9). Consequently, the relationship between heading name and heading number varies below product class level within individual volumes.

For each product class or subclass, a number of methods are described for synthesizing the general product type. Often there are variations on a method given. Both methods and variations contain experimental procedures with relevant background information and literature references. **Selected products and reactions** display the scope and limitations of the methods.

Science of Synthesis Reference Library

The **Science of Synthesis Reference Library** comprises volumes covering special topics of organic chemistry in a modular fashion, with six main classifications: (1) Classical, (2) Advances, (3) Transformations, (4) Applications, (5) Structures, and (6) Techniques. Volumes in the **Science of Synthesis Reference Library** focus on subjects of particular current interest with content that is evaluated by experts in their field. **Science of Synthesis**, including the **Knowledge Updates** and the **Reference Library**, is the complete information source for the modern synthetic chemist.

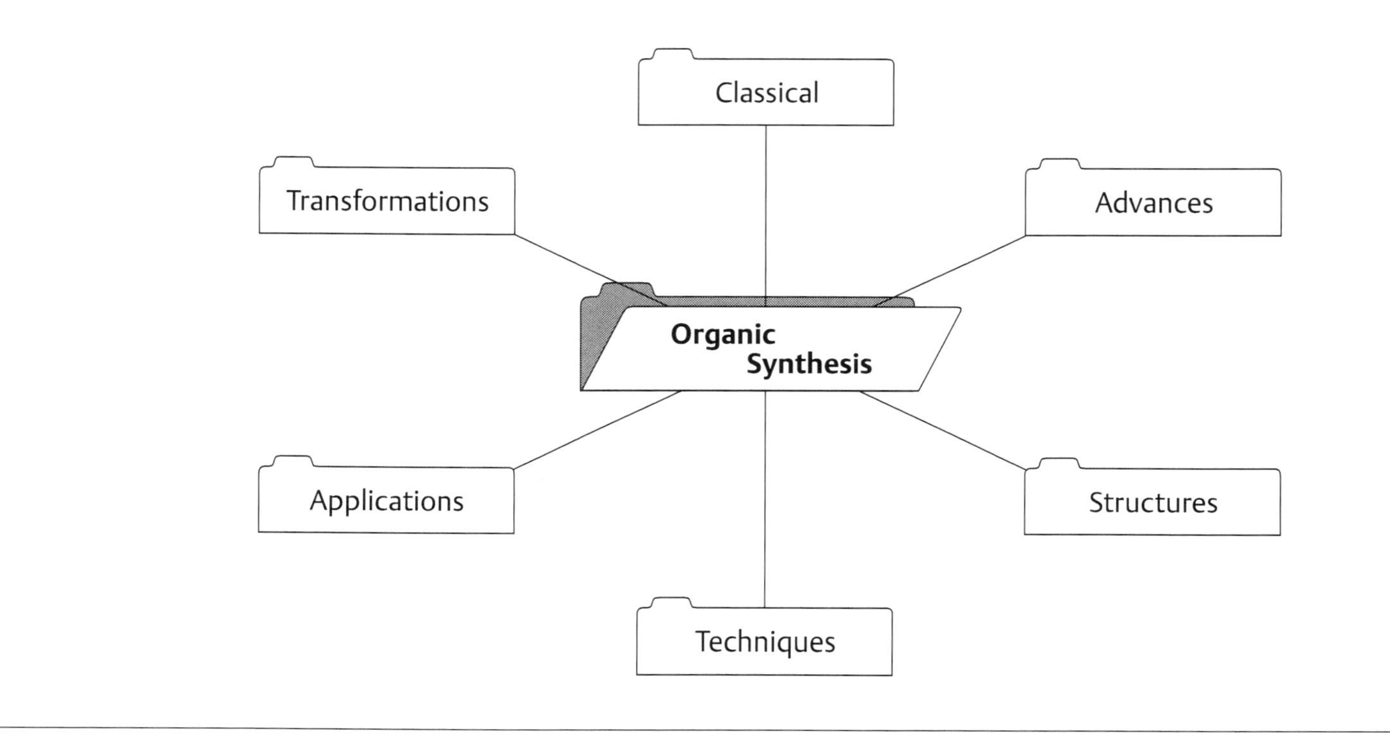

Science of Synthesis

Science of Synthesis is the authoritative and comprehensive reference work for the entire field of organic and organometallic synthesis.

Science of Synthesis presents the important synthetic methods for all classes of compounds and includes:
- Methods critically evaluated by leading scientists
- Background information and detailed experimental procedures
- Schemes and tables which illustrate the reaction scope

Science of Synthesis

Georg Thieme Verlag KG
Stuttgart · New York

Science of Synthesis

Applications of Domino Transformations in Organic Synthesis 2

Volume Editor S. A. Snyder

Responsible Member of the Editorial Board E. Schaumann

Authors

M. Bella
G. Blond
J. Boyce
I. Coldham
A. Dömling
M. Donnard
C. A. Guerrero
M. Gulea
E. Kroon
M. Moliterno
C. G. Neochoritis
A. V. Novikov
J. A. Porco, Jr.
P. Renzi
R. Salvio
N. S. Sheikh
A. Song
E. J. Sorensen
J. Suffert
W. Wang
J. G. West
Y.-Y. Yeung
Z. W. Yu
A. Zakarian
T. Zarganes Tzitzikas

2016
Georg Thieme Verlag KG
Stuttgart · New York

Rüdigerstrasse 14
D-70469 Stuttgart

Printed in Germany

Typesetting: Ziegler + Müller, Kirchentellinsfurt
Printing and Binding: AZ Druck und Datentechnik GmbH, Kempten

Bibliographic Information published by Die Deutsche Bibliothek

Die Deutsche Bibliothek lists this publication in the Deutsche Nationalbibliografie; detailed bibliographic data is available on the internet at <http://dnb.ddb.de>

Library of Congress Card No.: applied for

British Library Cataloguing in Publication Data

A catalogue record for this book is available from the British Library

ISBN 978-3-13-221151-3

Date of publication: May 11, 2016

Preface

As the pace and breadth of research intensifies, organic synthesis is playing an increasingly central role in the discovery process within all imaginable areas of science: from pharmaceuticals, agrochemicals, and materials science to areas of biology and physics, the most impactful investigations are becoming more and more molecular. As an enabling science, synthetic organic chemistry is uniquely poised to provide access to compounds with exciting and valuable new properties. Organic molecules of extreme complexity can, given expert knowledge, be prepared with exquisite efficiency and selectivity, allowing virtually any phenomenon to be probed at levels never before imagined. With ready access to materials of remarkable structural diversity, critical studies can be conducted that reveal the intimate workings of chemical, biological, or physical processes with stunning detail.

The sheer variety of chemical structural space required for these investigations and the design elements necessary to assemble molecular targets of increasing intricacy place extraordinary demands on the individual synthetic methods used. They must be robust and provide reliably high yields on both small and large scales, have broad applicability, and exhibit high selectivity. Increasingly, synthetic approaches to organic molecules must take into account environmental sustainability. Thus, atom economy and the overall environmental impact of the transformations are taking on increased importance.

The need to provide a dependable source of information on evaluated synthetic methods in organic chemistry embracing these characteristics was first acknowledged over 100 years ago, when the highly regarded reference source **Houben–Weyl Methoden der Organischen Chemie** was first introduced. Recognizing the necessity to provide a modernized, comprehensive, and critical assessment of synthetic organic chemistry, in 2000 Thieme launched **Science of Synthesis, Houben–Weyl Methods of Molecular Transformations**. This effort, assembled by almost 1000 leading experts from both industry and academia, provides a balanced and critical analysis of the entire literature from the early 1800s until the year of publication. The accompanying online version of **Science of Synthesis** provides text, structure, substructure, and reaction searching capabilities by a powerful, yet easy-to-use, intuitive interface.

From 2010 onward, **Science of Synthesis** is being updated quarterly with high-quality content via **Science of Synthesis Knowledge Updates**. The goal of the **Science of Synthesis Knowledge Updates** is to provide a continuous review of the field of synthetic organic chemistry, with an eye toward evaluating and analyzing significant new developments in synthetic methods. A list of stringent criteria for inclusion of each synthetic transformation ensures that only the best and most reliable synthetic methods are incorporated. These efforts guarantee that **Science of Synthesis** will continue to be the most up-to-date electronic database available for the documentation of validated synthetic methods.

Also from 2010, **Science of Synthesis** includes the **Science of Synthesis Reference Library**, comprising volumes covering special topics of organic chemistry in a modular fashion, with six main classifications: (1) Classical, (2) Advances, (3) Transformations, (4) Applications, (5) Structures, and (6) Techniques. Titles will include *Stereoselective Synthesis*, *Water in Organic Synthesis*, and *Asymmetric Organocatalysis*, among others. With expert-evaluated content focusing on subjects of particular current interest, the **Science of Synthesis Reference Library** complements the **Science of Synthesis Knowledge Updates**, to make **Science of Synthesis** the complete information source for the modern synthetic chemist.

The overarching goal of the **Science of Synthesis** Editorial Board is to make the suite of **Science of Synthesis** resources the first and foremost focal point for critically evaluated information on chemical transformations for those individuals involved in the design and construction of organic molecules.

Throughout the years, the chemical community has benefited tremendously from the outstanding contribution of hundreds of highly dedicated expert authors who have devoted their energies and intellectual capital to these projects. We thank all of these individuals for the heroic efforts they have made throughout the entire publication process to make **Science of Synthesis** a reference work of the highest integrity and quality.

The Editorial Board July 2010

Science of Synthesis Reference Library

Applications of Domino Transformations in Organic Synthesis (2 Vols.)
Catalytic Transformations via C—H Activation (2 Vols.)
Biocatalysis in Organic Synthesis (3 Vols.)
C-1 Building Blocks in Organic Synthesis (2 Vols.)
Multicomponent Reactions (2 Vols.)
Cross Coupling and Heck-Type Reactions (3 Vols.)
Water in Organic Synthesis
Asymmetric Organocatalysis (2 Vols.)
Stereoselective Synthesis (3 Vols.)

Volume Editor's Preface

Domino reactions have been a mainstay of synthetic chemistry for much of its history. Domino chemistry's roots trace to achievements such as the one-pot synthesis of tropinone in 1917 by Robinson and the generation of steroidal frameworks through polyene cyclizations, as originally predicted by the Stork–Eschenmoser hypothesis. In the ensuing decades, chemists have used these, and other inspiring precedents, to develop even more complicated domino sequences that rapidly and efficiently build molecular complexity, whether in the form of natural products, novel pharmaceuticals, or materials such as buckminsterfullerene.

Despite this body of achievements, however, the development of such processes remains a deeply challenging endeavor. Indeed, effective domino chemistry at the highest levels requires not only creativity and mechanistic acumen, but also careful planning at all stages of a typical experiment, from substrate design, to reagent and solvent choice, to timing of additions, and even the quench. Thus, if the frontiers are to be pushed even further, there is certainly much to master.

It was with these parameters in mind that the Editorial Board of **Science of Synthesis** decided to focus one of its Reference Library works on domino chemistry, covering the myriad ways that these sequences can be achieved with the full array of reactivity available, whether in the form of pericyclic reactions, radical transformations, anionic and cationic chemistry, metal-based cross couplings, and combinations thereof. In an effort to provide a unique approach in organizing and presenting such transformations relative to other texts and reviews on the subject, the sections within this book have been organized principally by the type of reaction that initiates the sequence. Importantly, only key and representative examples have been provided to highlight the best practices and procedures that have broad applicability. The hope is that this structure will afford a clear sense of current capabilities as well as highlight areas for future development and research.

A work on such a vibrant area of science would not have been possible, first and foremost, without a talented and distinguished author team. Each is mentioned in the introductory chapter, and I wish to thank all of them for their professionalism, dedication, and expertise. I am also grateful to all of the coaching, advice, and assistance provided by Ernst Schaumann, member of the Editorial Board of **Science of Synthesis**. Deep thanks also go, of course, to the entire editorial team at Thieme, particularly to Robin Padilla and Karen Muirhead-Hofmann who served as the scientific editors in charge of coordinating this reference work; Robin started the project, and Karen saw it through to the end. Their attention to detail and passion to produce an excellent final product made this project a true pleasure. Last, but not least, I also wish to thank my wife Cathy and my son Sebastian for their support of this project over the past two years.

Finally, I wish to dedicate this work, on behalf of the chapter authors and myself, to our scientific mentors. It was through their training that we learned how to better understand reactivity, propose novel chemistry, and identify the means to actually bring those ideas to fruition. Hopefully this text will serve the same role to those who study its contents, with even greater wisdom achieved as a result.

Scott A. Snyder — Chicago, October 2015

Abstracts

p 1

2.1.1 The Diels–Alder Cycloaddition Reaction in the Context of Domino Processes

J. G. West and E. J. Sorensen

The Diels–Alder cycloaddition has been a key component in innumerable, creative domino transformations in organic synthesis. This chapter provides examples of how this [4+2] cycloaddition has been incorporated into the said cascades, with particular attention to its interplay with the other reactions in the sequence. We hope that this review will assist the interested reader to approach the design of novel cascades involving the Diels–Alder reaction.

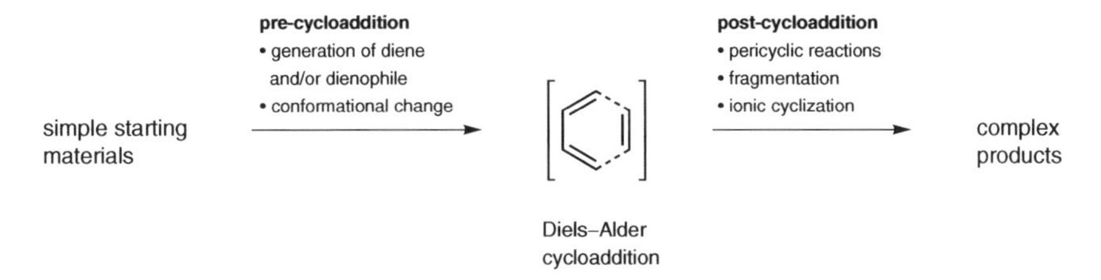

Keywords: Diels–Alder • cascade • domino reactions • pericyclic • [4+2] cycloaddition

p 47

2.1.2 Domino Reactions Including [2+2], [3+2], or [5+2] Cycloadditions

I. Coldham and N. S. Sheikh

This chapter covers examples of domino reactions that include a [2+2]-, [3+2]-, or [5+2]-cycloaddition reaction. The focus is on concerted reactions that occur in a tandem sequence in one pot, rather than overall "formal cycloadditions" or multicomponent couplings. The cycloaddition step typically involves an alkene or alkyne as one of the components in the ring-forming reaction. In addition to the key cycloaddition step, another bond-forming reaction will be involved that can precede or follow the cycloaddition. This other reaction is often an alkylation that generates the substrate for the cycloaddition, or is a ring-opening or rearrangement reaction that occurs after the cycloaddition. As the chemistry involves sequential reactions including at least one ring-forming reaction, unusual molecular structures or compounds that can be difficult to prepare by other means can be obtained. As a result, this strategy has been used for the regio- and stereoselective preparation of a vast array of polycyclic, complex compounds of interest to diverse scientific communities.

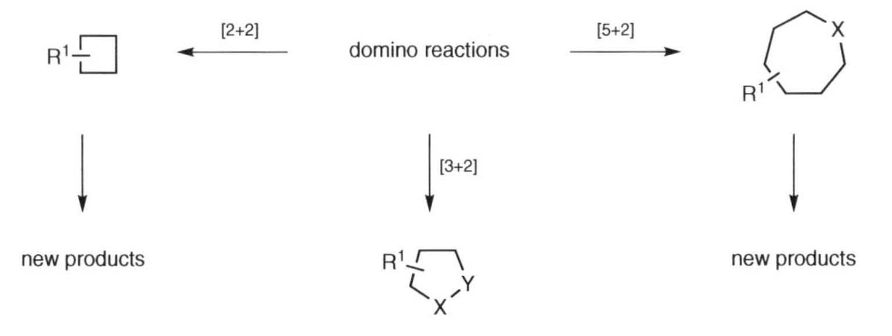

Keywords: alkylation • [2+2] cycloaddition • [3+2] cycloaddition • [5+2] cycloaddition • dipolar cycloaddition • domino reactions • Nazarov cyclization • ring formation • [3,3]-sigmatropic rearrangement • tandem reactions

p 93

2.1.3 Domino Transformations Involving an Electrocyclization Reaction

J. Suffert, M. Gulea, G. Blond, and M. Donnard

Electrocyclization processes represent a powerful and efficient way to produce carbo- or heterocycles stereoselectively. Moreover, when electrocyclizations are involved in domino processes, the overall transformation becomes highly atom and step economic, enabling access to structurally complex molecules. This chapter is devoted to significant contributions published in the last 15 years, focusing on synthetic methodologies using electrocyclization as a key step in a domino process.

Keywords: electrocyclization • hetero-electrocyclization • domino reactions • cascade reactions

p 159

2.1.4 Sigmatropic Shifts and Ene Reactions (Excluding [3,3])

A. V. Novikov and A. Zakarian

This chapter features a review and discussion of the domino transformations initiated by ene reactions and sigmatropic rearrangements, particularly focusing on [2,3]-sigmatropic shifts, such as Mislow–Evans and Wittig rearrangements, and [1,n] hydrogen shifts. A variety of examples of these domino processes are reviewed, featuring such follow-up processes to the initial reaction as additional ene reactions or sigmatropic shifts, Diels–Alder cycloaddition, [3+2] cycloaddition, electrocyclization, condensation, and radical cyclization. General practical considerations and specific features in the examples of the reported cascade transformation are highlighted. To complete the discussion, uses of these cascade processes in the synthesis of natural products are discussed, demonstrating the rapid assembly of structural complexity that is characteristic of domino processes. Overall, the domino transformations initiated by ene reactions and sigmatropic shifts represent an important subset of domino processes, the study of which is highly valuable for understanding key aspects of chemical reactivity and development of efficient synthetic methods.

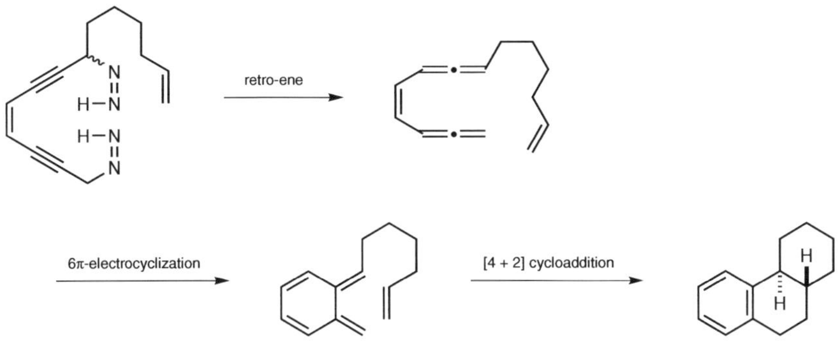

retro-ene

6π-electrocyclization

[4 + 2] cycloaddition

Keywords: ene reaction • sigmatropic shift • domino reactions • cascade reactions • hydrogen shift • [1,3]-shift • [1,5]-shift • [1,7]-shift • [2,3]-shift • [3,3]-shift • Mislow–Evans rearrangement • Wittig rearrangement • Diels–Alder cycloaddition • Claisen rearrangement • oxy-Cope rearrangement • electrocyclization • chloropupukeanolide D • isocedrene • steroids • mesembrine • joubertinamine • pinnatoxins • sterpurene • arteannuin M • pseudomonic acid A

p 195

2.1.5 Domino Transformations Initiated by or Proceeding Through [3,3]-Sigmatropic Rearrangements

C. A. Guerrero

This chapter concerns itself with domino transformations (i.e., cascade sequences and/or tandem reactions) that are either initiated by or proceed through at least one [3,3]-sigmatropic rearrangement. Excluded from this discussion are domino transformations that end with sigmatropy. The reactions included contain diverse forms of [3,3]-sigmatropic rearrangements and are followed by both polar chemistry or further concerted rearrangement.

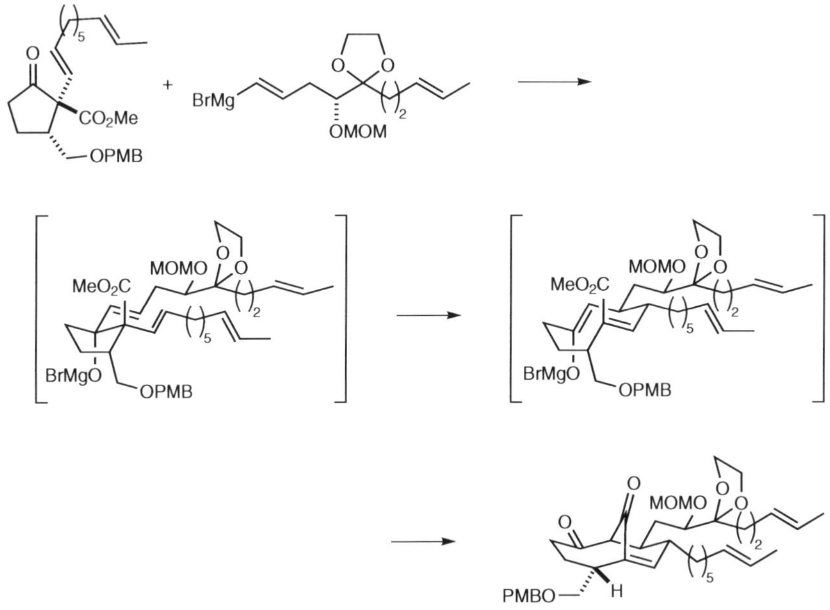

Keywords: rearrangement • sigmatropic • Bellus–Claisen • Cope • Overman • concerted • stereoselective • stereospecific • ene • trichloroacetimidate • Diels–Alder

p 229

2.2 Intermolecular Alkylative Dearomatizations of Phenolic Derivatives in Organic Synthesis

J. A. Porco, Jr., and J. Boyce

Intermolecular alkylative dearomatization products have shown promise as synthetic intermediates with diverse capabilities. This chapter describes the available methods for constructing these dearomatized molecules and demonstrates their value as synthetic intermediates for efficient total syntheses.

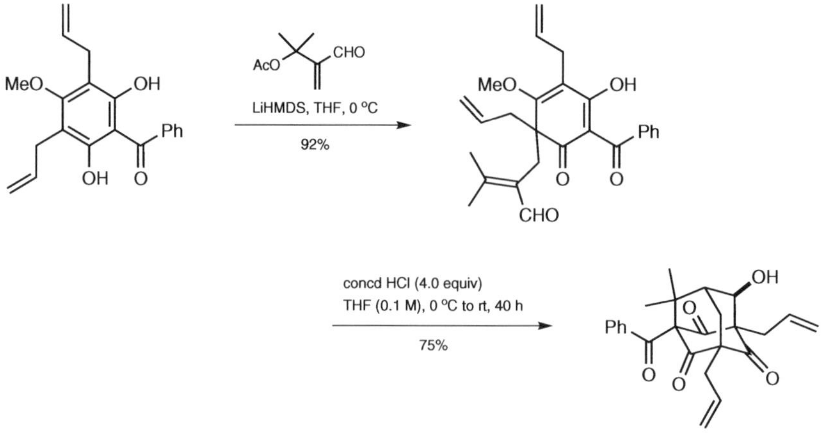

Keywords: alkylative dearomatization • dearomative alkylation • dearomative substitution • domino transformations • domino sequences • dearomative domino transformations • cationic cyclization • radical cyclization • alkylative dearomatization/annulation

p 293

2.3.1 Additions to Nonactivated C=C Bonds

Z. W. Yu and Y.-Y. Yeung

Electrophilic additions to nonactivated C=C bonds are one of the well-known classical reactions utilized by synthetic chemists as a starting point to construct useful complex organic molecules. This chapter covers a collection of electrophile-initiated domino transformations involving alkenes as the first reaction, followed by reaction with suitable nucleophiles in the succession and termination reactions under identical conditions. The discussion focuses on recent advances in catalysis, strategically designed alkenes, and new electrophilic reagents employed to improve reactivity and control of stereochemistry in the sequence of bond-forming steps.

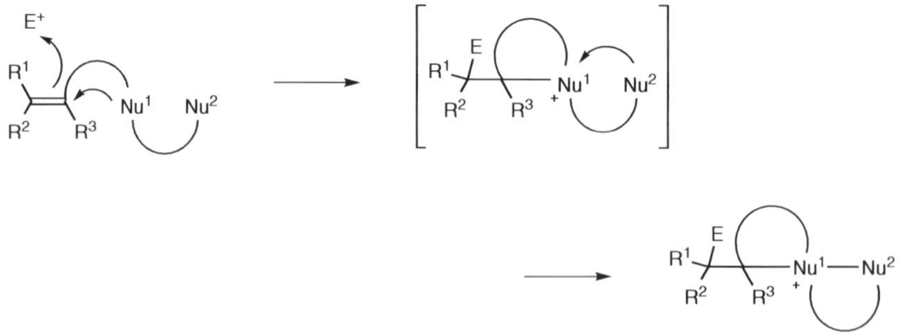

Keywords: nonactivated alkenes · addition · domino reactions · amination · etherification · carbonylation · polyenes · protons · halogens · transition metals · chalcogens

p 337

2.3.2 Organocatalyzed Addition to Activated C=C Bonds

P. Renzi, M. Moliterno, R. Salvio, and M. Bella

In this chapter, several examples of organocatalyzed additions to C=C bonds carried out through a domino approach are reviewed, from the early examples to recent applications of these strategies in industry.

organocatalyst
formal [4 + 2] cycloaddition

pharmaceuticals and
chiral diene ligands for
asymmetric catalysts

Keywords: organocatalysis · domino reactions · iminium ions · enamines · Michael/aldol reactions · nucleophilic/electrophilic addition · α,β-unsaturated carbonyl compounds · spirocyclic oxindoles · cinchona alkaloid derivatives · chiral secondary amines · Knoevenagel condensation · methyleneindolinones

p 387

2.3.3 Addition to Monofunctional C=O Bonds

A. Song and W. Wang

Catalytic asymmetric domino addition to monofunctional C=O bonds is a powerful group of methods for the rapid construction of valuable chiral building blocks from readily available substances. Impressive progress has been made on transition-metal-catalyzed and organocatalytic systems that promote such addition processes through reductive aldol, Michael/aldol, or Michael/Henry sequences. In addition, Lewis acid catalysis has also been developed in this area for the synthesis of optically active chiral molecules. This chapter covers the most impressive examples of these recent developments in domino chemistry.

Asymmetric Michael/Intermolecular Aldol or Henry Reaction

chiral transition-metal catalyst
or organocatalyst

$R^1 = R^2 = R^3 =$ H, alkyl, aryl; $R^2 = CO_2R^5$, NO_2

Asymmetric Michael/Intramolecular Aldol or Henry Reaction

chiral organocatalyst

n = 1, 2; EWG = electron-withdrawing group

chiral organocatalyst

$R^5 = H, CO_2R^6$

Keywords: aldol reactions • carbonyl ylides • chiral amine catalysis • domino reactions • epoxy alcohols • Lewis acid catalysis • Michael addition • organocatalysis • phosphoric acid catalysis • thiourea catalysis

p 419

2.3.4 Additions to C=N Bonds and Nitriles

E. Kroon, T. Zarganes Tzitzikas, C. G. Neochoritis, and A. Dömling

This chapter describes additions to imines and nitriles and their post-modifications within the context of domino reactions and multicomponent reaction chemistry.

1. $MeOH/H_2O$ (4:1) rt, 24–72 h
2. HCO_2H, rt, 16 h

87%

Keywords: multicomponent reactions • domino reactions • isocyanides • Ugi reaction • Pictet–Spengler reaction • Gewald reaction • isoindoles • benzodiazepines • cyanoacetamides • thiophenes

Applications of Domino Transformations in Organic Synthesis 2

Preface V

Volume Editor's Preface IX

Abstracts XI

Table of Contents XIX

2.1 **Pericyclic Reactions** 1

2.1.1 **The Diels–Alder Cycloaddition Reaction in the Context of Domino Processes**
J. G. West and E. J. Sorensen 1

2.1.2 **Domino Reactions Including [2+2], [3+2], or [5+2] Cycloadditions**
I. Coldham and N. S. Sheikh 47

2.1.3 **Domino Transformations Involving an Electrocyclization Reaction**
J. Suffert, M. Gulea, G. Blond, and M. Donnard 93

2.1.4 **Sigmatropic Shifts and Ene Reactions (Excluding [3,3])**
A. V. Novikov and A. Zakarian 159

2.1.5 **Domino Transformations Initiated by or Proceeding Through [3,3]-Sigmatropic Rearrangements**
C. A. Guerrero 195

2.2 **Intermolecular Alkylative Dearomatizations of Phenolic Derivatives in Organic Synthesis**
J. A. Porco, Jr., and J. Boyce 229

2.3 **Additions to Alkenes and C=O and C=N Bonds** 293

2.3.1 **Additions to Nonactivated C=C Bonds**
Z. W. Yu and Y.-Y. Yeung 293

2.3.2 **Organocatalyzed Addition to Activated C=C Bonds**
P. Renzi, M. Moliterno, R. Salvio, and M. Bella 337

2.3.3 **Addition to Monofunctional C=O Bonds**
A. Song and W. Wang 387

2.3.4 **Additions to C=N Bonds and Nitriles**
E. Kroon, T. Zarganes Tzitzikas, C. G. Neochoritis, and A. Dömling 419

Keyword Index 449

Author Index 481

Abbreviations 497

Table of Contents

2.1 **Pericyclic Reactions**

2.1.1 **The Diels–Alder Cycloaddition Reaction in the Context of Domino Processes**
J. G. West and E. J. Sorensen

2.1.1 **The Diels–Alder Cycloaddition Reaction in the Context of Domino Processes** 1
2.1.1.1 Cascades Not Initiated by Diels–Alder Reaction 2
2.1.1.1.1 Cascades Generating a Diene 2
2.1.1.1.1.1 Ionic Generation of a Diene 2
2.1.1.1.1.1.1 Through Wessely Oxidation of Phenols 2
2.1.1.1.1.1.2 Through Ionic Cyclization 6
2.1.1.1.1.1.3 Through Deprotonation of an Alkene 7
2.1.1.1.1.1.4 Through Elimination Reactions 8
2.1.1.1.1.1.5 Through Allylation 12
2.1.1.1.1.2 Pericyclic Generation of a Diene 12
2.1.1.1.1.2.1 Through Electrocyclization 13
2.1.1.1.1.2.1.1 Through Benzocyclobutene Ring Opening 13
2.1.1.1.1.2.1.2 Through Electrocyclic Ring Closure 14
2.1.1.1.1.2.2 Through Cycloaddition or Retrocycloaddition 17
2.1.1.1.1.2.3 Through Sigmatropic Reactions 18
2.1.1.1.1.3 Photochemical Generation of a Diene 19
2.1.1.1.1.4 Metal-Mediated Generation of a Diene 20
2.1.1.1.2 Cascades Generating a Dienophile 22
2.1.1.1.2.1 Ionic Generation of a Dienophile 22
2.1.1.1.2.1.1 Through Himbert Cycloadditions 22
2.1.1.1.2.1.2 Through Benzyne Formation 23
2.1.1.1.2.1.3 Through Wessely Oxidation 24
2.1.1.1.2.2 Pericyclic Generation of a Dienophile 27
2.1.1.1.2.2.1 Through Cycloaddition/Retrocycloaddition 27
2.1.1.1.2.2.2 Through Sigmatropic Rearrangement 27
2.1.1.1.2.2.3 Through Electrocyclization 29
2.1.1.1.3 Proximity-Induced Diels–Alder Reactions 29

2.1.1.2 Diels–Alder as the Initiator of a Cascade 31

2.1.1.2.1 Pericyclic Reactions Occurring in the Wake of a Diels–Alder Reaction 31

2.1.1.2.1.1 Cascades Featuring Diels–Alder/Diels–Alder Processes 31

2.1.1.2.1.2 Cascades Featuring Diels–Alder/Retro-Diels–Alder Processes 33

2.1.1.2.1.3 [4+2] Cycloaddition with Subsequent Desaturation 36

2.1.1.2.2 Diels–Alder Reactions with Concomitant Ionic Structural Rearrangements 36

2.1.1.2.2.1 Pairings of Diels–Alder Reactions with Structural Fragmentations 37

2.1.1.2.2.2 Combining a Diels–Alder Reaction with Ionic Cyclization 40

2.1.1.3 Conclusions 43

2.1.2 **Domino Reactions Including [2+2], [3+2], or [5+2] Cycloadditions**
I. Coldham and N. S. Sheikh

2.1.2 **Domino Reactions Including [2+2], [3+2], or [5+2] Cycloadditions** 47

2.1.2.1 Domino [2+2] Cycloadditions 47

2.1.2.1.1 Cycloaddition of an Enaminone and β-Diketone with Fragmentation 48

2.1.2.1.2 Cycloaddition of Ynolate Anions Followed by Dieckmann Condensation/Michael Reaction 48

2.1.2.1.3 Cycloaddition Cascade Involving Benzyne–Enamide Cycloaddition or a Fischer Carbene Complex 50

2.1.2.1.4 Cycloadditions with Rearrangement 51

2.1.2.1.4.1 Cycloaddition of an Azatriene Followed by Cope Rearrangement 51

2.1.2.1.4.2 Cycloaddition of a Propargylic Ether and Propargylic Thioether Followed by [3,3]-Sigmatropic Rearrangement 52

2.1.2.1.4.3 [3,3]-Sigmatropic Rearrangement of Propargylic Ester and Propargylic Acetate Followed by Cycloaddition 53

2.1.2.1.4.4 Cycloaddition of a Ketene Followed by Allylic Rearrangement 54

2.1.2.1.4.5 Allyl Migration in Ynamides Followed by Cycloaddition 55

2.1.2.1.4.6 1,3-Migration in Propargyl Benzoates Followed by Cycloaddition 56

2.1.2.2 Domino [3+2] Cycloadditions 57

2.1.2.2.1 Cycloadditions with Nitrones, Nitronates, and Nitrile Oxides 57

2.1.2.2.1.1 Reaction To Give a Nitrone Followed by Cycloaddition 58

2.1.2.2.1.2 Cycloaddition with a Nitrone and Subsequent Reaction 62

2.1.2.2.1.3 Reaction To Give a Nitronate Followed by Cycloaddition 63

2.1.2.2.1.4 Reaction To Give a Nitrile Oxide Followed by Cycloaddition 64

2.1.2.2.1.5 Cycloaddition with a Nitrile Oxide and Subsequent Reaction 65

2.1.2.2.2 Cycloadditions with Carbonyl Ylides 66

2.1.2.2.2.1 Reaction of an α-Diazo Compound To Give a Carbonyl Ylide Followed by Cycloaddition 66

2.1.2.2.2.2 Reaction of an Alkyne To Give a Carbonyl Ylide Followed by Cycloaddition 72

2.1.2.2.3 Cycloadditions with Azomethine Ylides 73

2.1.2.2.4 Cycloadditions with Azomethine Imines 80

2.1.2.2.5 Cycloadditions with Azides 81

2.1.2.2.5.1 Reaction To Give an Azido-Substituted Alkyne Followed by Cycloaddition 81

2.1.2.2.5.2 Cycloaddition of an Azide and Subsequent Reaction 83

2.1.2.3 Domino [5+2] Cycloadditions 84

2.1.2.3.1 Cycloaddition of a Vinylic Oxirane Followed by Claisen Rearrangement 85

2.1.2.3.2 Cycloaddition of an Ynone Followed by Nazarov Cyclization 86

2.1.2.3.3 Cycloaddition of an Acetoxypyranone Followed by Conjugate Addition 86

2.1.2.3.4 Cycloaddition Cascade Involving γ-Pyranone and Quinone Systems 87

2.1.3 **Domino Transformations Involving an Electrocyclization Reaction**
J. Suffert, M. Gulea, G. Blond, and M. Donnard

2.1.3 **Domino Transformations Involving an Electrocyclization Reaction** 93

2.1.3.1 Metal-Mediated Cross Coupling Followed by Electrocyclization 93

2.1.3.1.1 Palladium-Mediated Cross Coupling/Electrocyclization Reactions 93

2.1.3.1.1.1 Cross Coupling/6π-Electrocyclization 93

2.1.3.1.1.2 Cross Coupling/8π-Electrocyclization 101

2.1.3.1.1.3 Cross Coupling/8π-Electrocyclization/6π-Electrocyclization 103

2.1.3.1.2 Copper-Catalyzed Tandem Reactions 109

2.1.3.1.3 Zinc-Catalyzed Tandem Reactions 109

2.1.3.1.4 Ruthenium-Catalyzed Formal [2+2+2] Cycloaddition Reactions 110

2.1.3.2 Alkyne Transformation Followed by Electrocyclization 111

2.1.3.3 Isomerization Followed by Electrocyclization 116

2.1.3.3.1 1,3-Hydrogen Shift/Electrocyclization 116

2.1.3.3.2 1,5-Hydrogen Shift/Electrocyclization 117

2.1.3.3.3 1,7-Hydrogen Shift/Electrocyclization 120

2.1.3.4 Consecutive Electrocyclization Reaction Cascades 121

2.1.3.5 Alkenation Followed by Electrocyclization 123

2.1.3.6 Electrocyclization Followed by Cycloaddition 126

2.1.3.7 Miscellaneous Reactions 127

2.1.3.7.1 Electrocyclization/Oxidation 127

2.1.3.7.2 Photochemical Elimination/Electrocyclization 128

2.1.3.7.3 Domino Retro-electrocyclization Reactions 130

2.1.3.8 Hetero-electrocyclization 130

2.1.3.8.1 Aza-electrocyclization 130

2.1.3.8.1.1 Metal-Mediated Reaction/Hetero-electrocyclization 131

2.1.3.8.1.2 Imine or Iminium Formation/Hetero-electrocyclization 139

2.1.3.8.1.3 Isomerization or Rearrangement/Hetero-electrocyclization 144

2.1.3.8.2 Oxa-electrocyclization 150

2.1.3.8.3 Thia-electrocyclization 154

2.1.4 **Sigmatropic Shifts and Ene Reactions (Excluding [3,3])**
A. V. Novikov and A. Zakarian

2.1.4 **Sigmatropic Shifts and Ene Reactions (Excluding [3,3])** 159

2.1.4.1 Practical Considerations 159

2.1.4.2 Domino Processes Initiated by Ene Reactions 160

2.1.4.3 Domino Processes Initiated by [2,3]-Sigmatropic Rearrangements 168

2.1.4.4 Domino Processes Initiated by Other Sigmatropic Rearrangements 178

2.1.4.5 Domino Processes in the Synthesis of Natural Products 183

2.1.4.6 Conclusions 191

2.1.5 **Domino Transformations Initiated by or Proceeding Through [3,3]-Sigmatropic Rearrangements**
C. A. Guerrero

2.1.5 **Domino Transformations Initiated by or Proceeding Through [3,3]-Sigmatropic Rearrangements** 195

2.1.5.1 Cope Rearrangement Followed by Enolate Functionalization 196

2.1.5.1.1 Anionic Oxy-Cope Rearrangement Followed by Intermolecular Enolate Alkylation with Alkyl Halides 196

2.1.5.1.2 Anionic Oxy-Cope Rearrangement Followed by Enolate Alkylation by Pendant Allylic Ethers 198

2.1.5.1.3 Anionic Oxy-Cope Rearrangement Followed by Enolate Acylation 199

2.1.5.2 Aza- and Oxonia-Cope-Containing Domino Sequences 201

2.1.5.2.1 Ionization-Triggered Oxonia-Cope Rearrangement Followed by Intramolecular Nucleophilic Trapping by an Enol Silyl Ether 201

2.1.5.2.2 Intermolecular 1,4-Addition-Triggered Oxonia-Cope Rearrangement Followed by Intramolecular Nucleophilic Trapping by a Nascent Enolate 203

2.1.5.2.3 Iminium-Ion-Formation-Triggered Azonia-Cope Rearrangement Followed by Intramolecular Nucleophilic Trapping by a Nascent Enamine 204

2.1.5.3 Double, Tandem Hetero-Cope Rearrangement Processes 207

2.1.5.3.1 Double, Tandem [3,3]-Sigmatropic Rearrangement of Allylic, Homoallylic Bis(trichloroacetimidates) 207

2.1.5.4 Neutral Claisen Rearrangement Followed by Further (Non-Claisen) Processes 209

2.1.5.4.1 Oxy-Cope Rearrangement/Ene Reaction Domino Sequences 209

2.1.5.4.2 Oxy-Cope Rearrangement/Ene Reaction/Claisen Rearrangement and Oxy-Cope Rearrangement/Claisen Rearrangement/Ene Reaction Domino Sequences 211

2.1.5.5 Claisen Rearrangement Followed by Another Pericyclic Process 213

2.1.5.5.1 Double, Tandem Bellus–Claisen Rearrangement Reactions 213

2.1.5.5.2 Claisen Rearrangement Followed by [2,3]-Sigmatropic Rearrangement 215

2.1.5.5.3 Claisen Rearrangement/Diels–Alder Cycloaddition Domino Sequences 217

2.1.5.5.4 Claisen Rearrangement/[1,5]-H-Shift/6π-Electrocyclization Domino Sequences 220

2.1.5.6 Claisen Rearrangement Followed by Multiple Processes 222

2.1.5.6.1 Propargyl Claisen Rearrangement Followed by Tautomerization, Acylketene Generation, 6π-Electrocyclization, and Aromatization 222

2.1.5.6.2 Propargyl Claisen Rearrangement Followed by Imine Formation, Tautomerization, and 6π-Electrocyclization 223

2.2 **Intermolecular Alkylative Dearomatizations of Phenolic Derivatives in Organic Synthesis**
J. A. Porco, Jr., and J. Boyce

2.2 **Intermolecular Alkylative Dearomatizations of Phenolic Derivatives in Organic Synthesis** 229

2.2.1 Metal-Mediated Intermolecular Alkylative Dearomatization 232

2.2.1.1 Osmium(II)-Mediated Intermolecular Alkylative Dearomatization 232

2.2.1.2 Palladium-Catalyzed Intermolecular Alkylative Dearomatization 236

2.2.1.3 Tandem Palladium-Catalyzed Intermolecular Alkylative Dearomatization/Annulation 237

2.2.2 Non-Metal-Mediated Intermolecular Alkylative Dearomatization 240

2.2.2.1 Alkylative Dearomatizations of Phenolic Derivatives with Activated Electrophiles 240

2.2.2.2 Alkylative Dearomatizations of Phenolic Derivatives with Unactivated Electrophiles 248

2.2.3 Tandem Intermolecular Alkylative Dearomatization/Annulation 252

2.2.3.1 Tandem Alkylative Dearomatization/[4+2] Cycloaddition 252

2.2.3.2 Tandem Alkylative Dearomatization/Hydrogenation Followed by Lewis Acid Catalyzed Cyclization 252

2.2.3.3 Tandem Alkylative Dearomatization/Annulation To Access Type A and B Polyprenylated Acylphloroglucinol Derivatives 254

2.2.3.4 Enantioselective, Tandem Alkylative Dearomatization/Annulation 260

2.2.3.5 Tandem Alkylative Dearomatization/Radical Cyclization 263

2.2.4 Recent Methods for Alkylative Dearomatization of Phenolic Derivatives 268

2.2.4.1 Recent Applications to Intermolecular Alkylative Dearomatization of Naphthols 268

2.2.4.2 Dearomatization Reactions as Domino Transformations To Access Type A and B Polyprenylated Acylphloroglucinol Analogues 281

2.3 **Additions to Alkenes and C=O and C=N Bonds**

2.3.1 **Additions to Nonactivated C=C Bonds**
Z. W. Yu and Y.-Y. Yeung

2.3.1 **Additions to Nonactivated C=C Bonds** 293

2.3.1.1 Domino Amination 294

2.3.1.1.1 Proton-Initiated Events 294

2.3.1.1.2 Transition-Metal-Initiated Events 295

2.3.1.1.3 Halogen-Initiated Events 298

2.3.1.2 Domino Etherification 305

2.3.1.2.1 Halogen-Initiated Events 306

2.3.1.3 Domino Carbonylation 317

2.3.1.3.1 Transition-Metal-Initiated Events 317

2.3.1.3.2 Halogen-Initiated Events 320

2.3.1.4 Domino Polyene Cyclization 322

2.3.1.4.1 Transition-Metal-Initiated Events 323

2.3.1.4.2 Halogen-Initiated Events 325

2.3.1.4.3 Chalcogen-Initiated Events 329

2.3.2 Organocatalyzed Addition to Activated C=C Bonds
P. Renzi, M. Moliterno, R. Salvio, and M. Bella

2.3.2	**Organocatalyzed Addition to Activated C=C Bonds**	337
2.3.2.1	Organocatalyzed Domino Reactions with Activated Alkenes: The First Examples	337
2.3.2.1.1	Prolinol Trimethylsilyl Ethers as Privileged Catalysts for Enamine and Iminium Ion Activation	344
2.3.2.1.2	Increasing Complexity in Organocatalyzed Domino Reactions	347
2.3.2.2	Domino Organocatalyzed Reactions of Oxindole Derivatives	349
2.3.2.2.1	From Enders' Domino Reactions to Melchiorre's Methylene Oxindole	350
2.3.2.2.2	Michael Addition to Oxindoles	357
2.3.2.3	Synthesis of Tamiflu: The Hayashi Approach	365
2.3.2.4	One-Pot Synthesis of ABT-341, a DPP4-Selective Inhibitor	372
2.3.2.5	Large-Scale Industrial Application of Organocatalytic Domino Reactions: A Case Study	376
2.3.2.5.1	Transferring Organocatalytic Reactions from Academia to Industry: Not Straightforward	376
2.3.2.5.2	The Reaction Developed in the Academic Environment	377
2.3.2.5.3	The Reaction Developed in the Industrial Environment	379

2.3.3 Addition to Monofunctional C=O Bonds
A. Song and W. Wang

2.3.3	**Addition to Monofunctional C=O Bonds**	387
2.3.3.1	Transition-Metal-Catalyzed Domino Addition to C=O Bonds	387
2.3.3.1.1	Domino Reactions Involving Carbonyl Ylides	387
2.3.3.1.2	Reductive Aldol Reactions	389
2.3.3.1.3	Michael/Aldol Reactions	393
2.3.3.1.4	Other Domino Addition Reactions	394
2.3.3.2	Organocatalytic Domino Addition to C=O Bonds	395
2.3.3.2.1	Amine-Catalyzed Domino Addition to C=O Bonds	395
2.3.3.2.1.1	Enamine-Catalyzed Aldol/Aldol Reactions	395
2.3.3.2.1.2	Enamine-Catalyzed Aldol/Michael Reactions	396
2.3.3.2.1.3	Enamine-Catalyzed Diels–Alder Reactions	397
2.3.3.2.1.4	Enamine-Catalyzed Michael/Henry Reactions	399
2.3.3.2.1.5	Enamine-Catalyzed Michael/Aldol Reactions	400
2.3.3.2.1.6	Enamine-Catalyzed Michael/Hemiacetalization Reactions	400
2.3.3.2.1.7	Iminium-Catalyzed Michael/Aldol Reactions	402

2.3.3.2.1.8 Iminium-Catalyzed Michael/Henry Reactions 404

2.3.3.2.1.9 Iminium-Catalyzed Michael/Morita–Baylis–Hillman Reactions 404

2.3.3.2.1.10 Iminium-Catalyzed Michael/Hemiacetalization Reactions 405

2.3.3.2.2 Thiourea-Catalyzed Domino Addition to C=O Bonds 405

2.3.3.2.2.1 Aldol/Cyclization Reactions 405

2.3.3.2.2.2 Michael/Aldol Reactions 406

2.3.3.2.2.3 Michael/Henry Reactions 407

2.3.3.2.2.4 Michael/Hemiacetalization Reactions 408

2.3.3.2.3 Phosphoric Acid Catalyzed Domino Addition to C=O Bonds 410

2.3.3.3 Lewis Acid Catalyzed Domino Addition to C=O Bonds 411

2.3.3.4 Conclusions 414

2.3.4 **Additions to C=N Bonds and Nitriles**
E. Kroon, T. Zarganes Tzitzikas, C. G. Neochoritis, and A. Dömling

2.3.4 **Additions to C=N Bonds and Nitriles** 419

2.3.4.1 Addition to C=N Bonds and the Pictet–Spengler Strategy 422

2.3.4.2 Ugi Five-Center Four-Component Reaction Followed by Postcondensations 428

2.3.4.3 Addition to Nitriles 439

Keyword Index 449

Author Index 481

Abbreviations 497

2.1 Pericyclic Reactions

2.1.1 The Diels–Alder Cycloaddition Reaction in the Context of Domino Processes

J. G. West and E. J. Sorensen

General Introduction

The title that Otto Diels and Kurt Alder chose for their 1928 publication, "Syntheses in the Hydroaromatic Series", in *Annalen*[1] did not signal the revolution that their new insights would bring to the field of organic chemistry. Their pioneering paper described cycloadditions of 4π-electron systems (dienes) with 2π-electron systems (dienophiles), and captured the significance that [4+2] cycloadditions would hold for the field of organic chemical synthesis: "Thus, it appears to us that the possibility of synthesis of complex compounds related to or identical with natural products such as terpenes, sesquiterpenes, perhaps also alkaloids, has been moved to the near prospect." The very next sentence, "We explicitly reserve for ourselves the application of the reaction discovered by us to the solution of such problems", is even more colorful, but, in reality, nearly a quarter of a century would pass before the power of the "Diels–Alder" reaction was demonstrated in the context of natural product synthesis. In the year following the awarding of the 1950 Nobel Prize in Chemistry to Diels and Alder "for their discovery and development of the diene synthesis", R. B. Woodward and co-workers described their non-obvious use of a Diels–Alder construction to contend with the *trans*-fused C–D ring junction in cortisone,[2] and Gilbert Stork and his co-workers reported their stereospecific synthesis of cantharidin featuring a creative twofold Diels–Alder strategy.[3] Woodward's landmark 1956 synthesis of reserpine[4] and Eschenmoser's synthesis of colchicine by way of pericyclic reactions[5] provided further, powerful, demonstrations of the value of the Diels–Alder reaction as a structure-building process. On the foundation of these early achievements, the Diels–Alder reaction took its place beside the most reliable bond- and ring-forming methods in organic chemistry.

Today, nearly 90 years after that pioneering report by Diels and Alder, the cycloaddition reaction bearing their names has not lost its vitality. Indeed, few processes have captured the imagination of the practicing synthetic organic chemist as the Diels–Alder cycloaddition has. The ubiquity of six-membered rings in molecules of interest, from natural products to commodity chemicals, has brought this reaction to prominence not only as a singular operation, but also as a component in domino reaction sequences. This last capacity in particular has enabled highly original and powerful cascades to be designed and executed, leveraging the considerable risk inherent in such schemes into breathtaking advances for the field.

Due to the richness of precedent for cascade sequences featuring the Diels–Alder reaction, it is unavoidable that many inspiring examples will be omitted; the interested reader is encouraged to use this chapter and other reference materials[6–21] as a jumping-off point for entry into this fascinating body of literature. A similarly impressive collection of cascades involving formal Diels–Alder reactions permeates the chemical literature; however, these reactions are beyond the scope of this chapter and have been omitted. The examples in this work have been curated with the goal of presenting a broad survey of how Diels–Alder domino sequences can be used, with particular attention to how the aforementioned reaction is involved in the sequence.

It is the hope of the authors that students of organic chemistry will gain some appreciation for the myriad opportunities to advance a synthesis through the strategic application of Diels–Alder cascade processes and, in so doing, be able to advance not only their own project but, also, the field as a whole.

2.1.1.1 Cascades Not Initiated by Diels–Alder Reaction

A logical division one could imagine making in the presentation of this vast body of literature is whether the Diels–Alder reaction occurs at the beginning of the cascade or not. If the Diels–Alder is not the first step, it follows to consider three cases where it can be invoked: through the in situ generation of the diene or dienophile, or through the union of two pre-existing components promoted by earlier transformations. Each of these strategies presents certain advantages that may prove of high value in a target-oriented campaign.

2.1.1.1.1 Cascades Generating a Diene

The 4π component of the Diels–Alder reaction has been the target of extensive investigation for in situ generation. Presenting significant unsaturation, one can run into issues of chemoselectivity in a synthesis if one wishes to carry a diene through multiple operations. Numerous innovative methods to access dienes have thus been developed, of which many can be described as ionic, pericyclic, or radical in nature.

2.1.1.1.1.1 Ionic Generation of a Diene

The Diels–Alder reaction, a pericyclic process, proceeds through a mechanism that can be considered orthogonal to ionic reactivity modes. This exclusivity makes the Diels–Alder reaction inherently compatible with many two-electron processes, allowing for ionic components in domino reaction sequences. One area that has benefitted from significant exploration has been the in situ generation of dienes using ionic reactivity.

2.1.1.1.1.1.1 Through Wessely Oxidation of Phenols

The pioneering work of Wessely[22] revealed that electron-rich aromatic compounds can function as masked dienes when treated with a strong oxidant. Originally a two-step protocol, Liao and co-workers[23] showed that phenols could be oxidized in the presence of dienophiles to access Diels–Alder adducts in a cascade process. The use of arenes as diene surrogates provides several advantages in a synthetic design, most notably the relative stability imparted by the aromaticity of the arene. Indeed, it can improve the durability of the eventual diene in prior chemical steps and lead to heightened reactivity of the oxidatively generated diene in comparison to other classes.

This high utility is exemplified by the propensity of many Wessely oxidation products to undergo self-dimerization via a Diels–Alder reaction, a process that appears to have relevance in a biosynthetic sense. An example of this importance is provided by retrosynthetic analysis of the structure of the bis-sesquiterpene aquaticol by the lab of Quideau,[24] suggesting that its congested architecture might be accessible by Diels–Alder dimerization of orthoquinol **2**. Subjecting enantiopure phenol **1** to stabilized 1-hydroxy-1,2-benziodoxol-3(1*H*)-one 1-oxide [SIBX; a mixture of IBX (49%), benzoic acid (22%), and isophthalic acid (29%)] results in a 1:1 mixture of Diels–Alder adducts **3A** [(−)-aquaticol] and the closely related **3B**, in addition to catechol oxidation side product **4** (Scheme 1).

Scheme 1 Synthesis of Aquaticol through the Oxidative Dimerization of a Phenol[24]

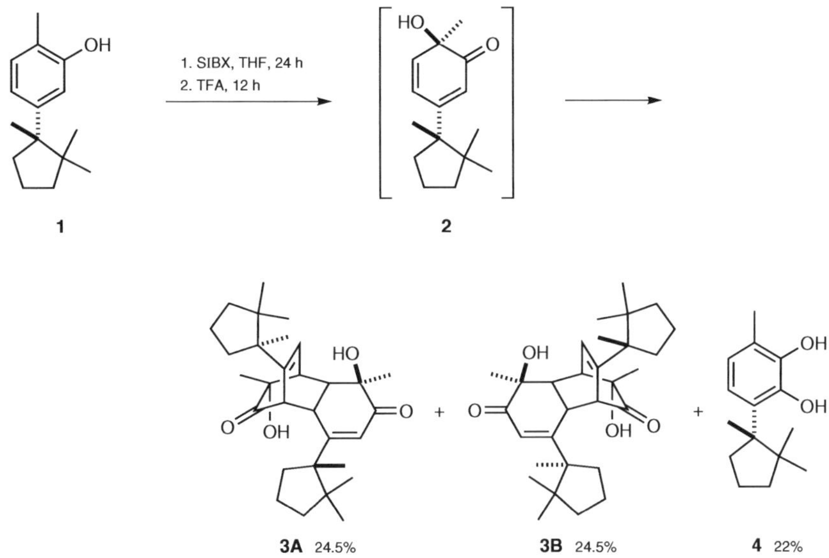

1. SIBX, THF, 24 h
2. TFA, 12 h

1 2

3A 24.5% 3B 24.5% 4 22%

Subsequent studies by Porco and co-workers[25] found that an enantioselective oxidation/dimerization cascade of the lithium phenoxide **5** of phenol *ent*-**1** can be effected using a copper/sparteine catalyst with oxygen serving as the terminal oxidant, furnishing (+)-aquaticol (**3C**) as a single diastereomer (Scheme 2).

Scheme 2 Asymmetric Oxidation/Dimerization Procedure Developed by Porco[25]

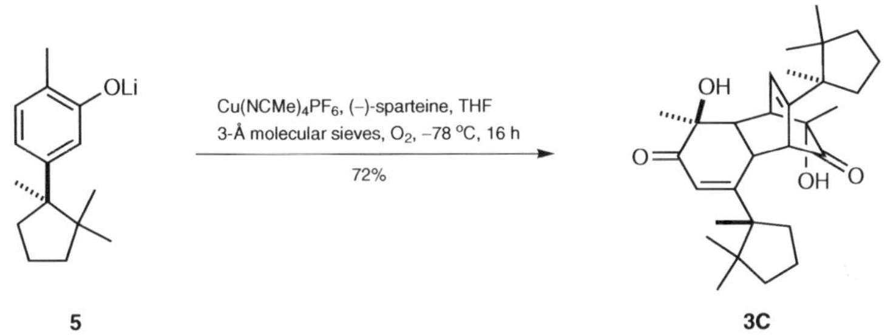

$Cu(NCMe)_4PF_6$, (–)-sparteine, THF
3-Å molecular sieves, O_2, –78 °C, 16 h
72%

5 3C

The group of Wood[26] found that an oxidation/intramolecular Diels–Alder cascade could be realized with the dienophilic partner already appended to the aromatic nucleus as a phenolic ether. Here, oxidation of compound **6** with (diacetoxyiodo)benzene in the presence of a variety of alcohols smoothly furnishes polycyclic products **8**, which serve as a convenient entry to the CP-263,114 architecture (Scheme 3). This method provides an added advantage as the alcohol present is incorporated into the final structure through the necessity of an *ortho*-quinone acetal intermediate **7**, allowing for this group to be widely varied through a simple change of solvent.

for references see p 44

Scheme 3 Access to the CP-263,114 Core Architecture through the Use of an Oxidation/Intramolecular Diels–Alder Cascade Sequence[26]

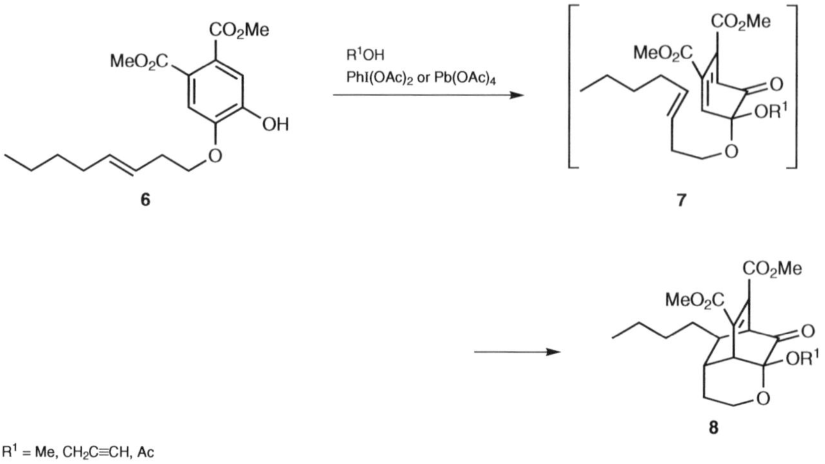

R^1 = Me, $CH_2C{\equiv}CH$, Ac

A variation on the strategy of Wood is found in the course of Rodrigo's synthesis of halenaquinone,[27] where oxidation of phenol **9** in the presence of alcohol-appended dienophile **10** produces masked *ortho*-quinone **11**. This intermediate smoothly undergoes a proximity-induced Diels–Alder reaction to furnish a key annulated bicyclo[2.2.2]octane product **12**. This diene substrate then undergoes a Cope rearrangement to form naphthofuranone product **13**, which has the core connectivity of halenaquinone (Scheme 4). This strategy is notable as it not only presents a union of both the nascent diene and dienophile, but also remodels the Diels–Alder cycloaddition product in a highly productive fashion.

Scheme 4 Wessely Oxidation in the Presence of an Unsaturated Alcohol Followed by Intramolecular Diels–Alder Reaction[27]

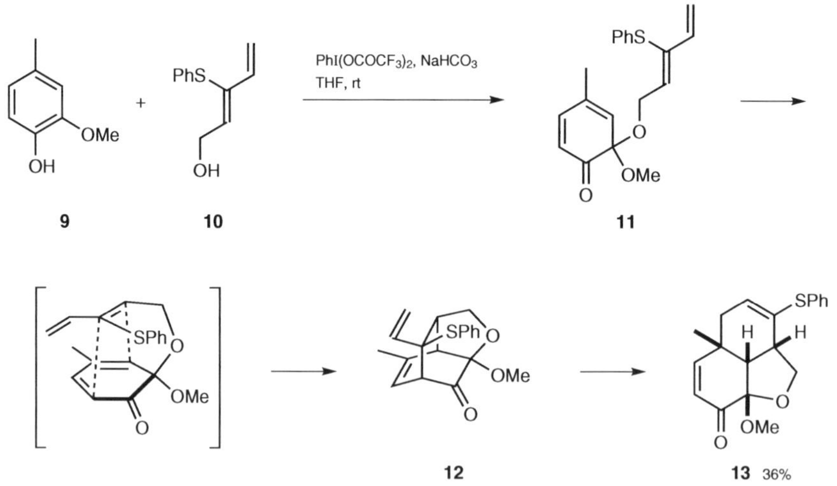

(1*S*,4*S*,8*R*,10*R*)-8,10-Dihydroxy-8,10-dimethyl-3,5-bis[(*S*)-1,2,2-trimethylcyclopentyl]-4,4a,8,8a-tetrahydro-1,4-ethanonaphthalene-7,9(1*H*)-dione (3A):[24]

To a soln of (*S*)-2-methyl-5-(1,2,2-trimethylcyclopentyl)phenol [(−)-**1**; 85 mg, 0.39 mmol, 1 equiv] in THF (4 mL) was added stabilized IBX [a mixture of IBX (49%), benzoic acid (22%), and isophthalic acid (29%); 365 mg, 0.58 mmol, 1.5 equiv] as a solid in one portion.

The resulting suspension was stirred at rt for 24 h, after which time TFA (30 μL, 0.394 mmol, 1 equiv) was added, and the mixture was stirred for a further 12 h. The mixture was then diluted with CH_2Cl_2 (10 mL) and H_2O (10 mL). 1 M aq NaOH (5 mL) was added dropwise (until pH 8). The aqueous phase was extracted with CH_2Cl_2 (3 × 10 mL). The combined organic phases were washed with 1 M aq NaOH (15 mL) and brine (2 × 15 mL), and then shaken vigorously with sat. aq $Na_2S_2O_4$ (40 mL), washed again with brine (40 mL), dried (Na_2SO_4), filtered, and concentrated at rt to give a crude pale-brown oily residue (95 mg). This residue was purified by flash chromatography (silica gel, CH_2Cl_2 then CH_2Cl_2/MeOH 100:1) to give two residues, which were again purified separately by flash chromatography (silica gel, hexanes/acetone 6:1), to furnish, respectively, (*S*)-benzene-1,2-diol (−)-**4** (20 mg; 22% yield) and a 1:1 mixture of (1*S*,4*S*,8*R*,10*R*)-product **3A** and the all-*S* dimer (−)-**3B** as white powders; yield: 45 mg (49%). The diastereomeric mixture of **3A** and **3B** was separated by semi-preparative reverse-phase HPLC [Thermo Spectra system; Deltapak C-18 column (7.8 × 300 mm, 15 μm); gradient elution; flow rate: 3 mL • min^{-1}; mobile phases: solvent A (H_2O/TFA 99:1) and solvent B (MeCN/TFA 99:1); 0–55 min: solvent (A/B) 70:30 to 0:100].

(1*R*,4*R*,8*S*,10*S*)-8,10-Dihydroxy-8,10-dimethyl-3,5-bis[(*R*)-1,2,2-trimethylcyclopentyl]-4,4a,8,8a-tetrahydro-1,4-ethanonaphthalene-7,9(1*H*)-dione [3C; **(+)-Aquaticol]:**[25]
A soln of Li phenolate **5** {derived from (*R*)-2-methyl-5-(1,2,2-trimethylcyclopentyl)phenol [(+)-**1**; 35.0 mg, 0.16 mmol, 1 equiv]} in anhyd THF (160 μL) was added to the Cu complex [prepared from $Cu(NCMe)_4PF_6$ (131.0 mg, 0.35 mmol, 2.2 equiv) and (−)-sparteine (84.8 μL, 0.37 mmol, 2.3 equiv) in anhyd THF (2.0 mL)] under an O_2 atmosphere at −78 °C. The mixture was stirred at −78 °C for 16 h and then the reaction was quenched with 5% aq H_2SO_4 (1.6 mL) at −78 °C. The mixture was extracted with EtOAc (3 ×), and the combined extracts were washed with 5% aq H_2SO_4, H_2O, and brine, dried ($MgSO_4$), and concentrated under reduced pressure. The crude residue was purified by flash chromatography (silica gel, hexanes/EtOAc 4:1) to afford (+)-aquaticol (**3C**) as a light yellow solid; yield: 27.1 mg (72%).

Dimethyl 9-Butyl-8a-methoxy-8-oxo-3,4,4a,7,8,8a-hexahydro-2*H*-4,7-methanobenzopyran-5,6-dicarboxylate (8, **R^1 = Me); Typical Procedure:**[26]
To a stirred soln of phenol **6** (47 mg, 0.14 mmol, 1 equiv) in MeOH (1.5 mL) was added $PhI(OAc)_2$ (54 mg, 0.17 mmol, 1.2 equiv). Upon addition of $PhI(OAc)_2$, the mixture changed immediately from colorless to clear yellow; upon stirring at rt for 2 h, it became clear again. The mixture was concentrated under reduced pressure and passed through a plug of silica gel to furnish analytically pure **8**; yield: 36 mg (70%).

(2a*S*,2a^1*S*,5a*S*,8a*S*)-8a-Methoxy-5a-methyl-3-(phenylsulfanyl)-2,2a,2a^1,5,5a,8a-hexahydro-8*H*-naphtho[1,8-*bc*]furan-8-one (13)**:**[27]
To a soln of 2-methoxy-4-methylphenol (**11**; 100 mg, 0.72 mmol, 1 equiv), 3-(phenylsulfanyl)penta-2,4-dien-1-ol (**12**; 500 mg, 2.60 mmol, 3.6 equiv), and 2,6-di-*tert*-butyl-4-methylphenol (1 crystal; ca. 2 mg) in THF (15 mL) at 0 °C was added [bis(trifluoroacetoxy)iodo]benzene (375 mg, 0.87 mmol, 1.2 equiv). The resulting soln was stirred for 5 min, after which time solid $NaHCO_3$ (150 mg, 1.79 mmol, 2.5 equiv) was added. The mixture was allowed to warm to rt and stirred overnight, and was then partitioned between H_2O and Et_2O. The aqueous phase was extracted twice more with Et_2O, and the combined organic layers were dried ($MgSO_4$) and filtered through a plug of silica gel. After removal of the solvent under reduced pressure, the resulting dark orange oil was dissolved in 1,2,4-trimethylbenzene and refluxed for 2 d. Removal of the solvent under reduced pressure followed by flash chromatography (Et_2O/hexane 3:7) gave a light yellow oil; yield: 86 mg (36%).

for references see p 44

2.1.1.1.1.1.2 Through Ionic Cyclization

Heathcock and co-workers[28] have provided a remarkable example of the utility of the Diels–Alder cycloaddition in polycyclization cascades. Treating a dihydrosqualene dialdehyde, obtained by Swern oxidation of diol **14**, with methylamine leads to the formation of dihydropyridinium species **15** through the conjugate addition/condensation of the methylamine enamine on the enal functionality. This fleeting intermediate can then undergo an intramolecular Diels–Alder reaction to form tetracycle **16**, which can then engage the pendent prenyl alkene in an aza-Prins cyclization to afford carbocation **17**. Next, a proximity-induced hydride transfer from the methyl group of the amine provides iminium species **18**, which, after hydrolysis, completes the preparation of 1,2-dihydro-*proto*-daphniphylline (**19**) in an impressive polycyclization cascade (Scheme 5).

Scheme 5 Synthesis of 1,2-Dihydro-*proto*-daphniphylline via a Cyclization/Diels–Alder/Hydride Transfer Cascade from an Acyclic Precursor[28]

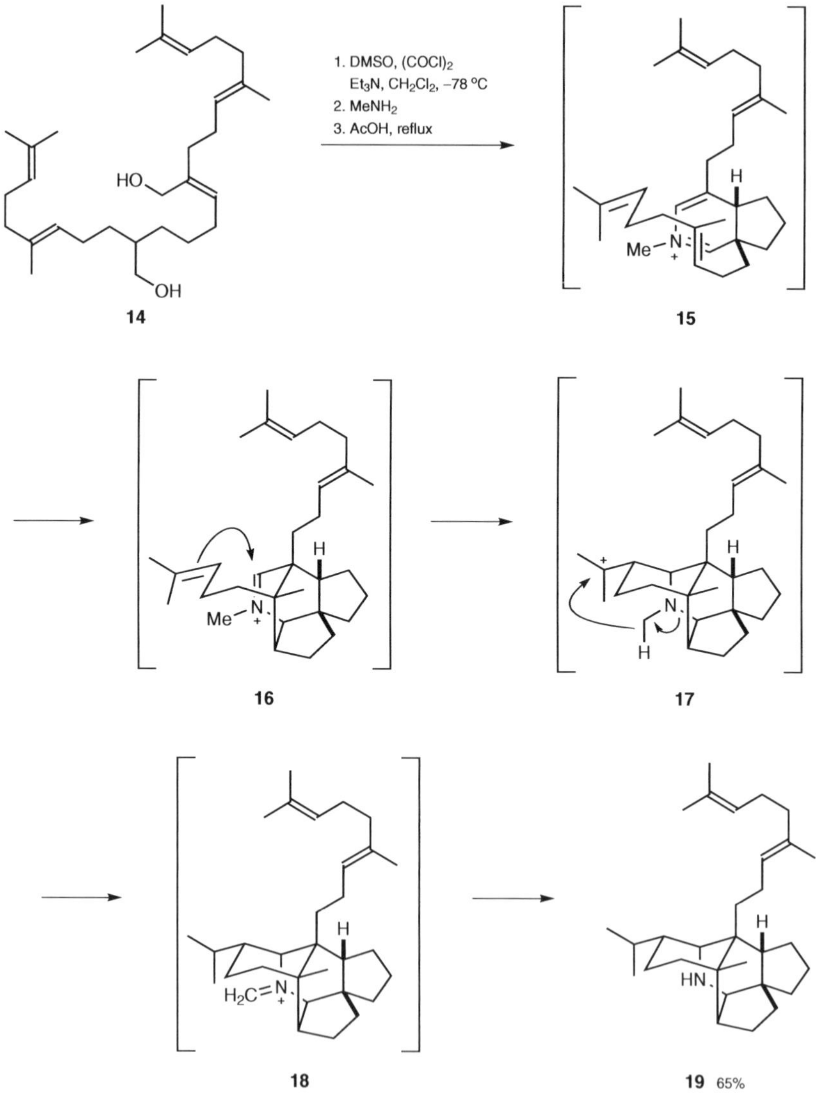

1. DMSO, $(COCl)_2$, Et_3N, CH_2Cl_2, −78 °C
2. $MeNH_2$
3. AcOH, reflux

14 **15** **16** **17** **18** **19** 65%

1,2-Dihydro-*proto*-daphniphylline (19):[28]
To a soln of DMSO (88 μL, 1.2 mmol, 9 equiv) in CH_2Cl_2 (1 mL) at −78 °C was added a 2.0 M soln of oxalyl chloride in CH_2Cl_2 (276 μL, 0.552 mmol, 4 equiv). After 20 min, diol **14** (61.3 mg, 0.138 mmol, 1 equiv) was added via cannula as a soln in CH_2Cl_2 (1 mL, followed by a 1-mL rinse). The resulting cloudy soln was stirred at −78 °C for 20 min and then treated with Et_3N (0.14 mL, 1.0 mmol). The dry ice bath was removed and the soln was allowed to warm to rt over 50 min. After cooling to 0 °C, a stream of anhyd $MeNH_2$ was then passed over the soln for 3 min. The flask was then sealed tightly and allowed to warm to rt over 5 h. The clear soln was concentrated by passing a stream of dry N_2 over it for 10 min. The resulting white, oily solid was triturated with Et_2O, filtered, and concentrated (high-vacuum pump, 4 h) to provide a clear yellow oil, which was utilized immediately in the next step; yield: 84.0 mg.

The crude bisimine was taken up in AcOH (1 mL) and placed in an 80 °C oil bath for 11 h. After cooling to 0 °C, the mixture was partitioned between CH_2Cl_2 (5 mL) and 2 M NaOH (5 mL) and stirred vigorously for 15 min. The layers were separated, and the aqueous phase was extracted with three portions of CH_2Cl_2. The combined organic phases were washed with brine and dried ($MgSO_4$). Filtration and concentration provided 68.0 mg of a brown oil, which was purified by flash chromatography (silica gel, gradient elution with 10:1 to 5:1 hexanes/EtOAc) to provide **19** as a clear, pale yellow oil; yield: 38.2 mg (65%).

2.1.1.1.1.1.3 Through Deprotonation of an Alkene

A keen insight from Danishefsky[29] and co-workers has been reported in the construction of the anthraquinone fragment of enediyne antibiotic dynemicin A. Treatment of homophthalic anhydride [1*H*-2-benzopyran-1,3(4*H*)-dione] **20** with lithium hexamethyldisilazanide results in the transient generation of xylylene (quinodimethane) **21** which, in the presence of quinone imine **22**, results in the production of sophisticated anthrone **24** ostensibly through the intermediacy of Diels–Alder adduct **23** followed by the extrusion of carbon dioxide (Scheme 6). The non-isolable intermediate anthrone **24** was immediately oxidized in the next step. The complexity of the fragments in this example illustrates well the ability of Diels–Alder cascades to merge two late-stage intermediates.

for references see p 44

Scheme 6 Participation of a Reactive Xylylene, Generated through Deprotonation, in a Diels–Alder/Retro-Diels–Alder Cascade[29]

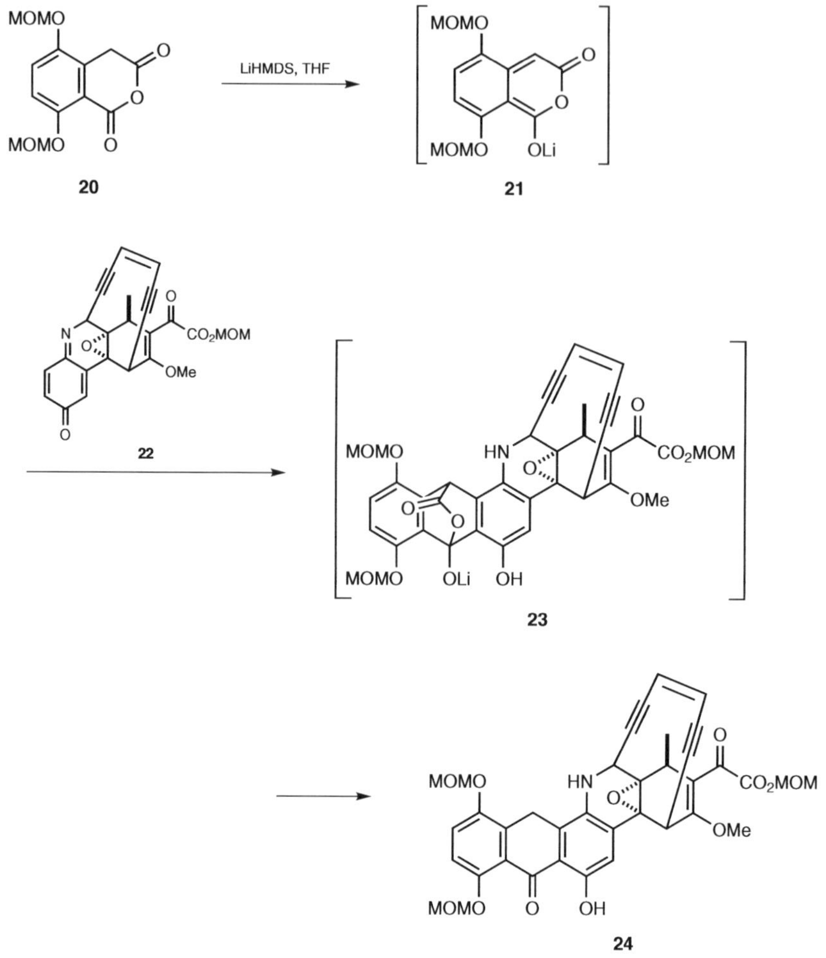

Anthrone 24:[29]

1*H*-2-Benzopyran-1,3(4*H*)-dione **20** (49 mg, 0.173 mmol, 6.0 equiv) was dissolved in THF (2.5 mL) and cooled to 0 °C. Then, a 1.0 M soln of LiHMDS (171 μL, 0.171 mmol, 5.9 equiv) was added dropwise, and the soln immediately became bright yellow. After 35 min, the quinone imine **22** (12 mg, 0.029 mmol, 1.0 equiv) was added as a soln in THF (1 mL). The mixture slowly became a dark red-brown color, and, after 35 min, TLC indicated that no starting material remained. The intermediate anthrone **24** was used directly in the next step.

2.1.1.1.1.1.4 Through Elimination Reactions

A classical method by which one could imagine generating unsaturation in a molecule is by elimination processes. An elimination reaction involving an allylic leaving group can be expected to lower the activation energy of the reaction, resulting in the convenient synthesis of a diene. Unsurprisingly, synthesis of a diene by this strategy has been successfully applied in several Diels–Alder cascade sequences.

A particularly striking example of an elimination process forming a diene is provided by the group of Grieco in a concise synthesis of pseudotabersonine.[30] Treatment of amino alcohol **25** with catalytic 4-toluenesulfonic acid in acetone/water leads to a vinylogous E1 reaction to furnish reactive diene intermediate **26**, which, upon heating, provides Diels–Alder adduct **27** (Scheme 7).

Scheme 7 In Situ Generation of a Diene by a Vinylogous E1 Reaction Prior to an Intramolecular Diels–Alder Reaction[30]

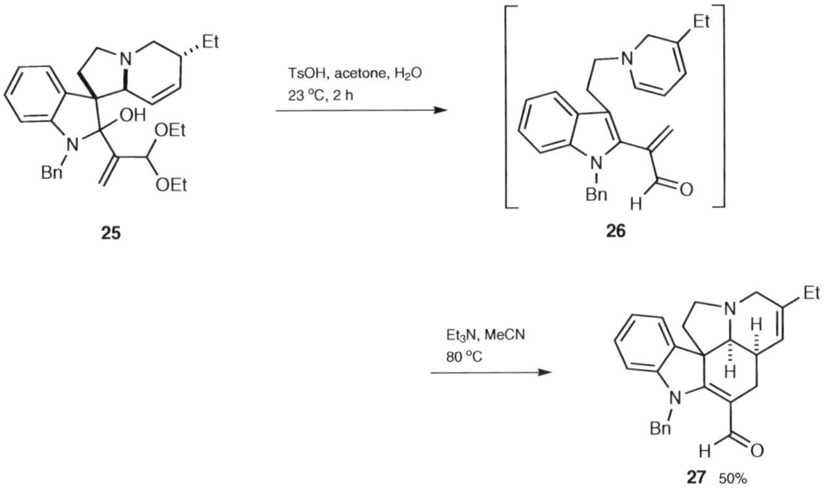

TsOH, acetone, H_2O
23 °C, 2 h

25

26

Et_3N, MeCN
80 °C

27 50%

A key Diels–Alder macrocyclization was envisioned by Sorensen and co-workers in the course of their synthesis of potent antibiotic abyssomycin C.[31] It was recognized that the requisite trienone diene partner in the desired Diels–Alder reaction would be exceptionally reactive, necessitating a design that introduces it directly prior to the cycloaddition. Treatment of silyl ether **28** with lanthanum(III) trifluoromethanesulfonate in hot toluene leads to the production of the desired trienone **29**. This transient intermediate then undergoes the Diels–Alder macrocyclization to furnish intermediate **30**, a process that is accelerated by heating (Scheme 8).

Scheme 8 A Lewis-Acid Catalyzed Elimination Reaction To Unveil a Reactive Trienone That Undergoes an Effective Diels–Alder Macrocyclization[31]

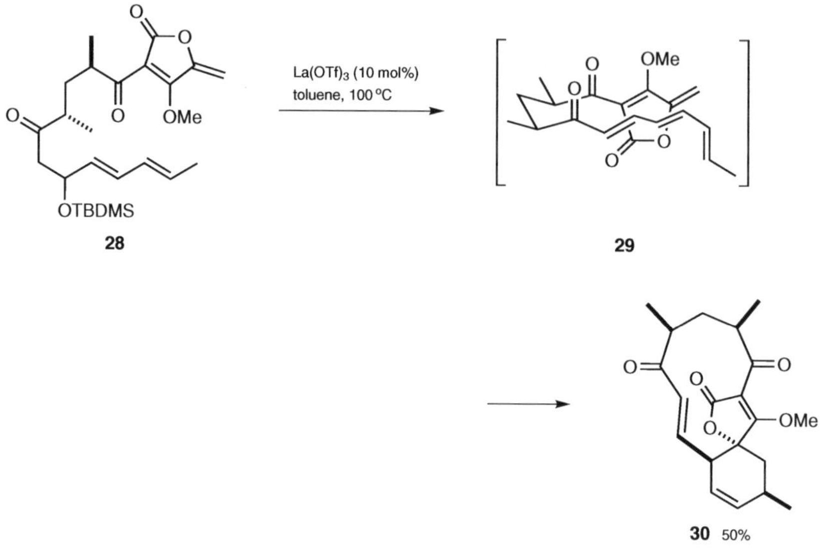

La(OTf)$_3$ (10 mol%)
toluene, 100 °C

OTBDMS
OMe
28
29
30 50%

Another strategy that can be described as highly reliant on a Diels–Alder cascade can be found in the Sorensen synthesis of the potent antitumor agent cyclostreptin (also known in early literature as FR182877).[32,33] Seeking to test the viability of a biogenesis involving a twofold transannular intramolecular Diels–Alder transformation, the Sorensen lab targeted macrocycle **31** as a starting material for the cascade. Oxidation of diastereomeric selenide **31** leads to a tandem elimination/double transannular Diels–Alder to furnish a sophisticated product **32** with the connectivity of cyclostreptin (Scheme 9). Evans and Starr completed a nearly contemporaneous, independent synthesis of cyclostreptin using a similar elimination/double Diels–Alder cascade initiated in a slightly different manner.[34]

Scheme 9 A Selenoxide Elimination Reaction To Enable a Bioinspired Diels–Alder/Diels–Alder Cascade[32,33]

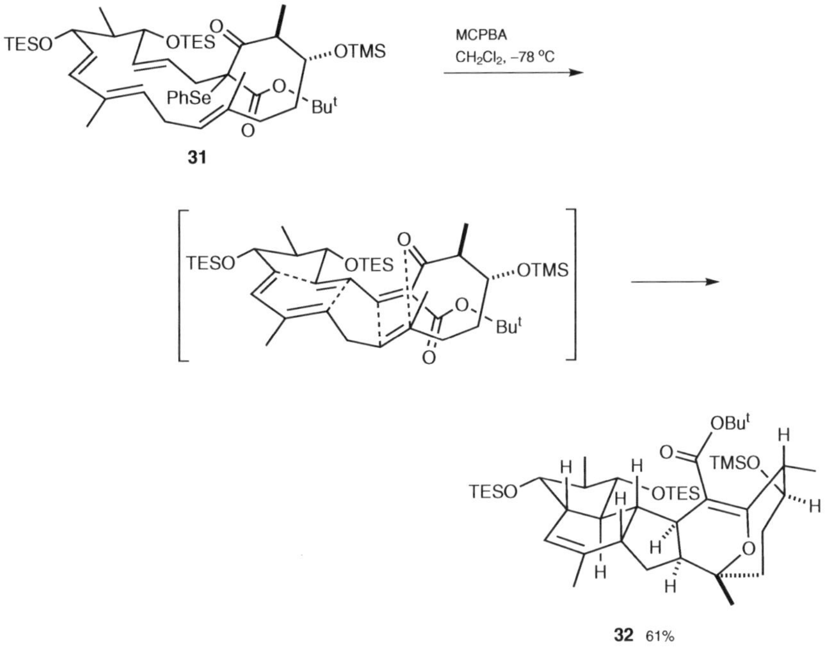

(3*E*,5*R*,7*S*,9*E*,10a*S*,13*R*,14a*R*)-15-Methoxy-5,7,13-trimethyl-6,7,13,14-tetrahydro-2*H*-3,14a-(metheno)benzo[*b*][1]oxacyclododecine-2,4,8(5*H*,10a*H*)-trione (30):[31]
To a soln of diene **28** (115 mg, 0.241 mmol, 1 equiv) in degassed toluene (24.1 mL) was added La(OTf)$_3$ (14 mg, 0.024 mmol, 0.1 equiv) and the mixture was stirred at 100 °C for 4 h. The mixture was concentrated under reduced pressure and rinsed through a plug of silica gel (EtOAc/hexanes 1:1) to remove the catalyst. The crude product was purified by column chromatography (EtOAc/hexanes 1:4). The target compound **30** was obtained as a colorless oil [yield: 42 mg (50%)] along with the *Z*-trienone isomer of **29** [yield: 5 mg (5%)]. The trienone was converted into the *E*-isomer and underwent the Diels–Alder macrocyclization in quantitative yield by heating in degassed toluene (0.01 M) in the presence of a single crystal of I_2.

Pentacycle 32:[33]
Selenide **31** [diastereomeric mixture (ratio 1:10); 19.0 g, 20.4 mmol, 1 equiv] was dissolved in CH_2Cl_2 (400 mL) at −78 °C, and a soln of MCPBA (70%; 6.04 g, 24.5 mmol, 1.2 equiv) in CH_2Cl_2 (50 mL) was slowly added via cannula. The reaction was complete immediately (determined by TLC analysis) and was quenched at −78 °C by addition of a mixture of sat. aq $Na_2S_2O_3$/sat. aq $NaHCO_3$ (1:1; 300 mL). The mixture was warmed to rt and extracted with hexanes (3 × 300 mL). The combined organic phases were washed with brine (500 mL), dried ($MgSO_4$), filtered, and concentrated. The crude product was dissolved in $CHCl_3$ (400 mL) and solid $NaHCO_3$ (5 g) was added. The flask was sealed under argon, and the mixture was protected from light and stirred at 45 °C for 4 h. The mixture was filtered through a plug of cotton and concentrated. Purification by chromatography (Et_2O/hexanes 1:99 to >5:95) afforded the desired pentacycle **32** as a clear, colorless, viscous oil; yield: 9.6 g (61%).

for references see p 44

2.1.1.1.1.1.5 Through Allylation

Allylation can be a viable precursor reaction to a Diels–Alder union by virtue of the unsaturation of the resultant product. An inspiring disconnection from the Nicolaou group is provided by their entry into the decahydrofluorenyl core of hirsutellone B.[35] Treatment of acyclic epoxide **33** with diethylaluminum chloride leads to a cascade sequence wherein a doubly vinylogous Sakurai-type alkenation provides dienyl intermediate **34**. Then, through the enhanced proximity provided by the nascent monocycle, this material smoothly undergoes an intramolecular Diels–Alder reaction to provide desired compound **35** (Scheme 10). Overall, this highly productive domino transformation results in the stereoselective formation of three rings in a single operation, allowing for the synthesis to be rapidly advanced.

Scheme 10 Synthesis of the Hirsutellone B Core through an Epoxide-Opening Allylation/Intramolecular Diels–Alder Cascade[35]

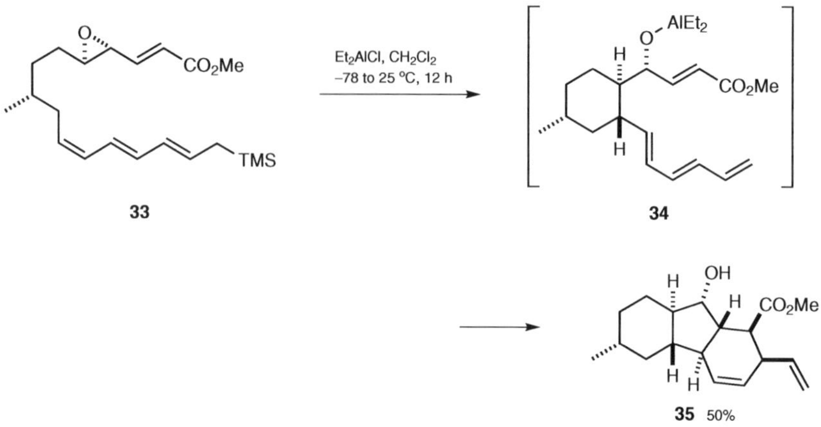

CO_2Me TMS **33**

Et_2AlCl, CH_2Cl_2, −78 to 25 °C, 12 h

$AlEt_2$ CO_2Me **34**

OH H CO_2Me **35** 50%

Methyl (1*S*,2*S*,4a*S*,4b*S*,6*R*,8a*R*,9*S*,9a*S*)-9-Hydroxy-6-methyl-2-vinyl-2,4a,4b,5,6,7,8,8a,9,9a-decahydro-1*H*-fluorene-1-carboxylate (35); Typical Procedure:[35]

To a stirred soln of epoxytetraene **33** (100 mg, 0.276 mmol, 1 equiv) in CH_2Cl_2 (10 mL) at −78 °C was added a 1.0 M soln of Et_2AlCl in hexanes (2.8 mL, 2.8 mmol, 10.0 equiv) dropwise. The resulting homogeneous soln was allowed to warm to −50 °C for 30 min, and then to rt slowly overnight (12 h). The reaction was then quenched with sat. aq $NaHCO_3$ (10 mL) and extracted with EtOAc (3 × 10 mL). The combined organic phases were washed with brine (20 mL), dried ($MgSO_4$), and concentrated. Purification of the residue by flash column chromatography (silica gel, EtOAc/hexanes 1:3) provided a white solid; yield: 40 mg (50%).

2.1.1.1.1.2 Pericyclic Generation of a Diene

The product of a pericyclic process, or a change in bonding resulting from the concerted rearrangement of electrons in a cyclic transition state, can sometimes participate in a subsequent pericyclic process, resulting in a so-called pericyclic cascade. The Diels–Alder cycloaddition, itself a pericyclic reaction, is no exception to this trend. The ability to significantly rearrange the bonding framework of molecules in a single chemical step has been recognized as exceptionally valuable for target-oriented synthesis and has resulted in many elegant examples of tandem pericyclic Diels–Alder reaction sequences being reported.

2.1.1.1.1.2.1 Through Electrocyclization

The electrocyclization of polyunsaturated substrates presents the attractive opportunity to alter the number of rings in a starting material. Moreover, one might consider using electrocyclizations to access reactive and otherwise inaccessible Diels–Alder partners from stable precursors.

2.1.1.1.1.2.1.1 Through Benzocyclobutene Ring Opening

ortho-Xylylenes (*ortho*-quinodimethanes) are well known as reactive dienes in Diels–Alder reactions. One method by which one might envision accessing such an intermediate is through the electrocyclic ring opening of a benzocyclobutene. First shown in the context of synthesis by Oppolzer and Keller's preparation of chelidonine,[36] it was soon recognized that many opportunities to streamline routes through this strategy exist in chemical space.

The concurrent development of a cobalt-mediated [2+2+2] cyclotrimerization served as a springboard for the Vollhardt group in their landmark synthesis of estrone. Vollhardt and Funk[37] envisioned forming both the B and C rings of the target through an intramolecular Diels–Alder annulation of a transiently formed *ortho*-xylylene **37** possessing an alkene attached to what is to become the D-ring. Thus, heating the benzocyclobutene intermediate **36** in benzene results in a conrotatory 4π-electrocyclic ring opening to unveil the highly activated *ortho*-dimethylene intermediate **37**, which smoothly engages the pendent alkene in an intramolecular Diels–Alder reaction to furnish intermediate **38** having the connectivity of estrone (Scheme 11). This application of the benzocyclobutene methodology is notable as it rapidly develops the architecture of its target through the simultaneous formation of two rings.

Scheme 11 Synthesis of the B and C Rings of Estrone through an Electrocyclization/Intramolecular Diels–Alder Cascade[37]

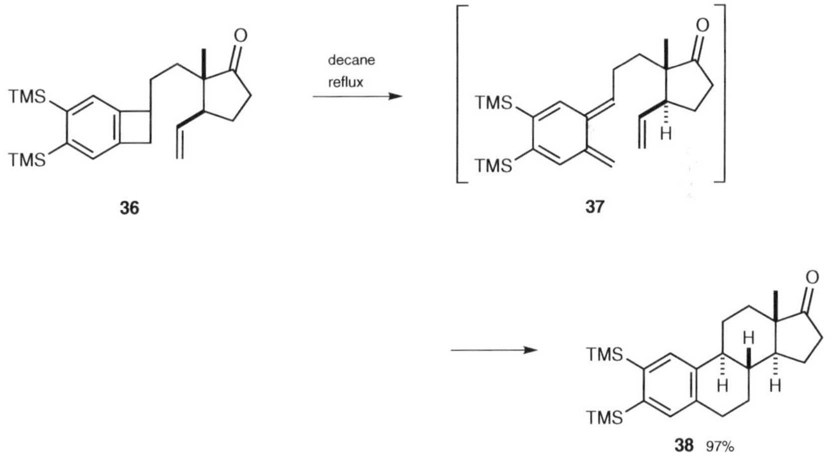

TMS TMS O **36** → decane, reflux → [TMS TMS O H **37**] → TMS TMS O H H H **38** 97%

A synthesis of rishirilide B by Danishefsky and co-workers also makes use of the benzocyclobutene diene surrogate to achieve a union of two partners, albeit in an intermolecular fashion.[38] Heating a mixture of benzocyclobutene **39** and enedione **40** in a sealed tube, followed by dehydration, results in the production of the desired tricycle **41** in a regioselective sense (Scheme 12). A notable aspect of this maneuver can be observed in the formulation of its dienophilic component **40**, namely in the hydroxy functionality α to the enone carbonyl. It is proposed that this hydroxy group is able to serve as an internal hy-

for references see p 44

drogen-bond donor, activating the enedione in a mode first recognized by Masamune.[39] This activation not only increases the ease with which the reaction can be effected, but also enhances the regioselectivity of the transformation through differential activation of one ketone. This factor was confirmed through the replacement of the hydroxy group with a silyl ether, resulting in lackluster reactivity with abysmal regioselectivity in the respective cascade.

Scheme 12 Regioselective, Intermolecular Benzocyclobutene/Diels–Alder Cascade Enabled by the Masamune Effect[38]

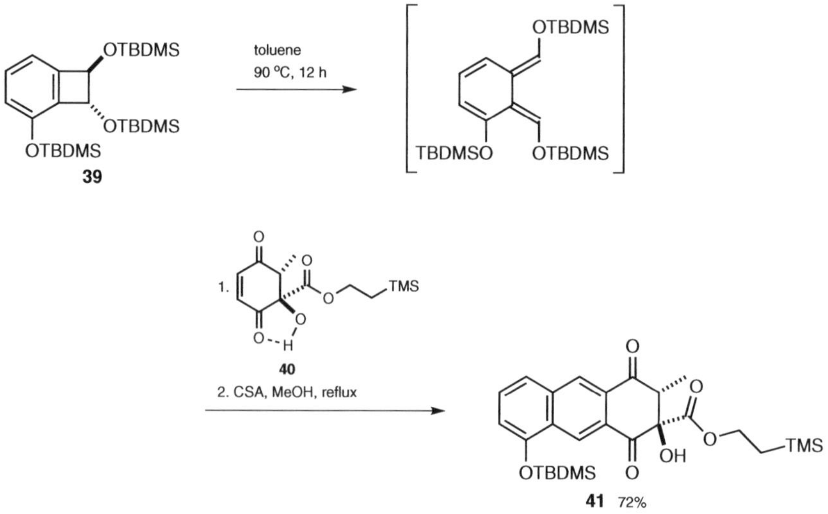

OTBDMS OTBDMS OTBDMS 39 toluene 90 °C, 12 h OTBDMS TBDMSO OTBDMS

1. 40 2. CSA, MeOH, reflux

41 72%

2,3-Bis(trimethylsilyl)estra-1,3,5(10)-triene-17-one (38):[37]
Benzocyclobutene **36** (142 mg, 0.36 mmol) was dissolved in degassed decane (35 mL) and refluxed for 20 h. The solvent was vacuum transferred and the residue was filtered through silica gel and crystallized (petroleum ether); yield: 135 mg (97%).

2-(Trimethylsilyl)ethyl (2*R*,3*R*)-8-(*tert*-Butyldimethylsiloxy)-2-hydroxy-3-methyl-1,4-dioxo-1,2,3,4-tetrahydroanthracene-2-carboxylate (41):[38]
Enedione **40** (50 mg, 85 μmol, 1 equiv) and benzocyclobutane **39** (20 mg, 70 μmol, 1 equiv) were dissolved in toluene-d_8 (1 mL) and heated at 90 °C for 12 h in a sealed tube. The solvent was evaporated, the residue was dissolved in MeOH (10 mL), and CSA (16 mg, 70 μmol, 1 equiv) and pyridine (6 μL, 70 μmol, 1 equiv) were added. After heating the mixture at reflux for 4 h, the mixture was concentrated and the residue was purified by column chromatography (silica gel) to afford a colorless oil; yield: 26 mg (72%).

2.1.1.1.1.2.1.2 Through Electrocyclic Ring Closure

Electrocyclic ring closure of arrays with six or more π-electrons generate products with conjugated alkenes, or functionalities able to serve as the diene component of a Diels–Alder cycloaddition. An advantage of producing polyenes via this method is that, by virtue of them being confined to a ring, they must conform to the prerequisite *s-cis* geometry needed for the cycloaddition to occur.

The postulated biosynthesis of torreyanic acid involves the dimerization of dihydropyran monomers through a Diels–Alder process.[40] Taking inspiration from this proposal, Porco and co-workers envisioned accessing the desired cycloaddition partners through oxidation of readily accessible allylic alcohol **42**, which, following aldehyde **43** formation, should be able to undergo a facile 6π-electrocyclization to provide the desired dihydropyran **44** (Scheme 13).[41] One can then imagine that this intermediate can dimerize via cycloaddition. Treatment of a racemic mixture of **42** with Dess–Martin periodinane at room temperature leads to the desired cascade, culminating in the formation of a 1:1 mixture of the *tert*-butyl esters of torreyanic acid (**45A**) and its close cousin isotorreyanic acid (**45B**).

Scheme 13 Oxidation/Electrocyclization/Diels–Alder Cascade in the Synthesis of Torreyanic Acid[41]

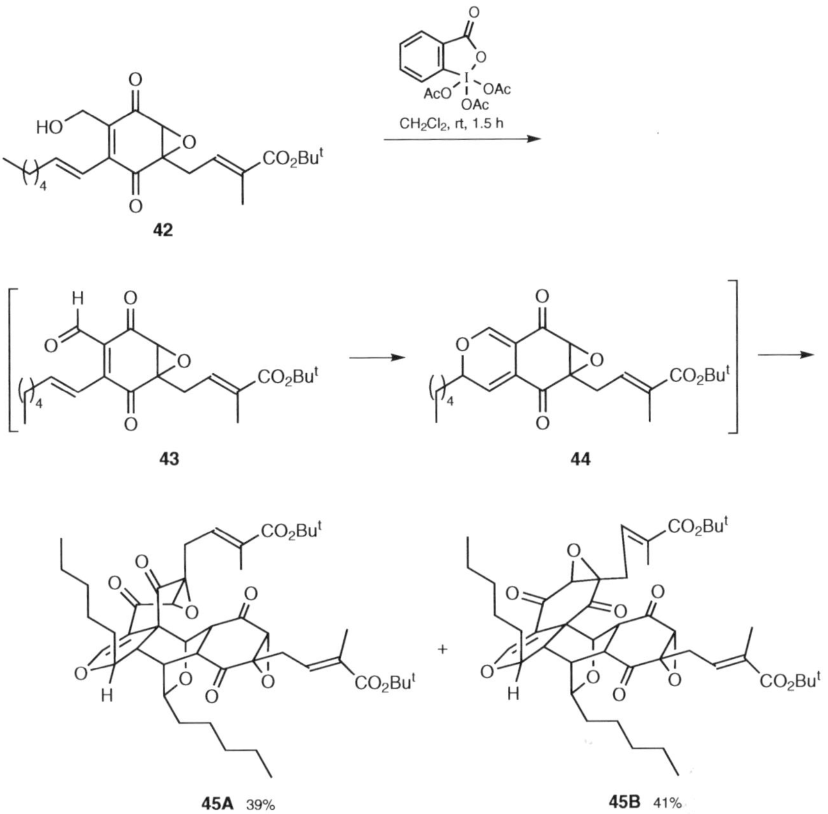

AcO, OAc, OAc; CH_2Cl_2, rt, 1.5 h

42 · **43** · **44**

45A 39% + **45B** 41%

There is no restriction on the number of electrocyclizations before the diene partner is formed, a truth that is exemplified by the work of Nicolaou. The daring biosynthetic proposal of Black for the racemic endiandric acids posits that their congested, polycyclic architecture could be achieved by a series of pericyclic reactions starting from a highly unsaturated, linear intermediate.[42] The group of Nicolaou was able to synthesize the methyl ester of this intermediate, compound **47**, through a twofold Lindlar hydrogenation of diyne **46**, and observed that racemic endiandric acid methyl esters B (**49**), C (**51**), F (**48**), and G (**50**) are formed after brief heating of the starting material.[43,44] Esters **49** and **51** are formed through intramolecular Diels–Alder reactions of esters **48** and **50**, themselves products of successive 8π- and 6π-electrocyclizations of linear precursor **47** (Scheme 14).

for references see p 44

Scheme 14 Pericyclic Cascades Interrelating Endiandric Acid Methyl Esters B, C, F, and G with an Acyclic Precursor[44]

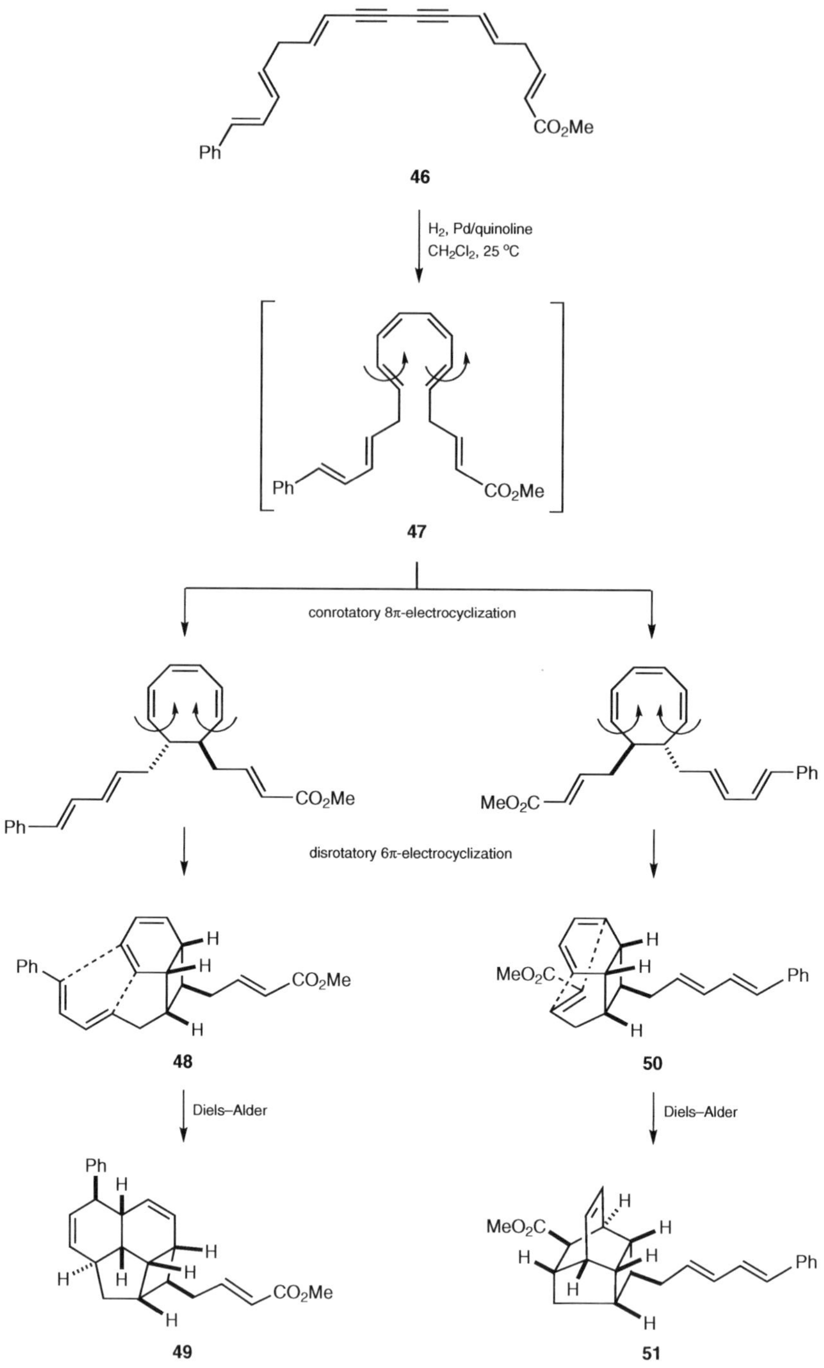

Torreyanic Acid *tert*-Butyl Ester (45A):[41]
Quinone monoepoxide **42** (10.8 mg, 0.027 mmol, 1 equiv) was dissolved in CH_2Cl_2 (2 mL) and Dess–Martin periodinane (18 mg, 0.042 mmol, 1.6 equiv) was added. After stirring at rt for 1.5 h, the mixture was neutralized with sat. $NaHCO_3/Na_2S_2O_3$ and extracted with Et_2O. The organic extracts were combined, washed with brine, dried ($MgSO_4$), filtered,

and concentrated under reduced pressure. ^{1}H NMR analysis of the crude product showed 2*H*-pyran monomer **44**, dimer **45A**, and dimer **45B** in a 1:1:1 ratio. The mixture was allowed to stand on a silica gel column for 1 h, and then eluted and purified by flash column chromatography to provide *tert*-butyl esters **45A** [yield: 4.3 mg (39%)] and **45B** [yield: 4.5 mg (41%)].

Endiandric Acid Methyl Esters 48–51:[44]
When the acetylenic precursor **46** was mildly hydrogenated (H_2, Lindlar catalyst, quinoline, CH_2Cl_2, 25 °C) followed by brief heating of the resulting mixture at 100 °C (toluene), endiandric acid methyl esters **48–51** were produced and chromatographically isolated.

2.1.1.1.1.2.2 Through Cycloaddition or Retrocycloaddition

The Diels–Alder reaction, itself a reversible cycloaddition, can be expected to have similar activation requirements to other reactions of this class. When combined with the inherent unsaturation of Diels–Alder substrates, it is unsurprising that cascades involving cycloadditions and retrocycloadditions can be implemented productively in synthetic designs.

One advantage of a retrocycloaddition-based approach to the 4π component of the Diels–Alder reaction is the ability to "protect" the unsaturated diene as a cycloadduct, allowing for operations usually incompatible with dienes to be performed with impunity. The synthetic strategy applied toward colombiasin A by the Nicolaou group called for a late-stage intramolecular Diels–Alder between cyclic alkene **53** and its appended diene.[45] In the course of their investigations, it was found that the diene introduced issues of chemoselectivity in the preceding oxidation step. Drawing from the work of Staudinger,[46] it was found that masking the diene as a sulfolene (2,5-dihydrothiophene 1,1-dioxide) group obviated much of this difficulty. Subjecting sulfolene intermediate **52** to heat and pressure unveils the diene **53**, which engages in an intramolecular Diels–Alder to furnish the desired intermediate **54** with the core connectivity of colombiasin A (Scheme 15).

Scheme 15 Use of a Sulfolene as a Diene Masking Group in the Nicolaou Synthesis of Colombiasin A[45]

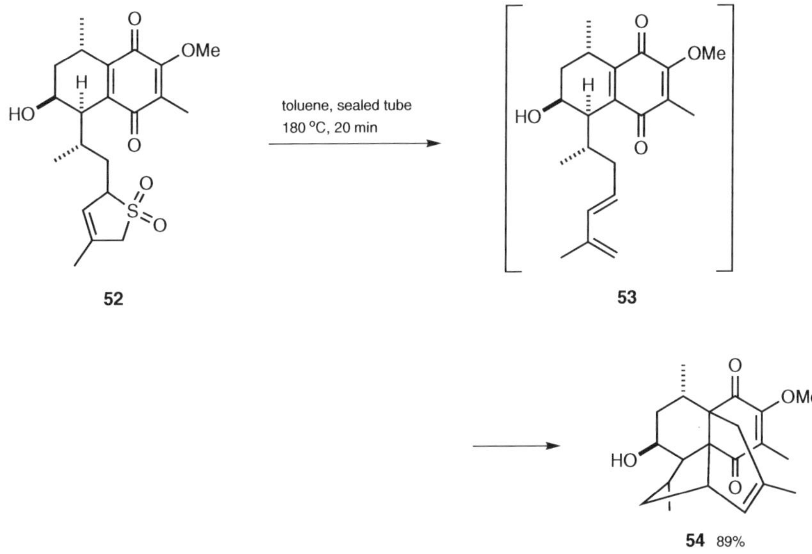

OMe HO H O O S O O **52** → toluene, sealed tube 180 °C, 20 min → [OMe HO H O O] **53** → OMe HO O O **54** 89%

for references see p 44

2.1.1.1.1.2.3 Through Sigmatropic Reactions

The rearrangement of σ-bonds across a π-framework presents an exciting opportunity to radically alter the connectivity of an intermediate in a controlled and predictable fashion. As with other pericyclic reactions, the product of this process can be engaged in a Diels–Alder reaction, allowing for the rapid generation of molecular complexity.

The aromatic Claisen rearrangement, the first reported [3,3]-sigmatropic rearrangement (in 1912),[47] transforms an allyl aryl ether into an allyl dienone. It can be recognized that this fleeting intermediate contains an exceptionally reactive diene that could be engaged in a Diels–Alder cycloaddition, provided that aromatization can be outcompeted. Quillinan and Scheinmann postulated just such a tandem transformation in their proposed biosynthesis of the *Garcinia* natural products.[48] While an elegant study by Nicolaou[49] provided a firm foundation for this hypothesis in his group's route to *O*-methylforbesione, it was not until the work of Theodorakis[50] that a bona fide *Garcinia* compound, forbesione, was synthesized using this strategy. Heating intermediate **55** results in the formation of dienone intermediate **56**, which can then be engaged by a pendent dienophile to provide the caged structure of forbesione (**57**) and isomer **58** through a Diels–Alder event (Scheme 16). Aside from providing validation for the biosynthetic hypothesis, the evolution of the intricate three-dimensional structure of forbesione from the planar starting material **59** represents another powerful demonstration of the Diels–Alder reaction in domino processes.

Scheme 16 Rapid Assembly of the Caged Architecture of Forbesione by an Aryl Claisen/Diels–Alder Domino Sequence[50]

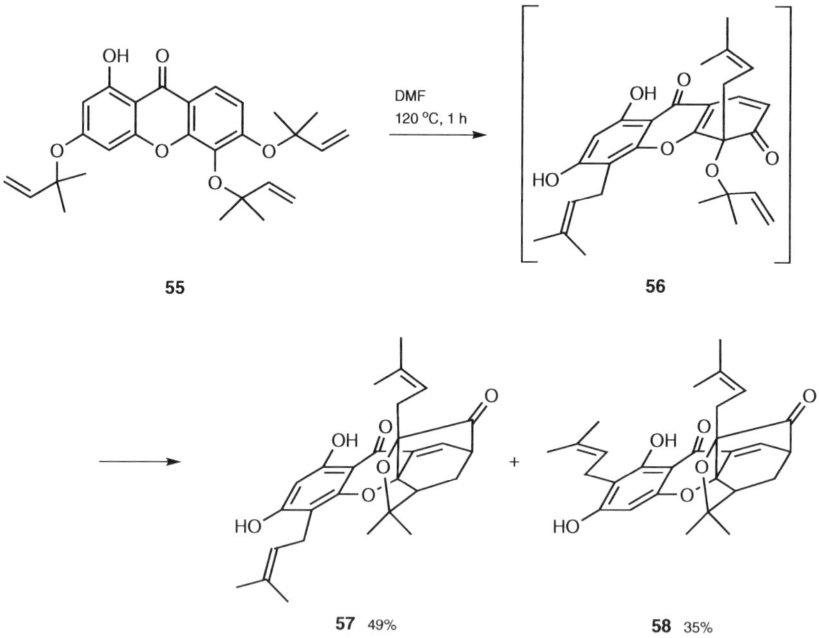

OH O
O O O
O
55
DMF
120 °C, 1 h
OH O
HO O O
O
56
OH O O O O HO
57 49%
+
OH O O O O HO
58 35%

Sigmatropic rearrangements provide another strategic factor for the in situ generation of dienes by virtue of the stereospecificity of their highly ordered transition states. Gibbs and Okamura[51,52] realized that asymmetry can be communicated in a pericyclic cascade through the conversion of axial into point chirality via a sigmatropic rearrangement. Sulfenate ester **60**, generated from chiral propargyl alcohol **59**, undergoes efficient [2,3]-rearrangement into transient allenyl sulfoxide **61**, the diene of which is smoothly captured by the pendent alkene to furnish tricycle **62**, a key intermediate in a synthesis of sterpurene (Scheme 17). This [2,3]-rearrangement/intramolecular Diels–Alder cascade provides a powerful method to translate stereochemical information. Critically, the sterpurene produced from **62** possessed an identical optical rotation to an authentic sample.

Scheme 17 Transfer of Axial to Point Chirality in a [2,3]-Rearrangement/Diels–Alder Cascade[51,52]

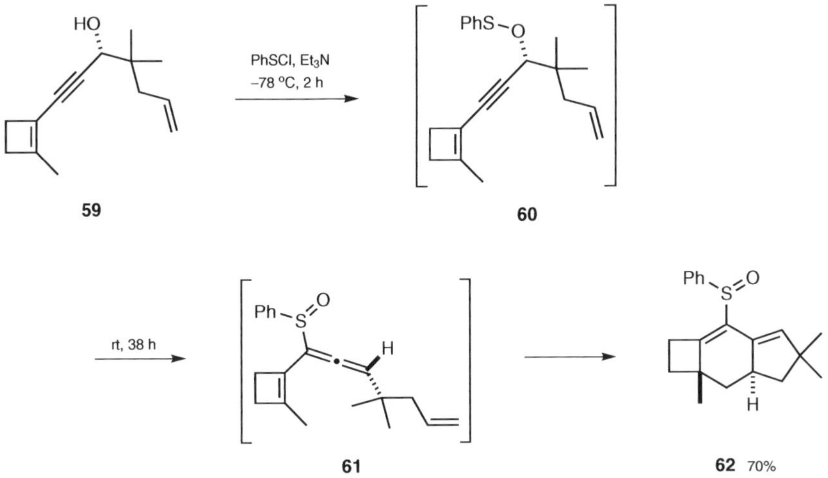

HO

PhSCl, Et_3N
−78 °C, 2 h

PhS−O

59 **60**

rt, 38 h

Ph−S=O

H

61 **62** 70%

(2aS,3aS)-2a,5,5-Trimethyl-7-(phenylsulfinyl)-2,2a,3,3a,4,5-hexahydro-1*H*-cyclobuta[*f*]indene (62):[52]
A 0.96 M soln of Cl_2 in CCl_4 (1.15 mL, 1.10 mmol) (**CAUTION:** *toxic*) was added to $(PhS)_2$ (240 mg, 1.10 mmol) in a 10-mL flask under N_2 at 0 °C. The resulting mixture was stirred for 10 min and then warmed to rt to give a 1.66 M orange-red soln of PhSCl (3.65 mL, 2.20 mmol).

A freshly prepared 1.66 M soln of PhSCl in CCl_4 (0.40 mL, 0.66 mmol, 1.2 equiv; prepared by the in situ method described above) was added to a stirred mixture of unsaturated alcohol **59** (113 mg, 0.55 mmol, 1 equiv) and Et_3N (0.19 mL, 1.39 mmol, 2.5 equiv; distilled from CaH) in CH_2Cl_2 (11.6 mL; distilled from CaH) under N_2 at −78 °C. After stirring for 2 h, the cooling bath was removed, and the mixture was stirred for 38 h at rt. The mixture was then worked up, and the product was obtained after purification by chromatography; yield: 120 mg (70%).

2.1.1.1.1.3 Photochemical Generation of a Diene

An alternative method for generating highly reactive *ortho*-xylylene (*ortho*-quinodimethane) dienes to the electrocyclic ring opening of benzocyclobutenes (see Section 2.1.1.1.1.2.1.1) has been developed by Pfau and co-workers.[53] Seeking a less thermally demanding route, it was found that photoexcitation of an aryl ketone possessing an *ortho* alkyl group could, reasonably, lead to a 1,5-hydrogen abstraction process to furnish an *ortho*-xylyl diradical which, through resonance, can be described as an *ortho*-xylylene. Performing this photolysis in the presence of an external dienophile leads to the smooth formation of the expected Diels–Alder adduct. This method is used to great effect in Nicolaou's[54] synthesis of hybocarpone. Here, 1,5-hydrogen abstraction from the *ortho* methyl group of the highly functionalized benzaldehyde **63** results in the production of ketyl–

for references see p 44

alkyl diradical **64**, a compound which can be described using *ortho*-xylylene resonance form **65**. This intermediate is trapped by a dienophilic acrylate to provide bicycle **66**, a useful building block in the synthesis of hybocarpone (Scheme 18).

Scheme 18 1,5-Hydrogen Abstraction as an Alternative Method To Generate *ortho*-Xylylenes in Nicolaou's Synthesis of Hybocarpone[54]

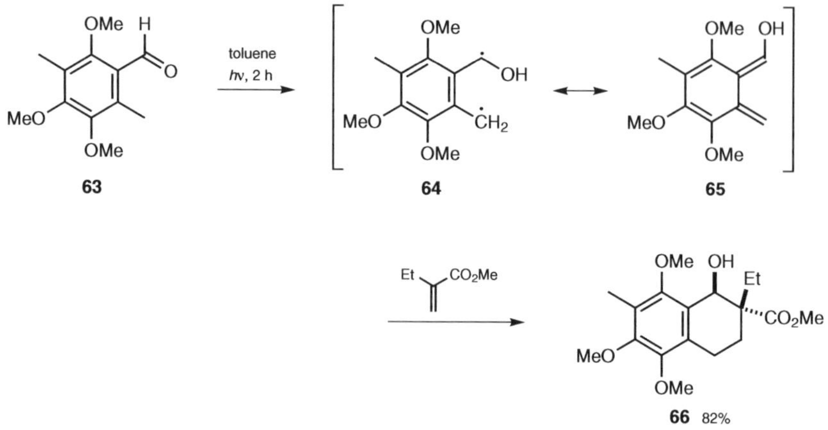

2.1.1.1.1.4 Metal-Mediated Generation of a Diene

Metal-mediated coupling reactions have taken an increasingly prominent place in the pantheon of useful synthetic transformations, a trend recently recognized by the 2010 Nobel Prize in Chemistry. The reasons for this attention are obvious: few other strategies allow for such versatility in the production of C—C bonds. Both cross coupling and oxidative coupling of fragments have been utilized to form dienes in Diels–Alder-containing domino processes.

The disconnection strategy utilized by Martin and co-workers in their synthesis of manzamine A involves a key intermolecular Diels–Alder reaction to assemble three of the four rings of the core subunit as intermediate **69**.[55] In the course of optimizing material flow to precursor **68**, it was recognized that the diene could be formed from a Stille cross coupling involving tributyl(vinyl)stannane and vinyl bromide **67**. Heating these two fragments in the presence of catalytic tetrakis(triphenylphosphine)palladium(0) leads to the smooth formation of the desired tricycle **69**, presumably through a Diels–Alder reaction of the in situ generated diene (Scheme 19).

Scheme 19 A Tandem Stille Coupling/Intramolecular Diels–Alder Sequence To Provide a Key Fragment of Manzamine A[55]

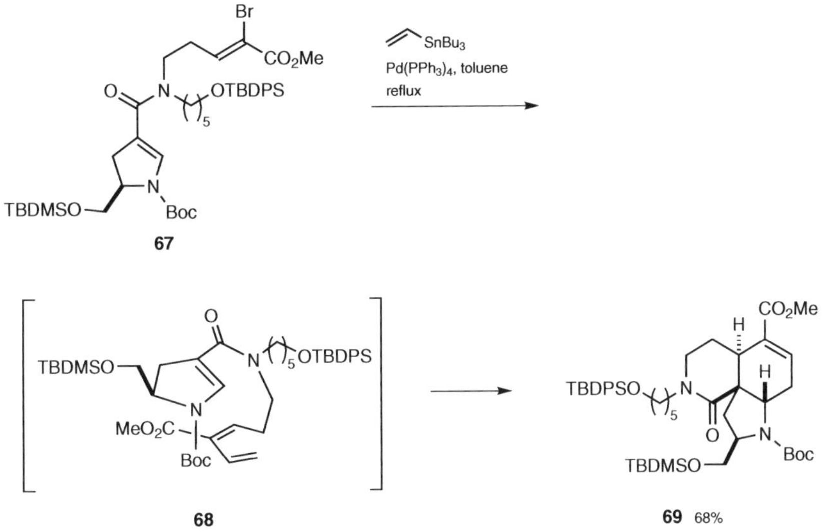

Oxidative coupling mediated by metals presents an opportunity to unite two fragments while also generating participants for a Diels–Alder reaction. The Chapman group envisioned accessing the hexacyclic natural product carpanone (**72**) through the oxidative dimerization of prop-1-enyl arene **70**.[56] Palladium-mediated oxidative coupling of two molecules of **70** results in the formation of bis(*ortho*-quinomethane) **71**, a highly reactive intermediate that can readily engage in an intramolecular Diels–Alder reaction to furnish carpanone (**72**) as the major product (Scheme 20).

Scheme 20 Palladium-Mediated Oxidative Dimerization of an Alkenic Phenol in an Exceptionally Direct Synthesis of Carpanone[56]

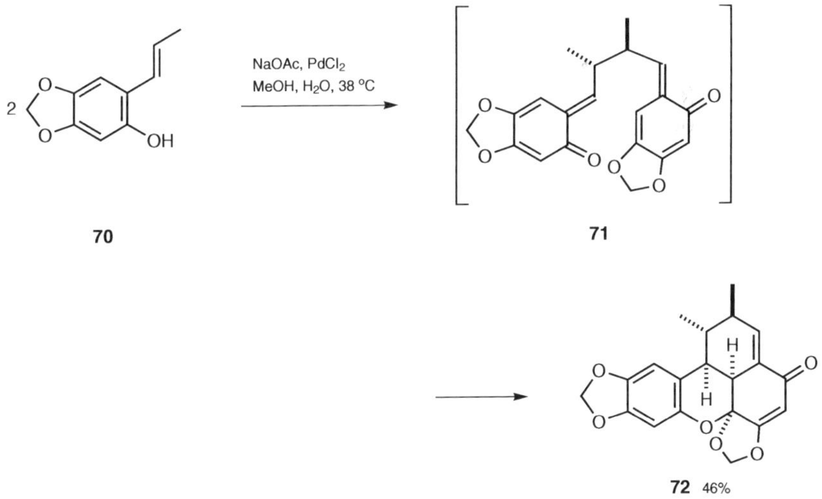

8-*tert*-Butyl 5-Methyl (4a*R*,7a*S*,9*R*,10a*S*)-9-[(*tert*-Butyldimethylsiloxy)methyl]-2-[5-(*tert*-butyldiphenylsiloxy)pentyl]-1-oxo-2,3,4,4a,7,7a,9,10-octahydropyrrolo[2,3-*i*]isoquinoline-5,8(1*H*)-dicarboxylate (69):[55]

A mixture of vinyl bromide **67** (27.1 g, 27.2 mmol, 1 equiv), tributyl(vinyl)stannane (9.49 g, 29.9 mmol, 1.1 equiv), and $Pd(PPh_3)_4$ (1.26 g, 1.09 mmol, 4 mol%) in freshly distilled toluene (270 mL; distilled from sodium/benzophenone) was heated under reflux for 30 h. The solvent was removed under reduced pressure at 40 °C and the residue was dissolved in Et_2O (100 mL) containing decolorizing carbon (5 g). The mixture was stirred for 20 min and filtered through Celite to give **69**; yield: 15.1 g (68%).

Carpanone (72):[56]

$PdCl_2$ (0.8 g, 4.5 mmol) was added to a rapidly stirred soln containing substrate **70** (1.6 g, 9.0 mmol) and NaOAc (6.0 g, 73 mmol) in MeOH (150 mL) and H_2O (30 mL). The soln was stirred for 2 h (38 °C) and then left to settle for 1 h at rt. After filtration and dilution with H_2O, the resultant suspension was extracted twice with Et_2O. The ethereal soln was washed with 10% aq NaOH and then H_2O, dried ($MgSO_4$), and then concentrated to give the crude product (1.0 g; 62%). CCl_4 (5 mL) (**CAUTION:** *toxic*) was added, and the product crystallized overnight. The crystals (0.98 g; mp 185 °C) thus obtained contained one molecule of CCl_4 per molecule of product **72** (X-ray crystal structure). Chromatography of the mother liquor gave additional **72**•CCl_4 (0.08 g); total yield of crystalline **72**: 0.733 g (46%).

2.1.1.1.2 Cascades Generating a Dienophile

While the extent to which the development of diene-generating reactions has been pursued is hinted at in the previous sections, it is fair to say that a significant amount of study has been devoted to the generation of their conjugates, dienophiles, in cascade processes. Some major advances in the production of dienophiles have come from exploring the same general strategies as have been applied to the generation of dienes, most notably ionic and pericyclic processes.

2.1.1.1.2.1 Ionic Generation of a Dienophile

As with the generation of dienes, many two-electron processes have been pursued to generate 2π partners in the Diels–Alder cycloaddition.

2.1.1.1.2.1.1 Through Himbert Cycloadditions

Himbert and Henn provided the first report of a thermal Diels–Alder reaction between anilines and tethered allenes in 1982.[57] This reaction is quite remarkable, as it not only uses a relatively unactivated arene as a diene, but also proceeds with no subsequent rearrangement to prevent reversibility. Despite some uncertainty as to the mechanism of the cycloaddition, recent computational and experimental study[58] has supported that this cyclization proceeds through a concerted [4+2]-Diels–Alder process (as opposed to a stepwise, radical pathway). The group of Vanderwal[59] found that treatment of skipped alkynamide starting materials, e.g. **73**, with base generates a reactive allene dienophile, e.g. **74**, which smoothly engages the pendent arene in the Himbert Diels–Alder cyclization, providing polycyclic lactam products such as **75** in fair to excellent yields (Scheme 21).

Scheme 21 A Complex Lactam Product from a Simple, Skipped Alkyne Precursor via the Himbert Arene/Allene Cycloaddition[59]

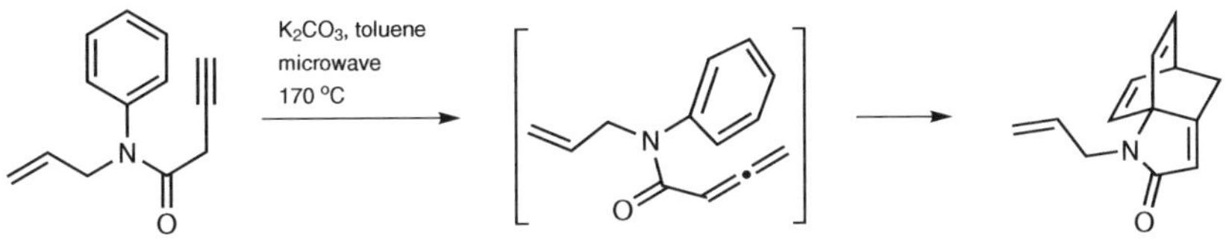

K_2CO_3, toluene
microwave
170 °C

73 74 75 74%

1-Allyl-4,5-dihydro-5,7a-ethenoindol-2(1*H*)-one 75; Typical Procedure:[59]
To a soln of amide **73** (2.00 g, 10.0 mmol, 1 equiv) in toluene (50 mL) was added K_2CO_3 (0.69 g, 5.00 mmol, 0.50 equiv) and the mixture was heated under microwave irradiation at 170 °C for 8 h. After cooling to rt, the solvent was evaporated. Chromatographic purification (silica gel, hexanes/EtOAc 2:1) of the residue afforded the cycloadduct **75** as a yellow solid; yield: 1.48 g (74%).

2.1.1.1.2.1.2 Through Benzyne Formation

Since their correct identification by Roberts in 1953,[60] benzynes have enjoyed a privileged position in the execution of concise molecule synthesis. Benzynes can be regarded as exceptionally labile intermediates and, as a result, are almost universally generated in situ, making them well suited to application in domino processes. One quality that has made these reactive intermediates so attractive is their ability to participate in Diels–Alder and/or formal Diels–Alder reactions as the 2π component, although it appears that a stepwise cycloaddition, as opposed to the concerted pericyclic pathway, may be the operative mechanism in some cases. Due to this ambiguity, this strategy will not be extensively treated in this chapter; however, the interested reader is directed to an excellent review on the use of benzynes in synthesis.[61]

An early, ingenious use of the benzyne/Diels–Alder disconnection can be found in the method developed in the Townsend synthesis of the aflatoxin biosynthetic precursor averufin.[62] Treatment of aryl lactone **76** with lithium tetramethylpiperidide at low temperature leads to the formation of the deprotonated species that can be represented as *ortho*-xylylene **77** (Scheme 22). Addition of benzyne precursor **78** to this basic mixture results in the highly regioselective formation of Diels–Alder adduct **79** which, upon workup, provides the desired asymmetrical anthraquinone **80**. This maneuver is particularly notable as Townsend and co-workers not only generate both the diene and dienophile in situ, but also make use of the decomposition of the Diels–Alder adduct to further advance their synthesis.

for references see p 44

Scheme 22 Regioselective Assembly of an Anthraquinone by Union of an In Situ Generated Diene and Dienophile[62]

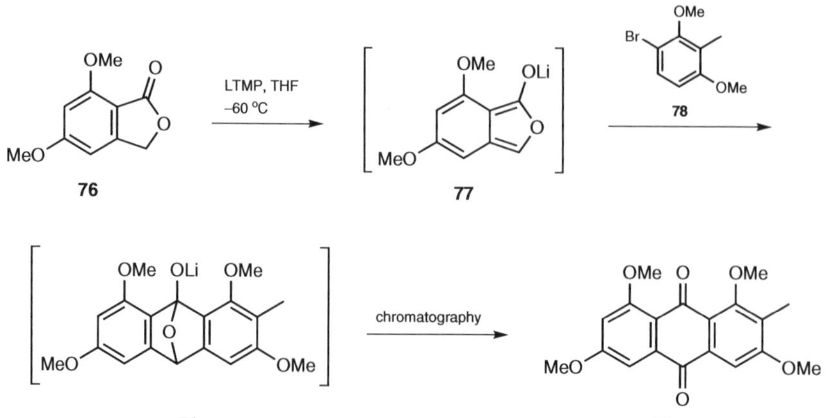

1,3,6,8-Tetramethoxy-2-methylanthracene-9,10-dione (80):[62]
2,2,6,6-Tetramethylpiperidine (135 μL, 113 mg, 0.80 mmol, 3.1 equiv) in THF (0.5 mL) was treated at −60 °C (CH_2Cl_2/dry ice bath) with a 1.4 M soln of BuLi in hexane (600 μL, 0.84 mmol, 3.3 equiv). After being stirred for 10 min, the soln was treated with 5,7-dimethoxybenzo[*c*]furan-1(3*H*)-one (**76**; 50.0 mg, 0.258 mmol, 1 equiv) in dry THF (4.5 mL). After 20 min, the orange soln was warmed to −40 °C (MeCN/dry ice bath) and a soln of 1-bromo-2,4-dimethoxy-3-methylbenzene (**78**; 125.9 mg, 0.545 mmol, 2.1 equiv) in THF (2 mL) was added. After being stirred for 15 min, the mixture was allowed to warm gradually to rt, the color becoming dark red to purple. After 1 h, the mixture was stirred open to the air for at least 4 h. Addition of H_2O (30 mL) and extraction of the resulting mixture with $CHCl_3$ (3 × 8 mL), washing of the combined organic extracts with brine, drying ($MgSO_4$), and removal of the solvents under reduced pressure gave the crude anthraquinone **80**. Purification of the yellow solid by column chromatography [silica gel (3.0 g), EtOAc/hexanes 1:1] gave the product; yield: 60.0 mg (66%).

2.1.1.1.2.1.3 Through Wessely Oxidation

The dimerization of in situ generated quinone acetals through a Diels–Alder process has been a popular area of inquiry. As this topic involves, by definition, the production of both diene and dienophile, it has already been treated in the section for diene generation (see Section 2.1.1.1.1.1.1). There are, however, several examples of the Wessely reaction where a dienophile is the product of the arene oxidation.

An interesting iteration of the tandem aromatic oxidation/Diels–Alder cascade has come from the group of Ciufolini in their efforts toward the polycyclic alkaloid himandrine.[63] It was found that treatment of aryl dienylsulfonamide **81** with (diacetoxyiodo)-benzene provides spirocycle **82** through an oxidative amidation (Scheme 23). Heating this intermediate with the addition of toluene provides a mixture of *cis*- and *trans*-decalone products **83A** and **83B**, presumably through an intramolecular Diels–Alder reaction followed by epimerization of the acidic bridgehead center α to the ketone.

Scheme 23 An Oxidative Amidation/Diels–Alder/Epimerization Cascade To Provide Decalin Sulfonamides from Simple Precursors[63]

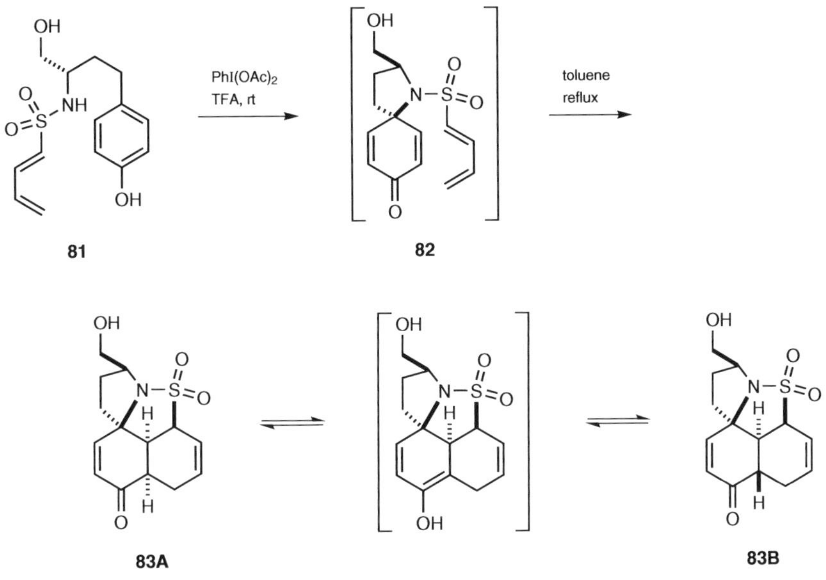

Taking inspiration from the insightful biosynthetic proposal of Schmitz,[64] Shair and coworkers have devised a route to the heptacyclic natural product longithorone A involving a transannular intramolecular Diels–Alder reaction between a quinone dienophile and a diene with an exocyclic double bond.[65] Treatment of cyclophane intermediate **84** with iodosylbenzene in a mixture of water and acetonitrile leads to the oxidation of both aromatic moieties with concomitant engagement of the desired quinone of oxidized intermediate **85** in an intramolecular Diels–Alder process (Scheme 24). This daring cascade sequence not only proceeds in excellent yield (90%), but also serves as the final step in the synthesis to furnish longithorone A (**86**) directly.

for references see p 44

Scheme 24 A Daring, Bioinspired Synthesis of Longithorone A with an Oxidation/Transannular Diels–Alder Cascade Deferred to the Final Step[65]

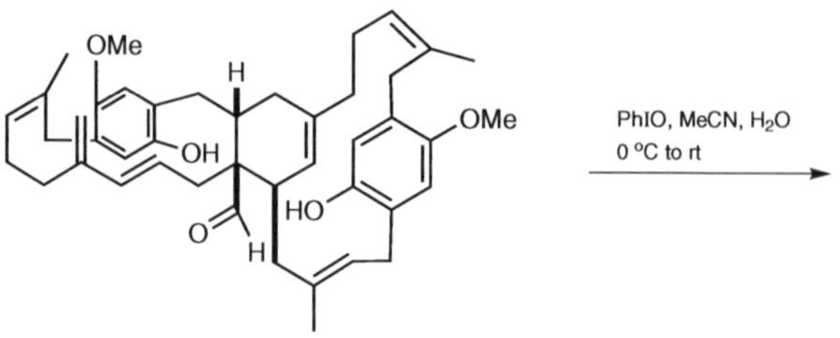

PhIO, MeCN, H_2O
0 °C to rt

84

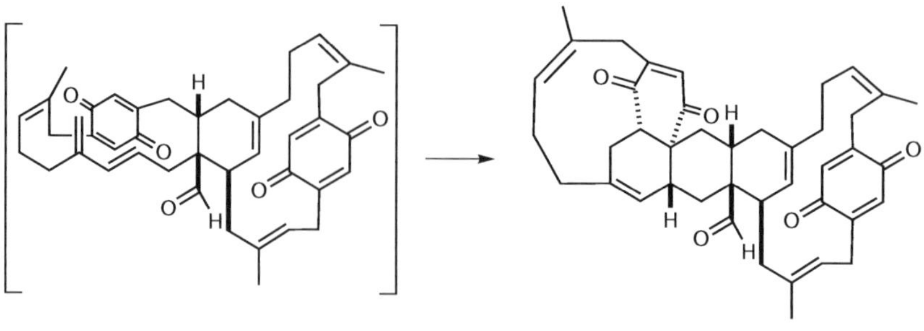

85

86 90%

(3aS,3a¹S,6S,8aR,11aR)- and (3aS,3a¹S,6S,8aR,11aS)-6-(Hydroxymethyl)-1,3a,3a¹,7,8,11a-hexahydro-6H,11H-naphtho[1,8-*cd*]pyrrolo[1,2-*b*]isothiazol-11-one 4,4-Dioxide (83A **and** 83B)**:**[63]

$PhI(OAc)_2$ (69.5 mg, 0.216 mmol, 1.2 equiv) was added slowly to a TFA (0.9 mL, 0.2 M) soln of phenolic compound **81** (54.0 mg, 0.180 mmol, 1 equiv) at rt. Toluene (2 mL) was added and the resulting mixture was heated to reflux overnight. Upon completion, the mixture was concentrated to dryness under reduced pressure. Chromatography (EtOAc/hexanes 3:1) of the residue gave an 8:1 mixture of diastereomers **83B** and **83A**; yield: 17.0 mg (32%).

Longithorone A (86)**:**[65]

To a soln of the bis(*tert*-butyldimethylsilyl)-protected derivative of **84** (4.2 mg, 0.0047 mmol, 1.0 equiv) in THF (0.5 mL) at 0 °C was added a 1.0 M soln of TBAF in THF (0.011 mL, 0.011 mmol, 2.2 equiv) with the clear, colorless soln turning yellow. After stirring for 15 min at 0 °C, the reaction was quenched with sat. aq $NaHCO_3$, and the mixture was warmed to rt, extracted with Et_2O, dried ($MgSO_4$), filtered, and concentrated to give the crude product **84**, which was used without further purification.

To a soln of crude bisphenol **84** (0.0047 mmol, theoretical; 1 equiv) in MeCN/H_2O (3:1; 0.5 mL) at 0 °C was added PhIO (7.0 mg, 0.032 mmol, 6 equiv) to give a canary yellow suspension, which was warmed to rt. After 10 min, TLC analysis showed no starting material and no product but another UV-active spot [R_f 0.34 (EtOAc/hexanes 1:3)], presumably the bisquinone **85**. After 40 min of stirring at rt, the reaction was quenched with sat. aq $NaHCO_3$ and extracted with CH_2Cl_2, and the extracts were dried ($MgSO_4$), filtered, and concentrated. ^1H NMR analysis of the crude mixture showed a trace of product **86**, along with two other aldehydes. The crude material was taken up in benzene-d_6 (1 mL) (**CAUTION:** *carcinogen*) and allowed to stand for 40 h, at which point ^1H NMR analysis revealed **86** to be the major aldehyde product. Purification by Et_3N-doped silica gel chromatography (EtOAc/hexane 1:19 to 2:18) afforded a pale yellow crystalline solid; yield: 2.6 mg (90%, over two steps).

2.1.1.1.2.2 Pericyclic Generation of a Dienophile

Pericyclic reactions are able to generate dienophiles in much the same fashion that they are able to generate dienes. Examples of such processes using retrocycloaddition and sigmatropic rearrangement are discussed in the following sections.

2.1.1.1.2.2.1 Through Cycloaddition/Retrocycloaddition

Cyclopentadiene has permeated the literature as a competent Diels–Alder partner since the first disclosure of the reaction in the original Diels–Alder report in 1928.[1] Grieco employed this classic diene in his synthesis of pseudotabersonine (already discussed for a subsequent cascade; see Section 2.1.1.1.1.1.4) to mask a reactive iminomethane dienophile.[30] Heating azanorbornene **87** in toluene in the presence of boron trifluoride–diethyl ether complex unveils dienophile **88**, which then engages the attached diene to provide a 1:1.5 mixture of spirocycles **89A** and **89B** (Scheme 25).

Scheme 25 Generation of a Reactive Imine Dienophile through Retrocycloaddition of an Azanorbornene[30]

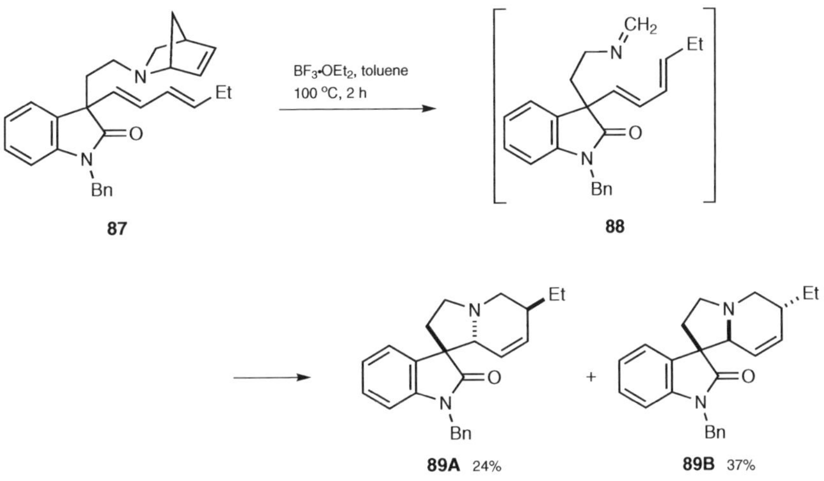

$BF_3•OEt_2$, toluene, 100 °C, 2 h

87 → [**88**] → **89A** 24% + **89B** 37%

2.1.1.1.2.2.2 Through Sigmatropic Rearrangement

The Claisen rearrangement, a [3,3]-sigmatropic rearrangement, allows for unsaturation to be transposed across a cyclic array via a well-defined transition state. This predictability allows for the strategic communication of stereochemical information during the transformation, the product of which can be envisioned serving as the dienophile of a subsequent Diels–Alder reaction.

A possible biogenesis of the *Thapsia garganica* transtaganolide terpenes, as proposed by Sterner and collaborators,[66] involves a rearrangement/intramolecular Diels–Alder cascade to furnish the caged architecture of the natural products. Toward a bioinspired synthesis, both Johansson[67] and Stoltz[68] have pursued an Ireland–Claisen rearrangement/intramolecular Diels–Alder strategy; this ultimately led to a total synthesis of transtaganolides A–D by Nelson and Stoltz.[69] Treating monocyclic 2-pyrone intermediate **90** with *N,O*-bis(trimethylsilyl)acetamide and catalytic triethylamine leads to the formation of tricycles **91A** and **91B** (through an alternative Claisen transition state to that illustrated) through a tandem Ireland–Claisen rearrangement/Diels–Alder cascade (Scheme 26). An

for references see p 44

impressive aspect of this domino process is that the relatively planar substrate **90** is able to assemble into topologically complex products with high selectivity [dr 3:1; 90% ee for (major) **91A** and 81% ee for (minor) **91B**].

Scheme 26 An Ireland–Claisen/Intramolecular Diels–Alder Cascade To Provide the Cage-Like Core of Transtaganolides A and B[69]

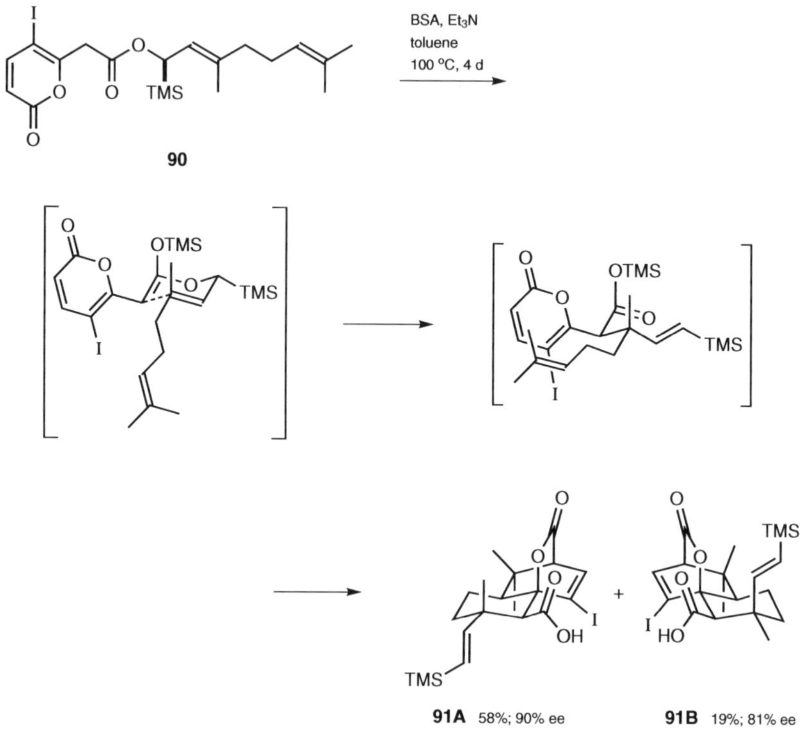

BSA, Et_3N
toluene
100 °C, 4 d

90

91A 58%; 90% ee

91B 19%; 81% ee

(2*S*,4a*R*,5*S*,6*R*,8a*S*)- and (2*R*,4a*S*,5*R*,6*R*,8a*R*)-4-Iodo-1,1,6-trimethyl-10-oxo-6-[(*E*)-2-(trimethylsilyl)vinyl]-1,5,6,7,8,8a-hexahydro-2*H*-4a,2-(epoxymethano)naphthalene-5-carboxylic Acids (91A and 91B); Typical Procedure:[69]

To a soln of pyrone ester **90** (384 mg, 0.79 mmol, 1 equiv) in toluene (4 mL), in a 500-mL sealed tube at 23 °C, was added BSA (384 μL, 1.58 mmol, 2 equiv) and Et_3N (11 μL, 0.08 mmol, 0.1 equiv). The mixture was heated to 110 °C and stirred for 20 min. The soln was then cooled to 23 °C and diluted with toluene (450 mL), leaving ample headspace in the sealed tube to allow for solvent expansion. The mixture was then reheated to 100 °C and stirred for 4 d until the reaction was complete, as determined by NMR analysis. The mixture was cooled to 23 °C, and then 0.02% aq HCl (20 mL) was added and the mixture was stirred vigorously for 1 min. The organic phase was separated and washed with 0.02% aq HCl (3 × 25 mL), making sure that the aqueous phase remained acidic. The aqueous phases were then combined and back-extracted with EtOAc (3 × 30 mL), and all organic phases were combined, dried (Na_2SO_4), and concentrated by rotary evaporation. The crude oil was purified by column chromatography (silica gel, Et_2O/hexanes 1:9 to 2:8 containing 0.1% AcOH) to yield a mixture of **91A** and **91B** as a white solid; yield: 294 mg (77%); ratio (**91A**/**91B**) 3:1.

2.1.1.1.2.2.3 Through Electrocyclization

The synthesis of torreyanic acid by the group of Porco[41] involves the Diels–Alder dimerization of an in situ generated dihydropyran. Because this domino process, by necessity, involves the generation of both diene and dienophile, it has already been discussed under the pericyclic generation of dienes (see Section 2.1.1.1.1.2.1.2).

2.1.1.1.3 Proximity-Induced Diels–Alder Reactions

The successful execution of the Diels–Alder reaction hinges on adherence to strict orbital overlap requirements between the diene and dienophile. Stemming from this restriction is the opportunity to control the Diels–Alder reaction through spatial means, providing a method to trigger a Diels–Alder event at a predetermined time through altering the relative position of the two partners.

The bold synthesis of dynemicin A[70] by the Schreiber group is a shining example of the marked effect of conformational change on the execution of Diels–Alder cycloadditions in conjunction with the great value that such an effect can have on the process. In this case, subjecting acid alcohol **92** to the macrocylization conditions of Yamaguchi[71] leads to the formation of reactive intermediate **93**, a compound that places the Diels–Alder components in a highly favorable conformation for a transannular cycloaddition to furnish the desired product **94** (Scheme 27). A Sonogashira-based cyclization that assembled the same Diels–Alder substrate was also successful, albeit with diminished efficiency in relation to the Yamaguchi approach.

Scheme 27 A Yamaguchi Macrolactonization of a Highly Unsaturated Intermediate To Trigger a Proximity-Induced Intramolecular Diels–Alder Reaction[70]

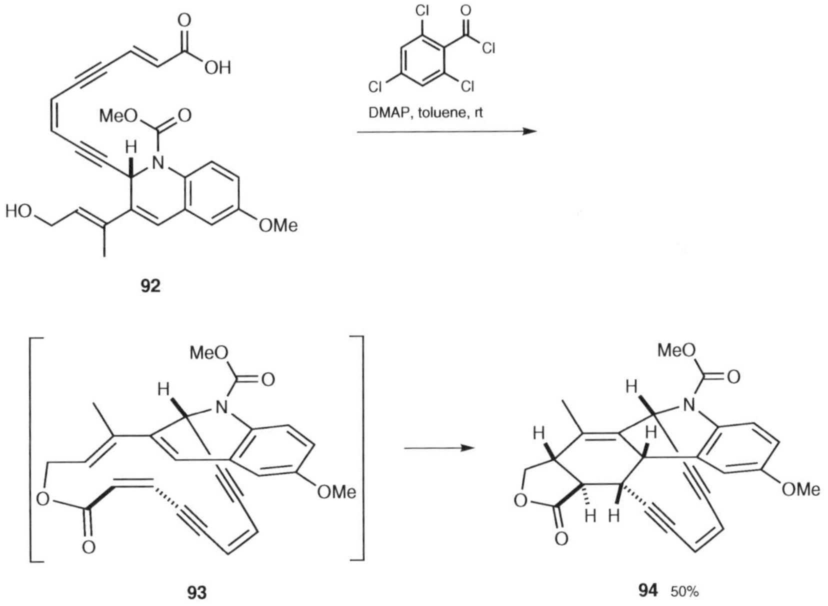

O
OH
MeO
O
H
N
HO
OMe
92

Cl O
Cl
Cl Cl
DMAP, toluene, rt

MeO
O
H
N
O
O
OMe
93

MeO
O
H
N
H
H
O
O
H H
OMe
94 50%

In Sorensen and co-workers' route to hirsutellone B,[72] like the Nicolaou synthesis of the same molecule (i.e., generation of a diene through allylation; see Section 2.1.1.1.1.1.5),[35] it was recognized that the sophisticated decahydrofluorene subunit of the target can be accessed from a Diels–Alder union of a trienic diene with its pendent α,β-unsaturated ketone. Unlike the Nicolaou synthesis, however, this event was envisioned by the Sorensen

for references see p 44

team to take place during a domino process whereby the unique architecture of the target would be assembled from a trisubstituted cyclohexane starting material. As shown in Scheme 28, thermalizing dioxinone **95** in benzene leads to the production of cyclophane product **97** through a cascade process whereby: (1) a cycloreversion of the dioxinone furnishes a high-energy α-oxo ketene intermediate **96**; (2) the ketene functionality engages the pendent amine to furnish the desired cyclophane; and (3) the unveiled α,β-unsaturated ketone undergoes a Diels–Alder cycloaddition with the pendent triene.

Scheme 28 An Acyl Ketene Addition/Intramolecular Diels–Alder Double Cyclization To Rapidly Access the Unique Architecture of Hirsutellone B[72]

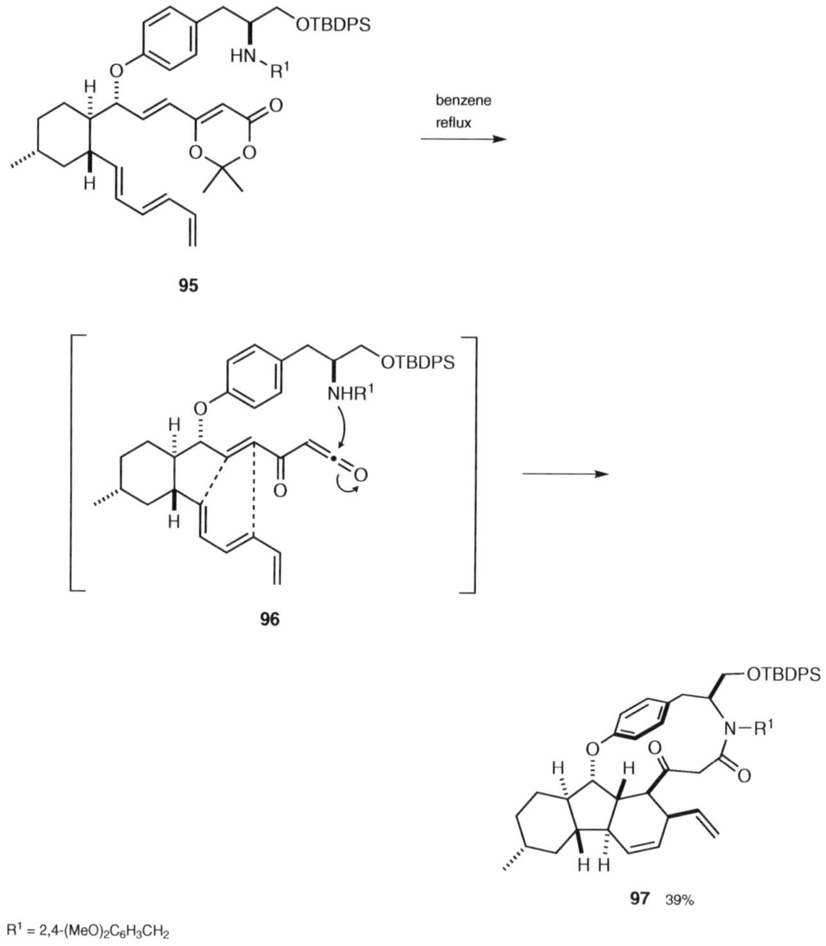

benzene reflux

95 → [**96**] → **97** 39%

$R^1 = 2,4\text{-}(MeO)_2C_6H_3CH_2$

Macrolactam 97:[72]

A two-necked, 500-mL flask equipped with a magnetic stirrer bar, Dean–Stark trap, and a reflux condenser was flame-dried under high vacuum and cooled to rt under argon. A soln of triene **95** (93.8 mg, 0.11 mmol) in freshly distilled benzene (350 mL; distilled from sodium/benzophenone) (**CAUTION:** *carcinogen*) was added, and the mixture was heated to reflux in an oil bath (105 °C) overnight. After cooling to rt, TLC analysis (hexanes/EtOAc 2:1; UV/4-methoxybenzaldehyde) revealed complete consumption of the starting material and formation of the product (R_f 0.63). The solvent was removed under reduced pressure, and the residue was purified by column chromatography (hexanes/EtOAc 3:1 to 2:1 to 1:1) to give macrolactam **97** as a colorless oil; yield: 34.1 mg (39%).

2.1.1.2 Diels–Alder as the Initiator of a Cascade

While the previous examples in this chapter have focused on the participation of the Diels–Alder reaction in cascades initiated by various chemical reactions, it is by no means limited to this role. Indeed, a great wealth of literature has focused on the use of the Diels–Alder reaction as an initiator of domino reaction sequences, a small selection of which follows.

2.1.1.2.1 Pericyclic Reactions Occurring in the Wake of a Diels–Alder Reaction

Pericyclic cascades can readily be initiated by the Diels–Alder cycloaddition. Two common classes of Diels–Alder-initiated domino sequences are Diels–Alder/Diels–Alder and Diels–Alder/retro-Diels–Alder strategies. Diels–Alder reactions with a concomitant group transfer are a recent addition to this class of transformations.

2.1.1.2.1.1 Cascades Featuring Diels–Alder/Diels–Alder Processes

The heptacyclic skeleton of bolivianine (**101**) presents an imposing challenge to target-oriented synthesis. The original isolation paper[73] proposes that this structure arises from an intramolecular Diels–Alder event involving intermediate **100**, presumably arising from a geranylation of a tetracyclic precursor followed by an ene reaction. Liu and co-workers considered whether the product of the hypothetical ene reaction might also be accessed from an all-carbon, intermolecular Diels–Alder reaction between tetracycle **98** and skipped triene **99** (ocimene); a subsequent intramolecular hetero-Diels–Alder reaction would then terminate the cascade and complete the synthesis of bolivianine (**101**) (Scheme 29).[74] Heating the two putative intermediates in a sealed tube directly furnishes bolivianine in an impressive example of an intermolecular–intramolecular Diels–Alder cascade.

Scheme 29 Bioinspired Synthesis of Bolivianine via an Intermolecular/Intramolecular Diels–Alder Cascade[74]

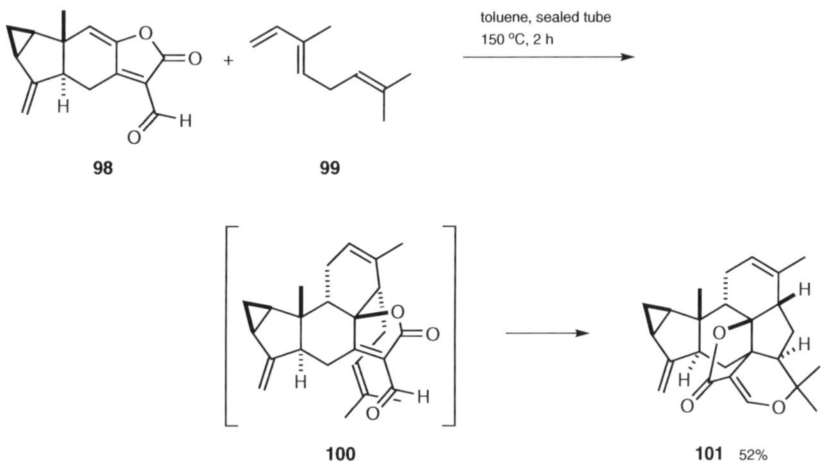

toluene, sealed tube
150 °C, 2 h

98 99 100 101 52%

An example of a twofold Diels–Alder event in a highly sophisticated context comes from the group of Roush[75] in the course of their creative synthesis of chlorothricolide. It was envisioned that an asymmetric intermolecular Diels–Alder reaction with chiral dienophile **103** could be carried out concurrently with an intramolecular Diels–Alder event of the proposed hexaenoate **102**. Heating hexaenoate **102** in the presence of 2 equivalents of

for references see p 44

chiral dienophile **103** allows for a tandem intermolecular–intramolecular Diels–Alder event to take place, furnishing advanced intermediate **104** and recyclable intramolecular Diels–Alder product **105** in yields of ca. 45 and 30%, respectively (Scheme 30). After one recycling of **105**, it was possible to obtain product **104** in a final yield of ca. 59%.

Scheme 30 Simultaneous Intramolecular and Intermolecular Diels–Alder Reactions in the Synthesis of Chlorothricolide[75]

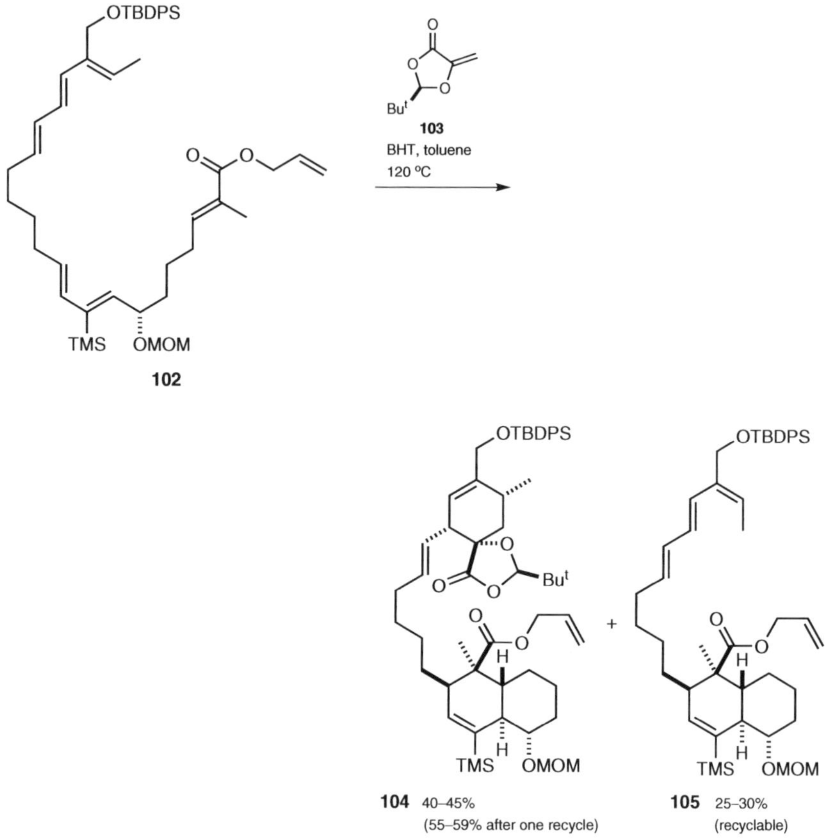

BHT = 2,6-di-*tert*-butyl-4-methylphenol

Bolivianine (101):[74]

A soln of compounds **98** (12.1 mg, 0.05 mmol, 1 equiv) and **99** (136.1 mg, 0.50 mmol, 10 equiv) in toluene (1 mL) was heated at 150 °C in a sealed tube for 2 h. After cooling to rt, the mixture was concentrated and purified by column chromatography (silica gel, petroleum ether/EtOAc 20:1) to give a white solid; yield: 9.8 mg (52%).

Double Diels–Alder Product 104:[75]

A soln of hexaenoate **102** (410 mg, 0.52 mmol, 1 equiv) in degassed toluene (350 μL, 1.5 M) was added to a resealable Carius tube that had been pre-silylated with BSA. Dienophile **103** (165 mg, 1.05 mmol, 2 equiv) and a crystal of 2,6-di-*tert*-butyl-4-methylphenol (radical inhibitor) were then added to the Carius tube, which was sealed and heated at 120 °C for 20 h. The mixture was concentrated under reduced pressure and the crude product was purified by flash column chromatography (silica gel, hexanes/Et_2O 7:1), producing the desired cycloadduct **104**; yield: 220 mg (45%; ca. 90% isomeric purity); and a mixture of isomeric double Diels–Alder adducts (91 mg; 19% yield) and adduct **105**; yield: 107 mg (26%).

2.1.1.2.1.2 Cascades Featuring Diels–Alder/Retro-Diels–Alder Processes

Diels–Alder/retro-Diels–Alder domino processes provide the opportunity to drive forward the normally reversible Diels–Alder cycloaddition through the loss of a small molecule, commonly in the form of a gas. Several successful retrocycloaddition byproducts are known, among them carbon dioxide, molecular nitrogen, and acetonitrile.

Haouamine A presents an interesting challenge to synthetic chemists in the form of its strained phenolic cyclophane. Burns and Baran addressed this challenging structural element through the use of a Diels–Alder/retro-Diels–Alder strategy.[76] In this example, heating pyrone–alkyne intermediate **106** in a microwave reactor leads to the desired cyclophane **108** through the intermediacy of the transient adduct **107**, thought to mimic the conformation of the putative "bent benzene" in the natural product (Scheme 31). Treatment of this product with potassium carbonate provides haouamine A (**108**) in 21% yield (30% starting material also recovered).

Scheme 31 Use of a Diels–Alder/Retro-Diels–Alder Strategy To Access the Strained Cyclophane of Haouamine A[76]

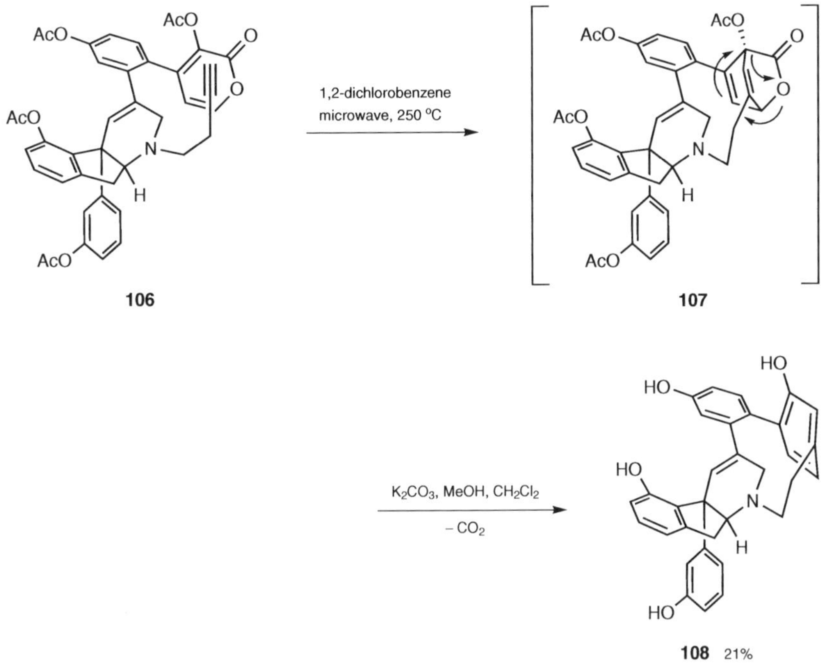

An approach that can be described as highly reliant on a Diels–Alder cascade sequence has been provided by the Boger group in their synthesis of the potent DNA cross-linking agent isochrysohermidin (Scheme 32).[77] The construction of the isochrysohermidin bis(1,5-dihydro-2*H*-pyrrol-2-one) architecture begins with a twofold hetero-Diels–Alder reaction of diene **109** with 2 equivalents of 1,2,4,5-tetrazine **110**. The resultant cycloadduct (**111**) can then undergo a double retro-hetero-Diels–Alder reaction with release of 2 equivalents of nitrogen (**111** to **112**) followed by 2 equivalents of methanol to provide bis(pyridazine) intermediate **113**. Overall, by utilizing four [4+2] operations in a single chemical step, the Boger group was able to elaborate a complex bis(pyridazine) intermediate from exceptionally simple, symmetric starting materials.

for references see p 44

Scheme 32 A Twofold Diels–Alder/Retro-Diels–Alder Sequence in the Synthesis of Isochrysohermidin[77]

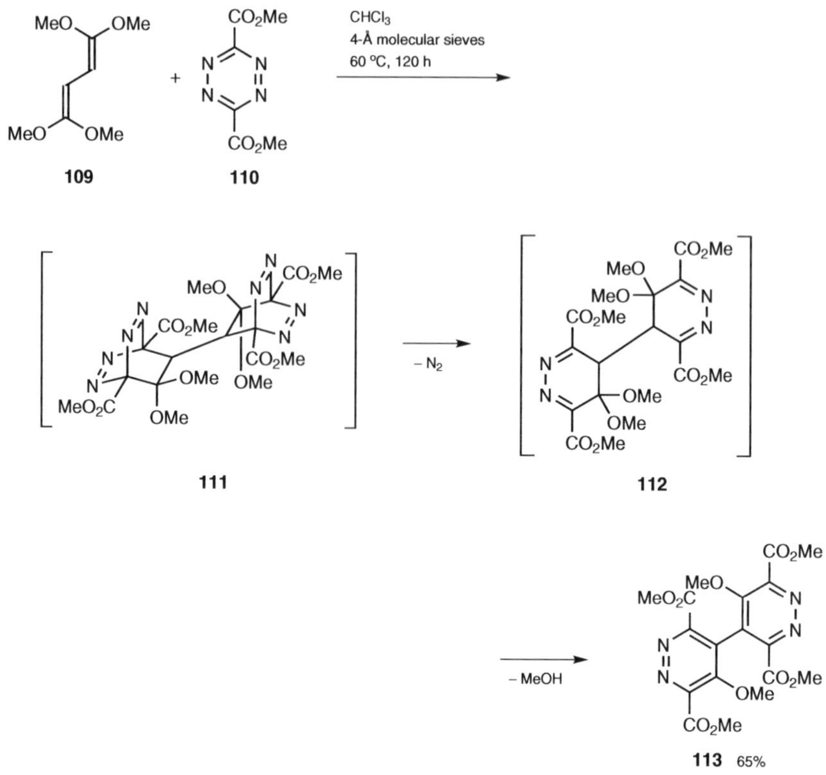

MeO
OMe
MeO
OMe
109
CO2Me
N
N
N
N
CO2Me
110
CHCl3
4-Å molecular sieves
60 °C, 120 h
111
– N2
112
– MeOH
113 65%

The byproducts of a Diels–Alder/retro-Diels–Alder sequence are not limited to nitrogen and carbon dioxide; indeed, an excellent example of an alternative may be drawn from research by the Jacobi group.[78] Thermolysis of oxazole–alkyne **114** in ethylbenzene produces the desired bicycle **116A** and its epimer **116B** (Scheme 33). This transformation proceeds through the intermediacy of bicyclic heterocycle **115**, which can undergo retrocycloaddition to liberate an equivalent of acetonitrile in addition to the ultimate product **116A**. It is notable that the undesired epimeric furanoketone **116B** can be epimerized through treatment with basic methanol, allowing for complete utilization of the cascade products.

Scheme 33 An Annulative Oxazole–Alkyne Diels–Alder/Retro-Diels–Alder Cascade in a Synthesis of Norsecurinine[78]

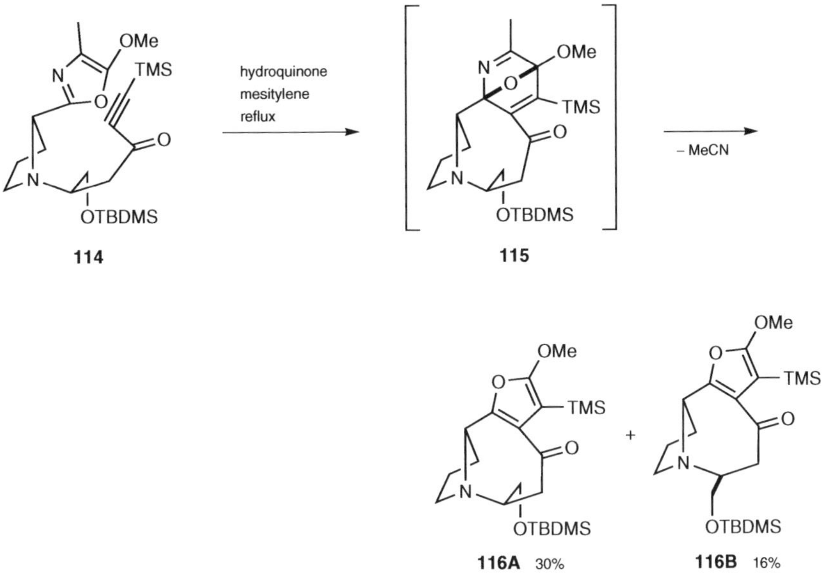

OMe
TMS
N
O
O
N
OTBDMS
114

hydroquinone
mesitylene
reflux

OMe
N
O
TMS
O
N
OTBDMS
115

– MeCN

OMe
O
TMS
O
N
OTBDMS
116A 30%

+

OMe
O
TMS
O
N
OTBDMS
116B 16%

Haouamine A (108):[76]
A portion of substrate **106** (5.0 mg, 0.0071 mmol, 1 equiv) was dissolved in 1,2-dichlorobenzene (7.1 mL) and 2,5-di-*tert*-butyl-4-methylphenol (12.1 mg, 0.055 mmol, 7.7 equiv) was added. The clear soln was then irradiated in a microwave at 250 °C for 10 h. After cooling to rt, the soln was passed through silica gel, eluting first with hexanes to remove 1,2-dichlorobenzene and subsequently with EtOAc. The crude mixture was concentrated and purified by preparatory TLC (silica gel, hexanes/EtOAc 2:1, run twice) to give recovered **106** (1.5 mg; 30%) as well as peracetyl haouamine A contaminated with its atropisomer (1.0 mg; 21% yield). This mixture was dissolved in $MeOH/CH_2Cl_2$ (1:1; 0.6 mL) and treated with K_2CO_3 (1.0 mg, 0.0072 mmol, 4.7 equiv). Stirring at rt was continued for 30 min, and 0.5 M aq KH_2PO_4/K_2PO_4 (1.0 mL; pH 7.0) was added. The aqueous layer was extracted with EtOAc (3 × 10 mL) and the combined organics were dried ($MgSO_4$), filtered, and concentrated to give **108**; yield: 0.7 mg (21%).

Tetramethyl 5,5′-Dimethoxy-[4,4′-bipyridazine]-3,3′,6,6′-tetracarboxylate (113):[77]
A soln of diene **109** (84 mg, 0.48 mmol, 1 equiv) in anhyd $CHCl_3$ (1.0 mL, 0.5 M) containing activated 4-Å molecular sieves (5 wt equiv) was stirred under argon for 5 min. The soln was treated with dimethyl 1,2,4,5-tetrazine-3,6-dicarboxylate (**110**; 428 mg, 2.16 mmol, 4.5 equiv) and stirred for 20 min at 22 °C under argon. After the vigorous evolution of N_2 subsided, the soln was stirred at 60 °C under argon for 5 d. The mixture was concentrated under reduced pressure. Chromatography [silica gel (4 × 12 cm), Et_2O] afforded a pale-yellow solid; yield: 140 mg (65%).

(6*R*,10a*R*)-6-[(*tert*-Butyldimethylsiloxy)methyl]-2-methoxy-3-(trimethylsilyl)-5,6,8,9,10,10a-hexahydro-4*H*-furo[2,3-*c*]pyrrolo[1,2-*a*]azepin-4-one (116A):[78]
A soln of ynone **114** (6.3 g, 13 mmol, 1 equiv) in degassed mesitylene (1200 mL) containing hydroquinone (68 mg, 0.6 mmol, 0.046 equiv) was heated at reflux for 30 min under N_2. The resulting brown soln was cooled to rt, concentrated under reduced pressure, and chromatographed (silica gel, Et_2O/hexanes 30:70) to afford **116A** and **116B** in 46% combined yield as yellow oils; yield (**116A**): 1.60 g (30%); yield (**116B**): 0.90 g (16%).

for references see p 44

2.1.1.2.1.3 [4+2] Cycloaddition with Subsequent Desaturation

A recent extension of the capabilities of the Diels–Alder reaction in the context of domino processes comes from the studies of the hexadehydro variant by Hoye and co-workers.[79] It was found that the cyclization of triyne intermediates of the general formulation **117** in cyclooctane as solvent results in the formation of arenes such as **119** along with a stoichiometric quantity of dehydrogenated solvent (cyclooctene) (Scheme 34). This process is believed to proceed through an initially formed benzyne intermediate which can then accept vicinal hydrogen atoms from a solvent molecule in a concerted, six-centered transition state **118** (analogous to the mechanism of diimide reduction).

Scheme 34 A Diels–Alder/Desaturation Cascade Involving a Hexadehydro-Diels–Alder Variant[79]

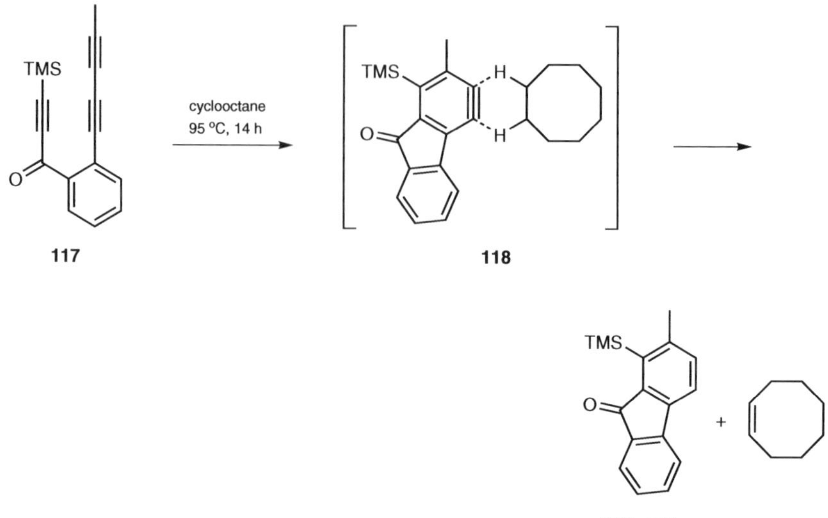

TMS
O
117
cyclooctane
95 °C, 14 h
TMS
O
H
H
118
TMS
O
+
119 97%

2-Methyl-1-(trimethylsilyl)-9*H*-fluoren-9-one (119); Typical Procedure:[79]
A typical double hydrogen-atom-transfer reaction comprised heating a soln of triyne precursor **117** (38 mg, 0.144 mmol, 1 equiv) in cyclooctane (14 mL, ca. 0.01 M) in a closed glass reaction vessel (for example, a screw-capped vial or culture tube). After heating at 95 °C for 14 h, the mixture was loaded directly onto a bed of silica gel and eluted first with hexanes to remove the excess cyclooctane, and then with EtOAc to capture the reduced benzenoid product **119**. The EtOAc fraction was concentrated and the residue was further purified by chromatography (silica gel, hexanes/EtOAc 19:1) to afford a yellow solid; yield: 37 mg (97%).

2.1.1.2.2 Diels–Alder Reactions with Concomitant Ionic Structural Rearrangements

As has been stated before, the large body of ionic reactivity is often compatible with the pericyclic nature of the Diels–Alder reaction. Indeed, it is notable that the same Lewis acid activation of some Diels–Alder reactions also promotes various intramolecular ionic reactions, presenting an opportunity to design highly sophisticated cascade transformations.

2.1.1.2.2.1 Pairings of Diels–Alder Reactions with Structural Fragmentations

Fragmentation reactions have long been an essential component of the synthetic organic chemist's toolkit, allowing for bonding frameworks to be predictably revised toward achieving a target.

A lesson in the stereochemical implications of a Diels–Alder cycloaddition, particularly on subsequent transformations in the course of a domino process, is provided by Aubé and co-workers in their synthesis of stenine (Scheme 35).[80] Subjecting triene azide **120** to methylaluminum dichloride leads to a mixture of tricyclic products **122A**, **122B**, and **123**. Stereoelectronic analysis shows that products **122A** and **123** both originate from an *endo*-Diels–Alder Process (cycloadduct **121A**), differing on whether the diazo substituent of the transient bicycle is equatorially (**122A**) or axially (**123**) disposed. Interestingly, only one product arising from the *exo*-Diels–Alder process (cycloadduct **121B**) is observed, as a Schmidt rearrangement (to give **122B**) is only feasible from an equatorially oriented diazo compound.

for references see p 44

Scheme 35 A Diels–Alder/Schmidt Rearrangement Cascade To Provide Tricyclic Lactam Products from a Linear Precursor[80]

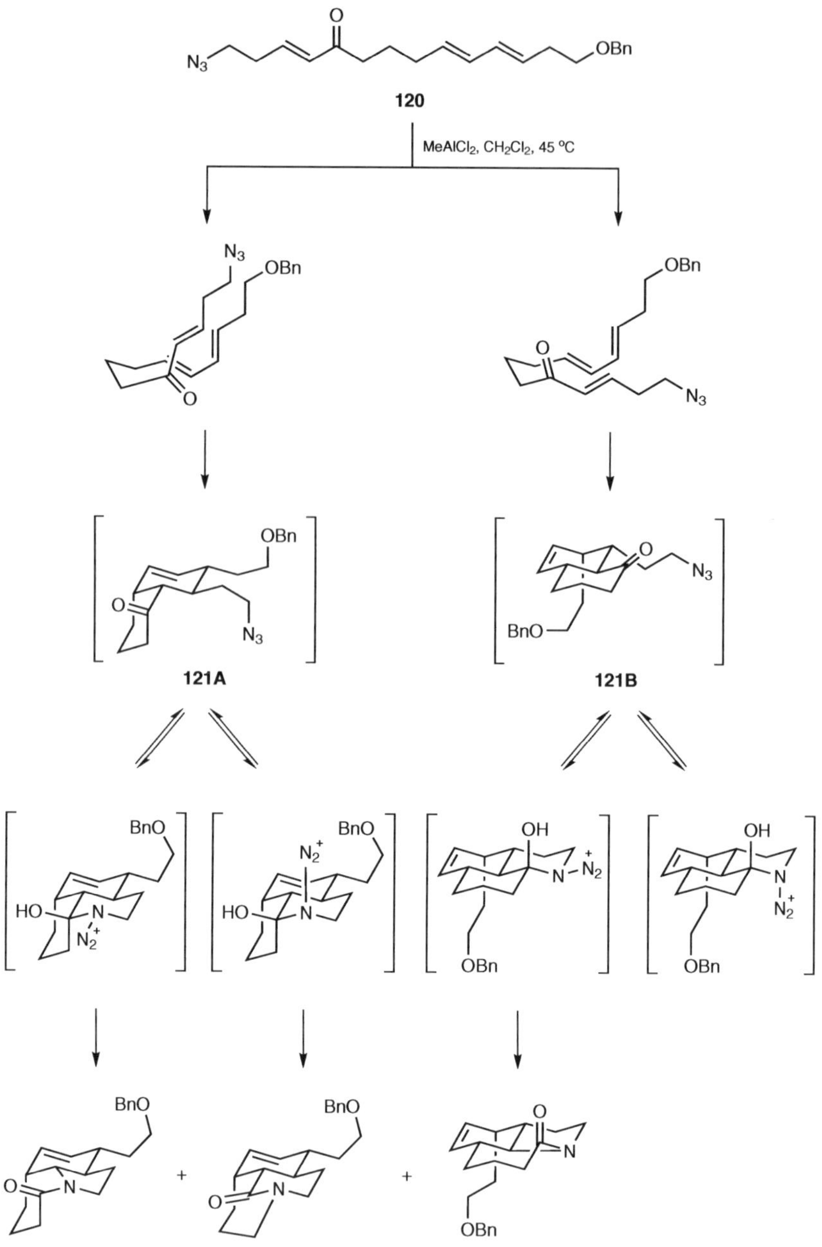

The challenge of assembling vicinal quaternary centers, as required for the synthesis of perophoramidine, served as inspiration for a creative application of a Diels–Alder/fragmentation cascade by Fuchs and Funk (Scheme 36).[81] It was envisioned that 3*H*-indole–oxindole **127** might be accessible through a hetero-Diels–Alder cycloaddition between 1*H*-indole **125** and transiently generated 2*H*-indol-2-one **124** followed by elimination of the resultant adduct (**126** to **127**). Thus, combination of **125** and 2*H*-indol-2-one **124** (generated from the corresponding 3-bromooxindole under the conditions of Kitagawa[82]) provides the desired product **127** in excellent yield. Further study of the reaction could not establish definitively whether it proceeds through the proposed Diels–Alder/elimination cascade or via a conjugate addition of 1*H*-indole **125** to 2*H*-indol-2-one **124**.

Scheme 36 Access to an Intermediate with Vicinal Quaternary Centers through an Ostensible Diels–Alder/Fragmentation Process[81]

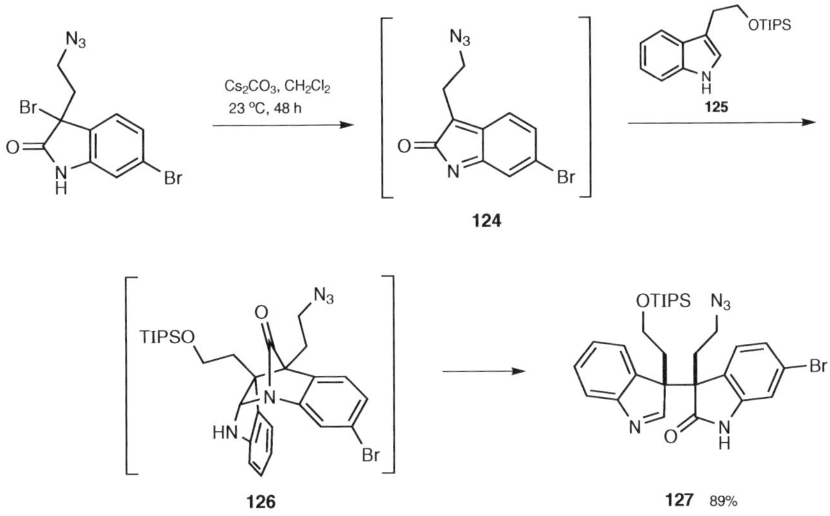

Cs_2CO_3, CH_2Cl_2, 23 °C, 48 h

124

125

126

127 89%

A rapid construction of the intricate 6,5,5 architecture of dendrobine has been achieved by Padwa and co-workers (Scheme 37).[83] Prior studies established the viability of a furanylcarbamate intramolecular Diels–Alder rearrangement to rapidly produce complex heterocyclic products bearing a striking resemblance to the core structure of dendrobine. When heated to 165 °C, furanylcarbamate **128** engages its pendent alkene in an intramolecular Diels–Alder reaction to provide rearranged, epimeric products **130**, presumably through the intermediacy of Diels–Alder adduct **129**, rupture of the oxacycle, and a subsequent semi-pinacol rearrangement.

Scheme 37 Use of a Simple Furanylcarbamate Intermediate To Provide a Complex Tricycle through a Diels–Alder/Fragmentation Reaction in a Synthesis of Dendrobine[83]

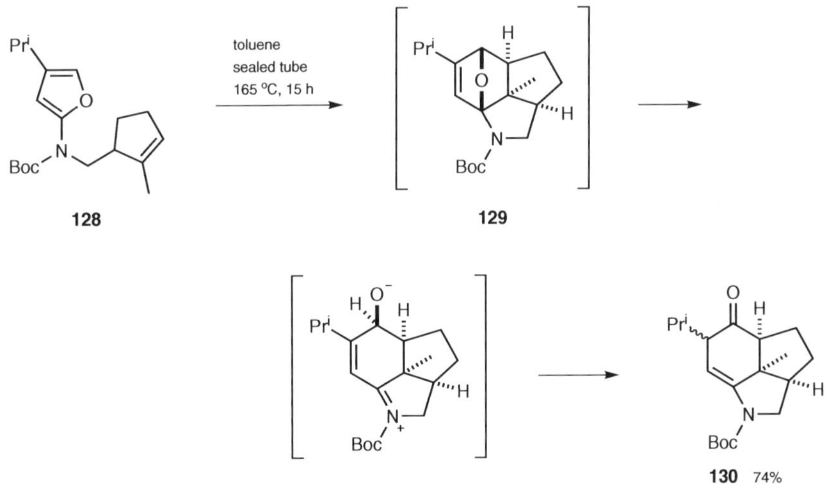

toluene, sealed tube, 165 °C, 15 h

128

129

130 74%

for references see p 44

Lactam Products 122A, 122B, **and** 123:[80]
To a flame-dried 50-mL flask was added azidotriene **120** (0.25 g, 0.71 mmol, 1 equiv) in CH_2Cl_2 (35 mL) and a 1.0 M soln of $MeAlCl_2$ in toluene (0.71 mL, 0.71 mmol, 1 equiv). Refluxing the yellow soln for 48 h produced a dark greenish soln, which was cooled to rt and then poured over sat. aq $NaHCO_3$ (100 mL). Upon shaking, the dark soln turned yellow. The mixture was extracted with EtOAc (3 × 100 mL) and the combined organic layers were dried (Na_2SO_4), filtered, and concentrated to afford a yellow oil. Flash chromatography (EtOAc) afforded the major lactam isomer **122A** as a viscous oil [yield: 100 mg (43%)], the bridged lactam **123** as a viscous oil [yield: 56 mg (24%)], and the minor lactam isomer **122B** as a white solid [yield: 27 mg (12%)].

(*S)-3-(2-Azidoethyl)-6-bromo-3-{(*R**)-3-[2-(triisopropylsiloxy)ethyl]-3*H*-indol-3-yl}indolin-2-one** (127):[81]
3-(2-Azidoethyl)-3,6-dibromoindolin-2-one (4.08 g, 11.34 mmol, 1 equiv) was dissolved in CH_2Cl_2 (58 mL). The resulting soln was then added via an addition funnel to a stirring soln of 3-[2-(triisopropylsiloxy)ethyl]-1*H*-indole (**125**; 5.39 g, 17.02 mmol, 1.5 equiv) and Cs_2CO_3 (12.94 g, 39.70 mmol) in CH_2Cl_2 (58 mL) at rt over a period of 30 min. The soln was stirred for 48 h at rt. The mixture was then filtered through a pad of Celite, washed thoroughly with CH_2Cl_2, and concentrated. The crude material was purified by column chromatography (silica gel, EtOAc/hexanes 1:3) to give a yellow foam; yield: 6.00 g (89%).

(2a*R,2a^1*R**,4a*S**)-1-(*tert*-Butoxycarbonyl)-6-isopropyl-2a^1-methyl-1,2,2a,2a^1,3,4,4a,6-octahydro-5*H*-cyclopenta[*cd*]indol-5-one** (130):[83]
A sample of carbamate **128** (1.0 g, 3.1 mmol) in toluene (10 mL) was heated in a sealed tube under argon at 165 °C for 15 h. After cooling to rt, the soln was concentrated under reduced pressure and the residue was purified by flash chromatography (silica gel) to a pale yellow oil consisting of a 2:1 mixture of diastereomers that proved to be inseparable; yield: 0.74 g (74%).

2.1.1.2.2.2 Combining a Diels–Alder Reaction with Ionic Cyclization

The Diels–Alder reaction can be followed by ionic cyclization in some cases. This ability is most pronounced in Diels–Alder reactions catalyzed by Lewis acids, as the electrophilic activation lends itself to subsequent attack by an appended nucleophile.

A Lewis acid catalyzed Diels–Alder reaction between tricyclic quinone **131** and doubly skipped triene **132** begins an impressive cascade toward the synthesis of perovskone by the Majetich lab (Scheme 38).[84,85] Subsequent isomerization of the pendent alkene followed by an intramolecular Prins cyclization and etherification provides polycycle **134**, which, upon treatment with an acid catalyst, adopts the form of perovskone. This domino reaction sequence is instructive as it rapidly assembles an imposing cage-like structure of perovskone through the recognition that the Diels–Alder reaction, the Prins cyclization of the resultant cycloadduct **133**, and the final etherification will be promoted by the same reaction conditions. Such synergy should be a goal of target-directed synthesis as it allows for the marked streamlining of routes.

Scheme 38 An Intermolecular Diels–Alder/Alkene Isomerization/Prins Cyclization/Etherification Cascade To Generate the Imposing Structure of Perovskone from Relatively Simple Starting Materials[84,85]

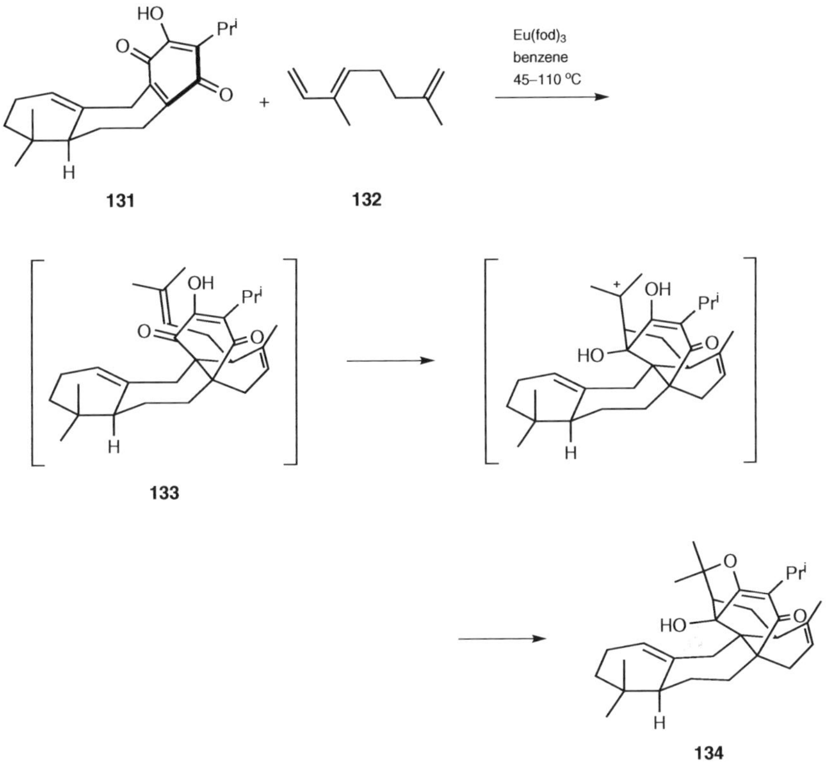

HO O Pr^i O H **131** + **132** Eu(fod)$_3$ benzene 45–110 °C

OH Pr^i O O H **133** → + OH Pr^i HO O H

→ O Pr^i HO O H **134**

fod = 6,6,7,7,8,8,8-heptafluoro-2,2-dimethyloctane-3,5-dionato

An interesting Diels–Alder/protection domino sequence is provided by the Baldwin group[86] in the course of a research effort directed at the synthesis of himgravine (Scheme 39). Treatment of butenolide–aldehyde **135** with 1,2-bis(trimethylsiloxy)ethane and trimethylsilyl trifluoromethanesulfonate results in the formation of tetracycle **138** in 53% yield. Control experiments point to the intermediacy of acetal-derived oxocarbenium ion **136** in the key cycloaddition step, a species whose LUMO should be significantly lowered when compared to the parent enal, promoting the desired Diels–Alder addition. The interception of the oxocarbenium intermediate **137** by the tethered alcohol allows for the direct formation of an acetal.

for references see p 44

Scheme 39 Lewis Acid Activation of an Intramolecular Diels–Alder Reaction Followed by Acetal Formation in Baldwin's Synthesis of Himgravine[86]

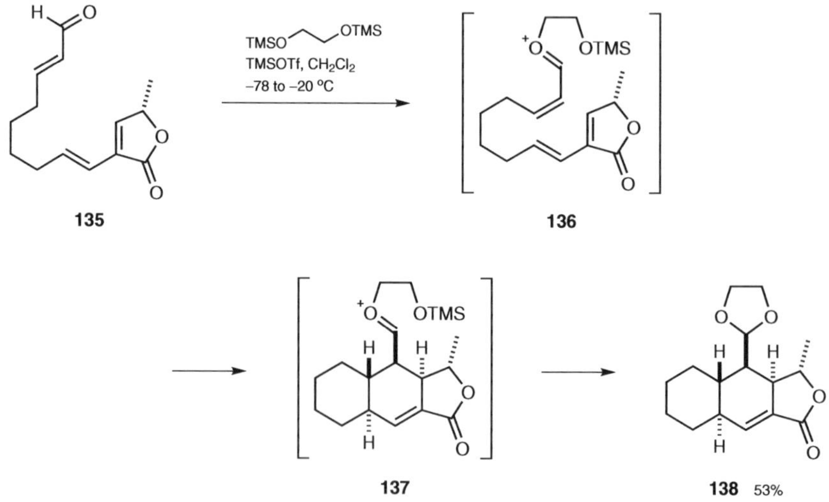

TMSO⌒OTMS, TMSOTf, CH_2Cl_2, −78 to −20 °C

135 → [**136**] → [**137**] → **138** 53%

A creative extension of the use of the Diels–Alder reaction in a cascade sequence comes from the group of Hoye[87] in their investigations of the hexadehydro-Diels–Alder transformation (Scheme 40) for the synthesis of diverse polycycles. Starting from tetrayne **139**, a hexadehydro-Diels–Alder reaction furnishes intermediate benzyne **140**, an electrophilic species, which can then be trapped by the pendent silyl ether to give zwitterion (or betaine) **141**. This intermediate then undergoes a rapid Brook rearrangement to provide tricyclic product **142**.

Scheme 40 Assembly of a Polycycle through a Diels–Alder/Cyclization/Brook Rearrangement Cascade[87]

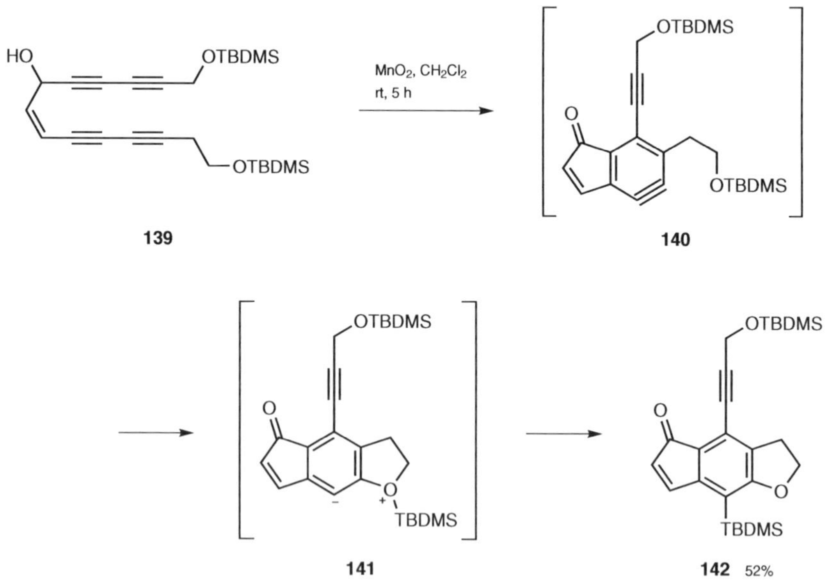

MnO_2, CH_2Cl_2, rt, 5 h

139 → [**140**] → [**141**] → **142** 52%

Polycyclic Ether 134:[84]
A mixture of quinone **131** (450 mg, 1.43 mmol, 1 equiv), triene **132** (1.50 g, 11.0 mmol, 7.7 equiv), $Eu(fod)_3$ (50 mg, 47 μmol, 0.03 equiv), and dry benzene (2.0 mL) (**CAUTION:** *carcinogen*) was stored at rt in a sealed tube for 3 d. The sealed reaction vessel was warmed to 45 °C and heated at that temperature for 48 h. Finally, the mixture was warmed to 100–110 °C and heated at that temperature for an additional 48 h. The mixture was concentrated under reduced pressure and the residue was purified by chromatography (silica gel, hexanes/EtOAc 10:1) to provide crystalline alcohol **134**, which was homogeneous by TLC analysis; yield: 519 mg (82%).

8-(*tert*-Butyldimethylsilyl)-4-[3-(*tert*-butyldimethylsiloxy)prop-1-ynyl]-2,3-dihydro-5*H*-indeno[5,6-*b*]furan-5-one (142)**; Typical Procedure:**[87]
MnO_2 (195 mg, 2.2 mmol, 25 equiv) was added to a soln of alcohol **139** (40 mg, 0.088 mmol, 1 equiv) in CH_2Cl_2 (1 mL) and the resulting black heterogeneous mixture was vigorously stirred at rt. After 5 h, the mixture was filtered through a plug of silica gel (CH_2Cl_2). Purification by flash chromatography (hexanes/EtOAc 9:1) gave a golden yellow oil; yield: 21 mg (52%).

2.1.1.3 Conclusions

In the nearly 90 years since it was initially described, the Diels–Alder reaction has evolved into one of organic chemistry's true power reactions. The vision for capitalizing on this process in the design and execution of a chemical synthesis is especially clear when the objective possesses the cyclohexene retron for a Diels–Alder reaction in its structure.[88] And yet, it is also possible to make much less obvious structural connections in a synthesis by coupling a Diels–Alder cycloaddition(s) with other types of bond formations in sophisticated cascades of reactions. This chapter has recounted several outstanding domino processes in which the Diels–Alder reaction has contributed to rapid formations of intricate molecular frameworks. Our sincere hope is that the examples described herein will inspire the reader to continue the search for those less obvious and especially powerful uses of Diels–Alder reactivity, and develop even more powerful examples.

References

[1] Diels, O.; Alder, K., *Justus Liebigs Ann. Chem.*, (1928) **460**, 98.
[2] Woodward, R. B.; Sondheimer, F.; Taub, D., *J. Am. Chem. Soc.*, (1951) **73**, 4057.
[3] Stork, G.; van Tamelen, E. E.; Friedman, L. J.; Burgstahler, A. W., *J. Am. Chem. Soc.*, (1951) **73**, 4501.
[4] Woodward, R. B.; Bader, F. E.; Bickel, H.; Frey, A. J.; Kierstead, R. W., *J. Am. Chem. Soc.*, (1956) **78**, 2023.
[5] Schreiber, J.; Leimgruber, W.; Pesaro, M.; Schudel, P.; Eschenmoser, A., *Angew. Chem.*, (1959) **71**, 637.
[6] Sauer, J., *Angew. Chem.*, (1966) **78**, 233; *Angew. Chem. Int. Ed.*, (1966) **5**, 211.
[7] Sauer, J., *Angew. Chem.*, (1967) **79**, 76; *Angew. Chem. Int. Ed.*, (1967) **6**, 16.
[8] Martin, J. G.; Hill, R. K., *Chem. Rev.*, (1961) **61**, 537.
[9] Oppolzer, W., In *Comprehensive Organic Synthesis*, Trost, B. M.; Fleming, I., Eds.; Pergamon: Oxford, (1991); Vol. 5, p 315.
[10] Carruthers, W., *Cycloaddition Reactions in Organic Synthesis*; Pergamon: Oxford, (1990).
[11] Kahn, S. D.; Pau, C. F.; Overman, L. E.; Hehre, W. J., *J. Am. Chem. Soc.*, (1986) **108**, 7381.
[12] Carlson, R. G., *Annu. Rep. Med. Chem.*, (1974) **9**, 270.
[13] Oppolzer, W., *Angew. Chem.*, (1977) **89**, 10; *Angew. Chem. Int. Ed. Engl.*, (1977) **16**, 10.
[14] Brieger, G.; Bennett, J. N., *Chem. Rev.*, (1980) **80**, 63.
[15] Taber, D. F., *Intramolecular Diels–Alder and Alder Ene Reactions*; Springer: Berlin, (1984).
[16] Fallis, A. G., *Can. J. Chem.*, (1984) **62**, 183.
[17] Craig, D., *Chem. Soc. Rev.*, (1987) **16**, 187.
[18] Roush, W. R., In *Advances in Cycloaddition*, Curran, D. P., Ed.; Jai: Greenwich, CT, (1990); Vol. 2, p 91.
[19] Corey, E. J., *Angew. Chem.*, (2002) **114**, 1724; *Angew. Chem. Int. Ed.*, (2002) **41**, 1650.
[20] Nicolaou, K. C.; Snyder, S. A.; Montagnon, T.; Vassilikogiannakis, G., *Angew. Chem.*, (2002) **114**, 1742; *Angew. Chem. Int. Ed.*, (2002) **41**, 1668.
[21] Stocking, E. M.; Williams, R. M., *Angew. Chem.*, (2003) **115**, 3186; *Angew. Chem. Int. Ed.*, (2003) **42**, 3078.
[22] Wessely, F.; Lauterbach-Keil, G.; Sinwel, F., *Monatsh. Chem.*, (1950) **81**, 811.
[23] Liao, C.-C.; Wei, C.-P., *Tetrahedron Lett.*, (1989) **30**, 2255.
[24] Gagnepain, J.; Castet, F.; Quideau, S., *Angew. Chem.*, (2007) **119**, 1555; *Angew. Chem. Int. Ed.*, (2007) **46**, 1533.
[25] Dong, S.; Zhu, J.; Porco, J. A., Jr., *J. Am. Chem. Soc.*, (2008) **130**, 2738.
[26] Njardarson, J. T.; McDonald, I. M.; Spiegel, D. A.; Inoue, M.; Wood, J. L., *Org. Lett.*, (2001) **3**, 2435.
[27] Sutherland, H. S.; Souza, F. E. S.; Rodrigo, R. G. A., *J. Org. Chem.*, (2001) **66**, 3639.
[28] Heathcock, C. H.; Piettre, S.; Ruggeri, R. B.; Ragan, J. A.; Kath, J. C., *J. Org. Chem.*, (1992) **57**, 2554.
[29] Shair, M. D.; Yoon, T. Y.; Mosny, K. K.; Chou, T. C.; Danishefsky, S. J., *J. Am. Chem. Soc.*, (1996) **118**, 9509.
[30] Carroll, W. A.; Grieco, P. A., *J. Am. Chem. Soc.*, (1993) **115**, 1164.
[31] Zapf, C. W.; Harrison, B. A.; Drahl, C.; Sorensen, E. J., *Angew. Chem.*, (2005) **117**, 6691; *Angew. Chem. Int. Ed.*, (2005) **44**, 6533.
[32] Vosburg, D. A.; Vanderwal, C. D.; Sorensen, E. J., *J. Am. Chem. Soc.*, (2002) **124**, 4552.
[33] Vanderwal, C. D.; Vosburg, D. A.; Weiler, S.; Sorensen, E. J., *J. Am. Chem. Soc.*, (2003) **125**, 5393.
[34] Evans, D. A.; Starr, J. T., *Angew. Chem.*, (2002) **114**, 1865; *Angew. Chem. Int. Ed.*, (2002) **41**, 1789.
[35] Nicolaou, K. C.; Sarlah, D.; Wu, T. R.; Zhan, W., *Angew. Chem.*, (2009) **121**, 7002; *Angew. Chem. Int. Ed.*, (2009) **48**, 6870.
[36] Oppolzer, W.; Keller, K., *J. Am. Chem. Soc.*, (1971) **93**, 3836.
[37] Funk, R. L.; Vollhardt, K. P. C., *J. Am. Chem. Soc.*, (1980) **102**, 5253.
[38] Allen, J. G.; Danishefsky, S. J., *J. Am. Chem. Soc.*, (2001) **123**, 351.
[39] Choy, W.; Reed, L. A., III; Masamune, S., *J. Org. Chem.*, (1983) **48**, 1137.
[40] Lee, J. C.; Strobel, G. A.; Lobkovsky, E.; Clardy, J., *J. Org. Chem.*, (1996) **61**, 3232.
[41] Li, C.; Lobkovsky, E.; Porco, J. A., Jr., *J. Am. Chem. Soc.*, (2000) **122**, 10484.
[42] Bandaranayake, W. M.; Banfield, J. E.; Black, D. St. C., *J. Chem. Soc., Chem. Commun.*, (1980), 902.
[43] Nicolaou, K. C.; Petasis, N. A.; Zipkin, R. E.; Uenishi, J., *J. Am. Chem. Soc.*, (1982) **104**, 5555.
[44] Nicolaou, K. C.; Petasis, N. A.; Zipkin, R. E., *J. Am. Chem. Soc.*, (1982) **104**, 5560.
[45] Nicolaou, K. C.; Vassilikogiannakis, G.; Magerlein, W.; Kranich, R., *Angew. Chem.*, (2001) **113**, 2543; *Angew. Chem. Int. Ed.*, (2001) **40**, 2482.

[46] Staudinger, H., DRP 506 839; *Chem. Abstr.*, (1913) **25**, 522.
[47] Claisen, L., *Ber. Dtsch. Chem. Ges.*, (1912) **45**, 3157.
[48] Quillinan, A. J.; Scheinmann, F., *J. Chem. Soc. D*, (1971), 966.
[49] Nicolaou, K. C.; Li, J., *Angew. Chem.*, (2001) **113**, 4394; *Angew. Chem. Int. Ed.*, (2001) **40**, 4264.
[50] Tisdale, E. J.; Slobodov, I.; Theodorakis, E. A., *Org. Biomol. Chem.*, (2003) **1**, 4418.
[51] Gibbs, R. A.; Okamura, W. H., *J. Am. Chem. Soc.*, (1988) **110**, 4062.
[52] Gibbs, R. A.; Bartels, K.; Lee, R. W. K.; Okamura, W. H., *J. Am. Chem. Soc.*, (1989) **111**, 3717.
[53] Pfau, M.; Combrisson, S.; Rowe, J. E., Jr.; Heindel, N. D., *Tetrahedron*, (1978) **34**, 3459.
[54] Nicolaou, K. C.; Gray, D., *Angew. Chem.*, (2001) **113**, 783; *Angew. Chem. Int. Ed.*, (2001) **40**, 761.
[55] Humphrey, J. M.; Liao, Y.; Ali, A.; Rein, T.; Wong, Y.-L.; Chen, H.-J.; Courtney, A. K.; Martin, S. F., *J. Am. Chem. Soc.*, (2002) **124**, 8584.
[56] Chapman, O. L.; Engel, M. R.; Springer, J. P.; Clardy, J. C., *J. Am. Chem. Soc.*, (1971) **93**, 6696.
[57] Himbert, G.; Henn, L., *Angew. Chem.*, (1982) **94**, 631; *Angew. Chem. Int. Ed.*, (1982) **21**, 620.
[58] Schmidt, Y.; Lam, J. K.; Pham, H. V.; Houk, K. N.; Vanderwal, C. D., *J. Am. Chem. Soc.*, (2013) **135**, 7339.
[59] Lam, J. K.; Schmidt, Y.; Vanderwal, C. D., *Org. Lett.*, (2012) **14**, 5566.
[60] Roberts, J. D.; Simmons, H. E., Jr.; Carlsmith, L. A.; Vaughan, C. W., *J. Am. Chem. Soc.*, (1953) **75**, 3290.
[61] Tadross, P. M.; Stoltz, B. M., *Chem. Rev.*, (2012) **112**, 3550.
[62] Townsend, C. A.; Davis, S. G.; Christensen, S. B.; Link, J. C.; Lewis, C. P., *J. Am. Chem. Soc.*, (1981) **103**, 6885.
[63] Liang, H.; Ciufolini, M. A., *Org. Lett.*, (2010) **12**, 1760.
[64] Fu, X.; Hossain, M. B.; van der Helm, D.; Schmitz, F. J., *J. Am. Chem. Soc.*, (1994) **116**, 12 125.
[65] Layton, M. E.; Morales, C. A.; Shair, M. D., *J. Am. Chem. Soc.*, (2002) **124**, 773.
[66] Appendino, G.; Prosperini, S.; Valdivia, C.; Ballero, M.; Colombano, G.; Billington, R. A.; Genazzani, A. A.; Sterner, O., *J. Nat. Prod.*, (2005) **68**, 1213.
[67] Larsson, R.; Sterner, O.; Johansson, M., *Org. Lett.*, (2009) **11**, 657.
[68] Nelson, H. M.; Stoltz, B. M., *Tetrahedron Lett.*, (2009) **50**, 1699.
[69] Nelson, H. M.; Gordon, J. R.; Virgil, S. C.; Stoltz, B. M., *Angew. Chem.*, (2013) **125**, 6831; *Angew. Chem. Int. Ed.*, (2013) **52**, 6699.
[70] Taunton, J.; Wood, J. L.; Schreiber, S. L., *J. Am. Chem. Soc.*, (1993) **115**, 10 378.
[71] Inanaga, J.; Hirata, K.; Saeki, H.; Katsuki, T.; Yamaguchi, M., *Bull. Chem. Soc. Jpn.*, (1979) **52**, 1989.
[72] Reber, K. P.; Tilley, S. D.; Carson, C. A.; Sorensen, E. J., *J. Org. Chem.*, (2013) **78**, 9584.
[73] Acebey, L.; Sauvain, M.; Beck, S.; Moulis, C.; Gimenez, A.; Jullian, V., *Org. Lett.*, (2007) **9**, 4693.
[74] Yuan, C.; Du, B.; Yang, L.; Liu, B., *J. Am. Chem. Soc.*, (2013) **135**, 9291.
[75] Roush, W. R.; Sciotti, R. J., *J. Am. Chem. Soc.*, (1994) **116**, 6457.
[76] Baran, P. S.; Burns, N. Z., *J. Am. Chem. Soc.*, (2006) **128**, 3908.
[77] Boger, D. L.; Baldino, C. M., *J. Am. Chem. Soc.*, (1993) **115**, 11 418.
[78] Jacobi, P. A.; Blum, C. A.; DeSimone, R. W.; Udodong, U. E. S., *Tetrahedron Lett.*, (1989) **30**, 7173.
[79] Niu, D.; Willoughby, P. H.; Woods, B. P.; Baire, B.; Hoye, T. R., *Nature (London)*, (2013) **501**, 531.
[80] Frankowski, K. J.; Golden, J. E.; Zeng, Y.; Lei, Y.; Aubé, J., *J. Am. Chem. Soc.*, (2008) **130**, 6018.
[81] Fuchs, J. R.; Funk, R. L., *J. Am. Chem. Soc.*, (2004) **126**, 5068.
[82] Kobayashi, M.; Aoki, S.; Gato, K.; Matsunami, K.; Kurosu, M.; Kitagawa, I., *Chem. Pharm. Bull.*, (1994) **42**, 2449.
[83] Padwa, A.; Dimitroff, M.; Liu, B., *Org. Lett.*, (2000) **2**, 3233.
[84] Majetich, G.; Zhang, Y.; Tian, X.; Britton, J. E.; Li, Y.; Phillips, R., *Tetrahedron*, (2011) **67**, 10 129.
[85] Majetich, G.; Zhang, Y., *J. Am. Chem. Soc.*, (1994) **116**, 4979.
[86] Baldwin, J. E.; Chesworth, R.; Parker, J. S.; Russell, A. T., *Tetrahedron Lett.*, (1995) **36**, 9551.
[87] Hoye, T. R.; Baire, B.; Niu, D.; Willoughby, P. H.; Woods, B. P., *Nature (London)*, (2012) **490**, 208.
[88] Corey, E. J.; Cheng, X.-M., *The Logic of Chemical Synthesis*; Wiley-Interscience: Hoboken, NJ, (1995).

2.1.2 Domino Reactions Including [2+2], [3+2], or [5+2] Cycloadditions

I. Coldham and N. S. Sheikh

General Introduction

Cycloaddition reactions lend themselves well to domino (tandem or cascade) processes. Section 2.1.1 outlines such reactions that include a [4+2]-cycloaddition step. In this chapter, we will focus on [2+2]-, [3+2]-, and [5+2]-cycloaddition reactions. One could argue that many cycloaddition reactions involve more than one step, as the formation of one of the π-components is often accomplished in the same pot as the subsequent cycloaddition. This is particularly the case for 1,3-dipoles that are unstable and not normally isolated but instead prepared in situ. For example, condensation of an amine and an aldehyde can provide an imine and hence a dipole (such as an azomethine ylide, azomethine imine, or nitrone) that is amenable to cycloaddition with an alkene. To provide some focus, this review will describe transformations that include an [m+n]-type cycloaddition reaction within a sequence of steps, other than condensations or other simple reactions that provide one of the components for the cycloaddition. This will discount formation of ylide dipoles by reaction of an aldehyde with an amine as one of the steps; otherwise, most dipolar cycloaddition reactions would then be termed domino reactions. We will also not describe examples in which a [4+2] step precedes the other [m+n] cycloaddition, as these are classified within Section 2.1.1. Our focus will be on cascade processes rather than multicomponent reactions, such that one transformation leads on to another (one of which involves a cycloaddition). Overall, at least three new σ-bonds will be formed, as two new σ-bonds are necessarily generated in the cycloaddition step.

This review aims to describe reliable and synthetically useful domino processes that incorporate a [2+2]-, [3+2]-, or [5+2]-cycloaddition reaction. We have selected what we consider to be some of the most instructive examples in the various sections, with emphasis on recent work. These efforts have given rise to the synthesis of a wide selection of interesting target molecules with diverse structures, often of biological relevance. As such domino reactions provide an efficient means to construct multiple bonds, we hope that this chapter will be a useful source of information and inspire further development of additional domino reactions using cycloadditions.

2.1.2.1 Domino [2+2] Cycloadditions

Domino [2+2]-cycloaddition reactions are instrumental in constructing both simple and complex skeletal architectures of many natural products and derivatives of synthetic utility. Depending upon the initialization modes and the nature of the starting substrates, the mechanism of [2+2] cycloaddition can either be concerted or follow a radical pathway, and it allows convenient access to four-membered rings. The regiochemistry and diastereofacial selectivity of such transformations are highly predictable owing to extensive research performed in this area. The prime focus of this section is to delineate the most synthetically useful domino [2+2]-cycloaddition protocols to provide insight into this class of reactions and possible directions toward new leads.

2.1.2.1.1 Cycloaddition of an Enaminone and β-Diketone with Fragmentation

Functionalized enaminones are prone to intramolecular [2+2] photocycloaddition, which results in constrained bicyclic systems, analogous to the classical de Mayo reaction.[1,2] The ring strain can be relieved in situ by fragmentation through a retro-Mannich reaction, and this leads to spiropyrrolenines, present in oxindole, β-carboline, 3*H*-indole (indolenine), and *strychnos* alkaloids. This reaction has been elaborated by White and co-workers, who subjected enaminone **1** to a [2+2]-cycloaddition/retro-Mannich tandem sequence to afford spiroimine **3** via cyclobutane **2** (Scheme 1).[1] Tryptamine-based substrate **1** is highly sensitive to the energy of incident radiation, and the best results are obtained by using a 450-W medium-pressure mercury lamp with a Corex filter (50% transmission at 290 nm). The method has been applied to various substituted enaminones, and its synthetic utility has been illustrated through the total syntheses of (±)-coerulescine, (±)-horsfiline, (±)-elacomine, and (±)-6-deoxyelacomine, in addition to the synthesis of the core structural motif of koumine.

Scheme 1 Intramolecular Photocycloaddition/Retro-Mannich Tandem Sequence[1]

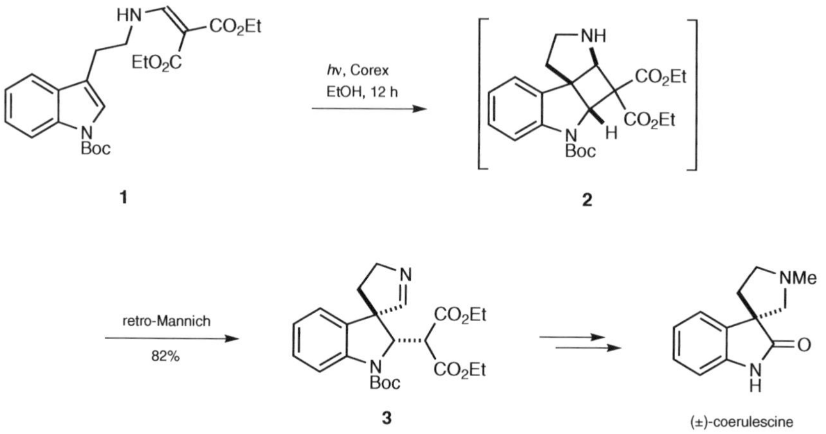

Another useful application of intramolecular [2+2] cycloaddition followed by fragmentation involves irradiation of a β-dicarbonyl compound and its reaction with an alkene.[3] The resultant cycloadduct, a cyclobutanol, undergoes spontaneous ring opening to a δ-dicarbonyl species, which upon treatment with alcoholic hydrogen chloride under aldol conditions cyclizes to afford the galanthan skeleton, a fundamental unit of the *amaryllidaceae* alkaloids.

Diethyl 2-{1-(*tert*-Butoxycarbonyl)-4′,5′-dihydrospiro[2,3-dihydroindole-3,3′-pyrrol]-2-yl}malonate (3); Typical Procedure:[1]
A degassed soln of enaminone **1** (200 mg, 0.47 mmol) in EtOH (100 mL) was irradiated with a 450-W medium-pressure mercury lamp through a Corex filter for 12 h. The soln was concentrated, and the resultant yellow oil was purified by flash chromatography (silica gel, hexanes/EtOAc 1:1) to give the title compound as an oil; yield: 164 mg (82%).

2.1.2.1.2 Cycloaddition of Ynolate Anions Followed by Dieckmann Condensation/Michael Reaction

Ynolates act as ketene anion equivalents and have a negatively charged oxygen atom attached to an alkyne functionality. These species react with γ- or δ-dicarbonyl compounds such as **4** through [2+2] cycloaddition to generate β-lactone enolates such as **5** that have an inherent tendency to cyclize, for example to give bicyclic β-lactone **6** (Scheme 2).[4,5] Not surprisingly, acid-catalyzed decarboxylation of the β-lactone yields 2,3-disubstituted cyclohex-2-enone **7**. An extension of this synthetically useful protocol involves aromatic substrates that ultimately afford a range of sterically demanding polysubstituted naphthalenes.[5] In this regard, the method has been applied to gain rapid access to dihydrojasmone, a flavoring and fragrance agent. It has also been used to prepare α-cuparenone, one of the most suitable candidates for showcasing synthetic routes, and to prepare cucumin E, a bioactive triquinane natural product.[5,6]

Scheme 2 [2+2] Cycloaddition of an Ynolate Anion Followed by Dieckmann Condensation[5]

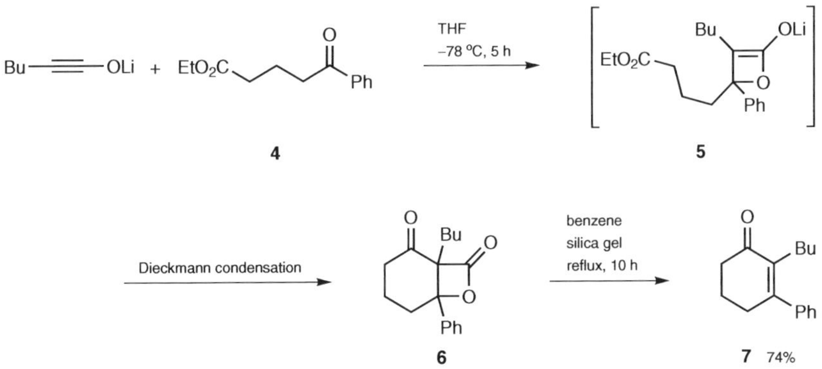

Bu—≡—OLi + EtO_2C…(O)Ph **4** → (THF, −78 °C, 5 h) → [EtO_2C … Bu, OLi, O, Ph] **5** → (Dieckmann condensation) → **6** (O, Bu, O, O, Ph) → (benzene, silica gel, reflux, 10 h) → **7** (O, Bu, Ph) 74%

A tandem sequence comprising [2+2] cycloaddition followed by a Michael reaction has been reported to provide a short and expedient way to construct five-, six-, and seven-membered 2-(cycloalkenyl)acetates after decarboxylation.[7] Incorporation of the Michael reaction has a particular advantage over the [2+2]-cycloaddition/Dieckmann condensation variant, as the latter fails to furnish seven-membered carbocycles.

2-Butyl-3-phenylcyclohex-2-enone (7); Typical Procedure:[5]

CAUTION: *Solutions of* tert-*butyllithium react explosively with water and may ignite in moist air.*

A 1.48 M soln of *t*-BuLi in pentane (2.70 mL, 4.0 mmol) was added dropwise to a soln of ethyl 2,2-dibromohexanoate (302 mg, 1.0 mmol) in anhyd THF (6 mL) at −78 °C under an atmosphere of argon, and the mixture was stirred for 3 h before it was warmed to 0 °C. After 30 min, the resultant colorless mixture was cooled to −78 °C, and a soln of ethyl 5-oxo-5-phenylpentanoate (**4**; 176 mg, 0.8 mmol) in THF (2 mL) was added dropwise. After stirring for 5 h at −78 °C, the mixture was diluted with sat. aq NH_4Cl and extracted with EtOAc. The organic layer was separated, washed successively with sat. aq $NaHCO_3$ and sat. aq NaCl, dried ($MgSO_4$), filtered, and concentrated to afford bicyclic β-lactone **6** as an oil. The crude mixture was dissolved in benzene (10 mL) (**CAUTION:** *carcinogen*) and 100–270 mesh chromatographic silica gel (100 mg) was added. The mixture was heated under reflux. After 10 h, the mixture was cooled to rt, filtered, and concentrated to give an oil. Purification by column chromatography (silica gel, EtOAc/hexanes 2:98 to 5:95) provided the title compound as an oil; yield: 136 mg (74%).

for references see p 89

2.1.2.1.3 Cycloaddition Cascade Involving Benzyne–Enamide Cycloaddition or a Fischer Carbene Complex

Hsung and co-workers have explored an unprecedented domino cycloaddition involving [2+2] benzyne–enamide cycloaddition, ring opening, then intramolecular [4+2] cycloaddition to access nitrogen-based heterocycles.[8] For example, treatment of benzyne precursor **8** with enamide **9** provides alkene-tethered enamide **10** through [2+2] cycloaddition. Compound **10** then undergoes ring opening and intramolecular [4+2] cycloaddition to give azatricycle **11** in excellent yield as a single diastereomer (Scheme 3). The method is well tolerated by various substituted enamides and can also be applied to access azatetracycles, albeit in lower yields. The stereochemistry present in the tethered alkene portion remains intact during the transformation, as confirmed by NOESY experiments. This chemistry can be extended to alkyne-tethered enamides, and the one-pot, three-component synthesis of azatetracycles can be achieved in good yields as a result.[8] However, with an alkyne motif attached to an enamide, this transformation requires an increase in the reaction temperature and removal of the solvent. The synthetic viability of the process is highlighted by application in the efficient total syntheses of (±)-chelidonine and (±)-norchelidonine.[9]

Scheme 3 Benzyne–Enamide [2+2] Cycloaddition/Electrocyclic Ring Opening/Intramolecular [4+2] Cycloaddition[8]

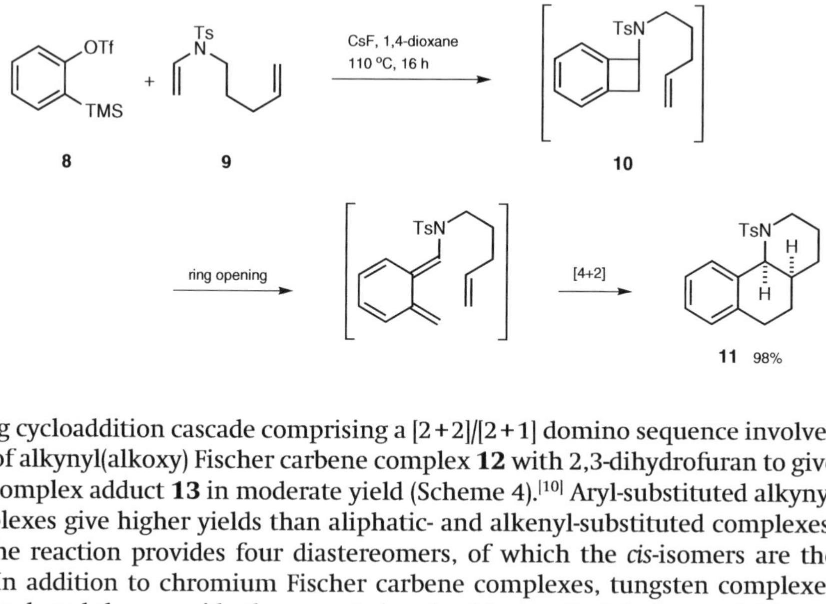

OTf, TMS, **8**; Ts, N, **9**; CsF, 1,4-dioxane, 110 °C, 16 h; TsN, **10**; ring opening; TsN; [4+2]; TsN, H, H, **11** 98%

An interesting cycloaddition cascade comprising a [2+2]/[2+1] domino sequence involves the reaction of alkynyl(alkoxy) Fischer carbene complex **12** with 2,3-dihydrofuran to give structurally complex adduct **13** in moderate yield (Scheme 4).[10] Aryl-substituted alkynyl Fischer complexes give higher yields than aliphatic- and alkenyl-substituted complexes. In general, the reaction provides four diastereomers, of which the *cis*-isomers are the major ones. In addition to chromium Fischer carbene complexes, tungsten complexes can also be used, and they provide the expected cycloadducts, albeit in lower yields and with poor stereocontrol.

Scheme 4 [2+2]/[2+1]-Cycloaddition Cascade involving a Chromium Fischer Carbene Complex[10]

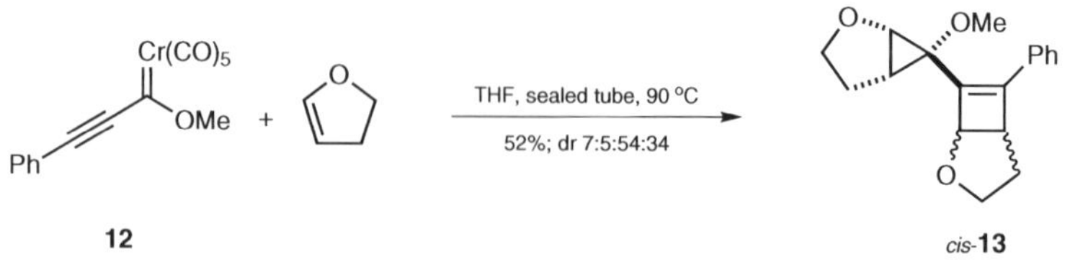

$Cr(CO)_5$, OMe, Ph, **12**; O; THF, sealed tube, 90 °C; 52%; dr 7:5:54:34; O, OMe, Ph, O, *cis*-**13**

(4a*R*,10b*S***)-1,2,3,4,4a,5,6,10b-Octahydro-1-tosylbenzo[*h*]quinoline (11); Typical Procedure:**[8]

2-(Trimethylsilyl)phenyl trifluoromethanesulfonate (**8**; 59.1 mg, 0.198 mmol, 3.0 equiv) was added dropwise to a soln of enamide **9** (17.5 mg, 0.066 mmol) and CsF (40.1 mg, 0.264 mmol, 4.0 equiv) in anhyd. 1,4-dioxane (0.05 M), and the resultant mixture was heated at 110 °C. After 16 h, the mixture was cooled to rt and filtered, and the filter residue was washed with EtOAc. The filtrate was concentrated, and the resultant crude product was purified by flash chromatography (silica gel, EtOAc/hexanes 1:4) to give the title compound as a solid; yield: 22.2 mg (98%); mp 180 °C (dec.).

2.1.2.1.4 Cycloadditions with Rearrangement

The tandem [2+2]-cycloaddition/rearrangement cascade (or vice versa) is a well-studied domino protocol. Intramolecular rearrangements play a pivotal role in creating complicated structural building blocks from simple substrates through relocation of σ- and π-electronic systems. This section highlights significant contributions in this area with a particular emphasis on the synthetic applicability of the reported procedures.

2.1.2.1.4.1 Cycloaddition of an Azatriene Followed by Cope Rearrangement

Conjugated imines are useful synthetic building blocks that can be used to construct nitrogen-containing heterocycles including azocinones. For example, Mahajan and coworkers have demonstrated the application of a domino [2+2] cycloaddition followed by in situ [3,3]-sigmatropic rearrangement to obtain azocinone **16** from the reaction of a conjugated imine **14** with a ketene (Scheme 5).[11] The system has been tested for a range of substituted imines and gives good yields of various azocinones. The proposed tandem sequence has been confirmed by energy-minimization calculations, which indicate that rearranged product **16** is more stable than [2+2]-cycloadduct **15** by 28.9 kcal·mol^{-1}.

Scheme 5 Cycloaddition of a Conjugated Imine and Ketene Followed by Cope Rearrangement[11]

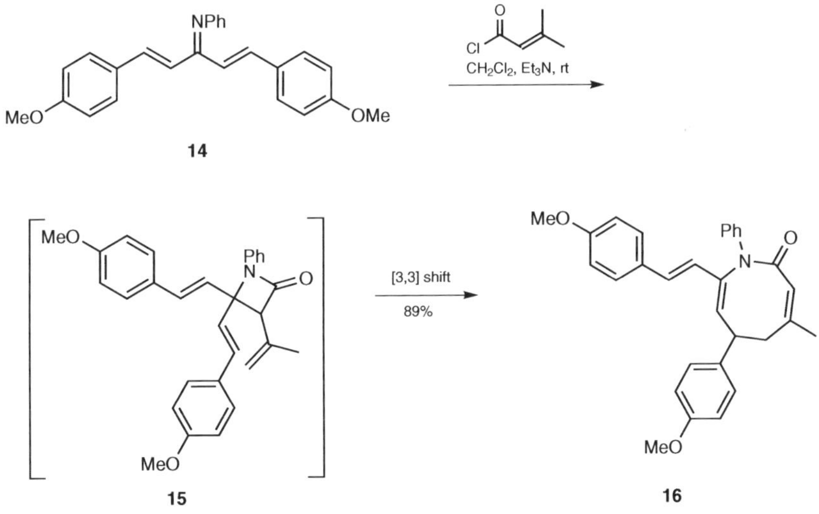

for references see p 89

(3Z,7Z)-6-(4-Methoxyphenyl)-8-[(E)-4-methoxystyryl]-4-methyl-1-phenyl-5,6-dihydroazocin-2(1H)-one (16); Typical Procedure:[11]
A soln of 3-methylbut-2-enoyl chloride in dry CH_2Cl_2 (30 mL) was added dropwise to a mixture of imine **14** (10 mmol) and Et_3N (15 mmol) in dry CH_2Cl_2 (30 mL) at rt. The reaction was monitored by TLC and, upon completion, the mixture was washed with sat. aq $NaHCO_3$ (50 mL) and H_2O (100 mL), dried (Na_2SO_4), filtered, and concentrated. Purification by column chromatography (silica gel, EtOAc/hexanes 1:10) gave the title compound as a colorless prismatic solid; yield: 4.02 g (89%); mp 196–197 °C.

2.1.2.1.4.2 Cycloaddition of a Propargylic Ether and Propargylic Thioether Followed by [3,3]-Sigmatropic Rearrangement

Allenes are generally responsive toward cycloadditions and have been applied to build polycyclic compounds. Inspired by this reactivity, Kanematsu and colleagues have demonstrated an intramolecular [2+2] cycloaddition of propargylic ether **17** under base-catalyzed conditions to give **18**. This event proceeds via an allenyl ether that immediately undergoes [3,3]-sigmatropic rearrangement to give tricyclic ring system **19** in excellent yield (Scheme 6).[12] The reaction has a competing pathway leading to [4+2] adducts. The fate of the cycloaddition, either [2+2] or [4+2], is remarkably controlled by the presence of a substituent at the C2 position, which, if present, directs the reaction toward [2+2] cycloaddition followed by the [3,3]-sigmatropic rearrangement sequence. In addition, substitution at C6 sways the reaction pathway such that 6-*trans*-substituted substrates lead to [2+2] products, whereas the similar reaction of 6-*cis*-substituted compounds gives a mixture of [2+2] and [4+2] cycloadducts. This unique tandem sequence has been applied to the rapid and effective asymmetric synthesis of the oxataxane skeleton, a potential precursor of numerous taxane derivatives, by using propargylic ether **21** via bicyclic allenyl ether **22**. Ether **21** can be readily obtained from optically pure (−)-Wieland–Miescher ketone (**20**) (Scheme 6).[13] An example of this tandem process has been described with a propargyl thioether, albeit with the [2+2] adduct obtained in a lower yield (34%) along with the [4+2] cycloadduct (19% yield).[14]

Scheme 6 Intramolecular [2+2] Cycloaddition of a Propargylic Ether Followed by [3,3]-Sigmatropic Rearrangement[12,13]

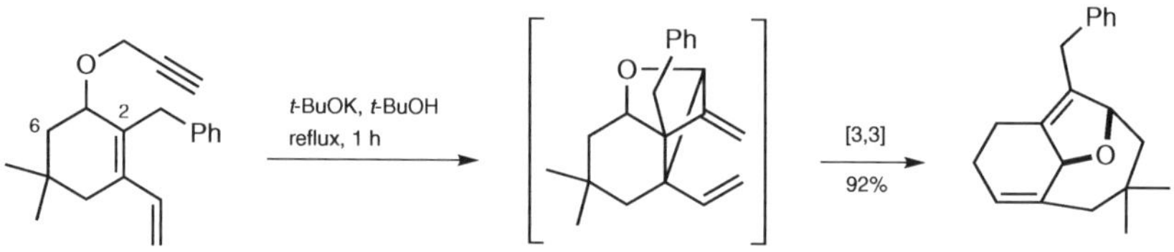

t-BuOK, t-BuOH, reflux, 1 h; [3,3] 92%

17 18 19

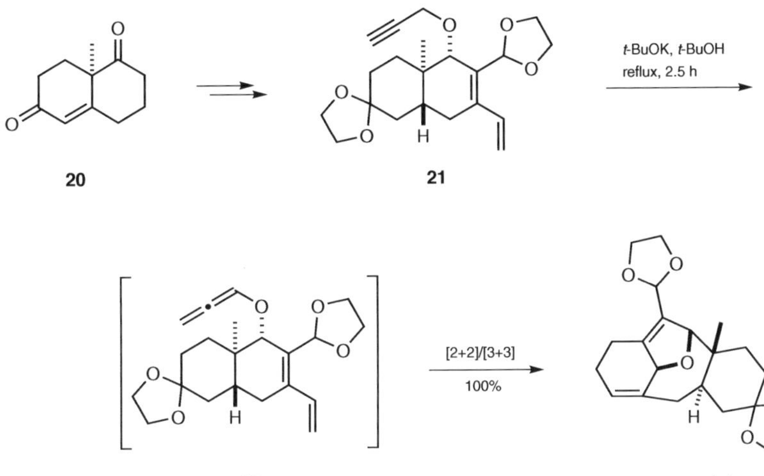

10-Benzyl-4,4-dimethyl-3,4,5,7,8,9a-hexahydro-2*H*-2,9-(metheno)benzo[*b*]oxepin (19); Typical Procedure:[12]

A soln of propargylic ether **17** (1.11 mmol) in *t*-BuOH (10 mL) was added dropwise to a soln of *t*-BuOK (8.91 mmol) in *t*-BuOH (10 mL) at 83 °C. The resulting mixture was heated at reflux and, after 1 h, was cooled and poured into a mixture of ice/water (20 mL). The aqueous phase was washed with Et_2O (3 × 45 mL) and the combined organic extracts were washed with brine (20 mL), dried (Na_2SO_4), filtered, and concentrated. Purification by medium-pressure liquid chromatography using a C.I.G. column system (EtOAc/hexane 1:50) gave the title compound as an oil; yield: 286 mg (92%).

2.1.2.1.4.3 [3,3]-Sigmatropic Rearrangement of Propargylic Ester and Propargylic Acetate Followed by Cycloaddition

A mild, metal-catalyzed [3,3]-sigmatropic rearrangement followed by a [2+2]-cycloaddition sequence has been demonstrated by Zhang to synthesize tetracyclic cyclobutanes with fused γ-lactones, 2,3-dihydroindoles, and an exocyclic double bond.[15] The [3,3]-sigmatropic rearrangement of propargylic ester **23** activated by a cationic gold(I) complex provides allenylic ester **24**, a material that undergoes sequential [2+2] cycloaddition to give highly functionalized cyclobutane **25** (Scheme 7). The scope of the reaction has been studied and proceeds equally well with alkyl substitution at the indole nitrogen atom, aliphatic or aromatic substitution at the propargylic position, as well as substitution at the alkyne terminus. However, longer reaction times (up to 12 h) and higher catalyst loadings are required for cases in which the starting materials have bulky substituents at either the propargylic position or the alkyne terminus.

for references see p 89

Scheme 7 Gold-Catalyzed [3,3]-Sigmatropic Rearrangement/[2+2]-Cycloaddition Sequence[15]

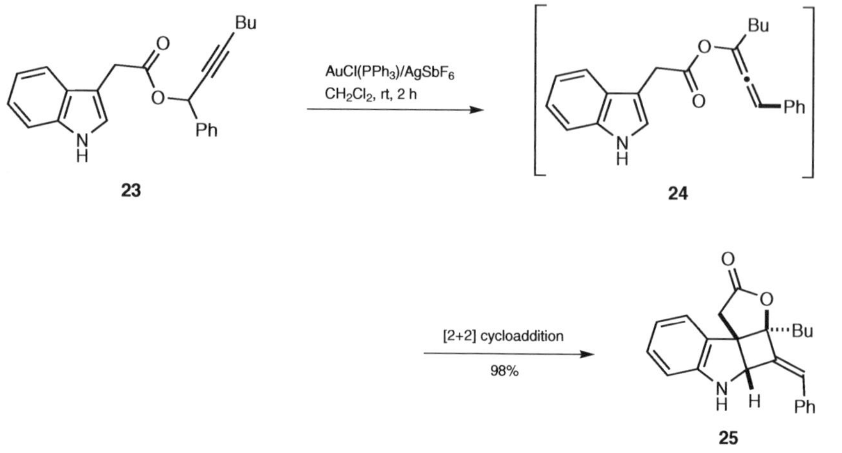

AuCl(PPh3)/AgSbF6
CH2Cl2, rt, 2 h

23 24

[2+2] cycloaddition
98%

25

A metal-free, thermally induced variant of this [3,3]-sigmatropic rearrangement/[2+2]-cycloaddition sequence has been reported by Brummond and Osbourn to achieve the synthesis of carbocyclic spiroxindoles in moderate yields (up to 61%).[16] The reaction is quite general and tolerant of various protecting groups on the oxindole nitrogen atom as well as to substitution of the alkyne terminus on the starting propargylic acetates. Interestingly, this reaction fails to provide the desired spirooxindole if performed in the presence of gold(III) catalysts (in that case, hydrolysis of the intermediate allenyl acetate is observed instead).

(3a*S,4a*S**,9b*S**,*E*)-4-Benzylidene-3a-butyl-3a,4,4a,5-tetrahydrofuro[3′,2′:2,3]cyclobuta-[1,2-*b*]indol-2(1*H*)-one (25); Typical Procedure:**[15]

$[Au(PPh_3)]^+SbF_6^-$ [generated as a 0.01 M soln in CH_2Cl_2 by treating $AuCl(PPh_3)$ with $AgSbF_6$ (1 mol%)] was added to a 0.05 M soln of indole **23** in CH_2Cl_2. The mixture was stirred at rt for 2 h, and the solvent was then evaporated. Purification by column chromatography (silica gel, hexanes/EtOAc 3:1) gave the title compound as an oil; yield: 98%.

2.1.2.1.4.4 Cycloaddition of a Ketene Followed by Allylic Rearrangement

A tandem sequence involving [2+2] cycloaddition of ketenes to α,β-unsaturated aldehydes followed by allylic rearrangement has been demonstrated by Hattori and co-workers.[17,18] The process is catalyzed by a cationic palladium(II) complex acting as a Lewis acid during [2+2]-cycloadduct formation and facilitating allylic rearrangement via a π–allylpalladium intermediate. For example, treatment of ketene with (*E*)-but-2-enal (**26**) gives δ-lactone **28** through allylic rearrangement of β-lactone **27**, formed as a result of initial [2+2] cycloaddition (Scheme 8).[17] Hydrolysis and subsequent saponification of lactone **28** provides a convenient route to prepare sorbic acid. The system has been evaluated for α,β-unsaturated ketones, which give low yields of the products. In addition, an asymmetric version, with the use of chiral aldehydes, has been reported, though it proceeds with poor to moderate levels of diastereocontrol.[18] The reaction of ketenes with aldehydes by employing [2+2] cycloaddition/cycloreversion to afford substituted alkenes has also been documented.[19]

Scheme 8 Palladium(II)-Catalyzed [2 + 2] Cycloaddition/[3,3]-Sigmatropic Rearrangement Involving Ketene and an α,β-Unsaturated Aldehyde[17]

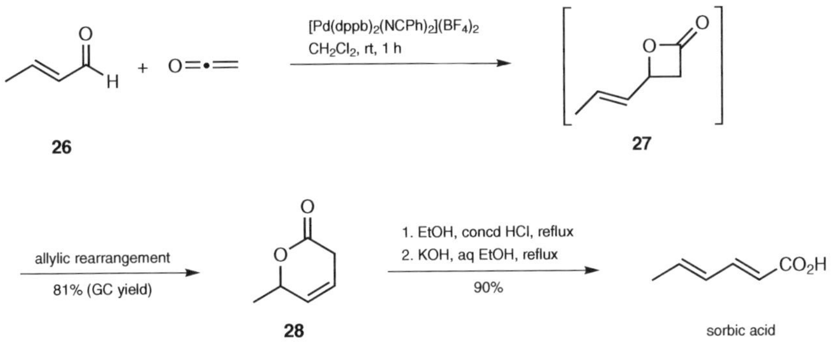

$[Pd(dppb)_2(NCPh)_2](BF_4)_2$, CH_2Cl_2, rt, 1 h; 26 → [27]; allylic rearrangement, 81% (GC yield) → 28; 1. EtOH, concd HCl, reflux; 2. KOH, aq EtOH, reflux; 90% → sorbic acid

6-Methyl-3,6-dihydro-2*H*-pyran-2-one (28); Typical Procedure:[17]

(*E*)-But-2-enal (**26**; 0.20 mmol) was added to a mixture of $[Pd(dppb)_2(NCPh)_2](BF_4)_2$ (0.05 mmol) in CH_2Cl_2 (500 mL). Ketene (about 0.25 mmol) was bubbled into the mixture over a period of 1 min, and the mixture was then stirred for 5 min. This series of operations was repeated until added aldehyde **26** reached a total amount of 1.0 mmol. An additional amount of ketene (about 1.0 mmol) was added, and the resulting mixture was stirred at rt for 1 h. After workup, the crude mixture was subjected to GC analysis (Quadrex MPS-10 column); yield: 81%.

2.1.2.1.4.5 Allyl Migration in Ynamides Followed by Cycloaddition

Hsung and co-workers have described an effective palladium(0)-activated cascade of *N*-allylynamide **29** to furnish highly substituted imine **30** (Scheme 9).[20] The reaction comprises initial allyl transfer from the nitrogen atom to the carbon atom, which leads to the generation of a ketenimine, followed by [2 + 2] cycloaddition. The method is fairly general and a range of propargylic substituents and tethered alkenes are tolerated. Interestingly, changes in the alkene substitution pattern can switch the regioselectivity from bridged to fused [2 + 2]-cycloaddition products, as represented by the synthesis of fused cycloadduct **32** from ynamide **31** (Scheme 9). Both cycloaddition pathways are highly diastereoselective and the products are obtained in ≥20:1 diastereomeric ratios, as determined by ^{1}H NMR spectroscopy.

Scheme 9 Bridged and Fused Intramolecular Ketenimine [2 + 2] Cycloadditions[20]

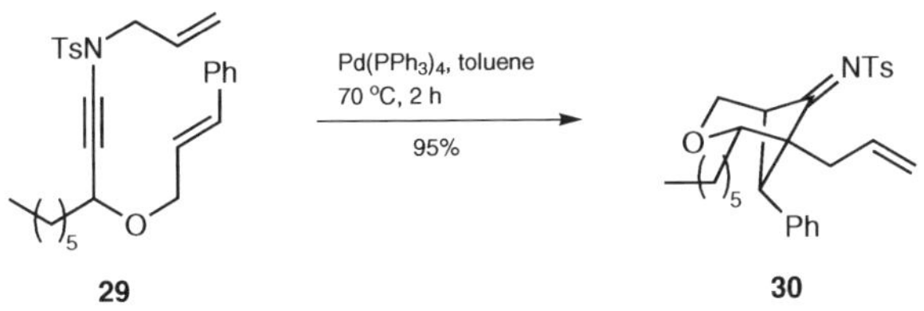

29 → 30: $Pd(PPh_3)_4$, toluene, 70 °C, 2 h; 95%

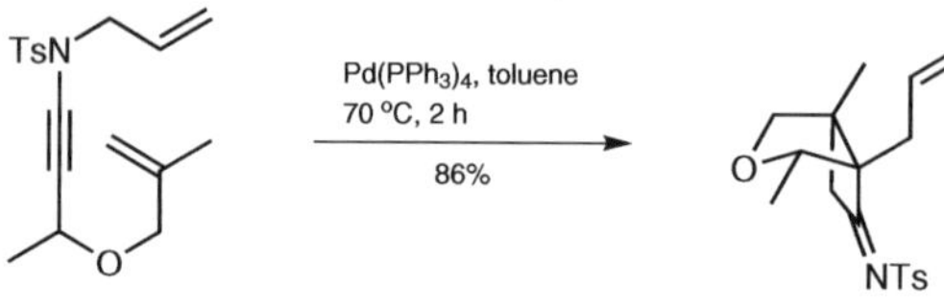

2-Hexyl-7-phenyl-1-(prop-2-enyl)-3-oxabicyclo[3.1.1]heptan-6-imine (30); **Typical Procedure:**[20]

A soln of ynamide **29** (107.5 mg, 0.23 mmol), $Pd(PPh_3)_4$ (0.0059 mmol), and toluene (1.2 mL) was added to a flame-dried, screw-capped vial. The vial was sealed under an atmosphere of N_2 and heated to 70 °C for 2 h. The solvent was then evaporated and the crude mixture was purified by column chromatography (silica gel, hexanes/EtOAc 4:1) to give the title compound as an oil; yield: 104.5 mg (95%).

2.1.2.1.4.6 1,3-Migration in Propargyl Benzoates Followed by Cycloaddition

Chan and co-workers have demonstrated the utility of gold catalysis in developing novel synthetic methodology toward the synthesis of azabicyclo[4.2.0]oct-5-ene **35**. This process relies on 1,3-migration followed by [2+2] cycloaddition of 1,7-enyne benzoate **34** using catalyst **33** (Scheme 10).[21] The substrate scope is broad and an array of functional groups is tolerated throughout the course of the reaction. In addition to this feature, a single diastereomeric product is obtained irrespective of whether the starting substrate is a single isomer or a mixture of diastereomers. As such, the reaction highlights an effective transfer of chirality from enantiopure starting materials to the cyclobutane-fused piperidine products containing up to four stereogenic centers.

Scheme 10 1,3-Migration/[2+2] Cycloaddition of a 1,7-Enyne Benzoate[21]

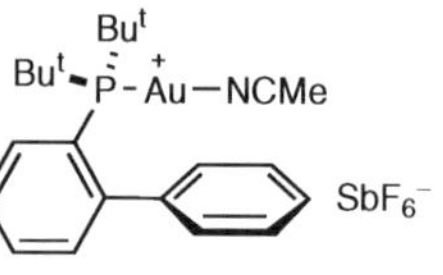

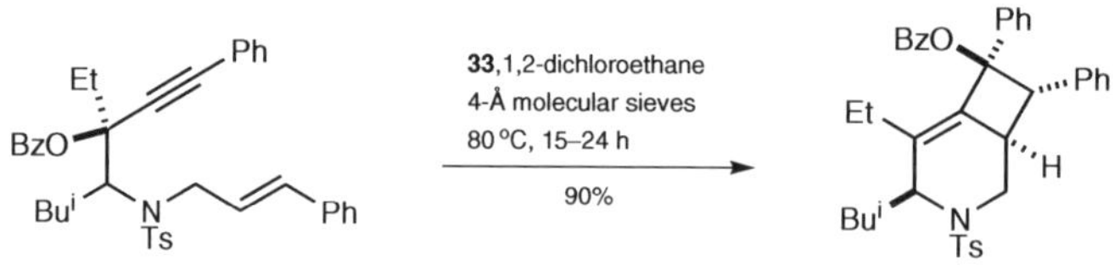

(1*R*,4*S*,7*R*,8*S*)-5-Ethyl-4-isobutyl-7,8-diphenyl-3-tosyl-3-azabicyclo[4.2.0]oct-5-en-7-yl Benzoate (35); **Typical Procedure:**[21]

Au(I) complex **33** (0.0075 mmol) was added to a soln of benzoate **34** (0.15 mmol) and 4-Å molecular sieves in dry 1,2-dichloroethane (15 mL) under an argon atmosphere. The resultant mixture was stirred at 80 °C for 15–24 h, cooled to rt, and filtered. The solid was washed with CH_2Cl_2, and the filtrate was concentrated. Purification by column chromatography (silica gel, EtOAc/hexane 1:9) gave the title compound as a white solid; yield: 83.7 mg (90%); mp 90–93 °C.

2.1.2.2 Domino [3+2] Cycloadditions

Cycloaddition reactions of 1,3-dipoles provide an efficient approach to the synthesis of five-membered heterocyclic compounds by formation of two new σ-bonds in the cycloaddition step.[22] Because at least one of the atoms must be a heteroatom, intermolecular examples can provide five-membered rings with up to four new stereogenic centers in the ring (especially if carbonyl ylides or azomethine ylides are used). Moreover, intramolecular examples allow access to products of considerable complexity, as two new rings are formed in the cycloaddition reaction.[23] If used as part of a domino process, more than two new σ-bonds are formed, and the cycloaddition step can precede or follow the other bond-forming reaction(s). Because the ylide dipole is typically prepared in situ by condensation (or another reaction) and then undergoes cycloaddition directly to give the new five-membered ring, it is possible to classify all such reactions as involving several steps, and therefore to be domino processes. However, given that this definition affords a broad array of potential choices, we will focus herein on reactions that involve more than ylide formation by condensation, deprotonation, oxidation, or other such simple methods. Indeed, the emphasis will be on reactions such as alkylations (e.g., by S_N2 reaction, conjugate addition, or carbene capture) that lead to the formation of the ylide that then undergoes cycloaddition (Scheme 11). Alternatively, reactions that involve dipolar cycloaddition to give the five-membered-ring products followed by subsequent in situ reaction(s) to give more complex products will be described (Scheme 11). As such, the following sections will be divided into two main types: (1) examples that involve the use of alkylation to prepare the ylide, followed by cycloaddition and (2) examples that involve [3+2] cycloaddition, followed by a subsequent reaction such as a ring-opening or rearrangement process. In each case, a range of different types of ylide dipole can be used, and this section will cover examples of 1,3-dipolar cycloaddition that involve domino reactions performed with the use of nitrones, nitronates, nitrile oxides, carbonyl ylides, azomethine ylides, azomethine imines, and azides.

Scheme 11 Domino Reactions with 1,3-Dipoles

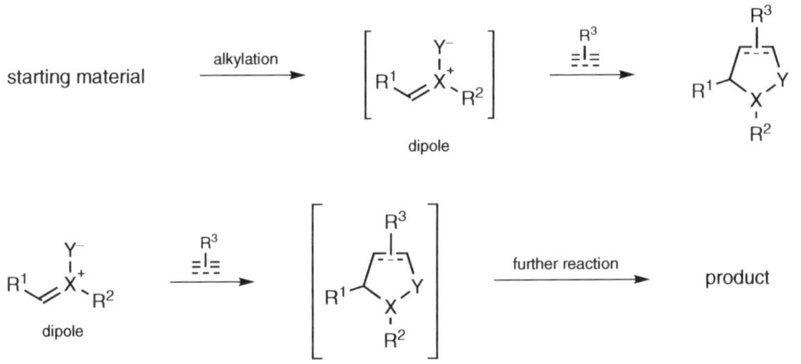

2.1.2.2.1 Cycloadditions with Nitrones, Nitronates, and Nitrile Oxides

Cycloaddition reactions of nitrones with alkynes or alkenes give access to dihydroisoxazole and tetrahydroisoxazole (isoxazolidine) products, respectively.[24] Isoxazolidines contain a weak N—O bond that can then be cleaved to afford a 1,3-amino alcohol functionality, which is present in many natural products and pharmaceutically active compounds. The wide range of different substituted alkynes and alkenes that are amenable to the cycloaddition process with a nitrone makes this chemistry particularly useful in chemical synthesis. In this section, some domino reactions involving nitronates and nitrile oxides that give rise to the same type of five-membered-ring products are also discussed.

for references see p 89

2.1.2.2.1.1 Reaction To Give a Nitrone Followed by Cycloaddition

The most common method to prepare a nitrone is by condensation of an *N*-alkyl or *N*-aryl hydroxylamine with an aldehyde or ketone. This process affords a nitrone directly that can then undergo cycloaddition. To promote domino transformations, events in which several bond-forming steps are conducted prior to cycloaddition, it is possible to simply use hydroxylamine itself (to give an intermediate oxime) that is combined with alkylation to give the nitrone. Such a cascade of condensation, alkylation, and cycloaddition provides a way to prepare complex polycyclic products. The alkylation step can consist of various types of reaction pathway, but it most typically involves reaction of the nitrogen atom of the intermediate oxime through a simple S_N2 reaction or conjugate addition.

For example, intramolecular alkylation of the oxime derived from aldehyde **36** and hydroxylamine occurs by nucleophilic substitution of chloride. The resultant nitrone undergoes intramolecular cycloaddition with the alkene to give tricyclic product **37** as a single stereoisomer (Scheme 12).[25] Product **37** can then be converted into the natural product (±)-myrioxazine A through two simple steps (reductive cleavage of the N—O bond and *N,O*-acetal formation with paraformaldehyde). Modification of this route to (±)-myrioxazine A has been reported by using a sulfonylated intermediate.[26] Critically, this type of chemistry has been extended from fused products to bridged products by using aldehydes such as **38**, in which branching of the alkyl halide and the alkene dipolarophile tether is now at the β-position rather than the α-position relative to the aldehyde (Scheme 12).[27,28] The intermediate oxime cyclizes (with displacement of bromide) to give nitrone **39**, which undergoes cycloaddition to afford final tricyclic product **40** containing the core ring system of an important subclass of the *daphniphyllum* alkaloids.

Scheme 12 Alkylation To Give a Nitrone Followed by Cycloaddition[25,27]

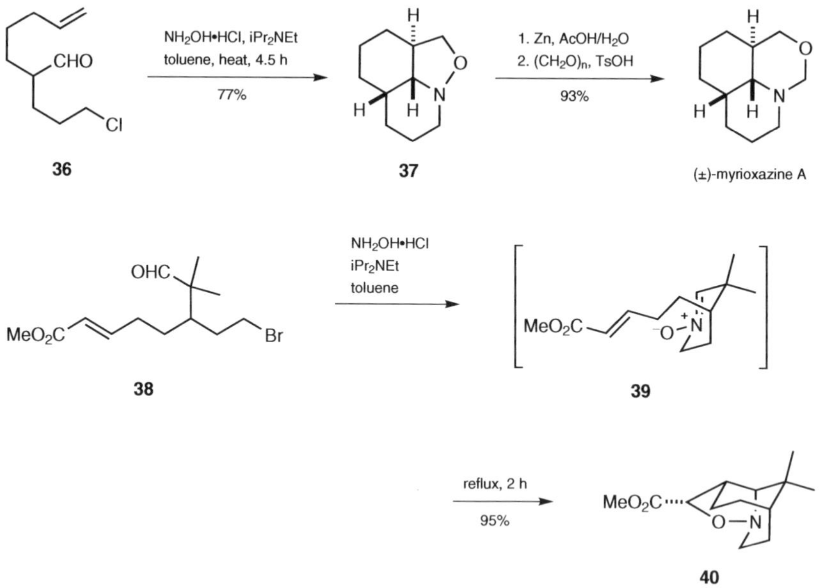

36: CHO, Cl → NH2OH•HCl, iPr2NEt, toluene, heat, 4.5 h, 77% → 37 → 1. Zn, AcOH/H2O 2. (CH2O)n, TsOH, 93% → (±)-myrioxazine A

38: OHC, MeO2C, Br → NH2OH•HCl, iPr2NEt, toluene → [39: MeO2C] → reflux, 2 h, 95% → 40: MeO2C

Another common electrophile for reaction with the intermediate oxime is an activated π-system. For example, conjugate/Michael addition of the oxime onto an α,β-unsaturated carbonyl or nitrile containing compound provides the desired nitrone for subsequent dipolar cycloaddition onto another alkene. Grigg and co-workers have reported several examples of this type of chemistry.[29,30] The groups of Holmes and Stockman have reported the reaction of hydroxylamine with symmetrical ketone **41** (Scheme 13).[31,32] Conjugate addition of the oxime with one of the unsaturated nitrile groups followed by cycloaddition with the other unsaturated nitrile group gives product **42**, which has been used in the syntheses of various histrionicotoxin alkaloids. Importantly, the kinetic product is actually the fused tricyclic ring system, but high reaction temperatures allow thermodynamic control to effect conversion into the final product **42**.[33]

Scheme 13 Intramolecular Conjugate Addition To Give a Nitrone Followed by Intramolecular Cycloaddition[31,32]

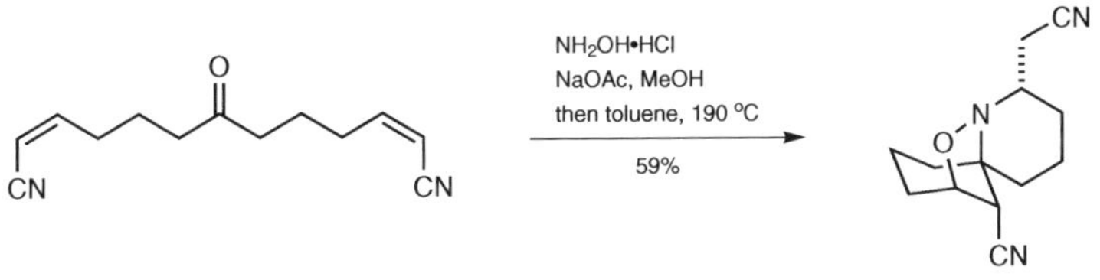

A related strategy has been reported by Padwa and co-workers, who have exploited the conjugate addition of oximes to 2,3-bis(phenylsulfonyl)buta-1,3-diene (**43**).[34,35] For example, in the synthesis of (±)-yohimbenone, conjugate addition of oxime **44** to unsaturated sulfone **43** gives an intermediate nitrone that undergoes dipolar cycloaddition with the other unsaturated sulfone to afford a new bridged product in the form of **45** (Scheme 14).[36] Reduction of the N—O bond provides ketone **46**, a material that can then be converted into the alkaloid yohimbenone through several additional steps.

Scheme 14 Intermolecular Conjugate Addition To Give a Nitrone Followed by Intramolecular Cycloaddition[36]

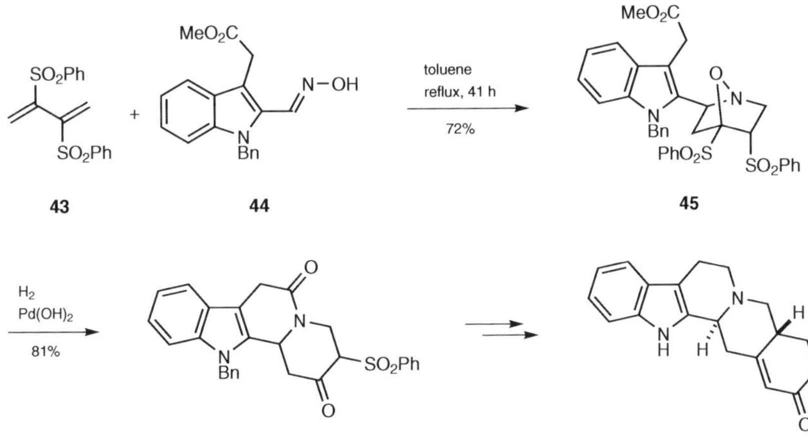

The reaction of an oxime through the nitrogen atom with a carbon-based electrophile provides the basis for the formation of a nitrone intermediate that can undergo subsequent dipolar cycloaddition. Another example is provided by the cyclization of an oxime onto an alkyne, an event that can be promoted by the presence of bromine to give an intermediate nitrone,[37] or alternatively by [2,3]-sigmatropic rearrangement of an *O*-propargylic oxime to give an *N*-allenylnitrone.[38] The nitrone can undergo domino cycloaddition with an activated π-system, followed by rearrangement.

Cyclization of an oxime onto a styrene group through a hydroamination reaction (retro-Cope-type reaction) also gives rise to a nitrone, and this species can undergo intramolecular cycloaddition if an alkene tether is present. For example, tricyclic product **47** is formed as a single stereoisomer through a domino intramolecular hydroamination/dipolar cycloaddition cascade (Scheme 15).[38]

Scheme 15 Hydroamination Reaction To Give a Nitrone Followed by Intramolecular Cycloaddition[38]

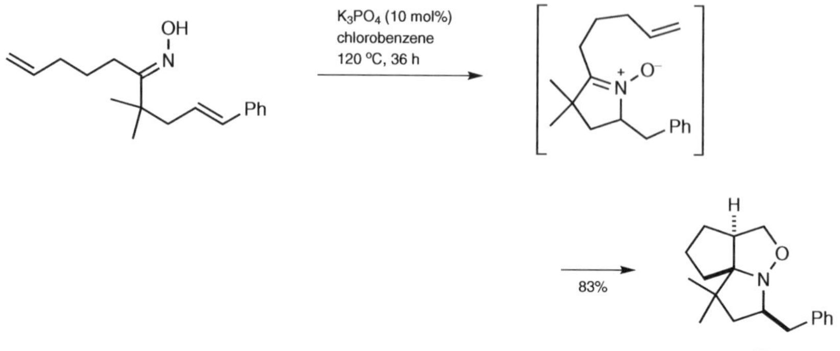

K_3PO_4 (10 mol%)
chlorobenzene
120 °C, 36 h

83%

47

An alternative approach to the formation of the required nitrone is through a domino ene reaction and then dehydrogenation. For example, the ene reaction of an allylbenzene with nitrosobenzene in the presence of a copper catalyst provides hydroxylamine **48**, which can then be oxidized in situ to give nitrone **49** (Scheme 16).[39] If the reaction is performed in the presence of an N-substituted maleimide dipolarophile, bicyclic product **50** is formed with high selectivity for the *trans*-isomer. Importantly, the reaction tolerates a range of electron-rich and electron-poor aromatic groups and various N-substituted maleimides.

Scheme 16 Ene Reaction Followed by Oxidation To Give a Nitrone and Subsequent Intermolecular Cycloaddition[39]

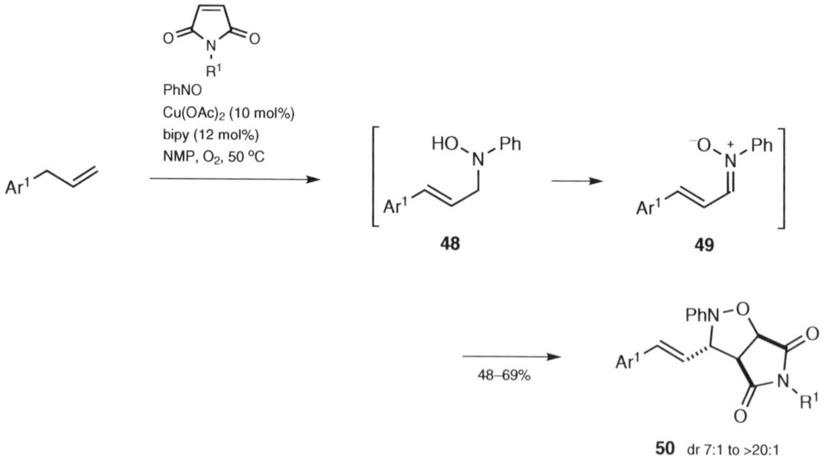

PhNO
$Cu(OAc)_2$ (10 mol%)
bipy (12 mol%)
NMP, O_2, 50 °C

48 **49**

48–69%

50 dr 7:1 to >20:1

Ar^1 = Ph, 4-FC_6H_4, 4-$MeOC_6H_4$, 4-$F_3CC_6H_4$, 2-Tol, 2-$TsOC_6H_4$, 3-pyridyl, 4-NCC_6H_4; R^1 = Me, Ph, Bn

Oxidation of a hydroxylamine can give a nitrone, and hydroxylamines can be formed by a variety of methods, such as the ene reaction shown in Scheme 16. Another method is the elimination of an amine *N*-oxide. Li and co-workers have shown that oxidation of amine **51** with 3-chloroperoxybenzoic acid promotes a cascade of reactions involving N-oxidation to the amine *N*-oxide, elimination (aided by a β-carbonyl group), oxidation of the hydroxylamine to nitrone **52**, and dipolar cycloaddition onto the newly formed alkene (Scheme 17).[40] This chemistry gives access to tetracyclic products such as **53**, which are amenable to N—O bond reduction with zinc in tetrahydrofuran/water to promote lactamization and formation of tricyclic products **54**.

Scheme 17 Oxidation, Elimination, and Oxidation To Give a Nitrone Followed by Intramolecular Cycloaddition[40]

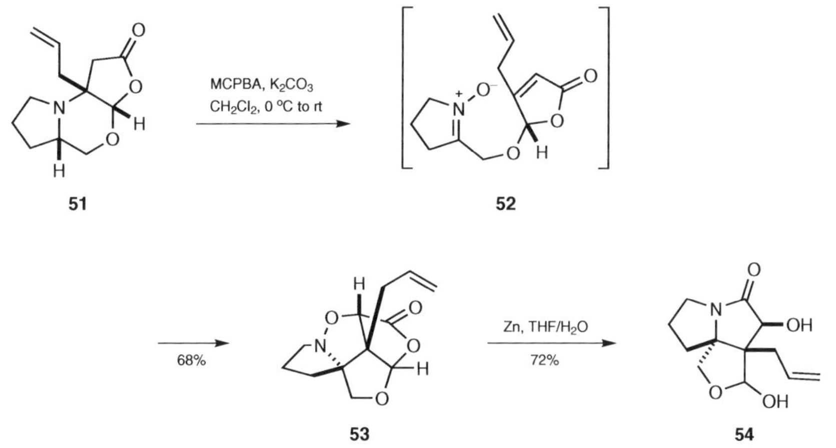

MCPBA, K_2CO_3, CH_2Cl_2, 0 °C to rt; 68%; Zn, THF/H_2O, 72%

51 52 53 54

Methyl (1*R,6*R**,7*S**,8*S**)-9,9-Dimethyl-5-oxa-4-azatricyclo[5.2.2.0^{4,8}]undecane-6-carboxylate (40); Typical Procedure:**[27]

A mixture of aldehyde **38** (250 mg, 0.82 mmol), NH_2OH•HCl (68 mg, 0.98 mmol), and iPr_2NEt (0.34 mL, 1.97 mmol) in toluene (8 mL) was heated under reflux. After 2 h, the solvent was evaporated. Purification by column chromatography (silica gel, CH_2Cl_2/MeOH/NH_3 98:2:0.2) gave the title compound as an oil; yield: 186 mg (95%).

Methyl 2-{1-Benzyl-2-[4,5-bis(benzenesulfonyl)-7-oxa-1-azabicyclo[2.2.1]heptan-2-yl]-1*H*-indol-3-yl}acetate (45); Typical Procedure:[36]

A mixture of oxime **44** (5.5 g, 17 mmol) and bis(sulfone) **43** (6.2 g, 19 mmol) in toluene (200 mL) was heated under reflux. After 41 h, the mixture was cooled to rt and a small amount of silica gel was added. The slurry was concentrated under reduced pressure. Purification by column chromatography (silica gel) gave the title compound as a white solid; yield: 7.9 g (72%); mp 142–143 °C.

(3*S,3a*S**,6a*R**)-5-Methyl-2-phenyl-3-[(*E*)-styryl]tetrahydro-4*H*-pyrrolo[3,4-*d*]isoxazole-4,6(5*H*)-dione (50, R^1 = Me; Ar^1 = Ph); Typical Procedure:**[39]

A mixture of $Cu(OAc)_2$ (9.1 mg, 0.05 mmol) and bipyridine (9.4 mg, 0.06 mmol) in 1-methylpyrrolidin-2-one (3 mL) was stirred in air for 1 h (until the copper salt had dissolved). Allylbenzene (0.75 mmol, 1.5 equiv), nitrosobenzene (2 mmol, 4.0 equiv), and *N*-methylmaleimide (0.5 mmol, 1.0 equiv) were then added. The vessel was capped with a rubber septum, degassed, and flushed with O_2 (3 ×); the mixture was then stirred at 50 °C under an atmosphere of O_2 (balloon). After 16 h, the mixture was cooled to rt and diluted with H_2O (10 mL) and EtOAc (10 mL). The organic layer was separated, and the aqueous phase was extracted with EtOAc (2 ×). The combined organic extracts were washed with H_2O (10 mL) and brine (10 mL), dried (Na_2SO_4), filtered, and concentrated. Purification by column chromatography (silica gel, petroleum ether/EtOAc) gave the title compound as a yellow solid; yield: 0.109 g (65%); dr 7:1; mp 152–154 °C.

for references see p 89

2.1.2.2.1.2 Cycloaddition with a Nitrone and Subsequent Reaction

The reaction of a nitrone with a π-system gives a five-membered ring that can be prone to subsequent in situ transformation; however, the use of an alkene π-system gives a stable isoxazolidine so, for a domino sequence, another reactive functionality must be present. An example of this has been demonstrated by Doyle and co-workers by using diazo(vinyl)-acetates.[41] Treating a mixture of a diarylnitrone, such as nitrone **55**, and rhodium(II) octanoate with methyl 2-diazobut-3-enoate (**56**) gives unusual bridged product **58** (Scheme 18). Its formation involves a domino process with several steps, including formation of the rhodium carbenoid, dipolar cycloaddition of the nitrone with the alkene (to give *trans*-3,4-disubstituted product **57**), cyclopropanation of the metallocarbene with the *N*-aryl ring, electrocyclic reaction with opening of the cyclopropane, and finally N—O bond breaking and rearrangement.

Scheme 18 Nitrone Cycloaddition with a Vinylated Metallocarbene Followed by a Pericyclic Cascade[41]

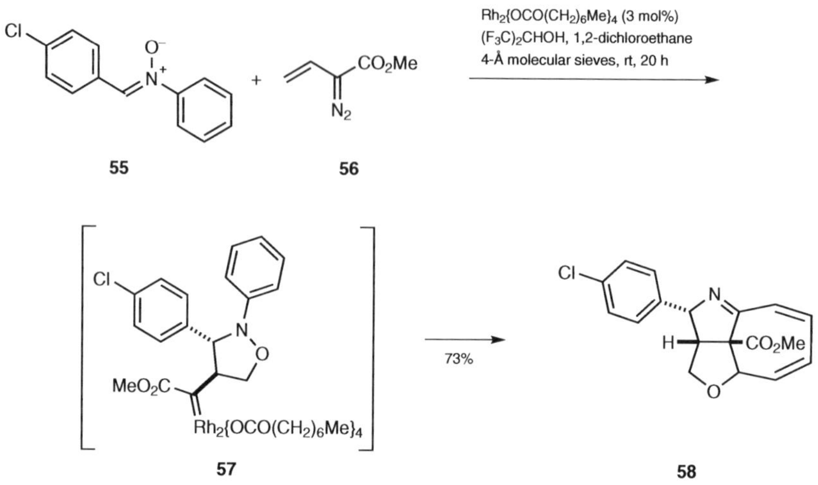

$Rh_2\{OCO(CH_2)_6Me\}_4$ (3 mol%)
$(F_3C)_2CHOH$, 1,2-dichloroethane
4-Å molecular sieves, rt, 20 h

55 + **56** → [**57**] —73%→ **58**

Nitrones are good dipoles for cycloaddition reactions with alkenes. Cycloaddition with a ketene can then lead to subsequent [3,3]-sigmatropic rearrangement.[42] By using disubstituted ketenes **60**, cycloaddition occurs at the carbonyl π-system rather than at the alkene through a regioselective process that furnishes five-membered products such as **61** (Scheme 19).[43] If an *N*-aryl nitrone is part of the reactant (as in **59**), subsequent [3,3]-sigmatropic rearrangement is promoted to give compound **62**. This event is followed by aromatization, ring opening, and subsequent conversion into oxindoles **63**. The use of a serine-derived aldehyde as a precursor to the nitrone provides a stereogenic center in the nitrone that is relayed to the new stereogenic center in the product. Importantly, in these cases, high levels of enantioselectivity can be achieved, particularly if the ring-nitrogen substituent in nitrone **59** is a (2,4,6-triisopropylphenyl)sulfonyl group. Cycloaddition of an *N*-phenylnitrone onto an alkyne also gives a product that is amenable to [3,3]-sigmatropic rearrangement and can be used to prepare indole derivatives.[44]

Scheme 19 Nitrone Cycloaddition with a Ketene Followed by Rearrangement[43]

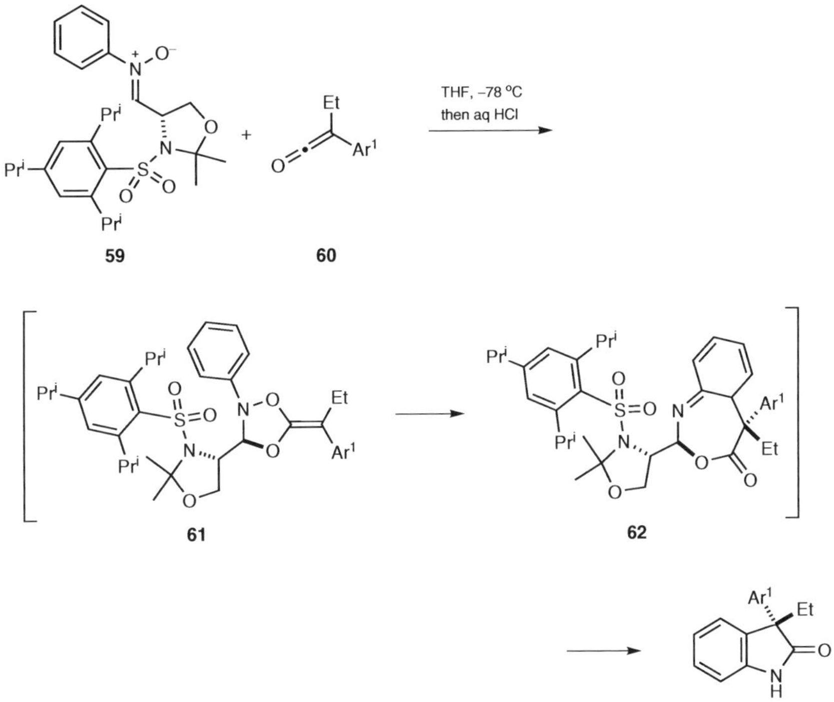

59 60

THF, –78 °C then aq HCl

61 62

63

Ar^1	er	Yield (%)	Ref
Ph	98:2	84	[43]
4-$MeOC_6H_4$	99:1	81	[43]

3-(4-Chlorophenyl)-8b-(methoxycarbonyl)-2,2a,3,8a-tetrahydro-8b*H*-1-oxa-4-azacyclopenta[*cd*]azulene (58); General Procedure:[41]

$Rh_2\{OCO(CH_2)_6Me\}_4$ (6.0 mg, 3 mol%), *C,N*-diarylnitrone **55** (0.25 mmol), 1,1,1,3,3,3-hexafluoropropan-2-ol (HFIP; 0.027 mL, 0.25 mmol), and 1,2-dichloroethane (1.5 mL) were added to a dry Schlenk flask containing 4-Å molecular sieves (100 mg) and a stirrer bar at rt under an atmosphere of N_2. The green soln was stirred for 5 min, and then the flask was wrapped with aluminum foil. A freshly prepared soln of methyl 2-diazobut-3-enoate (**56**; 95 mg, 0.75 mmol) in 1,2-dichloroethane (1 mL) was added to the flask over 1 h using a syringe pump. The mixture was stirred at rt for 20 h, the solvent was evaporated, and the residue was dissolved in a minimal amount of CH_2Cl_2. Purification by column chromatography (silica gel, hexane/EtOAc 3:1 with 5% Et_3N) gave the title compound; yield: 60 mg (73%).

2.1.2.2.1.3 Reaction To Give a Nitronate Followed by Cycloaddition

A domino process involving nitronate cycloaddition has been reported by Xiao and coworkers (Scheme 20).[45] Addition of sulfur ylide **64** to nitroalkene **65** gives nitronate **66**, which undergoes intramolecular dipolar cycloaddition with the tethered alkene. Product **67** is generated with very high selectivity (dr >95:5) for a range of aryl groups. It is of note that asymmetric versions of this reaction have been developed in the presence of a chiral C_2-symmetric urea catalyst or by using a chiral sulfur ylide.[46]

for references see p 89

Scheme 20 Formation of a Nitronate Followed by Cycloaddition[45]

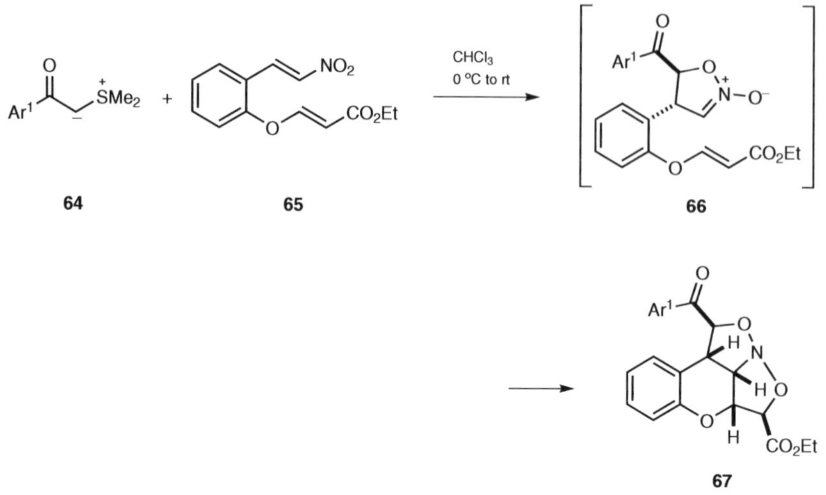

Ar^1	Yield (%)	Ref
Ph	89	[45]
4-$MeOC_6H_4$	92	[45]
2-$MeOC_6H_4$	90	[45]
4-Tol	93	[45]
4-FC_6H_4	90	[45]
4-ClC_6H_4	90	[45]
4-BrC_6H_4	90	[45]
3-BrC_6H_4	94	[45]
4-$O_2NC_6H_4$	90	[45]
2-furyl	97	[45]

Ethyl 1-Benzoyl-4,4a,9b,9c-tetrahydro-1*H*-2,3,5-trioxa-2a-azapentaleno[1,6-*ab*]naphthalene-4-carboxylate (67, Ar^1 = Ph); Typical Procedure:[45]

A soln of sulfur ylide **64** (0.55 mmol) in $CHCl_3$ (15 mL) was added slowly to a soln of nitroalkene **65** (0.5 mmol) in $CHCl_3$ (10 mL) at 0 °C using a syringe pump. After 12 h, the mixture was warmed to rt and stirred for another 12 h. The solvent was evaporated, and the residue was purified by column chromatography (silica gel, hexane/EtOAc 10:1 to 5:1) to give the title compound; yield: 170 mg (89%).

2.1.2.2.1.4 Reaction To Give a Nitrile Oxide Followed by Cycloaddition

In Section 2.1.2.2.1.3 (Scheme 20), it was shown that the addition of a carbanion species to a nitroalkene can lead to a nitronate dipole. Alternatively, ring opening of nitronates is known to give nitrile oxides, and these materials can then undergo cycloaddition with a π-system (Scheme 21).[47] For example, treatment of a nitroalkene with an isocyanide, generated from trimethylsilyl cyanide and an epoxide, gives intermediate nitronate **68** that ring opens to give nitrile oxide **69**. In the presence of an alkene dipolarophile, such as methyl acrylate, new cycloaddition product **70** is generated. The whole process is aided by the Lewis acids used in the different portions of the sequence [palladium(II) cyanide and lithium perchlorate], although the yields are not very high and the process has no diastereocontrol.

Scheme 21 Formation of a Nitrile Oxide Followed by Cycloaddition[47]

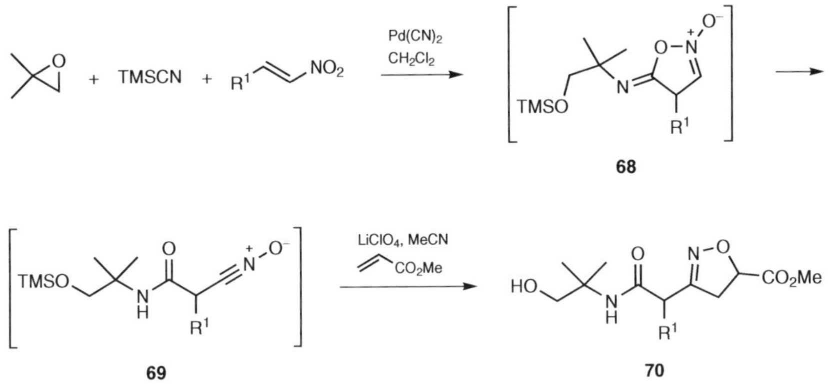

R^1	Yield (%)	Ref
Ph	37	[47]
4-$MeOC_6H_4$	31	[47]
$(CH_2)_5Me$	33	[47]

Methyl 3-{2-[(1-Hydroxy-2-methylpropan-2-yl)amino]-2-oxo-1-phenylethyl}-4,5-dihydroisoxazole-5-carboxylate (70, R^1 = Ph); Typical Procedure:[47]

CAUTION: *Trimethylsilyl cyanide and its hydrolysis products are extremely toxic.*

CAUTION: *Cyanide salts can be absorbed through the skin and are extremely toxic.*

$Pd(CN)_2$ (27 mg, 0.17 mmol) and 2,2-dimethyloxirane (0.328 mL, 3.69 mmol) were added to a soln of TMSCN (0.491 mL, 3.69 mmol) in CH_2Cl_2 (2 mL) in a sealed tube under an atmosphere of N_2. The mixture was heated to 60 °C and, after 18 h, allowed to cool to rt. A soln of (*E*)-(2-nitrovinyl)benzene (500 mg, 3.35 mmol) and methyl acrylate (3.02 mL, 33.5 mmol) in MeCN (7 mL) was added, which was followed by the addition of $LiClO_4$ (3.72 g, 35 mmol). The mixture was heated to 80 °C and, after 60 h, CH_2Cl_2 (20 mL) was added and the mixture was washed with sat. aq NH_4Cl (2 × 50 mL). The aqueous layer was extracted with CH_2Cl_2 (2 × 20 mL) and the combined organic layers were dried ($MgSO_4$), the solvent was evaporated, and the residue was purified by column chromatography (silica gel, CH_2Cl_2/MeOH 49:1 to 24:1) to give the title compound as an oil; yield: 419 mg (37%).

2.1.2.2.1.5 Cycloaddition with a Nitrile Oxide and Subsequent Reaction

Domino transformations performed with the use of nitrile oxides are also possible by initial cycloaddition followed by subsequent reaction; an example of this strategy is shown in Scheme 22.[48] Treatment of alkyne **71** with a nitrile oxide (prepared in situ from the α-chloro oxime) promotes dipolar cycloaddition, which occurs regioselectively to give intermediate isoxazole **72**. The presence of the *ortho*-amino group in the starting material then leads to cyclization to give isoxazolo[4,5-*c*]quinoline product **73**. The α-chloro oxime starting material is also amenable to formation in situ from the precursor oxime with N-chlorosuccinimide. An alternative isoxazole synthesis uses a domino dipolar cycloaddition of a nitrile oxide with a cyclopropene, followed by ring opening of the cyclopropane ring.[49]

Scheme 22 Cycloaddition with a Nitrile Oxide Followed by Cyclization[48]

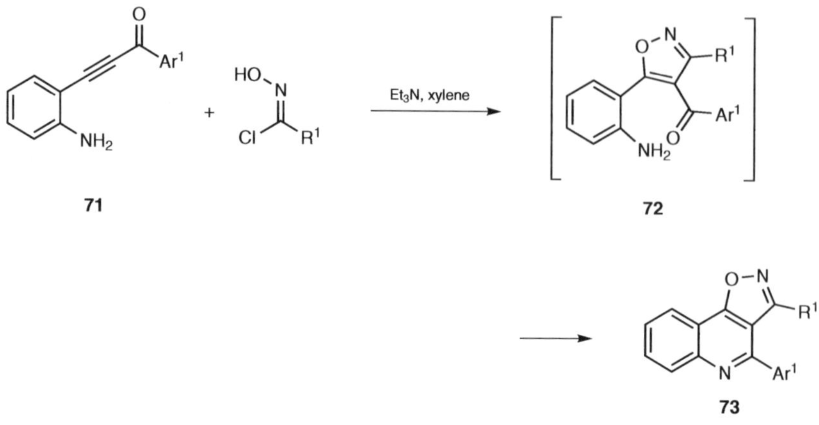

Ar^1	R^1	Yield (%)	Ref
4-ClC_6H_4	Ph	71	[48]
3-$F_3CC_6H_4$	CO_2Et	48	[48]

4-(4-Chlorophenyl)-3-phenylisoxazolo[4,5-c]quinoline (73, Ar^1 = 4-ClC_6H_4; R^1 = Ph); Typical Procedure:[48]

A soln of the α-chloro oxime (1.2 mmol) in dry xylene (5 mL) was added slowly to a hot soln of alkyne **71** (1.0 mmol) and Et_3N (1.5 mmol) in dry xylene (5 mL). After heating for 6 h, the solvent was evaporated under reduced pressure. The residue was taken up in EtOAc and was washed with 0.1 M HCl in H_2O. The organic layer was dried (Na_2SO_4), concentrated, and purified by column chromatography (silica gel, petroleum ether/EtOAc 49:1) to give the title compound as a solid; yield: 253 mg (71%); mp 186–188 °C.

2.1.2.2.2 Cycloadditions with Carbonyl Ylides

Cycloaddition reactions of carbonyl ylides have found considerable use in the construction of polycyclic natural products.[50,51] These ylides tend to be made either by cyclization of a metallocarbene (generated from an α-diazocarbonyl group) with another internal carbonyl or by cyclization of a carbonyl onto an alkyne activated by a transition metal such as platinum. The tetrahydrofuran or tetrahydropyran ring system that is typically generated from such protocols is itself commonly found in target molecules, or it can be ring opened to give the desired end product. Enantio- and diastereoselective versions of this chemistry, including reactions under continuous-flow conditions, are feasible.[52,53]

2.1.2.2.2.1 Reaction of an α-Diazo Compound To Give a Carbonyl Ylide Followed by Cycloaddition

The most common method to prepare a carbonyl ylide is by the rhodium-catalyzed reaction of an α-diazo compound in which the substrate has a suitably located carbonyl group to form the carbonyl ylide within a five- or six-membered ring. The carbonyl group used to cyclize onto the metallocarbene is typically a ketone or carboxylic amide, although examples with an aldehyde or a carboxylic ester are also known.

An example of cyclization with a ketone comes from Lam and Chiu, who have utilized such an approach to achieve the total synthesis of the diterpenoid (−)-indicol (Scheme 23).[54] In the key domino step, catalytic rhodium(II) promotes the formation of the metallocarbene from diazo compound **74**, with cyclization onto the ketone group to give rise to carbonyl ylide **75**. Dipolar cycloaddition with the terminal alkene then completes the sequence, which sets up the core bicyclo[5.4.0]undecane ring system as a 3.1:1 ratio of diastereomers **76A** and **76B**. Major isomer **76A** can then be converted into (−)-indicol (**77**) through 12 additional steps.

Scheme 23 Cyclization of a Linear Substrate with a Ketone To Give a Carbonyl Ylide Followed by Cycloaddition[54]

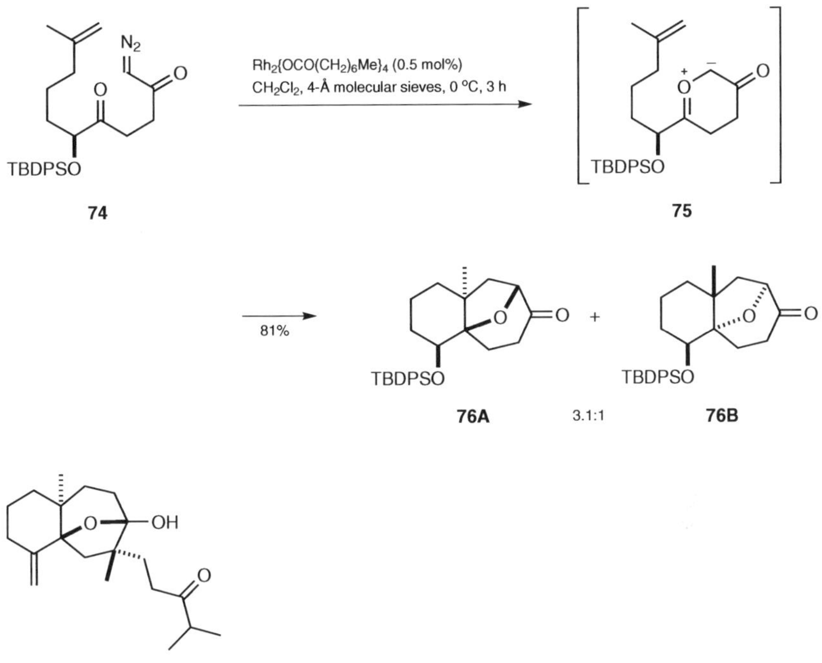

N_2
TBDPSO
74
$Rh_2\{OCO(CH_2)_6Me\}_4$ (0.5 mol%)
CH_2Cl_2, 4-Å molecular sieves, 0 °C, 3 h
TBDPSO
75
81%
TBDPSO
TBDPSO
76A
3.1:1
76B
OH

77 (−)-indicol

A similar strategy has been used for the synthesis of (+)-polygalolides A and B, which can be prepared using the domino cyclization/cycloaddition of carbonyl ylide **79**, generated from diazocarbonyl compound **78** (Scheme 24), as a key step.[55] In this sequence, rhodium(II) acetate is used to form the metallocarbene, which cyclizes onto the ketone to give carbonyl ylide **79**. Intramolecular cycloaddition onto the tethered alkene requires heating and gives tricyclic product **80** as a single stereoisomer. Significantly, the catalyst loading can be decreased to 2 mol% on a larger (500-mg) scale without a dramatic effect on the yield (70% yield for **80**).

for references see p 89

Scheme 24 Cyclization of a Branched Substrate with a Ketone To Give a Carbonyl Ylide Followed by Cycloaddition[55]

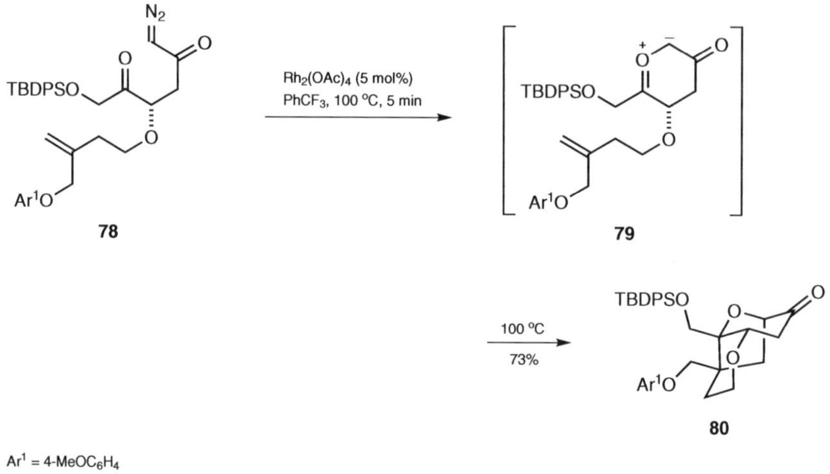

$Rh_2(OAc)_4$ (5 mol%)
$PhCF_3$, 100 °C, 5 min

100 °C
73%

78 79 80

Ar^1 = 4-$MeOC_6H_4$

The domino cyclization/dipolar cycloaddition reaction can also be extended to alkyne dipolarophiles.[56] For example, heating diazocarbonyl compound **81** with rhodium(II) acetate in toluene at 130 °C for 5 hours gives cycloadduct **82** smoothly (Scheme 25) in 72% yield as a single diastereomer.[57] This material can subsequently be advanced into analogues of the well-known natural product colchicine.

Scheme 25 Cyclization with a Ketone To Give a Carbonyl Ylide Followed by Cycloaddition onto an Alkyne[57]

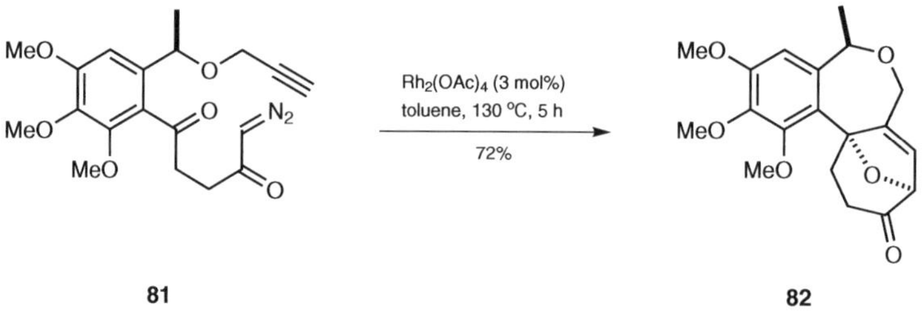

$Rh_2(OAc)_4$ (3 mol%)
toluene, 130 °C, 5 h
72%

81 82

The reactivity of carbonyl ylides also makes them amenable to intermolecular dipolar cycloaddition with external dipolarophiles. For example, heating diazocarbonyl compound **83** with rhodium(II) acetate and alkene **84** (a γ-alkylidenebutenolide) gives cycloadduct **85** (Scheme 26).[58] Computational studies have suggested that the rhodium metal is associated with the ylide during the domino process, a finding that can be of value to other applications and/or extensions of this process.

Scheme 26 Cyclization with a Ketone To Give a Carbonyl Ylide Followed by Intermolecular Cycloaddition[58]

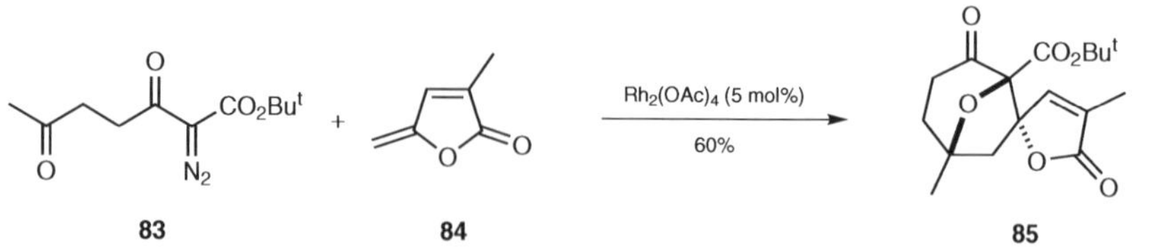

$Rh_2(OAc)_4$ (5 mol%)
60%

83 84 85

A useful approach to the synthesis of alkaloids using carbonyl ylides, one that has been particularly championed by Padwa and co-workers, is the cyclization of a metallocarbene onto a carboxylic amide group. In this design, following dipolar cycloaddition of the resulting carbonyl ylide, the newly formed tetrahydrofuran ring bears an amino substituent at the 2-position. This motif allows ready opening of the tetrahydrofuran ring to give an iminium ion that can undergo subsequent reaction. Overall, this process leads to two new rings with control of regio- and stereoselectivity, and the chemistry has found considerable use in synthesis. For example, treatment of diazocarbonyl compound **86** with rhodium(II) acetate as the catalyst leads to the formation of cycloadduct **87** in excellent yield (Scheme 27).[59] The diastereoselectivity arises from a preference for the *endo* transition state. Ring opening of the oxabicyclic moiety aided by the 2-amino substituent and subsequent transformations lead, in this case, to the synthesis of the *tacaman* alkaloid (±)-tacamonine.

Scheme 27 Cyclization with an Amide To Give a Carbonyl Ylide Followed by Cycloaddition onto an Alkene[59]

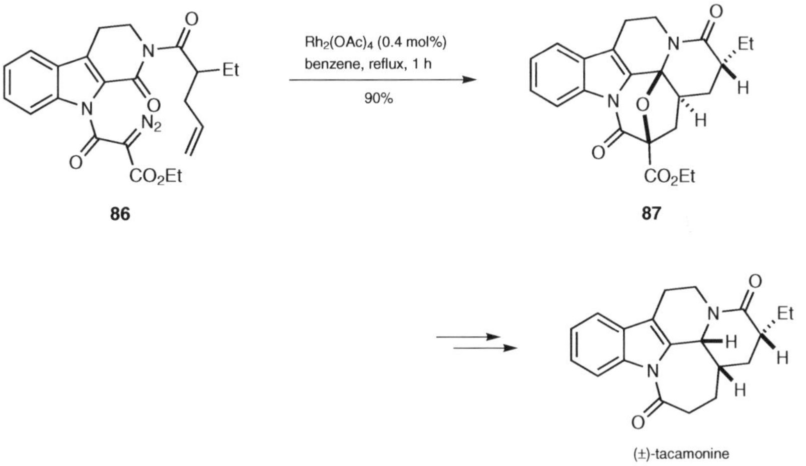

$Rh_2(OAc)_4$ (0.4 mol%), benzene, reflux, 1 h; 90%

86 → **87** → (±)-tacamonine

Another example of this type of chemistry is found in the synthesis of the complex alkaloid (±)-aspidophytine (Scheme 28).[60] Once again, rhodium(II) acetate is used to decompose diazocarbonyl compound **88** and to generate the metallocarbene that cyclizes onto the lactam carbonyl. The resulting carbonyl ylide then undergoes cycloaddition onto the indole π-system to yield polycyclic product **89** as a single stereoisomer. This outcome arises from a preference for the *exo* transition state owing to the presence of the bulky *tert*-butyl ester group. Finally, treatment of cycloadduct **89** with a Lewis acid ($BF_3•OEt_2$) promotes opening of the tetrahydrofuran ring and trapping of the iminium ion by the *tert*-butyl ester to give lactone **90**. Only a few additional steps are necessary to complete the total synthesis of (±)-aspidophytine (**91**).

for references see p 89

Scheme 28 Cyclization with an Amide To Give a Carbonyl Ylide Followed by Cycloaddition onto an Indole[60]

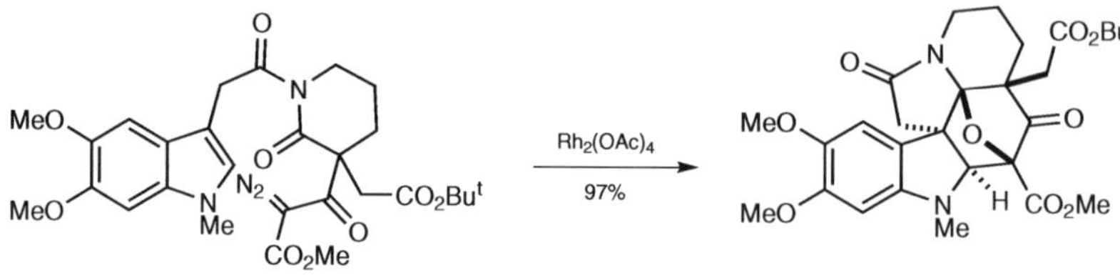

88 89

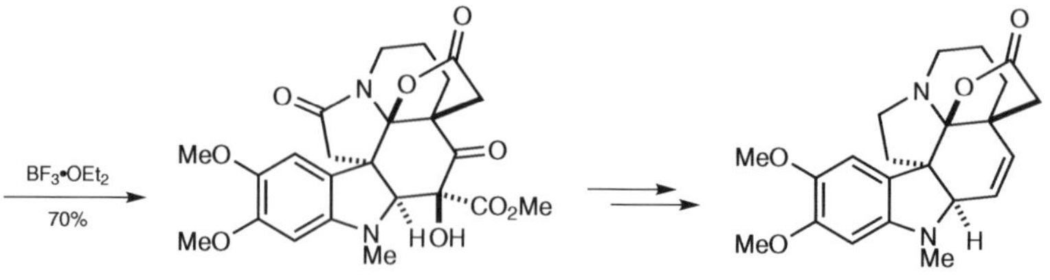

90 91 (±)-aspidophytine

If the starting carbonyl ylide substrate is an imide, cyclization of the metallocarbene creates a five-membered ring known as an isomünchnone. This class of ylide is capable of [3+2] cycloaddition with π-systems to generate new bicyclic ring systems.[61] For example, Padwa and co-workers have reported the formal synthesis of (±)-lycopodine by using, as a key step, the intramolecular cycloaddition of isomünchnone **93** generated by treatment of diazoimide **92** with dirhodium(II) tetrakis(perfluorobutanoate) [$Rh_2(pfb)_4$; pfb = heptafluorobutanoate] (Scheme 29).[62] A mixture of diastereomers **94A** and **94B** is obtained (dr 3:2), and this mixture can be used in subsequent chemistry toward the natural product.

Scheme 29 Cyclization To Give an Isomünchnone Followed by Cycloaddition onto an Unactivated Alkene[62]

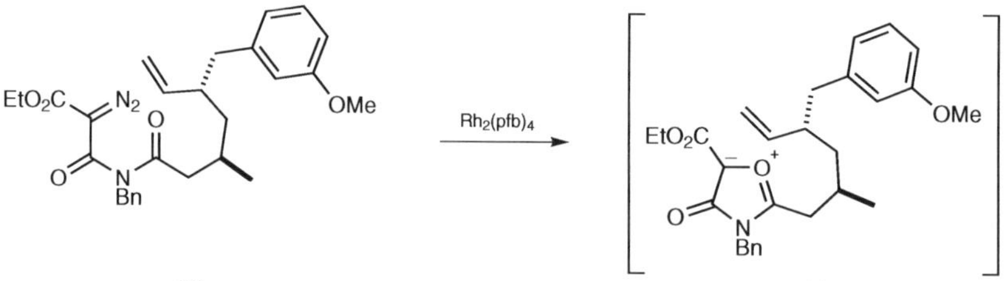

92 93

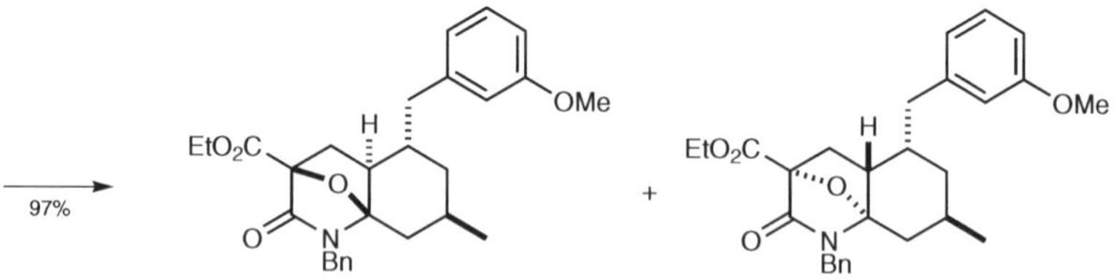

94A 3:2 94B

This type of chemistry has been applied to the formal synthesis of the antitumor agent camptothecin (Scheme 30).[63] Heating diazoimide **95** with rhodium(II) acetate in benzene gives a 5:1 mixture of cycloadducts; tetracycle **96** is the major isomer, and it can be isolated cleanly by recrystallization.

Scheme 30 Cyclization To Give an Isomünchnone Followed by Cycloaddition onto an Activated Alkene[63]

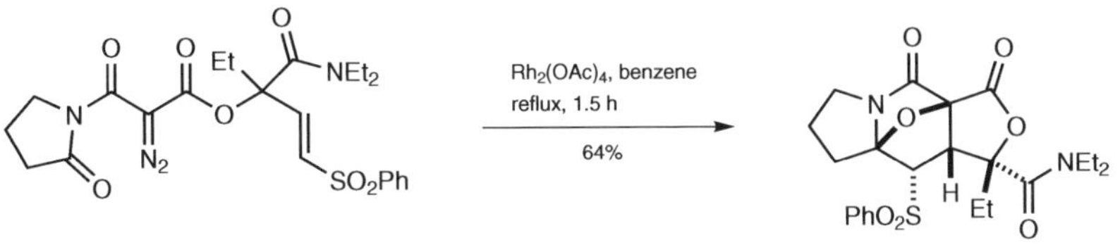

(4*S*,4a*R*,8*R*,9a*S*)- and (4*S*,4a*S*,8*S*,9a*R*)-4-(*tert*-Butyldiphenylsiloxy)-9a-methyloctahydro-4a,8-epoxybenzo[7]annulen-7(2*H*)-ones (76A **and** 76B)**; Typical Procedure:**[54]
Powdered, predried 4-Å molecular sieves (491 mg) were added to a soln of diazo ketone **74** (498 mg, 1.05 mmol) in CH_2Cl_2 (105 mL). The mixture was stirred for 5 min at 0 °C, $Rh_2\{OCO(CH_2)_6Me\}_4$ (4.7 mg, 0.006 mmol) was added, and the resulting mixture was stirred for 3 h at 0 °C and then filtered through sintered glass. The solvent was removed under reduced pressure, and the residue was purified by column chromatography (EtOAc/hexane 1:49 to 1:24) to give ketone **76A** as a white solid; yield: 286 mg (61%); mp 109–112 °C; and ketone **76B** as a white solid; yield: 94 mg (20%); mp 119–122 °C.

(1*R*,4*S*,8*R*,9*R*)-9-[(*tert*-Butyldiphenylsiloxy)methyl]-8-[(4-methoxyphenoxy)methyl]-5,10-dioxatricyclo[6.2.1.0^{4,9}]undecan-2-one (80)**; Typical Procedure:**[55]
A soln of diazo ketone **78** (33.8 mg, 0.056 mmol) in (trifluoromethyl)benzene (1.1 mL) was added over 5 min to a soln of $Rh_2(OAc)_4$ (1.4 mg, 0.003 mmol) in (trifluoromethyl)benzene (4.5 mL) at 100 °C. The resulting mixture was allowed to cool to rt, and the solvent was evaporated. The residue was purified by column chromatography (silica gel, hexane/EtOAc 12:1 to 9:1) to give the title compound as a white amorphous solid; yield: 23.4 mg (73%).

(2*R,3a^1*S**,11*R**,12a*S**)-2-Ethyl-11-(ethoxycarbonyl)-1,4,5,11,12,12a-hexahydro-3*H*-13-oxa-3a,9b-diaza-3a^1,11-methanobenzo[*a*]naphtho[2,1,8-*cde*]azulene-3,10(2*H*)-dione** (87)**; Typical Procedure:**[59]
$Rh_2(OAc)_4$ (2 mg, 0.0045 mmol) was added to a soln of diazocarbonyl compound **86** (500 mg, 1.15 mmol) in benzene (25 mL) (**CAUTION:** *carcinogen*), and the mixture was heated under reflux. After 1 h, the mixture was allowed to cool to rt, and the solvent was evaporated. The residue was purified by column chromatography (silica gel) to give the title compound as a white solid; yield: 420 mg (90%); mp 151–152 °C.

(1*R,3a*R**,8a*S**,9*R**,9a*S**)-*N*,*N*,1-Triethyl-3,4-dioxo-9-(phenylsulfonyl)tetrahydro-1*H*,3*H*,4*H*,6*H*-3a,8a-epoxyfuro[3,4-*f*]indolizine-1-carboxamide** (96)**; Typical Procedure:**[63]
$Rh_2(OAc)_4$ (7 mg, 0.016 mmol) and diazo compound **95** (400 mg, 0.8 mmol) were dried under high vacuum. After 1 h, dry benzene (13 mL) (**CAUTION:** *carcinogen*) was added, and the mixture was heated under reflux. After 1.5 h, the mixture was allowed to cool to rt and then filtered through Celite, washing with CH_2Cl_2. The solvent was evaporated, and the residue was purified by recrystallization (EtOAc) to give the title compound as a white solid; yield: 240 mg (64%); mp 212–214 °C.

for references see p 89

2.1.2.2.2.2 Reaction of an Alkyne To Give a Carbonyl Ylide Followed by Cycloaddition

An alternative and increasingly popular method to access carbonyl ylides is by activation of an alkyne with a transition metal to promote cyclization by a tethered carbonyl group. Cyclization of an aldehyde or ketone onto an alkyne activated by platinum generates a carbonyl ylide as part of a pyrylium ring bearing a platinum(II) moiety. Oh and co-workers have shown that these ylides, such as **98** generated from aldehyde **97**, can undergo dipolar cycloaddition with an alkene to generate a new platinum carbene, such as **99** (Scheme 31).[64] This metallocarbene can then undergo reactions such as C—H insertion.[65,66] For example, with compound **99** there is a suitably located benzyl group for insertion, and this gives rise to product **100**. Alternatively, in the presence of water, the metallocarbene can be trapped and the tetrahydrofuran ring can be opened to give a keto alcohol product.[67] Similar approaches with the use of gold catalysis have been reported,[68] as has a method avoiding metal catalysts altogether by using an iodine-mediated domino cyclization/cycloaddition sequence.[69]

Scheme 31 Cyclization Using Platinum Followed by Cycloaddition and C–H Insertion[64]

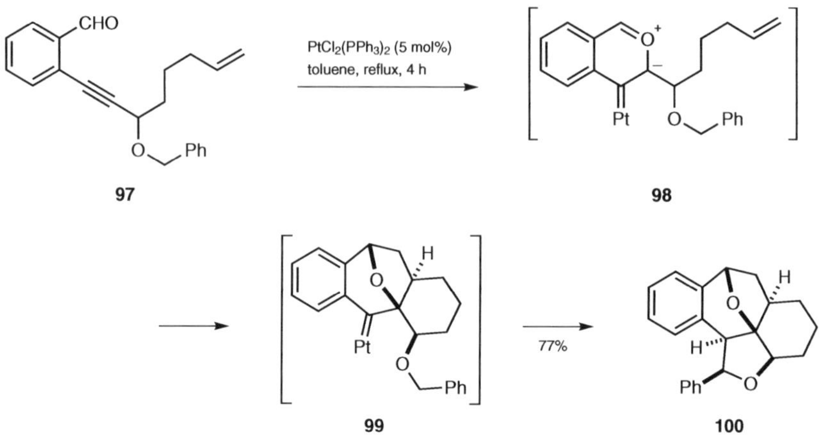

In related chemistry, She and co-workers have reported that oxygen-containing heterocycles can be prepared by cyclization of a carboxylic ester group onto an alkyne activated by platinum.[70] To prepare nitrogen-containing heterocycles, they have found that gold catalysis is preferable (Scheme 32).[71] Activation of alkyne **101** with a gold(I) species, followed by cyclization of the carbonyl group and rearrangement, gives (metallo)carbonyl ylide **102**. Intramolecular dipolar cycloaddition then provides initial product **103**, which undergoes an additional in situ transformation by hydrolysis in the wet dichloromethane solvent to give the final product **104**.

Scheme 32 Cyclization Using Gold(I) Followed by Cycloaddition and Hydrolysis[71]

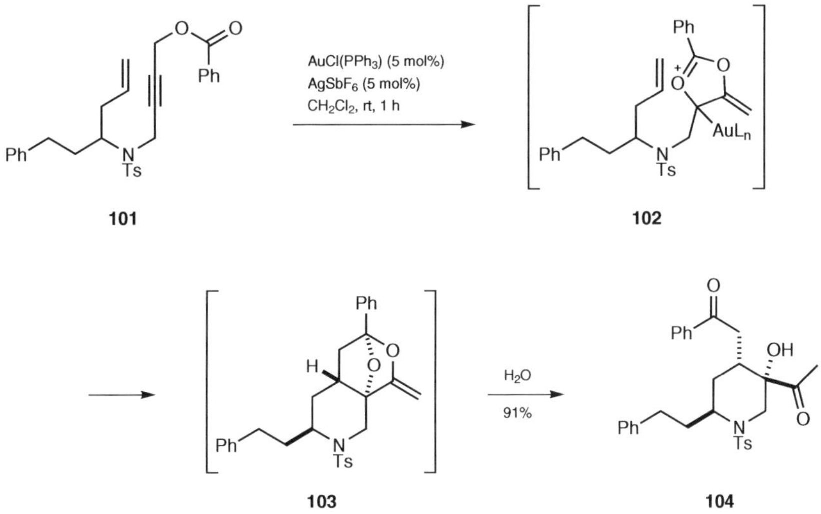

AuCl(PPh3) (5 mol%)
AgSbF6 (5 mol%)
CH2Cl2, rt, 1 h

101 102

H2O
91%

103 104

(1*R,2a*R**,2a¹*R**,5a*R**,7*S**,11b*R**)-1-Phenyl-2a,3,4,5,5a,6,7,11b-octahydro-1*H*-2a¹,7-epoxy-benzo[6,7]cyclohepta[1,2,3-*cd*]benzofuran (100); Typical Procedure:**[64]
A mixture of aldehyde **97** (0.1 mmol) and $PtCl_2(PPh_3)_2$ (5 mol%) in dry toluene (0.5 mL) was heated under reflux. After 4 h, the mixture was allowed to cool to rt and the solvent was evaporated. The residue was purified by column chromatography (silica gel) to give the title compound; yield: 77%.

2-[(2*S,4*R**,5*S**)-5-Acetyl-5-hydroxy-2-phenethyl-1-tosylpiperidin-4-yl]-1-phenylethan-1-one (104); Typical Procedure:**[71]
$AgSbF_6$ (0.05 equiv) was added to a suspension of propargylic ester **101** (0.15 mmol) and $AuCl(PPh_3)$ (0.05 equiv) in CH_2Cl_2 (1.5 mL) at rt. After 1 h, the mixture was loaded directly onto a silica gel column and purified (petroleum ether/EtOAc) to give the title compound as a solid; yield: 91%; mp 108–110 °C.

2.1.2.2.3 Cycloadditions with Azomethine Ylides

A very useful reaction in synthesis is the dipolar cycloaddition of an azomethine ylide to a π-bond. This process typically generates a pyrrolidine (or dihydropyrrole) ring, which is a common structural feature in alkaloid natural products. The azomethine ylide dipole cannot normally be isolated, and it is thus formed in situ by one of a number of methods, such as the condensation of an aldehyde and amine (followed by deprotonation, desilylation, or decarboxylation), addition of a metal salt to an imine (followed by deprotonation), ring opening of an aziridine, or addition of a nucleophile to an oxazolium salt (followed by ring opening). These methods are followed by dipolar cycloaddition, so the overall process is formally a domino sequence of reactions. Several new rings can be formed in the same pot if the dipolarophile is tethered to the azomethine ylide and an intramolecular dipolar cycloaddition takes place.[72]

One interesting method to access azomethine ylides is the ring opening of oxazolium salts, and this event has been exploited by Vedejs and co-workers for the preparation of aziridinomitosene analogues. Internal alkylation of oxazole **105** promoted by silver(I) trifluoromethanesulfonate gives oxazolium salt **106** (Scheme 33).[73] Treatment of this salt with cyanide results in addition and then ring opening to give intermediate azomethine ylide **107**. Intramolecular cycloaddition followed by loss of hydrogen cyanide gives final product **108**. This compound can ultimately be converted into a quinone aziridinomitosene that has been found to modify DNA at guanine sites.

Scheme 33 Ring Opening of an Oxazolium Salt Followed by Cycloaddition[73]

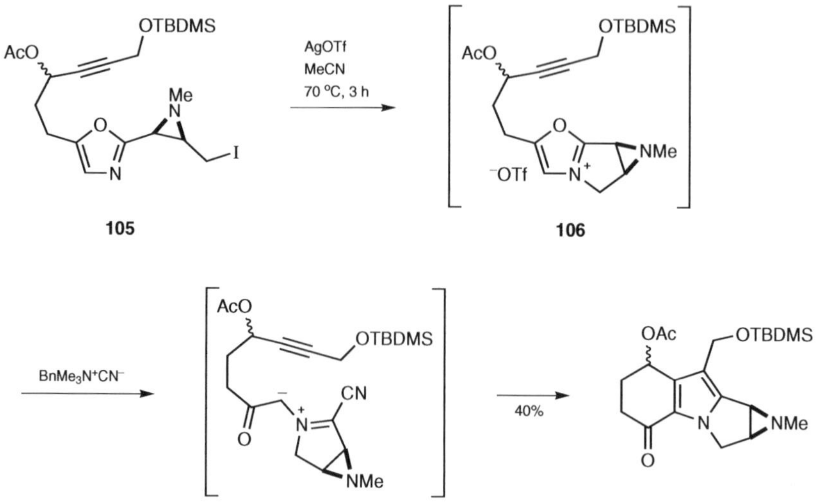

AgOTf, MeCN, 70 °C, 3 h

105 → 106

$BnMe_3N^+CN^-$

107 → (40%) 108

The addition of a secondary amino ester or amino acid to an aldehyde or ketone generates an iminium ion that is prone to deprotonation or decarboxylation to give an azomethine ylide. If a primary amine is used, an imine is formed; however, if the aldehyde or ketone bears an electrophilic center, such as an alkyl halide, at a suitable distance from the imine, then cyclization can serve as an alternate means to generate the iminium ion needed for azomethine ylide formation. This strategy has been exploited by the groups of Pearson and Coldham. An example from Pearson and co-workers is shown in Scheme 34 starting from ketone **109**.[74] Heating this ketone in toluene at 85 °C for 2 hours with (tributylstannyl)methanamine gives product **111**. In this event, the intermediate imine cyclizes with loss of iodide to give a seven-membered ring, and this is followed by destannylation to give azomethine ylide **110**. Intramolecular cycloaddition then gives product **111**, and final reductive desulfonylation gives (±)-demethoxyschelhammericine (**112**). If the ethyl sulfoxide is used rather than the ethyl sulfone, in situ elimination occurs after the cycloaddition to give the diene demethoxyschelhammeridine.

Scheme 34 Cyclization, Destannylation, and Subsequent Cycloaddition in the Synthesis of (±)-Demethoxyschelhammericine[74]

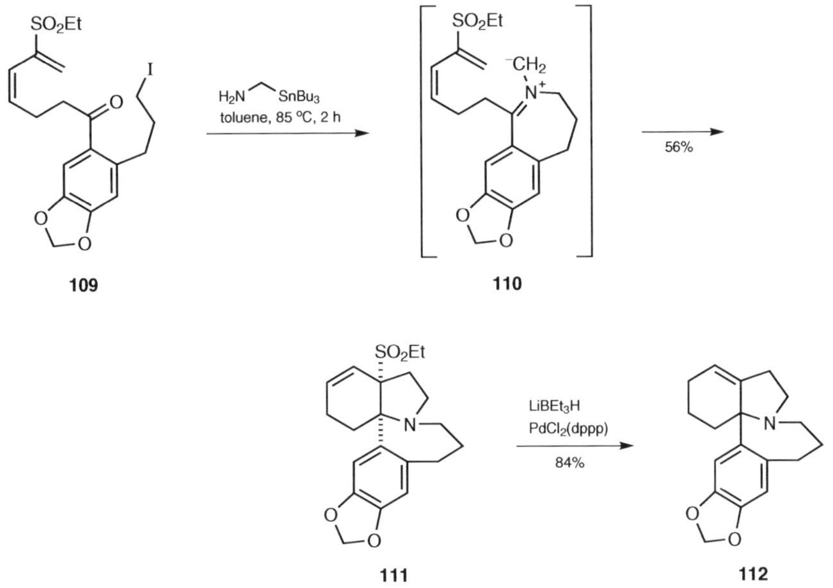

H_2N⌒$SnBu_3$
toluene, 85 °C, 2 h

56%

109 110

LiBEt$_3$H
PdCl$_2$(dppp)
84%

111 112

This strategy sets up three rings in a domino process involving condensation, cyclization, ylide formation, and then dipolar cycloaddition. Critically, it is possible to avoid the use of tin compounds and to make use of decarboxylation or deprotonation to form the azomethine ylide, particularly for non-enolizable substrates.[75] Thus, Coldham and co-workers have developed an efficient synthesis of some aspidosperma alkaloids starting from aldehyde **113** (Scheme 35).[76] This material is heated with glycine and a catalytic amount of 10-camphorsulfonic acid to give a single stereoisomer of product **114**. This reaction is thought to proceed by imine formation, cyclization, and then loss of carbon dioxide to give the intermediate azomethine ylide. The dipolar cycloaddition step is enhanced by the presence of the acid to give product **114**. Hydrolysis of **114** causes epimerization α to the ketone to give the desired tricyclic compound **115**. This material can ultimately be converted into the alkaloids (±)-aspidospermidine, (±)-aspidospermine, and (±)-quebrachamine. Similar domino cascade chemistry has been employed for the synthesis of the alkaloid (±)-crispine A[77] and for the core ring system found in stenine and neostenine[78] as well as scandine and meloscine.[79]

Scheme 35 Cyclization, Decarboxylation, and Subsequent Cycloaddition in the Synthesis of Aspidosperma Alkaloids[76]

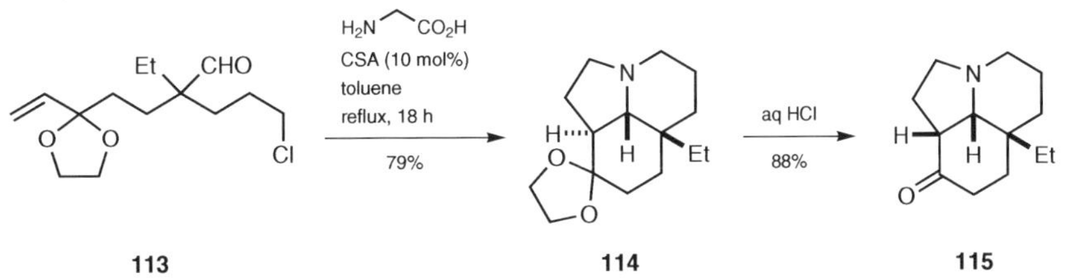

H_2N⌒CO_2H
CSA (10 mol%)
toluene
reflux, 18 h
79%

aq HCl
88%

113 114 115

for references see p 89

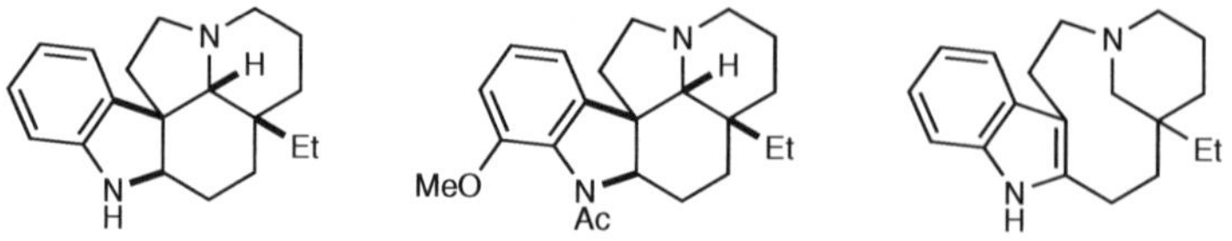

aspidospermidine aspidospermine quebrachamine

In Section 2.1.2.2.2, several methods for preparing carbonyl ylides are described, and these approaches have found some use with imines to generate azomethine ylides. For example, Martin and co-workers have described the formation of diazo compound **116** and its reaction with rhodium(II) acetate to form azomethine ylide **117A** (Scheme 36).[80,81] This intermediate undergoes dipolar cycloaddition to give a mixture of regioisomeric products; however, if the crude diazo compound is used, in which some acid (trifluoroacetic acid–triethylamine complex) is present from the previous reaction steps, then ylide **117A** is believed to isomerize to ylide **117B**, which leads to the desired regioisomeric cycloadduct **118** only. Compound **118** can be used to complete the formal synthesis of didehydrostemofoline (**119**).

Scheme 36 Cyclization with a Diazocarbonyl Compound Followed by Cycloaddition[80]

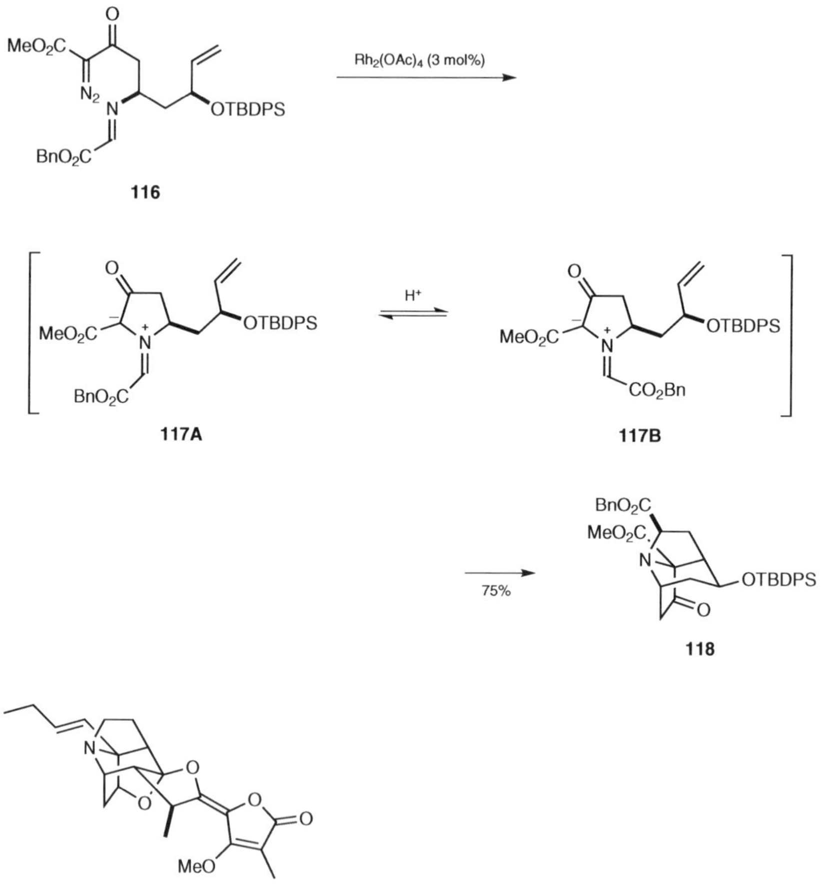

116 → $Rh_2(OAc)_4$ (3 mol%) → [117A ⇌ (H^+) 117B] → 75% → 118

119 didehydrostemofoline

Another method related to that presented in Section 2.1.2.2.2.2 is cyclization onto an alkyne activated by a transition metal, particularly platinum or gold. For example, the use of an imine and platinum(II) chloride for cyclization leads to the direct formation of an azomethine ylide, and this process has been used for the synthesis of tricyclic indole derivatives.[82] An alternative strategy is to use a nitrone for cyclization. For instance, in efforts with nitrone **120**, gold(III) chloride has been found to be preferable for activation (Scheme 37).[83] Cyclization first generates intermediate **121** and this is followed by cleavage of the N—O bond and cyclization of the imine onto the metallocarbene to furnish azomethine ylide **122**. This compound can then undergo intramolecular dipolar cycloaddition to give product **123**, the structure of which has been confirmed by single-crystal X-ray diffraction analysis.

Scheme 37 Cyclization of a Nitrone To Give an Azomethine Ylide Followed by Cycloaddition[83]

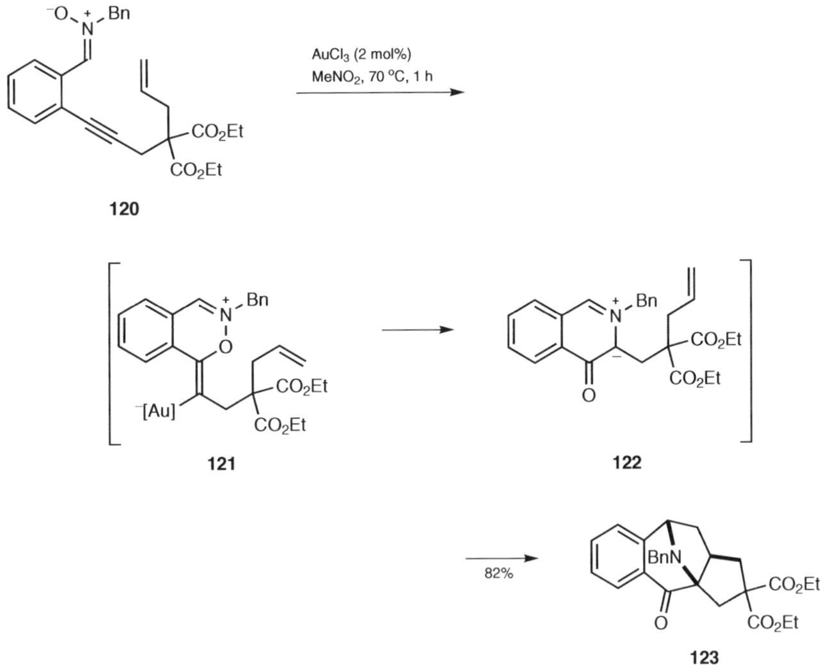

$AuCl_3$ (2 mol%)
$MeNO_2$, 70 °C, 1 h

120 → [**121** → **122**] → 82% **123**

The oxidation of a benzylamine can lead to an azomethine ylide and, hence, after cycloaddition, to a pyrrolidine product that is typically oxidized further under the reaction conditions to give a pyrrole.[84,85] This chemistry has been combined with ring-closing metathesis in which ruthenium is thought to aid the oxidation processes (and/or mediated by the quinone). Thus, treatment of diene **124** with Grubbs second-generation initiator **125** gives dihydroquinoline **126** (Scheme 38).[86] This material is oxidized to azomethine ylide **127** under the reaction conditions and undergoes dipolar cycloaddition with benzoquinone (added in excess amount). The resulting pyrrolidine product is then oxidized in situ to give pyrrole **128** (an isoindolo[2,1-*a*]quinoline).

for references see p 89

Scheme 38 Ring-Closing Metathesis and Oxidation To Give an Azomethine Ylide Followed by Cycloaddition and Oxidation[86]

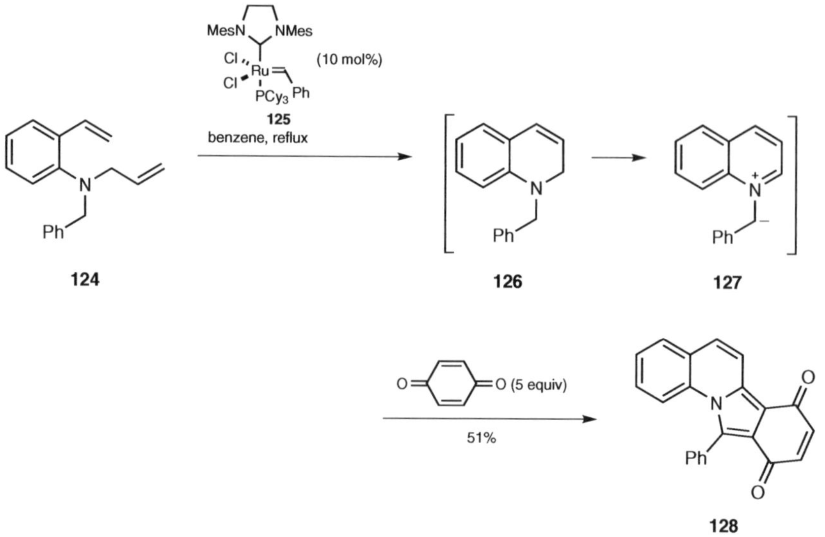

Most dipolar cycloaddition reactions reported to date involve alkene dipolarophiles. However, it is also possible to use a π-system containing one or two heteroatoms. With carbonyl dipolarophiles, in particular, aldehydes, ketones, and lactones can sometimes be successful. For example, isatoic anhydride (**129**, $R^1 = H$) reacts (at the more reactive carbonyl group) with the "nonstabilized" azomethine ylide generated from *N*-(methoxymethyl)-*N*-[(trimethylsilyl)methyl]benzylamine and trifluoroacetic acid (Scheme 39).[87] Cycloaddition product **130** ($R^1 = H$) is not isolated but undergoes domino opening of the oxazolidine ring to give an iminium ion. Loss of carbon dioxide and ring closure gives product **131** ($R^1 = H$), the structure of which has been confirmed by single-crystal X-ray diffraction analysis. Higher yields of the products **131** can be achieved if *N*-alkylisatoic anhydrides are used.

Scheme 39 Cycloaddition with a Carbonyl Group Followed by Ring Opening, Decarboxylation, and Cyclization[87]

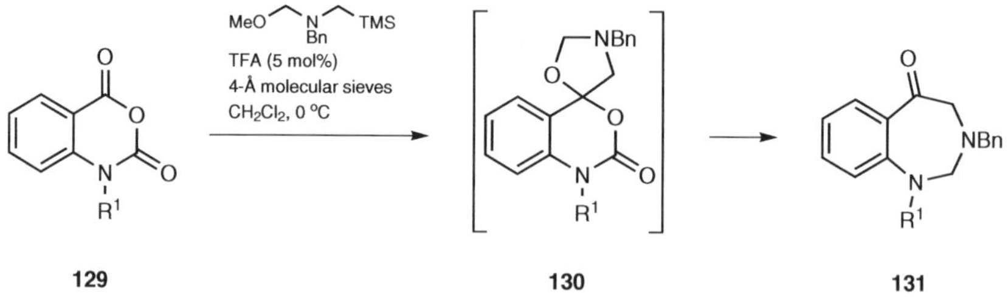

R^1	Yield (%) of **131**	Ref
H	42	[87]
Me	92	[87]
$CH_2CH{=}CH_2$	71	[87]

(1a*S*,8b*S*)-8-[(*tert*-Butyldimethylsiloxy)methyl]-1-methyl-4-oxo-1,1a,2,4,5,6,7,8b-octahydroazirino[2′,3′:3,4]pyrrolo[1,2-*a*]indol-7-yl Acetate (108); Typical Procedure:[73]
A soln of iodide **105** (429 mg, 0.806 mmol) in MeCN (4 mL) was added to AgOTf (336 mg, 1.31 mmol) and the mixture was heated under an atmosphere of N_2 at 70 °C. After 3 h, the mixture was transferred dropwise by cannula over 5 min to a suspension of $BnMe_3N^+CN^-$ (719 mg, 4.08 mmol) in MeCN (15 mL). The remaining AgI precipitate in the first flask was suspended in MeCN (4 mL) and the supernatant soln was transferred by cannula to the mixture at rt. After 30 min, the mixture was poured into EtOAc (100 mL) and washed with brine (3 × 50 mL). The organic phase was dried (Na_2SO_4), the solvent was evaporated, and the residue was purified by column chromatography (silica gel, hexane/EtOAc/Et_3N 100:100:1) to give the title compound as an oil; yield: 131 mg (40%).

6a-Ethyldecahydrospiro[pyrrolo[3,2,1-*ij*]quinoline-9,2′-[1,3]dioxolane] (114); Typical Procedure:[76]
A mixture of aldehyde **113** (0.27 g, 0.98 mmol), CSA (25 mg, 0.1 mmol), and glycine (0.22 g, 2.9 mmol) was heated under reflux in toluene (7.5 mL). After 18 h, the mixture was cooled to rt and the solvent was evaporated. Purification by column chromatography (silica gel, CH_2Cl_2/MeOH/NH_3 95:5:0.1) gave the title compound, which was then recrystallized (petroleum ether) as plates; yield: 194 mg (79%); mp 94–96 °C.

Diethyl (3a*S,9*R**,10a*S**)-11-Benzyl-4-oxo-4,9,10,10a-tetrahydro-1*H*-3a,9-epiminobenzo[*f*]azulene-2,2(3*H*)-dicarboxylate (123); Typical Procedure:**[83]
$AuCl_3$ (0.5 mg, 0.0018 mmol) was added to a soln of nitrone **120** (40 mg, 0.089 mmol) in $MeNO_2$ (0.8 mL), and the mixture was heated to 70 °C. After 1 h, the solvent was evaporated under reduced pressure. Purification by column chromatography (silica gel, hexane/EtOAc 2:1) gave the title compound as a solid; yield: 33 mg (82%); mp 98 °C.

11-Phenylisoindolo[2,1-*a*]quinoline-7,10-dione (128); Typical Procedure:[86]
Second-generation Grubbs initiator **125** (10 mol%) was added to a 0.01 M soln of aniline **124** in benzene (0.1 mmol) (**CAUTION:** *carcinogen*) under an atmosphere of argon, and the mixture was heated under reflux. After 30 min, benzo-1,4-quinone (0.5 mmol) was added and heating was continued. After 1 h, sat. aq NH_4Cl was added, the solvent was partially evaporated, and the mixture was extracted with EtOAc. The solvent was partially evaporated and the organic layer was washed with brine, dried (Na_2SO_4), and concentrated under reduced pressure. The residue was purified by column chromatography (silica gel, toluene/EtOAc 24:1) to give the title compound as a red solid; yield: 51%; mp 268–269 °C.

3-Benzyl-1-methyl-1,2,3,4-tetrahydro-5*H*-benzo[*d*][1,3]diazepin-5-one (131, R^1 = Me); Typical Procedure:[87]
N-(Methoxymethyl)-*N*-[(trimethylsilyl)methyl]benzylamine (0.85 g, 3.6 mmol) was added to a mixture of anhydride **129** (R^1 = Me; 0.35 g, 2.0 mmol) and powdered 4-Å molecular sieves (0.05 g) in CH_2Cl_2 (5 mL) at 0 °C under an atmosphere of N_2. A 1 M soln of TFA in CH_2Cl_2 (0.1 mL, 0.1 mmol) was added dropwise and the resulting soln was allowed to warm to rt. After 24 h, the solvent was evaporated and the residue was purified by flash chromatography (silica gel, hexane/Et_2O 1:1) to give the title compound as a solid; yield: 0.49 g (92%); mp 96–97 °C.

for references see p 89

2.1.2.2.4 Cycloadditions with Azomethine Imines

Cycloaddition reactions of azomethine imines with alkynes or alkenes give access to dihydropyrazole or tetrahydropyrazole (pyrazolidine) products. These reactions are typically induced by condensation of an aldehyde with a hydrazine to generate a hydrazone. This hydrazone then undergoes cyclization and loss of a proton to generate the intermediate azomethine imine dipole, which is then set up for [3+2] cycloaddition with a π-system, typically an alkene dipolarophile. The resulting products are prone to oxidation to yield a pyrazole, particularly if the starting hydrazine bears an arylsulfonyl group. For example, Wu and co-workers have reported the silver-catalyzed three-component reaction of aldehyde **132** with tosylhydrazine and the dipolarophile ethyl acrylate (Scheme 40).[88] Initial condensation gives hydrazone **133**, which undergoes cyclization onto the alkyne, activated by silver(I) trifluoromethanesulfonate, to give ylide **134**. The subsequent dipolar cycloaddition is notably regioselective for isomer **135**. Final loss of the tosyl group and aromatization then gives pyrazolo[5,1-*a*]isoquinoline **136**. Significantly, the reaction is successful with a variety of substituted alkynylaldehydes and alkene dipolarophiles. A similar process catalyzed by silver(I) has been reported by Liu and co-workers using an *exo-dig* cyclization rather than an *endo-dig* cyclization.[89] The same hydrazone **133**, in the absence of a silver(I) salt, can be induced to cyclize by activation of the alkyne with bromine.[90] Using ethyl acrylate as the dipolarophile then leads to cycloaddition and, depending on the conditions, brominated analogues of **135** or **136** can be isolated.

Scheme 40 Cyclization of a Hydrazone by Silver Catalysis Followed by Cycloaddition and Aromatization[88]

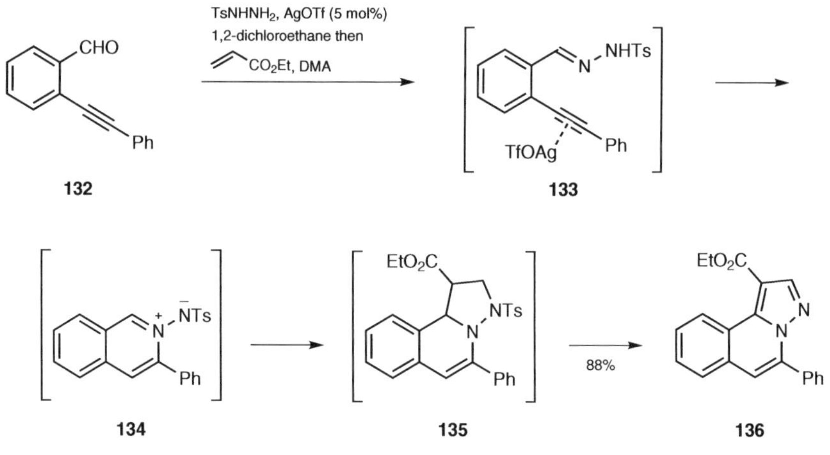

Coldham and co-workers have found that addition of hydrazine to aldehyde **137** gives an intermediate hydrazone that cyclizes with loss of bromide to give azomethine imine **138** (Scheme 41).[91] Intermolecular dipolar cycloaddition with, for example, dimethyl fumarate, then gives pyrazolidine products **139A** and **139B** in a 2.7:1 ratio. The structure and stereochemistry of the major isomer has been confirmed by single-crystal X-ray diffraction analysis. Examples of related domino reactions involving cyclization with loss of bromide followed by intramolecular dipolar cycloaddition to give tricyclic products have also been reported.

Scheme 41 Cyclization of a Hydrazone onto an Alkyl Halide Followed by Cycloaddition[91]

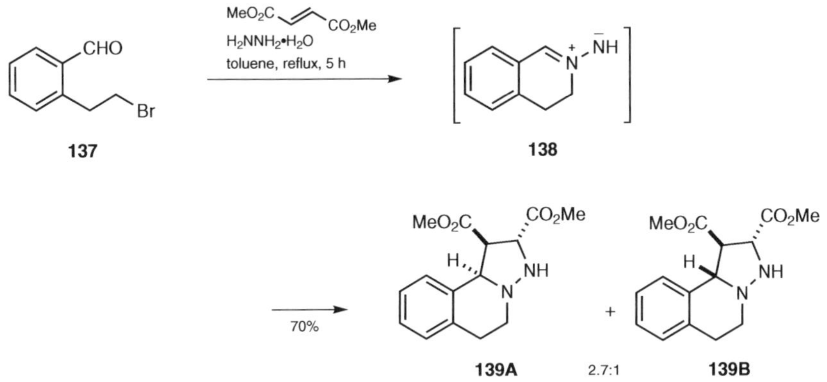

Ethyl 5-Phenylpyrazolo[5,1-*a*]isoquinoline-1-carboxylate (136); Typical Procedure:[88]
Aldehyde **132** (0.2 mmol) was added to a soln of tosylhydrazine (0.2 mmol) in 1,2-dichloroethane (0.5 mL) at rt. After 30 min, AgOTf (0.01 mmol) was added, and the mixture was heated to 60 °C. After 1 h, DMA (2 mL) and ethyl acrylate (0.4 mmol) were added, and the mixture was stirred at 60 °C. After completion of the reaction, as indicated by TLC, 1.0 M NH_4Cl in H_2O (10 mL) was added. The mixture was extracted with EtOAc (2 × 10 mL) and dried (Na_2SO_4), and the solvent was evaporated. Purification by column chromatography (silica gel) gave the title compound; yield: 56 mg (88%).

Dimethyl 1,2,3,5,6,10*b*-Hexahydropyrazolo[5,1-*a*]isoquinoline-1,2-dicarboxylates (139A and 139B); Typical Procedure:[91]
A mixture of aldehyde **137** (264 mg, 1.24 mmol) and H_2NNH_2•H_2O (0.12 mL, 2.48 mmol) in toluene (16 mL) was heated to reflux. After 30 min, dimethyl fumarate (268 mg, 1.86 mmol) was added, and the mixture was further heated under reflux. After 5 h, the mixture was cooled to rt and the solvent was evaporated. Purification by column chromatography (CH_2Cl_2/MeOH 98:2) gave cycloadduct **139A** as an amorphous solid; yield: 182 mg (51%); and cycloadduct **139B** as an oil; yield: 69 mg (19%).

2.1.2.2.5 Cycloadditions with Azides

The cycloaddition of an azide with an alkyne gives rise to a 1,2,3-triazole ring system. It is a common reaction in synthetic chemistry that has gained further widespread use by the advent of transition-metal-promoted variants, particularly as catalyzed by copper(I) species. There are two main strategies for domino reactions that include dipolar cycloaddition of an azide and an alkyne. First, a bond-forming reaction of some sort, from simple azide displacement of a leaving group to more complex C—C bond-forming reactions in which an azide is present in one component, is followed by in situ intramolecular cycloaddition.[92] Second, initial dipolar cycloaddition can be followed by an in situ reaction, such as cyclization or loss of nitrogen from the triazole, followed by subsequent chemistry. Some selected examples of these two approaches will be discussed in the following sections.

2.1.2.2.5.1 Reaction To Give an Azido-Substituted Alkyne Followed by Cycloaddition

This section will focus on domino C—C bond formation followed by dipolar cycloaddition. We begin with an example from Chowdhury and co-workers, who have found that treatment of iodide **140** with a terminal alkyne under palladium and copper catalysis (Sonogashira conditions) provides triazole **142** directly (Scheme 42).[93] Presumably, intermediate alkyne **141** undergoes intramolecular dipolar cycloaddition in situ with the tethered azide. Related couplings of aryl iodides bearing an azide functional group followed by dipolar cycloaddition are known.[94] Alternatively, an alkyne and an azide component can be coupled together by using a Michael reaction[95] or N-alkylation.[96] As before, this approach gives an azido-substituted alkyne that undergoes in situ dipolar cycloaddition.

Scheme 42 Alkyne Coupling Followed by Cycloaddition[93]

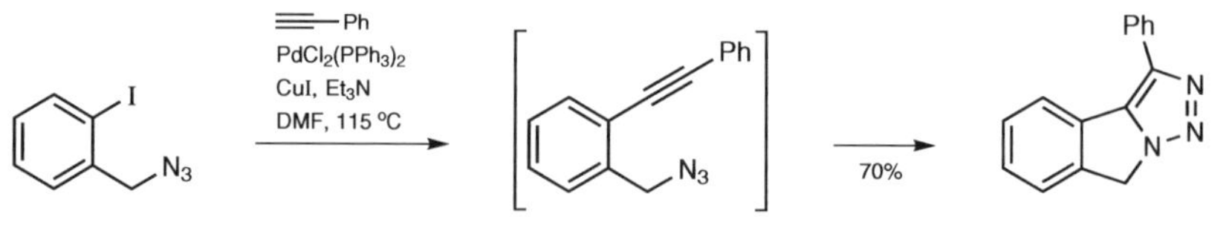

Another approach is to prepare the alkyne itself, such as by the reaction of an aldehyde with the Ohira–Bestmann reagent, which is followed by in situ cycloaddition of a tethered azide. For example, treatment of aldehyde **143** with the Ohira–Bestmann reagent **144** gives tricyclic product **146** directly via intermediate alkyne **145** (Scheme 43).[97]

Scheme 43 Alkyne Formation Followed by Cycloaddition[97]

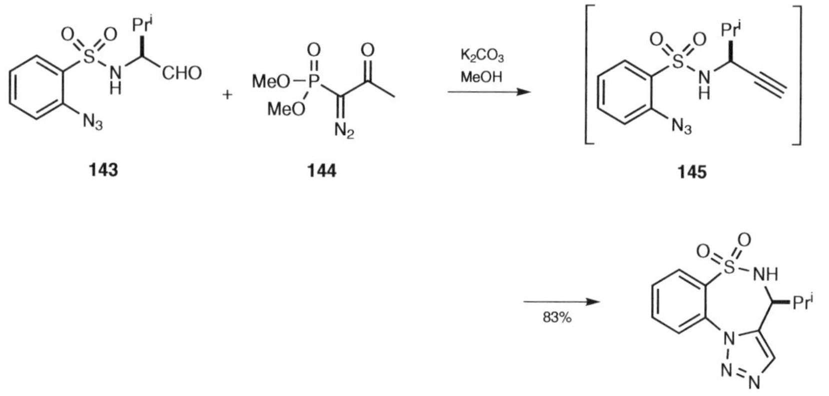

3-Phenyl-8*H*-[1,2,3]triazolo[5,1-*a*]isoindole (142); **Typical Procedure:**[93]
$PdCl_2(PPh_3)_2$ (24.6 mg, 0.035 mmol), CuI (13.3 mg, 0.07 mmol), and Et_3N (0.71 mL, 5.0 mmol) were added to a soln of iodo azide **140** (1.0 mmol) in dry DMF (8 mL) at rt under an atmosphere of argon. After 20 min, phenylacetylene (1.25 mmol) in dry DMF (1 mL) was added, and the mixture was heated to 115 °C. After 2 h, the mixture was cooled to rt and the solvent was evaporated. H_2O (30 mL) was added, and the mixture was extracted with EtOAc (2 × 25 mL). The organic layer was dried (Na_2SO_4), the solvent was evaporated, and the residue was purified by column chromatography (silica gel, petroleum ether/EtOAc ~5:1) to give the title compound as a light-brown solid; yield: 163 mg (70%); mp 152–154 °C.

2.1.2.2.5.2 Cycloaddition of an Azide and Subsequent Reaction

Dipolar cycloaddition of an azide with an alkyne gives a 1,2,3-triazole. This product is amenable to subsequent in situ reaction to provide an overall domino transformation. For example, warming alkyne **147** with sodium azide and copper(I) iodide in the presence of L-proline promotes dipolar cycloaddition followed by N-arylation to give triazolo[1,5-*a*]quinoxaline product **148** (Scheme 44).[98] Similar chemistry has been used for the synthesis of other heterocycles containing a triazole group such as triazolothiadiazepine derivatives[99] and triazolylquinolin-2(1*H*)-ones.[100]

Scheme 44 Cycloaddition of an Azide and an Alkyne Followed by C—N Bond Formation[98]

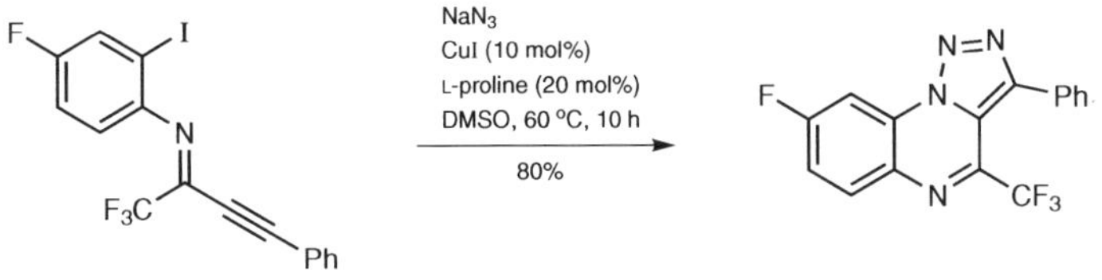

NaN_3, CuI (10 mol%), L-proline (20 mol%), DMSO, 60 °C, 10 h; 80%

147 → **148**

In certain cases, the 1,2,3-triazole products formed by dipolar cycloaddition can fragment and lose nitrogen. This leads to imine/enamine or aziridine formation.[101] For example, Tu and co-workers have reported that enamine **152** can be formed by heating chloride **149** with sodium azide (Scheme 45).[102] Initial displacement of chloride gives azide **150**, which undergoes dipolar cycloaddition to give intermediate triazole **151** which can itself be isolated under basic conditions in a mixture of triethylamine/dichloromethane (the structure has been confirmed by single-crystal X-ray diffraction analysis). Loss of nitrogen and rearrangement gives an imine that loses a proton to give enamine **152**.

Scheme 45 Cycloaddition Followed by Loss of Nitrogen and Rearrangement[102]

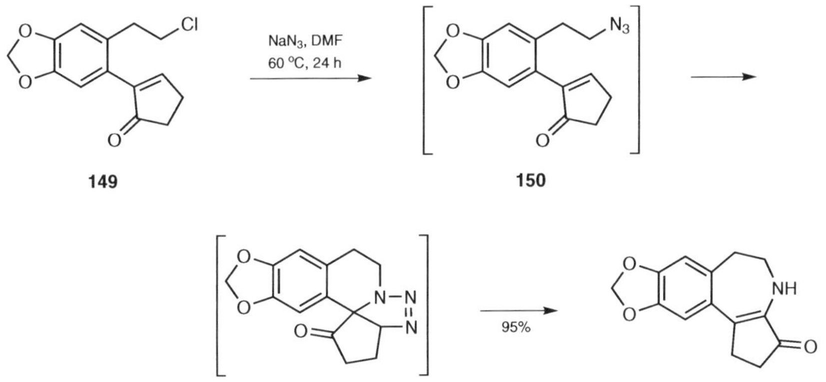

149 → (NaN_3, DMF, 60 °C, 24 h) → [**150**] → [**151**] → (95%) → **152**

Mann and co-workers have described the formation of enamines from 1,2,3-triazoles and their subsequent cyclization chemistry.[103,104] For example, heating azide **153** using microwave irradiation results in the formation of bicyclic imine **156** (Scheme 46).[104] The transformation involves a domino dipolar cycloaddition that gives triazole **154** followed

for references see p 89

by loss of nitrogen to generate imine **155**, which is in equilibrium with its enamine counterparts. Final cyclization of one of these enamines with the unsaturated carbonyl component then gives ultimate product **156**, formed as a single stereoisomer.

Scheme 46 Cycloaddition Followed by Loss of Nitrogen and Cyclization[104]

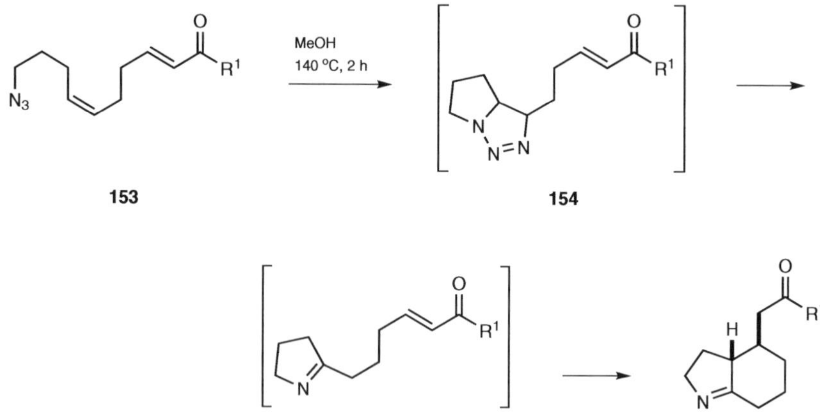

MeOH
140 °C, 2 h

153 154 155 156

R^1	Yield (%)	Ref
Ph	55	[104]
Me	64	[104]
OMe	68	[104]

8-Fluoro-3-phenyl-4-(trifluoromethyl)[1,2,3]triazolo[1,5-*a*]quinoxaline (148); **Typical Procedure:**[98]

CAUTION: *Sodium azide can explode on heating and is highly toxic. Contact of metal azides with acids liberates the highly toxic and explosive hydrazoic acid.*

Alkyne **147** (0.3 mmol) was added to a soln of CuI (10 mol%), L-proline (20 mol%), and NaN_3 (0.33 mmol) in DMSO (2 mL), and the mixture was warmed to 60 °C. After 10 h, the mixture was partitioned between EtOAc and H_2O. The organic layer was washed with brine, dried ($MgSO_4$), and concentrated. Purification by column chromatography (silica gel, petroleum ether/EtOAc) gave the title compound as a white solid; yield: 80 mg (80%); mp 178–181 °C.

Methyl 2-[(3a*R,4*R**)-3,3a,4,5,6,7-Hexahydro-2*H*-indol-4-yl]acetate** (156, **R^1 = OMe); Typical Procedure:**[104]

A soln of azide **153** (R^1 = OMe; 60 mg, 0.27 mmol) in dry MeOH (12 mL) was heated in a microwave reactor at 140 °C. After 2 h, the mixture was cooled to rt, and the solvent was evaporated. Purification by column chromatography (silica gel, CH_2Cl_2/MeOH 100:0 to 97:3) gave the title compound as an oil; yield: 36 mg (68%).

2.1.2.3 Domino [5+2] Cycloadditions

The [5+2] cycloaddition reaction involving pericyclic or transition-metal-promoted processes has been studied as a way to access natural products and to establish novel synthetic methodologies for the preparation of seven-membered rings.[105] Despite the fact that they have been known for more than a century, domino transformations containing [5+2] cycloadditions are rare. There are, however, some synthetically significant approaches that provide functionalized polycyclic systems with excellent stereocontrol, and this section highlights such pertinent reactions.

2.1.2.3.1 Cycloaddition of a Vinylic Oxirane Followed by Claisen Rearrangement

Zhang and Feng have reported an impressive use of an intramolecular [5+2] cycloaddition/Claisen rearrangement domino sequence to construct bicyclo[3.1.0] products in a regioselective and highly diastereoselective manner.[106] Treatment of vinylic oxiranes having an internally tethered alkyne motif, such as **157**, with the rhodium carbene complex [1,3-bis(2,6-diisopropylphenyl)imidazol-2-ylidene]chloro(cyclooctadiene)rhodium(I) [RhCl(IPr)(cod); IPr = 1,3-bis(2,6-diisopropylphenyl)imidazol-2-ylidene] gives bicyclic compounds in the form of **159** (Scheme 47). Mechanistically, initial [5+2] cycloaddition results in cycloadduct **158**, which subsequently undergoes thermally induced Claisen rearrangement to give the desired product without the need for any additional reagents or catalysts. A range of functionalities including sulfonamide, ether, and geminal diester groups attached to the tethered segment in combination with both aliphatic and aromatic substituents are tolerated. This tandem process has been explored with an enantioenriched vinylic oxirane, and complete transfer of chirality from the starting substrate to the final product has been reported.

Scheme 47 Intramolecular [5+2] Cycloaddition/Claisen Rearrangement Domino Sequence Applied to Vinylic Oxirane Substrates[106]

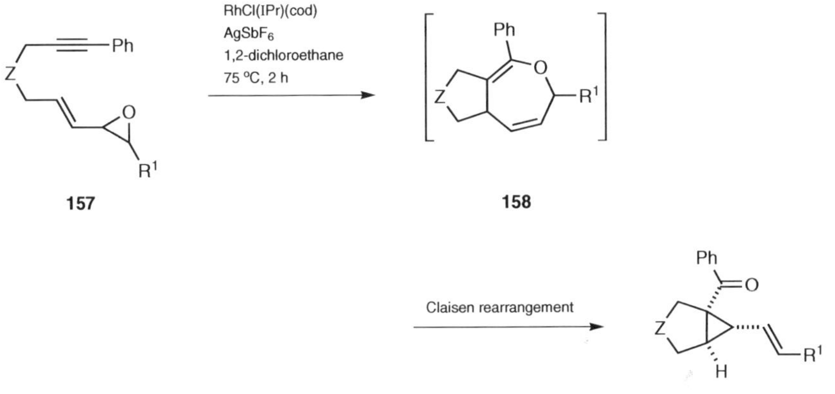

RhCl(IPr)(cod), $AgSbF_6$, 1,2-dichloroethane, 75 °C, 2 h: **157** → [**158**]

Claisen rearrangement: → **159**

R^1	Z	Yield (%)	Ref
Me	NTs	94	[106]
Ph	O	85	[106]
Ph	$C(CO_2Me)_2$	92	[106]

Phenyl{(1*R,5*R**,6*R**)-6-[(*E*)-prop-1-enyl]-3-tosyl-3-azabicyclo[3.1.0]hexan-1-yl}methanone (159, R¹ = Me; Z = NTs); Typical Procedure:**[106]

A mixture of RhCl(IPr)(cod) (0.01 mmol, 5 mol%) and $AgSbF_6$ (0.01 mmol, 5 mol%) in 1,2-dichloroethane (1 mL) was stirred under an atmosphere of N_2 at rt for 30 min. A soln of vinylic oxirane **157** (R^1 = Me; Z = NTs; 0.2 mmol) in 1,2-dichloroethane (1.5 mL) was added, and the resulting mixture was stirred at 75 °C for 2 h. The soln was concentrated, and the resulting crude residue was purified by column chromatography (silica gel, hexanes/EtOAc 5:1) to give the title compound as a white solid; yield: 94%; mp 105–107 °C.

for references see p 89

2.1.2.3.2 Cycloaddition of an Ynone Followed by Nazarov Cyclization

Bicyclo[5.3.0]decanes are found in natural products and compounds of medicinal importance, including prostratin and resiniferatoxin. Wender and co-workers have devised a unique strategy to access bicyclo[5.3.0]decane derivative **162** by employing a chemoselective reaction of ynone **160** with readily available vinylcyclopropane **161** using a cationic rhodium-catalyzed [5+2] cycloaddition/Nazarov cyclization sequence (Scheme 48).[107] A range of enynones and arylynones show remarkable reactivity with good diastereoselectivity, and the reaction can be performed using various rhodium-based catalyst systems.

Scheme 48 Rhodium-Catalyzed [5+2] Cycloaddition/Nazarov Cyclization Sequence To Access Bicyclo[5.3.0]decane Derivatives[107]

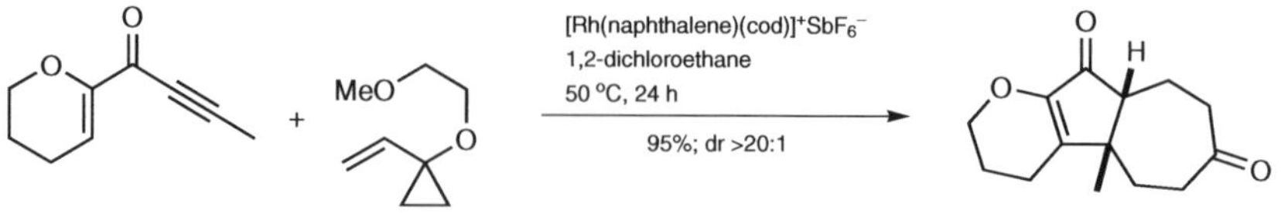

[Rh(naphthalene)(cod)]$^+$SbF$_6^-$
1,2-dichloroethane
50 °C, 24 h
95%; dr >20:1

160 161 162

(4b*S,9a*R**)-4b-Methyl-3,4,4b,5,6,8,9,9a-octahydro-2*H*-azuleno[2,1-*b*]pyran-7,10-dione (162); Typical Procedure:**[107]

Enynone **160** (14 mg, 0.093 mmol) and 1-(2-methoxyethoxy)-1-vinylcyclopropane (**161**; 17 μL, 16 mg, 0.112 mmol) were added to a mixture of [Rh(naphthalene)(cod)]$^+$SbF$_6^-$ (1.1 mg, 1.9 μmol, 2 mol%) in 1,2-dichloroethane (0.6 mL), and the resulting yellow soln was heated at 50 °C for 24 h. The mixture was then cooled to rt and the solvent was evaporated. Purification of the resulting residue by column chromatography (silica gel, pentane/EtOAc 1:3) gave the title compound as an oil; yield: 20.8 mg (95%).

2.1.2.3.3 Cycloaddition of an Acetoxypyranone Followed by Conjugate Addition

In continuation of their investigations toward establishing the reactivity of *anti*- and *syn*-acetoxypyranones for [5+2] cycloaddition reactions, Mitchell and co-workers have observed a tandem cascade process that leads to a unique caged lactol.[108] Heating *syn*-acetoxypyranone **163** in the presence of quinuclidine furnishes lactol **165** via bridged, tricyclic ether **164** with high diastereofacial selectivity (Scheme 49). The sequence is believed to proceed through [5+2] cycloaddition followed by attack of a water molecule on the aldehyde in concert with the conjugate addition. The corresponding *anti*-acetoxypyranone undergoes the same cascade with the same stereoselectivity, though in moderate yield.

Scheme 49 [5+2] Cycloaddition of an Acetoxypyranone Followed by Conjugate Addition[108]

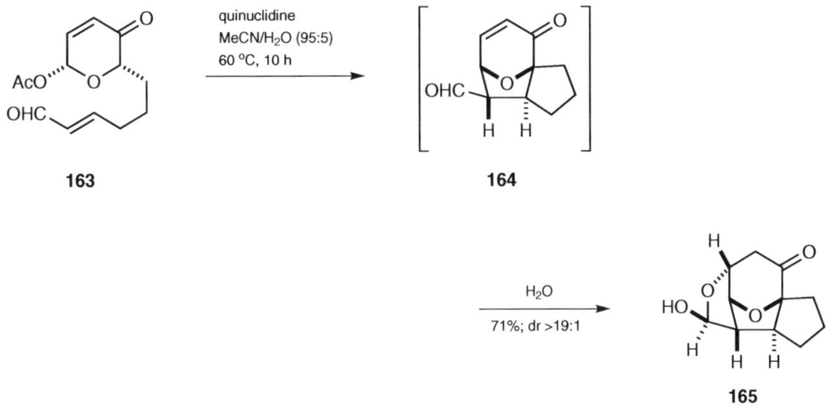

quinuclidine
$MeCN/H_2O$ (95:5)
60 °C, 10 h

163 164

H_2O
71%; dr >19:1

165

Lactol 165; Typical Procedure:[108]

Quinuclidine (178 mg, 1.60 mmol) was added to a soln of acetoxypyranone **163** (100 mg, 0.4 mmol) in $MeCN/H_2O$ (4 mL; 95:5). The mixture was stirred at 60 °C for 10 h, and then the solvent was evaporated. Purification by column chromatography (silica gel, hexanes/EtOAc 7:3) gave the title compound as an oil; yield: 51 mg (71%).

2.1.2.3.4 Cycloaddition Cascade Involving γ-Pyranone and Quinone Systems

The sequential domino [5+2]- and [4+2]-cycloaddition cascade utilizing γ-pyranone **166** and diene precursor **167** is an elegant approach to tetracyclic compound **168** (Scheme 50).[109] This material is a commonly encountered skeleton that is found in several natural products of terpenoid origin. This one-pot intramolecular/intermolecular cycloaddition cascade is efficient and leads to the formation of four new C—C bonds and five stereogenic centers with high stereocontrol. Furthermore, the methodology has been successfully applied to γ-pyranone-bearing tethered alkyne **169** and 2,3-dimethylbuta-1,3-diene to give tetracyclic adduct **170** in excellent yield (Scheme 50).[110] Theoretical calculations validate the selectivity obtained during this thermal cycloaddition cascade.[111] Lewis acid promoted domino sequential cycloaddition reactions involving [5+2]/[3+2] and [5+2]/[3+3] cycloadditions of functionalized quinones with aromatic substituted alkenes have also been reported, albeit in lower yields.[112]

Scheme 50 Intramolecular [5+2] Cycloaddition Followed by Intermolecular [4+2] Cycloaddition in a Domino Cascade[109,110]

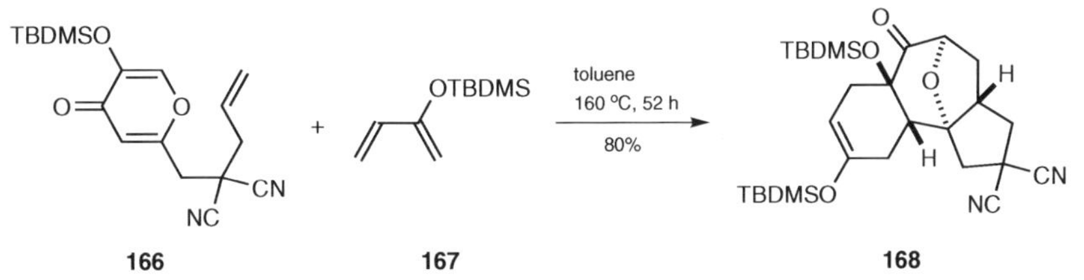

TBDMSO NC CN OTBDMS
toluene
160 °C, 52 h
80%

166 167 168

for references see p 89

TBDMSO PhO$_2$S SO$_2$Ph O O + toluene 160 °C, 12 h 93% TBDMSO O PhO$_2$S SO$_2$Ph O H

169 170

(1*R,3*R**,8*S**,9*S**)-3-(*tert*-Butyldimethylsiloxy)-5,6-dimethyl-13,13-bis(phenylsulfonyl)-15-oxatetracyclo[7.5.1.0^{1,11}.0^{3,8}]pentadeca-5,10-dien-2-one** (170); **Typical Procedure:**[110]

A soln of γ-pyranone **169** (425 mg, 0.74 mmol) and 2,3-dimethylbuta-1,3-diene (0.41 mL, 3.7 mmol) in toluene (25 mL) was heated in a sealed tube at 160 °C for 12 h. The solvent was then evaporated and the residue was purified by column chromatography (silica gel, EtOAc/hexanes 5:95 to 10:90) to give the title compound as a white solid; yield: 452 mg (93%); mp 206 °C.

References

[1] White, J. D.; Li, Y.; Ihle, D. C., *J. Org. Chem.*, (2010) **75**, 3569.
[2] White, J. D.; Ihle, D. C., *Org. Lett.*, (2006) **8**, 1081.
[3] Minter, D. E.; Winslow, C. D., *J. Org. Chem.*, (2004) **69**, 1603.
[4] Shindo, M.; Sato, Y.; Shishido, K., *J. Am. Chem. Soc.*, (1999) **121**, 6507.
[5] Shindo, M.; Sato, Y.; Shishido, K., *J. Org. Chem.*, (2001) **66**, 7818.
[6] Shindo, M.; Sato, Y.; Shishido, K., *Tetrahedron Lett.*, (2002) **43**, 5039.
[7] Shindo, M.; Matsumoto, K.; Sato, Y.; Shishido, K., *Org. Lett.*, (2001) **3**, 2029.
[8] Feltenberger, J. B.; Hayashi, R.; Tang, Y.; Babiash, E. S. C.; Hsung, R. P., *Org. Lett.*, (2009) **11**, 3666.
[9] Ma, Z.-X.; Feltenberger, J. B.; Hsung, R. P., *Org. Lett.*, (2012) **14**, 2742.
[10] Pérez-Anes, A.; García-García, P.; Suárez-Sobrino, Á. L.; Aguilar, E., *Eur. J. Org. Chem.*, (2007), 3480.
[11] Singh, P.; Bhargava, G.; Mahajan, M. P., *Tetrahedron*, (2006) **62**, 11 267.
[12] Hayakawa, K.; Aso, K.; Shiro, M.; Kanematsu, K., *J. Am. Chem. Soc.*, (1989) **111**, 5312.
[13] Yeo, S.-K.; Hatae, N.; Seki, M.; Kanematsu, K., *Tetrahedron*, (1995) **51**, 3499.
[14] Yeo, S.-K.; Shiro, M.; Kanematsu, K., *J. Org. Chem.*, (1994) **59**, 1621.
[15] Zhang, L., *J. Am. Chem. Soc.*, (2005) **127**, 16 804.
[16] Brummond, K. M.; Osbourn, J. M., *Beilstein J. Org. Chem.*, (2010) **6**, 1.
[17] Hattori, T.; Suzuki, Y.; Uesugi, O.; Oi, S.; Miyano, S., *Chem. Commun. (Cambridge)*, (2000), 73.
[18] Hattori, T.; Suzuki, Y.; Ito, Y.; Hotta, D.; Miyano, S., *Tetrahedron*, (2002) **58**, 5215.
[19] Arrastia, I.; Cossio, F. P., *Tetrahedron Lett.*, (1996) **37**, 7143.
[20] Dekorver, K. A.; Hsung, R. P.; Song, W.-Z.; Wang, X.-N.; Walton, M. C., *Org. Lett.*, (2012) **14**, 3214.
[21] Rao, W.; Susanti, D.; Chan, P. W. H., *J. Am. Chem. Soc.*, (2011) **133**, 15 248.
[22] *Synthetic Applications of 1,3-Dipolar Cycloaddition Chemistry Toward Heterocycles and Natural Products*, Padwa, A.; Pearson, W. H., Eds.; Wiley: New York, (2003).
[23] Burrell, A. J. M.; Coldham, I., *Curr. Org. Synth.*, (2010) **7**, 312.
[24] Banerji, A.; Bandyopadhyay, D., *J. Indian Chem. Soc.*, (2004) **81**, 817.
[25] Burrell, A. J. M.; Coldham, I.; Oram, N., *Org. Lett.*, (2009) **11**, 1515.
[26] Coldham, I.; Burrell, A. J. M.; Watson, L.; Oram, N.; Martin, N. G., *Heterocycles*, (2012) **84**, 597.
[27] Coldham, I.; Burrell, A. J. M.; Guerrand, H. D. S.; Oram, N., *Org. Lett.*, (2011) **13**, 1267.
[28] Coldham, I.; Watson, L.; Adams, H.; Martin, N. G., *J. Org. Chem.*, (2011) **76**, 2360.
[29] Grigg, R.; Markandu, J.; Surendrakumar, S.; Thornton-Pett, M.; Warnock, W. J., *Tetrahedron*, (1992) **48**, 10 399.
[30] Armstrong, P.; Grigg, R.; Heaney, F.; Surendrakumar, S.; Warnock, W. J., *Tetrahedron*, (1991) **47**, 4495.
[31] Davison, E. C.; Fow, M. E.; Holmes, A. B.; Roughley, S. D.; Smith, C. J.; Williams, G. M.; Davies, J. E.; Raithby, P. R.; Adams, J. P.; Forbes, I. T.; Press, N. J.; Thompson, M. J., *J. Chem. Soc., Perkin Trans. 1*, (2002), 1494.
[32] Stockman, R. A.; Sinclair, A.; Arini, L. G.; Szeto, P.; Hughes, D. L., *J. Org. Chem.*, (2004) **69**, 1598.
[33] Krenske, E. H.; Agopcan, S.; Aviyente, V.; Houk, K. N.; Johnson, B. A.; Holmes, A. B., *J. Am. Chem. Soc.*, (2012) **134**, 12 010.
[34] Wilson, M. S.; Padwa, A., *J. Org. Chem.*, (2008) **73**, 9601.
[35] Flick, A. C.; Caballero, M. J. A.; Padwa, A., *Tetrahedron*, (2010) **66**, 3643.
[36] Stearman, C. J.; Wilson, M.; Padwa, A., *J. Org. Chem.*, (2009) **74**, 3491.
[37] Ding, Q.; Wang, Z.; Wu, J., *J. Org. Chem.*, (2009) **74**, 921.
[38] Nakamura, I.; Kudo, Y.; Terada, M., *Angew. Chem.*, (2013) **125**, 7684; *Angew. Chem. Int. Ed.*, (2013) **52**, 7536.
[39] Chen, H.; Wang, Z.; Zhang, Y.; Huang, Y., *J. Org. Chem.*, (2013) **78**, 3503.
[40] Li, Y.-J.; Chuang, H.-Y.; Yeh, S.-M.; Huang, W.-S., *Eur. J. Org. Chem.*, (2011), 1932.
[41] Wang, X.; Abrahams, Q. M.; Zavalij, P. Y.; Doyle, M. P., *Angew. Chem.*, (2012) **124**, 6009; *Angew. Chem. Int. Ed.*, (2012) **51**, 5907.
[42] Liu, H.; Liu, G.; Pu, S.; Wang, Z., *Org. Biomol. Chem.*, (2013) **11**, 2898.
[43] Çelebi-Ölçüm, N.; Lam, Y.; Richmond, E.; Ling, B. L.; Smith, A. D.; Houk, K. N., *Angew. Chem.*, (2011) **123**, 11 680; *Angew. Chem. Int. Ed.*, (2011) **50**, 11 478.
[44] Yang, H.-B.; Wei, Y.; Shi, M., *Tetrahedron*, (2013) **69**, 4088.
[45] Lu, L.-Q.; Li, F.; An, J.; Zhang, J.-J.; An, X.-L.; Hua, Q.-L.; Xiao, W.-J., *Angew. Chem.*, (2009) **121**, 9706; *Angew. Chem. Int. Ed.*, (2009) **48**, 9542.
[46] An, J.; Lu, L.-Q.; Yang, Q.-Q.; Wang, T.; Xiao, W.-J., *Org. Lett.*, (2013) **15**, 542.

[47] Fédou, N. M.; Parsons, P. J.; Viseux, E. M. E.; Whittle, A. J., *Org. Lett.*, (2005) **7**, 3179.
[48] Abbiati, G.; Arcadi, A.; Marinelli, F.; Rossi, E., *Eur. J. Org. Chem.*, (2003), 1423.
[49] Chen, S.; Ren, J.; Wang, Z., *Tetrahedron*, (2009) **65**, 9146.
[50] Padwa, A., *Chem. Soc. Rev.*, (2009) **38**, 3072.
[51] Padwa, A., *Tetrahedron*, (2011) **67**, 8057.
[52] France, S.; Phun, L. H., *Curr. Org. Synth.*, (2010) **7**, 332.
[53] Takeda, K.; Oohara, T.; Shimada, N.; Nambu, H.; Hashimoto, S., *Chem.–Eur. J.*, (2011) **17**, 13 992.
[54] Lam, S. K.; Chiu, P., *Chem.–Eur. J.*, (2007) **13**, 9589.
[55] Sugano, Y.; Kikuchi, F.; Toita, A.; Nakamura, S.; Hashimoto, S., *Chem.–Eur. J.*, (2012) **18**, 9682.
[56] Shimada, N.; Hanari, T.; Kurosaki, Y.; Anada, M.; Nambu, H.; Hashimoto, S., *Tetrahedron Lett.*, (2010) **51**, 6572.
[57] Termath, A. O.; Ritter, S.; König, M.; Kranz, D. P.; Neudörfl, J.-M.; Prokop, A.; Schmalz, H.-G., *Eur. J. Org. Chem.*, (2012), 4501.
[58] Rodier, F.; Rajzmann, M.; Parrain, J.-L.; Chouraqui, G.; Commeiras, L., *Chem.–Eur. J.*, (2013) **19**, 2467.
[59] England, D. B.; Padwa, A., *J. Org. Chem.*, (2008) **73**, 2792.
[60] Mejía-Oneto, J. M.; Padwa, A., *Helv. Chim. Acta*, (2008) **91**, 285.
[61] Li, H.; Bonderoff, S. A.; Cheng, B.; Padwa, A., *J. Org. Chem.*, (2014) **79**, 392.
[62] Padwa, A.; Brodney, M. A.; Marino, J. P., Jr.; Sheehan, S. M., *J. Org. Chem.*, (1997) **62**, 78.
[63] Grillet, F.; Sabot, C.; Anderson, R.; Babjak, M.; Greene, A. E.; Kanazawa, A., *Tetrahedron*, (2011) **67**, 2579.
[64] Oh, C. H.; Lee, J. H.; Lee, S. J.; Kim, J. I.; Hong, C. S., *Angew. Chem.*, (2008) **120**, 7615; *Angew. Chem. Int. Ed.*, (2008) **47**, 7505.
[65] Oh, C. H.; Lee, J. H.; Lee, S. M.; Yi, H. J.; Hong, C. S., *Chem.–Eur. J.*, (2009) **15**, 71.
[66] Kim, J. H.; Ray, D.; Hong, C. S.; Han, J. W.; Oh, C. H., *Chem. Commun. (Cambridge)*, (2013) **49**, 5690.
[67] Oh, C. H.; Lee, S. M.; Hong, C. S., *Org. Lett.*, (2010) **12**, 1308.
[68] Gross, T.; Metz, P., *Chem.–Eur. J.*, (2013) **19**, 14 787.
[69] Xie, Y.-X.; Yan, Z.-Y.; Qian, B.; Deng, W.-Y.; Wang, D.-Z.; Wu, L.-Y.; Liu, X.-Y.; Liang, Y.-M., *Chem. Commun. (Cambridge)*, (2009), 5451.
[70] Zheng, H.; Zheng, J.; Yu, B.; Chen, Q.; Wang, X.; He, Y.; Yang, Z.; She, X., *J. Am. Chem. Soc.*, (2010) **132**, 1788.
[71] Zheng, H.; Huo, X.; Zhao, C.; Jing, P.; Yang, J.; Fang, B.; She, X., *Org. Lett.*, (2011) **13**, 6448.
[72] Coldham, I.; Hufton, R., *Chem. Rev.*, (2005) **105**, 2765.
[73] Vedejs, E.; Naidu, B. N.; Klapars, A.; Warner, D. L.; Li, V.-S.; Na, Y.; Kohn, H., *J. Am. Chem. Soc.*, (2003) **125**, 15 796.
[74] Pearson, W. H.; Kropf, J. E.; Choy, A. L.; Lee, I. Y.; Kampf, J. W., *J. Org. Chem.*, (2007) **72**, 4135.
[75] Burrell, A. J. M.; Coldham, I.; Watson, L.; Oram, N.; Pilgram, C. D.; Martin, N. G., *J. Org. Chem.*, (2009) **74**, 2290.
[76] Coldham, I.; Burrell, A. J. M.; White, L. E.; Adams, H.; Oram, N., *Angew. Chem.*, (2007) **119**, 6271; *Angew. Chem. Int. Ed.*, (2007) **46**, 6159.
[77] Coldham, I.; Jana, S.; Watson, L.; Martin, N. G., *Org. Biomol. Chem.*, (2009) **7**, 1674.
[78] Burrell, A. J. M.; Watson, L.; Martin, N. G.; Oram, N.; Coldham, I., *Org. Biomol. Chem.*, (2010) **8**, 4530.
[79] Coldham, I.; Burrell, A. J. M.; Guerrand, H. D. S.; Watson, L.; Martin, N. G.; Oram, N., *Beilstein J. Org. Chem.*, (2012) **8**, 107.
[80] Fang, C.; Shanahan, C. S.; Paull, D. H.; Martin, S. F., *Angew. Chem.*, (2012) **124**, 10 748; *Angew. Chem. Int. Ed.*, (2012) **51**, 10 596.
[81] Shanahan, C. S.; Fang, C.; Paull, D. H.; Martin, S. F., *Tetrahedron*, (2013) **69**, 7592.
[82] Takaya, J.; Miyashita, Y.; Kusama, H.; Iwasawa, N., *Tetrahedron*, (2011) **67**, 4455.
[83] Yeom, H.-S.; Lee, J.-E.; Shin, S., *Angew. Chem.*, (2008) **120**, 7148; *Angew. Chem. Int. Ed.*, (2008) **47**, 7040.
[84] Yu, C.; Zhang, Y.; Zhang, S.; Wang, W., *Chem. Commun. (Cambridge)*, (2011) **47**, 1036.
[85] Guo, S.; Zhang, H.; Huang, L.; Guo, Z.; Xiong, G.; Zhao, J., *Chem. Commun. (Cambridge)*, (2013) **49**, 8689.
[86] Arisawa, M.; Fujii, Y.; Kato, H.; Fukuda, H.; Matsumoto, T.; Ito, M.; Abe, H.; Ito, Y.; Shuto, S., *Angew. Chem.*, (2013) **125**, 1037; *Angew. Chem. Int. Ed.*, (2013) **52**, 1003.
[87] D'Souza, A. M.; Spiccia, N.; Basutto, J.; Jokisz, P.; Wong, L. S.-M.; Meyer, A. G.; Holmes, A. B.; White, J. M.; Ryan, J. H., *Org. Lett.*, (2011) **13**, 486.

[88] Ye, S.; Yang, X.; Wu, J., *Chem. Commun. (Cambridge)*, (2010) **46**, 5238.
[89] Huple, D. B.; Chen, C.-H.; Das, A.; Liu, R.-S., *Adv. Synth. Catal.*, (2011) **353**, 1877.
[90] Ren, H.; Ye, S.; Liu, F.; Wu, J., *Tetrahedron*, (2010) **66**, 8242.
[91] Guerrand, H. D. S.; Adams, H.; Coldham, I., *Org. Biomol. Chem.*, (2011) **9**, 7921.
[92] Majumdar, K. C.; Ray, K., *Synthesis*, (2011), 3767.
[93] Brahma, K.; Achari, B.; Chowdhury, C., *Synthesis*, (2013) **45**, 545.
[94] Hu, Y.-Y.; Hu, J.; Wang, X.-C.; Guo, L.-N.; Shu, X.-Z.; Niu, Y.-N.; Liang, Y.-M., *Tetrahedron*, (2010) **66**, 80.
[95] Arigela, R. K.; Mandadapu, A. K.; Sharma, S. K.; Kumar, B.; Kundu, B., *Org. Lett.*, (2012) **14**, 1804.
[96] Majumdar, K. C.; Ganai, S., *Synthesis*, (2013) **45**, 2619.
[97] Chambers, C. S.; Patel, N.; Hemming, K., *Tetrahedron Lett.*, (2010) **51**, 4859.
[98] Chen, X.; Zhu, J.; Xie, H.; Li, S.; Wu, Y.; Gong, Y., *Adv. Synth. Catal.*, (2010) **352**, 1296.
[99] Barange, D. K.; Tu, Y.-C.; Kavala, V.; Kuo, C.-W.; Yao, C.-F., *Adv. Synth. Catal.*, (2011) **353**, 41.
[100] Qian, W.; Wang, H.; Allen, J., *Angew. Chem.*, (2013) **125**, 11 198; *Angew. Chem. Int. Ed.*, (2013) **52**, 10 992.
[101] Bhattacharya, D.; Ghorai, A.; Pal, U.; Maiti, N. C.; Chattopadhyay, P., *RSC Adv.*, (2014) **4**, 4155.
[102] Zhao, Y.-M.; Gu, P.; Tu, Y.-Q.; Zhang, H.-J.; Zhang, Q.-W.; Fan, C.-A., *J. Org. Chem.*, (2010) **75**, 5289.
[103] de Miguel, I.; Herradón, B.; Mann, E., *Adv. Synth. Catal.*, (2012) **354**, 1731.
[104] de Miguel, I.; Velado, M.; Herradón, B.; Mann, E., *Adv. Synth. Catal.*, (2013) **355**, 1237.
[105] Ylijoki, K. E. O.; Stryker, J. M., *Chem. Rev.*, (2013) **113**, 2244.
[106] Feng, J.-J.; Zhang, J., *J. Am. Chem. Soc.*, (2011) **133**, 7304.
[107] Wender, P. A.; Stemmler, R. T.; Sirois, L. E., *J. Am. Chem. Soc.*, (2010) **132**, 2532.
[108] Woodall, E. L.; Simanis, J. A.; Hamaker, C. G.; Goodell, J. R.; Mitchell, T. A., *Org. Lett.*, (2013) **15**, 3270.
[109] Rodríguez, J. R.; Rumbo, A.; Castedo, L.; Mascareñas, J. L., *J. Org. Chem.*, (1999) **64**, 966.
[110] Rodríguez, J. R.; Castedo, L.; Mascareñas, J. L., *J. Org. Chem.*, (2000) **65**, 2528.
[111] Domingo, L. R.; Zaragozá, R. J., *Tetrahedron*, (2001) **57**, 5597.
[112] Engler, T. A.; Scheibe, C. M.; Iyengar, R., *J. Org. Chem.*, (1997) **62**, 8274.

2.1.3 Domino Transformations Involving an Electrocyclization Reaction

J. Suffert, M. Gulea, G. Blond, and M. Donnard

General Introduction

In addition to molecular complexity, the challenge of today's chemist is also synthetic efficiency. Among the impressive number of reactions that have been proposed and developed by chemists, electrocyclization processes are one of the most powerful and efficient ways to produce polycyclic molecules. Mainly in the form of 6π- and 8π-electrocyclizations, these transformations were, are, and will continue to be extremely useful for the stereospecific preparation of the carbo- and heterocycles present in many structurally complex molecules. Critically, 6π- and 8π-electrocyclization reactions can serve as the start of a much more sophisticated process that involves several reactions in one pot, namely, domino and cascade-type reactions. These domino/cascade reactions can be initiated by the use of a catalytic amount of a metal, acid, or base, or by heat or light. Atom economy, step economy, efficiency, complexity, and stereoselectivity drive the processes presented in this chapter.

This overview will show that easy access to complex polycyclic molecules can result from readily available simple starting materials through 6π-, 8π-, or successive 8π–6π-electrocyclization reactions. We will focus our attention on the most representative examples published in the last 15 years, with particular emphasis on methodologies that allow for the formation of nonnatural and natural products by using a 6π- or 8π-electrocyclization reaction as the key step.

2.1.3.1 Metal-Mediated Cross Coupling Followed by Electrocyclization

2.1.3.1.1 Palladium-Mediated Cross Coupling/Electrocyclization Reactions

Domino-type combination of consecutive palladium-catalyzed cross-coupling reactions with subsequent pericyclic transformations offers rapid access to various oligocyclic skeletons.

2.1.3.1.1.1 Cross Coupling/6π-Electrocyclization

Houk, Kwon, and co-workers recently reported their experimental and computational results on the origin of 1,6-stereoinduction in the torquoselective 6π-electrocyclization reaction.[1,2] In their study, they employ a tandem Suzuki/6π-electrocyclization reaction to reach reserpine-type skeleton **2** (Scheme 1). They show that the diastereoselectivity of the process is sensitive to the substituent at the C3-position of triene **1**. Indeed, moderately bulky substituents such as methyl and methyl ester groups at this position allow the system to undergo 6π-electrocyclization with high levels of diastereoselectivity. By contrast, bulky groups such as *tert*-butyl and trimethylsilyl groups lead to much lower levels of stereoselectivity in the ring closure.

for references see p 156

Scheme 1 Suzuki Cross Coupling/6π-Electrocyclization[1,2]

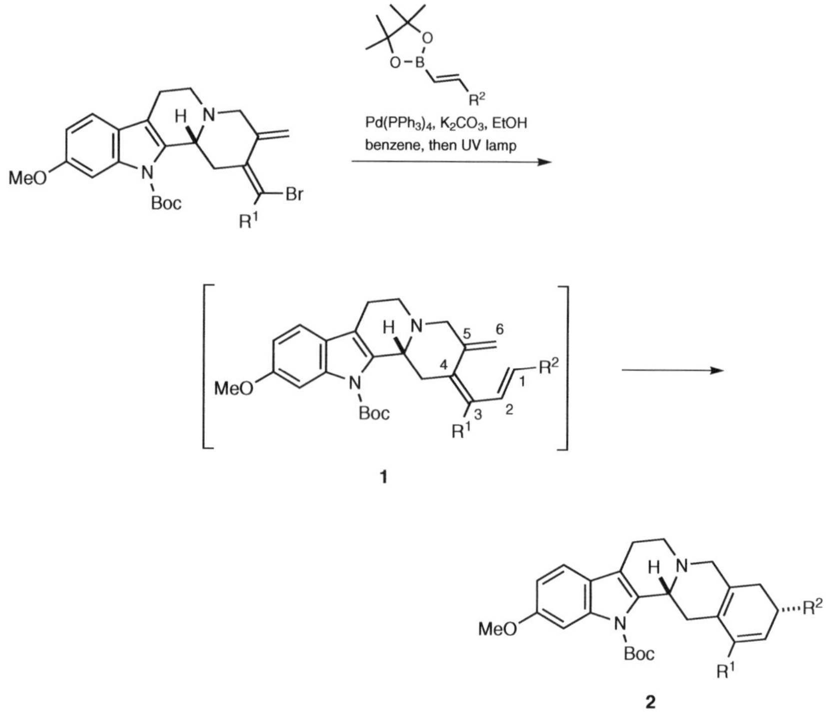

R^1	R^2	Yield (%)	Ref
CO_2Me	Me	23	[2]
CO_2Me	OBn	54	[1]
TMS	Me	83	[2]

Zhang and co-workers have developed a palladium-catalyzed tandem reaction of yne–propargylic carbonates **3** with boronic acids for the synthesis of fused aromatic rings **4** through allene-mediated electrocyclization (Scheme 2).[3]

Scheme 2 Suzuki Cross Coupling/6π-Electrocyclization/Aromatization[3]

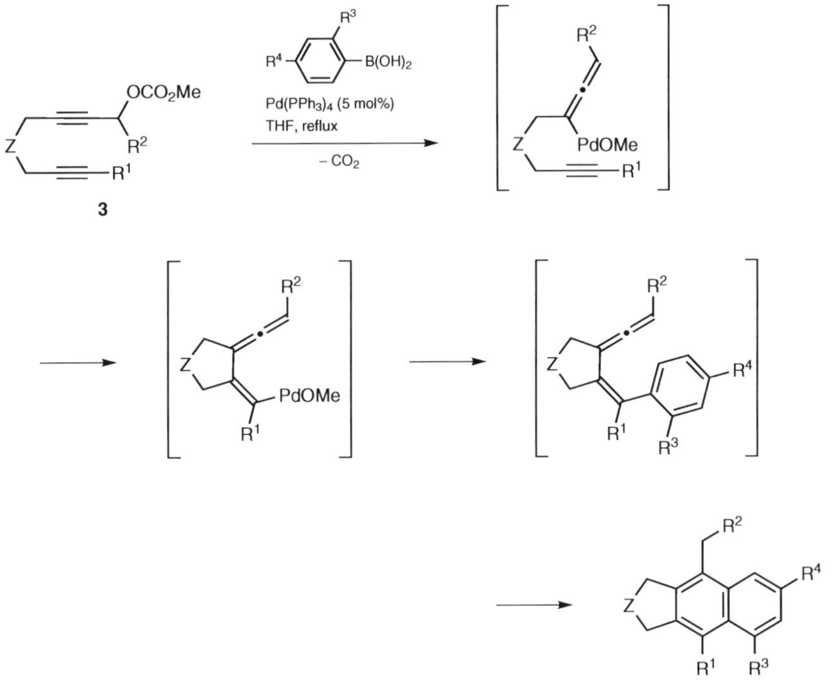

R^1	R^2	R^3	R^4	Z	Time (h)	Yield (%)	Ref
Me	H	H	H	O	1	65	[3]
H	H	H	Cl	O	2	61	[3]
Me	H	H	Cl	O	2	61	[3]
H	H	H	Me	O	1	60	[3]
Me	H	H	Me	O	1	69	[3]
H	H	H	OMe	O	2	61	[3]
H	H	H	H	NTs	1	55	[3]
H	H	H	Me	NTs	1	64	[3]
H	H	Me	H	NTs	2	59	[3]
H	Ph	H	H	O	1	76	[3]

The same kind of strategy can be applied by starting the domino sequence with a Heck cross-coupling reaction. The outcome and the mechanistic pathway of the palladium-catalyzed Heck-type cascade oligocyclization of various 2-bromoalkenediynes and 2-bromoalkadienynes have been explored with respect to the length of the tether between the multiple bonds and the nature of the substituent at the acetylene terminus.[4–6] For example, substrate **5** with an electron-donating substituent yields bisannulated benzene derivative **6**, whereas substrate **7** with an electron-withdrawing substituent yields a tetracyclic system (e.g., **8**) containing a three-membered ring (Scheme 3).[5] The process to bisannulated product **6** involves two consecutive 6-*exo-dig* carbopalladation reactions followed by 6π-electrocyclization and β-hydride elimination, whereas the process to tetracyclic system **8** involves a 6-*exo-dig*/5-*exo-trig*/3-*exo-trig*/β-hydride elimination sequence.

for references see p 156

Scheme 3 Palladium-Catalyzed Cyclization of 2-Bromoalkenediynes[5]

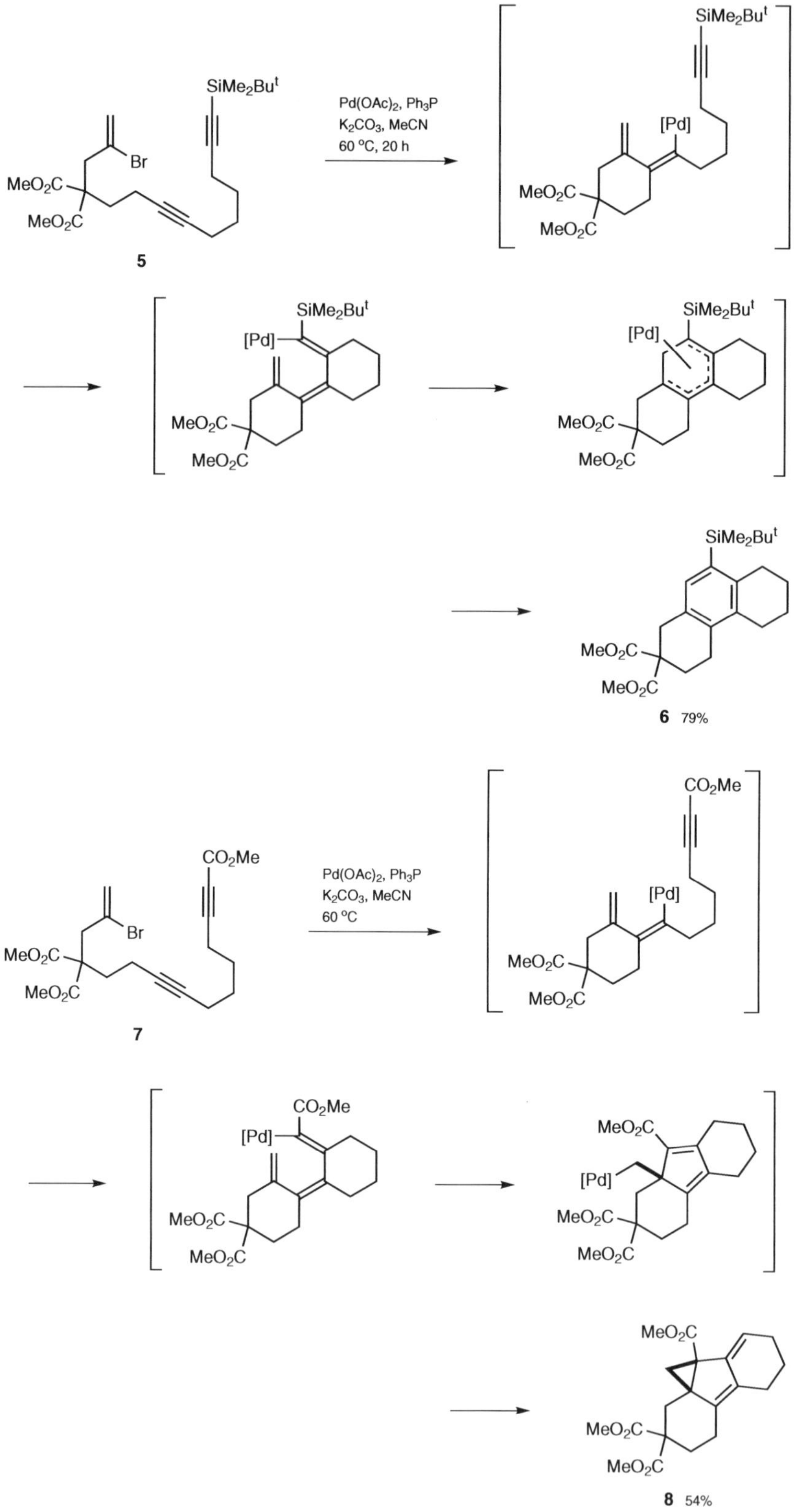

$Pd(OAc)_2$, Ph_3P
K_2CO_3, MeCN
60 °C, 20 h

5

6 79%

$Pd(OAc)_2$, Ph_3P
K_2CO_3, MeCN
60 °C

7

8 54%

The same type of cascade has been developed by Suffert and co-workers, and it involves 4-*exo-dig* cyclocarbopalladation. These processes are shown to convert bromoenediynes **9** and bromodienyne **11** into strained aromatic compounds **10** and **12** in a single step (Scheme 4).[7]

Scheme 4 Synthesis of Strained Aromatic Polycycles[7]

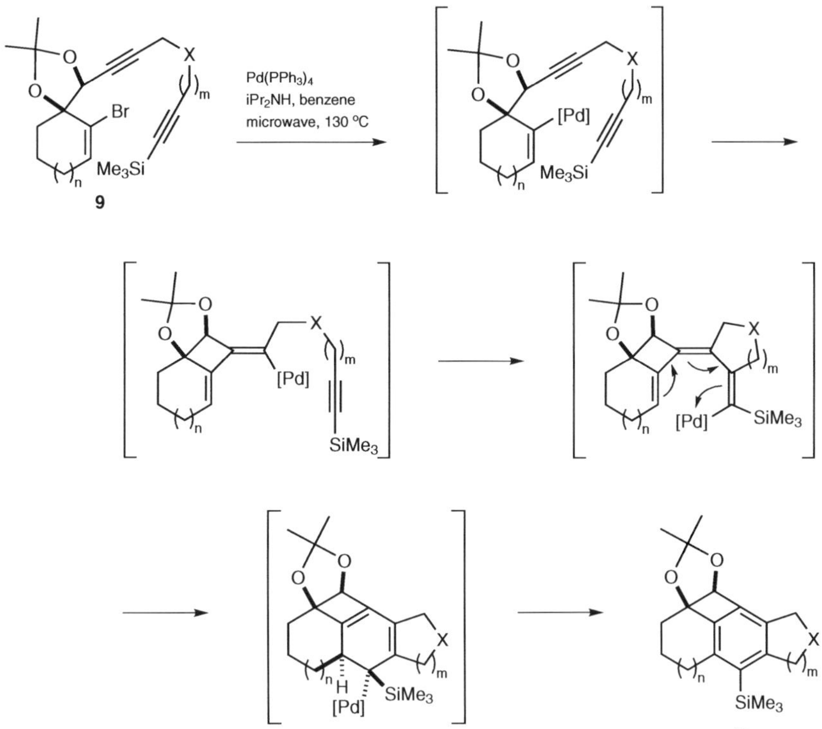

Pd(PPh$_3$)$_4$
iPr$_2$NH, benzene
microwave, 130 °C

X	m	n	Yield (%)	Ref
O	1	1	89	[7]
O	1	2	89	[7]
O	2	1	62	[7]
O	3	1	72	[7]
NBoc	1	1	50	[7]
CH$_2$	1	1	69	[7]

for references see p 156

By using palladium(II) acetate, dibenzothiophenes **14** (Z = S)[8] and dibenzofurans **14** (Z = O)[9] can be prepared from 2,3-dibromobenzo[*b*]thiophene (**13**, Z = S) and 2,3-dibromobenzo[*b*]furan (**13**, Z = O) using SPhos as the ligand. Furthermore, carbazoles **16** can be obtained from 2,3-dibromo-1-methyl-1*H*-indole (**15**) if SPhos is used,[10] and functionalized anthraquinones[11] **17** and diarylfluorenones[12] **18** can be obtained from the corresponding dibromo compounds using XPhos or *t*-BuXPhos as the ligands. All of these compounds are prepared through a domino two-fold Heck/6π-electrocyclization reaction, as outlined by Langer and co-workers (Scheme 5).

Scheme 5 Synthesis of Aromatic Polycycles[8–12]

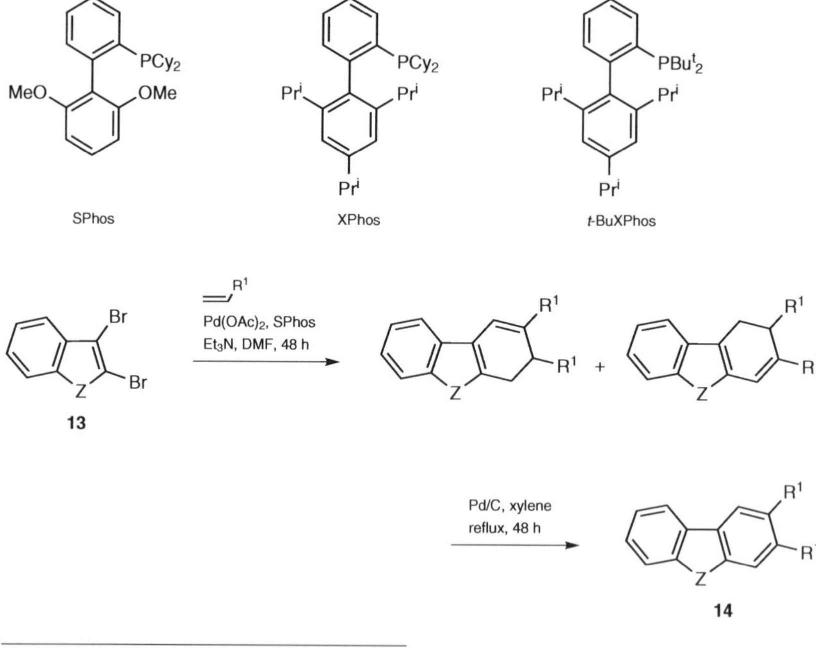

R^1	Z	Temp (°C)	Yield[a] (%)	Ref
CO_2Et	O	100	73	[9]
CO_2Bu	O	100	76	[9]
CO_2iBu	O	100	84	[9]

R^1	Z	Temp (°C)	Yield[a] (%)	Ref
CO_2t-Bu	O	100	75	[9]
CO_2Et	S	130	76	[8]
CO_2Bu	S	130	81	[8]
CO_2iBu	S	130	74	[8]
$CO_2(CH_2)_5Me$	S	130	77	[8]
Ph	S	130	79	[8]
4-ClC_6H_4	S	130	83[b]	[8]

[a] Yield of the isolated product based on starting material **13**.
[b] Product was directly formed from **13** (X = S) in one step.

$Pd(OAc)_2$, SPhos, Et_3N
DMF, 120 °C, 48 h

15 → **16**

R^1	Yield (%)	Ref
CO_2Me	69	[10]
CO_2Et	93	[10]
CO_2Bu	77	[10]
CO_2iBu	66	[10]
$CO_2(CH_2)_5Me$	81	[10]
CO_2t-Bu	85	[10]
$(CH_2)_2NMe_2$	79	[10]

$Pd(OAc)_2$, XPhos
Et_3N, DMF, 90 °C, 8 h

17

R^1	Yield (%)	Ref
CO_2iBu	83	[11]
CO_2t-Bu	82	[11]
Ph	82	[11]
4-t-BuC_6H_4	80	[11]
4-$MeOC_6H_4$	76	[11]

for references see p 156

R^1	Yield (%)	Ref
4-*t*-BuOC$_6$H$_4$	95	[12]
4-*t*-BuC$_6$H$_4$	92	[12]
Ph	80	[12]
4-BrC$_6$H$_4$	75	[12]
4-AcC$_6$H$_4$	65	[12]

4-Benzyl-1,3-dihydronaphtho[2,3-*c*]furan (4, $R^1 = R^3 = R^4 = H$; $R^2 = Ph$; Z = O); Typical Procedure:[3]
A Schlenk tube was charged with a soln of phenylboronic acid ($R^3 = R^4 = H$; 55 mg, 0.45 mmol, 1.5 equiv) and Pd(PPh$_3$)$_4$ (17.3 mg, 0.015 mmol, 0.05 equiv) in THF (2.0 mL) under an argon atmosphere. A soln of yne–propargylic carbonate **3** ($R^1 = H$; $R^2 = Ph$; Z = O; 77 mg, 0.3 mmol, 1 equiv) in THF (1.0 mL) was then introduced into the tube by syringe in one portion with stirring. The mixture was heated at reflux until the carbonate was consumed, as monitored by TLC. The mixture was cooled to rt, one drop of TFA was added, and the mixture was then stirred for ~30 min. After evaporation of the solvent under reduced pressure, the mixture was purified by chromatography (silica gel, hexane/EtOAc) to afford the title compound as a white solid; yield: 59 mg (76%).

Dimethyl 9-(*tert*-Butyldimethylsilyl)-3,4,5,6,7,8-hexahydrophenanthrene-2,2(1*H*)-dicarboxylate (6); Typical Procedure:[5]
Compound **5** (249 mg, 0.50 mmol, 1 equiv) was added to a sealed tube containing a degassed mixture of Pd(OAc)$_2$ (11 mg, 0.05 mmol, 0.1 equiv), Ph$_3$P (33 mg, 0.13 mmol, 0.25 equiv), and K$_2$CO$_3$ (207 mg, 1.5 mmol, 3 equiv) in MeCN (5 mL). After heating at 60 °C for 20 h, the mixture was cooled to rt, filtered through a layer each of Celite and charcoal, and then concentrated. The resulting residue was purified by column chromatography (hexane/Et$_2$O 10:1) to give the title compound as a colorless oil; yield: 163 mg (79%).

(3a*R,10c*S**)-2,2-Dimethyl-7-(trimethylsilyl)-5,6,10,10c-tetrahydro-4*H*,8*H*-furo[3″,4″:2′,3′]naphtha[8′,1′:1,4,3]cyclobuta[1,2-*d*][1,3]dioxole (10, X = O; m = n = 1); Typical Procedure:**[7]
Pd(PPh$_3$)$_4$ (26 mg, 0.023 mmol, 0.1 equiv) and iPr$_2$NH (3 mL) were added to a soln of compound **9** (100 mg, 0.243 mmol, 1 equiv) in anhyd benzene (7 mL) (**CAUTION:** *carcinogen*). The mixture was purged for 5 min with argon, heated at 130 °C under microwave irradiation for 20 min, concentrated under reduced pressure, and purified by chromatography (silica gel, heptane/Et$_2$O 9:1) to afford the title compound as a yellow solid; yield: 89%.

(3a*R,6a*S**)-5,5-Dimethyl-7-(trimethylsilyl)-2,3-dihydro-1*H*,6a*H*-naphtho[1′,8′:3,4,1]cyclobuta[1,2-*d*][1,3]dioxole (12); Typical Procedure:**[7]
Pd(PPh$_3$)$_4$ (8 mg, 0.007 mmol, 0.1 equiv) and iPr$_2$NH (0.2 mL) were added to a soln of compound **11** (26 mg, 0.07 mmol, 1 equiv) in benzene (0.6 mL) (**CAUTION:** *carcinogen*). The mixture was purged for 5 min with argon and heated at 130 °C under microwave irradiation for 20 min. Purification by chromatography (silica gel, heptane/EtOAc 19:1) afforded the title compound as a white solid; yield: 13 mg (66%, as reported).

Diethyl 9-Methyl-2,9-dihydro-1*H*-carbazole-2,3-dicarboxylate (16, $R^1 = CO_2Et$); Typical Procedure:[10]

In a pressure tube (glass bomb), a suspension of $Pd(OAc)_2$ (11 mg, 0.05 mmol, 0.05 equiv) and SPhos (41 mg, 0.10 mmol, 0.10 equiv) in DMF (5 mL) was purged with argon and stirred at 20 °C to afford a yellowish or brownish transparent soln. 2,3-Dibromo-1-methyl-1*H*-indole (**15**; 289 mg, 1.0 mmol, 1 equiv), Et_3N (1.1 mL, 8.0 mmol), and the acrylate ($R^1 = CO_2Et$; 0.27 mL, 2.5 mmol, 2.5 equiv) were added with stirring. The mixture was stirred at 120 °C for 48 h. The soln was cooled to 20 °C, poured into H_2O and CH_2Cl_2 (25 mL each), and the organic and aqueous layers were separated. The latter was extracted with CH_2Cl_2 (3 × 25 mL) and the combined organic layers were washed with H_2O (3 × 20 mL), dried (Na_2SO_4), and concentrated under reduced pressure. The residue was purified by flash chromatography (silica gel, heptane/EtOAc) to afford the title compound as a yellow solid; yield: 303 mg (93%).

2.1.3.1.1.2 Cross Coupling/8π-Electrocyclization

Suffert and co-workers have reported an efficient synthesis of 5,8,5- and 6,4,8,5-polycyclic systems in a one-pot, domino operation.[13,14] Although the yields are modest, the outcome is certainly compensated for by the structural complexity of the final products. The strategy for the preparation of 6,4,8,5-polycyclic system **21** is based on a cascade reaction starting from vinyl bromide **19** and consists of a 4-*exo-dig* cyclocarbopalladation, a Stille cross-coupling reaction with stannane **20**, and a terminating conrotatory 8π-electrocyclization reaction (Scheme 6).

Scheme 6 Synthesis of Eight-Membered Ring Systems by a 4-*exo-dig* Cyclocarbopalladation/Stille Cross-Coupling/8π-Electrocyclization Sequence[13,14]

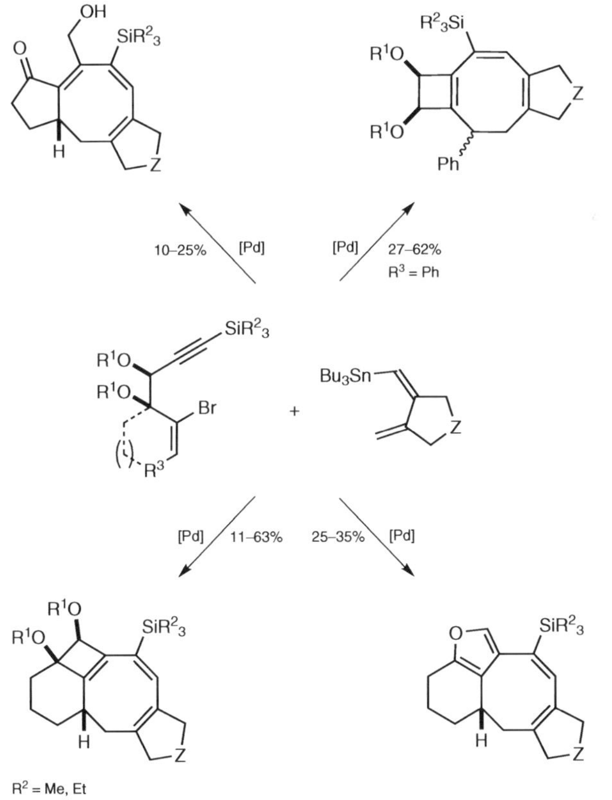

R^2 = Me, Et

for references see p 156

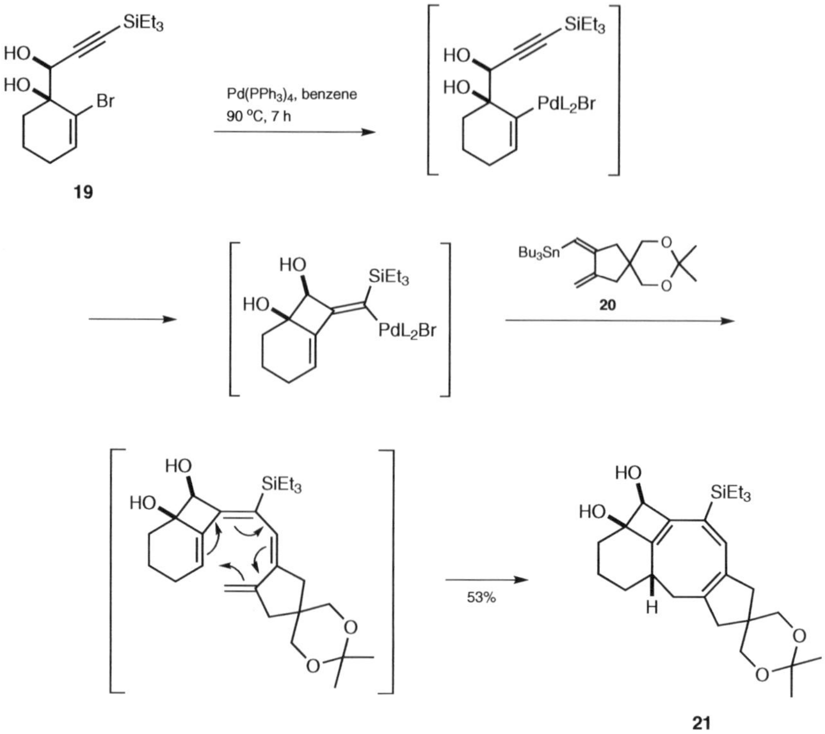

Following the same strategy, Anderson and co-workers have described a palladium-catalyzed cascade for the synthesis of fused bi- and tricyclic ring systems by finishing the cascade with a 6π-electrocyclization reaction. If it is ended with an 8π-electrocyclization reaction instead, then eight-membered rings are obtained.[15]

Interestingly, the electrocyclization step can be subsequently followed by an additional pericyclic reaction. For example, the use of a Stille cross-coupling/8π-electrocyclization/Diels–Alder cascade to assemble the core of (−)-PF-1018 using copper(I) thiophene-2-carboxylate (CuTC) has been reported by Trauner and co-workers (Scheme 7).[16] Considering the number of pericyclic reaction modes available from intermediate polyene **22**, the fact that tricycle **23** can be reproducibly isolated as a single diastereomer is remarkable.

Scheme 7 Synthesis of (−)-PF-1018 Using a Cascade Approach[16]

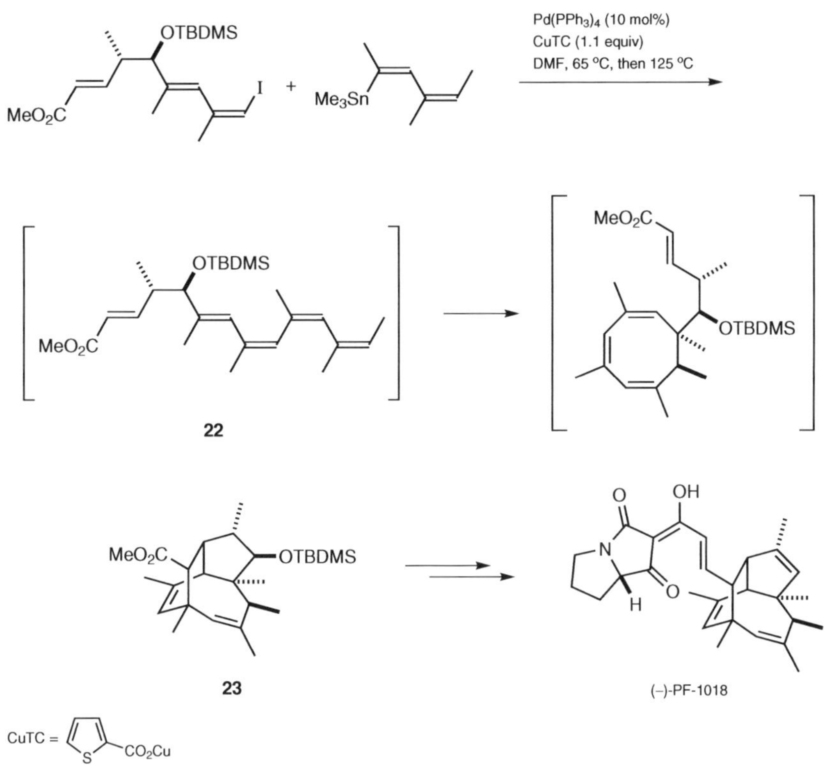

$Pd(PPh_3)_4$ (10 mol%)
CuTC (1.1 equiv)
DMF, 65 °C, then 125 °C

22

23

(−)-PF-1018

A one-pot approach developed by Ma and co-workers allows the installation of eight-membered rings onto indoles. This reaction is highly efficient and proceeds through palladium-catalyzed C—C coupling, double-bond migration, and 8π-electrocyclization.[17]

(7*R*,10a*S*,12*S*,*E*)-2′,2′-Dimethyl-5-(triethylsilyl)-3,8,9,10,10a,11-hexahydrospiro[[6,7]methanobenzo[*a*]cyclopenta[*d*][8]annulene-2,5′-[1,3]dioxane]-7,12(1*H*)-diol (21);
Typical Procedure:[13]
An oven-dried, 25-mL, two-necked flask equipped with a reflux condenser was charged with a soln of vinyl bromide **19** (100 mg, 0.29 mmol, 1 equiv) in dry benzene (10 mL) (**CAUTION:** *carcinogen*) under an argon atmosphere. $Pd(PPh_3)_4$ (33 mg, 0,03 mmol, 0.1 equiv) was added, followed by stannane **20** (182 mg, 0.38 mmol, 1.3 equiv). The mixture was stirred in a preheated 90 °C oil bath for 7 h, and the progress of the reaction was monitored by TLC. Upon completion of the reaction, the mixture was concentrated under reduced pressure and immediately purified by flash chromatography (EtOAc/hexane 10:90 to 35:65) to give the title compound; yield: 70 mg (53%).

2.1.3.1.1.3 **Cross Coupling/8π-Electrocyclization/6π-Electrocyclization**

The groups of Trauner and Parker have independently investigated the structures of SNF4435 C and D (**26A** and **26B**) because of their significant biological activity and to understand their biosynthetic origins.[18–22] SNF4435 C and D are immunosuppressant and multidrug-resistance reversal agents that were isolated from the culture broth of the Okinawan strain of *Streptomyces spectabilis* in 2001.

Both groups have reported synthetic strategies to access these materials by using a Stille cross-coupling/8π-electrocyclization/6π-electrocyclization domino sequence. The

for references see p 156

retrosynthetic analysis is identical, although it should be noted that whereas bond breaking is projected in the same places, the two Stille cross-coupling partners (e.g., **24**/**25** in the Trauner strategy and **27**/**28** in the Parker strategy) are inverted (Scheme 8).

Scheme 8 Synthetic Strategies by Trauner and Parker for the Synthesis of SNF4435 C and D[18–22]

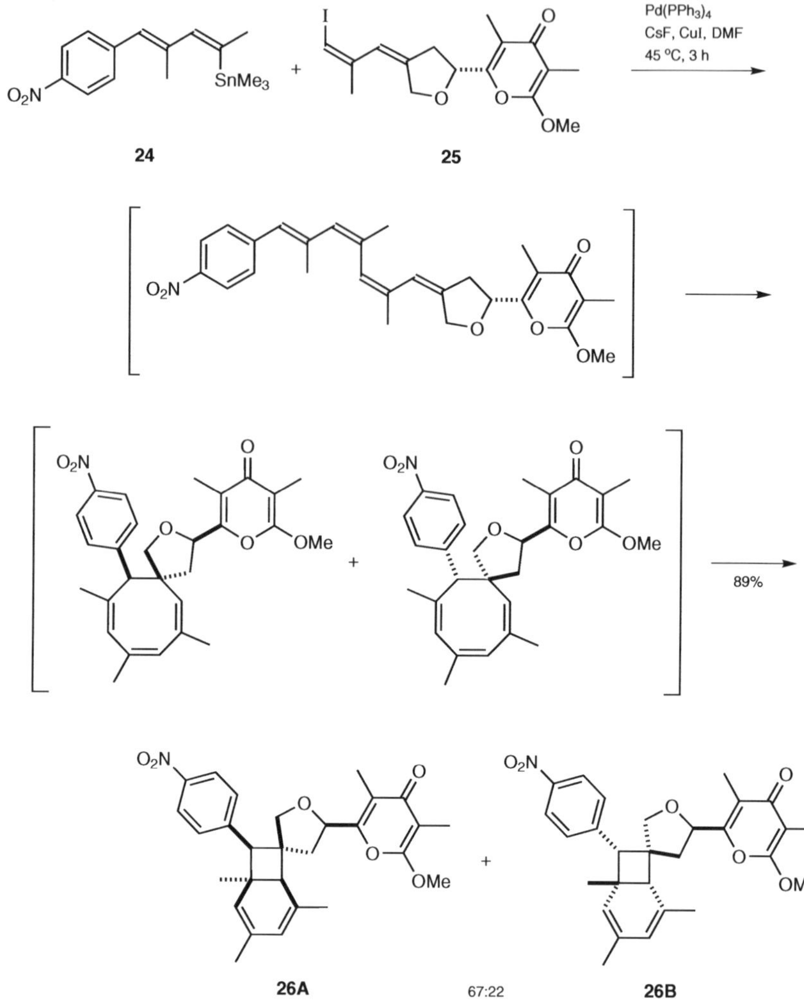

$Pd(PPh_3)_4$, CsF, CuI, DMF, 45 °C, 3 h

24 + **25**

89%

26A 67:22 **26B**

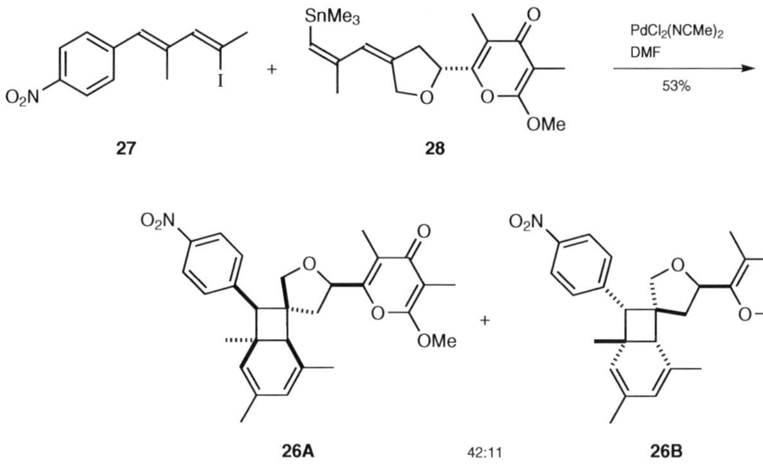

SnMe3 O PdCl2(NCMe)2 DMF 53% O2N I O O OMe 27 28

O2N O O2N O O O O O OMe + OMe 26A 42:11 26B

Critically, Trauner and co-workers have used an identical cascade reaction to realize the biomimetic synthesis of several different natural products, including elysiapyrone A and B,[23] ocellapyrone A and B,[24] and shimalactone A and B (Scheme 9).[25]

Scheme 9 Biomimetic Synthesis of Elysiapyrones, Ocellapyrones, and Shimalactones[23–25]

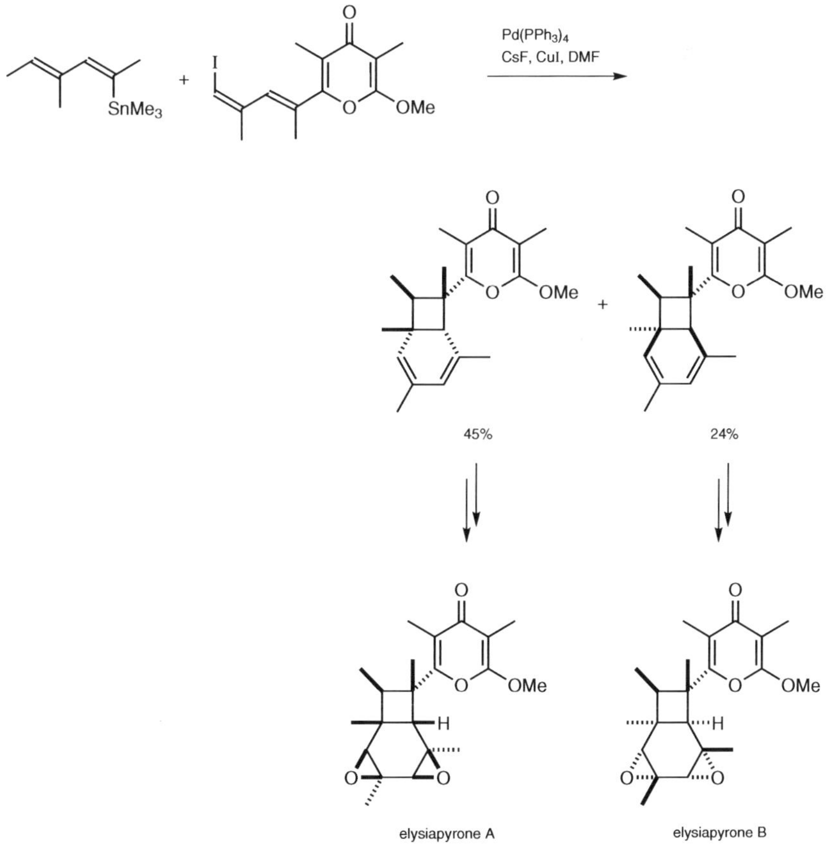

O Pd(PPh3)4 CsF, CuI, DMF I SnMe3 O OMe

O O O OMe + O OMe 45% 24%

O O O OMe O OMe H H O O O O elysiapyrone A elysiapyrone B

for references see p 156

Pd(PPh3)4
CsF, CuI, DMF

ocellapyrone A 8%

78%

O_2
89%

ocellapyrone B

Pd(PPh3)4
CuTC

shimalactone A 55%

shimalactone B 11%

CuTC = CO_2Cu

Moses and co-workers have used the same reaction cascade in a synthetic approach to kingianin A based on biosynthetic speculation (Scheme 10).[26] Their method highlights both the power and the generality of the highlighted sequence of reactions.

Scheme 10 Synthetic Approach to Kingianin A Involving a Stille Cross-Coupling/8π-Electrocyclization/6π-Electrocyclization Domino Sequence[26]

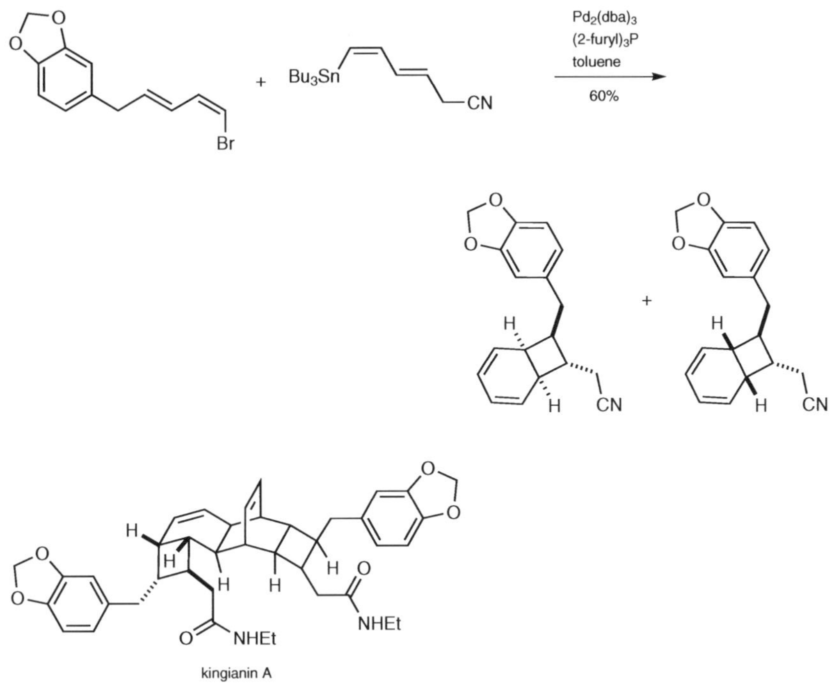

Suffert and co-workers have also developed a remarkable domino reaction that involves 4-*exo-dig* cyclocarbopalladation, Sonogashira-type coupling, regioselective alkynylation of a disubstituted triple bond, and 8π-electrocyclization/6π-electrocyclization (Black's cascade sequence) for the preparation of [4.6.4.6]fenestradienes **29** (Scheme 11).[27,28] The exceptional efficiency of this process is highlighted by the limited number of steps that is required to synthesize these structurally complex products. The resulting compounds, known as fenestranes, are unusual in part due to their highly strained bond angles.[29]

Scheme 11 Synthesis of [4.6.4.6]Fenestradienes[27]

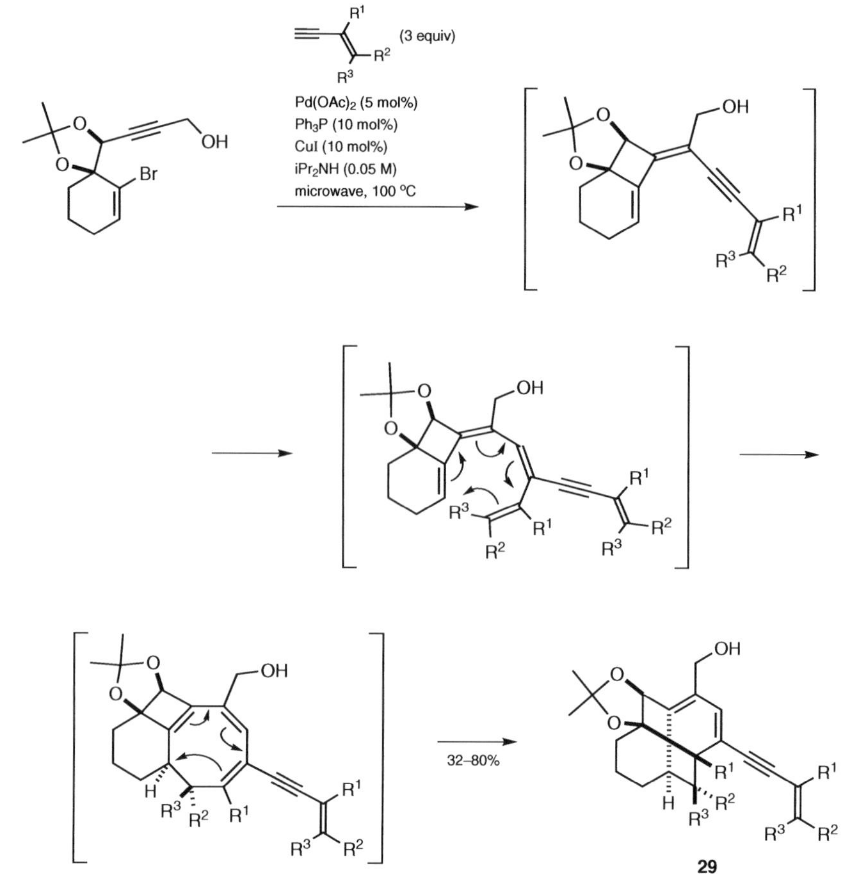

(3 equiv)
$Pd(OAc)_2$ (5 mol%)
Ph_3P (10 mol%)
CuI (10 mol%)
iPr_2NH (0.05 M)
microwave, 100 °C

32–80%

29

R^1	R^2	R^3	Time (min)	Yield (%)	Ref
H	CH_2NHBoc	H	60	43	[27]
Me	H	H	60	73	[27]
H	 1-hydroxycyclohexyl	H	120	39	[27]
H	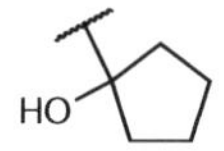 1-hydroxycyclopentyl	H	60	59	[27]
Me	CH_2OH	H	60	68	[27]
Me	$CH_2OTBDMS$	H	60	72	[27]
Me	H	$CH_2OTBDMS$	120	40	[27]
H	$(CH_2)_2Ph$	H	30	46	[27]

SNF4435 C and D (26A and 26B); Typical Procedure:[22]

CsF (16 mg, 0.10 mmol, 2 equiv), CuI (2 mg, 0.01 mmol, 0.2 equiv), and $Pd(PPh_3)_4$ (6 mg, 0.005 mmol, 0.1 equiv) were added to a soln of iodide **25** (21 mg, 0.052 mmol, 1 equiv) and stannane **24** (29 mg, 0.078 mmol, 1.5 equiv) in DMF (1 mL) at rt. The mixture was heated to 45 °C for 3 h, cooled to rt, and diluted with EtOAc (15 mL). The organic layer was washed with sat. aq NH_4Cl (3 × 10 mL) and the combined aqueous layers were extracted with EtOAc (3 × 15 mL). The organic layers were combined, dried ($MgSO_4$), filtered, and concentrated. Purification by chromatography (silica gel, hexane/EtOAc 1:1) gave the title compounds; yield: 22 mg (89%); ratio (**26A/26B**) 67:22 (~3:1). The mixture was separated by HPLC to afford **26A** and **26B** as pale yellow solids. The enantiomeric excess was determined by HPLC on an analytical Chiralcel OD column (id = 4.6 mm, length = 250 mm).

2.1.3.1.2 Copper-Catalyzed Tandem Reactions

Kim and co-workers have developed an expedient synthetic approach to naphthalenes **31** by the reaction of Morita–Baylis–Hillman bromide derivatives **30** and arylacetylenes. The synthesis involves a tandem copper-catalyzed alkynylation/propargyl–allenyl isomerization/6π-electrocyclization process (Scheme 12).[30]

Scheme 12 Synthesis of Naphthalene Derivatives by Copper-Catalyzed Cross Coupling/6π-Electrocyclization[30]

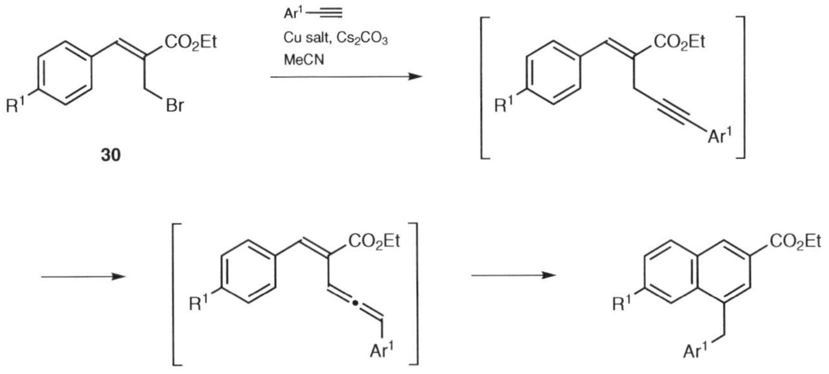

R^1	Ar^1	Cu Salt	Time (h)	Yield (%)	Ref
H	Ph	CuI	5	56	[30]
Cl	Ph	CuI	3	54	[30]
Me	Ph	CuBr	5	74	[30]
OMe	Ph	CuBr	18	81	[30]
H	4-$MeOC_6H_4$	CuBr	15	44	[30]
H	4-Tol	CuI	16	56	[30]
H	4-FC_6H_4	CuI	13	54	[30]
OMe	4-$MeOC_6H_4$	CuBr	19	56	[30]

2.1.3.1.3 Zinc-Catalyzed Tandem Reactions

Zinc-mediated arylation of 2-(bromomethyl)indoles **32** with arenes leads to the formation of carbazoles **33** through an unprecedented [1,5]-sigmatropic rearrangement/6π-electrocyclization/aromatization sequence (Scheme 13).[31]

Scheme 13 Synthesis of Carbazole Derivatives[31]

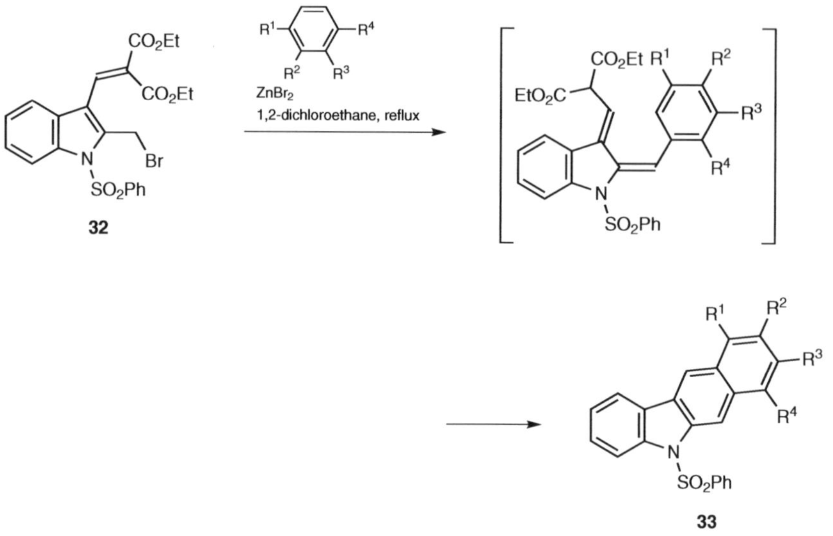

$ZnBr_2$
1,2-dichloroethane, reflux

R^1	R^2	R^3	R^4	Time (h)	Yield (%)	Ref
H	H	H	H	24	25	[31]
H	OMe	H	H	1	62	[31]
H	H	$(CH{=}CH)_2$		5	40	[31]
H	OMe	OMe	H	2	57	[31]
H	Me	H	Me	1	55	[31]
Me	H	H	Me	2	54	[31]

2.1.3.1.4 Ruthenium-Catalyzed Formal [2+2+2] Cycloaddition Reactions

Standard ruthenium(II)-catalyzed [2+2+2] cycloaddition of 1,6-diynes **34** with alkenes **35** gives bicyclic cyclohexa-1,3-dienes **37** in good yields. Mechanistically, ruthenium-catalyzed coupling of 1,6-diynes with acyclic alkenes occurs to give open 1,3,5-trienes **36** which, after thermal disrotatory 6π-electrocyclization, afford the final cyclohexa-1,3-dienes **37** (Scheme 14).[32,33]

Scheme 14 Synthesis of Polycyclic Cyclohexadienes by Formal [2+2+2] Cycloaddition[32,33]

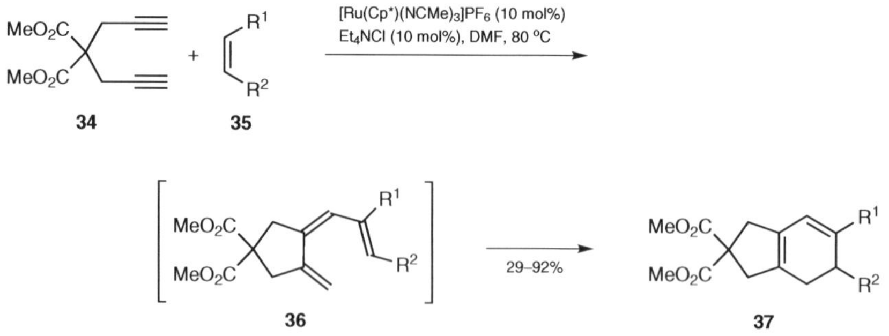

$[Ru(Cp^*)(NCMe)_3]PF_6$ (10 mol%)
Et_4NCl (10 mol%), DMF, 80 °C

29–92%

R^1	R^2	Yield (%)	Ref
CO_2Me	H	62	[33]
Ac	H	51	[33]
CHO	H	33	[33]
CN	H	46	[33]
CO_2Me	CO_2Me	61	[33]
CH_2OEt	H	52	[33]
CH_2OPh	H	66	[33]
CH_2OTMS	H	58	[33]

Schreiber and co-workers have described a cascade reaction that provides tricyclic diene **38**.[34] They hypothesize that this compound results from a consecutive ene–yne–yne ring-closing metathesis sequence followed by spontaneous 6π-electrocyclic ring closure and sigmatropic 1,5-hydride shift. During this domino process, three rings, three C—C bonds, and one stereocenter are formed (Scheme 15).

Scheme 15 Synthesis of a Polycyclic Cyclohexadiene[34]

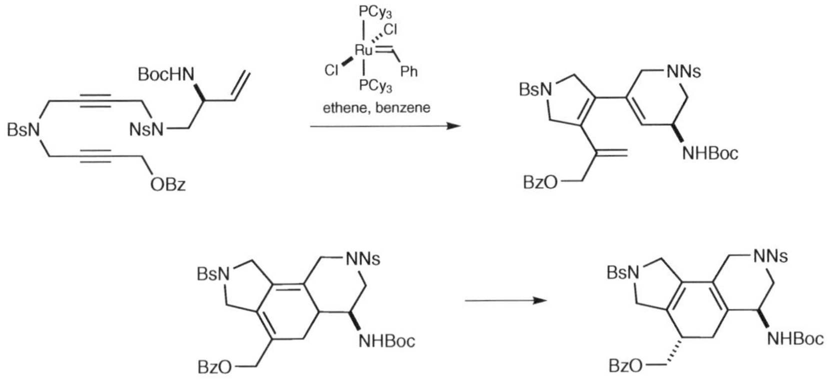

Bs = 4-bromophenylsulfonyl; Ns = 2-nitrophenylsulfonyl

2.1.3.2 Alkyne Transformation Followed by Electrocyclization

Partial reduction of the alkyne functionality generates polyenic compounds that can undergo electrocyclization. Most examples of this general approach in the literature describe the synthesis of cyclooctatrienes.

For example, Lindlar semihydrogenation of vitamin D type trienyne **39** leads spontaneously to **41**, which is a derivative of vitamin D3.[35] The tetraene intermediate **40** resulting from the reduction undergoes rapid stereoselective electrocyclization to afford the unique steroid framework (Scheme 16).

for references see p 156

Scheme 16 Synthesis of a Cyclooctatriene through a Semihydrogenation/8π-Electrocyclization Cascade[35]

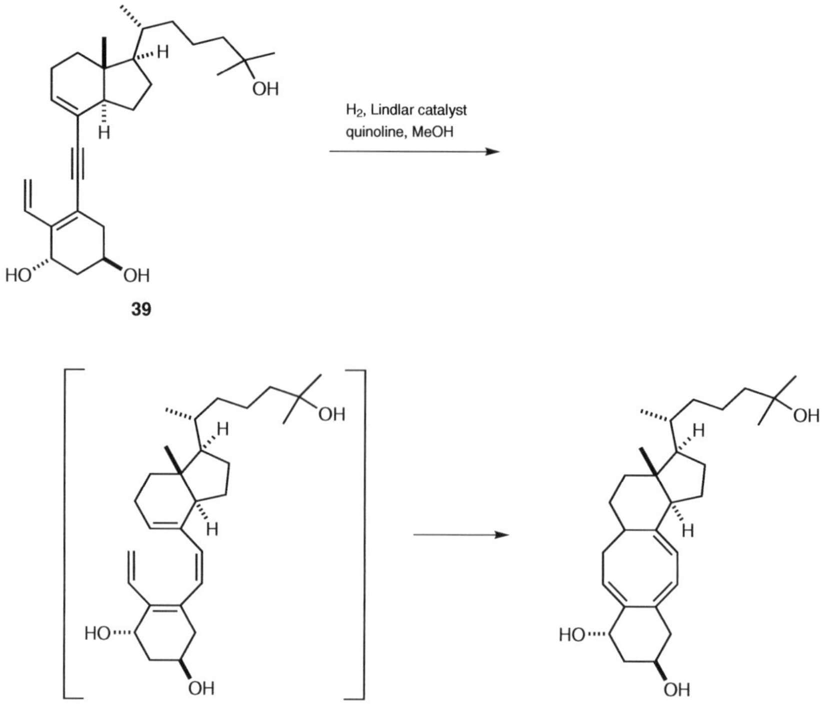

H_2, Lindlar catalyst
quinoline, MeOH

39

40

41

Similarly, twofold Lindlar hydrogenation of diyne **42** gives octatetraene **43**, which spontaneously undergoes 8π-conrotatory electrocyclization to form cyclooctatriene **44** as a racemic mixture. This compound cannot be isolated because it undergoes 6π-disrotatory electrocyclization to give endiandric esters D and E (Scheme 17). Heating endiandric ester E at 100 °C in toluene results in an intramolecular Diels–Alder reaction to give endiandric ester A. Notably, endiandric ester D cannot undergo such a cycloaddition but can equilibrate with endiandric ester E. Indeed, it has been shown that this entire cascade can be achieved in one operation by Lindlar hydrogenation of diyne **42** and subsequent heating of the resulting material to give endiandric ester A in 30% overall yield. Other members of the endiandric acid family (B, C, F, and G) have been prepared by the Nicolaou group by making use of the same strategy.[36–40]

Scheme 17 Synthesis of Endiandric Acids A, D, and E[36–40]

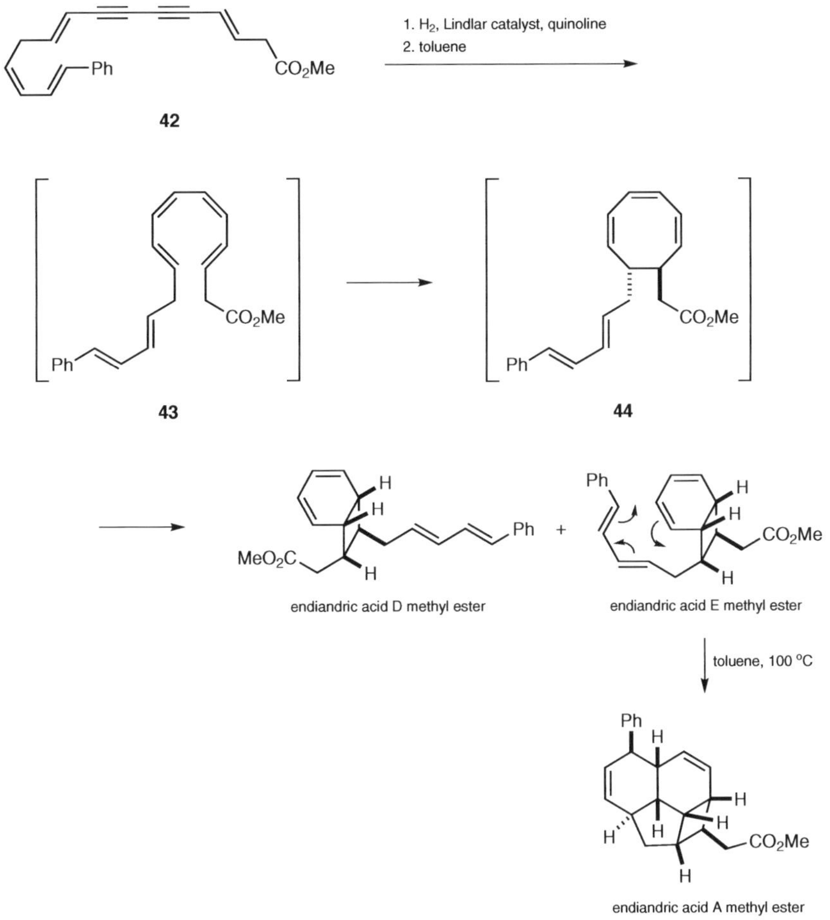

Ph
CO2Me
1. H2, Lindlar catalyst, quinoline
2. toluene
42
CO2Me
Ph
43
CO2Me
Ph
44
H
H
MeO2C
H
Ph
+
Ph
H
H
CO2Me
H
endiandric acid D methyl ester
endiandric acid E methyl ester
toluene, 100 °C
Ph
H
H
H
H
H
CO2Me
H
endiandric acid A methyl ester

Suffert and co-workers have reported a new family of [4.6.4.6]fenestradienes **46** and [4.6.4.6]fenestrenes **47** through a unique 8π-electrocyclization/6π-electrocyclization/oxidation domino sequence starting from substrates **45**.[41–43] The high yields and limited number of steps in the full reaction sequence make this process very attractive for the synthesis of new fenestranes in general terms. Furthermore, depending on the temperature [70 °C (for n = 1) or room temperature (for n = 2)], cyclooctatrienes **48** are also obtained in moderate to good yields. The high torquoselectivities observed experimentally and the complete diastereoselectivity for the 8π-electrocyclization products have been thoroughly studied by density functional computations (Scheme 18).[42]

Scheme 18 Synthesis of Fenestradienes, Fenestrenes, and Cyclooctatrienes through a Semihydrogenation/8π-Electrocyclization and Potentially a 6π-Electrocyclization Cascade[41,42]

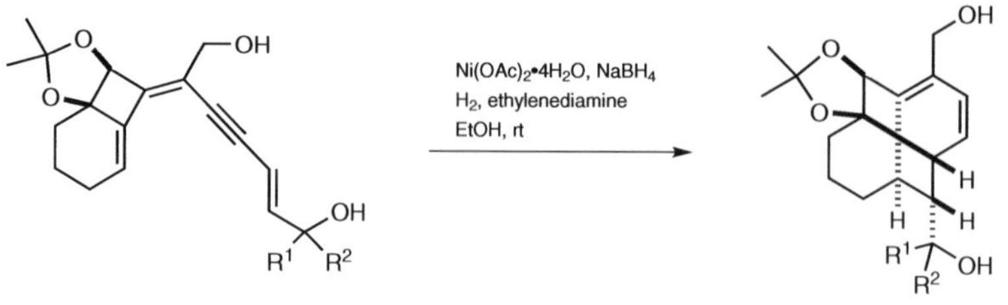

$Ni(OAc)_2 \cdot 4H_2O$, $NaBH_4$, H_2, ethylenediamine, EtOH, rt

45 → **46**

MCPBA, CH_2Cl_2, 0 °C, 1 h

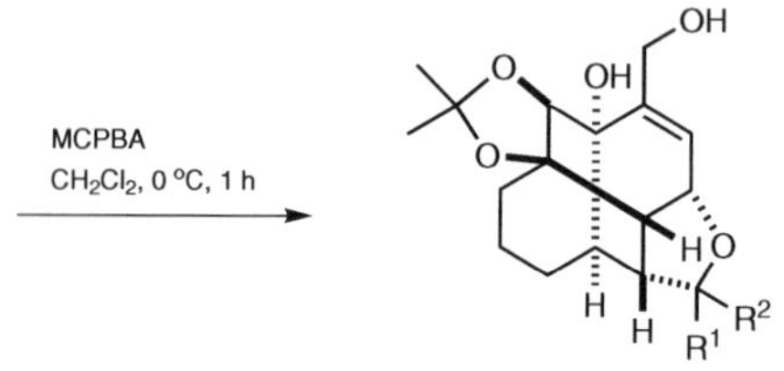

→ **47**

R^1	R^2	Yield (%) of **46**	Yield (%) of **47**	Ref
Me	Me	88	63	[41]
iBu	iBu	63	67	[41]
$(CH_2)_5$		86	68	[41]
$(CH_2)_4$		93	62	[41]
H	H	–[a]	35	[41]

[a] Yield not determined.

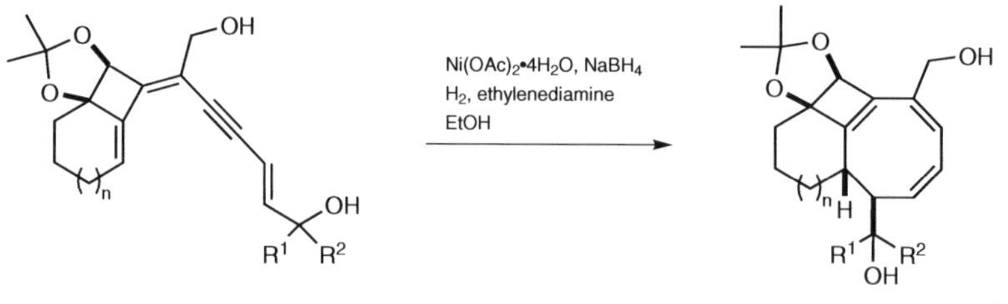

$Ni(OAc)_2 \cdot 4H_2O$, $NaBH_4$, H_2, ethylenediamine, EtOH

→ **48**

R^1	R^2	n	Temp	Yield (%)	Ref
Me	Me	1	70 °C	43	[42]
Me	Me	2	rt	77	[42]
$(CH_2)_5$		2	rt	81	[42]
$(CH_2)_4$		2	rt	81	[42]
iBu	iBu	2	rt	88	[42]

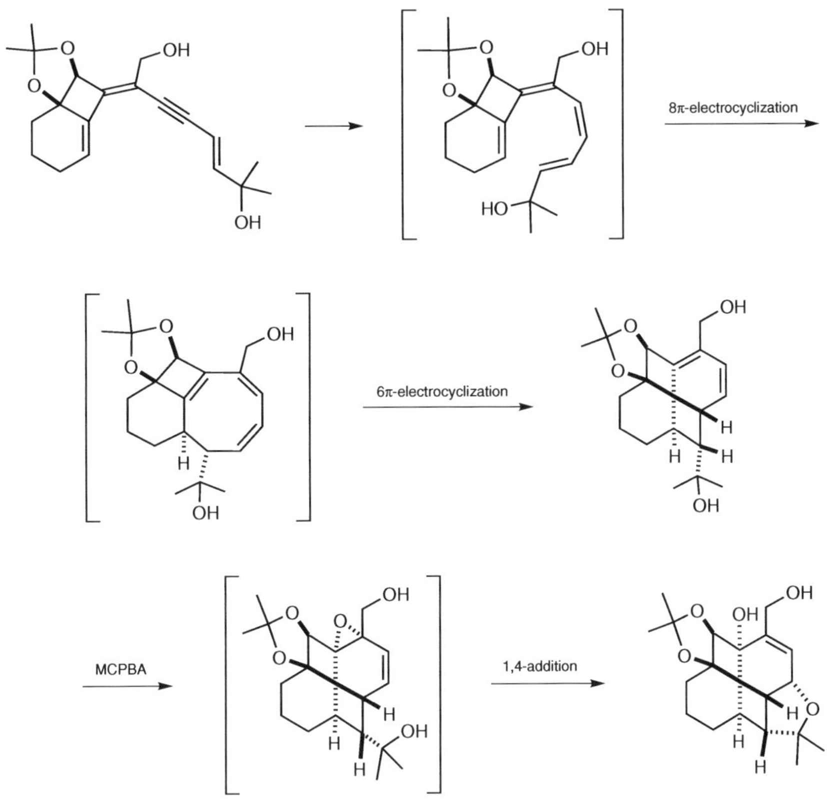

In another example, conjugated hexa-1,3,5-trienes **50** based on a bridged bicyclic skeleton are prepared by half-reduction of dienynes **49**.[44] Trienes **50** undergo 6π-electrocyclization at ambient or elevated temperature to furnish complex, polycyclic cyclohexadienes **51**. In all cases, complete selectivity in favor of cyclization from the *exo* face of the bridged bicyclic system is observed (Scheme 19).

Scheme 19 Cascade Semireduction/6π-Electrocyclization[44]

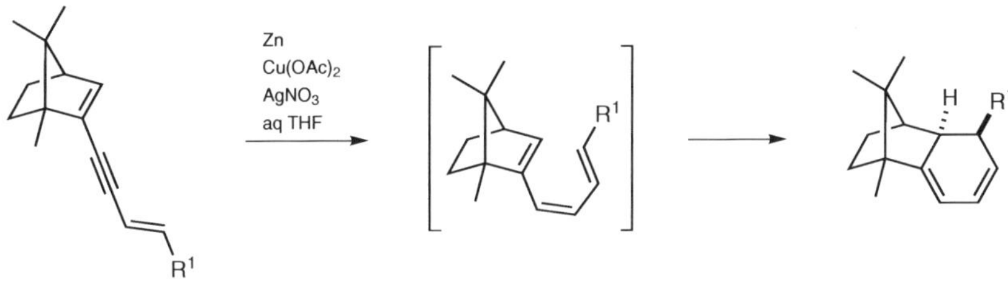

R¹	Yield (%)	Ref
Ph	67	[44]
TMS	90	[44]

Tricycles **54** can be obtained by the addition of vinylmagnesium chloride to propargyl alcohols **52** to generate trienes **53** as magnesium chelates that can then undergo 6π-electrocyclization (Scheme 20).[45]

Scheme 20 Cascade Carbomagnesiation/6π-Electrocyclization[45]

R^1	Solvent	Yield (%)	Ref
H	toluene	43	[45]
Me	cyclohexane	81	[45]

2-{(1*R,4a*R**,7a*S**,11*R**,12*R**)-8-(Hydroxymethyl)-6,6-dimethyl-1,2,3,4,7a,11-hexahydro-1,11-methanodibenzo[1,4:3,4]cyclobuta[1,2-*d*][1,3-dioxol-12-yl]}propan-2-ol (46, $R^1 = R^2 =$ Me); Typical Procedure:**[41]

A soln of $NaBH_4$ (12 mg, 0.33 mmol, 1 equiv) in EtOH (0.3 mL) was added to a soln of $Ni(OAc)_2{\bullet}4H_2O$ (83 mg, 0.33 mmol, 1 equiv) in EtOH (1 mL) at rt under an atmosphere of argon. After stirring for 1 h under an atmosphere of H_2, a soln of trienyne **45** (110 mg, 0.33 mmol) and ethylenediamine (80 μL, 1.16 mmol, 3.5 equiv) in EtOH (0.7 mL) was added. The black mixture was stirred at rt overnight and then filtered through silica gel treated with 5% Et_3N (EtOAc). The filtrate was concentrated to afford the title compound as a yellow oil; yield: 97 mg (88%).

2.1.3.3 Isomerization Followed by Electrocyclization

2.1.3.3.1 1,3-Hydrogen Shift/Electrocyclization

In 2010, Zhou and co-workers reported a sulfur-assisted propargyl–allenyl isomerization/6π-electrocyclization sequence that produces polyfunctionalized benzene and naphthalene derivatives (e.g., **56**) starting from fully acyclic precursors **55** (Scheme 21).[46] On the basis of this preliminary work, the methodology has been extended to other heteroatoms such as nitrogen and oxygen.[47] In 2012, Spencer and Frontier efficiently applied this methodology to the synthesis of polysubstituted phenols.[48]

Scheme 21 Access to Benzene Derivatives by a 1,3-Tautomerization/Electrocyclization Process[46,47]

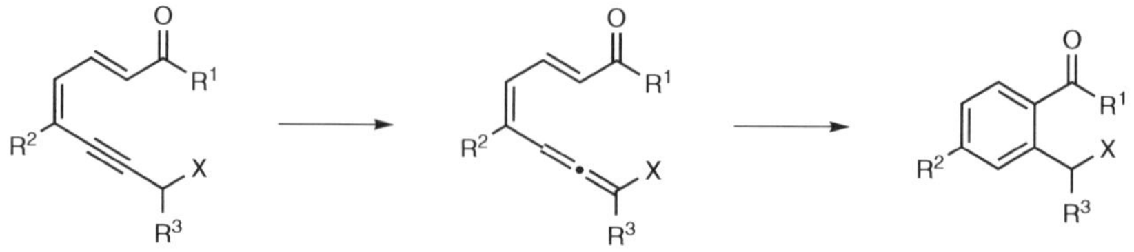

R^1	R^2	R^3	X	Conditions	Yield (%)	Ref
OEt	Ph	H	SPh	Et_3N, MeCN, 60 °C, 12 h	80	[46]
NMePh	$(CH_2)_5Me$	H	SBu	DBU, MeCN, rt, 3 h	88	[46]
OEt	$(CH_2)_5Me$	H	SBu	DBU, MeCN, rt, 3 h	76	[46]
OEt	Ph	Me	SPh	Et_3N, MeCN, 60 °C, 12 h	70	[46]
OEt	Ph	H	4-TolO	Et_3N, MeCN, 60 °C, 12 h	91	[47]
Ph	Ph	H	4-TolO	Et_3N, MeCN, 60 °C, 12 h	68	[47]
Me	Ph	H	4-TolO	Et_3N, MeCN, 60 °C, 12 h	70	[47]
Me	$(CH_2)_2iPr$	H	4-TolO	Et_3N, MeCN, 60 °C, 12 h	77	[47]

Ethyl 3-[(Phenylsulfanyl)methyl]biphenyl-4-carboxylate (56, **R^1= OEt; R^2= Ph; R^3= H; X = SPh); Typical Procedure:**[46]

Et_3N (139 μL, 1 mmol, 2 equiv) was added to a soln of ethyl (2*E*,4*Z*)-5-phenyl-8-(phenylsulfanyl)octa-2,4-dien-6-ynoate (175 mg, 0.5 mmol) in MeCN (3 mL) under an atmosphere of N_2 at 60 °C. The reaction was monitored by TLC until completion (~12 h). After evaporation of the volatiles, chromatography (silica gel, hexane/EtOAc 50:1) of the mixture afforded the title compound; yield: 139 mg (80%).

2.1.3.3.2 1,5-Hydrogen Shift/Electrocyclization

In 2006, Ma and co-workers described a straightforward method to access fused bicyclic compounds containing an eight-membered ring (e.g., **59** and **62**) in moderate to good yield from (4*Z*)-1,2,4,7-tetraenes **57** and **60** via enones **58** and **61** (Scheme 22).[49,50] The reaction is thought to proceed by a 1,5-hydrogen shift followed by an 8π-electrocyclization reaction. Interestingly, if a dienophile is added to the reaction mixture, the triene product undergoes Diels–Alder cycloaddition in good yields.

for references see p 156

Scheme 22 Rearrangement of (4*Z*)-1,2,4,7-Tetraenes Involving 1,5-Hydrogen Shift and Electrocyclization[49,50]

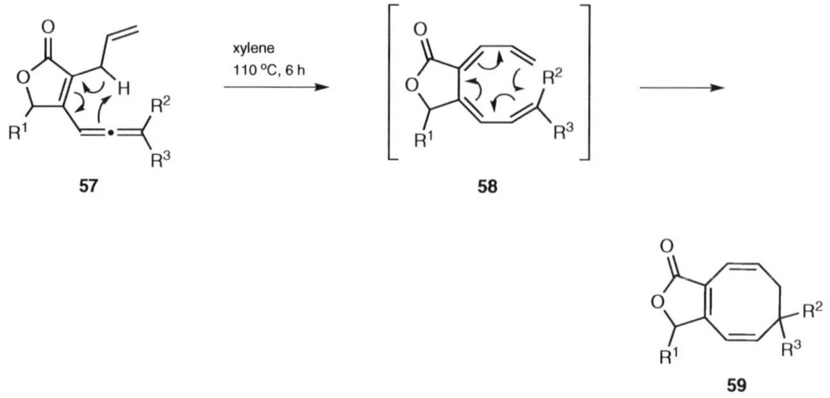

xylene
110 °C, 6 h

57 58 59

R^1	R^2	R^3	Yield (%)	Ref
Ph	Et	Et	89	[49]
Ph	Me	Me	84	[49]
1-naphthyl	Et	Et	97	[49]
Et	Et	Et	84	[49]
Et	$(CH_2)_5$		89	[49]
4-BrC_6H_4	Et	Et	95	[49]
4-BrC_6H_4	Me	Me	84	[49]
4-BrC_6H_4	$(CH_2)_4$		66	[49]

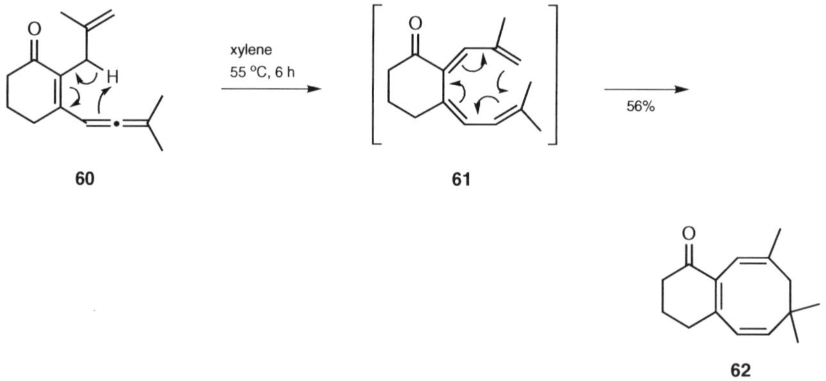

xylene
55 °C, 6 h

56%

60 61 62

In 2013, Alajarin, Sanchez-Andrada, Vidal, and co-workers reported a domino sequence based on a 1,5-hydrogen shift of an acetal proton followed by 6π-electrocyclization. An allene is first formed from benzyl alkyne **63**. A cyclic or acyclic acetal functional group located at the *ortho* position of the resulting phenylallene **64** activates the proton on the carbon atom attached to the acetal group. This results in transfer of the proton to the central carbon atom of allene **64** through a thermally induced 1,5-shift. Newly formed trienic intermediate **65** then spontaneously undergoes 6π-electrocyclization to form polyfunctionalized derivative **66**, which can undergo aromatization (R^2 = H) to give naphthalene derivative **67** (Scheme 23).[51] These cascade reactions are facilitated by electron-withdrawing groups at the terminal cumulene carbon atom (i.e., if R^3 is an electron-withdrawing group).

Scheme 23 1,5-Hydrogen Shift/Electrocyclization Reaction of 2-(1,3-Dioxolan-2-yl)phenylallenes[51]

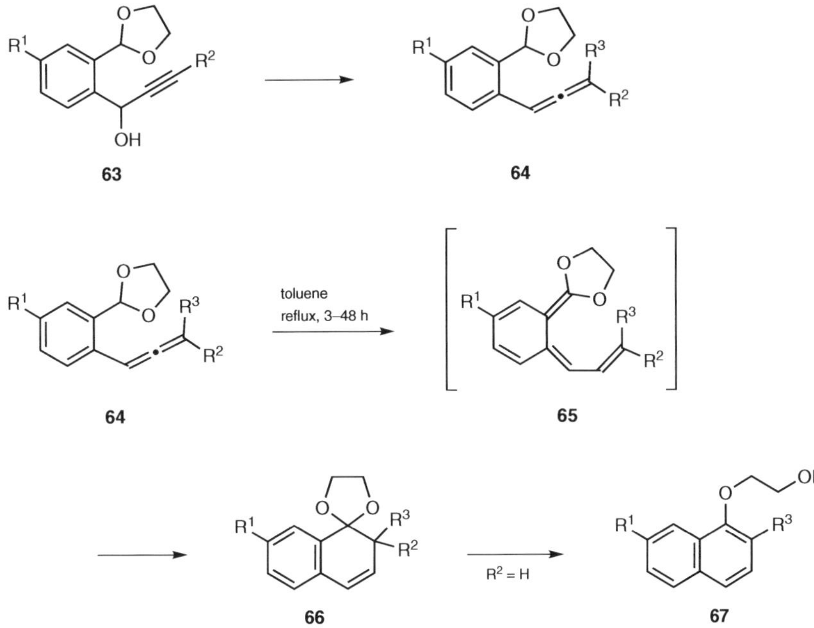

toluene
reflux, 3–48 h

R^2 = H

R^1	R^2	R^3	Yield (%) of **67**	Ref
H	H	Me	30	[51]
H	H	*t*-Bu	50	[51]
H	H	Ph	70	[51]
H	H	4-Tol	58	[51]
H	H	4-$MeOC_6H_4$	65	[51]
H	H	$P(O)(OEt)_2$	93	[51]
H	H	$P(O)Ph_2$	80	[51]
OMe	H	$P(O)Ph_2$	89	[51]
H	Ph	$P(O)(OEt)_2$	65[a]	[51]
H	Ph	$P(O)Ph_2$	79[a]	[51]

[a] Yield (%) of **66**.

(4*Z*,8*Z*)-6,6-Diethyl-3-phenyl-6,7-dihydrocycloocta[*c*]furan-1(3*H*)-one (59, R^1 = Ph; R^2 = R^3 = Et)**; Typical Procedure:**[49]

Under an argon atmosphere, a soln of 3-allyl-4-(3-ethylpenta-1,2-dienyl)-5-phenylfuran-2(5*H*)-one (56 mg, 0.19 mmol) in xylene (4 mL) was stirred at 110 °C for 6 h. After complete consumption of the starting material, as monitored by TLC, the mixture was directly purified by flash chromatography (silica gel, petroleum ether/EtOAc) to afford the title compound; yield: 50 mg (89%).

2-{[2-(Diphenylphosphoryl)-7-methoxynaphthalen-1-yl]oxy}ethanol [**67**, $R^1 = OMe$; $R^3 = P(O)Ph_2$]; **Typical Procedure:**[51]

A soln of allene **64** (418 mg, 1 mmol) in anhyd toluene (15 mL) was heated at reflux for 3–48 h. After cooling, the solvent was removed under reduced pressure, and the crude material was purified by column chromatography (silica gel, EtOAc) to afford the title compound; yield: 376 mg (89%).

2.1.3.3.3 1,7-Hydrogen Shift/Electrocyclization

In 2004, Flynn and co-workers reported an elegant multicomponent coupling reaction to generate triene **68** that efficiently undergoes a 1,7-hydrogen shift/8π-electrocyclization sequence to afford fused tetracyclic cyclooctatriene **69**. The efficiency of this domino reaction can be attributed to a decrease in the degrees of freedom of the molecule, as induced by the benzofuran and cyclohexyl rings, and to the ultimate restoration of aromaticity at the end of the sequence (Scheme 24).[52]

Scheme 24 Synthesis of a Cyclooctatriene Derivative through a 1,7-Hydrogen Shift/8π-Electrocyclization Process[52]

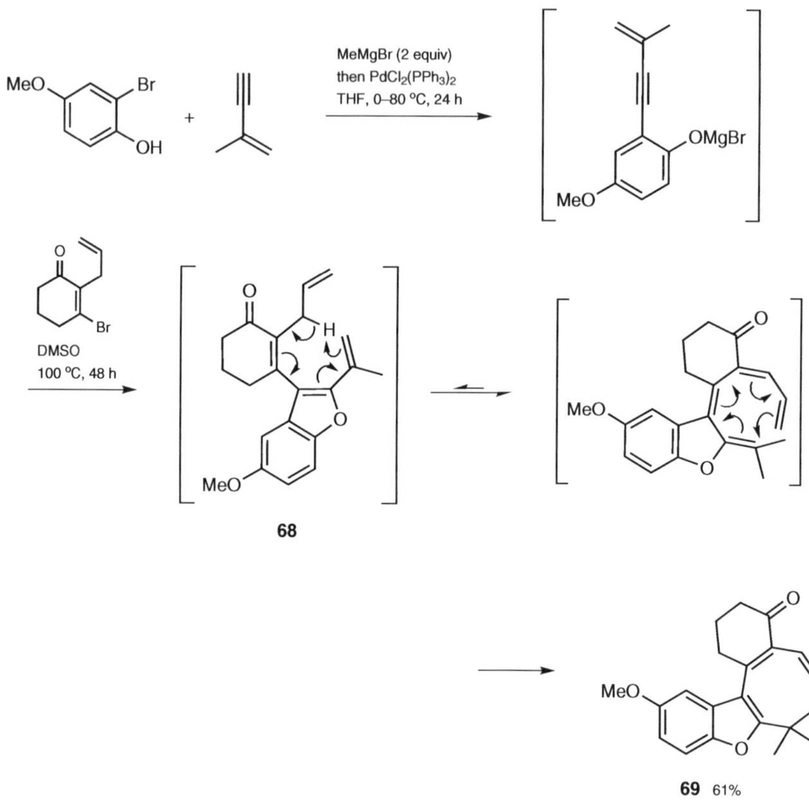

MeMgBr (2 equiv)
then $PdCl_2(PPh_3)_2$
THF, 0–80 °C, 24 h

DMSO
100 °C, 48 h

68

69 61%

(*Z*)-12-Methoxy-8,8-dimethyl-2,3,7,8-tetrahydrobenzo[3,4]cycloocta[1,2-*b*]benzofuran-4(1*H*)-one (**69**); **Typical Procedure:**[52]

In a round-bottomed flask, a 2.35 M soln of MeMgBr in Et_2O (1.47 mL, 3.45 mmol, 2.1 equiv) was added dropwise to a soln of 2-bromo-4-methoxyphenol (0.326 g, 1.61 mmol, 1 equiv) and 2-methylbut-1-en-3-yne (0.17 mL, 1.8 mmol, 1.1 equiv) in dry THF (3.5 mL) at 0 °C. The mixture was then warmed to rt and $PdCl_2(PPh_3)_2$ (56 mg, 0.08 mmol, 0.002 equiv) was added. This mixture was heated to 80 °C for 24 h and subsequently cooled to rt. DMSO (4.4 mL) and 2-allyl-3-bromocyclohex-2-enone (0.345 g, 1.61 mmol, 1 equiv) were added. The mixture was next stirred at 100 °C under a flow of N_2 for 1 h to remove THF, and then under a stationary N_2 atmosphere at 100 °C for another 48 h. The mixture was then cooled to rt, diluted with Et_2O (40 mL), washed with aq NH_4Cl (2 × 40 mL), dried ($MgSO_4$), and concentrated onto silica gel (ca. 1.5 g). Purification by flash chromatography (silica gel, hexane/Et_2O 5:2) afforded a colorless solid; yield: 314 mg (61%).

2.1.3.4 Consecutive Electrocyclization Reaction Cascades

Thermal electrocyclization reactions of benzooctatetraenes **70** and benzodecapentaenes **72** have been studied experimentally and computationally (Scheme 25). Compounds **70** and **72** (where R^1 = H or Me) give [4.2.0]bicyclooctadienes (**71** and **73**, respectively) by an 8π-electrocyclization/6π-electrocyclization sequence known as the Black cascade. The mechanism of this reaction has been studied by DFT quantum-chemical calculations.[53,54]

Scheme 25 Thermal 8π-Electrocyclization/6π-Electrocyclization Cascade[53,54]

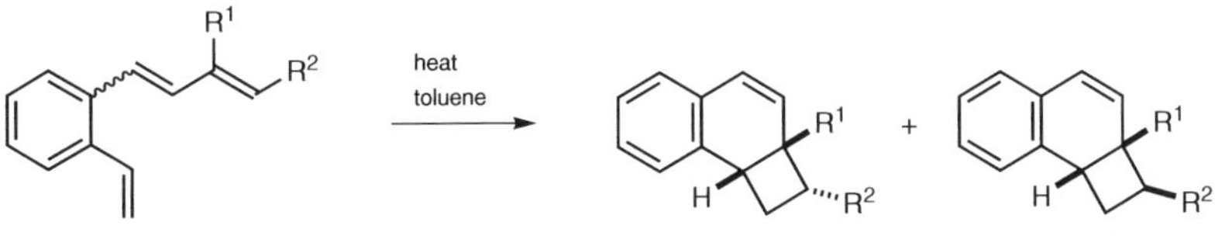

R^1	R^2	Yield (%) of *endo*-**71**	Yield (%) of *exo*-**71**	Ref
H	Me	33	11	[53]
H	Ph	73	–[a]	[53]
H	2-furyl	56	–[a]	[53]
Me	Ph	90	–[a]	[53]

[a] Not detected.

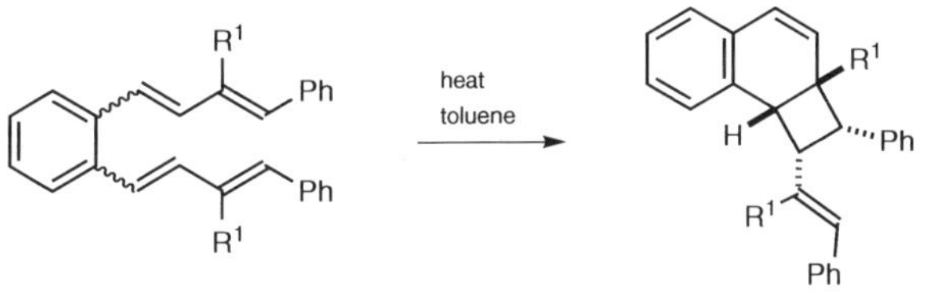

R^1	Yield (%) of *endo*-**73**	Ref
H	20	[54]
Me	40	[54]

The electrocyclization of trienones **74** has been investigated as a model system to determine possible access to (−)-coprinolone. In polar and/or hydrogen-bond-acceptor solvents the enol form of trienone **74** (i.e., **75**) undergoes an 8π-electrocyclization/6π-electrocyclization sequence directly to afford compound **76** in moderate to good yield (Scheme 26).[55]

Scheme 26 Thermal Enolization/8π-Electrocyclization/6π-Electrocyclization Cascade, and Structure of (−)-Coprinolone[55]

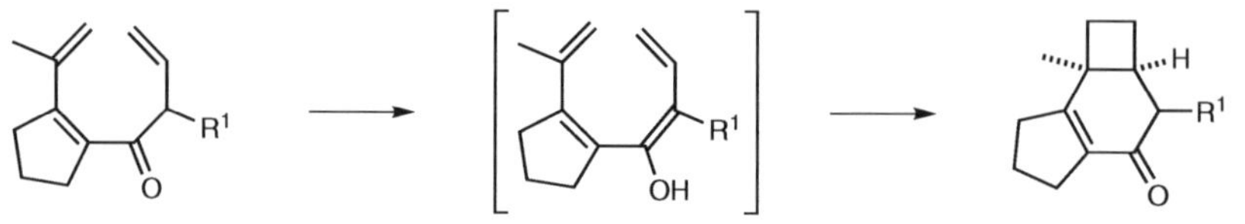

R^1	Solvent	Temp (°C)	Yield (%)	Ref
H	MeOH	110	24	[55]
Me	MeOH	110	33	[55]
Ph	MeOH	120	8	[55]
CH=CH_2	DMSO	100	59	[55]

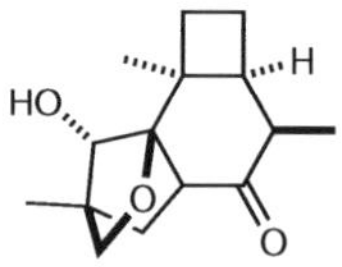

Another example of a thermal enolization/electrocyclization cascade is described by Snider and Harvey.[56] Here, a base-catalyzed conversion provides a very simple route to highly substituted bicyclo[4.2.0]octenones **77A**, **77B**, and **77C** as a mixture of isomers. A 4π-electrocyclization/retro-4π-electrocyclization/enolization/8π-electrocyclization/6π-electrocyclization cascade is likely involved in this impressive domino transformation (Scheme 27).

Scheme 27 Thermal Enolization/Electrocyclization Cascade[56]

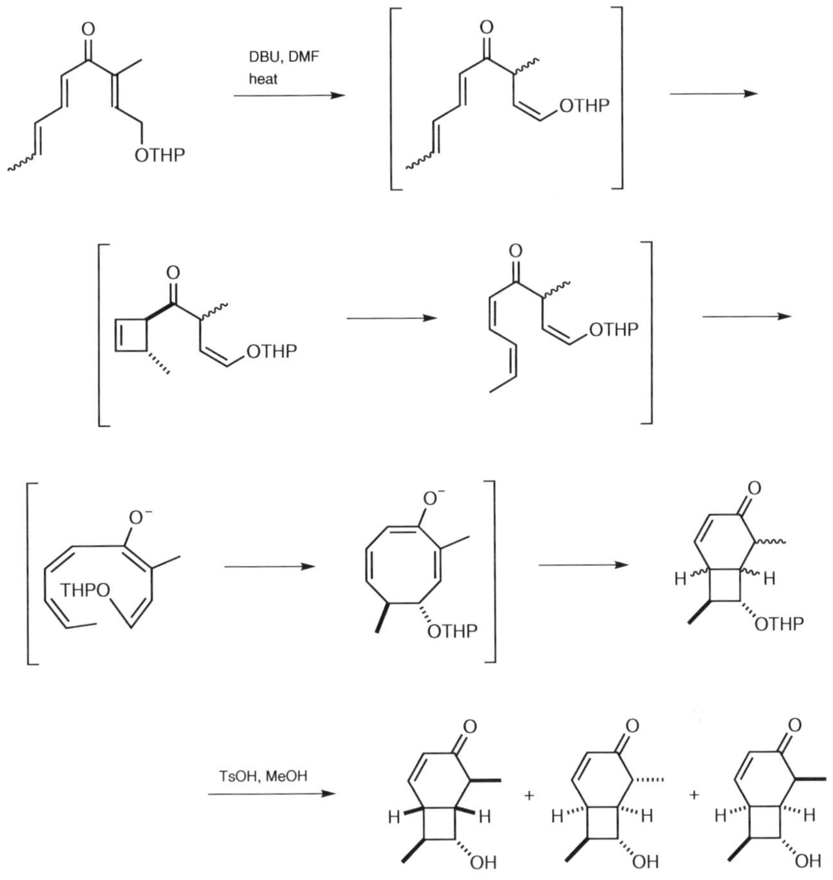

DBU, DMF heat

TsOH, MeOH

77A 10% **77B** 10% **77C** 9%

2.1.3.5 Alkenation Followed by Electrocyclization

As electrocyclic reactions are mainly based on the movement of conjugated π-electrons to create one new σ-bond that closes a ring, a major concern in using this type of approach in synthesis is the way in which the required polyene system will be installed. One of the most direct ways to generate this series of C=C bonds is to start from a carbonyl group bearing a polyunsaturated side chain and to perform a classical alkenation reaction, such as a Wittig, Julia–Kocienski, or Horner–Wadsworth–Emmons reaction. Logically, the reaction can be performed in the opposite way by using a simple carbonyl group and an alkenation reagent bearing several conjugated C=C bonds. Surprisingly though, this latter strategy has not been widely used, and only a small number of examples have been reported. Because of the very rare use of this methodology, only two recently published approaches will be presented. These examples have been selected as they involve two different types of alkenation and are followed by two distinct types of electrocyclization.

The first method reported by Kim and co-workers[57] is based on Wittig alkenation of a benzaldehyde derivative with a phosphonium salt bearing two C=C bonds (synthesized in situ from the Morita–Baylis–Hillman adducts **78** of cinnamaldehyde analogues[58]). This sequence leads to triene **79**, which then undergoes 6π-electrocyclization to form various *ortho*-terphenyl derivatives **80** under an oxidative aerobic atmosphere (Scheme 28).

Scheme 28 Synthesis of *ortho*-Terphenyl Derivatives through a Wittig Alkenation/6π-Electrocyclization/Oxidation Sequence[57]

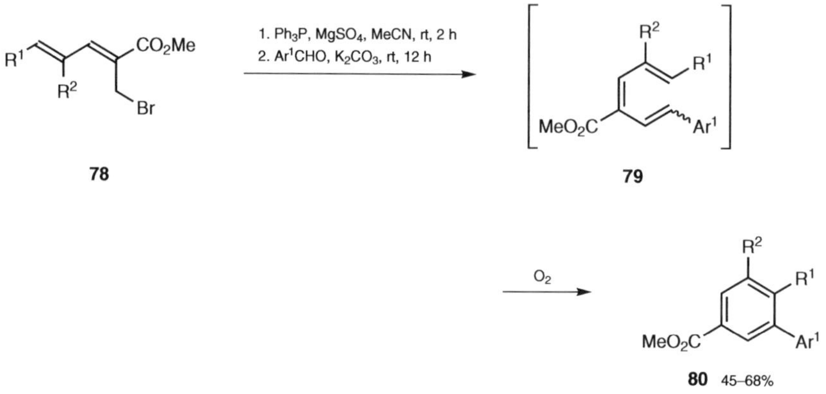
CO2Me
R1
R2
Br
78
1. Ph3P, MgSO4, MeCN, rt, 2 h
2. Ar1CHO, K2CO3, rt, 12 h
R2
R1
MeO2C
Ar1
79
O2
R2
R1
MeO2C
Ar1
80 45–68%

R^1 = Ph, 4-$MeOC_6H_4$, 4-FC_6H_4, Me; R^2 = H, Me;
Ar^1 = Ph, 4-ClC_6H_4, 4-$MeOC_6H_4$, 4-$O_2NC_6H_4$, 4-PhC_6H_4, 4-pyridyl, 2-furyl, 5-methyl-2-thienyl

The second method is characterized by an aldol condensation/alkene migration/8π-electrocyclization/6π-electrocyclization sequence starting from reaction of *o*-allylbenzaldehydes **81** with sulfones **82** (Scheme 29). This elegant domino reaction, reported by Chang and co-workers,[59] leads to benzo-fused tricyclic 1,2,2a,8b-tetrahydrocyclobuta[*a*]naphthalene derivatives **83** in good yields, and the products can be further functionalized. Extensive studies of the sequence have revealed that the benzaldehyde component **81** must be electron rich, and the sulfonyl substituent on **82**, serving as an electron-withdrawing group, also plays a key factor in initiating the formation of tricyclic product **83** through a delocalized electron push–pull conjugated relationship. By using this domino strategy with cinnamaldehyde-type substrates, this group has also reported the synthesis of *para*-terphenyl derivatives.[60]

Scheme 29 Synthesis of Tetrahydrocyclobuta[*a*]naphthalenes through an Aldol Condensation/Alkene Migration/Electrocyclization Sequence[59]

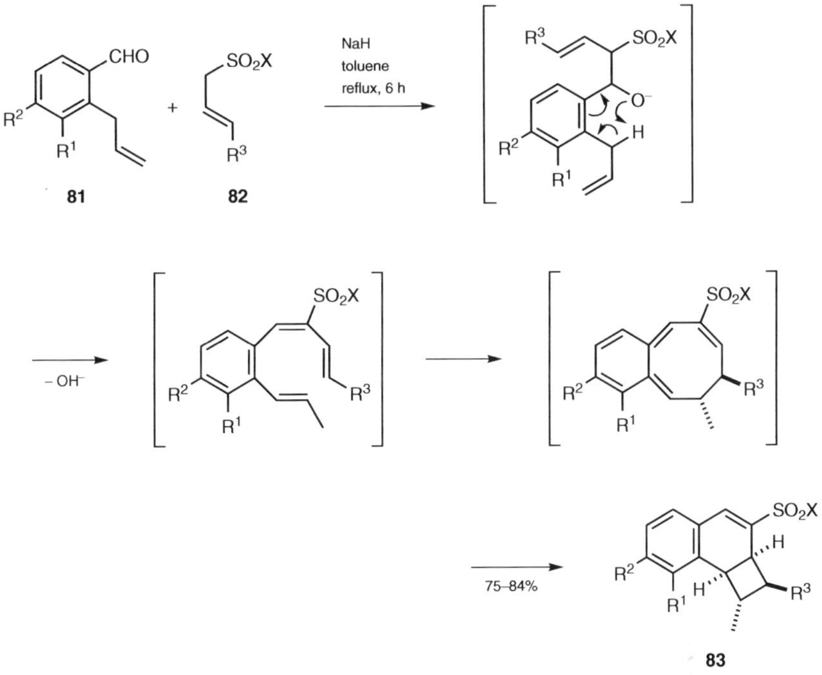

R^1 = OMe, OBn, cyclopentyloxy; R^2 = H, OMe; R^3 = H, Me, Ph, 4-FC_6H_4, 4-$MeOC_6H_4$, benzo-1,3-dioxol-5-yl; X = Me, 4-Tol

Methyl 1,1′:2′,1″-Terphenyl-4′-carboxylate (80, **R^1 = Ar^1 = Ph; R^2 = H); Typical Procedure:**[57]
A mixture of methyl (2*Z*,4*E*)-2-(bromomethyl)-5-phenylpenta-2,4-dienoate (**78**, R^1 = Ph; R^2 = H; 281 mg, 1.0 mmol, 1 equiv),[58] Ph_3P (288 mg, 1.1 mmol, 1.1 equiv), and $MgSO_4$ (1.0 g) in MeCN (3 mL) was stirred at rt for 2 h to form the phosphonium salt. Then, benzaldehyde (117 mg, 1.1 mmol, 1.1 equiv) and K_2CO_3 (276 mg, 2.0 mmol, 2 equiv) were added sequentially, and the resulting mixture was stirred at rt for an additional 12 h to form the target triene. The mixture was then heated at reflux for 15 h under an O_2 atmosphere (balloon). The usual aqueous extractive workup and purification by column chromatography (silica gel, hexane/Et_2O 100:1) gave the title compound as a white solid; yield: 159 mg (55%).

(1*S,2*S**,2a*S**,8b*R**)-7,8-Dimethoxy-1-methyl-2-phenyl-3-tosyl-1,2,2a,8b-tetrahydrocyclobuta[*a*]naphthalene** (83, **R^1 = R^2 = OMe; R^3 = Ph; X = 4-Tol); Typical Procedure:**[59]
A 60% dispersion of NaH in mineral oil (120 mg, 3.0 mmol, 10 equiv) was added to a soln of [3-(4-toluenesulfonyl)prop-1-enyl]benzene (**82**, R^3 = Ph; X = 4-Tol; 82 mg, 0.3 mmol, 1 equiv) in toluene (8 mL) at rt over 10 min. Then, a soln of 2-allyl-3,4-dimethoxybenzaldehyde (**81**, R^1 = R^2 = OMe; 62 mg, 0.3 mmol, 1 equiv) in toluene (2 mL) was added at rt, and the resulting mixture was stirred at reflux for 6 h. Upon completion of the reaction, the mixture was cooled to rt, the solvent was evaporated under reduced pressure, the residue was diluted with H_2O (10 mL), and the mixture was extracted with EtOAc (3 × 20 mL). The combined organic layers were washed with brine, dried, filtered, and evaporated to afford the crude product. Purification (silica gel, hexane/EtOAc 10:1 to 6:1) afforded the title compound; yield: 104 mg (75%).

for references see p 156

2.1.3.6 Electrocyclization Followed by Cycloaddition

The transformation of (1*E*,3*Z*,5*E*)-hexa-1,3,5-trienes into ring-annulated cyclohexa-1,3-dienes can be followed by intermolecular [4+2] cycloaddition with dienophiles in a domino process, thereby giving an exceptional elevation of molecular complexity. With this process, a wide variety of functionally substituted tricyclic skeletons (e.g., **84**–**87**) are accessible (Scheme 30) simply by varying the starting materials.[61]

Scheme 30 Domino 6π-Electrocyclization/Diels–Alder Reactions of 1,6-Disubstituted (1*E*,3*Z*,5*E*)-Hexa-1,3,5-trienes[61]

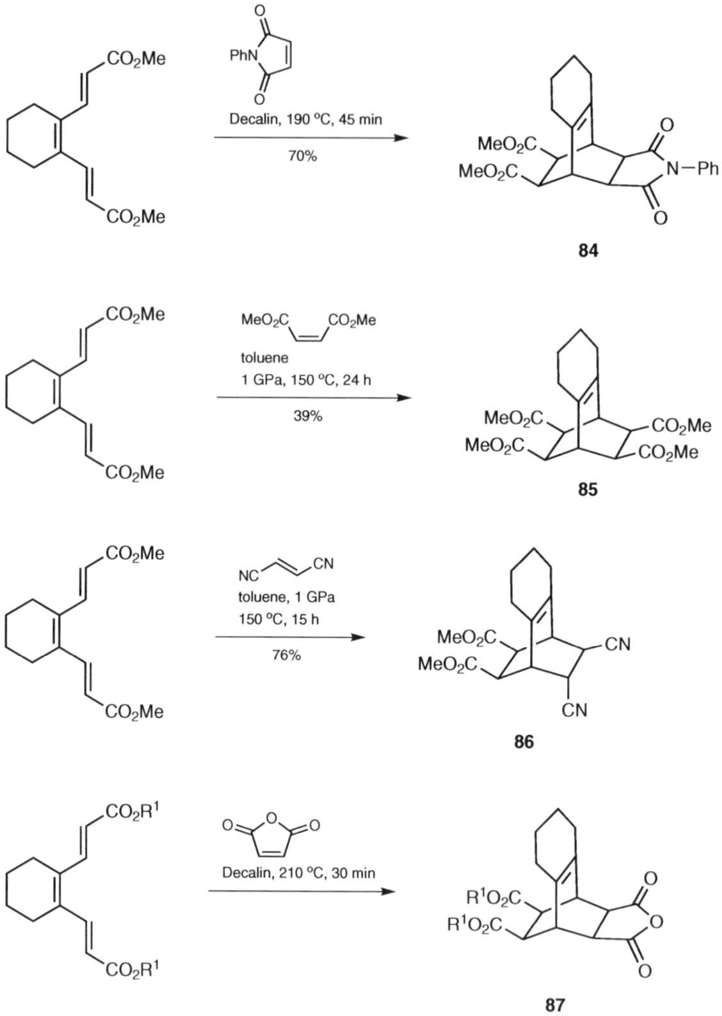

Decalin, 190 °C, 45 min; 70%; **84**

toluene, 1 GPa, 150 °C, 24 h; 39%; **85**

toluene, 1 GPa, 150 °C, 15 h; 76%; **86**

Decalin, 210 °C, 30 min; **87**

R^1	Yield (%)	Ref
t-Bu	19	[61]
Me	32	[61]

Similarly, even higher molecular complexity can be generated if the dienophile partner is directly linked to the diene moiety to perform the terminating [4+2] cycloaddition intramolecularly. Acyclic conjugated (*E*,*Z*,*E*,*E*)-tetraenes **88** can, upon thermolysis, undergo a domino pericyclic process involving a 6π-electrocyclization/intramolecular Diels–Alder sequence to give tricyclo[3.2.1.$0^{2,7}$]oct-3-enes **89** in reasonable yields (Scheme 31).[62] Notably, equivalent acyclic conjugated tetraenes bearing an *E*,*Z*,*Z*,*E* configuration typically undergo conrotatory 8π-electrocyclization to give a cyclooctatriene, and this is usually then followed by disrotatory 6π-electrocyclization to ultimately afford a bicyclo[4.2.0]octadiene.[63]

Scheme 31 Domino 6π-Electrocyclization/Intramolecular Diels–Alder Reaction of Acyclic Conjugated (*E*,*Z*,*E*,*E*)-Tetraenes[62]

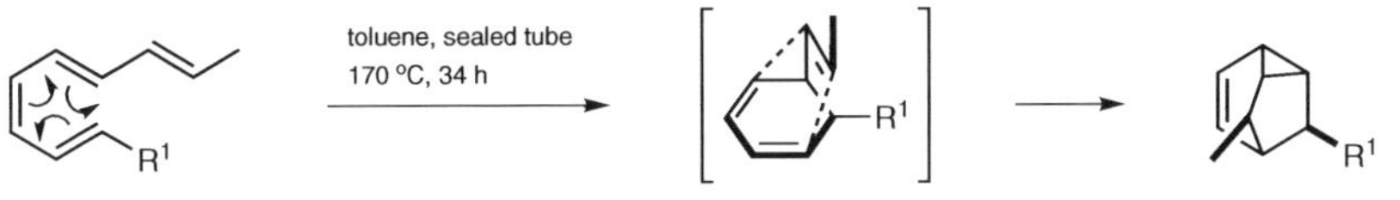

toluene, sealed tube
170 °C, 34 h

88 → [intermediate] → 89

R^1	Yield[a] (%)	Ref
(*E*)-CH=CHMe	60	[62]
Me	87	[62]
Ph	17	[62]

[a] Yield determined by GC.

Dimethyl 9,10-Dicyano-1,2,3,4,5,6,7,8-octahydro-1,4-ethanonaphthalene-2,3-dicarboxylate (86); Typical Procedure:[61]
A soln of dimethyl (3,3′-cyclohex-1-ene-1,2-diyl)(2*E*,2′*E*)-diacrylate (150 mg, 0.60 mmol, 1 equiv) and fumaronitrile (258 mg, 3.30 mmol, 5.5 equiv) in deoxygenated anhyd toluene (2 mL) in a sealed Teflon tube was heated at 150 °C under a pressure of 1 GPa for 15 h. After removal of the solvent under reduced pressure, purification (silica gel, Et_2O/pentane 1:1) afforded a colorless solid; yield: 149 mg (76%).

(6*R,8*S**)-6-Methyl-8-[(*E*)-prop-1-enyl]tricyclo[3.2.1.$0^{2,7}$]oct-3-ene [89, R^1 = (*E*)-CH=CHMe]; Typical Procedure:**[62]
A dilute soln of (2*E*,4*E*,6*Z*,8*E*,10*E*)-dodeca-2,4,6,8,10-pentaene [**88**, R^1 = (*E*)-CH=CHMe; 126 mg, 0.78 mmol, containing 30% of the all-*E* isomer] in toluene (5 mL) was placed in a tube with a few crystals of 2,6-di-*tert*-butyl-4-methylphenol to inhibit polymerization. The resulting mixture was degassed using the freeze–pump–thaw technique. The tube was then sealed and heated at 170 °C in an oil bath. After 34 h, the sealed tube was cooled to −78 °C and opened, and the contents (yield: 60% by GC) were removed and concentrated (MeOH was added to assist the removal of toluene). The residue was purified by chromatography (silica gel, hexanes) to afford the title compound as a colorless oil; yield: 45 mg (36%; 51% based on the presence of 30% of the all-*E* isomer in the starting material that could not productively undergo such a process).

2.1.3.7 Miscellaneous Reactions

2.1.3.7.1 Electrocyclization/Oxidation

In 2012, Pinto, Silva, and colleagues reported a photoinduced electrocyclization/oxidation process starting from (*E*)-3-styrylflavones **90**, performed using a high-pressure mercury UV lamp. The sequence ultimately leads to 5-arylbenzo[*c*]xanthone derivatives **91** in moderate yields (Scheme 32).[64]

for references see p 156

Scheme 32 Synthesis of 5-Arylbenzo[c]xanthone Derivatives[64]

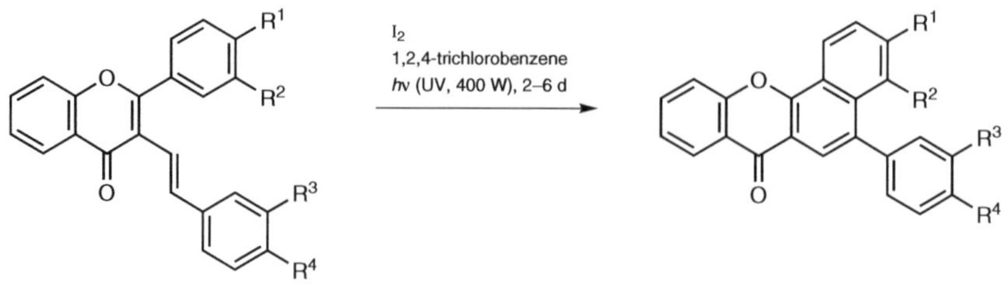

I_2
1,2,4-trichlorobenzene
$h\nu$ (UV, 400 W), 2–6 d

90 → **91**

R^1	R^2	R^3	R^4	Yield (%)	Ref
H	H	H	H	70	[64]
Me	H	H	H	45	[64]
Cl	H	H	H	73	[64]
NO_2	H	H	H	30	[64]
OMe	H	H	H	74	[64]
OMe	OMe	H	H	20	[64]
H	H	OMe	OMe	60	[64]

5-Phenyl-7*H*-benzo[*c*]xanthen-7-one (91, **$R^1 = R^2 = R^3 = R^4 = H$**)**; Typical Procedure:**[64]
A mixture of (*E*)-2-phenyl-3-styryl-4*H*-1-benzopyran-4-one (**90**, $R^1 = R^2 = R^3 = R^4 = H$; 49 mg, 0.15 mmol) and a catalytic amount of I_2 (4 mg, 0.015 mmol, 0.1 equiv) in 1,2,4-trichlorobenzene (20 mL) was irradiated for 2 to 6 d using a high-pressure mercury UV lamp (400 W) until complete consumption of the starting material, as observed by TLC. Purification of the mixture by chromatography [silica gel, petroleum ether (to remove I_2 and 1,2,4-trichlorobenzene) and then EtOAc/petroleum ether 1:9] and recrystallization (EtOH) afforded the title compound; yield: 34 mg (70%).

2.1.3.7.2 Photochemical Elimination/Electrocyclization

A photochemical route to benzo[*a*]carbazoles **93** through a domino reaction from 2-aryl-3-(1-tosylalkyl)-1*H*-indoles **92** was reported by Protti, Palmieri, and co-workers in 2013.[65] Irradiation of these substrates at a wavelength of 366 nm for up to 30 hours in polar, aprotic solvents, such as acetone or tetrahydrofuran, selectively affords target products **93** in moderate to acceptable yields though a homolytic cleavage/elimination/6π-electrocyclization/aromatization sequence (Scheme 33).

Scheme 33 Photochemical Synthesis of Benzo[*a*]carbazole Derivatives[65]

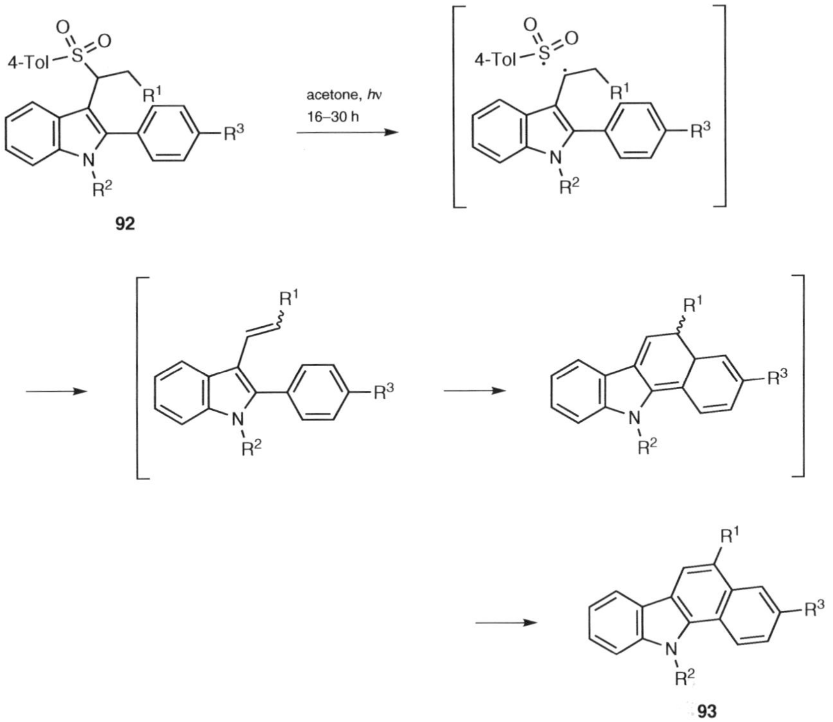

R^1	R^2	R^3	Time (h)	Yield (%)	Ref
Me	H	H	16	49	[65]
Bu	H	H	16	55	[65]
$(CH_2)_5Me$	H	H	16	57	[65]
Bn	H	H	30	56	[65]
Et	H	Me	24	46	[65]
$(CH_2)_5Me$	H	Me	24	54	[65]
Pr	H	CF_3	30	61	[65]
Et	H	OMe	24	62	[65]
$(CH_2)_6Me$	Me	H	30	54	[65]

5-Methyl-11*H*-benzo[*a*]carbazole (93, **R^1 = Me; R^2 = R^3 = H); Typical Procedure:**[65]
Under an atmosphere of N_2, a soln of 2-phenyl-3-[1-(4-toluenesulfonyl)propyl]-1*H*-indole (**92**, R^1 = Me; R^2 = R^3 = H; 156 mg, 0.4 mmol) in acetone (40 mL) was irradiated in a multilamp reactor equipped with 10 phosphor-coated lamps (emission centered at $\lambda = 366$ nm) until complete consumption of the substrate. The mixture was then concentrated under reduced pressure and purified by chromatography (silica gel, cyclohexane/$CHCl_3$) to afford the title compound; yield: 45 mg (49%).

2.1.3.7.3 Domino Retro-electrocyclization Reactions

Diederich and co-workers recently described a cascade process to access various electron-rich molecules (e.g., **94**) having optical properties of general interest, especially as chromophores. The general method is based on a domino sequence of [2+2] cycloadditions of donor-substituted alkynes to alkenenitriles followed by thermally promoted retro-electrocyclization reactions (Scheme 34).[66–68]

Scheme 34 Double [2+2]-Cycloaddition/Retro-electrocyclization Reactions between Tetracyanoethene and Various Aniline-Capped Buta-1,3-diynes[66]

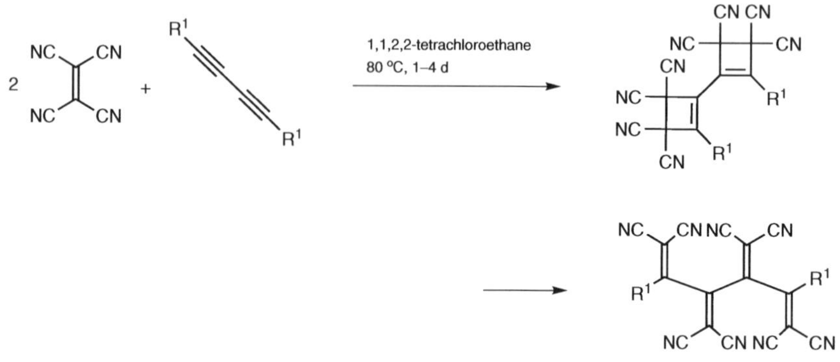

1,1,2,2-tetrachloroethane
80 °C, 1–4 d

94 79–87%

R^1 = 4-iPrHNC$_6$H$_4$, 4-BuHNC$_6$H$_4$, 4-Me$_2$NC$_6$H$_4$, 4-Me(iPr)NC$_6$H$_4$, 4-iPr$_2$NC$_6$H$_4$, 4-[Me(CH$_2$)$_5$]$_2$NC$_6$H$_4$

3,4-Bis(dicyanomethylene)-2,5-bis[4-(diisopropylamino)phenyl]hexa-1,5-diene-1,1,6,6-tetracarbonitrile (94, R^1 = 4-iPr$_2$NC$_6$H$_4$); Typical Procedure:[66]
Tetracyanoethene (522 mg, 4.0 mmol, 8 equiv) was added to a soln of 4,4′-(buta-1,3-diyne-1,4-diyl)bis(*N,N*-diisopropylaniline) (200 mg, 0.49 mmol, 1 equiv) in 1,1,2,2-tetrachloroethane (25 mL), and the mixture was stirred under an argon atmosphere at 80 °C for 1 d. Upon completion of the reaction, the solvent was evaporated under reduced pressure, and the residue was purified by flash chromatography (silica gel, CH_2Cl_2) to give the title compound as a green solid; yield: 266 mg (81%).

2.1.3.8 Hetero-electrocyclization

Domino reactions involving hetero-electrocyclizations represent an atom- and step-economic route to access variously substituted heterocycles. In this section, the reactions are classified as aza-, oxa-, or thia-electrocyclizations based entirely on the nature of the heteroatom (nitrogen, oxygen, or sulfur) involved in the electrocyclization process of the domino sequence.

2.1.3.8.1 Aza-electrocyclization

In most of the reactions selected for this chapter, an azatriene is formed in situ and then undergoes aza-electrocyclization through C—N or C—C bond formation. The methods are organized into three parts according to whether the electrocyclization precursor is obtained via a metal-mediated reaction (i.e., C—N or C—C cross coupling), imine or iminium formation (mainly from a starting α,β-unsaturated aldehyde or ketone), or a rearrangement or isomerization.

2.1.3.8.1.1 Metal-Mediated Reaction/Hetero-electrocyclization

Various polysubstituted pyridines can be synthesized from α,β-unsaturated *O*-pentafluorobenzoyl ketoximes **95** and alkenylboronic acids. The method consists of a domino reaction involving copper-catalyzed C—N cross coupling, electrocyclization of the resulting 3-azatrienes **96**, and aromatization under air oxidation into pyridine derivatives **97** (Scheme 35).[69]

Scheme 35 Synthesis of Substituted Pyridines through a Copper-Catalyzed C–N Cross-Coupling/Electrocyclization Cascade[69]

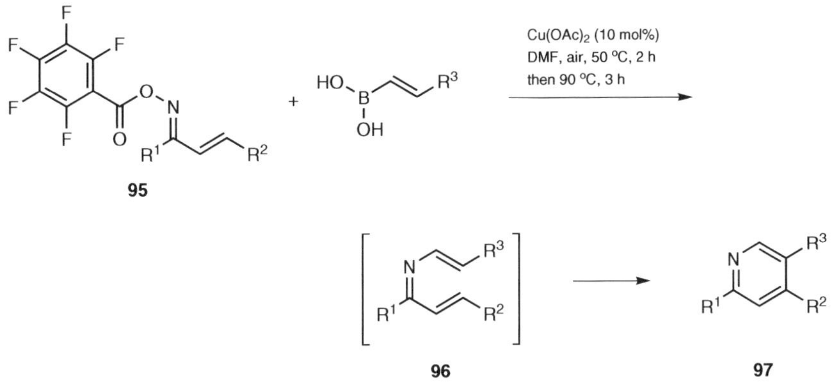

Cu(OAc)2 (10 mol%)
DMF, air, 50 °C, 2 h
then 90 °C, 3 h

95 96 97

R^1	R^2	R^3	Yield (%)	Ref
Ph	Ph	Bu	87	[69]
t-Bu	4-MeOC$_6$H$_4$	Bu	86	[69]
Ph	Bu	4-ClC$_6$H$_4$	58	[69]
4-pyridyl	Ph	Bu	62	[69]
Ph	2-furyl	Bu	78	[69]
(*E*)-CH=CHPh	Ph	Bu	74	[69]
2-ClC$_6$H$_4$	Ph	2-O$_2$NC$_6$H$_4$	70	[69]
4-NCC$_6$H$_4$	3-BrC$_6$H$_4$	2-thienyl	73	[69]

Alternatively, substituted pyridines can be synthesized by performing the reaction under palladium catalysis starting from sulfonamides, α,β-unsaturated β-iodo aldehydes, and alkenyl stannanes. The transformation sequence involves imine formation, a Stille cross-coupling reaction, and 6π-aza-electrocyclization followed by aromatization.[70]

Highly substituted dihydropyridines **100** can be prepared by a one-pot C—H alkenylation/electrocyclization sequence starting from α,β-unsaturated imines **99** and alkynes (Scheme 36).[71] Chlorobis(cyclooctene)rhodium(I) dimer [Rh_2Cl_2(coe)$_4$; coe = cyclooctene] is used and 4-(diethylphosphino)-*N*,*N*-dimethylaniline (**98**) is the most efficient ligand for this process. Aromatization of the products obtained into the corresponding pyridines **101** can be achieved in one pot by oxidation/debenzylation. The same research group has extended this methodology to the synthesis of tetrahydropyridines and other nitrogen-based heterocycles by subsequent reduction or other reaction of the dihydropyridines resulting after electrocyclization.[72,73]

Scheme 36 Synthesis of Dihydropyridines and Pyridines through a Rhodium-Catalyzed C—H Alkenylation/Electrocyclization Cascade[71]

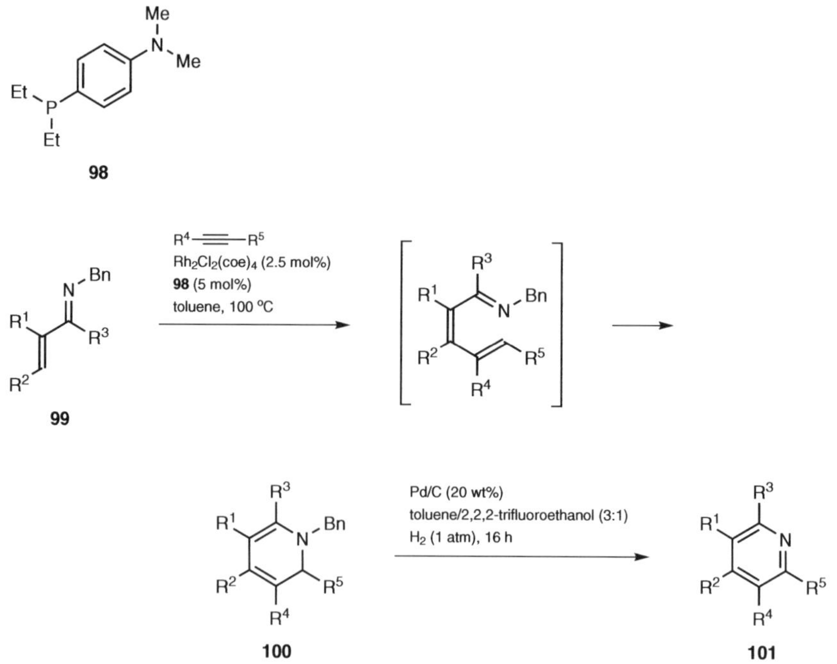

R^1	R^2	R^3	R^4	R^5	Yield (%)	Ref
Me	H	H	Et	Et	80	[71]
Me	H	H	Me	iPr	69	[71]
Me	H	H	Bn	H	66	[71]
H	Me	H	Et	Et	54	[71]
Me	Me	H	Et	Et	61	[71]
Me	Me	Me	Et	Et	75	[71]
Me	Me	Me	Pr	Me	59	[71]
$(CH_2)_4$		Me	Et	Et	32	[71]

Two research groups have independently reported the rhodium-catalyzed synthesis of variously polysubstituted pyridines (e.g., **102–104**) through electrocyclization starting from α,β-unsaturated ketoximes and alkynes (Scheme 37).[74–76] For instance, Cheng's group describes the use of disubstituted alkynes as substrates and chlorotris(triphenylphosphine)rhodium(I) as the catalyst,[74] whereas Ellman's group extends the scope of the reaction to terminal alkynes by using chlorobis(cyclooctene)rhodium(I) dimer/triisopropyl phosphite as the catalytic promoter.[76] In the latter case, the regioselectivity depends on the substituents at the positions α and β to the ketoxime, and the results range from modest (1.6:1) to full (1:0) control.[76]

Scheme 37 Rhodium-Catalyzed Synthesis of Pyridines from α,β-Unsaturated Ketoximes and Alkynes[74–76]

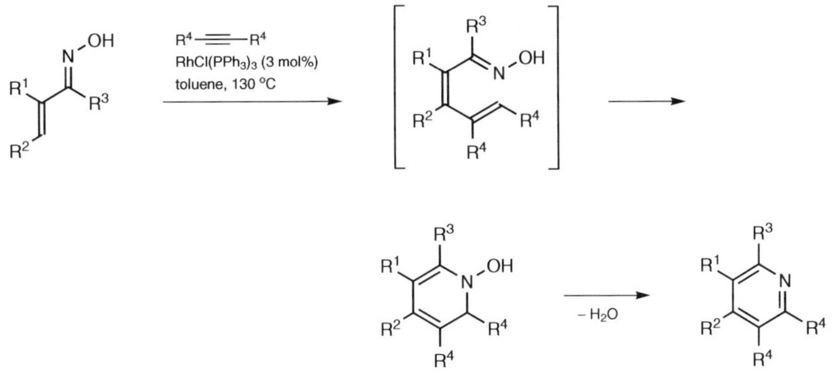

R^1	R^2	R^3	R^4	Yield (%)	Ref
Me	Me	Me	Ph	92	[74]
Me	H	Me	Ph	80	[74]
H	H	Pr	Ph	62	[74]
$(CH_2)_3$		Me	Ph	94	[74]
Me	Me	Me	Me	76	[74]
Ph	H	Me	Et	75	[74]
$(CH_2)_4$		Me	Pr	81	[74]
$(CH_2)_3$		Me	2-thienyl	89	[74]

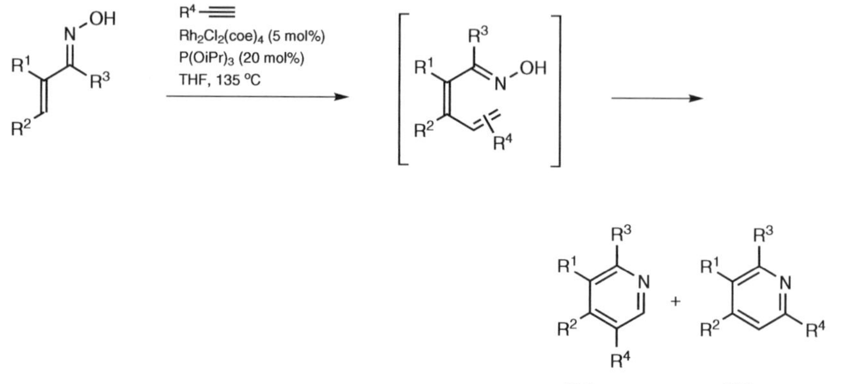

R^4—≡

$Rh_2Cl_2(coe)_4$ (5 mol%)
$P(OiPr)_3$ (20 mol%)
THF, 135 °C

R^1 R^2 R^3 N OH R^4

103 major + **104** minor

R^1	R^2	R^3	R^4	Ratio (**103**/**104**)	Yield (%)	Ref
Me	Me	Me	Bu	3.2:1	73	[76]
H	H	Me	Bu	1:0	39	[76]
iPr	H	Me	Bu	6:1	72	[76]
Ph	H	Me	Bu	2.3:1	52	[76]
H	Me	Me	Bu	4.8:1	52	[76]
H	Me	Et	Bu	2.5:1	69	[76]
Me	H	Me	Ph	1:0	63	[76]
Me	H	Me	Bn	1:0	61	[76]

As an alternative to the previous methods, a cobalt-catalyzed annulation reaction can be used to synthesize polysubstituted dihydropyridines **107** from α,β-unsaturated *N*-(4-methoxyphenyl) imines **105** and internal alkynes under mild conditions. The reaction involves cobalt-mediated C—H bond activation, migratory insertion into the triple bond, and 6π-electrocyclization of the resulting azatriene intermediate **106** as shown in Scheme 38.[77]

Scheme 38 Cobalt-Catalyzed Synthesis of Substituted Dihydropyridines[77]

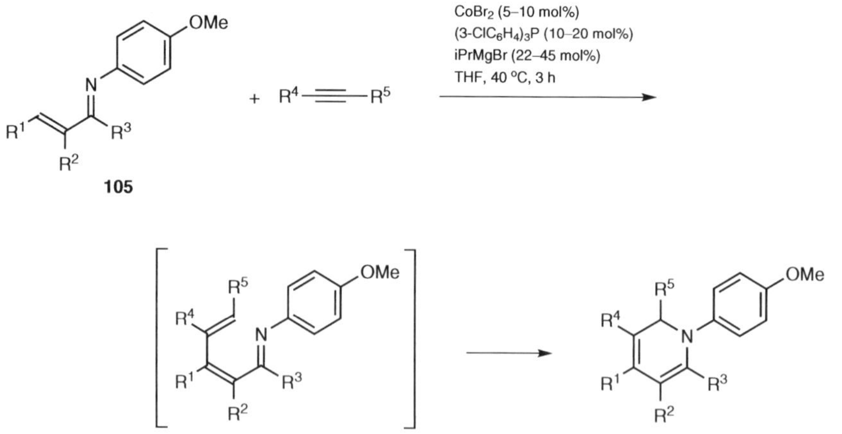

OMe N R^1 R^2 R^3 **105** + R^4—≡—R^5

$CoBr_2$ (5–10 mol%)
$(3\text{-}ClC_6H_4)_3P$ (10–20 mol%)
iPrMgBr (22–45 mol%)
THF, 40 °C, 3 h

106 → **107**

R^1	R^2	R^3	R^4	R^5	Yield (%)	Ref
Ph	H	Me	Pr	Pr	59	[77]
Ph	H	Me	Ph	Ph	91	[77]
4-$MeOC_6H_4$	H	Me	Ph	Ph	95	[77]
4-NCC_6H_4	H	Me	Ph	Ph	84	[77]
4-ClC_6H_4	H	Ph	Ph	Ph	96	[77]
Ph	Me	Me	Ph	Ph	77	[77]
Ph	H	H	Ph	Ph	79	[77]
Ph	H	Me	4-$F_3CC_6H_4$	4-$MeOC_6H_4$	86	[77]
$(CH_2)_4$		Me	Ph	Ph	79	[77]

The synthesis of δ-carbolines from 2-iodoanilines **108** and *N*-tosyl enynamines **109** is achieved by a palladium-catalyzed sequential reaction. The process consists of Larock heteroannulation and elimination of 4-toluenesulfinic acid, and this event is followed by electrocyclization of the resulting 2-azatrienes **110** and oxidative aromatization to give δ-carbolines **111** in a domino fashion (Scheme 39).[78]

Scheme 39 Palladium-Catalyzed Synthesis of δ-Carbolines[78]

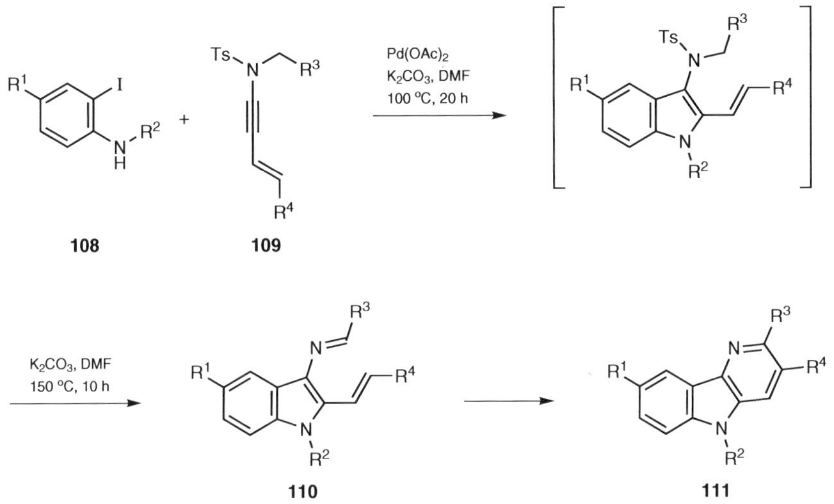

R^1	R^2	R^3	R^4	Yield (%)	Ref
H	H	Ph	Ph	74	[78]
H	Me	Ph	Ph	66	[78]
Me	Bn	Ph	Ph	72	[78]
Cl	Bn	Ph	Ph	43	[78]
H	Bn	4-Tol	Ph	69	[78]
H	Bn	4-FC_6H_4	Ph	60	[78]
H	Bn	Ph	4-Tol	65	[78]
H	Bn	Ph	3-ClC_6H_4	58	[78]

N-Phenyl alkynyl imines **112** react with sulfonyl azides **113** in a domino reaction consisting of 1,3-dipolar cycloaddition, ketenimine formation, 6π-electrocyclization, and 1,3-hydrogen shift (Scheme 40). The process is catalyzed by a copper salt [e.g., copper(I) iodide] at room temperature and this leads to various 4-sulfonamidoquinolines **114** in moderate to good yields.[79]

Scheme 40 Copper(I)-Catalyzed Synthesis of 4-Sulfonamidoquinolines[79]

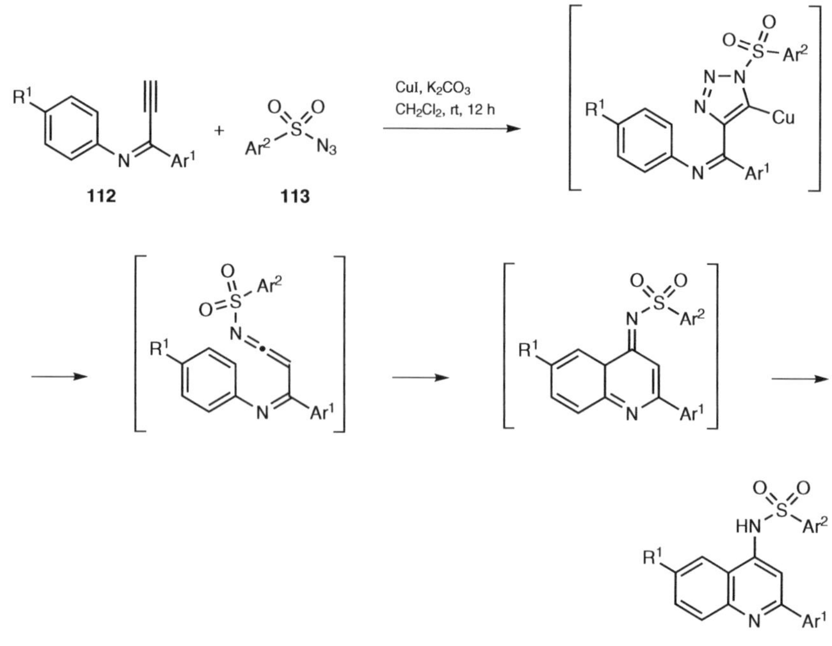

R^1	Ar^1	Ar^2	Yield (%)	Ref
H	Ph	4-Tol	78	[79]
H	Ph	Ph	68	[79]
H	Ph	$4\text{-}O_2NC_6H_4$	47	[79]
Me	Ph	4-Tol	79	[79]
OMe	Ph	4-Tol	84	[79]
F	Ph	4-Tol	37	[79]
H	$4\text{-}t\text{-}BuC_6H_4$	4-Tol	65	[79]
H	2-thienyl	4-Tol	56	[79]

3,4,5-Trisubstituted pyrazoles **117** are synthesized by palladium-catalyzed tandem cross coupling/electrocyclization of enol trifluoromethanesulfonates **115** and diazoacetates **116** as outlined in Scheme 41.[80]

Scheme 41 Palladium-Catalyzed Tandem Synthesis of Trisubstituted Pyrazoles[80]

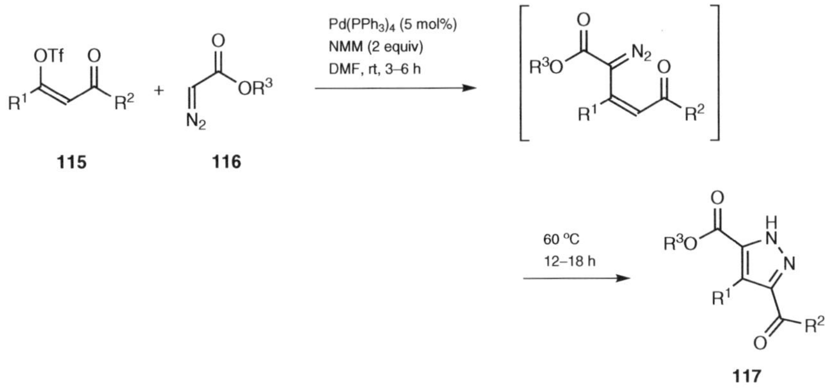

115 + 116 → [Pd(PPh$_3$)$_4$ (5 mol%), NMM (2 equiv), DMF, rt, 3–6 h] → [intermediate] → [60 °C, 12–18 h] → 117

R^1	R^2	R^3	Yield (%)	Ref
Me	O*t*-Bu	Et	84	[80]
$(CH_2)_2Ph$	O*t*-Bu	Et	79	[80]
Me	OBn	Et	80	[80]
Ph	O*t*-Bu	Et	40	[80]
iPr	OEt	*t*-Bu	83	[80]
CH_2OEt	OEt	*t*-Bu	52	[80]
Me	Ph	Et	73	[80]
Me	Me	Et	81	[80]

α,β-Unsaturated (*E*)-*O*-propargylic oximes **118** lead to polysubstituted pyridine *N*-oxides through a copper-catalyzed [2,3]-sigmatropic rearrangement/6π-electrocyclization tandem reaction.[81] The proposed mechanism involves nucleophilic attack by the oxime nitrogen atom onto the electrophilic C≡C bond activated by the π-acidic copper catalyst, ionic cleavage of the C—O bond of the cyclic intermediate obtained, and subsequent elimination of the copper catalyst to give *N*-allenyl nitrone intermediate **119**. This is followed by 6π-electrocyclization of the 3-azatriene to afford dihydropyridine **120**, which then isomerizes in situ to pyridine *N*-oxide **121** (Scheme 42). The same authors have applied this strategy to the synthesis of four-membered cyclic nitrones from (*E*)-*O*-propargylic aryl aldoximes.[82]

Scheme 42 Synthesis of Polysubstituted Pyridine *N*-Oxides by Copper-Catalyzed Tandem [2,3]-Sigmatropic Rearrangement/6π-Electrocyclization[82]

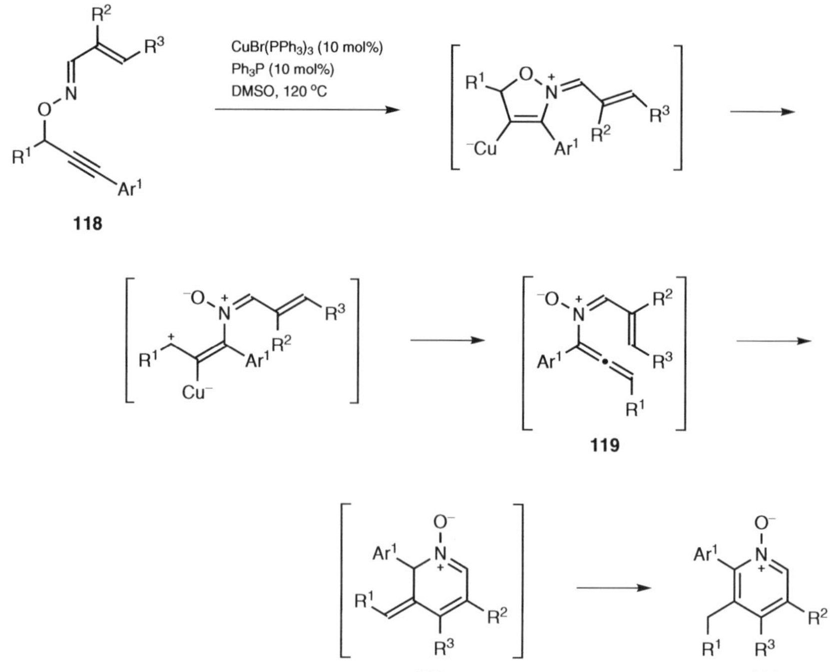

CuBr(PPh_3)$_3$ (10 mol%)
Ph_3P (10 mol%)
DMSO, 120 °C

118 119 120 121

Ar^1	R^1	R^2	R^3	Yield (%)	Ref
Ph	Ph	H	H	84	[82]
Ph	Ph	H	Me	64	[82]
Ph	Ph	Me	H	71	[82]
Ph	Ph	$(CH_2)_4$		53	[82]
4-MeOC$_6$H$_4$	Ph	H	H	75	[82]
4-F$_3$CC$_6$H$_4$	Ph	H	H	52	[82]
Ph	Pr	H	H	47	[82]
Ph	4-MeOC$_6$H$_4$	H	H	86	[82]

5-Butyl-2,4-diphenylpyridine (97, $R^1 = R^2 = Ph$; $R^3 = Bu$); Typical Procedure:[69]
Dry DMF (2 mL) was added to a Schlenk tube containing *O*-pentafluorobenzoyl 1,3-diphenylprop-2-en-1-one oxime (**95**, $R^1 = R^2 = Ph$; 41.7 mg, 0.1 mmol, 1 equiv), (*E*)-hex-1-enylboronic acid (15.4 mg, 0.12 mmol, 1.2 equiv), 4-Å molecular sieves (100 mg), and Cu(OAc)$_2$ (1.8 mg, 0.01 mmol, 0.1 equiv). The mixture was stirred under air at 50 °C for 2 h and then at 90 °C for 3 h. The mixture was extracted with Et_2O, and the organic phase was washed with brine and then dried ($MgSO_4$). After evaporation of the solvent, the residue was subjected to flash chromatography (hexane/EtOAc 20:1) to give a colorless oil; yield: 25.1 mg (87%).

1-(4-Methoxyphenyl)-6-methyl-4-phenyl-2,3-dipropyl-1,2-dihydropyridine (107, $R^1 = Ph$; $R^2 = H$; $R^3 = Me$; $R^4 = R^5 = Pr$); Typical Procedure:[77]
A Schlenk tube equipped with a stirrer bar was charged with (*E*)-4-methoxy-*N*-[(*E*)-4-phenylbut-3-en-2-ylidene]aniline (**105**, $R^1 = Ph$; $R^2 = H$; $R^3 = Me$; 50.3 mg, 0.20 mmol, 1 equiv), oct-4-yne (35.3 μL, 0.24 mmol, 1.2 equiv), (3-ClC$_6$H$_4$)$_3$P (7.3 mg, 0.020 mmol, 0.1 equiv), a

soln of 0.067 M $CoBr_2$ in THF (0.15 mL, 0.010 mmol, 0.05 equiv), and THF (0.47 mL). A soln of 0.89 M iPrMgBr in THF (51 μL, 0.045 mmol, 0.225 equiv) was added at rt and the resulting mixture was stirred at 40 °C for 3 h before the reaction was quenched with H_2O. The aqueous layer was extracted with EtOAc (3 × 4 mL) and the combined organic layers were dried (Na_2SO_4) and concentrated under reduced pressure. The crude product was purified by chromatography (silica gel, hexane/Et_3N 50:1) to afford a red oil; yield: 42.3 mg (59%, as reported).

2.1.3.8.1.2 Imine or Iminium Formation/Hetero-electrocyclization

Cyclic α,β,γ,δ-unsaturated aldehydes **123** prepared by ruthenium-catalyzed cycloisomerization of diynes **122** are involved in a tandem imine formation/electrocyclization/dehydration process to give pyridine-fused carbocycles or heterocycles **124** (Scheme 43).[83]

Scheme 43 Synthesis of Pyridine-Fused Carbocycles or Heterocycles[83]

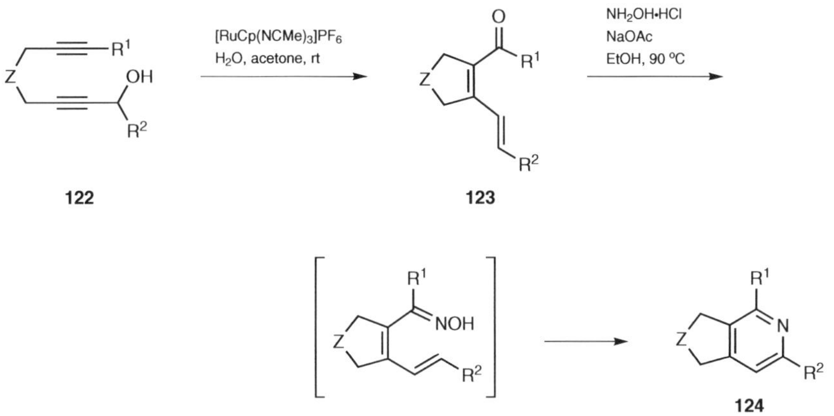

R^1	R^2	Z	Yield (%) of **124** from **123**	Ref
H	iPr	$C(CO_2Me)_2$	95	[83]
H	Ph	$C(CO_2Me)_2$	99	[83]
H	iPr	NTs	40	[83]
Me	H	O	75	[83]
Me	Ph	O	85	[83]
H	Et	$C(CO_2Me)_2$	97	[83]
H	Et	$C(CH_2OMe)_2$	73	[83]
H	Ph	$C(CH_2OMe)_2$	87	[83]

A series of 2,4-disubstituted pyridines **126** have been synthesized from acyclic α,β,γ,δ-unsaturated aldehydes **125** through a tandem transformation involving three reactions: (1) imine formation, (2) electrocyclization, and (3) aromatization by oxidation in air (Scheme 44).[84] For substrates bearing an electron-withdrawing substituent (e.g., **127**), pyrroles **129** are obtained in varying amounts along with the desired pyridines **128**.

Scheme 44 Synthesis of 2,4-Disubstituted Pyridines[84]

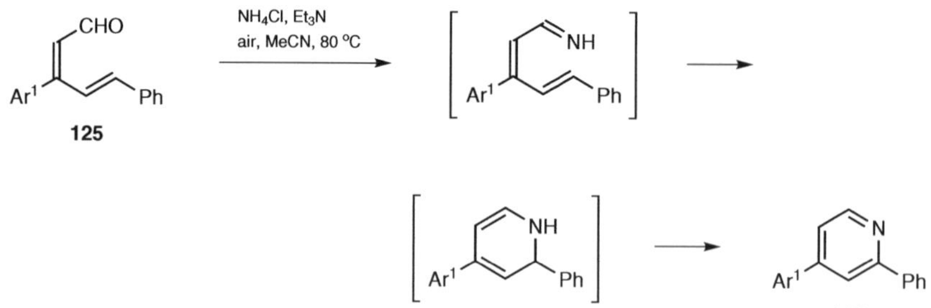

NH_4Cl, Et_3N
air, MeCN, 80 °C

125 → 126

Ar^1	Yield (%)	Ref
Ph	82	[84]
4-Tol	84	[84]
4-$MeOC_6H_4$	71	[84]
1-naphthyl	73	[84]
2-naphthyl	81	[84]
3,4-$(MeO)_2C_6H_3$	78	[84]

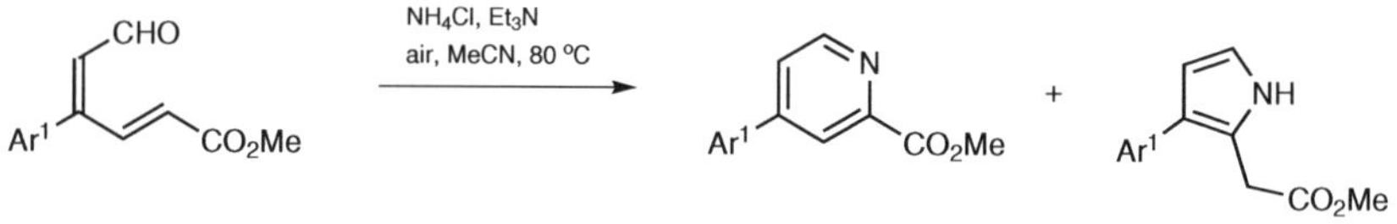

NH_4Cl, Et_3N
air, MeCN, 80 °C

127 → **128** 25–43% + **129** 24–38% (40:60 to 62:38)

Ar^1 = Ph, 4-$MeOC_6H_4$, 3,4-$(MeO)_2C_6H_3$, 3-Tol, 2-naphthyl, benzo-1,3-dioxol-5-yl

Baudoin and co-workers have developed a strategy for the synthesis of 3-aryl-3,4-dihydroisoquinolines **131** from iminobenzocyclobutenes **130**. The key step of the strategy is based on tandem electrocyclic ring opening/6π-electrocyclization. The utility of this methodology has been demonstrated by the synthesis of a natural product, the alkaloid coralydine, in three steps from the corresponding 3,4-dihydroisoquinoline **131** (Scheme 45).[85]

Scheme 45 Synthesis of Isoquinolines by an Electrocyclic Ring Opening/Aza-electrocyclization Tandem Reaction[85]

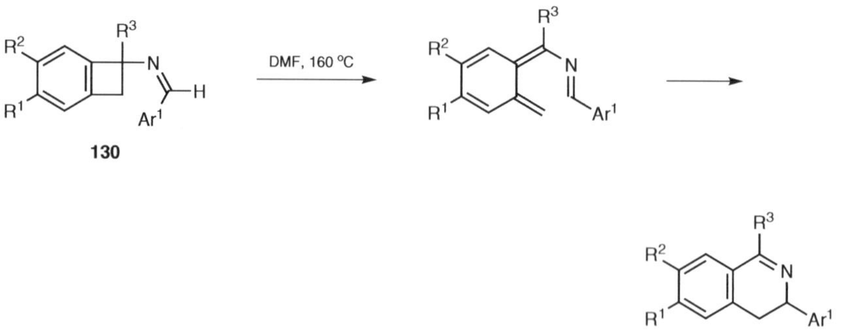

DMF, 160 °C

130 → 131

R^1	R^2	R^3	Ar^1	Yield (%)	Ref
H	H	Me	Ph	73	[85]
H	H	Me	4-$MeOC_6H_4$	65	[85]
H	H	Me	4-$F_3CC_6H_4$	52	[85]
H	H	Me	3-pyridyl	68	[85]
H	H	Me	2-furyl	60	[85]
OMe	OMe	Me	Ph	61	[85]
H	H	Pr	Ph	61	[85]
H	CF_3	Me	4-ClC_6H_4	59	[85]

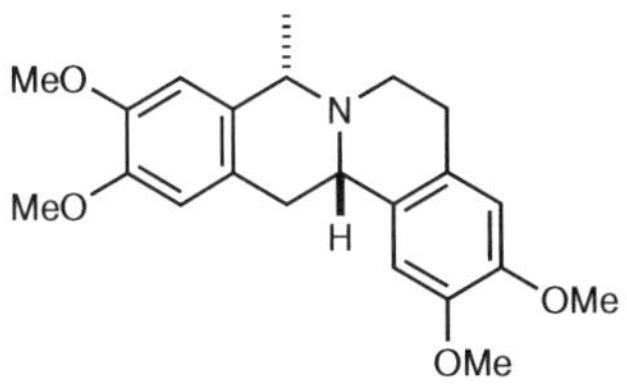

MeO MeO N H OMe OMe

coralydine

Isothiocyanates **133**, formed by treatment of 2-isocyanostyrenes **132** with sulfur in the presence of a catalytic amount of selenium, lead to quinoline-2(1*H*)-thiones **134** through spontaneous 6π-electrocyclization (Scheme 46).[86]

Scheme 46 Synthesis of Quinoline-2(1*H*)-thiones[86]

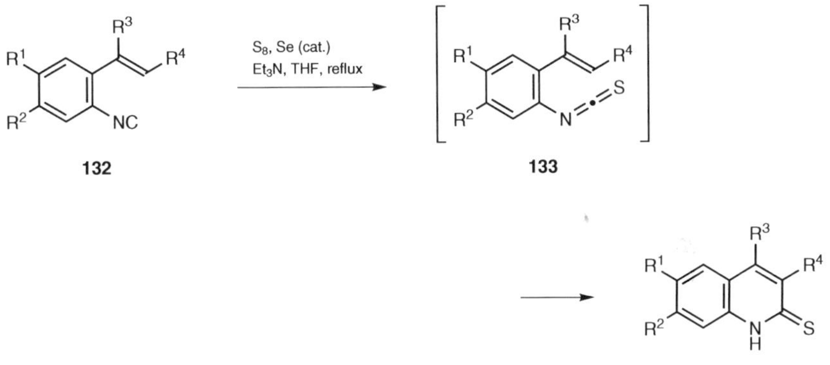

S_8, Se (cat.)
Et_3N, THF, reflux

132 133 134

R^1	R^2	R^3	R^4	Yield (%)	Ref
H	H	Ph	H	81	[86]
H	H	H	4-Tol	89	[86]
H	H	4-$MeOC_6H_4$	H	91	[86]
H	H	Me	H	37	[86]
Cl	H	4-Tol	H	45	[86]
OMe	OMe	Ph	H	58	[86]

Asymmetric, organocatalytic iminium formation/6π-electrocyclization can be used for the synthesis of optically active 1,4-dihydropyridazines **137** (Scheme 47).[87] Enantiopure phosphonic acid **135**, used as the Brønsted acid catalyst, represents the chiral counteranion of iminium species **136**, which induces chirality during the 6π-disrotatory electrocyclization process.

Scheme 47 Asymmetric Organocatalytic Synthesis of 1,4-Dihydropyridazines[87]

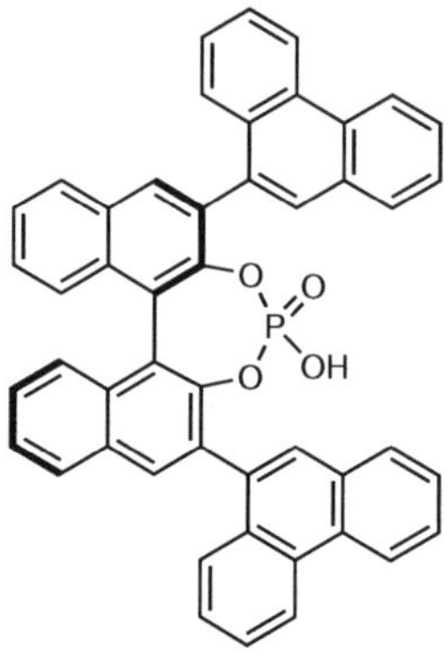

135

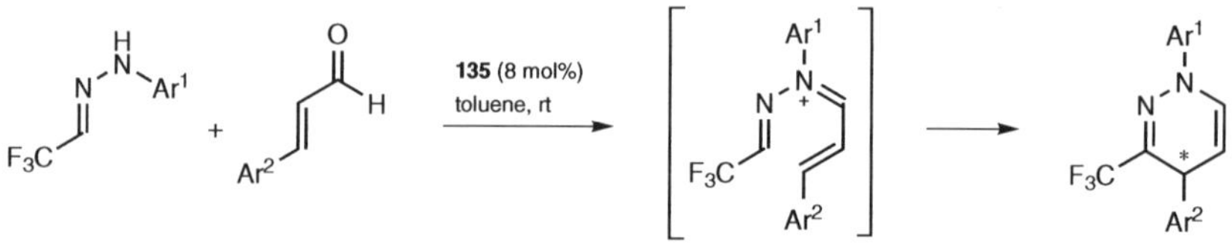

135 (8 mol%)
toluene, rt

136

137

Ar^1	Ar^2	ee (%)	Yield (%)	Ref
Ph	Ph	80	85	[87]
1-naphthyl	Ph	90	87	[87]
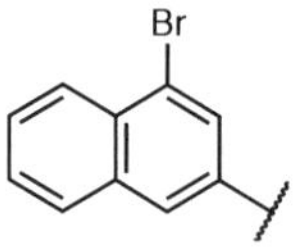 Br	Ph	90	84	[87]
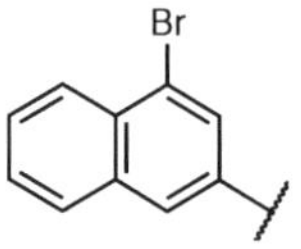 Br	2-$MeOC_6H_4$	96	81	[87]
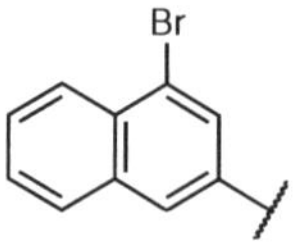 Br	2-$O_2NC_6H_4$	98	79	[87]

Ar^1	Ar^2	ee (%)	Yield (%)	Ref
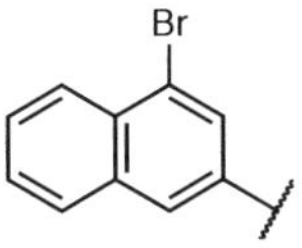 Br	2-BrC_6H_4	97	84	[87]
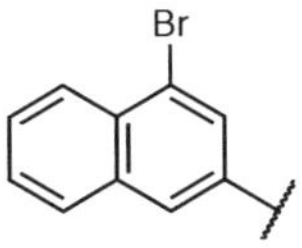 Br	4-FC_6H_4	95	85	[87]
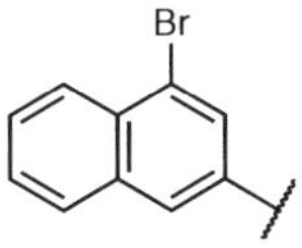 Br	4-NCC_6H_4	94	79	[87]

A tandem reaction involving the generation of an imine by an aza-Wittig reaction combined with electrocyclization affords nitrogen heterocycles.[88–90] As shown in Scheme 48, iminophosphorane **138** reacts initially with α,β-unsaturated aldehyde or ketone **139** to form aza-hexa-1,3,5-triene intermediate **140**, a material that undergoes thermal electrocyclization to give substituted pyridine **141**.[88] Good yields are obtained with the use of enones bearing an electron-withdrawing substituent (e.g., a methoxycarbonyl group) on the β-ethene carbon atom.

Scheme 48 Synthesis of Pyridines by a Tandem Aza-Wittig/Electrocyclization Reaction[88]

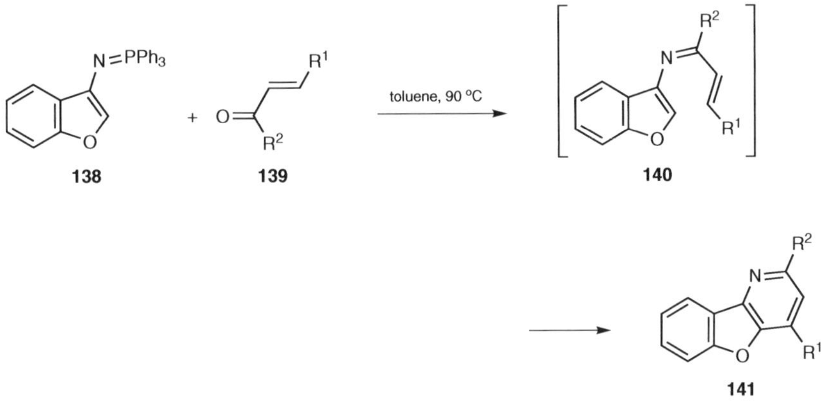

R^1	R^2	Yield (%)	Ref
H	H	34	[88]
Me	H	20	[88]
Ph	H	21	[88]
H	Me	28	[88]
CO_2Me	Me	80	[88]
CO_2Me	4-Tol	85	[88]

for references see p 156

The reaction of dibromo-substituted *O*-methyl oximes **142** with alkyl or aryl Grignard reagents leads to pyrimidines **144**. The mechanism is thought to consist of a domino reaction that initially forms a 1,5-diazahexa-1,3,5-triene intermediates **143**, and this event is followed by electrocyclization and elimination of methanol (Scheme 49)[91] to effect the final aromatization.

Scheme 49 Synthesis of Pyrimidines from Dibromo-Substituted *O*-Methyl Oximes and Grignard Reagents[91]

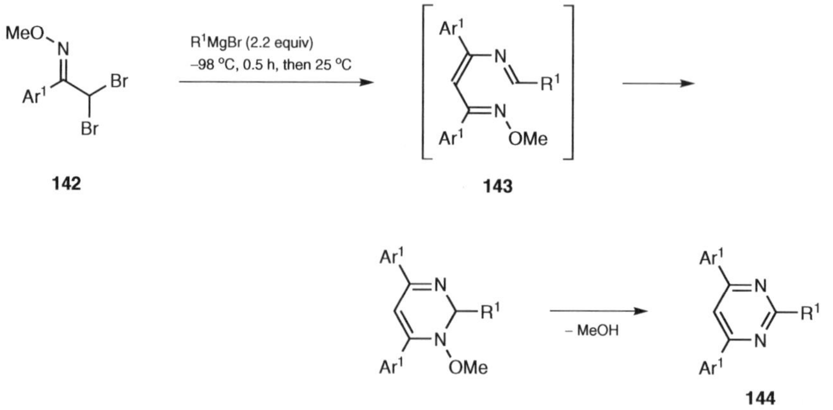

Ar^1	R^1	Yield (%)	Ref
Ph	Pr	64	[91]
Ph	Cy	55	[91]
Ph	4-$MeOC_6H_4$	75	[91]
Ph	4-BrC_6H_4	80	[91]
4-Tol	Bu	70	[91]
4-ClC_6H_4	Bu	72	[91]
4-BrC_6H_4	Bu	75	[91]
2-furyl	Bu	65	[91]

Dimethyl 3-Ethyl-5,7-dihydro-6*H*-cyclopenta[*c*]pyridine-6,6-dicarboxylate [124, **R^1 = H; R^2 = Et; Z = $C(CO_2Me)_2$]; Typical Procedure:**[83]

A capped vial containing cyclopentene **123** [R^1 = H; R^2 = Et; Z = $C(CO_2Me)_2$; 29 mg, 0.110 mmol, 1 equiv], EtOH (1.1 mL), NH_2OH•HCl (9 mg, 0.132 mmol, 1.2 equiv), and NaOAc (5 mg, 0.066 mmol, 0.6 equiv) was heated in an oil bath at 90 °C for 24 h. The vial was cooled to rt and the contents were diluted with $CHCl_3$, washed with sat. aq $NaHCO_3$, and extracted with 1 M HCl. The aqueous phases were then combined and basified (pH >7) by the slow addition of sat. aq $NaHCO_3$. The aqueous phase was then extracted with $CHCl_3$ and these extracts were dried (Na_2SO_4) and the solvent was removed under reduced pressure to yield a light oil; yield: 28 mg (97%).

2.1.3.8.1.3 Isomerization or Rearrangement/Hetero-electrocyclization

An efficient tandem one-pot reaction has been developed for the synthesis of polysubstituted pyridines **149** from nitriles **145**, Reformatsky reagents, and 1,3-enynes with complete control of the substitution pattern.[92,93] The reaction proceeds through the regio- and chemoselective addition of the Blaise reaction intermediate **146** to the 1,3-enyne followed by an isomerization/cyclization/aromatization cascade. The proposed mechanism involves isomerization of β-enaminozincate **147** into N-zincated 1-azatriene **148**, which undergoes 6π-electrocyclization to produce pyridine **149** after elimination of zinc hydrogen bromide (Scheme 50).

Scheme 50 Synthesis of Pyridines from Nitriles Using 1,3-Enynes[92]

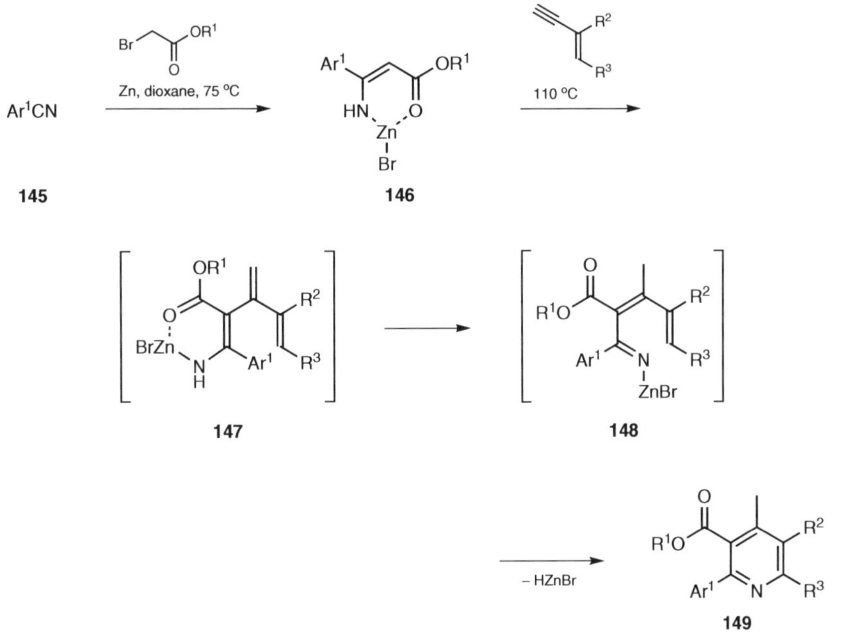

Ar^1CN **145** → ($BrCH_2CO_2R^1$; Zn, dioxane, 75 °C) → **146** → (1,3-enyne, 110 °C) → [**147**] → [**148**] → (– HZnBr) → **149**

Ar^1	R^1	R^2	R^3	Yield (%)	Ref
Ph	Et	$(CH_2)_4$		90	[92]
2-Tol	Et	$(CH_2)_4$		89	[92]
$4\text{-}MeOC_6H_4$	Et	$(CH_2)_4$		72	[92]
Ph	iPr	$(CH_2)_4$		92	[92]
Ph	Et	Ph	Ph	40	[92]
Ph	Et	Ph	H	53	[92]
Ph	Et	Pr	Pr	49	[92]
Ph	Et	$(CH_2)_5$		82	[92]

Successive anion-accelerated 4π-electrocyclization/8π-electrocyclization reactions enable the facile transformation of benzocyclobutenones **150** into 2,3-benzodiazepines **151** through reaction with a lithiated diazo compound.[94] In this reaction, the oxy anion formed by nucleophilic addition facilitates the first electrocyclic ring opening, and the reconstruction of the stable aromatic system favors the second 8π-electrocyclization reaction that completes the targeted structure (Scheme 51).

Scheme 51 Synthesis of 2,3-Benzodiazepines from Benzocyclobutenones[94]

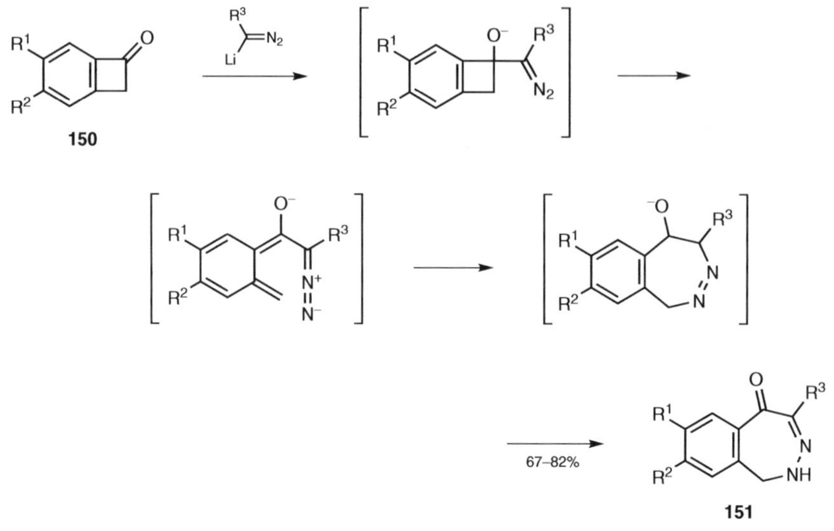

$R^1 = R^2 = H, OMe; R^3 = TMS, CO_2Et$

Equally significantly, 2-(ketenimino)thiobenzoates **152** can be transformed into quinolin-4-ones **153** by a domino sequence involving 1,5-migration of the sulfanyl group from the carbonyl carbon atom to the central carbon atom of the ketenimine moiety followed by 6π-electrocyclic ring closure of the formed 3-azatriene (Scheme 52). Experimental and theoretical studies show that these two processes are pseudopericyclic.[95,96] The same authors have developed similar methodology using substrates **154** bearing a monosubstituted 1,3-dioxolane or 1,3-dithiolane hydride releasing fragment (Scheme 52). The tandem reaction consists of a 1,5-hydride shift, followed by in situ 6π-electrocyclization of the azatriene derived from the ketenimine, to give quinolin-4-ones **155**.[97] The reaction has also been performed using carbodiimides via diazatriene intermediates.

Scheme 52 Synthesis of Quinolines and Quinazolines by 1,5-Migration/6π-Electrocyclization Domino Reaction[95–97]

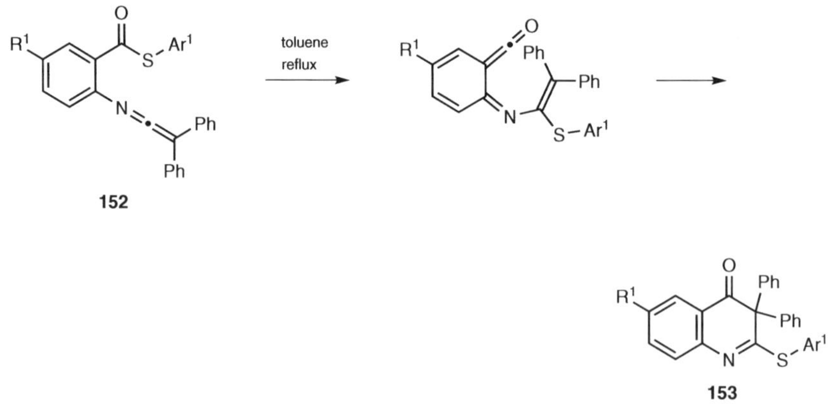

R^1	Ar^1	Yield (%)	Ref
H	4-Tol	74	[95]
H	4-MeOC$_6$H$_4$	89	[95]
H	4-pyridyl	79	[95]
Cl	4-Tol	97	[95]

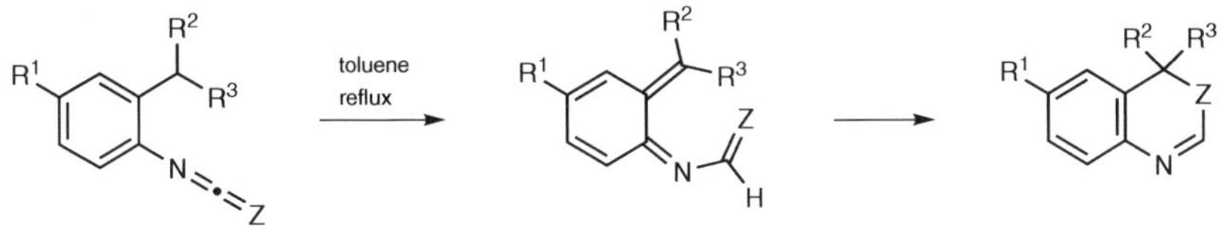

R^1	R^2 R^3	Z	Yield (%)	Ref
Me	O(CH$_2$)$_2$O	CPh$_2$	68	[97]
H	S(CH$_2$)$_2$S	C(Ph)Me	52	[97]
Cl	S(CH$_2$)$_2$S	CPh$_2$	77	[97]
H	O(CH$_2$)$_2$O	C(Ph)Me	58	[97]

3-Aza-1,5-enynes **156** lead selectively to 1,2-dihydropyridines **158** by an aza-Claisen rearrangement/electrocyclization domino reaction (Scheme 53) that requires only simple heating in methanol to proceed. The enamine is in equilibrium with the rearranged imine **157** via 1,3-hydrogen shift, and this leads ultimately to an azatriene which generates the 1,2-dihydropyridine **158** by subsequent 6π-electrocyclization. Aromatization of 1,2-dihydropyridine **158** is not observed.[98]

Scheme 53 Synthesis of 1,2-Dihydropyridines by Tandem Aza-Claisen/6π-Electrocyclization Reaction[98]

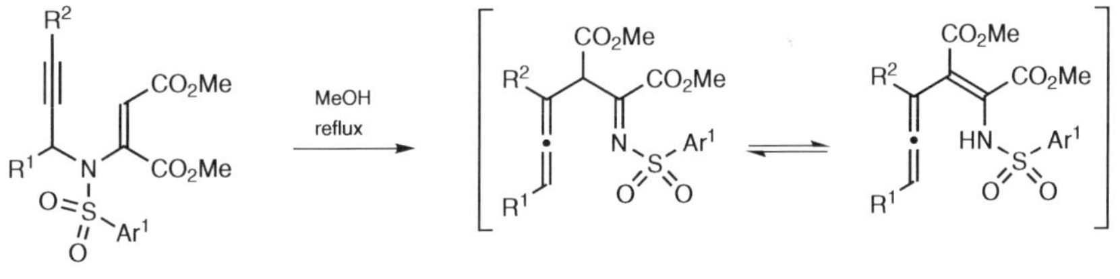

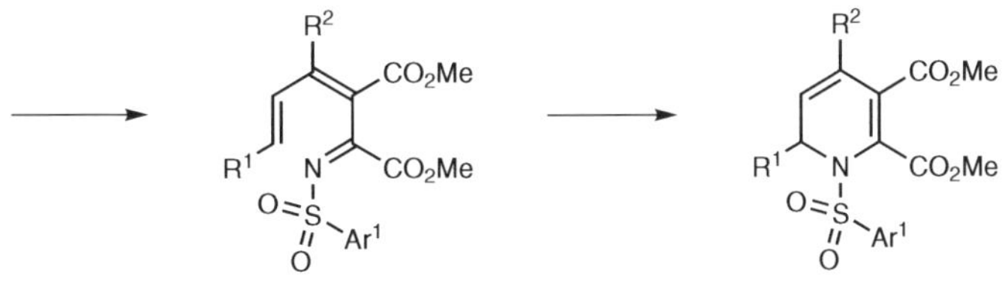

158 72–97%

R^1 = Ph, iPr, cyclopropyl, 2-Tol, 2-ClC$_6$H$_4$, 2-F$_3$CC$_6$H$_4$, 4-FC$_6$H$_4$, 1-naphthyl; R^2 = Ph, iPr, cyclopropyl, 4-Tol, 4-FC$_6$H$_4$;
Ar^1 = Ph, 4-Tol, 2-ClC$_6$H$_4$

In another approach, polyfunctionalized quinolines **160** can be synthesized from aza-enynes **159** through a propargyl–allenyl isomerization/aza-electrocyclization domino reaction promoted by a catalytic amount of 1,8-diazabicyclo[5.4.0]undec-7-ene (Scheme 54).[99] This method has been extended by other research groups to include trifluoromethylated substrates.[100]

Scheme 54 Synthesis of Quinolines through a Propargyl–Allenyl Isomerization/Aza-electrocyclization Domino Reaction[99]

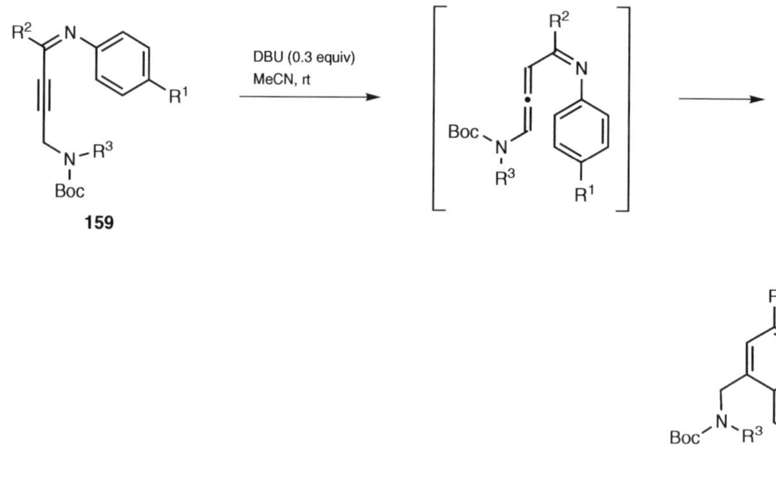

DBU (0.3 equiv)
MeCN, rt

159

160

R^1	R^2	R^3	Yield (%)	Ref
H	Ph	4-Tol	65	[99]
Me	Ph	4-Tol	69	[99]
OMe	Ph	4-ClC_6H_4	87	[99]
H	*t*-Bu	4-ClC_6H_4	88	[99]
Br	Ph	4-ClC_6H_4	65	[99]
H	Ph	Ph	85	[99]
H	Ph	Bn	80	[99]
H	Ph	4-AcC_6H_4	92	[99]

In a recent total synthesis of (−)-lyconadin C, the key step involves a tandem Curtius rearrangement/6π-electrocyclization reaction to form the central pyridone ring in intermediate **161** (Scheme 55).[101] This process represents the first example of the involvement of a vinyl isocyanate in this type of transformation for the synthesis of an alkaloid.

Scheme 55 Synthesis of a Pyridone Ring through a Curtius Rearrangement/Aza-electrocyclization Domino Reaction[101]

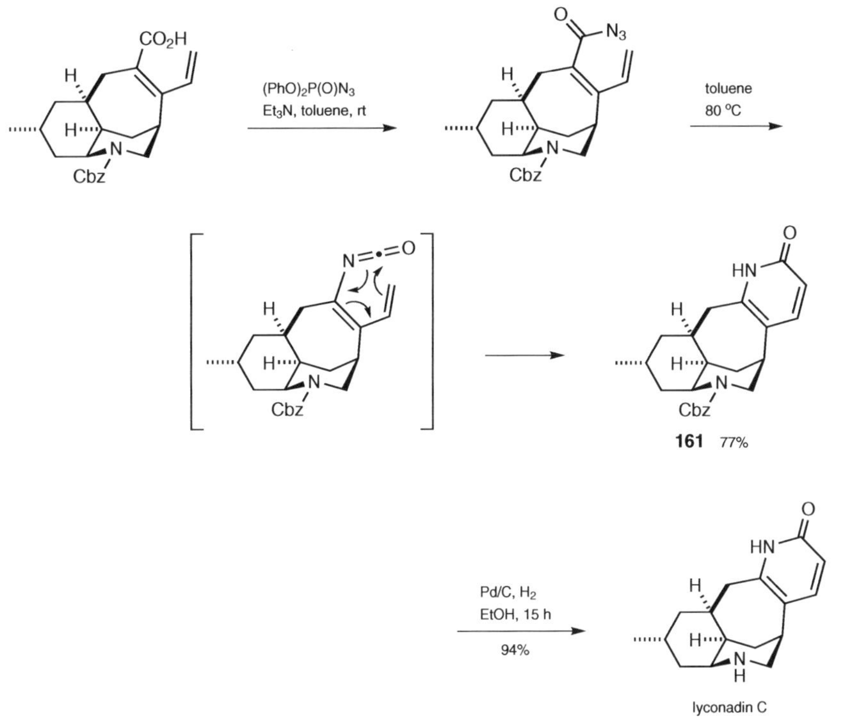

CO_2H H H Cbz N; $(PhO)_2P(O)N_3$, Et_3N, toluene, rt; O N_3 H H N Cbz; toluene 80 °C; N=•=O H H N Cbz; HN O H H N Cbz **161** 77%; Pd/C, H_2 EtOH, 15 h 94%; HN O H H N H lyconadin C

An interesting domino reaction enabling the transformation of *N*-vinyl-β-lactams into aminocyclobutanes has been developed by Cheung and Yudin.[102] The reaction takes place in the presence of a copper source, such as copper(I) iodide, and cesium carbonate under microwave irradiation. Under these conditions, *N*-vinyl-β-lactam precursor **162** undergoes ring expansion through [3,3]-sigmatropic rearrangement, which leads to an eight-membered enamide-ring intermediate. This intermediate then affords cyclobutane δ-lactam **163** by 6π-electrocyclization (Scheme 56).

Scheme 56 Synthesis of Aminocyclobutanes by a [3,3]-Sigmatropic Rearrangement/Electrocyclization Domino Reaction[102]

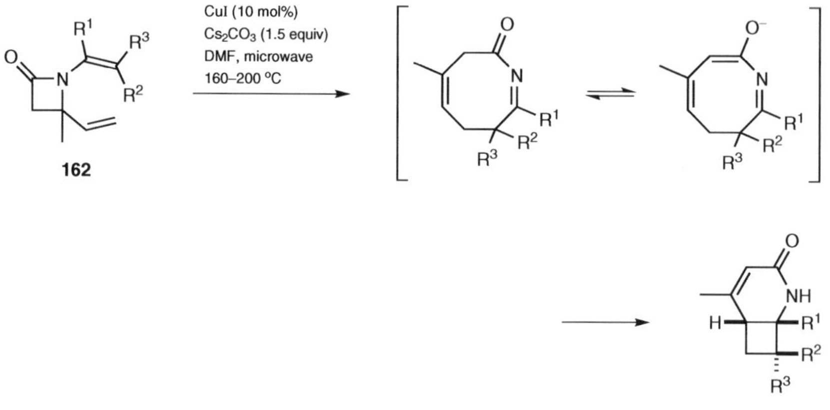

CuI (10 mol%)
Cs_2CO_3 (1.5 equiv)
DMF, microwave
160–200 °C

162 → 163

R^1	R^2	R^3	Yield (%)	Ref
Ph	H	H	42	[102]
Cy	H	H	84	[102]
cyclopropyl	H	H	89	[102]
Me	$(CH_2)_3$		93	[102]

Ethyl 4-Methyl-2-phenyl-5,6,7,8-tetrahydroquinoline-3-carboxylate [149, Ar^1 = Ph; R^1 = Et; R^2,R^3 = $(CH_2)_4$]; Typical Procedure:[92]
A 1 M soln of MsOH in 1,4-dioxane (0.3 mL, 0.3 mmol, 0.1 equiv) was added to a stirred suspension of commercial Zn dust (Aldrich 10 μm; 392 mg, 6.0 mmol, 2 equiv) in 1,4-dioxane (1.2 mL) at 75 °C. After 10 min, benzonitrile (**145**, Ar^1 = Ph; 306 μL, 3.0 mmol, 1 equiv) was added in a single portion. At the same temperature, ethyl bromoacetate (0.5 mL, 4.5 mmol, 1.5 equiv) was added slowly over 1 h using a syringe pump, and the mixture was stirred for an additional 1 h. 1-Ethynylcyclohexene (388 μL, 3.3 mmol, 1.1 equiv) and 1,4-dioxane (2.0 mL) were added to this mixture, and the bath temperature was increased to 110 °C. After 3 h, the mixture was cooled to rt and the reaction was quenched with sat. aq NH_4Cl. This mixture was neutralized with sat. aq Na_2CO_3 and extracted with EtOAc (3 × 40 mL). The combined organic layers were dried ($MgSO_4$), filtered, and concentrated under reduced pressure. The residue was purified by column chromatography (silica gel, hexane/EtOAc 4:1) to afford the title compound as a pale yellow solid; yield: 796 mg (90%).

2.1.3.8.2 Oxa-electrocyclization

Generally, oxa-electrocyclizations involve an oxatriene. In the examples below, the oxatriene is generated by Knoevenagel condensation or by hydroxy-to-carbonyl oxidation of a 2,4-dien-1-ol. A few less general methods are also outlined, however, due to their interesting design and potential to inspire additional future solutions.

Firstly, Knoevenagel condensation of 2-alkylalk-2-enals **164** and β-oxo esters **165**, followed by 6π-electrocyclization of the resulting adducts, results in the synthesis of polysubstituted 2*H*-pyran-5-carboxylates **166**. The reaction is performed in the presence of piperidine and acetic acid (Scheme 57).[103]

Scheme 57 Synthesis of Substituted 2*H*-Pyrans through a Knoevenagel/Electrocyclization Domino Reaction[103]

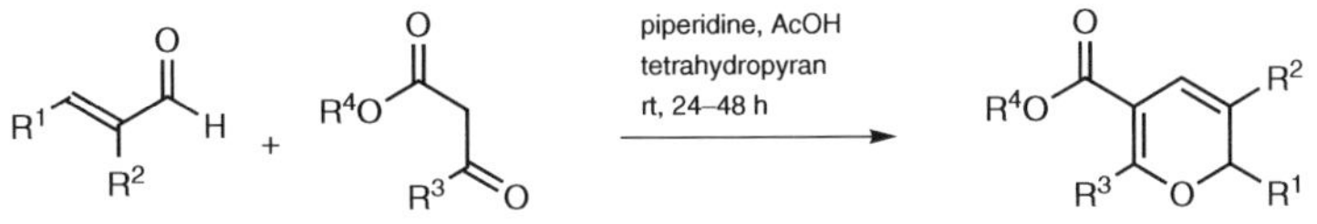

piperidine, AcOH
tetrahydropyran
rt, 24–48 h

164 + **165** → **166**

R^1	R^2	R^3	R^4	Yield (%)	Ref
Pr	Et	Me	Me	71	[103]
Pr	Et	Cy	Me	70	[103]
Pr	Et	CH_2SPh	Me	66	[103]
Me	Me	Me	Me	76	[103]
$(CH_2)_4$		Me	Me	72	[103]
Ph	Me	Me	Me	60	[103]
Pr	Et	Cy	Me	70	[103]
Pr	Et	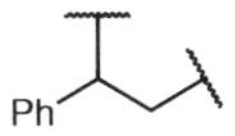 Ph		69	[103]
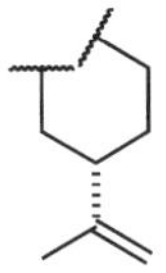		Me	Me	44	[103]

Some other versions of this type of reaction leading to 2*H*-pyrans[104] have been reported, but they use different reactions conditions (either solvent free and/or catalyst free under microwave irradiation and/or in water).[105–108]

Tandem Knoevenagel/6π-electrocyclization sequences have also been used efficiently to produce highly substituted pyrano[3,2-*c*]pyridones **167** (Scheme 58).[109] The authors show that some of the prepared pyrano[3,2-*c*]pyridones can serve as precursors for the synthesis of structurally diverse natural-product-like materials.

for references see p 156

Scheme 58 Synthesis of Pyrano[3,2-c]pyridones by a Knoevenagel/Oxa-electrocyclization Domino Reaction[109]

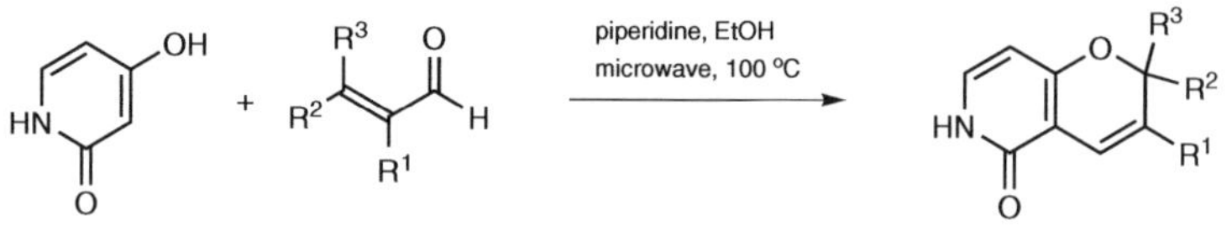

R^1	R^2	R^3	Yield (%)	Ref
H	$(CH_2)_2CH{=}CMe_2$	Me	87	[109]
Me	Me	H	52	[109]
H	Ph	Me	42	[109]
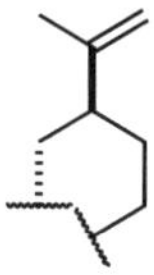		H	60	[109]

The total synthesis of the interleukin-1β converting enzyme inhibitor EI-1941–2 (**170**) can be achieved starting from epoxide **168** by an oxidation/oxa-electrocyclization/oxidation domino sequence via α-pyrone intermediate **169** (Scheme 59). An oxoammonium salt (Bobbitt's reagent) is used for the terminating oxidation, which leads to **170**.[110]

Scheme 59 Synthesis of the Enzyme Inhibitor EI-1941–2 through an Oxidation/Oxa-electrocyclization/Oxidation Cascade[110]

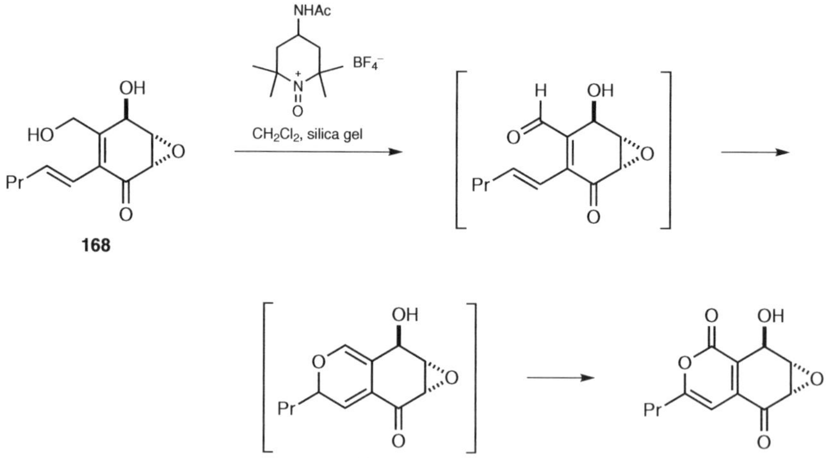

An oxidation/oxa-electrocyclization/Diels–Alder dimerization cascade has been used in the total synthesis of (+)-torreyanic acid by Porco and co-workers[111] and was later applied to the synthesis of epoxyquinols A, B, and C and epoxytwinol A.[112]

The synthesis of pyranocarbazoles can be achieved by the reaction of α,β-unsaturated aldehydes with 2- or 4-hydroxycarbazoles in the presence of ethylenediamine diacetate. The transformation involves an aldol-type reaction/6π-electrocyclization domino process. One example involving 2-hydroxycarbazole **171** and 3-methylbut-2-enal to give corresponding pyranocarbazole **172** is shown in Scheme 60.[113] This methodology has been used by others for the synthesis of several natural pyranocarbazole alkaloids.[114,115]

Scheme 60 Synthesis of 3,3-Dimethyldihydropyranocarbazole by an Aldol Reaction/Oxa-electrocyclization Cascade[113]

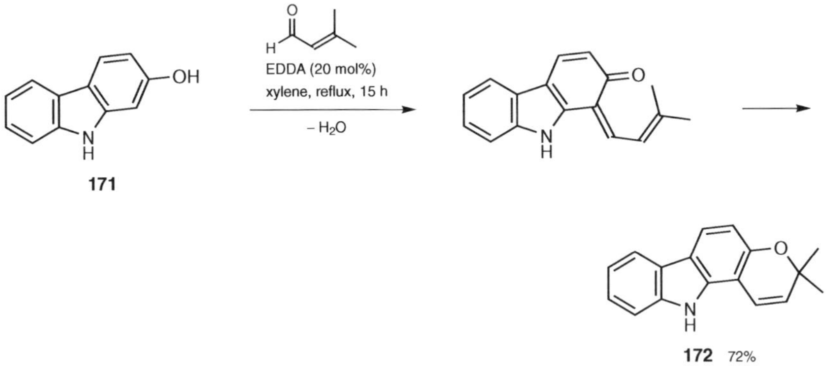

EDDA (20 mol%)
xylene, reflux, 15 h
$- H_2O$

171

172 72%

EDDA = ethylenediamine diacetate

A practical synthesis of 2*H*-1-benzopyrans **174** has been developed starting from 2-silylated aryl trifluoromethanesulfonates (as aryne precursors) and α,β-unsaturated aldehydes **173** in the presence of cesium fluoride. The transformation proceeds by a tandem [2+2] cycloaddition/thermal electrocyclic ring opening/6π-electrocyclization sequence (Scheme 61).[116]

Scheme 61 Synthesis of 2*H*-1-Benzopyrans by a Tandem [2+2] Cycloaddition/Electrocyclic Ring Opening/Oxa-electrocyclization Sequence and the Proposed Mechanism[116]

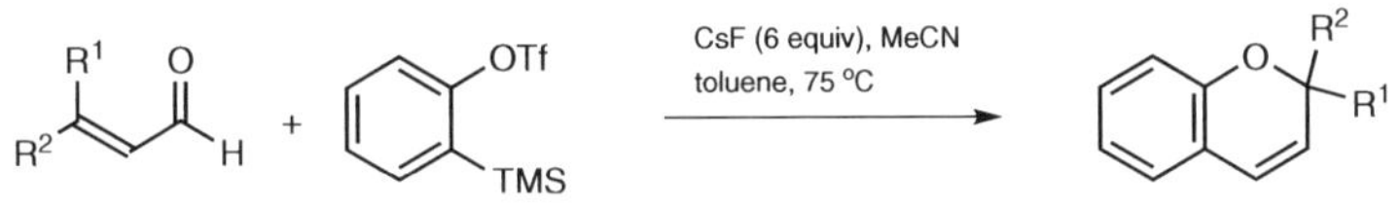

CsF (6 equiv), MeCN
toluene, 75 °C

173

174

R^1	R^2	Yield (%)	Ref
H	2-$MeOC_6H_4$	70	[116]
H	4-Tol	66	[116]
H	4-BrC_6H_4	34	[116]
4-$MeOC_6H_4$	4-$MeOC_6H_4$	61	[116]
Me	3-$MeOC_6H_4$	52	[116]
Me	iBu	31	[116]
Ph	4-$MeOC_6H_4$	61	[116]
Ph	Ph	80	[116]

for references see p 156

[2 + 2]
ring opening
6π-electrocyclization

Methyl 3-Ethyl-6-methyl-2-propyl-2*H*-pyran-5-carboxylate (166, **R^1 = Pr; R^2 = Et; R^3 = R^4 = Me); Typical Procedure:**[103]

Piperidine (170 mg, 2.0 mmol, 2 equiv) and AcOH (12 mg, 0.2 mmol, 0.2 equiv) were added consecutively at rt to a soln of 2-ethylhex-2-enal (**164**, R^1 = Pr; R^2 = Et; 126 mg, 1.0 mmol, 1 equiv) and methyl acetoacetate (**165**, R^3 = R^4 = Me; 175 mg, 1.5 mmol, 1.5 equiv) in tetrahydropyran (2 mL). The resulting mixture was stirred until the aldehyde was no longer detectable by TLC (~24 h). The reaction was then quenched by the addition of an aqueous soln of NH_4Cl and this mixture was extracted with EtOAc. The crude product was purified by column chromatography (silica gel, hexane/EtOAc 30:1) to give the title compound as a colorless oil; yield: 160 mg (71%); R_f 0.70 (hexane/EtOAc 7:1).

2-Methyl-2-(4-methylpent-3-enyl)-2,6-dihydro-5*H*-pyrano[3,2-*c*]pyridin-5-one [167, **R^1 = H; R^2 = $(CH_2)_2CH{=}CMe_2$; R^3 = Me]; Typical Procedure:**[109]

A soln of 4-hydroxypyridin-2(1*H*)-one (18 mg, 0.16 mmol, 1 equiv), citral (25 mg, 0.16 mmol, 1 equiv), and piperidine (4.86 μL, 0.049 mmol, 0.3 equiv) in EtOH (0.4 mL) was irradiated under microwave conditions at 100 °C for 15 min. The reaction was quenched by the addition of sat. aq NH_4Cl (4 mL), and the aqueous layer was extracted with EtOAc. The combined organic extracts were dried (Na_2SO_4), filtered, and concentrated under reduced pressure. Purification by flash column chromatography (silica gel, CH_2Cl_2/acetone 1:1) afforded the title compound as a white solid; yield: 35 mg (87%, as reported).

2.1.3.8.3 Thia-electrocyclization

Domino reactions involving thia-electrocyclization with the formation of a C—S bond are very rare. One example is the Knoevenagel/thia-electrocyclization domino reaction between phosphonodithioacetate **175** and an α,β-unsaturated aldehyde such as cinnamaldehyde or crotonaldehyde. This process leads directly to 5-phosphono-substituted 2*H*-thiopyrans **176** (Scheme 62). The reaction takes place slowly, presumably on the basis of the rate of the electrocyclization of the 1-thioxatriene intermediate obtained under standard Knoevenagel reaction conditions.[117]

Scheme 62 Synthesis of Substituted 2*H*-Thiopyrans by a Knoevenagel/Thia-electrocyclization Domino Reaction[117]

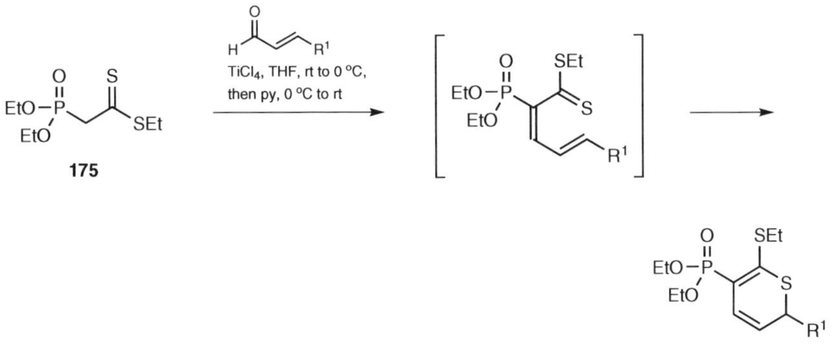

R1	Yield (%)	Ref
Me	67	[117]
Ph	74	[117]

Diethyl [6-(Ethylsulfanyl)-2-phenyl-2*H*-thiopyran-5-yl]phosphonate (176, **R^1 = Ph); Typical Procedure:**[117]

A soln of cinnamaldehyde (0.5 mmol, 1 equiv) in dry THF (1 mL) and a soln of phosphonodithioacetate **175** (128 mg, 0.5 mmol, 1 equiv) in dry THF (5 mL) were added successively to a soln of $TiCl_4$ [1 mmol in 250 μL CCl_4 (**CAUTION:** *carcinogen*), 2 equiv] in dry THF (4 mL). The mixture was stirred at rt for 15 min and then cooled to 0 °C. A soln of pyridine (162 μL, 2 mmol, 4 equiv) in dry THF (3 mL) was added dropwise over 1 h, and the resulting mixture was stirred at rt until completion of the reaction. The mixture was poured at 0 °C into 1 M aq HCl (10 mL) and vigorously stirred for 1 h. The organic compounds were extracted with CH_2Cl_2 (3 × 20 mL) and the combined organic layers were dried ($MgSO_4$), filtered, and concentrated. The residue was purified by flash chromatography (silica gel, EtOAc/cyclohexane 3:2) to afford a beige solid; yield: 137 mg (74%).

for references see p 156

References

[1] Patel, A.; Barcan, G. A.; Kwon, O.; Houk, K. N., *J. Am. Chem. Soc.*, (2013) **135**, 4878.
[2] Barcan, G. A.; Patel, A.; Houk, K. N.; Kwon, O., *Org. Lett.*, (2012) **14**, 5388.
[3] Wang, F.; Tong, X.; Cheng, J.; Zhang, Z., *Chem.–Eur. J.*, (2004) **10**, 5338.
[4] de Meijere, A.; von Zezschwitz, P.; Bräse, S., *Acc. Chem. Res.*, (2005) **38**, 413.
[5] Tokan, W. M.; Meyer, F. E.; Schweizer, S.; Parsons, P. J.; de Meijere, A., *Eur. J. Org. Chem.*, (2008), 6152.
[6] Schweizer, S.; Tokan, W. M.; Parsons, P. J.; de Meijere, A., *Eur. J. Org. Chem.*, (2010), 4687.
[7] Blond, G.; Bour, C.; Salem, B.; Suffert, J., *Org. Lett.*, (2008) **10**, 1075.
[8] Toguem, S.-M. T.; Malik, I.; Hussain, M.; Iqbal, J.; Villinger, A.; Langer, P., *Tetrahedron*, (2013) **69**, 160.
[9] Hussain, M.; Hung, N. T.; Langer, P., *Tetrahedron Lett.*, (2009) **50**, 3929.
[10] Hussain, M.; Toguem, S.-M. T.; Ahmad, R.; Tung, T. D.; Knepper, I.; Villinger, A.; Langer, P., *Tetrahedron*, (2011) **67**, 5304.
[11] Hussain, M.; Zinad, D. S.; Salman, G. A.; Sharif, M.; Villinger, A.; Langer, P., *Synlett*, (2010), 276.
[12] Akrawi, O. A.; Khan, A.; Hussain, M.; Mohammed, H. H.; Villinger, A.; Langer, P., *Tetrahedron Lett.*, (2013) **54**, 3037.
[13] Salem, B.; Suffert, J., *Angew. Chem. Int. Ed.*, (2004) **21**, 2826.
[14] Bour, C.; Blond, G.; Salem, B.; Suffert, J., *Tetrahedron*, (2006) **62**, 10567.
[15] Kan, S. B. J.; Anderson, E. A., *Org. Lett.*, (2008) **10**, 2323.
[16] Webster, R.; Gaspar, B.; Mayer, P.; Trauner, D., *Org. Lett.*, (2013) **15**, 1866.
[17] Zhu, C.; Zhang, X.; Lian, X.; Ma, S., *Angew. Chem. Int. Ed.*, (2012) **51**, 7817.
[18] Parker, K. A.; Lim, Y.-H., *J. Am. Chem. Soc.*, (2004) **126**, 15968.
[19] Kim, K.; Lauher, J. W.; Parker, K. A., *Org. Lett.*, (2012) **14**, 138.
[20] Parker, K. A.; Lim, Y.-H., *Org. Lett.*, (2004) **6**, 161.
[21] Parker, K. A.; Wang, Z., *Org. Lett.*, (2006) **8**, 3553.
[22] Beaudry, C. M.; Trauner, D., *Org. Lett.*, (2005) **7**, 4475.
[23] Barbarow, J. E.; Miller, A. K.; Trauner, D., *Org. Lett.*, (2005) **7**, 2901.
[24] Miller, A. K.; Trauner, D., *Angew. Chem. Int. Ed.*, (2005) **44**, 4602.
[25] Sofiyev, V.; Navarro, G.; Trauner, D., *Org. Lett.*, (2008) **10**, 149.
[26] Sharma, P.; Ritson, D. J.; Burnley, J.; Moses, J. E., *Chem. Commun. (Cambridge)*, (2011) **47**, 10605.
[27] Charpenay, M.; Boudhar, A.; Blond, G.; Suffert, J., *Angew. Chem. Int. Ed.*, (2012) **51**, 4379.
[28] Charpenay, M.; Boudhar, A.; Hulot, C.; Blond, G.; Suffert, J., *Tetrahedron*, (2013) **69**, 7568.
[29] Boudhar, A.; Charpenay, M.; Blond, G.; Suffert, J., *Angew. Chem. Int. Ed.*, (2013) **52**, 12786.
[30] Lim, J. W.; Kim, K. H.; Kim, S. H.; Kim, J. N., *Tetrahedron Lett.*, (2012) **53**, 5449.
[31] Mohanakrishnan, A. K.; Dhayalan, V.; Clement, J. A.; Sureshbabu, R. B. R.; Kumar, N. S., *Tetrahedron Lett.*, (2008) **49**, 5850.
[32] Rubín, S. G.; Varela, J. A.; Castedo, L.; Saá, C., *Chem.–Eur. J.*, (2008) **14**, 9772.
[33] Varela, J. A.; Rubín, S. G.; González-Rodríguez, C.; Castedo, L.; Saá, C., *J. Am. Chem. Soc.*, (2006) **128**, 9262.
[34] Spiegel, D. A.; Schroeder, F. C.; Duvall, J. R.; Schreiber, S. L., *J. Am. Chem. Soc.*, (2006) **128**, 14766.
[35] Hayashi, R.; Fernández, S.; Okamura, W. H., *Org. Lett.*, (2002) **4**, 851.
[36] Nicolaou, K. C.; Montagnon, T.; Snyder, S. A., *J. Chem. Soc., Chem. Commun.*, (2003), 551.
[37] Nicolaou, K. C.; Petasis, N. A.; Uenishi, J.; Zipkin, R. E., *J. Am. Chem. Soc.*, (1982) **104**, 5557.
[38] Nicolaou, K. C.; Petasis, N. A.; Zipkin, R. E., *J. Am. Chem. Soc.*, (1982) **104**, 5560.
[39] Nicolaou, K. C.; Petasis, N. A.; Zipkin, R. E.; Uenishi, J., *J. Am. Chem. Soc.*, (1982) **104**, 5555.
[40] Nicolaou, K. C.; Zipkin, R. E.; Petasis, N. A., *J. Am. Chem. Soc.*, (1982) **104**, 5558.
[41] Hulot, C.; Blond, G.; Suffert, J., *J. Am. Chem. Soc.*, (2008) **130**, 5046.
[42] Hulot, C.; Amiri, S.; Blond, G.; Schreiner, P.; Suffert, J., *J. Am. Chem. Soc.*, (2009) **131**, 13387.
[43] Hulot, C.; Peluso, J.; Blond, G.; Muller, C. D.; Suffert, J., *Bioorg. Med. Chem. Lett.*, (2010) **20**, 6836.
[44] Benson, C. L.; West, F. G., *Org. Lett.*, (2007) **9**, 2545.
[45] Tessier, P. E.; Nguyen, N.; Clay, M. D.; Fallis, A. G., *Org. Lett.*, (2005) **7**, 767.
[46] Zhou, H.; Xing, Y.; Yao, J.; Chen, J., *Org. Lett.*, (2010) **12**, 3674.
[47] Zhou, H.; Xing, Y.; Yao, J.; Lu, Y., *J. Org. Chem.*, (2011) **76**, 4582.
[48] Spencer, W. T.; Frontier, A. J., *J. Org. Chem.*, (2012) **77**, 7730.
[49] Ma, S.; Gu, Z., *J. Am. Chem. Soc.*, (2006) **128**, 4942.
[50] Gu, Z.; Ma, S., *Chem.–Eur. J.*, (2008) **14**, 2453.

[51] Alajarin, M.; Bonillo, B.; Marin-Luna, M.; Sanchez-Andrada, P.; Vidal, A., *Chem.–Eur. J.*, (2013) **19**, 16093.
[52] Kerr, D. J.; Willis, A. C.; Flynn, B. L., *Org. Lett.*, (2004) **6**, 457.
[53] Škorić, I.; Pavošević, F.; Vazdar, M.; Marinić, Ž.; Šindler-Kulyk, M.; Eckert-Maksić, M.; Margetić, D., *Org. Biomol. Chem.*, (2011) **9**, 6771.
[54] Vuk, D.; Marinić, Ž.; Molčanov, K.; Margetić, D.; Škorić, I., *Tetrahedron*, (2014) **70**, 886.
[55] Lawrence, A. L.; Wegner, H. A.; Jacobsen, M. F.; Adlington, R. M.; Baldwin, J. E., *Tetrahedron Lett.*, (2006) **47**, 8717.
[56] Snider, B. B.; Harvey, T. C., *J. Org. Chem.*, (1994) **59**, 504.
[57] Lim, C. H.; Kim, S. H.; Kim, K. H.; Kim, J. N., *Tetrahedron Lett.*, (2013) **54**, 2476.
[58] Kim, K. H.; Kim, S. H.; Park, S.; Kim, J. N., *Tetrahedron*, (2011) **67**, 3328.
[59] Chang, M.-Y.; Wu, M.-H.; Chen, Y.-L., *Org. Lett.*, (2013) **15**, 2822.
[60] Chang, M.-Y.; Chan, C.-K.; Wu, M.-H., *Tetrahedron*, (2013) **69**, 7916.
[61] von Essen, R.; Frank, D.; Sünnemann, H. W.; Vidović, D.; Magull, J.; de Meijere, A., *Chem.–Eur. J.*, (2005) **11**, 6583.
[62] Skropeta, D.; Rickards, R. W., *Tetrahedron Lett.*, (2007) **48**, 3281.
[63] Huisgen, R.; Dahmen, A.; Huber, H., *J. Am. Chem. Soc.*, (1967) **89**, 7130.
[64] Rocha, D. H. A.; Pinto, D. C. G. A.; Silva, A. M. S.; Patonay, T.; Cavaleiro, J. A. S., *Synlett*, (2012) **23**, 559.
[65] Protti, S.; Palmieri, A.; Petrini, M.; Fagnoni, M.; Ballini, R.; Albini, A., *Adv. Synth. Catal.*, (2013) **355**, 643.
[66] Jarowski, P. D.; Wu, Y.-L.; Boudon, C.; Gisselbrecht, J.-P.; Gross, M.; Schweizer, W. B.; Diederich, F., *Org. Biomol. Chem.*, (2009) **7**, 1312.
[67] Breiten, B.; Wu, Y.-L.; Jarowski, P. D.; Gisselbrecht, J.-P.; Boudon, C.; Griesser, M.; Onitsch, C.; Gescheidt, G.; Schweizer, W. B.; Langer, N.; Lennartz, C.; Diederich, F., *Chem. Sci.*, (2011) **2**, 88.
[68] Kivala, M.; Boudon, C.; Gisselbrecht, J.-P.; Seiler, P.; Gross, M.; Diederich, F., *Angew. Chem. Int. Ed. Engl.*, (2007) **46**, 6357.
[69] Liu, S.; Liebeskind, L. S., *J. Am. Chem. Soc.*, (2008) **130**, 6918.
[70] Kobayashi, T.; Hatano, S.; Tsuchikawa, H.; Katsumura, S., *Tetrahedron Lett.*, (2008) **49**, 4349.
[71] Colby, D. A.; Bergman, R. G.; Ellman, J. A., *J. Am. Chem. Soc.*, (2008) **130**, 3645.
[72] Ischay, M. A.; Takase, M. K.; Bergman, R. G.; Ellman, J. A., *J. Am. Chem. Soc.*, (2013) **135**, 2478.
[73] Duttwyler, S.; Lu, C.; Rheingold, A. L.; Bergman, R. G.; Ellman, J. A., *J. Am. Chem. Soc.*, (2012) **134**, 4064.
[74] Parthasarathy, K.; Jeganmohan, M.; Cheng, C.-H., *Org. Lett.*, (2008) **10**, 325.
[75] Parthasarathy, K.; Cheng, C.-H., *J. Org. Chem.*, (2009) **74**, 9359.
[76] Martin, R. M.; Bergman, R. G.; Ellman, J. A., *J. Org. Chem.*, (2012) **77**, 2501.
[77] Yamakawa, T.; Yoshikai, N., *Org. Lett.*, (2013) **15**, 196.
[78] Cao, J.; Xu, Y.; Kong, Y.; Cui, Y.; Hu, Z.; Wang, G.; Deng, Y.; Lai, G., *Org. Lett.*, (2012) **14**, 38.
[79] Cheng, G.; Cui, X., *Org. Lett.*, (2013) **15**, 1480.
[80] Babinski, D. J.; Aguilar, H. R.; Still, R.; Frantz, D. E., *J. Org. Chem.*, (2011) **76**, 5915.
[81] Nakamura, I.; Zhang, D.; Terada, M., *J. Am. Chem. Soc.*, (2010) **132**, 7884.
[82] Nakamura, I.; Araki, T.; Zhang, D.; Kudo, Y.; Kwon, E.; Terada, M., *Org. Lett.*, (2011) **13**, 3616.
[83] Trost, B. M.; Gutierrez, A. C., *Org. Lett.*, (2007) **9**, 1473.
[84] Singha, R.; Dhara, S.; Ray, J. K., *Tetrahedron Lett.*, (2013) **54**, 4841.
[85] Chaumontet, M.; Piccardi, R.; Baudoin, O., *Angew. Chem. Int. Ed.*, (2009) **48**, 179.
[86] Kobayashi, K.; Fujita, S.; Fukamachi, S.; Konishi, H., *Synthesis*, (2009), 3378.
[87] Das, A.; Volla, C. M. R.; Atodiresei, I.; Bettray, W.; Rueping, M., *Angew. Chem. Int. Ed.*, (2013) **52**, 8008.
[88] Funicello, M.; Laboragine, V.; Pandolfo, R.; Spagnolo, P., *Synlett*, (2010), 77.
[89] Bonini, C.; Funicello, M.; Scialpi, R.; Spagnolo, P., *Tetrahedron*, (2003) **59**, 7515.
[90] Jayakumar, S.; Kumar, V.; Mahajan, M. P., *Tetrahedron Lett.*, (2001) **42**, 2235.
[91] Kakiya, H.; Yagi, K.; Shinokubo, H.; Oshima, K., *J. Am. Chem. Soc.*, (2002) **124**, 9032.
[92] Chun, Y. S.; Lee, J. H.; Kim, J. H.; Ko, Y. O.; Lee, S.-g., *Org. Lett.*, (2011) **13**, 6390.
[93] Kim, J. H.; Chun, Y. S.; Shin, H.; Lee, S.-g., *Synthesis*, (2012) **44**, 1809.
[94] Matsuya, Y.; Ohsawa, N.; Nemoto, H., *J. Am. Chem. Soc.*, (2006) **128**, 13072.
[95] Alajarin, M.; Ortin, M.-M.; Sanchez-Andrada, P.; Vidal, A.; Bautista, D., *Org. Lett.*, (2005) **7**, 5281.
[96] Alajarin, M.; Ortin, M.-M.; Sanchez-Andrada, P.; Vidal, A., *J. Org. Chem.*, (2006) **71**, 8126.
[97] Alajarin, M.; Bonillo, B.; Ortin, M.-M.; Sanchez-Andrada, P.; Vidal, A., *Org. Lett.*, (2006) **8**, 5645.

[98] Xin, X.; Wang, D.; Wu, F.; Li, X.; Wan, B., *J. Org. Chem.*, (2013) **78**, 4065.
[99] Zhou, H.; Liu, L.; Xu, S., *J. Org. Chem.*, (2012) **77**, 9418.
[100] Zhu, M.; Wang, Z.; Xu, F.; Yu, J.; Fu, W., *J. Fluorine Chem.*, (2013) **156**, 21.
[101] Cheng, X.; Waters, S. P., *Org. Lett.*, (2013) **15**, 4226.
[102] Cheung, L. L. W.; Yudin, A. K., *Org. Lett.*, (2009) **11**, 1281.
[103] Peng, W.; Hirabaru, T.; Kawafuchi, H.; Inokuchi, T., *Eur. J. Org. Chem.*, (2011), 5469.
[104] Hsung, R. P.; Kurdyumov, A. V.; Sydorenko, N., *Eur. J. Org. Chem.*, (2005), 23; and references cited therein.
[105] da Rocha, D. R.; Mota, K.; da Silva, I. M. C. B.; Ferreira, V. F.; Ferreira, S. B.; da Silva, F. de C., *Tetrahedron*, (2014) **70**, 3266.
[106] Edayadulla, N.; Lee, Y. R., *Bull. Korean Chem. Soc.*, (2013) **34**, 2963.
[107] Jung, E. J.; Park, B. H.; Lee, Y. R., *Green Chem.*, (2010) **12**, 2003.
[108] Peña, J.; Moro, R. F.; Basabe, P.; Marcos, I. S.; Díez, D., *RSC Adv.*, (2012) **2**, 8041.
[109] Fotiadou, A. D.; Zografos, A. L., *Org. Lett.*, (2012) **14**, 5664.
[110] Kleinke, A. S.; Li, C.; Rabasso, N.; Porco, J. A., Jr., *Org. Lett.*, (2006) **8**, 2847.
[111] Li, C.; Johnson, R. P.; Porco, J. A., Jr., *J. Am. Chem. Soc.*, (2003) **125**, 5095.
[112] Shoji, M.; Imai, H.; Mukaida, M.; Sakai, K.; Kakeya, H.; Osada, H.; Hayashi, Y., *J. Org. Chem.*, (2005) **70**, 79.
[113] Pandit, R. P.; Lee, Y. R., *Synthesis*, (2012) **44**, 2910.
[114] Kumar, V. P.; Gruner, K. K.; Kataeva, O.; Knölker, H.-J., *Angew. Chem. Int. Ed.*, (2013) **52**, 11 073.
[115] Hesse, R.; Gruner, K. K.; Kataeva, O.; Schmidt, A. W.; Knölker, H.-J., *Chem.–Eur. J.*, (2013) **19**, 14 098.
[116] Zhang, T.; Huang, X.; Wu, L., *Eur. J. Org. Chem.*, (2012), 3507.
[117] Riu, A.; Harrison-Marchand, A.; Maddaluno, J.; Gulea, M.; Albadri, H.; Masson, S., *Eur. J. Org. Chem.*, (2007), 4948.

2.1.4 Sigmatropic Shifts and Ene Reactions (Excluding [3,3])

A. V. Novikov and A. Zakarian

General Introduction

Pericyclic reactions are a powerful tool for organic synthesis. While processes such as the Diels–Alder cycloaddition[1] and the Claisen rearrangement[2,3] have enjoyed the most dramatic success, spectacular examples of the ene reaction[4,5] and [2,3]-sigmatropic shifts, such as the Wittig rearrangement,[6,7] have also been documented.

Pericyclic reactions naturally lend themselves to domino sequences, because they tend to occur under the same set of conditions and display a general tolerance to diverse functionalization. As a matter of fact, upon simple thermal activation, appropriately structured unsaturated compounds readily undergo cascades of pericyclic transformations.[8] However, in many cases, these processes require exceedingly high temperatures, resulting in multiple products formed by competing pathways. To achieve a synthetically useful transformation using a domino series of pericyclic reactions, the substrate usually has to be carefully designed and its structure finely tuned.

In this chapter, we will concentrate our discussion on two types of reactions that have commonly been used to initiate domino sequences: (1) ene reactions and (2) [2,3]-sigmatropic rearrangements (Scheme 1), although examples of domino sequences initiated by other sigmatropic rearrangements will also be demonstrated. The reactions that are formally equivalent to pericyclic processes by outcome, but in fact proceed via a completely different type of mechanism (e.g., transition-metal catalysis) will not be discussed.

Scheme 1 Ene Reaction and [2,3]-Sigmatropic Rearrangement

ene reaction

[2,3]-sigmatropic rearrangement

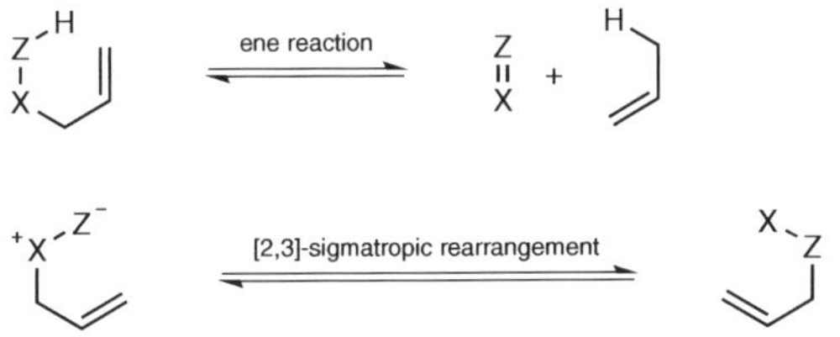

2.1.4.1 Practical Considerations

In most cases, domino pericyclic processes are performed by heating the substrate in an appropriate inert solvent. The choice of solvent is usually of low importance, as solvents typically have limited influence on pericyclic reactions, with the most important variable being its boiling point. Additives are also used occasionally. For instance, radical scavengers, such as 2,6-di-*tert*-butyl-4-methylphenol, are routinely employed to suppress competing radical processes, especially at higher temperatures. Frequently, the reactions are performed in sealed reaction vessels, in some cases under high pressure. The use of microwave heating is effective in many instances. Catalysis is also a possibility in isolated examples. In specific cases, other factors may play a role, such as the nature of the counterion

for references see p 192

in the case of rearrangements of anionic or cationic substrates, whereas in others, the generation of reactive precursors for some cascade transformations may require special reaction conditions.

Perhaps the most important factor for the success of a particular transformation is the structure of the substrate. The appropriate substrate can be selected depending on the reactions involved, using known structural predispositions and influences as associated with a specific reaction. Computational analysis is often a helpful tool for understanding the subtle effects governing the process. To achieve a successful tandem transformation, each step of the domino sequence is typically optimized independently, before combining them into a single-pot transformation.

Many of the above issues will be highlighted in the discussion of individual domino transformations in the examples that follow.

2.1.4.2 Domino Processes Initiated by Ene Reactions

The ene reaction has proven to be an effective transformation for chemical synthesis.[4] Several well-established synthetic methods involve an ene or a retro-ene reaction, including allylic oxidations with singlet oxygen (Schenck ene reaction),[9] allylic oxidations with selenium(IV) oxide,[10] reduction of the tosylhydrazones of α,β-unsaturated ketones,[11] and such classic reactions as the malonic ester syntheses in the decarboxylation step following functionalization of the 1,3-dicarbonyl group.

To effect a typical intermolecular ene reaction, the choice of an appropriate alkene substrate and/or the use of a highly reactive enophile are necessary. In some intramolecular processes, however, the reaction is quite facile and selective due to geometrical constraints within the substrate.[12] These factors play an even greater role in achieving effective domino processes involving ene reactions; conformational predisposition of the substrate toward the reaction has often proven key.

A number of effective domino processes originating with an ene reaction have been reported. In the examples in Scheme 2, the ene reaction precedes a Diels–Alder cyclization. The initially described transformation was later optimized via microwave heating.[13] The use of nonconjugated cyclic 1,4-dienes **1** is essential. It both facilitates the ene reaction, which is thermodynamically favorable as the double bond rearranges into conjugation, and sets up the 1,3-diene for the ensuing Diels–Alder cycloaddition. A large excess of the ene component is necessary for success. With equimolar amounts of the reagents, the product of the intermolecular Diels–Alder reaction between the intermediate 1,3-diene **3** and the acetylenic dienophile **2** predominates. Substitution in the substrates can be varied to a certain extent, allowing the use of methyl propynoate (**2**, $R^3 = CO_2Me$; $R^4 = H$), propynoic acid (**2**, $R^3 = CO_2H$; $R^4 = H$), but-3-yn-2-one (**2**, $R^3 = Ac$; $R^4 = H$), and dimethyl acetylenedicarboxylate (**2**, $R^3 = R^4 = CO_2Me$) as the enophile, as well as several substituted cyclohexadienes **1** in addition to the parent cyclohexa-1,3-diene (**1**, $R^1 = R^2 = H$) reactant. Notably, complete regioselectivity is observed when nonsymmetric alkynes are used. This outcome is thought to be the result of dynamic kinetic resolution, with the observed isomer **4** forming by a faster [4+2] cycloaddition.

Scheme 2 Ene/Diels–Alder Cascade[13]

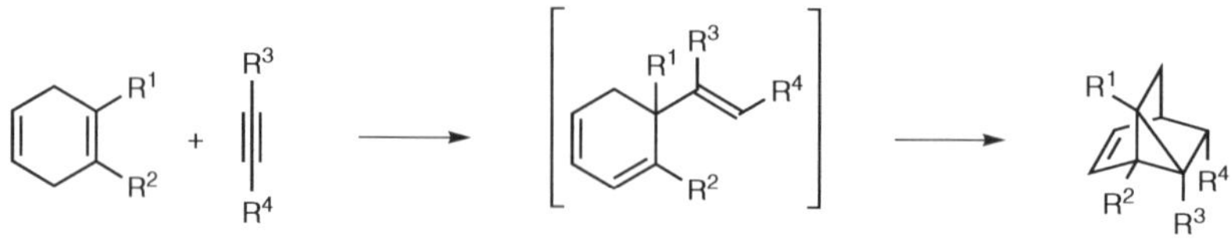

1 2 3 4

R^1	R^2	R^3	R^4	Ratio (1/2)	Conditions	Yield (%)	Ref
H	H	CO_2Me	CO_2Me	10:1	325–361 °C, microwave, 6 min	82	[13]
H	H	CO_2Me	CO_2Me	30:1	317–325 °C, microwave, 6 min	87	[13]
H	H	CO_2Me	H	10:1	$ZnCl_2$, 240 °C, 4 h	77	[13]
H	H	Ac	H	10:1	$ZnCl_2$, 220–230 °C, 18 h	68	[13]
H	H	CO_2H	H	10:1	neat, 300 °C, 21 h	26	[13]
$(CH_2)_3$		CO_2Me	CO_2Me	15:1	317–325 °C, microwave, 6 min	80	[13]
$CH_2CH{=}CHCH_2$		CO_2Me	CO_2Me	15:1	300–317 °C, microwave, 6 min	65	[13]
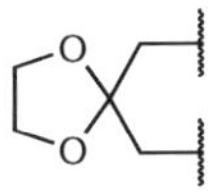		CO_2Me	CO_2Me	15:1	275–300 °C, microwave, 6 min	37	[13]

The seemingly unrelated transformation of diynes of type **5** also involves the ene reaction/Diels–Alder cascade (Scheme 3).[14] The construction of the diene for the Diels–Alder process is achieved via an intramolecular ene reaction of one alkyne onto another. The resulting 1,2,4-triene **6**, containing an allene fragment, is a highly reactive Diels–Alder substrate, readily cyclizing onto alkyne dienophiles to give the products **7**. Alternative radical mechanisms were ruled out based on experiments with added radical initiators or inhibitors and an EPR study. Critically, cyclohexa-1,4-diene additives, initially used to test for radical intermediates in the reaction, turned out to have a beneficial effect on the yield of the product, evidently by suppressing competing undesired radical processes. Using triazatriyne and triazaenediyne macrocycles **8** and **10**, this intramolecular ene reaction/Diels–Alder cascade gives the corresponding tetracyclic products **9** and **11**, respectively. Domino reactions of this type using different external dienophiles **13** and diynes **12** provide products **14** with good to excellent yields and good stereoselectivity (Scheme 3).[15]

Scheme 3 Propargyl Ene/Diels–Alder Cascade[14,15]

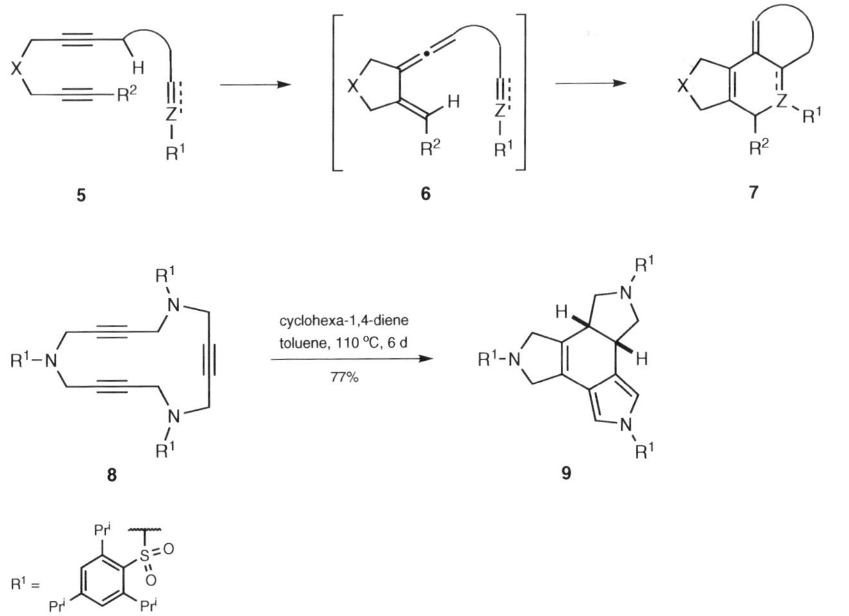

5 → **6** → **7**

8 → (cyclohexa-1,4-diene, toluene, 110 °C, 6 d; 77%) → **9**

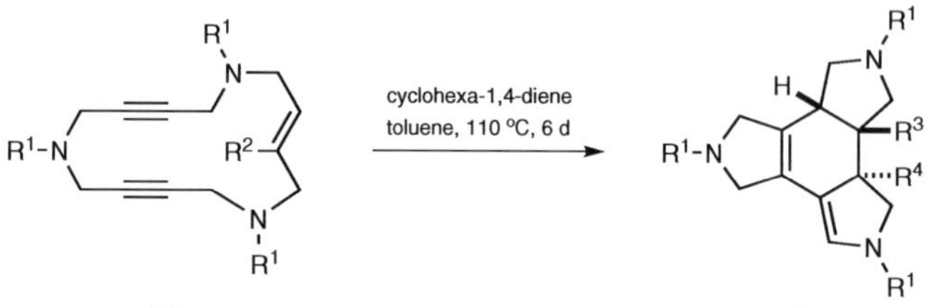

cyclohexa-1,4-diene
toluene, 110 °C, 6 d

10 → **11**

R^1	R^2	R^3	R^4	Yield (%)	Ref
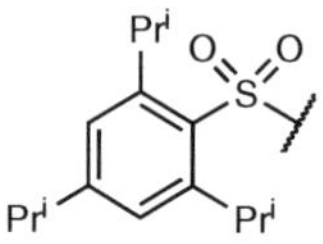 2,4,6-(Pr^i)$_3C_6H_2SO_2$	H	H	H	78	[14]
Ts	Ph	Ph	H	81[a]	[14]

[a] Obtained as a mixture of **11** (R^3 = Ph; R^4 = H) and **11** (R^3 = H; R^4 = Ph).

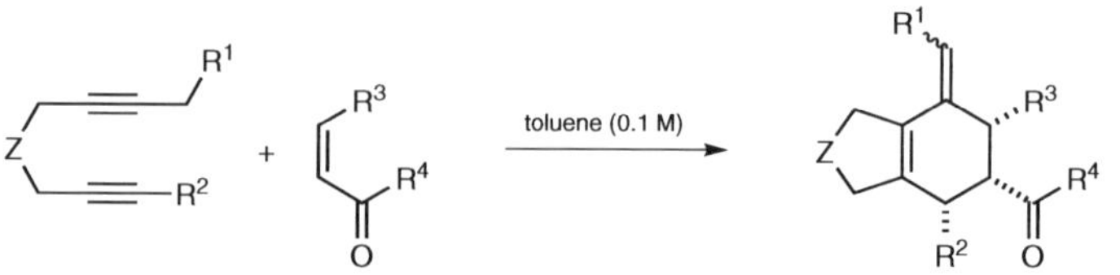

12 + **13** → toluene (0.1 M) → **14**

R^1	R^2	R^3	R^4	Z	Conditions	Ratio (Z/E)	Yield (%)	Ref
Pr	H	CON(Me)		O	160 °C, 21 h	91:9	94	[15]
H	H	CON(Me)		O	160 °C, 21 h	–	52	[15]
Pr	H	CON(Me)		NTs	160 °C, 11 h	74:26	75	[15]
Pr	H	CON(Me)		$C(SO_2Ph)_2$	150 °C, 8 h	69:31	73	[15]
Pr	H	H	Me	O	160 °C, 21 h	84:16	72[a]	[15]
Pr	CO_2Me	CON(Me)		O	160 °C, 1 h	93:7	93	[15]
Pr	C≡CTIPS	H	Me	O	110 °C, 21 h	90:10	68	[15]

[a] Diels–Alder regioisomer formed in 11% yield.

In the case of alkynyl dienophiles **16** and oxadiynes **15**, the products of the cascade can be aromatized via a base-catalyzed rearrangement, yielding benzene derivatives **17** (Scheme 4).[15] The formal overall result of the transformation is a [2+2+2] cycloaddition.

Scheme 4 Propargyl Ene/Diels–Alder Cascade with Alkyne Dienophiles[15]

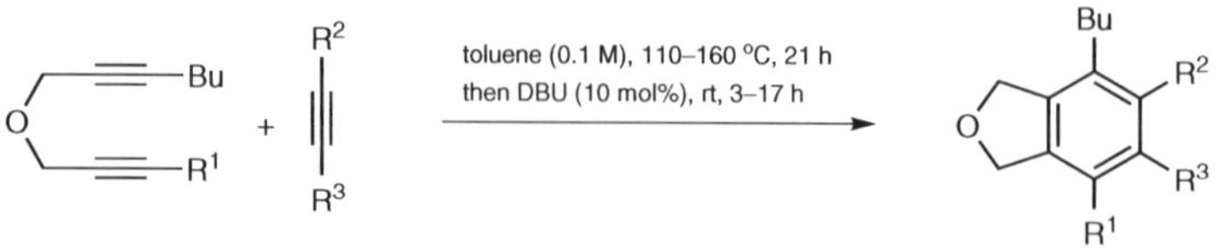

toluene (0.1 M), 110–160 °C, 21 h
then DBU (10 mol%), rt, 3–17 h

15 + **16** → **17**

R^1	R^2	R^3	Temp (°C)	Yield (%)	Ref
H	CO_2Me	CO_2Me	160	75	[15]
H	H	Ac	160	40	[15]
C≡CTIPS	CO_2Me	CO_2Me	110	81	[15]

This sequence is also accomplished with the cyano group as one of the triple bond components in diyne substrates **18** and **20**, producing substituted pyridines **19** and **21**, respectively (Scheme 5).[16] Variation in the substitution can be carried out at different positions. Other types of reactions following the initial ene reaction, such as condensation and hydrogen shift, have also been described, although with only one substrate in each case.

Scheme 5 Propargyl Ene/Diels–Alder Cascade with Nitriles[16]

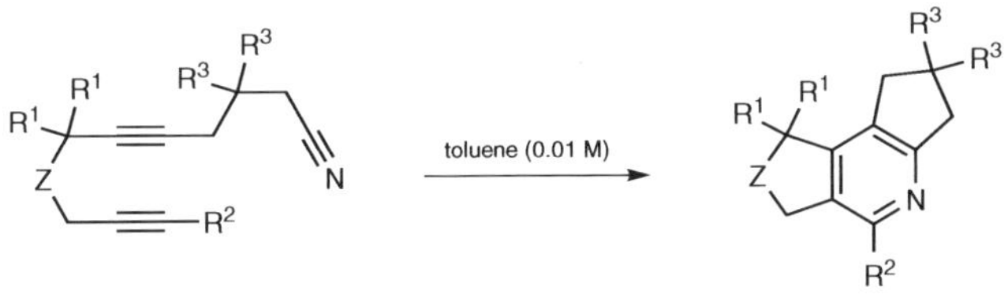

toluene (0.01 M)

18 → **19**

R^1	R^2	R^3	Z	Conditions	Yield (%)	Ref
H	H	H	O	160 °C, 21 h	71	[16]
Me	H	H	O	reflux, 66 h	96	[16]
H	C≡CTMS	H	O	reflux, 24 h	30	[16]
H	C≡CTIPS	Et	CH_2	200 °C, 16 h	51	[16]
H	C≡CTIPS	CO_2Et	CH_2	200 °C, 19 h	46	[16]
H	H	CO_2Et	CH_2	200 °C, 48 h	0	[16]

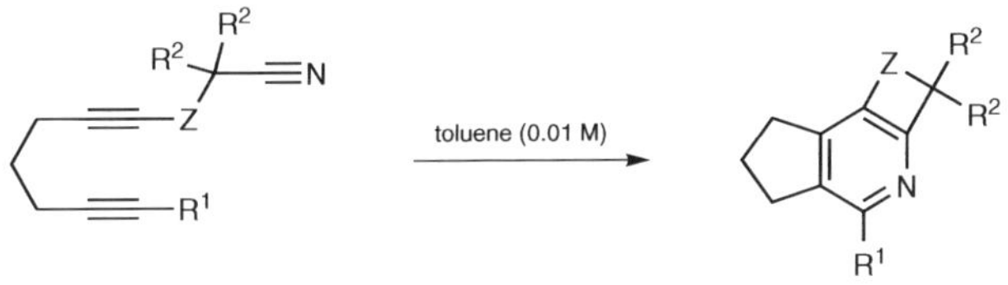

toluene (0.01 M)

20 → **21**

R^1	R^2	Z	Conditions	Yield (%)	Ref
C≡CTIPS	Me	CH=N	150 °C, 4 h	86	[16]
C≡CTIPS	Me	CON(Me)	115 °C, 41 h	74	[16]
C≡CTIPS	H	CON(Me)	160 °C, 36 h	58	[16]
C≡CTIPS	Me	C(O)O	150 °C, 20 h	62	[16]
C≡CTIPS	H	C(O)O	210 °C, 6 h	35	[16]
C≡CTIPS	Me	$C(O)CH_2$	170 °C, 20 h	0[a]	[16]

[a] A different product (resulting from ynone hetero-Diels–Alder addition) was obtained in 91% yield.

for references see p 192

Curiously, while several examples of this transformation were also reported earlier,[17–19] different mechanisms were proposed for the events, featuring either biradicals or a [2+2] cycloaddition. The isolation of byproducts supported those mechanisms.[19] The transformations in question were later reexamined to reveal that the structures of these isolated byproducts had been misassigned, and they were, in fact, consistent with the propargyl ene/Diels–Alder mechanistic pathway.[15]

Another interesting example of an ene reaction/Diels–Alder cascade features a hetero-Diels–Alder cycloaddition succeeding a Schenck ene reaction (Scheme 6).[20] In this process, the sequence begins with an ene reaction between diene **22** and singlet oxygen, with the substrate already containing the preinstalled diene. However, its diminished reactivity due to increased substitution suppresses the immediate [4+2] reaction with singlet oxygen. Instead, an ene reaction takes place and, as expected, it occurs at the more electron-rich site of the diene. The resultant less-hindered diene **23** readily reacts with another equivalent of singlet oxygen to give the final [4+2] cycloadduct **24**.

Scheme 6 Ene/Hetero-Diels–Alder Cascade[20]

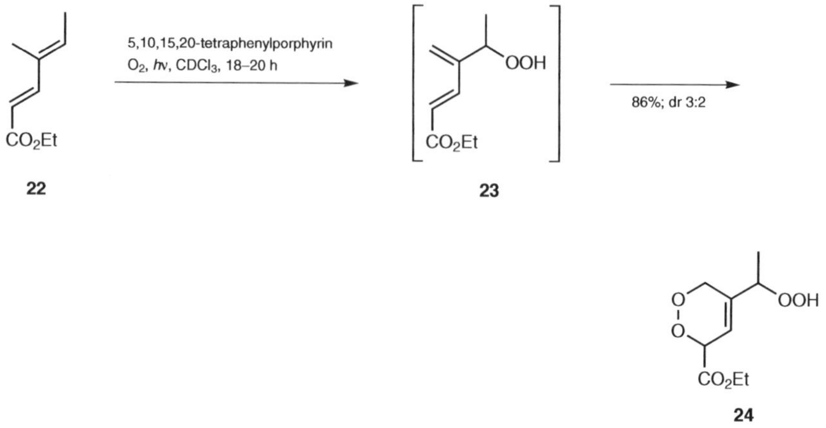

A domino sequence of an ene reaction followed by another ene reaction is the basis of an annulation method that relies on the use of a reactive enophile **26** along with a good ene substrate **25** (Scheme 7).[21–23] The use of this combination permits the first intermolecular ene reaction to proceed effectively. The subsequent ene reaction of intermediate **27** is facilitated by virtue of being intramolecular, leading to the formation of a six-membered ring **28**. This annulation reaction displays a certain degree of generality, tolerating substitution on the enophile alkene and the cyclohexane ring. As can be expected, the rates of the reaction are very sensitive to the substrate structure.

Alkylidenecyclopentane substrates **29** (n = 1) also effectively participate in this type of annulation with enophiles **30**. The preparation of bicyclic compounds **31** substituted at the bridgehead position is also feasible (Scheme 7).[22,23]

This domino reaction is also observed with nonsymmetrically substituted alkylidenecyclohexanes and -cyclopentanes, although, as might be expected, mixtures of regioisomers are formed, typically with low selectivity.[23]

Scheme 7 Ene/Ene Cascade Annulation[22,23]

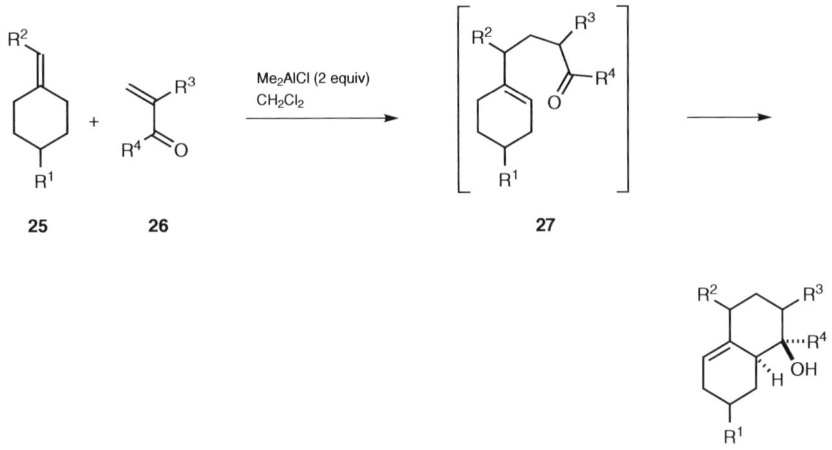

25 + 26 → (Me_2AlCl (2 equiv), CH_2Cl_2) → [27] → 28

R^1	R^2	R^3	R^4	Conditions	dr	Yield (%)	Ref
H	H	H	H	0 °C, 20 min	–	63	[22]
H	H	H	Me	−20 °C, 2 h	–	4 (39)[a]	[22]
H	H	H	Me	25 °C, 1 h	–	49 (9)[a]	[22]
H	H	Me	H	0–25 °C, 30 min	49:17	66	[22]
H	H	Br	H	−78 °C, 30 min	55:4	59	[22]
Me	H	H	H	−40 °C, 4 h	2:3	49	[22]
t-Bu	H	H	H	−40 °C, 4 h	33:38	71	[22]
H	Me	H	H	0 °C, 1.5 h	–	39	[22]
H	Me	H	Me	25 °C, 45 min	61:4	65	[22]
H	Me	Me	H	−30 °C, 4.5 h	–	35	[22]
H	Me	Br	H	−78 °C, 30 min	42:3:3	48	[22]

[a] Yield of noncyclized product of a single ene reaction.

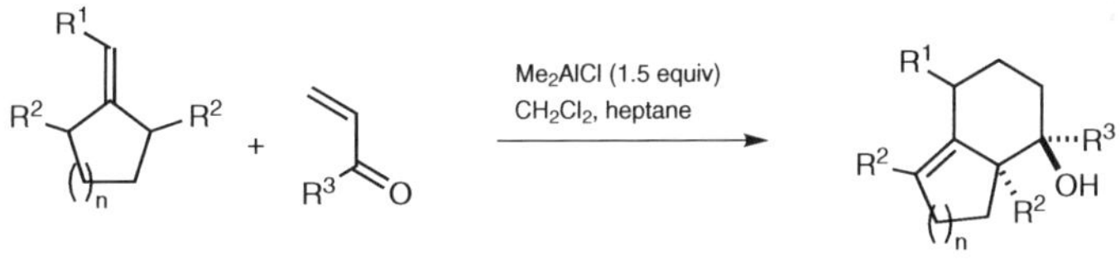

29 + 30 → (Me_2AlCl (1.5 equiv), CH_2Cl_2, heptane) → 31

R^1	R^2	R^3	n	Conditions	Yield (%)	Ref
H	H	H	1	0 °C, 20 min	38	[22]
Me	H	H	1	0 °C, 2 h	72	[22]
Me	H	Me	1	25 °C, 30 min	51	[22]
H	Me	H	1	0 °C, 45 min	69	[23]
H	Me	H	2	0 °C, 45 min	51	[23]
H	Me	H	2	0 °C, 45 min	57	[23]

A retro-ene decomposition of propargyl diimine **33** initiates a domino reaction sequence (Scheme 8).[24] The propargyl diimine **33** is generated by the Mitsunobu reaction of diol **32**. Once formed, it undergoes a retro-ene reaction, releasing molecular nitrogen and forming bis(allene) **34**, which undergoes subsequent 6π-electrocyclization to give intermediate **35** followed by [4+2] cycloaddition to produce 1,2,3,4,4a,9,10,10a-octahydrophenanthrene **36**. This cascade pericyclic reaction sequence has been employed in the synthesis of racemic estrone.

Scheme 8 Retro-Ene/Electrocyclization/Diels–Alder Cascade[24]

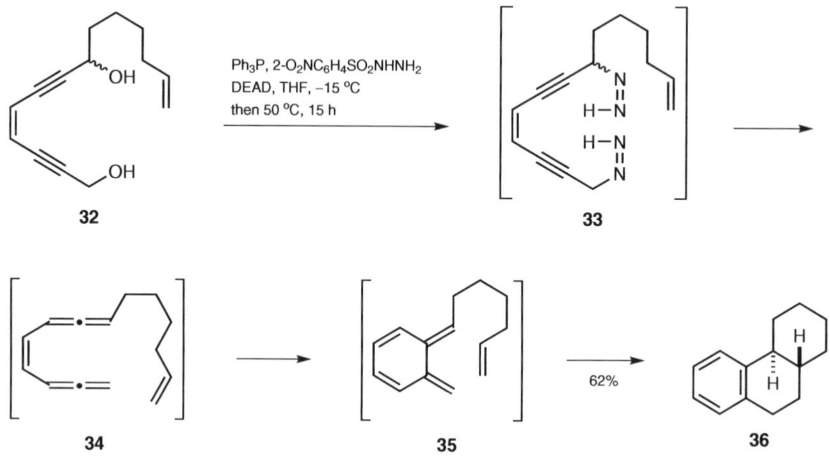

An intriguing example of a pericyclic domino reaction sequence is initiated by a nitroso-ene reaction (Scheme 9).[25] In this case, the initial ene reaction between an arylnitroso compound and a 3-arylprop-1-ene **37** results in the formation of allylic hydroxylamine **38**. The hydroxylamine is oxidized in situ to nitrone **39**, which undergoes a [3+2] cycloaddition with a maleimide **40** to produce bicyclic isoxazolidines **41**. Again, the initial ene reaction is facilitated by a highly reactive enophile along with a good ene substrate. Notably, the reaction produces much lower yields of the products when performed stepwise (i.e., first generating the nitrone and then subjecting it to [3+2] cycloaddition), evidently due to the poor stability of nitrones under the reaction conditions and during isolation.

Scheme 9 Nitroso Ene/Oxidation/[3+2] Cycloaddition Cascade[25]

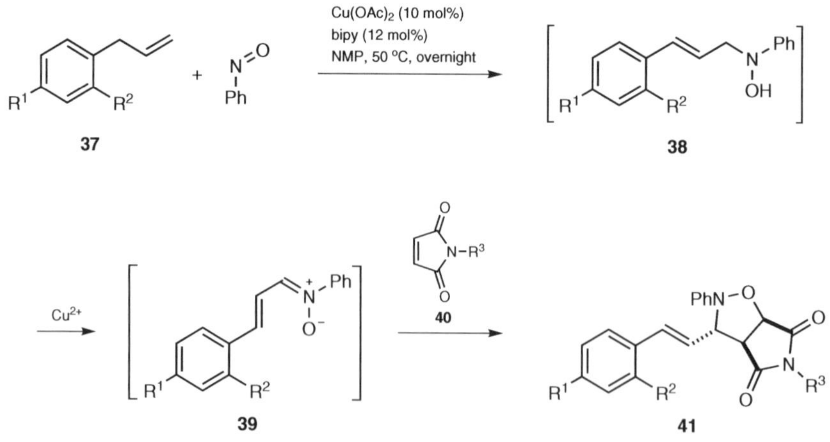

R^1	R^2	R^3	Ratio (*E*/*Z*)	Yield (%)	Ref
H	H	Me	7:1	61	[25]
F	H	Me	9:1	51	[25]
OMe	H	Ph	>20:1	54	[25]
CF_3	H	Ph	15:1	57	[25]
F	H	Ph	10:1	56	[25]
H	Me	Ph	10:1	63	[25]
CF_3	H	Bn	37:1	61	[25]
F	H	Bn	15:1	53	[25]
OMe	H	Bn	9:1	48	[25]
H	Me	Bn	9:1	54	[25]
H	H	Bn	17:1	55	[25]
OTs	H	Me	15:1	53	[25]
CN	H	Me	>20:1	46	[25]

(4a*R,7a*R**,*Z*)-4-Butylidene-6-methyl-1,3,4,4a,7a,8-hexahydro-5*H*-furo[3,4-*f*]isoindole-5,7(6*H*)-dione** [14, **R^1 = Pr; R^2 = H; R^3,R^4 = CON(Me); Z = O]; Typical Procedure:**[15]

A 10-cm threaded Pyrex tube (18 mm outer diameter; 14 mm inner diameter) equipped with a rubber septum fitted with an argon inlet needle was charged with a soln of 1-(prop-2-ynyloxy)hept-2-yne (**12**, R^1 = Pr; R^2 = H; Z = O; 100 mg, 0.67 mmol, 1.0 equiv) and *N*-methylmaleimide [**13**, R^3,R^4 = CON(Me); 74 mg, 0.67 mmol, 1.0 equiv] in toluene (6.7 mL). The soln was degassed (three freeze–pump–thaw cycles at −196 °C, <0.5 Torr) and then the tube was sealed with a threaded Teflon cap. The mixture was heated at 160 °C for 21 h, allowed to cool to rt, and then concentrated to afford a pale yellow oil. Column chromatography [silica gel (3 g), EtOAc/hexanes 3:7] afforded the product as a white solid; yield: 164 mg (94%); ratio (*Z*/*E*) 91:9 as determined by ^{1}H NMR analysis; mp 107–113 °C.

(3*S,3a*S**,6a*R**)-2-Phenyl-3-[(*E*)-2-phenylvinyl]tetrahydro-4*H*-pyrrolo[3,4-*d*]isoxazole-4,6(5*H*)-diones** 41; **General Procedure:**[25]

$Cu(OAc)_2$ (9.1 mg, 0.05 mmol, 0.1 equiv) and bipy (9.4 mg, 0.06 mmol, 0.12 equiv) were weighed in a test tube and NMP (3 mL) was added. The mixture was stirred under air until the copper salt completely dissolved (ca. 1 h). The allylbenzene **37** (0.75 mmol, 1.5 equiv), PhNO (2 mmol, 4.0 equiv), and maleimide **40** (0.5 mmol, 1.0 equiv) were sequentially added. The test tube was capped with a rubber septum, degassed, flushed with O_2 (3 ×), and stirred at 50 °C overnight under an O_2 balloon. The mixture was cooled to rt and diluted with H_2O (10 mL) and EtOAc (10 mL). The organic layer was separated, and the aqueous phase was extracted with EtOAc (2 ×). The combined extract was washed with H_2O (10 mL) and brine (10 mL), dried (Na_2SO_4), filtered, and concentrated to dryness. The residue was purified by flash column chromatography (silica gel, EtOAc/petroleum ether) to afford an analytically pure product.

for references see p 192

2.1.4.3 Domino Processes Initiated by [2,3]-Sigmatropic Rearrangements

[2,3]-Sigmatropic rearrangements are a well-known set of transformations that provide the basis for a number of powerful synthetic methods, such as the interconversion of allylic sulfoxides and allylic sulfenates (the Mislow–Evans rearrangement),[26] [2,3]-Wittig rearrangements,[6,7] allylic ylide rearrangements triggered by the addition of carbenoids,[27] and several other processes. Due to polarization of the key bonds, [2,3]-sigmatropic rearrangements tend to proceed under milder conditions than many other sigmatropic rearrangements, sometimes taking place at very low temperatures.[28]

One of the early domino processes initiated by a [2,3]-sigmatropic rearrangement was the Wittig rearrangement of diallylic ethers (e.g., **42**). For instance, the products (e.g., **44**) of a [2,3]-sigmatropic rearrangement of anions of diallyl ethers (e.g., **43**) easily undergo an oxy-Cope rearrangement under appropriate conditions to give δ,ε-unsaturated aldehydes (e.g., **45**) (Scheme 10). This cascade has found several synthetic applications.[29–31]

Scheme 10 Wittig Rearrangement/Oxy-Cope Cascade[30]

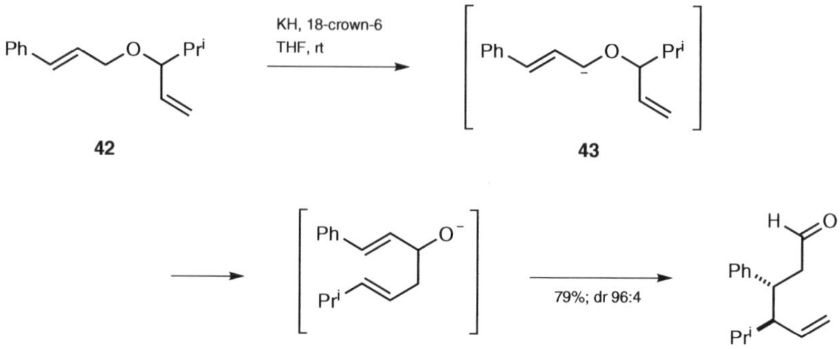

42 → (KH, 18-crown-6, THF, rt) → [43] → [44] → (79%; dr 96:4) → 45

A general and practical protocol for this transformation uses enolates of allyloxy esters. α,β-Unsaturated allyloxy esters **46** are easily prepared from the corresponding allylic alcohols, ethyl bromoacetate, and an aldehyde. They are treated with lithium diisopropylamide to generate enolate **47**, which then undergoes a [2,3]-sigmatropic rearrangement (Wittig rearrangement), producing a transient substrate for the oxy-Cope rearrangement. With structures of type **48**, having a *gem*-dimethyl substitution at the C3 position of the allylic group, the oxy-Cope rearrangement takes place at room temperature to give intermediates **49**. The α-oxo esters **50**, obtained in crude form after acidic workup, are directly subjected to an ene reaction by heating in decane to produce cyclic products **51** (Scheme 11).[32] The ene reaction under these conditions does not proceed to completion, leaving a certain amount of the α-oxo ester **50** due to the thermodynamically controlled equilibrium, according to the authors' hypothesis for the observed outcome.

In the case of other substitution patterns on the allylic fragment in esters **52** and **56**, the acidic workup is performed after the [2,3]-sigmatropic rearrangement to give α-hydroxy esters **53** and **57**, respectively. The oxy-Cope rearrangement and the ene reaction take place subsequently during the thermal activation in a separate step to give cyclopentanols **55** and cyclohexanols **59** via the intermediate α-oxo esters **54** and **58**, respectively (Scheme 11).[32]

Scheme 11 Enolate [2,3]-/[3,3]-Sigmatropic Rearrangement/Ene Reaction Cascade[32]

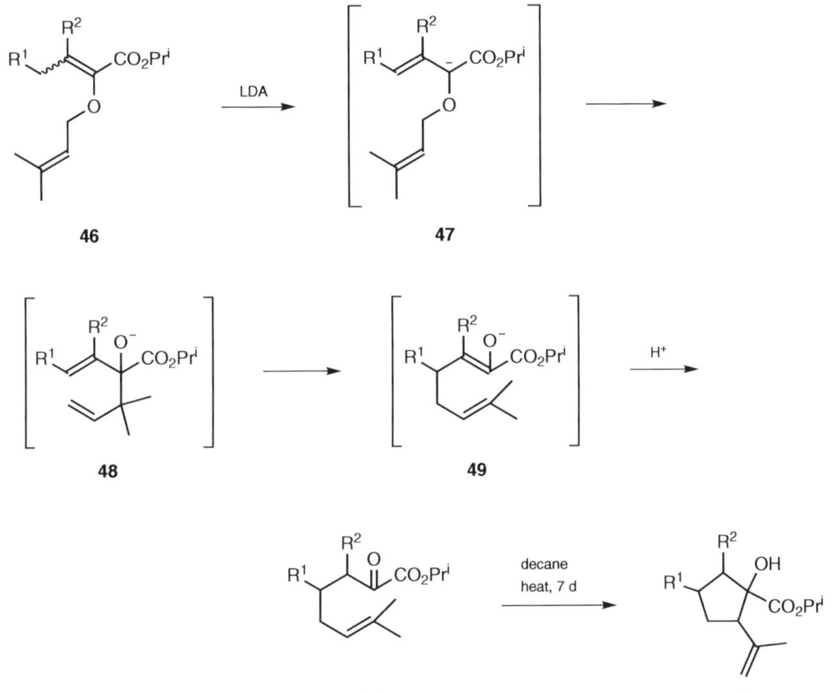

R^1	R^2	dr of **51**	Yield (%)		Ref
			50	**51**	
H	H	91:9	22	47	[32]
H	Me	74:13	7	56	[32]
Ph	H	53:35	12	71	[32]

for references see p 192

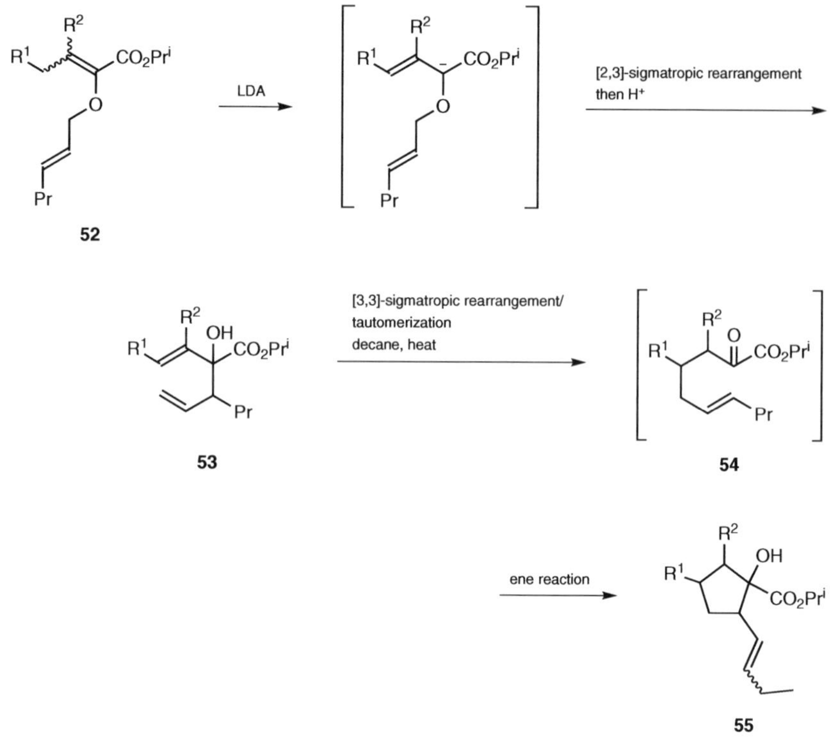

R^2
R^1 CO_2Pr^i
O
Pr
52

LDA

R^2
R^1 CO_2Pr^i
O
Pr

[2,3]-sigmatropic rearrangement
then H^+

R^2
OH
R^1 CO_2Pr^i
Pr
53

[3,3]-sigmatropic rearrangement/
tautomerization
decane, heat

R^2
O
R^1 CO_2Pr^i
Pr
54

ene reaction

R^2
OH
R^1 CO_2Pr^i
55

R^1	R^2	Yield (%)	Ref
H	H	74[a]	[32]
H	Me	80[b]	[32]
Ph	H	87	[32]

[a] α-Oxo ester **54** was isolated in 9% yield.
[b] α-Oxo ester **54** was isolated in 10% yield.

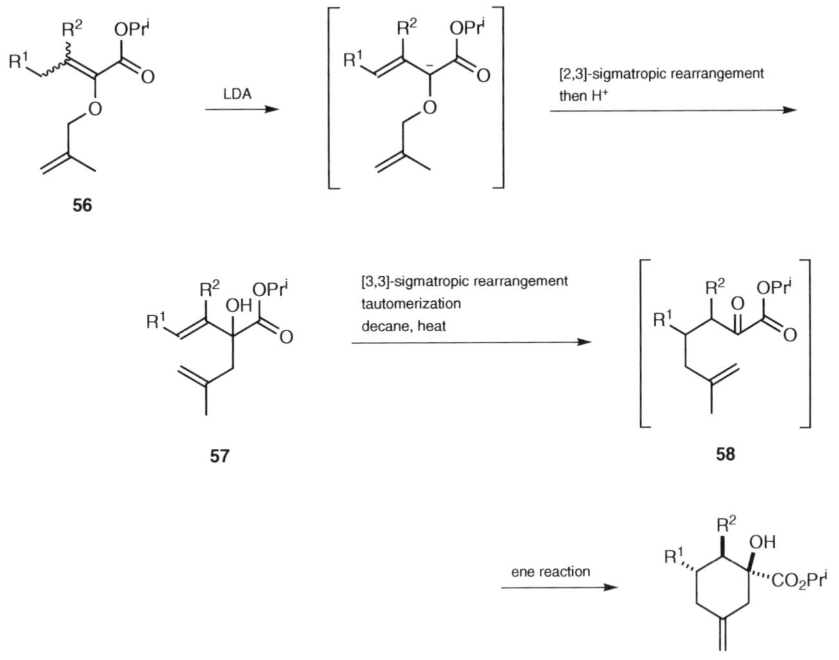

R^1	R^2	Yield (%)	Ref
H	H	59	[32]
H	Me	51	[32]
Ph	H	76	[32]

Several unexpected domino sequences initiated by the Mislow–Evans reaction and producing structurally unusual products have been reported on propargylic substrates. One such domino sequence involving [2,3]- and [3,3]-sigmatropic rearrangements is achieved for ethynyl propargyl sulfoxides (Scheme 12).[33] Here, highly reactive thioketenes **60** are generated which are either intercepted by amines in situ or dimerized. Several examples of this reaction with trimethylsilyl, *tert*-butyldimethylsilyl, and phenyl substituents at the alkynyl positions have been described.

Scheme 12 Mislow–Evans/[3,3]-Sigmatropic Rearrangement Cascade[33]

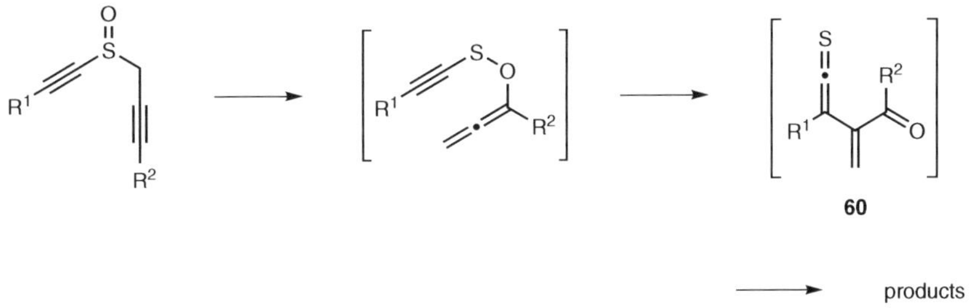

hexane, reflux, 24 h

R^1	Yield (%)	Ref
TMS	82	[33]
TBDMS	61	[33]

$BnNH_2$ (20 equiv), benzene, reflux, 6 h: 35% + 11%

$BnNH_2$ (20 equiv), benzene, reflux, 6 h: 33% + 21%

Et_2NH (20 equiv), benzene, reflux, 6 h: 26% + 42%

A related domino sequence on dipropargyl disulfides **61** is observed upon heating in chloroform. The substrates undergo a sequence of two [2,3]-sigmatropic rearrangements to form diallenic disulfides **63** via intermediacy of an unusual thiosulfoxide intermediate **62**; its intermediacy is supported by the detection of allenic propargylic sulfides when the reaction is carried out in the presence of triphenylphosphine. The resulting diallenic disulfide **63** undergoes a [3,3]-sigmatropic rearrangement followed by a double Michael addition and either aromatization to give 1*H*,3*H*-thieno[3,4-*c*]thiophene **64** or oxidation by oxygen to produce disulfide **65**. Performing the reaction under an inert atmosphere greatly diminishes the amount of oxidized products, whereas the use of dimethyl sulfoxide as a solvent leads to their exclusive formation (Scheme 13).[34]

Scheme 13 [2,3]-/[2,3]-/[3,3]-Sigmatropic Rearrangement Cascade on Disulfide Substrates[34]

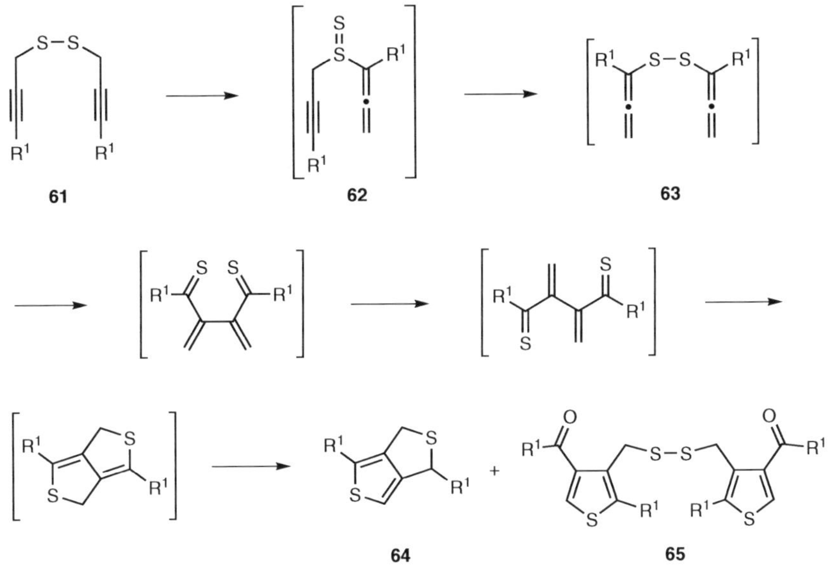

R^1	Conditions	Ratio (**64**/**65**)	Yield[a] (%)	Ref
H	$CDCl_3$, 60 °C, 1.5 h	1:2	40	[34]
H	$CDCl_3$, 60 °C, 2 h, argon	10:1	80[b]	[34]
Me	$CDCl_3$, 60 °C, 160 h	5:1	65	[34]
Me	DMSO, 70 °C, 24 h	0:100	45	[34]
Ph	$CDCl_3$, 60 °C, 16 h	6:1	53	[34]

[a] Total yield of **64** and **65** after column chromatography.
[b] Yield of product **64**.

Two more fascinating domino processes are initiated by a [2,3]-sigmatropic rearrangement on propargylic substrates. The first one includes three sequential [2,3]-sigmatropic rearrangements of propargylic disulfenates **66** followed by a [3,3]-sigmatropic rearrangement and a [2+2] cycloaddition to give the products **67** (Scheme 14).[35] Several examples are reported, but the yields and physical characterization data are only provided for two cases.

Scheme 14 Rearrangement of Propargylic Disulfenates[35]

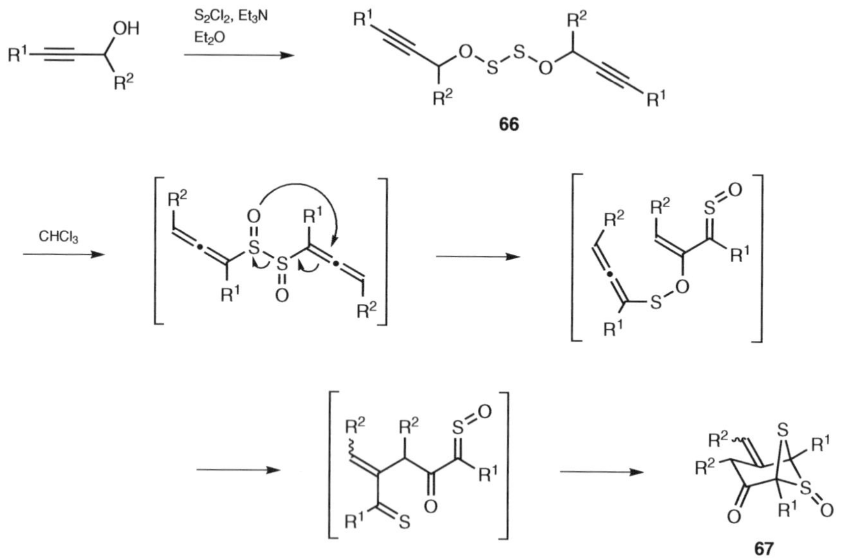

R^1	R^2	Yield (%)	Ref
H	H	69	[35]
H	Me	57	[35]

In the other cascade reaction, the dipropargylic alcohols **68** are first converted into diallenic sulfoxides **69** via sequential [2,3]-sigmatropic rearrangements. The resulting diallene is set up for the formation of a four-membered ring by 4π-electrocyclization, giving dimethylenecyclobutene **70** (Scheme 15).[36] Two examples of the transformation were reported (R^1 = H or Ph).

Scheme 15 [2,3]-Sigmatropic Rearrangement/4π-Electrocyclization Cascade[36]

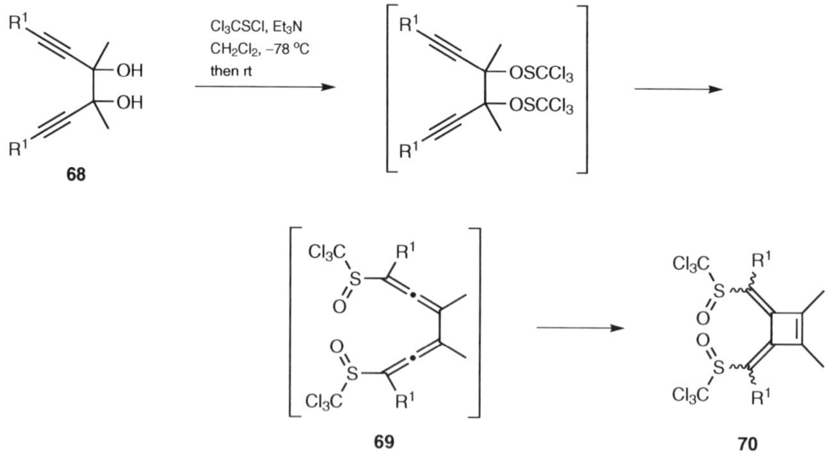

R^1 = H, Ph

A synthetic application of a similar cascade reaction uses dipropargylic alcohol **71** to produce diallene **72** that undergoes a 6π-electrocyclization to give the six-membered dimethylenecyclohexadiene **73**, which is a very reactive diene for a Diels–Alder reaction. Upon [4+2] cycloaddition onto an alkene, the aromatic benzene ring in **74** is produced (Scheme 16).[37] The latter transformation is used along with the ene reaction shown in Scheme 8 (Section 2.1.4.2) in the synthesis of steroids.

Scheme 16 Mislow–Evans Rearrangement/6π-Electrocyclization/[4+2] Cycloaddition Cascade[37]

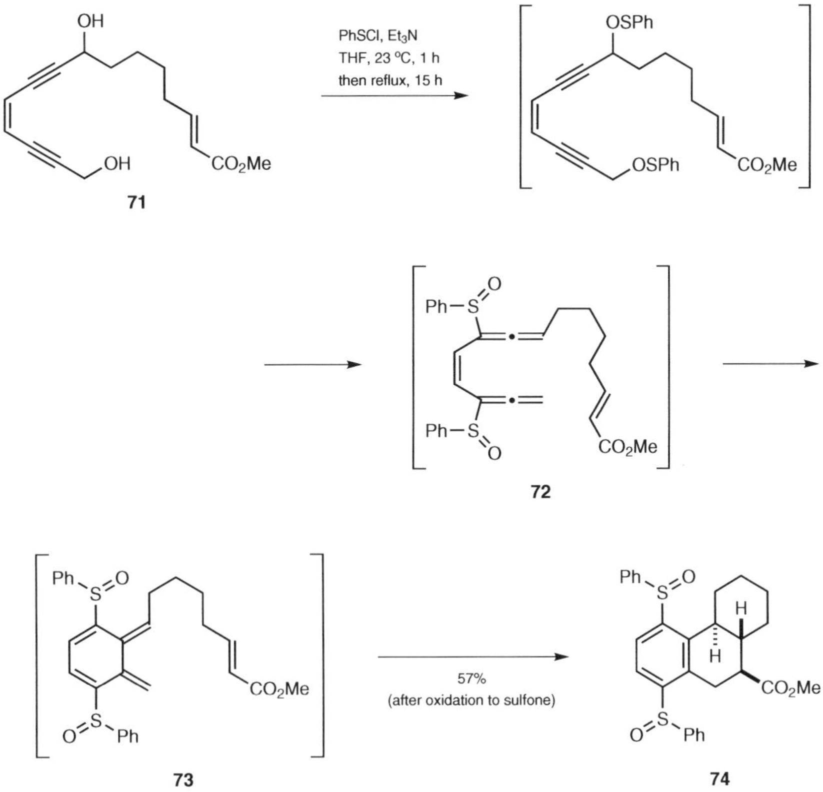

PhSCl, Et_3N
THF, 23 °C, 1 h
then reflux, 15 h

57%
(after oxidation to sulfone)

The example in Scheme 17 highlights [2,3]-sigmatropic rearrangements followed by a [1,5]-sigmatropic hydrogen shift, a process that has been used in the synthesis of the retinoids.[38,39] The cascade reaction is initiated by equilibration of the double-bond configuration in sulfenate intermediate **75** via a reversible [2,3]-sigmatropic rearrangement to give propargylic sulfenate intermediate **76**, followed by a [2,3]-sigmatropic rearrangement. It concludes with a [1,5]-sigmatropic hydrogen shift in vinylallenyl sulfoxide **77**, producing the conjugated pentaene product **78**. Labeling experiments were performed to confirm the origin of the shifting hydrogen. The involvement of the sulfoxide is responsible for rate acceleration, stereoselectivity, and the required rearrangement of the double-bond configuration.

for references see p 192

Scheme 17 [2,3]-Sigmatropic Rearrangement/[1,5]-Sigmatropic Hydrogen Shift Cascade[38,39]

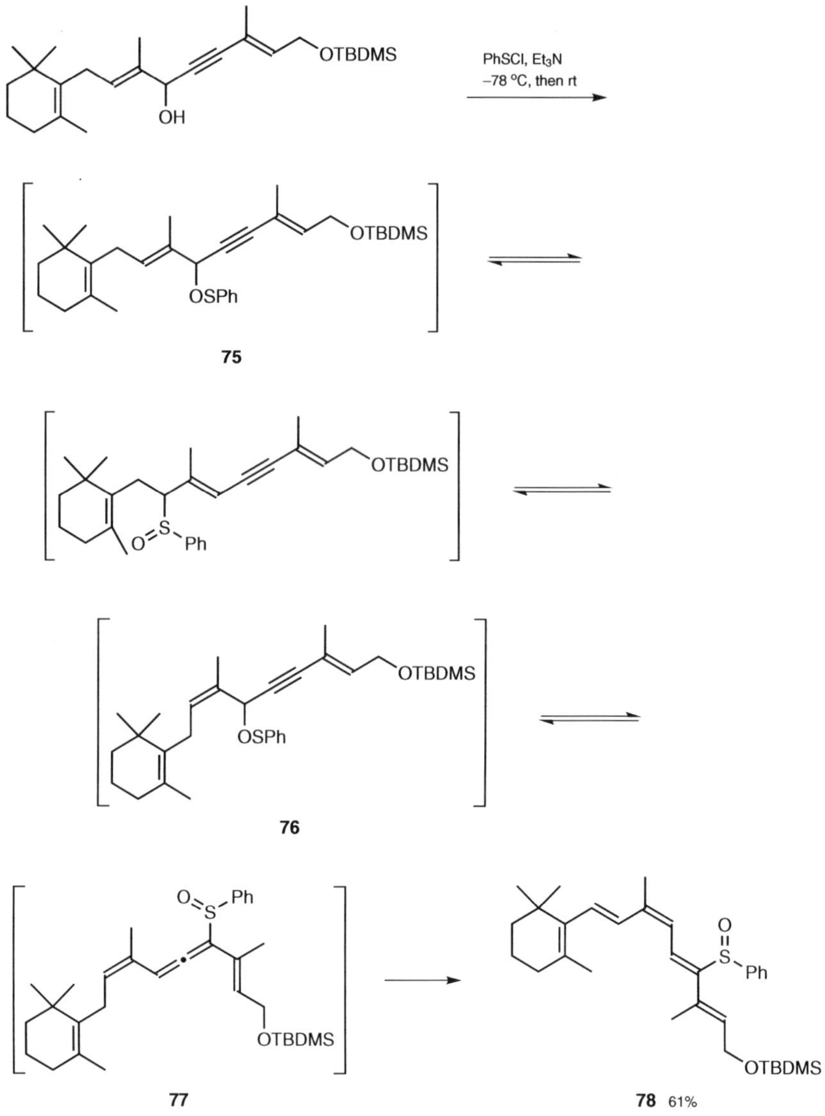

OTBDMS
OH
PhSCl, Et_3N
−78 °C, then rt

OTBDMS
OSPh
75

OTBDMS
O=S–Ph

OTBDMS
OSPh
76

O=S–Ph
OTBDMS
77

O=S–Ph
OTBDMS
78 61%

The classic Sommelet–Hauser rearrangement[40] can also be considered a combination of successive [2,3]- and [1,3]-sigmatropic rearrangements (Scheme 18).[40] However, the [1,3]-shift in this case likely proceeds through a stepwise deprotonation/reprotonation mechanism rather than a direct sigmatropic shift in the usual sense.

Scheme 18 Sommelet–Hauser Rearrangement as a Domino Sequence[40]

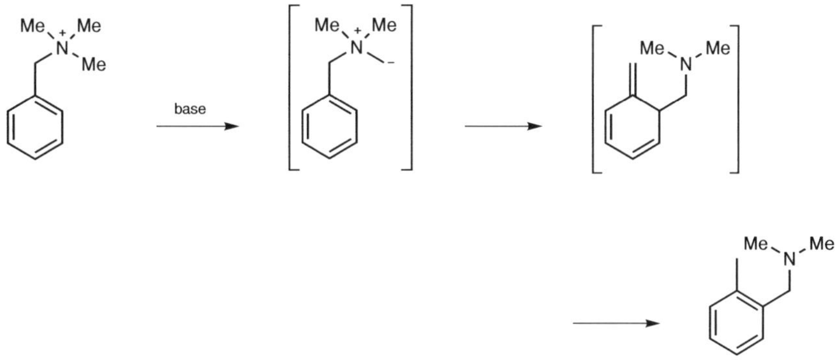

Me Me N+ Me base Me Me N+ − Me N Me Me N Me

In an atypical example of a domino sequence initiated by a [2,3]-sigmatropic rearrangement, a radical reaction follows rather than a pericyclic process (Table 1).[41] Equilibration between allylic sulfoxide **79** and allylic sulfenate **80** is the starting point of a radical process initiated by radical abstraction of the phenylsulfanyl group by a tributyltin radical. The resulting oxo radical **81** undergoes fragmentation, and the carbon-centered radical **82** formed in this manner cyclizes onto the α,β-unsaturated ketone generated upon the fragmentation to give a cyclohexanone. Ultimately, a net ring expansion is achieved (Table 1).

Table 1 [2,3]-Sigmatropic Rearrangement/Radical Cascade[41]

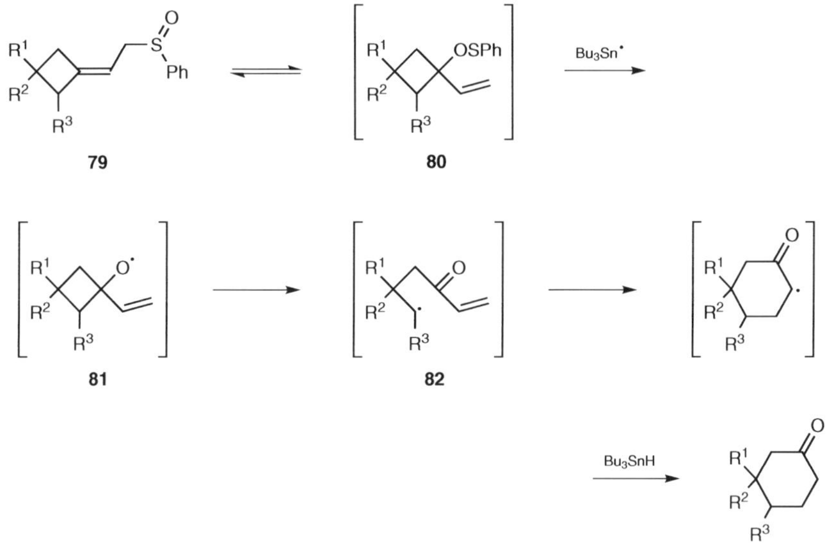

R1 R2 R3 S O Ph **79** R1 R2 R3 OSPh **80** $Bu_3Sn^•$ R1 R2 R3 O• **81** R1 R2 R3 O **82** R1 R2 R3 O Bu_3SnH R1 R2 R3 O

for references see p 192

Reactant	Product	Yield (%)	Ref
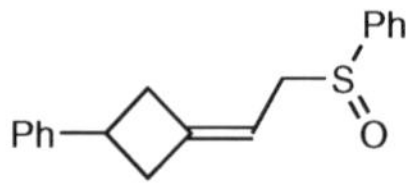	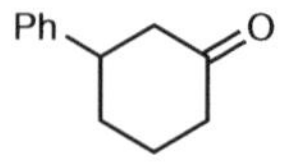	69[a]	[41]
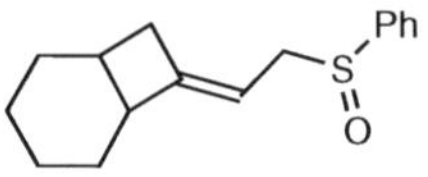	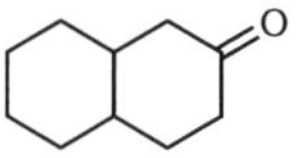	43	[41]
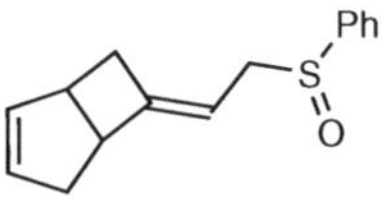	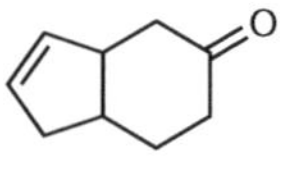	67	[41]
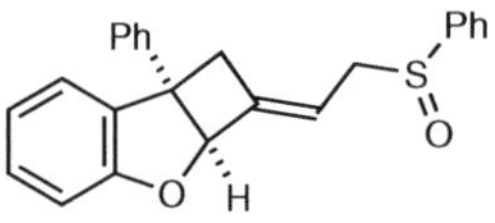	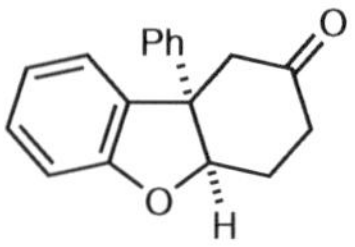	64[b]	[41]
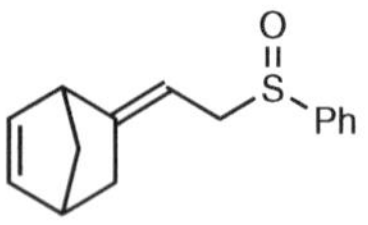	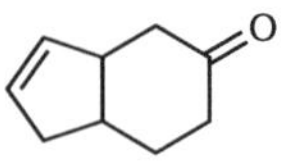	66	[41]

[a] The cyclopentanone isomer (5-*exo* cyclization) was obtained in 19% yield.
[b] The isomer resulting from fragmentation on the other side of the ring was formed in 10% yield.

1*H*,3*H*-Thieno[3,4-c]thiophenes 64 **and Bis(3-thienylmethyl) Disulfides** 65; **General Procedure:**[34]

Disulfide **61** (1 mmol) was dissolved in the appropriate anhyd solvent (5 mL) and heated under stirring at 60–70 °C until disappearance of the starting material as shown by TLC. After evaporation of the solvent ($CHCl_3$ or MeCN), the crude reaction mixture was subjected to column chromatography (silica gel). In the case of reactions performed in DMSO, the soln was diluted with CH_2Cl_2 (30 mL) and washed with H_2O (3 × 20 mL). The combined organic phase was dried ($MgSO_4$) and concentrated to give the mixture of compounds **64** and **65**, which were separated by column chromatography (silica gel).

2.1.4.4 Domino Processes Initiated by Other Sigmatropic Rearrangements

A number of other types of sigmatropic rearrangements are known. These reactions typically require harsher reaction conditions (higher temperatures or photochemical conditions) than those already described. When placed under these conditions, however, many compounds readily undergo cascades of pericyclic transformations.[8] On the other hand, these processes frequently give mixtures of products due to the harsh reaction conditions and multiple possible reaction trajectories. With constrained substrates, useful transformations are possible using this approach. For example, dienyl phenols undergo a [1,7]-sigmatropic hydrogen shift, followed by 6π-electrocyclization, to form 2*H*-1-benzopyrans upon heating (Scheme 19).[8,42]

Scheme 19 [1,7]-Sigmatropic Hydrogen Shift/6π-Electrocyclization Cascade[8,42]

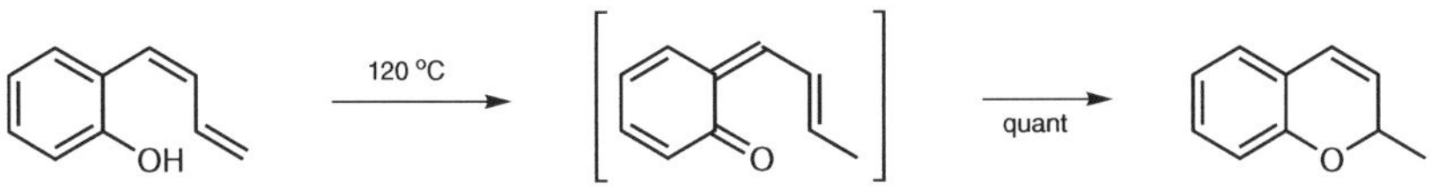

Through a judicious design of the substrate structure and choice of reaction conditions, synthetic applications of such transformations become feasible. One such case involves a [1,5]-sigmatropic rearrangement of aryl ketene imines **85** or aryl carbodiimides **88** to produce intermediates such as **86** or **89**, which undergo 6π-electrocyclization with simultaneous rearomatization to yield bicyclic products **87** and **90**, respectively (Scheme 20 and Scheme 21).[43–46] The [1,5]-sigmatropic rearrangement is facilitated in this case, according to the performed DFT calculations, due to the hydridic character of the migrating hydrogen and a complementary electrophilic character of the ketene imine, along with the decreased steric requirement of the allenic geometry of the ketene imine.

Scheme 20 [1,5]-Sigmatropic Rearrangement/6π-Electrocyclization Cascade of Aryl Ketene Imines[43–46]

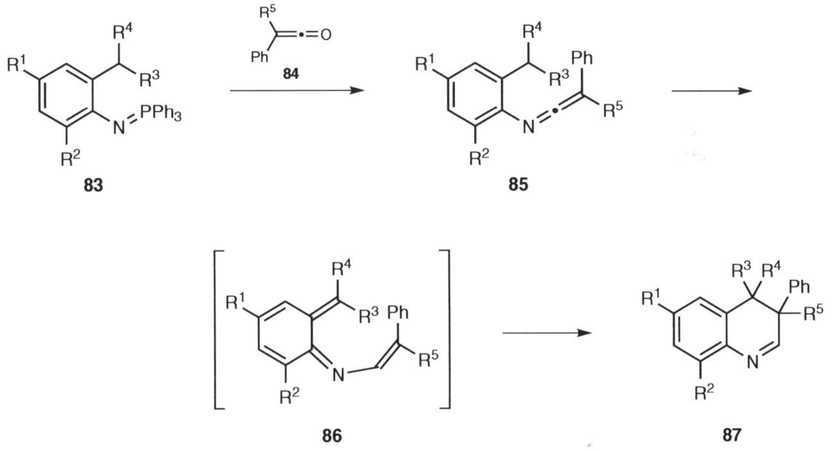

R^1	R^2	R^3	R^4	R^5	Yield (%)	Ref
H	H	$O(CH_2)_2O$		Ph	70	[43,46]
H	H	$O(CH_2)_2O$		Me	58	[43,46]
H	Me	$O(CH_2)_2O$		Ph	67	[43,46]
H	H	$S(CH_2)_2S$		Ph	74	[43,46]
Cl	H	$S(CH_2)_2S$		Ph	77	[43,46]
H	H	$O(CH_2)_3O$		Ph	90	[43,46]
H	H	$O(CH_2)_3O$		Me	81	[43,46]
H	H	$S(CH_2)_3S$		Ph	71	[43,46]
Cl	H	$S(CH_2)_3S$		Ph	88	[43,46]
H	H	$S(CH_2)_3O$		Ph	73	[45]
H	H	Ph	Ph	Me	60	[44]
H	H	Ph	Ph	Et	15	[44]
H	Me	Ph	Ph	Me	99	[44]
H	H	$4-Me_2NC_6H_4$	$4-Me_2NC_6H_4$	Et	80	[44]

for references see p 192

R^1	R^2	R^3	R^4	R^5	Yield (%)	Ref
H	Me	4-$Me_2NC_6H_4$	4-$Me_2NC_6H_4$	Me	78	[44]
H	H	4-$MeOC_6H_4$	4-$MeOC_6H_4$	Et	52	[44]
H	Me	4-$MeOC_6H_4$	4-$MeOC_6H_{4p}$	Et	99	[44]
H	H	Ph	Ph	Ph	0[a]	[44]

[a] A different product, 10-(2,2-diphenylethenyl)-9-phenyl-9,10-dihydroacridine, resulting from electrocyclization onto one of the aromatic rings, was isolated in 95% yield.

Scheme 21 [1,5]-Sigmatropic Rearrangement/6π-Electrocyclization Cascade of Aryl Carbodiimides[43,44,46]

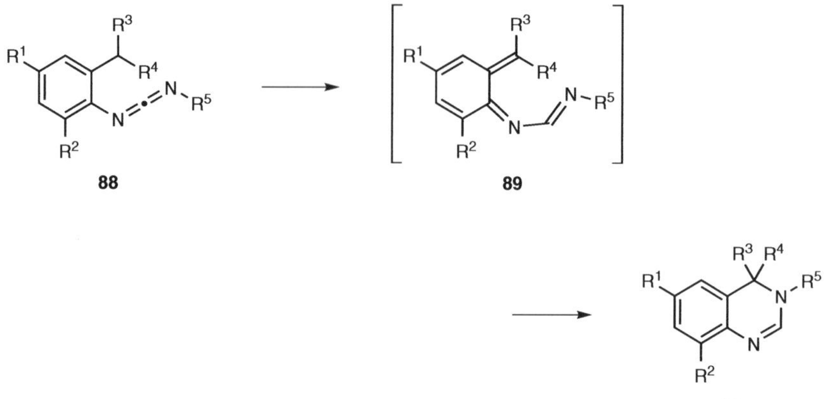

R^1	R^2	R^3	R^4	R^5	Yield (%)	Ref
H	H	$O(CH_2)_2O$		4-Tol	47	[46]
H	H	$O(CH_2)_2O$		4-$MeOC_6H_4$	81	[43,46]
H	Me	$O(CH_2)_2O$		4-Tol	99	[43,46]
H	Me	$O(CH_2)_2O$		4-BrC_6H_4	98	[43,46]
Cl	H	$O(CH_2)_2O$		4-Tol	49	[46]
Me	H	$O(CH_2)_2O$		4-ClC_6H_4	85	[43,46]
H	H	4-$Me_2NC_6H_4$	4-$Me_2NC_6H_4$	4-BrC_6H_4	82	[44]
H	Me	4-$Me_2NC_6H_4$	4-$Me_2NC_6H_4$	4-Tol	76	[44]
H	H	4-$MeOC_6H_4$	4-$MeOC_6H_4$	4-ClC_6H_4	78	[44]

A set of domino processes initiated by a rare [1,3]-sigmatropic rearrangement are demonstrated on an unusual class of substrates, namely allenyl amides.[47–49] The thermal [1,3]-sigmatropic rearrangement, although possible, is normally a very difficult process. It is unclear if the rearrangement in this case is concerted or occurs through some other type of mechanism. Regardless, the facile [1,3]-sigmatropic rearrangement of the allenyl amide can be followed by 6π-electrocyclization to give amide **91**, sequential [1,7]-sigmatropic rearrangements to produce amide **92** (Scheme 22),[47] [4+2] cycloaddition to yield aza heterocycles **93** (Scheme 23),[48] and, most spectacularly, a cascade of 6π-electrocyclizations and [4+2] cycloadditions to afford **96** (Scheme 24).[49]

Scheme 22 [1,3]-Sigmatropic Rearrangement Initiated Cascades on Allenyl Amides[47]

Conditions	Yield (%)	Ref
MeCN, 115 °C, 16 h	91	[47]
CSA, CH_2Cl_2, rt, 10 min	95	[47]

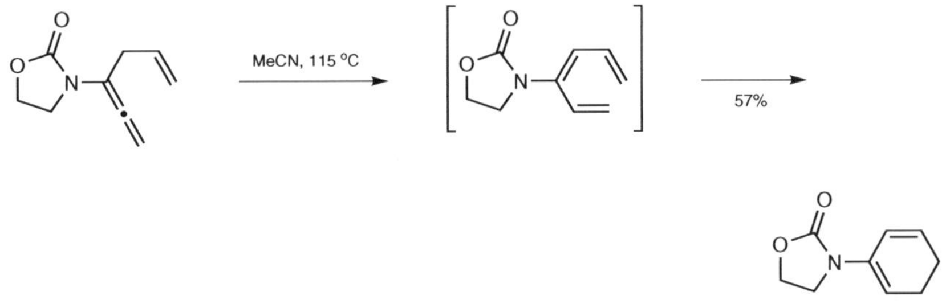

MeCN, 115 °C

57%

91

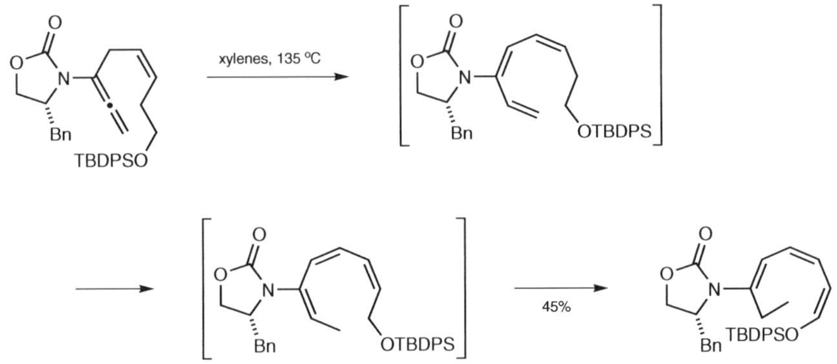

xylenes, 135 °C

Bn TBDPSO

Bn OTBDPS

Bn OTBDPS

45%

Bn TBDPSO

92

for references see p 192

Scheme 23 [1,3]-Sigmatropic Rearrangement/Diels–Alder Cascade[48]

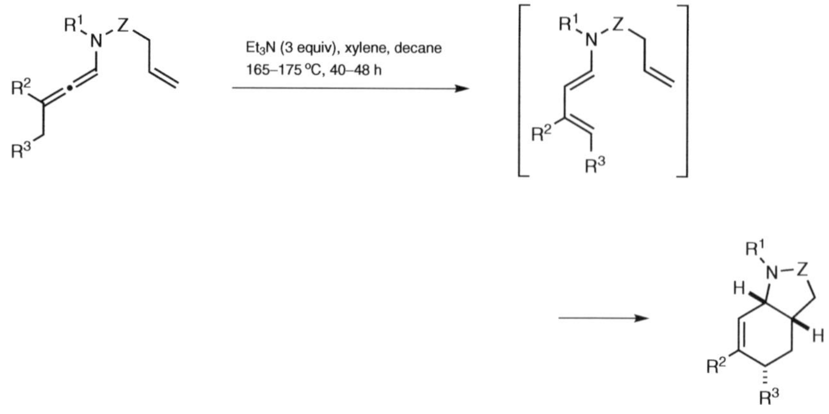

Et_3N (3 equiv), xylene, decane
165–175 °C, 40–48 h

93

R^1	R^2	R^3	Z	dr	Yield (%)	Ref
Ts	H	H	CH_2CH_2	7:1	46	[48]
$4\text{-}O_2NC_6H_4SO_2$	Me	H	CH_2	11:1	58	[48]
Ac	Me	H	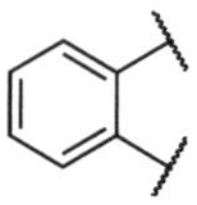	3:1	65	[48]
$4\text{-}O_2NC_6H_4SO_2$	$(CH_2)_4$		CH_2CH_2	12:1	75	[48]
Ac	$(CH_2)_4$		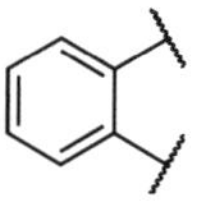	10:1	74	[48]

In the latter case, allenyl amide **94** is converted into triene **95**, which undergoes 6π-electrocyclization and then a [4+2] cycloaddition to form bridged tricyclic product **96** (Scheme 24), which constitutes the BCD ring system of the natural product atropurpuran. This transformation, however, could not be performed as a true domino reaction, and was only successful when performed in a stepwise fashion.

Scheme 24 [1,3]-Sigmatropic Rearrangement/6π-Electrocyclization/Diels–Alder Sequence[49]

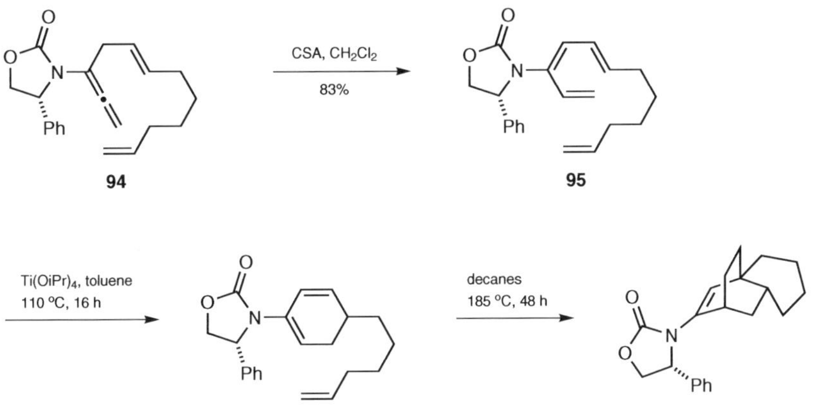

CSA, CH_2Cl_2
83%

94 **95**

$Ti(OiPr)_4$, toluene
110 °C, 16 h

decanes
185 °C, 48 h

96 73% (from **95**)

3*H*-Spiro[quinoline-4,2′-[1,3]dioxolanes] 87 [R^3,R^4 = $O(CH_2)_2O$]; General Procedure:[43]
To a soln of the corresponding 2-(triphenylphosphoranylideneamino)benzaldehyde acetal **83** [R^3,R^4 = $O(CH_2)_2O$; 1 mmol] in anhyd toluene (15 mL) was added methyl(phenyl)ketene (**84**, R^5 = Me; 0.13 g, 1 mmol) or diphenylketene (**84**, R^5 = Ph; 0.19 g, 1 mmol) in toluene (5 mL). The mixture was stirred at rt for 15 min, and the formation of the corresponding ketenimine **85** was established by IR spectroscopy (the IR spectra of the reaction mixtures show strong absorptions around 2000 cm^{-1}, characteristic of the N=C=C functional group). Next, the soln containing ketenimine **85** was heated at reflux temperature until the total disappearance of the cumulenic band in the IR spectra, (approximately 1 h). After cooling, the solvent was removed under reduced pressure. The resulting material was purified by column chromatography (silica gel, hexanes/Et_2O 7:3).

2.1.4.5 Domino Processes in the Synthesis of Natural Products

In this section, domino sequences of pericyclic transformations integrating ene reactions and sigmatropic rearrangements as key events in total synthesis endeavors are highlighted.

An effective cascade of ene and hetero-Diels–Alder reactions has been employed in the construction of the key structural fragment in the synthesis of pseudomonic acid A (**101**) (Scheme 25). The use of a highly reactive enophile, formaldehyde, and nonconjugated diene **97** enables the initial ene reaction to proceed. This ene reaction produces conjugated diene intermediate **98** and installs an alcohol handle that coordinates to a Lewis acid, delivering the second molecule of formaldehyde for a hetero-Diels–Alder cyclization via transition structure **99** that installs the tetrahydropyran ring. The resulting product **100** was utilized as an intermediate in the synthesis of pseudomonic acid A (**101**).[50,51]

Scheme 25 Ene/Hetero-Diels–Alder Cascade in the Synthesis of Pseudomonic Acid A[50,51]

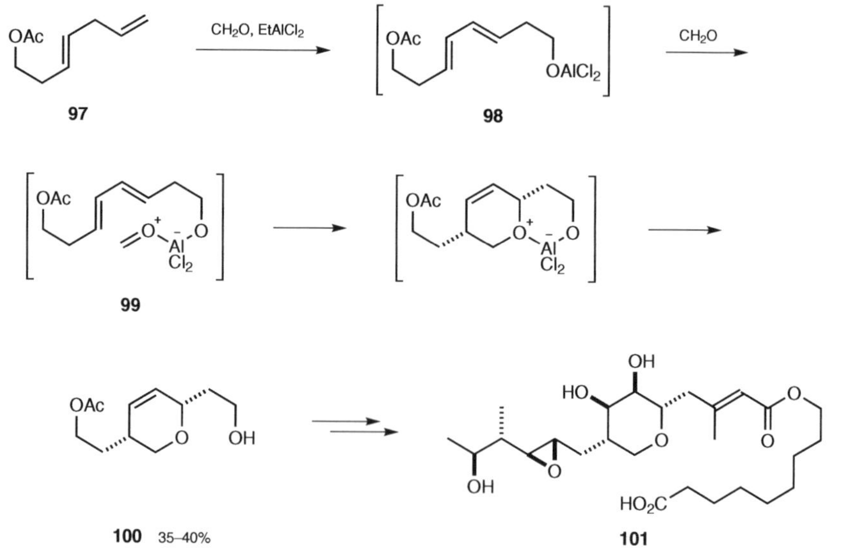

An ene reaction was used to construct the intricate skeleton of chloropupukeanolide D (**106**) within a cascade sequence using a Diels–Alder reaction (Scheme 26).[52] The reverse-electron-demand Diels–Alder reaction between vinylallene **103** and diene **102** affords a [4+2]-cycloaddition adduct under high pressure as a mixture of *exo-* and *endo-*isomers. The *endo-*isomer **104** further undergoes the ene reaction, producing the complex carbon skeleton of **105** in 70% yield. Intramolecular hydrogen bonding is proposed as a key factor in the preferential *endo* selectivity of the initial Diels–Alder reaction. The proximity and appropriate orientation of the allene group in **104**, in turn, is essential for the successful ene reaction.

Scheme 26 [4+2] Cycloaddition/Ene Reaction Cascade in an Approach to Chloropupukeanolide D[52]

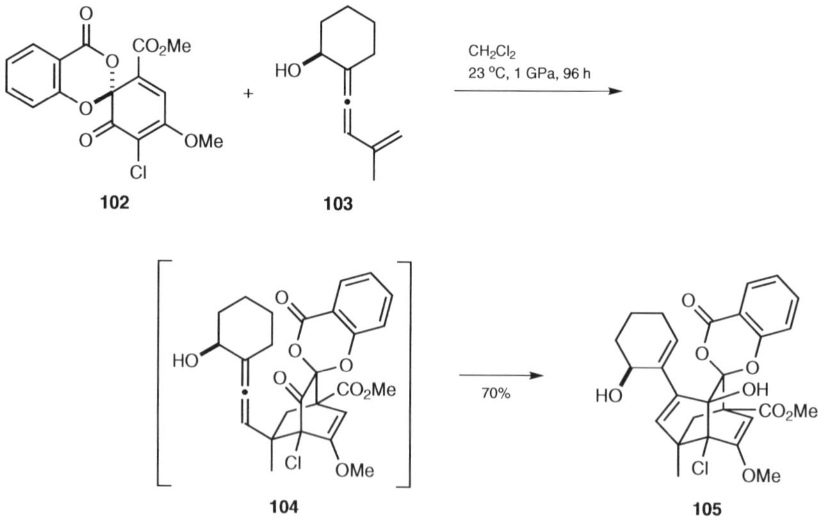

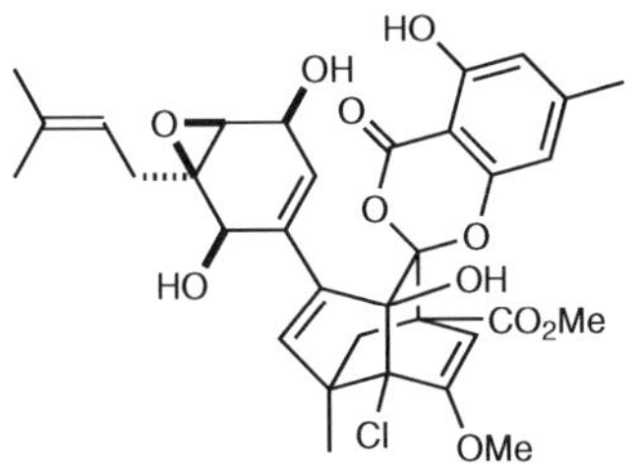

106

A combination of a [3,3]-sigmatropic rearrangement and an ene reaction has been effectively used in the synthesis of steroids (Scheme 27).[53] The Claisen rearrangement of allyl vinyl ether **107** produces the enyne substrate **108** for the ene reaction, which proceeds effectively to construct the D ring of steroid structure **109** with the required stereochemistry, setting up at the same time the functional group handles for the construction of the C ring.

Scheme 27 [3,3]-Sigmatropic Rearrangement/Ene Reaction Cascade in the Synthesis of Steroids[53]

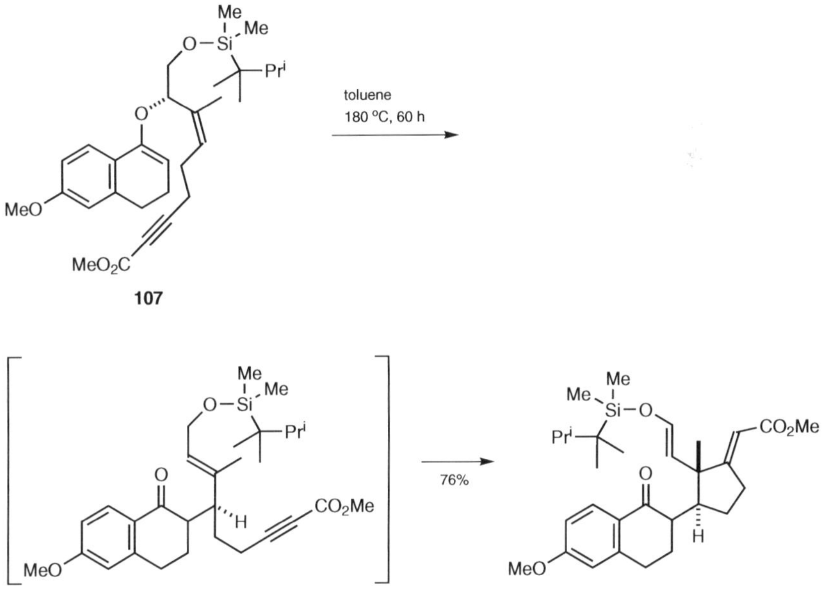

107

toluene
180 °C, 60 h

108

76%

109

Another example of an effective [3,3]-sigmatropic rearrangement/ene reaction cascade (see also Section 2.1.5.4.1) was applied to the synthesis of (+)-arteannuin M (**114**) (Scheme 28).[54] Here, the oxy-Cope rearrangement of alcohol **110** produces, after keto–enol tautomerization, ketone **112**. An intramolecular ene reaction then produces the bicyclic core **113** of arteannuin M (**114**). Notably, the ten-membered cyclic intermediate enol **111** does not have chiral centers. However, due to conformational constraints, it retains the chirality, and its subsequent conversion into the ketone occurs stereoselectively, resulting in an overall 89% retention of chirality.

for references see p 192

Scheme 28 [3,3]-Sigmatropic Rearrangement/Ene Reaction in the Synthesis of Arteannuin M[54]

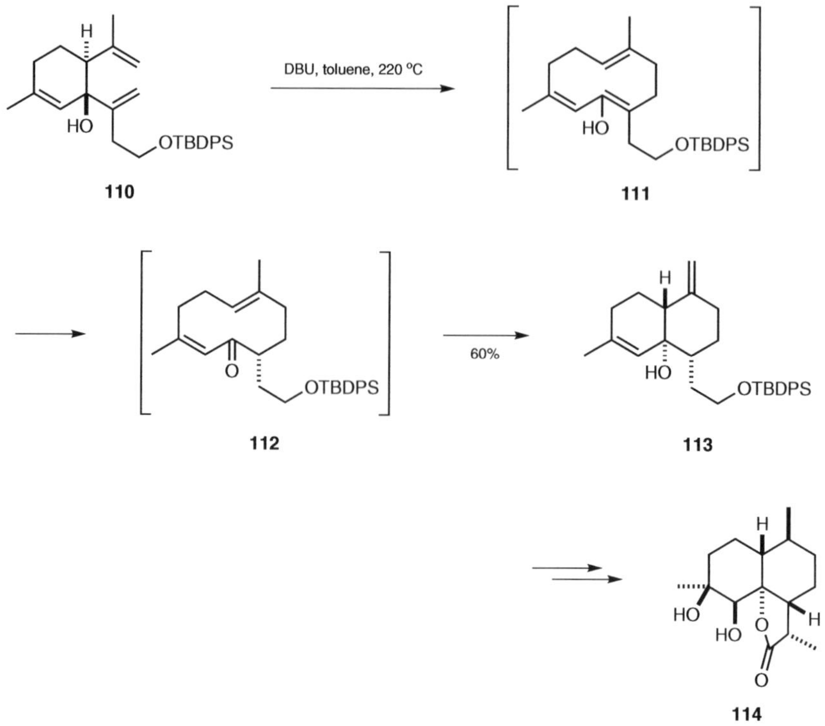

DBU, toluene, 220 °C

HO OTBDPS **110** → [HO OTBDPS **111**] → [O OTBDPS **112**] → (60%) HO OTBDPS **113** ⇉ HO HO O O **114**

The synthesis of (+)-sterpurene (**119**) was based on a domino sequence comprising a [2,3]-sigmatropic rearrangement of propargyl sulfenates and other pericyclic processes (Scheme 29).[55] In this case, the propargylic sulfenate **116** derived from alcohol **115** undergoes a [2,3]-sigmatropic rearrangement to give allenic sulfoxide **117**, generating the diene component for intramolecular Diels–Alder reaction. The final tricyclic sulfoxide product **118** is obtained in 70% yield and is subsequently converted into (+)-sterpurene (**119**) in two additional steps.

Scheme 29 [2,3]-Sigmatropic Rearrangement/Diels–Alder Cascade for the Synthesis of (+)-Sterpurene[55]

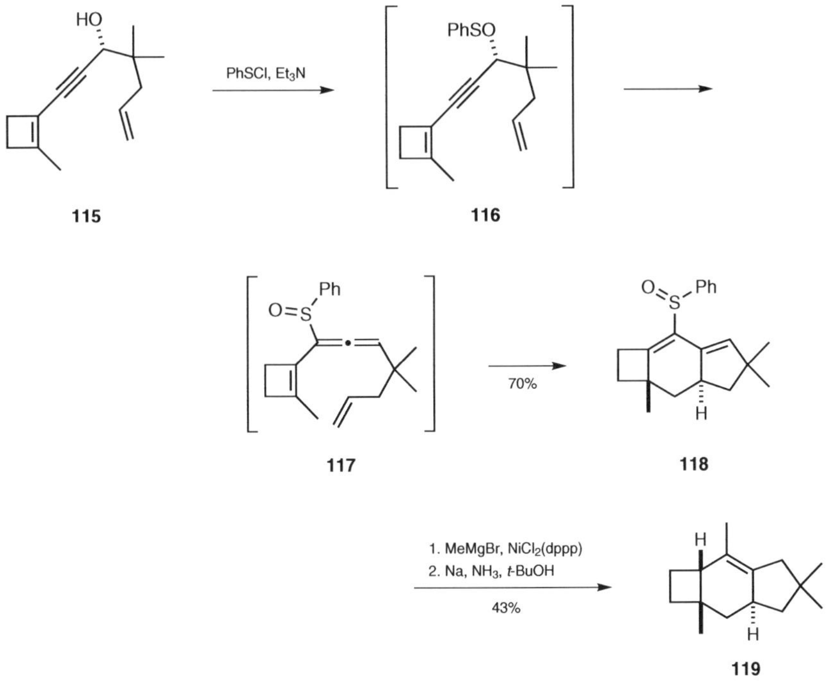

HO

PhSCl, Et_3N

PhSO

115

116

Ph

O=S

O=S-Ph

70%

H

117

118

1. MeMgBr, $NiCl_2(dppp)$
2. Na, NH_3, t-BuOH

43%

H

H

119

Another effective use of the [2,3]-sigmatropic rearrangement of a sulfoxide gives a precursor of the cyclic imine fragment of pinnatoxins [e.g., pinnatoxin A (**123**)] and pteriatoxins (Scheme 30).[56] In this case, allylic sulfoxide **121** is generated through a Claisen rearrangement of vinylic sulfoxide **120**, followed by a Mislow–Evans rearrangement producing allylic alcohol **122** (see also Section 2.1.5.5.2). Notably, adjacent quaternary and tertiary centers are stereoselectively generated as a result of this sequence.

Scheme 30 Claisen/[2,3]-Sigmatropic Rearrangement Cascade toward the Pinnatoxins[56]

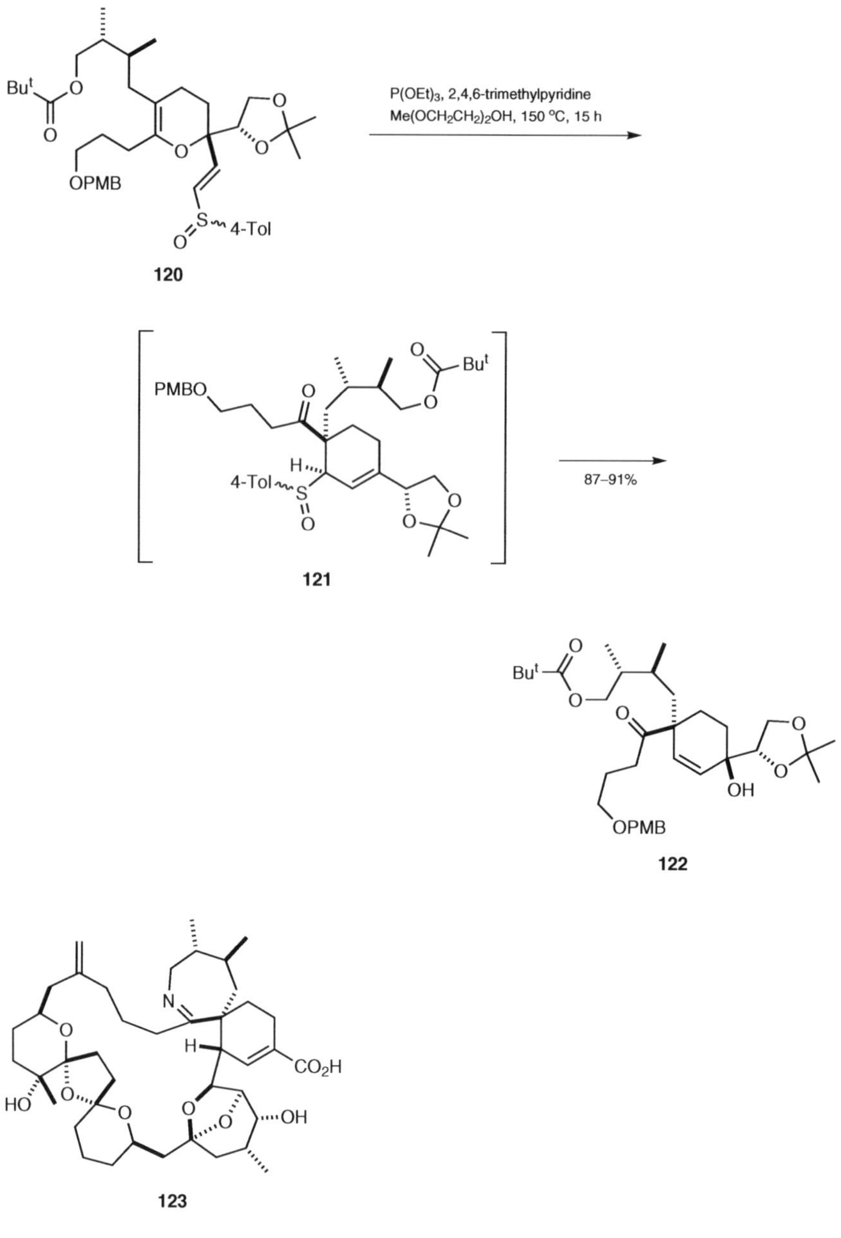

P(OEt)$_3$, 2,4,6-trimethylpyridine
Me(OCH$_2$CH$_2$)$_2$OH, 150 °C, 15 h

120

121

87–91%

122

123

A similar sequence of [3,3]- and [2,3]-sigmatropic rearrangements has been applied in the synthesis of the *Sceletium* alkaloids (−)-joubertinamine (**128**) and (−)-mesembrine (**129**).[57] In this case, however, the domino version of the transformation proved unsuccessful, and the transformation was performed in a stepwise fashion (Scheme 31). Although *Z*-vinyl sulfide (*Z*)-**124** is more accessible, the [3,3]-sigmatropic rearrangement reaction can be channeled through preferred *E*-isomer (*E*)-**124**, which is much more reactive, by in situ isomerization in the presence of benzenethiol. The isomerization presumably occurs by a radical addition/elimination pathway. The resultant allylic sulfide **125** is subjected to a Wittig reaction and then oxidized to sulfoxide **126**, and the subsequent [2,3]-rearrangement of the unpurified sulfoxide **126** under reductive conditions cleanly delivers allylic alcohol **127**.

Scheme 31 Application of the Claisen/[2,3]-Sigmatropic Rearrangement Cascade to the Synthesis of *Sceletium* Alkaloids[57]

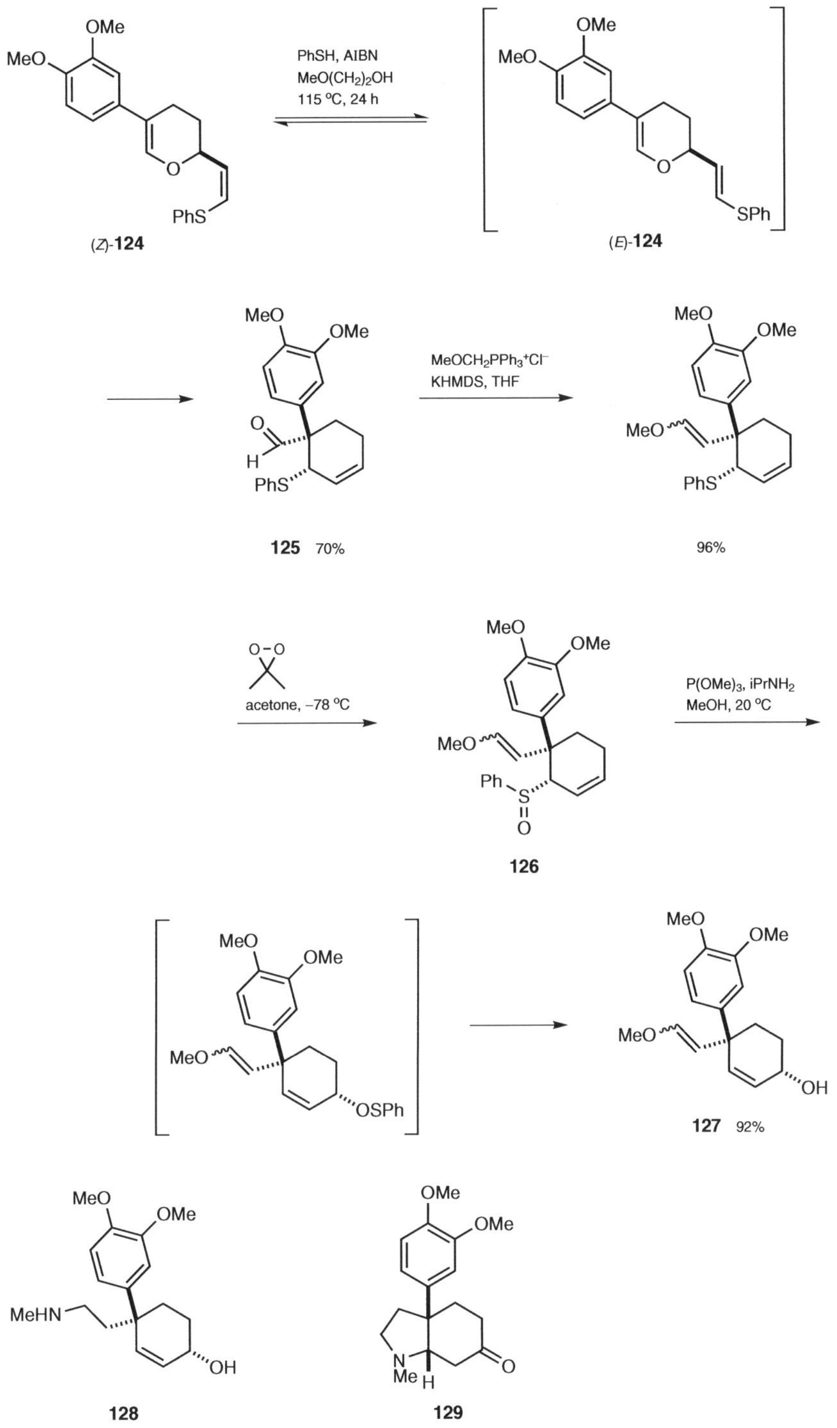

for references see p 192

The application of [1,5]-hydrogen sigmatropic rearrangements is demonstrated in the following two syntheses.

In the classic synthesis of isocedrene (**134**) and its derivatives, a [1,5]-sigmatropic rearrangement equilibrates the mixture of two regioisomeric cyclopentadienes **130** and **131** carried over from the previous step (Scheme 32).[58] Isomer **132** then preferentially undergoes a Diels–Alder cycloaddition, resulting in the complete conversion of the mixture into the single Diels–Alder adduct **133**, which is then elaborated to isocedrene (**134**). The [1,5]-sigmatropic rearrangement on the cyclopentadiene framework is particularly facile due to the conformational confinement of the structure in the geometry conducive for the [1,5]-sigmatropic rearrangement.

Scheme 32 [1,5]-Sigmatropic Rearrangement/[4+2] Cycloaddition Cascade in the Synthesis of Isocedrene[58]

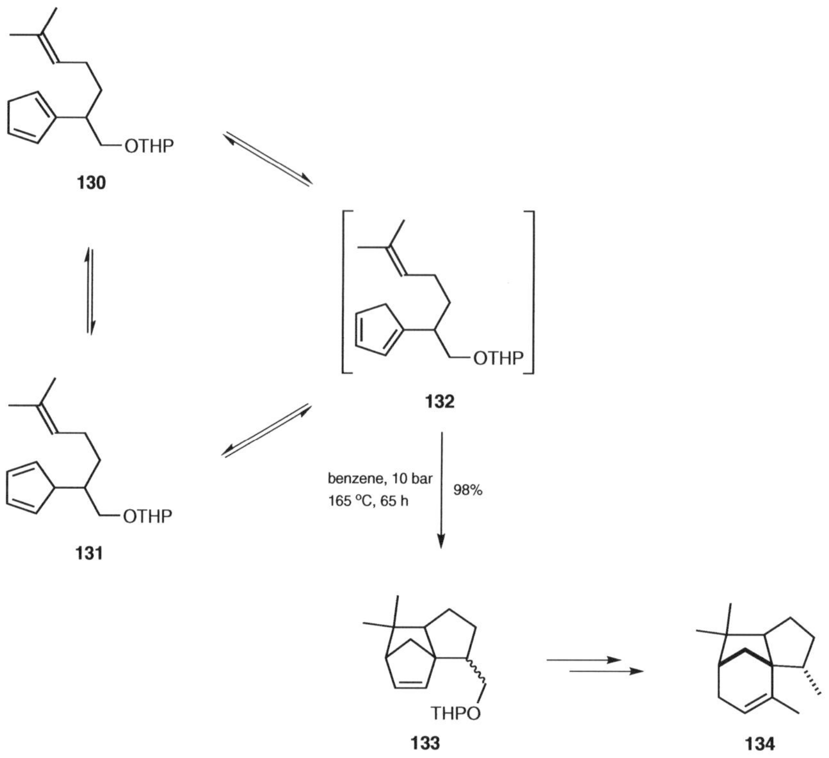

OTHP
130
OTHP
131
OTHP
132
benzene, 10 bar
165 °C, 65 h
98%
THPO
133
134

An effective photoinduced [1,5]-sigmatropic rearrangement has been used in the synthesis of steroids (Scheme 33).[59] Irradiation of aryl ketone **135** induces enolization by a [1,5]-sigmatropic rearrangement, which is believed to proceed via 1,5-hydrogen abstraction by the oxygen of the carbonyl group in the excited state rather than as a concerted pericyclic process. The outcome of the reaction, however, is equivalent to a [1,5]-sigmatropic rearrangement, producing dearomatized enol **136**, which rearomatizes by intramolecular [4+2] cycloaddition to form rings B and C of steroid structure **137**.

Scheme 33 Photoinduced [1,5]-Sigmatropic Rearrangement/[4+2] Cycloaddition Cascade in the Synthesis of Steroids[59]

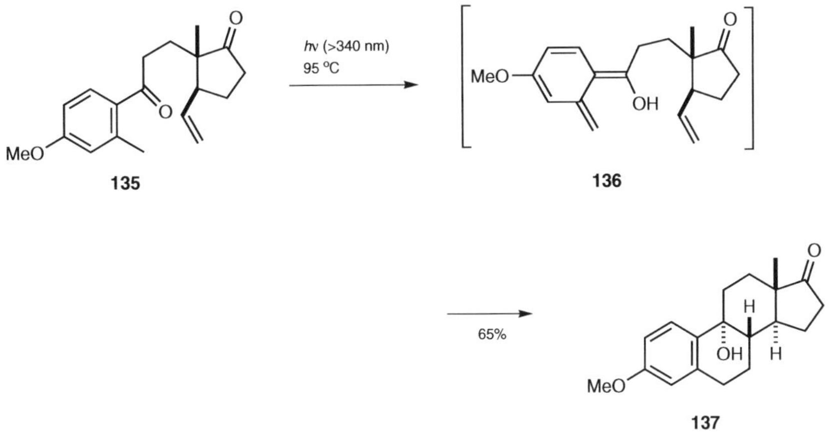

2.1.4.6 Conclusions

Ene reactions and sigmatropic rearrangements are powerful transformations in stereoselective organic synthesis. Arranging them into domino sequences amplifies their power, permitting the rapid assembly of complex structures in an atom-economical fashion from simple starting materials. Although it is relatively easy to initiate domino transformations by ene reactions and sigmatropic rearrangements, careful selection and fine-tuning of the substrate structure is often necessary to achieve an effective, selective, and high-yielding transformation.

The reviewed examples demonstrate powerful domino sequences that lead to core structures of natural products, as well as nonnatural compounds of unusual and intricate structure. The successful transformations presented highlight the principles of substrate selection and design to enable cascade transformations to proceed effectively. In the future, more emphasis will be placed on the brevity and efficiency of organic synthesis. Therefore, research into the design and mechanism of cascade pericyclic reactions and their application in synthesis is sure to continue.

References

[1] Nicolaou, K. C.; Snyder, S. A.; Montagnon, T.; Vassilikogiannakis, G., *Angew. Chem. Int. Ed.*, (2002) **41**, 1668.
[2] Majumdar, K. C.; Nandi, R. K., *Tetrahedron*, (2013) **69**, 6921.
[3] Rehbein, J.; Hiersemann, M., *Synthesis*, (2013) **45**, 1121.
[4] Clarke, M. L.; France, M. B., *Tetrahedron*, (2008) **64**, 9003.
[5] Mikami, K.; Shimizu, M., *Chem. Rev.*, (1992) **92**, 1021.
[6] Nakai, T.; Mikami, K., *Org. React. (N. Y.)*, (1994) **46**, 105.
[7] Nakai, T.; Mikami, K., *Chem. Rev.*, (1986) **86**, 885.
[8] Spangler, C. W., *Chem. Rev.*, (1976) **76**, 187.
[9] Prein, M.; Adam, W., *Angew. Chem. Int. Ed. Engl.*, (1996) **35**, 477.
[10] Rabjohn, N., *Org. React. (N. Y.)*, (1976) **24**, 261.
[11] Kabalka, G. W.; Hutchins, R.; Natale, N. R.; Yang, D. T. C.; Broach, V., *Org. Synth., Coll. Vol. VI*, (1988), 293.
[12] Oppolzer, W.; Snieckus, V., *Angew. Chem. Int. Ed. Engl.*, (1978) **17**, 476.
[13] Giguere, R. J.; Namen, A. M.; Lopez, B. O.; Arepally, A.; Ramos, D. E.; Majetich, G.; Defauw, J., *Tetrahedron Lett.*, (1987) **28**, 6553.
[14] González, I.; Pla-Quintana, A.; Roglans, A.; Dachs, A.; Solà, M.; Parella, T.; Farjas, J.; Roura, P.; Lloveras, V.; Vidal-Gancedo, J., *Chem. Commun. (Cambridge)*, (2010) **46**, 2944.
[15] Robinson, J. M.; Sakai, T.; Okano, K.; Kitawaki, T.; Danheiser, R. L., *J. Am. Chem. Soc.*, (2010) **132**, 11 039.
[16] Sakai, T.; Danheiser, R. L., *J. Am. Chem. Soc.*, (2010) **132**, 13 203.
[17] Kociolek, M. G.; Johnson, R. P., *Tetrahedron Lett.*, (1999) **40**, 4141.
[18] Parsons, P. J.; Waters, A. J.; Walter, D. S.; Board, J., *J. Org. Chem.*, (2007) **72**, 1395.
[19] Saaby, S.; Baxendale, I. R.; Ley, S. V., *Org. Biomol. Chem.*, (2005) **3**, 3365.
[20] Griesbeck, A. G.; de Kiff, A., *Org. Lett.*, (2013) **15**, 2073.
[21] Snider, B. B.; Deutsch, E. A., *J. Org. Chem.*, (1982) **47**, 745.
[22] Snider, B. B.; Deutsch, E. A., *J. Org. Chem.*, (1983) **48**, 1822.
[23] Snider, B. B.; Goldman, B. E., *Tetrahedron*, (1986) **42**, 2951.
[24] Hakuba, H.; Kitagaki, S.; Mukai, C., *Tetrahedron*, (2007) **63**, 12 639.
[25] Chen, H.; Wang, Z.; Zhang, Y.; Huang, Y., *J. Org. Chem.*, (2013) **78**, 3503.
[26] Evans, D. A.; Andrews, G. C., *Acc. Chem. Res.*, (1974) **7**, 147.
[27] Zhang, Y.; Wang, J., *Coord. Chem. Rev.*, (2010) **254**, 941.
[28] Lesuisse, D.; Canu, F.; Tric, B., *Tetrahedron*, (1994) **50**, 8491.
[29] Sayo, N.; Kimura, Y.; Nakai, T., *Tetrahedron Lett.*, (1982) **23**, 3931.
[30] Greeves, N.; Lee, W.-M.; Barkley, J. V., *Tetrahedron Lett.*, (1997) **38**, 6453.
[31] Greeves, N.; Lee, W.-M.; McLachlan, S. P.; Oakes, G. H.; Purdie, M.; Bickley, J. F., *Tetrahedron Lett.*, (2003) **44**, 9035.
[32] Hiersemann, M., *Eur. J. Org. Chem.*, (2001), 483.
[33] Aoyagi, S.; Makabe, M.; Shimada, K.; Takikawa, Y.; Kabuto, C., *Tetrahedron Lett.*, (2007) **48**, 4639.
[34] Braverman, S.; Cherkinsky, M.; Meridor, D.; Sprecher, M., *Tetrahedron*, (2010) **66**, 1925.
[35] Braverman, S.; Pechenick, T.; Gottlieb, H. E., *Tetrahedron Lett.*, (2003) **44**, 777.
[36] Braverman, S.; Kumar, E. V. K. S.; Cherkinsky, M.; Sprecher, M.; Goldberg, I., *Tetrahedron Lett.*, (2000) **41**, 6923.
[37] Kitagaki, S.; Ohdachi, K.; Katoh, K.; Mukai, C., *Org. Lett.*, (2006) **8**, 95.
[38] de Lera, A. R.; Castro, A.; Torrado, A.; López, S., *Tetrahedron Lett.*, (1998) **39**, 4575.
[39] Iglesias, B.; Torrado, A.; de Lera, A. R., *J. Org. Chem.*, (2000) **65**, 2696.
[40] Kantor, S. W.; Hauser, C. R., *J. Am. Chem. Soc.*, (1951) **73**, 4122.
[41] Chuard, R.; Giraud, A.; Renaud, P., *Angew. Chem. Int. Ed.*, (2002) **41**, 4323.
[42] Schweizer, E. E; Crouse, D. M.; Dalrymple, D. L., *J. Chem. Soc. D*, (1969), 354.
[43] Alajarin, M.; Bonillo, B.; Ortin, M.-M.; Sanchez-Andrada, P.; Vidal, A., *Org. Lett.*, (2006) **8**, 5645.
[44] Alajarin, M.; Bonillo, B.; Ortin, M.-M.; Sanchez-Andrada, P.; Vidal, A.; Orenes, R.-A., *Org. Biomol. Chem.*, (2010) **8**, 4690.
[45] Alajarin, M.; Bonillo, B.; Marin-Luna, M.; Sanchez-Andrada, P.; Vidal, A.; Orenes, R.-A., *Tetrahedron*, (2012) **68**, 4672.
[46] Alajarin, M.; Bonillo, B.; Ortin, M.-M.; Sanchez-Andrada, P.; Vidal, A., *Eur. J. Org. Chem.*, (2011), 1896.

[47] Hayashi, R.; Feltenberger, J. B.; Lohse, A. G.; Walton, M. C.; Hsung, R. P., *Beilstein J. Org. Chem.*, (2011) **7**, 410.
[48] Feltenberger, J. B.; Hsung, R. P., *Org. Lett.*, (2011) **13**, 3114.
[49] Hayashi, R.; Ma, Z.-X.; Hsung, R. P., *Org. Lett.*, (2012) **14**, 252.
[50] Snider, B. B.; Phillips, G. B., *J. Am. Chem. Soc.*, (1982) **104**, 1113.
[51] Snider, B. B.; Phillips, G. B.; Cordova, R., *J. Org. Chem.*, (1983) **48**, 3003.
[52] Suzuki, T.; Miyajima, Y.; Suzuki, K.; Iwakiri, K.; Koshimizu, M.; Hirai, G.; Sodeoka, M.; Kobayashi, S., *Org. Lett.*, (2013) **15**, 1748.
[53] Mikami, K.; Takahashi, K.; Nakai, T., *J. Am. Chem. Soc.*, (1990) **112**, 4035.
[54] Barriault, L.; Deon, D. H., *Org. Lett.*, (2001) **3**, 1925.
[55] Gibbs, R. A.; Okamura, W. H., *J. Am. Chem. Soc.*, (1988) **110**, 4062.
[56] Pelc, M. J.; Zakarian, A., *Org. Lett.*, (2005) **7**, 1629.
[57] Ilardi, E. A.; Isaacman, M. J.; Qin, Y.-c.; Shelly, S. A.; Zakarian, A., *Tetrahedron*, (2009) **65**, 3261.
[58] Steinmeyer, A.; Schwede, W.; Bohlmann, F., *Liebigs Ann. Chem.*, (1988), 925.
[59] Quinkert, G.; Schwartz, U.; Stark, H.; Weber, W.-D.; Adam, F.; Baier, H.; Frank, G.; Dürner, G., *Liebigs Ann. Chem.*, (1982), 1999.

2.1.5 Domino Transformations Initiated by or Proceeding Through [3,3]-Sigmatropic Rearrangements

C. A. Guerrero

General Introduction

The domino processes[1] discussed in this section are those that are either initiated by, or proceed through, [3,3]-sigmatropic rearrangements, primarily the Claisen[2] and Cope[3] [3,3]-sigmatropic rearrangements; in-depth reviews of these classic reactions are available.[4–7] Given that virtually all organic reactions may be classified as proceeding through one of four general classes of reaction mechanisms, i.e. polar, free radical, carbene, or pericyclic, it is only logical that domino reactions are found that include the Cope, Claisen, or other rearrangements as part of a more complex, domino sequence. Moreover, as there is no mechanistic prohibition on the amalgamation of mechanistic types in a domino reaction, there is no requirement that domino processes begin with, proceed through, or end with the same type of mechanism from this list of general classes of reaction mechanisms. Consequently, an enormous number of processes might well be contained in this section (and others in this volume)! However, due to space limitations and also for the convenience of the reader, certain restrictions have been placed on the material covered. First, only those domino sequences that are initiated by or proceed through an intermediate [3,3]-sigmatropic rearrangement are discussed. In other words, for a domino sequence to be included here, the [3,3]-sigmatropic rearrangement involved must lead to further and significant structural transformation. Consequently, domino sequences wherein the [3,3]-sigmatropic rearrangement is the terminating event (e.g., Davies' tandem C—H activation/divinylcyclopropane [3,3]-sigmatropic rearrangement[8]) are not covered. By contrast, double (or tandem) Claisen or Cope processes are discussed, since an initial [3,3]-sigmatropic event installs the functionality required for the second rearrangement. Additionally, several variants of both the Cope and Claisen rearrangements require the generation of a 1,5-diene (or heteroatom-bearing equivalent), necessitating flexible selection criteria for this discussion. For example, the widely used Ireland–Claisen, Johnson–Claisen, and oxy-Cope rearrangements, to name a few, involve the generation of reactive intermediates prior to the actual, concerted rearrangement. Again, although such considerations for inclusion (or exclusion) may seem arbitrary, they were adopted to maximize coverage of those domino processes containing Cope or Claisen rearrangements that would be of highest synthetic utility and thus of most interest, noting that some excellent work is not presented.

In a more global sense, the successful and often highly valuable incorporation of Cope and Claisen rearrangements into domino sequences stems largely from three factors. First, the requisite 1,5-dienes/heterodienes are intermediates that are easily prepared by myriad methods, either in advance of the domino transformation or in situ. Second, these sigmatropic rearrangements often proceed through transition states that are both easily predictable and highly organized, enabling firm predictions to be made regarding the configuration and constitution of the resulting reactive intermediates and/or products. Third, these nascent intermediates themselves often operate in exceedingly diverse reaction manifolds of their own (e.g., enolates are subject to myriad further chemistry); it is perhaps this final point that lends domino reactions involving these pericyclic processes to the dramatic structural diversifications discussed.

for references see p 225

SAFETY: The starting materials and products of the domino transformations described in this section should be treated with the same care as any other organic molecule with incompletely tested biological, physiological, or chemical properties. Additionally, any user who attempts experiments described therein (or similar ones or ones inspired by the chemical content of this section) should do so only if they are experienced in the handling of reactive organic and organometallic reagents, wearing appropriate personal protective equipment. Experimentalists should also use extreme caution when generating, handling, and quenching the bases used herein (e.g., potassium hydride, potassium *tert*-butoxide, potassium hexamethyldisilazanide, and others) as well as organometallic agents (e.g., Grignard reagents, organolithiums, organoceriums) as these classes of reagents are all pyrophoric and pose significant fire hazards in any laboratory. All personnel are strongly advised to use air-free handling techniques, seek the advice of more experienced researchers, and be aware of and trained to use appropriate personal protective equipment and fire extinguishers. A recent article on the storage, handling, and disposal of pyrophoric reagents provides highly relevant recommended practices.[9]

2.1.5.1 Cope Rearrangement Followed by Enolate Functionalization

2.1.5.1.1 Anionic Oxy-Cope Rearrangement Followed by Intermolecular Enolate Alkylation with Alkyl Halides

The functionalization of the nascent enolate generated as a consequence of an anionic oxy-Cope rearrangement[10,11] often serves as a more obvious method for accomplishing multiple transformations in a single reaction vessel. Although alkylation with reactive electrophiles introduced after rearrangement does not qualify as a domino reaction according to the strictest definition thereof,[1] this sequence is included for discussion because doing so illustrates the basic framework of a reaction that is initiated, or at least includes, an early [3,3]-sigmatropic rearrangement. Thus, it serves as a foundation for discussing more-elaborate sequences.

The simplest, and most common, method to initiate anionic oxy-Cope rearrangement/alkylation sequences is simple deprotonation of an alcohol suitably functionalized for oxy-Cope rearrangement by a potassium base, followed by alkylation of the resulting enolate after the rearrangement is complete. This tactic has frequently been employed by the Paquette group,[12,13] especially in studies aimed at a synthesis of taxol.[14–18] Thus, treatment of tertiary carbinol **1** with potassium hexamethyldisilazanide and 18-crown-6 in tetrahydrofuran followed by the addition of iodomethane gives methylated, ring-expanded ketone **2** in 73% isolated yield (Scheme 1).[17] This example is fairly representative of this "domino" sequence; however, it has more generality than might be immediately apparent. For example, it may be surprising to note that the anionic variant of the oxy-Cope rearrangement need not be accessed to perform this tandem reaction. For example, there is an instance of a tandem oxy-Cope rearrangement/alkylation sequence that proceeds under purely thermal agitation.[19]

Scheme 1 Anionic Oxy-Cope Rearrangement/Enolate Alkylation Domino Sequence[17]

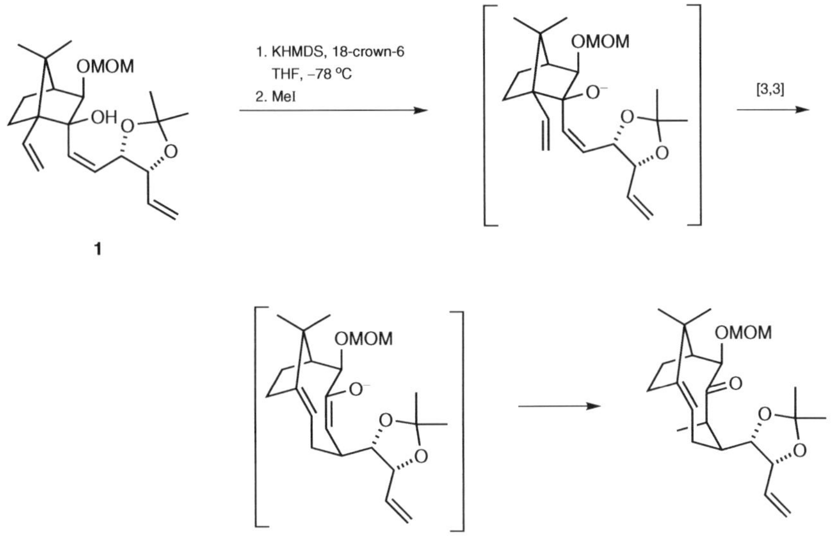

A logical, but powerful, extension of this tandem sequence is the initiation of the anionic oxy-Cope rearrangement by the addition of an organometallic nucleophile to a suitable carbonyl group, generating the requisite alkoxide. Somewhat remarkably, although potassium is the best metal cation in anionic oxy-Cope rearrangements, there are numerous examples of domino sequences involving lithium and halomagnesium enolates where rearrangement occurs well below ambient temperature. For example, treatment of ketone **3** with cyclopent-1-enylithium at –78 °C results in rapid consumption of substrate via carbonyl addition and spontaneous anion-assisted oxy-Cope rearrangement (at –78 °C). Subsequent addition of iodomethane at the same temperature and gradual warming to ambient temperature gives ketone **4** in an excellent 96% yield (Scheme 2).[20]

Scheme 2 Organometallic Carbonyl Addition/Anionic Oxy-Cope Rearrangement/Enolate Alkylation Domino Sequence[20]

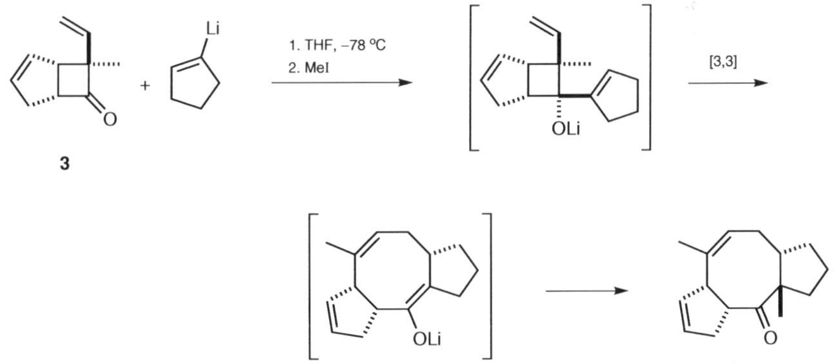

for references see p 225

(1*S*,2*R*,*E*)-5-[(4*S*,5*R*)-2,2-Dimethyl-5-vinyl-1,3-dioxolan-4-yl]-2-(methoxymethoxy)-4,11,11-trimethylbicyclo[6.2.1]undec-7-en-3-one (2); Typical Procedure:[17]

CAUTION: *Inhalation, ingestion, or skin absorption of iodomethane can be fatal.*

To a soln of alcohol **1** (905 mg, 2.39 mmol) and 18-crown-6 (695 mg, 2.63 mmol) in dry THF (50 mL) at −78 °C was added 0.5 M KHMDS in toluene (7.2 mL, 3.6 mmol). After 45 min at −78 °C, MeI (750 μL, 12.0 mmol) was introduced, and the mixture was warmed to 0 °C over 1 h. Sat. aq $NaHCO_3$ was added and the separated aqueous phase was extracted with EtOAc. The combined organic layers were washed with brine, dried, and concentrated. The residue was subjected to chromatography (silica gel, hexanes/EtOAc 8:1) to provide **2** as a colorless oil; yield: 683 mg (73%).

(3a*S,6a*S**,9a*R**,10a*S**,*Z*)-6,10a-Dimethyl-2,3,3a,4,6a,9,9a,10a-octahydrodicyclopenta[*a*,*d*]-[8]annulen-10(1*H*)-one (4); Typical Procedure:**[20]

CAUTION: *Inhalation, ingestion, or skin absorption of iodomethane can be fatal.*

Lithium wire (1.4 g, 0.2 mol) was flattened into a foil and scraped clean while submersed in petroleum ether. Following rapid transfer to a dry flask containing anhyd THF (20 mL), 1-bromocyclopent-1-ene (4.7 g, 32.2 mmol) in anhyd THF (10 mL) was added at a rate such that gentle reflux was maintained (ca. 1 h). After the olive-green soln had cooled to rt, transfer to a separate dry flask was accomplished via cannula. The residual lithium fragments were washed with dry THF (5 mL) and this rinse soln was also transferred. The organometallic soln was cooled to −78 °C and treated dropwise with a soln of cyclobutanone **3** (2.0 g, 13.5 mmol) in anhyd THF (5 mL). After 1 h at −78 °C, freshly filtered (basic alumina) MeI (9.0 g, 64 mmol) was added dropwise and stirring was maintained as the mixture was allowed to warm to rt over 2–3 h. H_2O (50 mL) was added and the aqueous layer was extracted with Et_2O (3 × 50 mL). The combined organic layers were dried and concentrated to provide a yellow oil, which was purified by flash chromatography (silica gel, petroleum ether) to give ketone **4** as a colorless oil; yield: 3.0 g (96%).

2.1.5.1.2 Anionic Oxy-Cope Rearrangement Followed by Enolate Alkylation by Pendant Allylic Ethers

Paquette has expressed the opinion[12] that "the most fascinating development so far is the discovery made independently by two research groups that medium ring enolates generated by anionic oxy-Cope rearrangement possess the latent ability to undergo transannular cyclization with concomitant S_N' displacement of a methoxy leaving group"; this discovery was made independently and simultaneously by Schreiber.[21] Thus, treatment of tertiary allylic alcohol **5** with potassium hydride in tetrahydrofuran at reflux leads to anionic oxy-Cope rearrangement followed immediately by S_N' ether displacement, giving ketones **6A** and **6B** in 47 and 24% yield, respectively (Scheme 3).[22] As would be expected, this process has geometrical requirements. The methoxy group of the initial anionic oxy-Cope product must be aligned with the π-electrons of the allyl fragment to which it is bonded if the methoxy group is to depart. If it is not then the allylic displacement does not occur and the products isolated are those deriving only from anionic oxy-Cope rearrangement.

Scheme 3 Anionic Oxy-Cope Rearrangement/Intramolecular S_N' Displacement Domino Sequence[21,22]

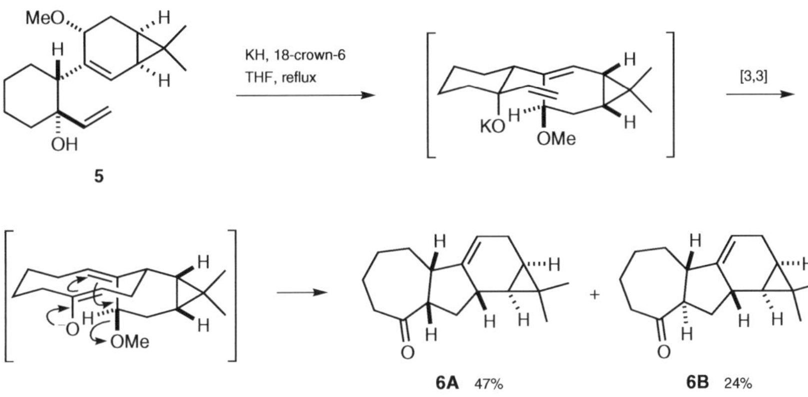

MeO H H H OH 5 KH, 18-crown-6 THF, reflux KO H OMe H H [3,3] O H OMe H H H H O H H H 6A 47% H H O H H H 6B 24%

(1a*R*,3b*S*,8a*S*,9a*S*,9b*R*)-1,1-Dimethyl-1,1a,2,3b,4,5,6,7,8a,9,9a,9b-dodecahydro-8*H*-cyclopropa[3,4]benzo[1,2-*a*]azulen-8-one (6A) and (1a*R*,3b*S*,8a*R*,9a*S*,9b*R*)-1,1-Dimethyl-1,1a,2,3b,4,5,6,7,8a,9,9a,9b-dodecahydro-8*H*-cyclopropa[3,4]benzo[1,2-*a*]azulen-8-one (6B); Typical Procedure:[22]

To a suspension of KH (24% in oil; 140 mg, 0.826 mmol) and 18-crown-6 (219 mg, 0.827 mmol) in dry THF (10 mL) was added a soln of alcohol **5** (45.8 mg, 0.165 mmol) in dry THF (5 mL). The mixture was stirred at rt for 45 min and at reflux for 2 h, cooled to −78 °C, and quenched with sat. aq NH_4Cl (3 mL). The resultant mixture was poured into 20% aq NH_4Cl (15 mL) and extracted with Et_2O (3 × 50 mL). The combined organic layers were washed with H_2O and brine, dried, and concentrated. Purification by flash column chromatography (silica gel, EtOAc/petroleum ether 2:98) gave **6A** [yield: 19.9 mg (47%)] and **6B** [yield: 9.6 mg (24%)] both as colorless oils.

2.1.5.1.3 Anionic Oxy-Cope Rearrangement Followed by Enolate Acylation

In experiments analogous to Paquette and Schreiber's transannular allylic displacement, the enolate ions stereospecifically generated in the wake of an anionic oxy-Cope rearrangement may also be trapped intramolecularly by ester electrophiles, comprising a terminating Dieckmann cyclization. Indeed, such process are somewhat general and lead to highly complex, polycyclic architectures.[23,24] The diastereoselective addition of substituted Grignard reagents **8** to the ketone function of α-vinyl-β-oxo esters **7** furnishes a transient alkoxide that undergoes anionic sigmatropy followed by enolate acylation with the pendant ester to give bicycle **9** (Scheme 4). There are three key points to this reaction:[23] First, despite the intricate nature of the domino sequence, involving the formation and fragmentation of several C—C bonds; the sequence is stereospecific, meaning that the alkene geometry of the Grignard reagent is faithfully transmitted to the alkyl groups at the β and γ positions relative to both carbonyls in the product diketone. Second, silyl-protected ethers are tolerated under the reaction conditions. Third, and perhaps most interestingly, ordinary Grignard reagents, which are generated from vinyllithiums by transmetalation with magnesium bromide in situ, were the only organometallic agents identified that underwent the full, triple reaction domino sequence. Organolithiums themselves or their organocerium derivatives are competent nucleophiles that, after facile carbonyl addition, undergo anionic sigmatropy. However, the newly formed enolates do not give products of transannular cyclization; instead, products of enolate quenching are isolated. Thus, it can be concluded that these lithium and cerium counterparts are not sufficiently reactive to undergo the terminating Dieckmann cyclization. Alternatively, it may be that

for references see p 225

the products of such acylations are unstable and revert to enolate-bearing materials. Regardless, easily obtained Grignard reagents are optimally suited to bring about the full domino transformation, the generality of which was demonstrated in a subsequent report.[24]

Scheme 4 Grignard Carbonyl Addition/Anionic Oxy-Cope Rearrangement/Dieckmann Cyclization Domino Sequence[23,24]

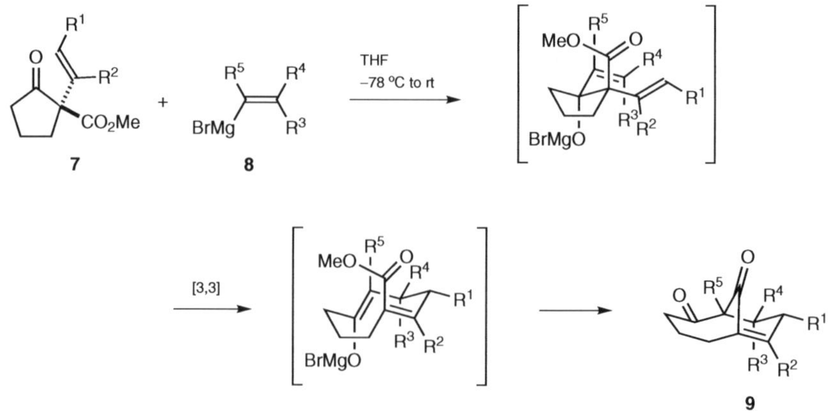

R^1	R^2	R^3	R^4	R^5	Yield (%)	Ref
$(CH_2)_5Me$	H	Me	H	H	65	[23]
$(CH_2)_5Me$	H	H	Me	H	30	[23]
$(CH_2)_5Me$	H	H	$CH_2CH_2OTBDMS$	H	64	[23]
$(CH_2)_5Me$	H	H	$(CH_2)_3$		82	[24]
$(CH_2)_5Me$	H	H	$(CH_2)_4$		80	[24]
$(CH_2)_3$		H	$(CH_2)_2OTBDMS$	H	73	[24]
$(CH_2)_4$		H	$(CH_2)_2OTBDMS$	H	84	[24]

The robustness of this domino sequence and its applicability to target-directed synthesis at the limits of molecular complexity are highlighted by the most efficient (step-economical) approach to any member of the phomoidride family of polyketide natural products. In the synthesis of CP-263,114,[25] carbonyl addition of Grignard **11** to β-oxo ester **10** initiates stereospecific anionic oxy-Cope rearrangement followed by spontaneous Dieckmann cyclization, ultimately giving diketone **12** (Scheme 5). Notably, a number of protecting groups for critical functionality are compatible, including 4-methoxybenzyl ethers, methoxymethyl acetals, and ketals. Overall, the application of this domino methodology to the total synthesis of these structures very strongly underlines its utility and reliability in complex molecular settings.

Scheme 5 Domino Sequence as Applied to the Total Synthesis of CP-263,114 (Phomoidride B)[25]

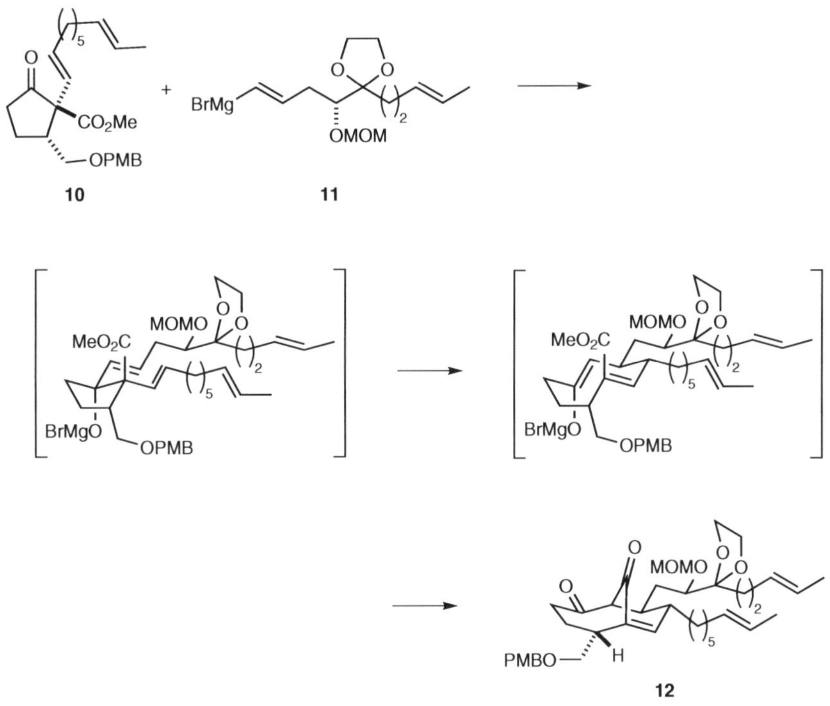

(1*S,8*S**,9*R**)-8-Hexyl-9-methylbicyclo[4.3.1]dec-6-ene-2,10-dione [9, $R^1 = (CH_2)_5Me$; $R^2 = R^4 = R^5 = H$; $R^3 = Me$]; Typical Procedure:**[23]

A soln of 0.49 M (Z)-prop-1-enylmagnesium bromide (**8**, $R^3 = Me$; $R^4 = R^5 = H$) in THF (3.7 mL, 1.8 mmol, 2.1 equiv) was added to a suspension of β-oxo ester **7** [$R^1 = (CH_2)_5Me$; $R^2 = H$; 0.214 g, 0.85 mmol, 1 equiv] and 4-Å molecular sieves (0.79 g) in THF (8 mL) at −78 °C. The suspension was stirred at −78 °C for 1.5 h, at 0 °C for 1 h, and at rt for 12 h. The mixture was quenched by the addition of glacial AcOH (0.12 mL, 1.8 mmol, 2.1 equiv) and stirred for 10 min. The mixture was filtered through a pad of Celite, which was washed extensively with Et_2O (100 mL), and the filtrate was concentrated. Purification of the residue by flash column chromatography (silica gel, CH_2Cl_2/hexanes 1:1 to 7:3 to 1:0) afforded the product as a colorless oil; yield: 0.145 g (65%).

2.1.5.2 Aza- and Oxonia-Cope-Containing Domino Sequences

2.1.5.2.1 Ionization-Triggered Oxonia-Cope Rearrangement Followed by Intramolecular Nucleophilic Trapping by an Enol Silyl Ether

The discussion in Section 2.1.5.1 has focused only on all-carbon 3-hydroxy-bearing hexa-1,5-dienes undergoing Cope rearrangement and further bond-forming processes. Only by considering similar structures bearing heteroatoms as part of the six-atom chain, however, can the full richness and diversity of domino reactions containing Cope rearrangements be fully appreciated. Although there are myriad examples of such reactivity, only a few selected examples will be discussed.

One such set of processes involves trapping of intermediates rapidly equilibrating via oxonia-Cope rearrangement[26] with a pendant nucleophilic alkene, generating differentially substituted complex tetrahydropyrans. One process is set into motion by an initial ionization–Prins reaction (Scheme 6).[27] In simpler substrates, such equilibrating rearrangements cannot be observed; however, when a pendant nucleophilic alkene is present in the substrate, the kinetic preference for 6-*endo-trig* versus 4-*exo-trig* capture of the equilibrating oxocarbenium ions leads to the formation of tetrahydropyranones, some bearing all-carbon quaternary centers. Thus, structures of type **13** are initially ionized using strong Lewis acids at low temperature. The oxocarbenium ion **14** so generated constitutes a 2-oxahexa-1,5-diene, and enters into a rapid equilibrium of oxonia-Cope rearrangements. However, only one of these isomers can undergo facile nucleophilic trapping to productively terminate the domino sequence, giving diastereomers **15A** and **15B**. Based on the products observed, it can be concluded that the nucleophilic trapping event proceeds through a chair-like transition state. The high stereochemical fidelity of this event, coupled to the expected and presumably operative chair-like transition state of the oxonia-Cope rearrangement, permits rigorous prediction of the stereochemical outcome of this domino sequence. Notably, this intricate, yet high-yielding, set of events is a beautiful and textbook illustration of the Curtin–Hammett principle being leveraged in a synthetically useful manner.

Scheme 6 Ionization/Oxonia-Cope Rearrangement/Oxocarbenium Trapping Domino Sequence for the Synthesis of Highly Substituted Tetrahydropyranones[27]

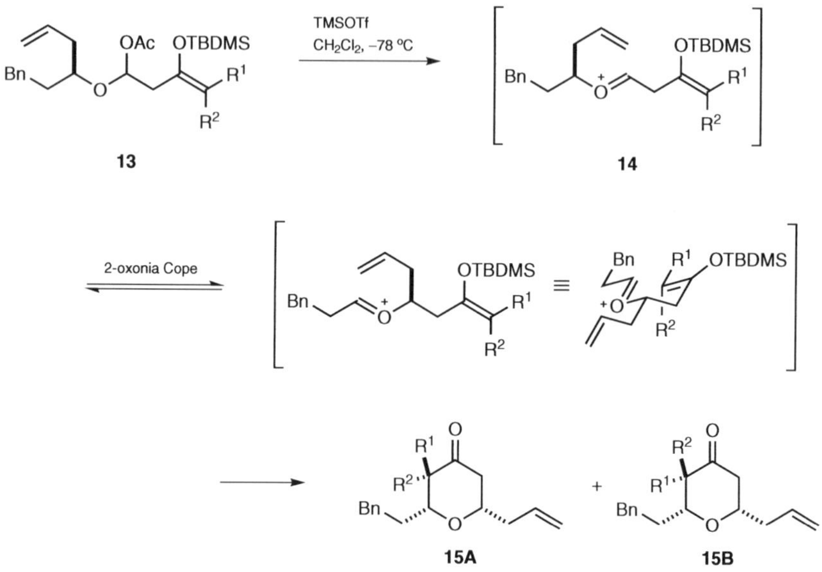

R^1	R^2	Ratio (**15A**/**15B**)	Combined Yield (%)	Ref
H	H	–	99	[27]
Et	H	1.6:1	99	[27]
CH=CH$_2$	H	2.1:1	84	[27]
Me	Me	–	92	[27]
(CH$_2$)$_5$		–	93	[27]
Ph	Me	–	88	[27]
CH$_2$OTBDPS	Me	1.2:1	77	[27]

(2*R,3*R**,6*S**)-6-Allyl-3-ethyl-2-(2-phenylethyl)tetrahydro-4*H*-pyran-4-one** (15A, **R^1 = Et; R^2 = H) and (2*R**,3*S**,6*S**)-6-Allyl-3-ethyl-2-(2-phenylethyl)tetrahydro-4*H*-pyran-4-one** (15B, **R^1 = Et; R^2 = H); Typical Procedure:**[27]
TMSOTf (36 μL, 201 μmol, 3.0 equiv) was added to a cooled (−78 °C) soln of α-acetoxy ether **13** (R^1 = Et; R^2 = H; 30 mg, 67 μmol, 1.0 equiv) in CH_2Cl_2 (1.3 mL). The mixture was stirred for 30 min at −78 °C until no starting acetoxy ether remained; the mixture was quenched with sat. aq $NaHCO_3$ (2 mL). The aqueous layer was extracted with Et_2O (3 × 10 mL), and the combined organic layers were washed with brine (10 mL), dried ($MgSO_4$), and concentrated. Purification [silica gel, Et_2O/hexanes 1:19 to 1:9 (R_f 0.20)] provided products **15A** [yield: 11 mg (61%)] and **15B** [yield: 7 mg (38%)].

2.1.5.2.2 Intermolecular 1,4-Addition-Triggered Oxonia-Cope Rearrangement Followed by Intramolecular Nucleophilic Trapping by a Nascent Enolate

In addition to the oxonia-Cope/oxocarbenium trapping sequence in described Section 2.1.5.2.1, a domino sequence that generates 3-acyltetrahydropyrans by a related process, wherein an equilibrating oxonia-Cope process is terminated by an in situ generated enol, has been developed. However, in contrast to the reaction reported in Section 2.1.5.2.1, the reaction in this section is initiated by the intermolecular attack of a Michael acceptor on a vinyl ether function in the substrate.[28] Thus, treatment of vinyl ether **16** with methyl vinyl ketone (**17**) and a strong Lewis acid elicits an initial intermolecular 1,4-addition (Scheme 7). The nascent oxocarbenium ion **18** formed then spontaneously enters into an oxonia-Cope equilibrium. However, and again in contrast to the reaction in Section 2.1.5.2.1,[27] the terminating nucleophile in this instance is the in situ generated titanium enolate that results from 1,4-addition, rather than a pre-generated silyl enol ether, ultimately furnishing 3-acyltetrahydrofuran **19**. Overall, this impressive set of events is formally equivalent to a Diels–Alder cycloaddition between an electron-poor alkene and an *O*-vinyl oxocarbenium ion. Various Michael acceptors are suitable co-substrates for this reaction. Thus, although methyl vinyl ketone is the most common Michael acceptor, ethyl acrylate and even cyclohexenone are also suitable Michael acceptors.

Scheme 7 Mukaiyama/Michael/Oxonia-Cope Rearrangement/Oxocarbenium Trapping Domino Sequence for the Synthesis of Highly Substituted 3-Acyltetrahydropyrans[28]

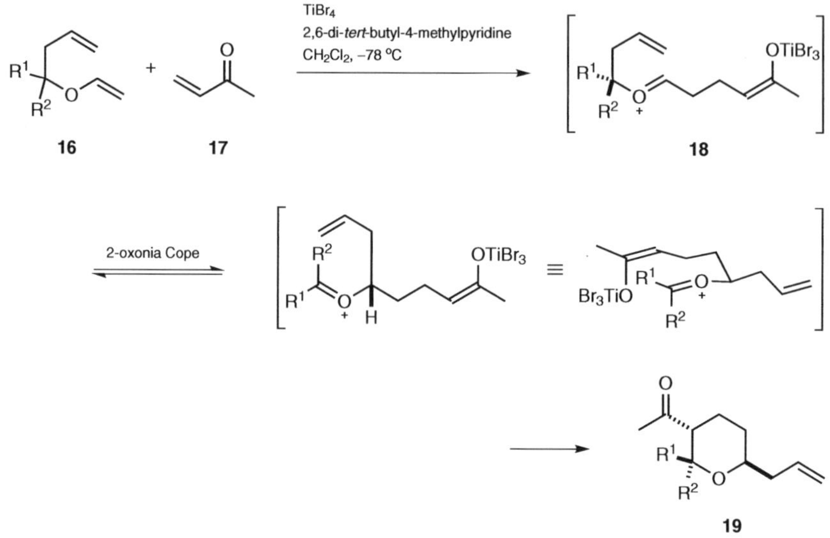

$TiBr_4$
2,6-di-*tert*-butyl-4-methylpyridine
CH_2Cl_2, −78 °C

16 + 17 → [18]

2-oxonia Cope

→ 19

R^1	R^2	Yield (%)	Ref
$(CH_2)_2Ph$	H	74	[28]
$(CH_2)_2OBn$	H	72	[28]
$(CH_2)_2OTIPS$	H	63	[28]
iBu	H	72	[28]
$(CH_2)_5Me$	H	74	[28]
Ph	H	65	[28]
$(CH_2)_5$		74	[28]

(2*S,3*R**,6*S**)-3-Acetyl-6-allyl-2-(2-phenylethyl)tetrahydro-2*H*-pyran [19, $R^1 = (CH_2)_2Ph$; $R^2 = H$]; Typical Procedure:**[28]

A soln of methyl vinyl ketone (**17**; 350 μL, 4.20 mmol, 2.0 equiv), vinyl ether **16** [$R^1 = (CH_2)_2Ph$; $R^2 = H$; 424 mg, 2.10 mmol, 1.0 equiv], and 2,6-di-*tert*-butyl-4-methylpyridine (647 mg, 3.15 mmol, 1.5 equiv) in CH_2Cl_2 (20.0 mL) was cooled to −78 °C. A soln of $TiBr_4$ (0.474 mL, 4.20 mmol, 2.0 equiv) was then added dropwise by syringe. The mixture was then stirred at −78 °C for 1 h. Sat. aq $NaHCO_3$ (15.0 mL) was added at −78 °C, and the mixture was allowed to warm to rt. The layers were separated and the aqueous layer was extracted with CH_2Cl_2 (3 × 10 mL). The combined organic layers were dried (Na_2SO_4) and concentrated under reduced pressure. The crude oil was then purified by flash chromatography (Et_2O/hexanes 1:19 to 1:4) to afford the product as a colorless oil; yield: 429 mg (74%).

2.1.5.2.3 Iminium-Ion-Formation-Triggered Azonia-Cope Rearrangement Followed by Intramolecular Nucleophilic Trapping by a Nascent Enamine

While the use of 2-oxonia-Cope rearrangements as part of an overall domino sequence is relatively undeveloped, the incorporation of the nitrogen analogue of this process, the 2-azonia-Cope rearrangement,[29] into domino processes has seen continuous, rigorous development and application over the past 35 years by the Overman group.[30] Although the inaugural report[31] of this now highly systematized methodology first appeared as a methodological study, it is worth noting that it was developed logically and rationally in response to the chemical problem posed by a laboratory synthesis of gephyrotoxin, a poison-dart alkaloid.[32,33]

As with other domino sequences discussed in this section and in the strictest sense, the aza-Cope/Mannich reaction does not commence with a [3,3]-sigmatropic rearrangement, but rather requires the in situ formation of a hexa-1,5-diene. Condensation of an aldehyde with a homoallylic amine **20** that also bears a hydroxy group at the allylic position gives an unsaturated iminium ion **21** that may then enter into a [3,3]-sigmatropic rearrangement manifold (Scheme 8). This process is greatly facilitated by the cationic nature of intermediate **21**, enabling rearrangement to occur 100–200 °C lower than a conventional all-carbon (non-oxy) Cope rearrangement. By design, the "forward" product of azonia-Cope rearrangement generates an enol **22** that is suitably positioned to engage the highly electrophilic iminium ion by 5-*endo-trig* nucleophilic capture. The final product is a 3-acylpyrrolidine **23** and, from a retrosynthetic design standpoint, this structural feature is the retrosynthetic keying element that signals the applicability of using Overman's aza-Cope/Mannich sequence to construct such systems.

Scheme 8 Iminium-Ion-Formation-Triggered Azonia-Cope Rearrangement Followed by Intramolecular Nucleophilic Trapping by a Nascent Enamine: The Overman Aza-Cope/Mannich Domino Sequence[31,33]

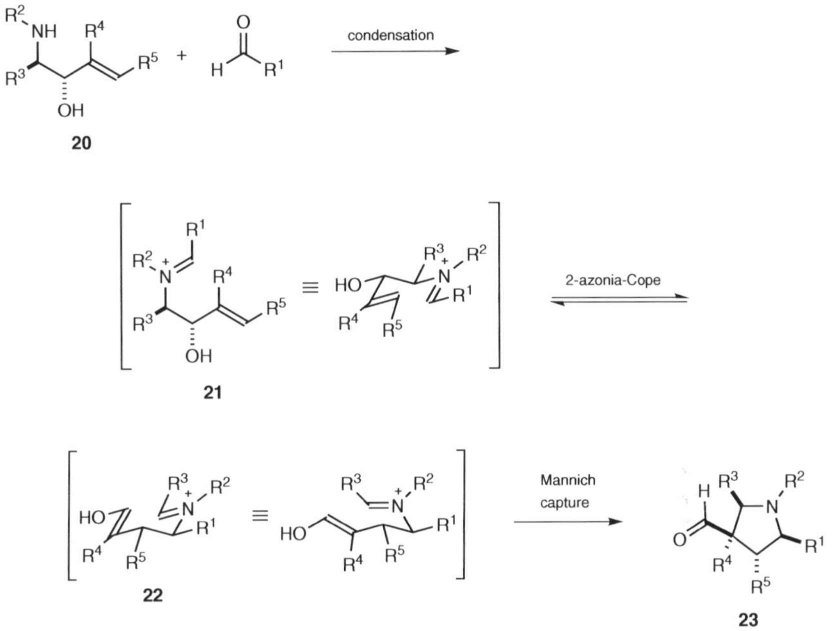

The scope of this domino sequence is notably broad. Understandably, it has been utilized, not only by the Overman group, but also several other groups, for the synthesis of numerous alkaloid natural products; a selection are shown in Scheme 9.[34–46] Scheme 10 illustrates a relatively simple Overman aza-Cope/Mannich reaction of **24** to give 2-substituted 4-acetylpyrrolidines **25**,[33] while Scheme 11 illustrates the strategic application of this powerful domino reaction for the reaction of **26** to give **27** in the enantioselective synthesis of strychnine.[40,41] Notably, the reaction works well in both simple acyclic and more complex polycyclic settings using relatively simple reactions and reaction conditions. Its popularity derives from its stereochemical predictability and the fact that it enables its users to generate a high degree of molecular complexity, particularly because the initial condensation serves as a robust fragment union event.

for references see p 225

Scheme 9 A Selection of Alkaloid Natural Products Synthesized Using the Aza-Cope/Mannich Domino Sequence[34–46]

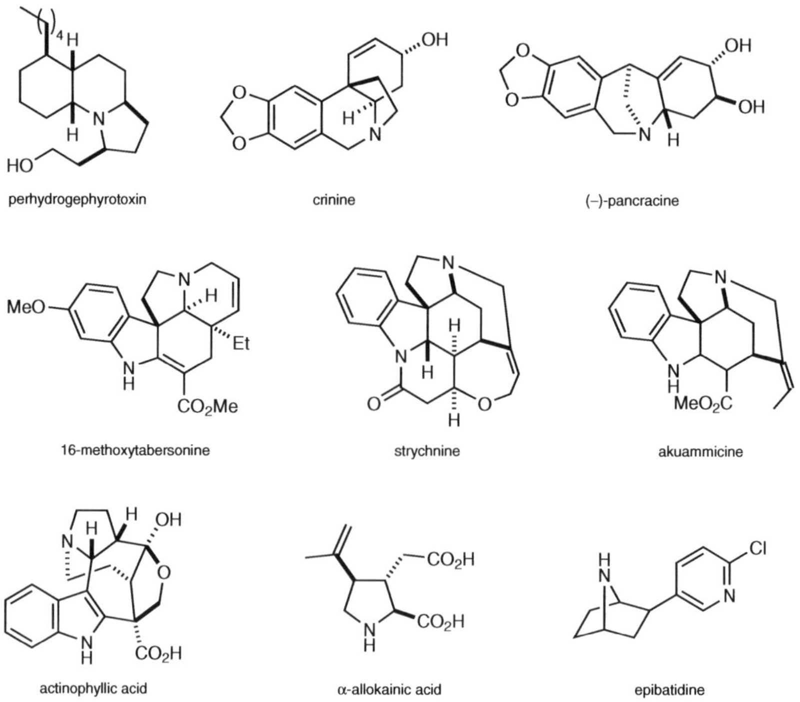

perhydrogephyrotoxin

crinine

(−)-pancracine

16-methoxytabersonine

strychnine

akuammicine

actinophyllic acid

α-allokainic acid

epibatidine

Scheme 10 Aza-Cope/Mannich Domino Sequence as Applied to a Simple, Acyclic System[31,33]

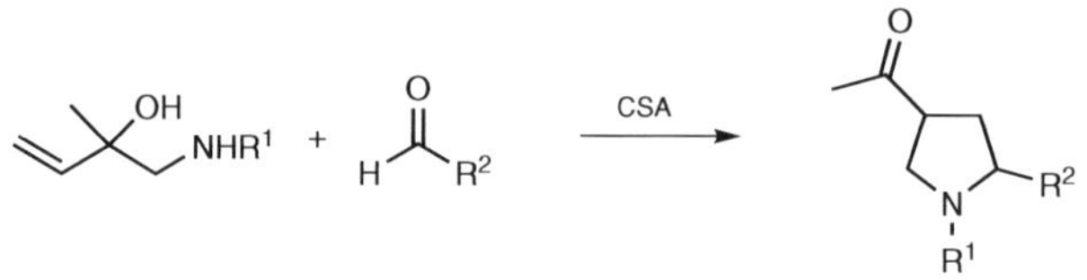

24 + $HC(O)R^2$ —CSA→ **25**

R^1	R^2	Temp (°C)	Yield (%)	Ref
Bn	H	80	89	[33]
Pr	H	79	85	[33]
Ph	H	110	66	[33]
2-ClC_6H_4	H	110	74	[33]
Bn	Ph	24	94	[33]
Ph	Ph	110	56	[33]
4-FC_6H_4	$(CH_2)_4Me$	110	83	[33]
Bn	2-furyl	80	95	[33]
Me	3-pyridyl	80	84	[33]

Scheme 11 Aza-Cope/Mannich Domino Sequence as Applied to the Synthesis of Strychnine[40,41]

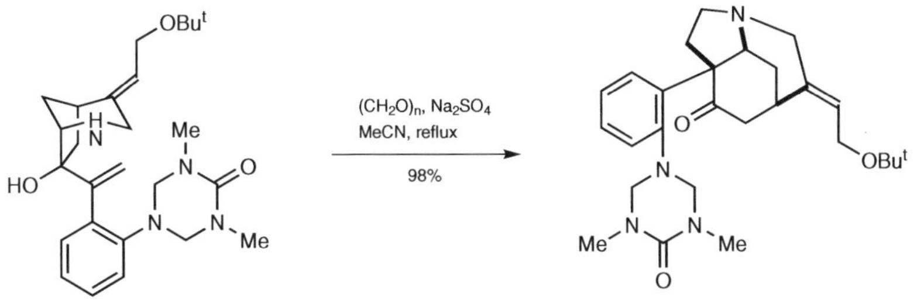

3-Acetyl-1-benzylpyrrolidine (25, R^1 = Bn; R^2 = H); Typical Procedure:[33]
A mixture of amino alcohol **24** (R^1 = Bn; 450 mg, 2.35 mmol), paraformaldehyde (78 mg, 2.6 mmol), CSA (495 mg, 2.13 mmol), and dry benzene (5 mL) (**CAUTION:** *carcinogen*) was heated at reflux for 10 h. After cooling to rt, 1 M NaOH (3 mL) was added, the organic layer was separated, and the aqueous layer was extracted with Et_2O (3 × 5 mL). The combined organic extracts were dried (Na_2SO_4) and concentrated, and the residue was purified by bulb-to-bulb distillation (oven temperature 95–105 °C/0.4 Torr) to give the product as a colorless oil; yield: 425 mg (89%).

(3a*R*,6*R*,7a*S*,*E*)-8-[2-(*tert*-Butoxy)ethylidene]-3a-[2-(3,5-dimethyl-4-oxo-1,3,5-triazinan-1-yl)phenyl]hexahydro-1,6-ethanoindol-4(2*H*)-one (27); Typical Procedure:[41]
A mixture of azabicyclooctane **26** (1.6 g, 3.5 mmol), Na_2SO_4 (4.9 g, 35 mmol), paraformaldehyde (320 mg, 11 mmol), and MeCN (40 mL) was heated at reflux for 10 min and then concentrated. Purification of the residue by flash chromatography ($CHCl_3$/MeOH 19:1) gave the product as a colorless oil; yield: 1.6 g (98%).

2.1.5.3 Double, Tandem Hetero-Cope Rearrangement Processes

2.1.5.3.1 Double, Tandem [3,3]-Sigmatropic Rearrangement of Allylic, Homoallylic Bis(trichloroacetimidates)

In the 1970s, Overman reported the [3,3]-sigmatropic rearrangement of allylic trichloroacetimidates to give allylic trichloroacetamides simply upon heating.[47,48] In a clever application of this process, the Chida group described double, tandem Overman rearrangements of allylic, homoallylic bis(trichloroacetimidate) structures for the synthesis of diastereomerically pure vicinal diamines.[49–51] This tactic for stereocontrolled synthesis of diamines capitalizes on the ease with which diols are prepared in stereochemically pure form and leverages this facility to generate stereochemically pure diamines, which are more typically prepared only with considerable difficulty. Thus, heating allylic bis(trichloroacetimidate) **28A** or **28B** in *tert*-butylbenzene in the presence of sodium carbonate under sealed tube conditions elicits an initial Overman trichloroacetimidate rearrangement (Scheme 12). However, this initial rearrangement generates a new allylic trichloroacetimidate, such as **29**, that then itself undergoes [3,3]-sigmatropic rearrangement, furnishing bis(trichloroacetamide) **30A** or **30B**.[51]

for references see p 225

Scheme 12 Double, Tandem [3,3]-Sigmatropic Rearrangement of Allylic, Homoallylic Bis(trichloroacetimidates)[51]

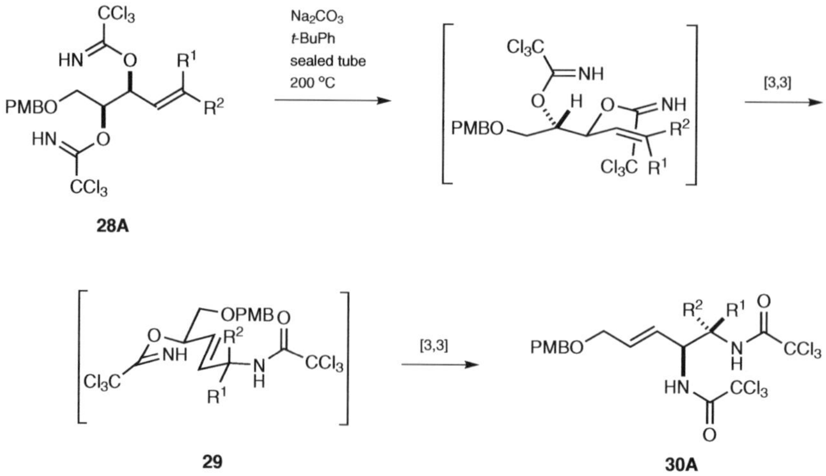

R^1	R^2	Yield (%)	Ref
CH_2OBn	H	95	[51]
H	CH_2OBn	89	[51]
Me	CH_2OBn	90	[51]

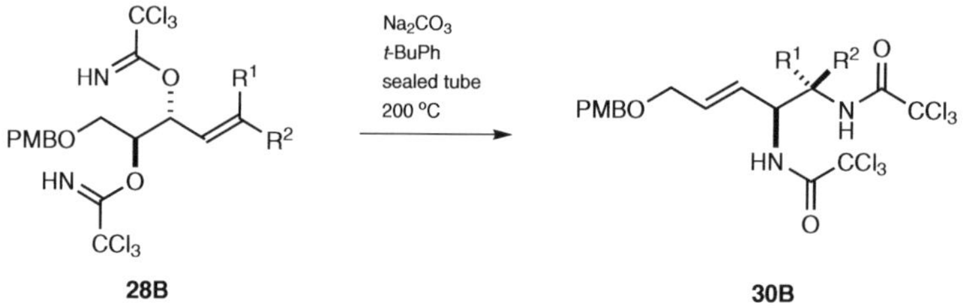

R^1	R^2	Yield (%)	Ref
CH_2OBn	H	73	[51]
H	CH_2OBn	95	[51]
Me	CH_2OBn	81	[51]

(2*S*,3*S*,*E*)-1-(Benzyloxy)-6-(4-methoxybenzyloxy)-*N*,*N*′-bis(trichloroacetyl)hex-4-ene-2,3-diamine (30A, **$R^1 = CH_2OBn$; $R^2 = H$); Typical Procedure:**[51]

A sealed tube was charged with bisimidate **28A** ($R^1 = CH_2OBn$; $R^2 = H$; 42.5 mg, 65.7 μmol), Na_2CO_3 (2.8 mg, 26.3 μmol), and *tert*-butylbenzene (2.2 mL). The soln was then purged with a flow of argon for 5 min, and heated to 200 °C for 40 min. After cooling to rt, the resulting mixture was directly purified by column chromatography (silica gel, EtOAc/hexane 1:9 to 1:2) to give the product as colorless crystals; yield: 40.2 mg (95%).

2.1.5.4 Neutral Claisen Rearrangement Followed by Further (Non-Claisen) Processes

2.1.5.4.1 Oxy-Cope Rearrangement/Ene Reaction Domino Sequences

An important variation on post-[3,3]-sigmatropic rearrangement functionalization is to engage the carbonyl group (rather than an enol/enolate) of the immediate product in further pericyclic processes. The groups of Sutherland[52] and Rajagopalan[53–55] made early contributions to the development of tandem, neutral oxy-Cope/ene domino processes. The Barriault group has also studied this process in detail,[56] identifying structural factors that can derail the domino sequence; these contributions have been reviewed.[57,58] When 3-hydroxyhexa-1,5-dienes such as **31** are heated at high temperatures (typically 220 °C), oxy-Cope rearrangement occurs followed by tautomerization and spontaneous transannular ene cyclization to give **32** (Scheme 13).[56] Further examples are illustrated in Table 1. The entire sequence is highly diastereoselective as a consequence of: (1) the faithful transmission of relative stereochemical information of the starting materials to the alkene geometry of the macrocycle, and (2) the tendency of transannular ene reactions to proceed through chair-like transition states. Barriault noted that this "strategy offers a simple highly diastereoselective method for the synthesis [of] polycyclic structures with a tertiary alcohol at [a] ring junction."[56] To underscore the utility of this domino process, a concise ten-step synthesis of sesquiterpene natural product arteannuin was completed.[59]

Scheme 13 Mechanism of the Thermal Oxy-Cope Rearrangement/Ene Reaction Domino Sequence[56]

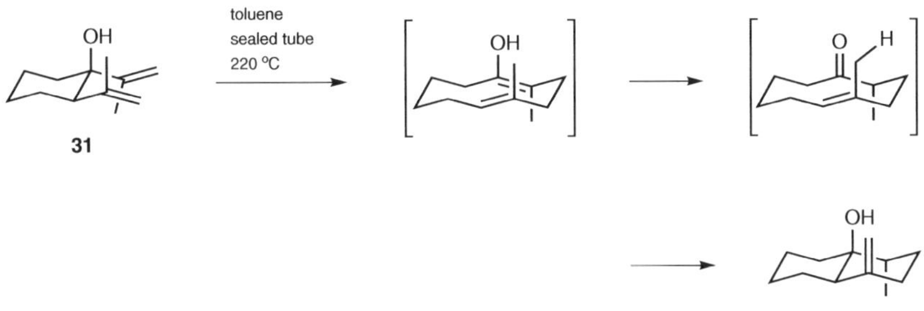

OH
toluene
sealed tube
220 °C
31
OH
O H
OH
32

for references see p 225

Table 1 Thermal Oxy-Cope Rearrangement/Ene Reaction Domino Sequence[56]

toluene
sealed tube
220 °C

Entry	Substrate	Product	Yield (%)	Ref
1			72	[56]
2			63	[56]
3			44	[56]
4			70	[56]
5	**33**	**34**	83	[56]

(4b*R*,8a*S***,10a*R***)-1,3,4b,5,6,7,8,9,10,10a-Decahydrophenanthren-8a(2*H*)-ol** (34); **Typical Procedure:**[56]

A soln of diene **33** (24 mg, 0.12 mmol) in dry toluene (5 mL) was heated in a pressure tube (previously washed with aq iPrOH/NaOH soln, H_2O, and acetone) for 5 h at 220 °C. The tube was cooled to rt, and the soln was transferred and concentrated. The residue was purified by flash chromatography (Et_2O/hexanes 1:9) to give the product as a colorless oil; yield: 20 mg (83%).

2.1.5.4.2 Oxy-Cope Rearrangement/Ene Reaction/Claisen Rearrangement and Oxy-Cope Rearrangement/Claisen Rearrangement/Ene Reaction Domino Sequences

Although the impressive oxy-Cope/ene domino sequence developed by the Barriault group (Section 2.1.5.4.1) builds an impressive level of complexity, a limitation was immediately identified.[57] Specifically, it was noted that quaternary stereocenters cannot be constructed at one of the bridgehead centers. To address this shortcoming, starting materials were constructed that contained an allyl group tethered to the core carbocyclic motif by an ether function. The logic behind incorporation of this design element is that after the initial oxy-Cope rearrangement/ene domino sequence, a Claisen rearrangement would ensue (Scheme 14). In one experiment, heating ether **35** in toluene at high temperature (220 °C) for several hours led smoothly to lactol **36**, the product of the desired oxy-Cope rearrangement/tautomerization/ene reaction/Claisen rearrangement sequence, in 75% isolated yield.[60] Equally impressive, this domino sequence has been generalized across a range of differentially substituted ethers **37**, demonstrating that diverse decahydronaphthalenes **38** with unique side chains can be synthesized very efficiently with this protocol (Scheme 15). The utility of this newly developed methodology in complex molecule synthesis has been demonstrated by preparing the core structures of tetradecomycin[61] and desdimethylambliol B.[62]

Scheme 14 Mechanism of the Oxy-Cope Rearrangement/Ene Reaction/Claisen Rearrangement Domino Sequence[60]

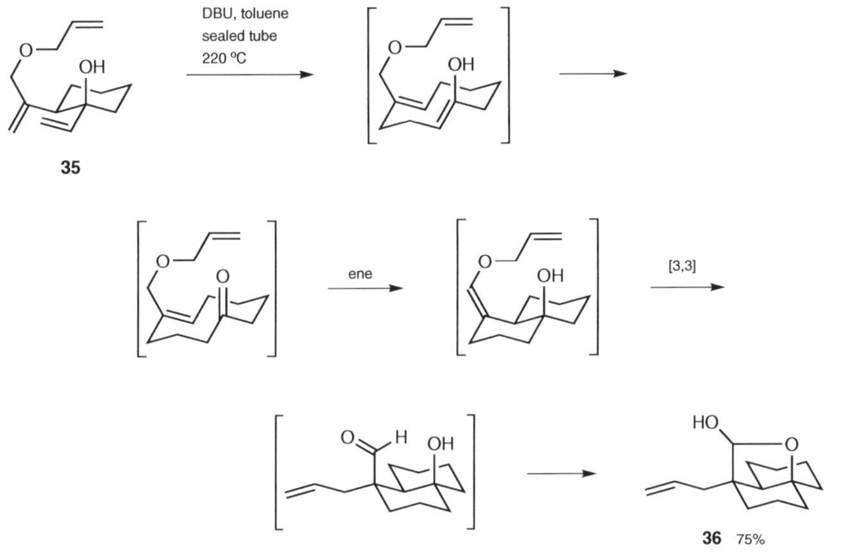

DBU, toluene
sealed tube
220 °C

35

ene

[3,3]

36 75%

for references see p 225

Scheme 15 Oxy-Cope Rearrangement/Ene Reaction/Claisen Rearrangement Domino Sequence[60]

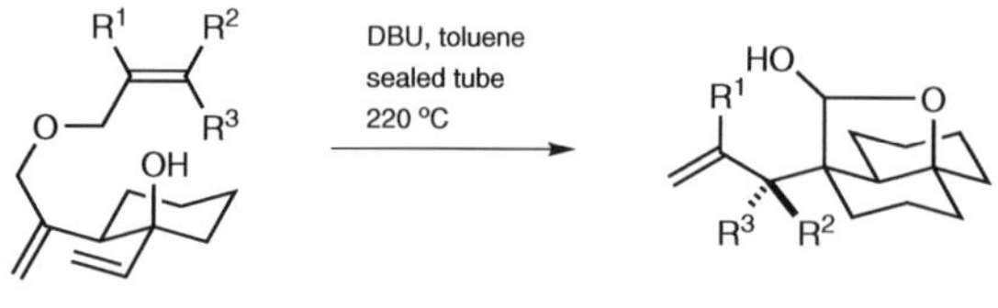

R^1	R^2	R^3	Yield (%)	Ref.
H	H	H	75	[60]
H	Ph	H	75	[60]
$(CH_2)_4$		H	60	[60]
H	Me	Me	76	[60]
H	Ph	Me	75	[60]

A related synthesis of decahydronaphthalenes involving the same three pericyclic reactions, although ordered differently, has also been reported. Thus, heating of allyl dienyl ether **39** at 200 °C in either dimethylformamide or toluene under microwave irradiation gives decalin **42** via the sequence of: (1) oxy-Cope rearrangement to give macrocyclic structure **40**, (2) Claisen rearrangement to reveal a macrocyclic ketone **41**, and (3) final ene reaction to give the desired product (Scheme 16);[63] this domino sequence has again been generalized. Moreover, this new methodology has been applied to the synthesis of a complex molecular target, the skeleton of the *neo*-clerodane diterpenoid teucrolivin A.[64] Finally, domino sequences are also available that involve four pericyclic processes for the synthesis of complex carbocycles, the fourth being an intramolecular Diels–Alder cycloaddition.[65]

Scheme 16 Oxy-Cope Rearrangement/Claisen Rearrangement/Ene Reaction Domino Sequence[63]

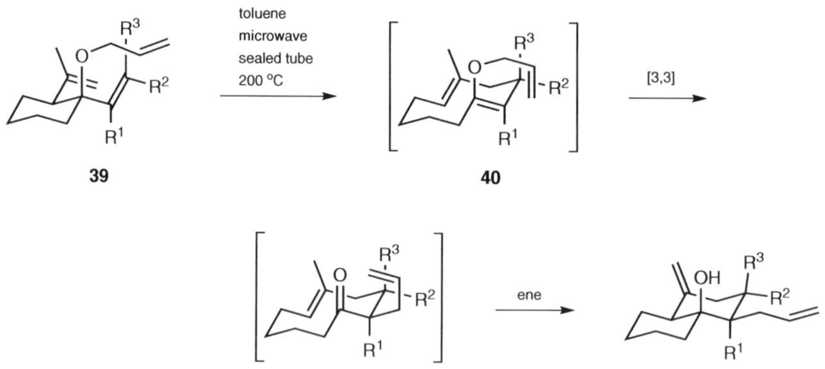

R^1	R^2	R^3	Temp (°C)	dr	Yield (%)	Ref
Me	H	H	200	2.5:1	87	[63]
OEt	H	H	200	>25:1	91	[63]
SEt	H	H	200	3:1	91	[63]

R^1	R^2	R^3	Temp (°C)	dr	Yield (%)	Ref
Me	H	Me	200	4:1	73	[63]
Me	Me	H	210	>25:1	98	[63]
Me	$CH_2OTBDMS$	H	180	17:1	90	[63]

(1*S,4a*S**,8a*R**)-Octahydro-2*H*-4a,1-(epoxymethano)naphthalen-10-ol** (38, $R^1 = R^2 = R^3 = H$); **Typical Procedure:**[60]
A soln of allyl ether **37** ($R^1 = R^2 = R^3 = H$; 80 mg, 0.36 mmol) in dry, deoxygenated toluene (10 mL) and DBU (110 mg, 0.72 mmol) was heated in a sealed quartz tube (previously washed with aq iPrOH/NaOH soln, H_2O, and acetone) for 60 min at 220 °C. The soln was cooled to rt, and then transferred and concentrated. The residue was purified by flash chromatography (EtOAc/hexanes 1:4) to afford the product as a colorless oil; yield: 60 mg (75%).

(4*R,4a*S**,8a*S**)-4-Allyl-4-ethoxy-1-methyleneoctahydronaphthalen-4a(2*H*)-ol** (42, $R^1 = OEt$; $R^2 = R^3 = H$); **Typical Procedure:**[63]
A soln of allyl ether **39** ($R^1 = OEt$; $R^2 = R^3 = H$; 38 mg, 0.152 mmol) in toluene (15 mL) was degassed with argon for 15 min and then heated in a microwave to 200 °C. After 2 h, the mixture was concentrated and purified by chromatography (silica gel, Et_2O/hexanes 4:96, basified with Et_3N) to give the product as a colorless oil; yield: 35 mg (91%).

2.1.5.5 Claisen Rearrangement Followed by Another Pericyclic Process

2.1.5.5.1 Double, Tandem Bellus–Claisen Rearrangement Reactions

In studies aimed toward the development of a catalytic, asymmetric Claisen rearrangement, it has been noted that Lewis acids catalyze the Bellus–Claisen[66,67] rearrangement between allylic amines and in situ generated ketenes (from carboxylic acid chlorides).[68] It has also been noted that suitably constructed allylic diamines undergo consecutive Bellus–Claisen rearrangements to furnish stereochemically rich pimelic acid diamides (heptanedioic acid diamides, Scheme 17).[69] In this very synthetically useful reaction, ketenes are generated in situ by dehydrochlorination of a carboxylic acid chloride through the action of a tertiary amine base in the presence of allylic diamines (e.g., **43**) and a Lewis acid. Subsequent reversible ketene trapping then leads to diastereomeric quaternary ammonium salts **44**, only one of which does not suffer from significant $A^{1,2}$ strain and undergoes Bellus–Claisen rearrangement. However, as the product of this rearrangement **45** still bears a tertiary allylic amine, it can and does enter into a second ketene-trapping/Bellus–Claisen sequence, with strain-minimization still serving as a diastereoselectivity determining element, giving product **46**. This reaction sequence has been generalized (Scheme 18) and it has proven to be a powerful method for the rapid establishment of acyclic stereochemical complexity.

for references see p 225

Scheme 17 Mechanism of the Double Bellus–Claisen Rearrangement[69]

$Yb(OTf)_3$
CH_2Cl_2

43

(*E*)-**44** (productive)

(*Z*)-**44** (non-productive)

45

"methyl over chair" (productive)

"carboxamide over chair" (non-productive)

46

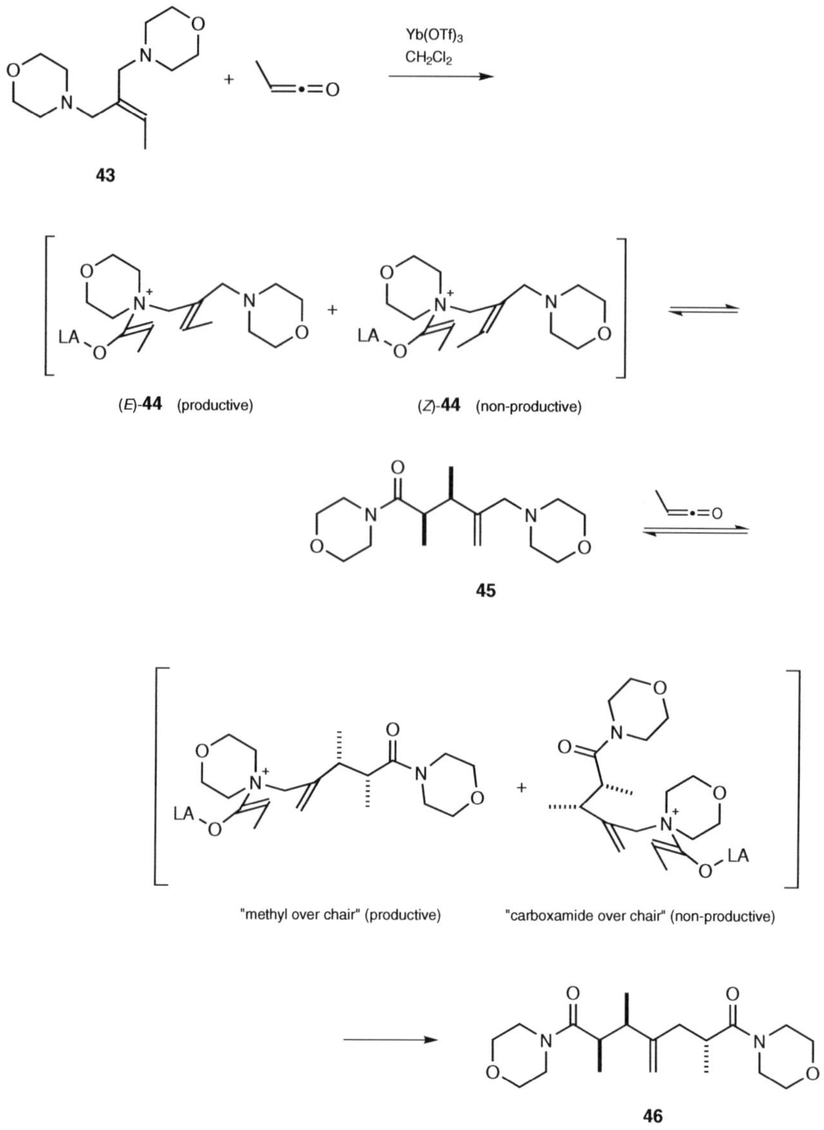

Scheme 18 Double Bellus–Claisen Rearrangement[69]

Lewis acid
iPr_2NEt
CH_2Cl_2, rt

47

48

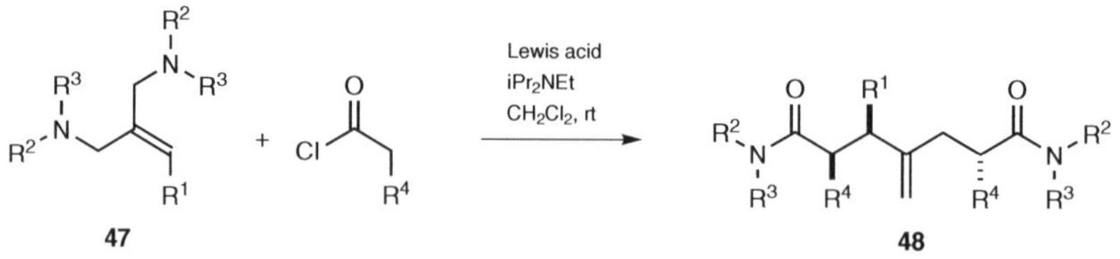

R^1	R^2	R^3	R^4	Lewis Acid	dr	Yield (%)	Ref
Me	$(CH_2)_2O(CH_2)_2$		Me	$Yb(OTf)_3$	98:2	97	[69]
Me	$(CH_2)_4$		Me	$Yb(OTf)_3$	95:5	90	[69]
Me	$(CH_2)_5$		Me	$Yb(OTf)_3$	96:4	99	[69]
Cl	$(CH_2)_2O(CH_2)_2$		Me	$Yb(OTf)_3$	99:1	98	[69]
OBz	$(CH_2)_2O(CH_2)_2$		Me	$Yb(OTf)_3$	91:9	86	[69]
CN	$(CH_2)_2O(CH_2)_2$		Me	$Yb(OTf)_3$	97:3	78	[69]
SPh	$(CH_2)_2O(CH_2)_2$		Me	$Yb(OTf)_3$	93:7	70	[69]
Me	$(CH_2)_2O(CH_2)_2$		$(CH_2)_2Ph$	$Yb(OTf)_3$	92:8	99	[69]
Me	$(CH_2)_2O(CH_2)_2$		CH_2NPhth	$Yb(OTf)_3$	95:5	98	[69]
Me	$(CH_2)_2O(CH_2)_2$		CH_2OCO*t*-Bu	$TiCl_4$•2THF	97:3	97	[69]
OBz	$(CH_2)_2O(CH_2)_2$		CH_2OCO*t*-Bu	$TiCl_4$•2THF	92:8	71	[69]

(2*R,3*R**,6*R**)-2,3,6-Trimethyl-4-methylene-1,7-dipiperidinoheptane-1,7-dione [48, $R^1 = R^4 = Me$; $R^2,R^3 = (CH_2)_5$]; Typical Procedure:**[69]
To a flask containing $Yb(OTf)_3$ (258 mg, 0.416 mmol) was added diamine **47** [$R^1 = Me$; $R^2,R^3 = (CH_2)_5$; 50.0 mg, 0.212 mmol] in CH_2Cl_2 (4.0 mL), followed by iPr_2NEt (0.15 mL, 0.85 mmol) at 23 °C. After 5 min, a 1 M soln of propanoyl chloride in CH_2Cl_2 (0.80 mL, 0.80 mmol) was added dropwise over 1 min. The resulting dark red soln was maintained at 23 °C until diamine **47** was consumed (TLC monitoring, EtOAc; 4–6 h). The mixture was then diluted with EtOAc (20 mL) and washed with 1 M aq NaOH (20 mL). The aqueous layer was then extracted with EtOAc (3 × 20 mL), and the combined organic layers were washed with brine, dried (Na_2SO_4), and concentrated. The resulting residue was purified by chromatography (silica gel, EtOAc) to provide the product as a colorless oil; 73.4 mg (99%); dr 96:4.

2.1.5.5.2 Claisen Rearrangement Followed by [2,3]-Sigmatropic Rearrangement

Among the most intricate domino cascades initiated by, or featuring, an early [3,3]-sigmatropic rearrangement are those that have another "non-[3,3]" sigmatropic rearrangement that follows. These types of sequences are often the most difficult to design during a retrosynthetic analysis; however, the structural complexity achieved can often outweigh the "setup cost" associated with generating a domino reaction substrate with appropriately placed functionality to execute the designed pathway. Given that two alkenes migrate during the [3,3]-sigmatropic rearrangement, at least one of those usually takes part in the second pericyclic process.

In studies aimed toward the purely chemical synthesis of pinnatoxin A,[70,71] a domino aliphatic Claisen/Mislow–Evans[72] sequence is used to address the constitutional and stereochemical problem posed by the spiroimine core functionality. Thus, heating dihydropyran **49** in diethylene glycol monomethyl ether at 150 °C in the presence of triethyl phosphite and 2,4,6-trimethylpyridine (as a mildly basic buffer for the sulfenic acid produced) leads efficiently to cyclohexenol **50** in 92% yield (Scheme 19).[73] If the reaction is performed using the sulfoxide diastereomer of **49** (not shown), the reaction is essentially equally efficient. Although this domino sequence was not ultimately used in the published synthesis of pinnatoxin A,[74–76] its application to a system of considerable molecular complexity bodes well for its future application in other settings.

Scheme 19 Claisen Rearrangement/Mislow–Evans Transposition Domino Sequence[73]

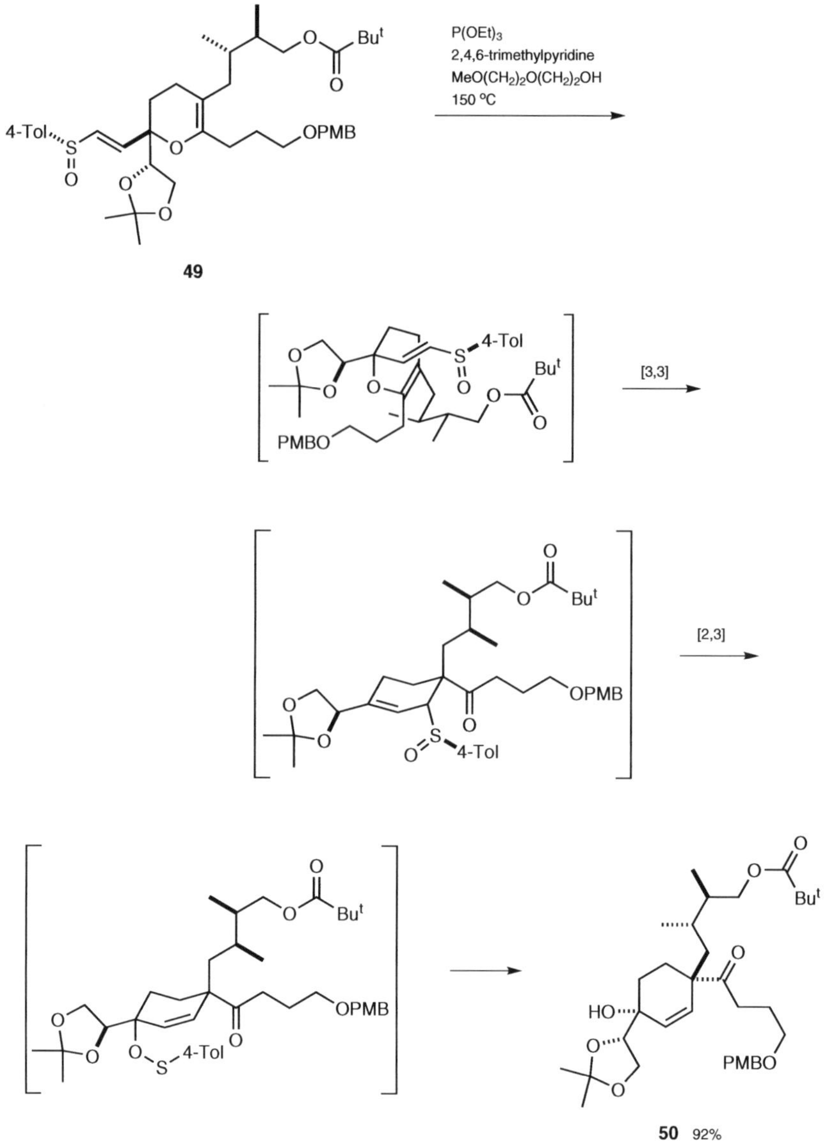

P(OEt)3
2,4,6-trimethylpyridine
MeO(CH2)2O(CH2)2OH
150 °C

49

[3,3]

[2,3]

50 92%

(1*R*,4*R*)-1-[(4*S*)-2,2-Dimethyl-1,3-dioxolan-4-yl]-4-[(2*S*,3*R*)-2,3-dimethyl-4-(pivaloyloxy)butyl]-4-[4-(4-methoxybenzyloxy)-1-oxobutyl]cyclohex-2-en-1-ol (50); Typical Procedure:[73]

A soln of vinylic sulfoxide **49** (0.154 g, 0.217 mmol), $P(OEt)_3$ (0.18 mL, 1.05 mmol), and 2,4,6-trimethylpyridine (0.14 mL, 1.05 mmol) in diethylene glycol monomethyl ether (2.7 mL) was heated at 150 °C for 15 h in a sealed vial under argon. After cooling, the mixture was diluted with EtOAc and washed with sat. aq NH_4Cl and H_2O (3 ×), and the combined aqueous layers were extracted with EtOAc. The combined organic layers were washed with brine, dried (Na_2SO_4), and concentrated. The residue was subjected to column chromatography (silica gel, EtOAc/hexanes 4:6) and dried under high vacuum (50 °C) to remove residual 2,4,6-trimethylpyridine to give the product; yield: 0.117 g (92%).

2.1.5.5.3 Claisen Rearrangement/Diels–Alder Cycloaddition Domino Sequences

Another powerful, complexity-generating set of tandem processes are those that are initiated by a Claisen rearrangement and lead immediately to a Diels–Alder cycloaddition. Within this framework, synthetic studies toward two families of natural products, the *Garcinia* caged xanthones[77] and transtaganolide[78]/basiliolide[79] family of terpenoids, constitute some of the best studied and best applied instances of domino Claisen rearrangement/Diels–Alder cycloaddition cascades, especially since their application rests on a biomimetic hypothesis.

The first report documenting the successful application of a tandem Claisen/Diels–Alder cascade toward the *Garcinia* caged xanthone family of natural products emanated from the Nicolaou laboratories.[80–82] In 2001, they reported the first total synthesis of 1-*O*-methylforbesione[79] by a pathway predicated on the probable biogenesis of this natural product put forth in 1971 by Quillinan and Scheinmann.[83] Thus, simple heating of xanthone **51** at 120 °C in dimethylformamide initiates a Claisen rearrangement wherein the catechol diether undergoes dearomatization to generate two constitutionally isomeric, conjugated cyclohexadienones **52** and **53** (Scheme 20). The existence of these highly electrophilic intermediates, however, is fleeting as they are immediately captured in a Diels–Alder cycloaddition by the pendant alkene of the remaining allylic ether, delivering the signature skeletal structure of the target natural products. While our attention has been focused on the right-hand side of these xanthone natural products, it should also be noted that Claisen rearrangement also takes place with the allylic ethers on the left side of the molecules. These other Claisen rearrangements, however, are not followed by a Diels–Alder cycloaddition as the intermediate Claisen products can and do tautomerize to restore aromaticity. The four reaction products generated reflect all four possible products one would expect from two independent Claisen rearrangements that each can give two regioisomeric products. Ultimately, Quillinan and Scheinmann's biogenetic proposal for the *Garcinia* xanthones proved too enticing for only the Nicolaou group to pursue it. Indeed, synthetic strategies incorporating their proposal in part or in full have also been studied and published by the Theodorakis[84–87] and You groups.[88–90] In the Theodorakis study,[87] use of an unprotected phenol in **54** enabled access to forbesione (**55**) itself with complete control over the regiochemistry of the Claisen/Diels–Alder domino sequence (Scheme 21).

for references see p 225

Scheme 20 Claisen Rearrangement/Diels–Alder Cycloaddition Domino Sequence for the Synthesis of the Caged *Garcinia* Xanthones[80]

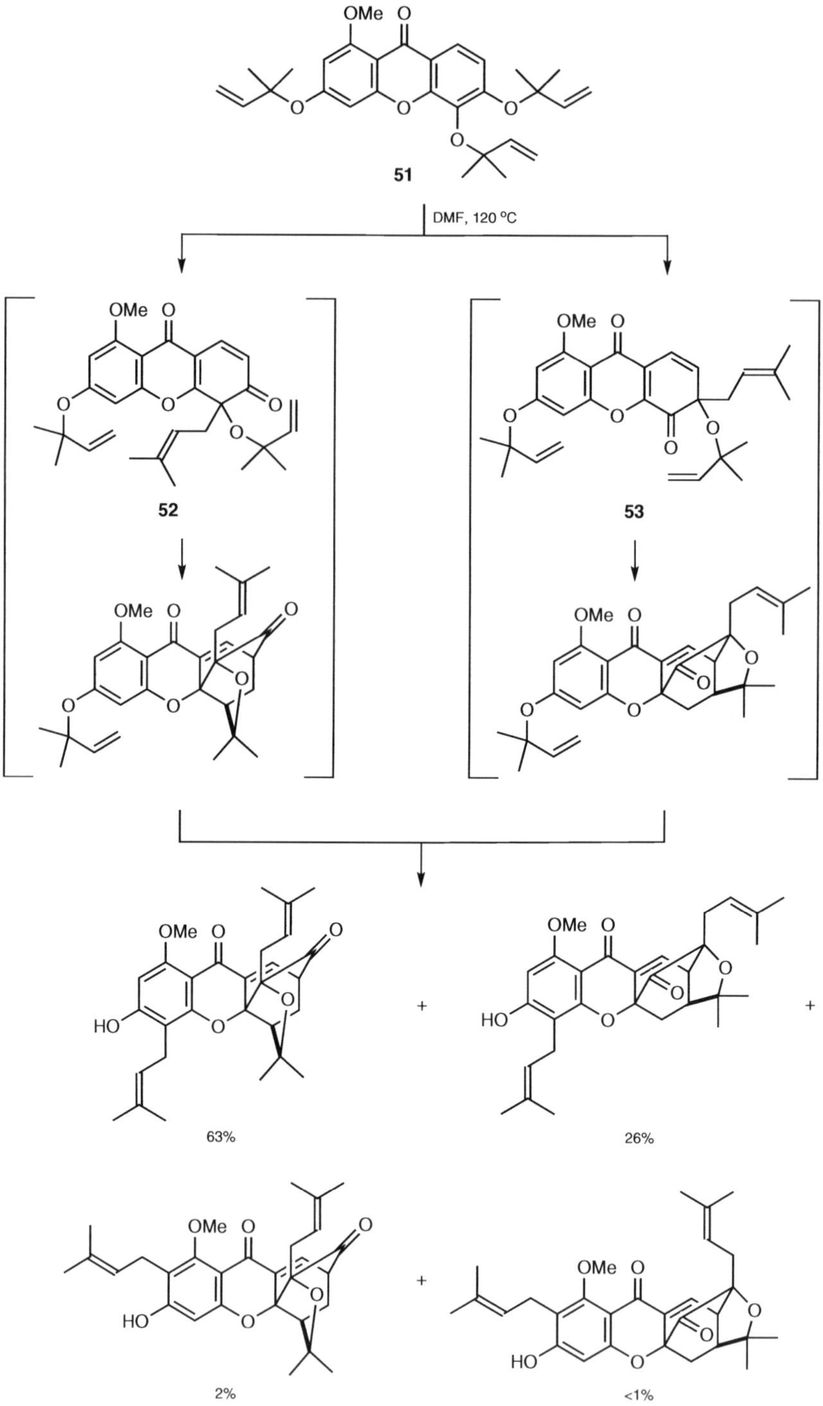

OMe O

51

DMF, 120 °C

OMe O

52

OMe O

OMe O

53

OMe O

OMe O

HO

63%

OMe O

HO

26%

OMe O

HO

2%

OMe O

HO

<1%

Scheme 21 Claisen Rearrangement/Diels–Alder Cycloaddition Domino Sequence for the Synthesis of Forbesione[87]

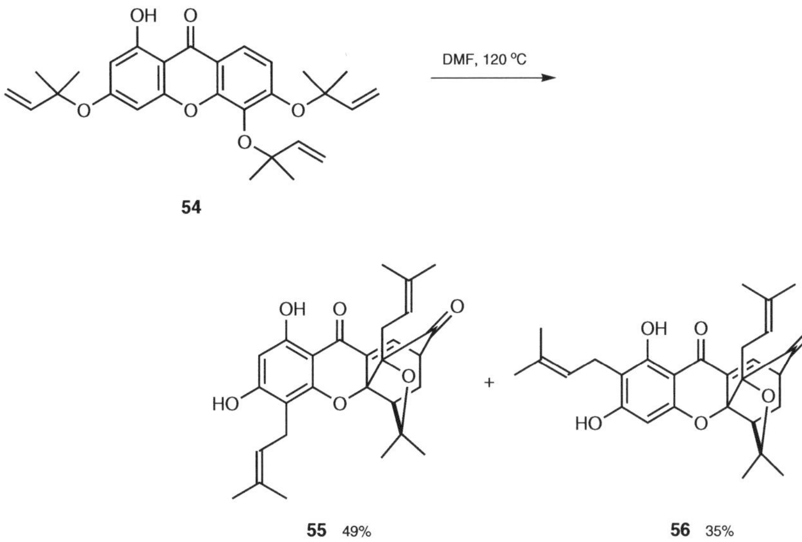

DMF, 120 °C

54

55 49%

56 35%

Another Claisen/Diels–Alder domino sequence appears in the synthesis of the basiliolide/ transtaganolide family of meroterpenoids (Scheme 22). This family of natural products has received considerable attention from the synthetic community, and includes published studies by the Lee,[91] Dudley,[92] Johansson/Sterner,[93,94] and Stoltz[95–98] research groups. However, it was Stoltz and co-workers who were the first to demonstrate that the Claisen and Diels–Alder events, thought to be operative in the biogenesis of these compounds, could be conducted in a single flask as a domino sequence.[97] Thus, treatment of pyrone **57** with *N*,*O*-bis(trimethylsilyl)acetamide and triethylamine at 100 °C for 2 days furnishes complex lactones **58A** and **58B**, the products of an impressive Ireland–Claisen/Diels–Alder cascade (Scheme 23).

Scheme 22 The Basiliolide/Transtaganolide Family of Natural Products

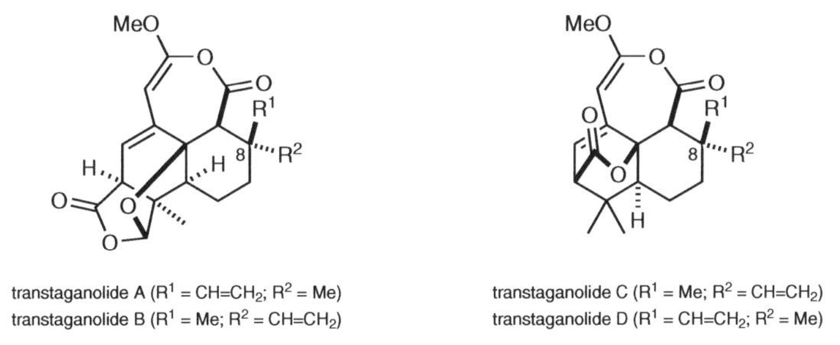

for references see p 225

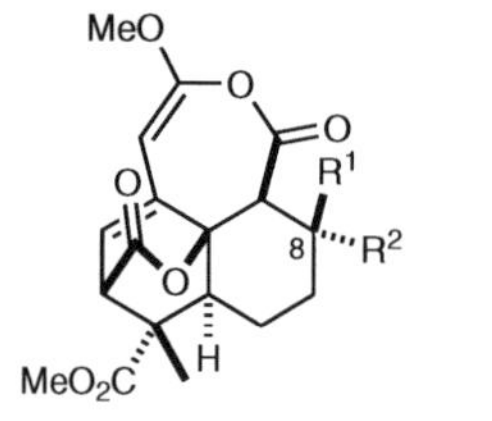

epi-8-basiliolide B (R^1 = Me; R^2 = CH=CH$_2$)
basiliolide B (R^1 = CH=CH$_2$; R^2 = Me)

Scheme 23 Ireland–Claisen Rearrangement/Diels–Alder Cycloaddition Domino Sequence[97]

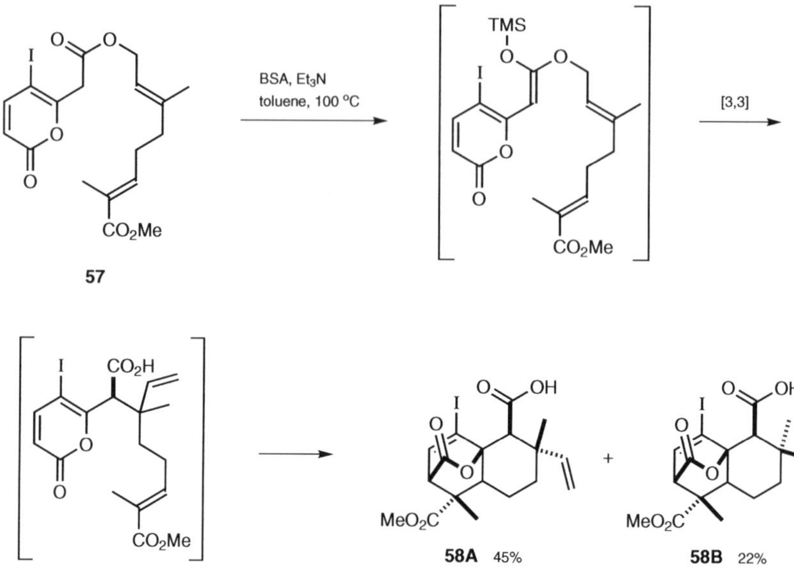

BSA, Et_3N, toluene, 100 °C; [3,3]

57

58A 45% + **58B** 22%

(1*S,3a*S**,5*S**,12a*S**)-8,10-Dihydroxy-2,2-dimethyl-3a,11-bis(3-methylbut-2-enyl)-1,2-dihydro-1,5-methanofuro[2,3-*d*]xanthene-4,7(3a*H*,5*H*)-dione (55) and (1*S**,3a*S**,5*S**,12a*S**)-8,10-Dihydroxy-2,2-dimethyl-3a,9-bis(3-methylbut-2-enyl)-1,2-dihydro-1,5-methanofuro[2,3-*d*]xanthene-4,7(3a*H*,5*H*)-dione (56); Typical Procedure:**[87]

A soln of substrate **54** (27.3 mg, 58.8 μmol) in DMF (1.00 mL) was heated at 120 °C and continuously monitored by TLC (Et_2O/hexanes 1:1). While heating, the mixture turned yellow, providing a qualitative indication of the progress of the reaction. After 1 h, the mixture was diluted with toluene and concentrated several times at 60 °C under reduced pressure to remove DMF by azeotropic distillation. The crude material was purified by preparative TLC (Et_2O/hexane 3:7; same plate several times) to give **55** [yield: 13.4 mg (49%)] and **56** [yield: 9.6 mg (35%)].

2.1.5.5.4 Claisen Rearrangement/[1,5]-H-Shift/6π-Electrocyclization Domino Sequences

Domino sequences initiated by Claisen rearrangement may be followed by other pericyclic processes beside cycloadditions. For example, a popular method for benzopyran formation starts from *O*-propargylphenols and involves heating substrates to moderate to high temperatures, sometimes under microwave irradiation, to elicit a domino sequence consisting of propargyl Claisen rearrangement,[99] [1,5]-hydrogen shift, and 6π-electrocyclization. This tactic is exemplified by the synthesis of benzopyran **63**, a key building block in the synthesis of avrainvillamide and stephacidin B (Scheme 24).[100] Here, heating substrate **59** at 140 °C induces an initial propargyl Claisen rearrangement, which temporarily compromises aromaticity and generates an β,γ,γ,δ-doubly unsaturated ketone **60**. The existence of this intermediate, however, is fleeting as it immediately undergoes tautomerization to restore aromaticity, generating a phenol **61**. This new intermediate **61** then undergoes a [1,5]-hydrogen shift to give a newly dearomatized α,β,γ,δ-doubly unsaturated ketone **62** that undergoes the final 6π-electrocyclization to restore aromaticity and furnish the ultimate product of elevated thermal treatment, benzopyran **63**.

Scheme 24 Claisen Rearrangement/[1,5]-Hydrogen Shift/6π-Electrocyclization Domino Sequence for Benzopyran Synthesis[100]

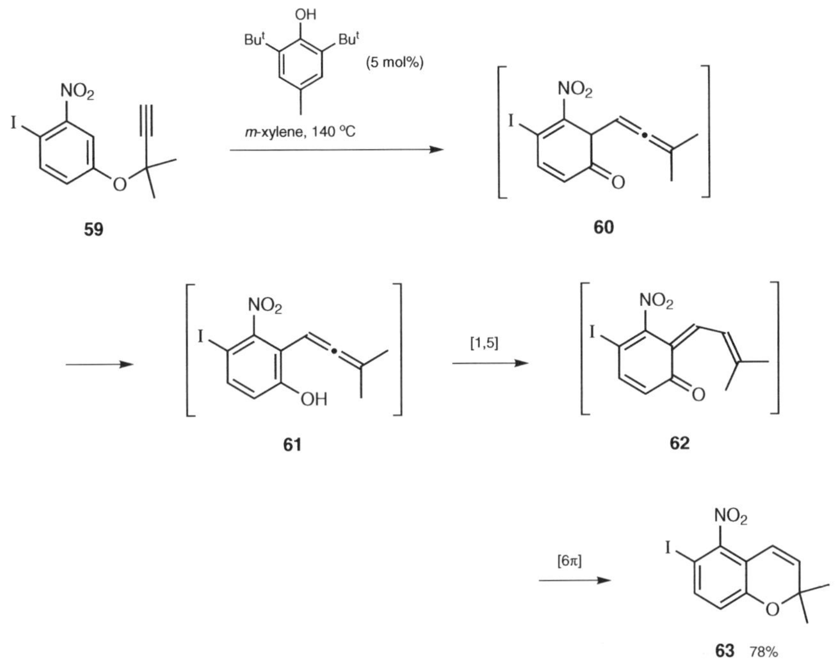

6-Iodo-2,2-dimethyl-5-nitro-2*H*-1-benzopyran (63); Typical Procedure:[100]

A soln of 1-iodo-4-(2-methylbut-3-yn-2-yloxy)-2-nitrobenzene (**59**; 526 mg, 1.59 mmol, 1 equiv) and 2,6-di-*tert*-butyl-4-methylphenol (17.6 mg, 79.4 μmol, 0.05 equiv) in *m*-xylene (15.8 mL) was heated to 140 °C for 15 h and then was allowed to cool to 23 °C. The cooled product soln was loaded onto a pad of silica gel (5 cm), eluting with hexanes (discarded) and then EtOAc/hexanes (1:4). The eluent was concentrated and the residue obtained was purified by flash column chromatography (EtOAc/hexanes 1:49) to furnish the product as a pale yellow oil; yield: 409 mg (78%).

for references see p 225

2.1.5.6 Claisen Rearrangement Followed by Multiple Processes

2.1.5.6.1 Propargyl Claisen Rearrangement Followed by Tautomerization, Acylketene Generation, 6π-Electrocyclization, and Aromatization

The domino transformation of propargyl vinyl ethers **64** gives highly substituted phenols **66** (Scheme 25).[101] The starting materials **64** are easily obtained via the conjugate addition of propargylic alcohols to propynoate esters. When heated, ethers **64** undergo a Claisen rearrangement to give allene intermediates **65**. These new species **65** then undergo numerous tautomerizations, funneling through hexatriene, cyclohexadiene, and acylketene intermediates[102] to eventually arrive at the phenol products **66**. An alternative working mechanism was presented that does not involve acylketenes; however, the intermediacy of acylketenes also explains the observed products, fits the mechanistic experimental data obtained, and more closely resembles the mechanism of the Danheiser benzannulation reaction, a process which also gives phenolic products.[103]

Scheme 25 Propargyl Claisen Rearrangement Followed by Tautomerization, Acylketene Generation, 6π-Electrocyclization, and Aromatization[101]

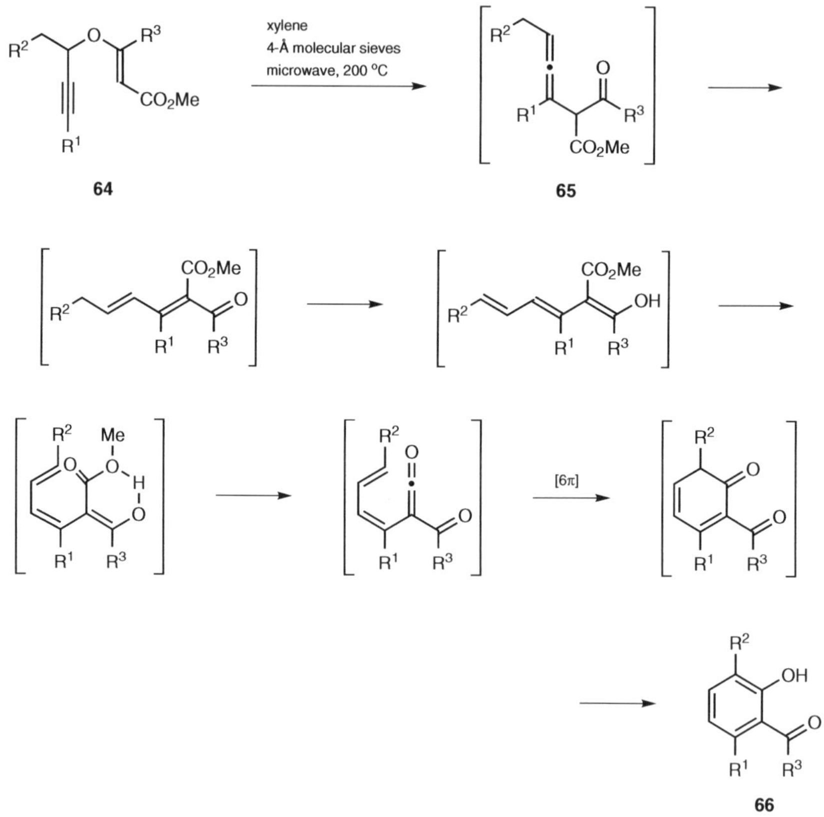

xylene
4-Å molecular sieves
microwave, 200 °C

64 65 [6π] 66

R^1	R^2	R^3	Yield (%)	Ref
Ph	Et	H	76	[101]
Ph	iPr	H	72	[101]
Ph	*t*-Bu	H	67	[101]
Ph	H	H	27	[101]
Ph	Ph	H	89	[101]

R^1	R^2	R^3	Yield (%)	Ref
Ph	Bn	H	61	[101]
Ph	F	H	63	[101]
4-$MeOC_6H_4$	Et	H	72	[101]
3,4-$Cl_2C_6H_3$	Et	H	77	[101]
4-$MeOC_6H_4$	Et	H	81	[101]
Ph	Et	Me	65	[101]
Ph	Et	$(CH_2)_4Me$	72	[101]
Ph	Et	Ph	75	[101]
Bu	Et	H	43	[101]
CO_2Me	Pr	H	70	[101]
CO_2Me	Ph	H	56	[101]
CO_2Me	H	H	29	[101]
TMS	Et	H	30	[101]

3′-Hydroxy-[1,1′;4′,1″-terphenyl]-2′-carbaldehyde (66, **R^1 = R^2 = Ph; R^3 = H); Typical Procedure:**[101]

Propargyl vinyl ether **64** (R^1 = R^2 = Ph; R^3 = H; 0.700 mmol), activated 4-Å molecular sieves (250 mg), and dry xylene (1 mL) were placed in a sealed microwave vial and the mixture was irradiated for 1 h in a single-mode microwave oven (300 W, 200 °C). The mixture was filtered through a pad of Celite using CH_2Cl_2 as solvent. After removing the solvent at reduced pressure, the products were purified by flash column chromatography (silica gel, hexane/EtOAc 19:1) to give the product: yield: 89%.

2.1.5.6.2 Propargyl Claisen Rearrangement Followed by Imine Formation, Tautomerization, and 6π-Electrocyclization

A related dihydropyridine synthesis uses the same starting materials as the reaction shown in Section 2.1.5.6.1, but in addition involves primary amines, including anilines.[104] Mechanistically, the process begins with the same propargyl Claisen rearrangement of propargyl vinyl ethers **67** (Scheme 26). However, this reaction gives heterocyclic dihydropyridines **68** under the standard reaction conditions, perhaps because the doubly unsaturated imine persists long enough (i.e., does not tautomerize) to undergo *trans–cis* isomerization across the central π-bond and electrocyclization.

for references see p 225

Scheme 26 Propargyl Claisen Rearrangement Followed by Imine Formation, Tautomerization, and 6π-Electrocyclization[104]

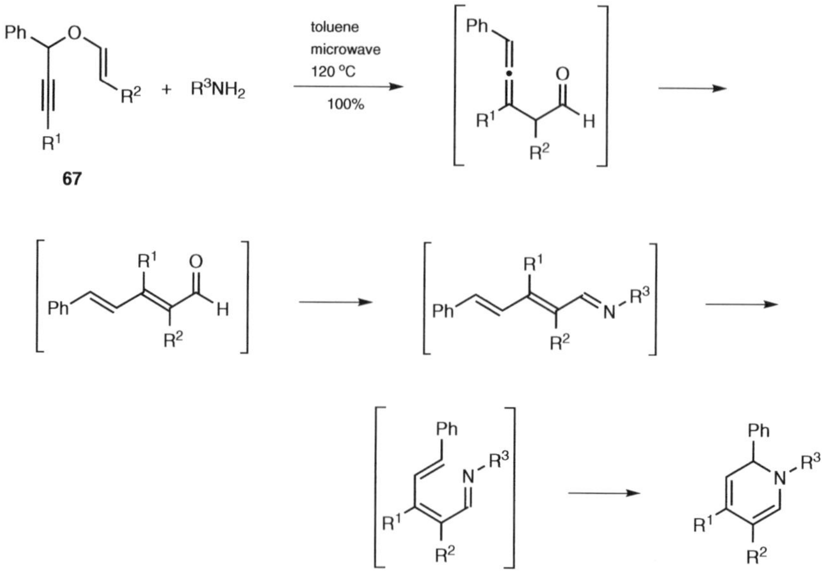

R^1	R^2	R^3	Yield (%)	Ref
H	CO_2Me	4-$MeOC_6H_4$	100	[104]
Ph	CO_2Me	4-$MeOC_6H_4$	95	[104]
H	Ts	4-$MeOC_6H_4$	80	[104]
H	CO_2Me	Bn	83	[104]
H	CO_2Me	$CH_2CH{=}CH_2$	72	[104]
H	CO_2Me	1-adamantyl	87	[104]
H	CO_2Me	(S)-CHMePh	83	[104]
H	CO_2Me	4-$MeOC_6H_4CH_2$	78	[104]
H	CO_2Me	Ph	93	[104]
H	CO_2Me	4-ClC_6H_4	88	[104]

Methyl 1-(4-Methoxyphenyl)-4,6-diphenyl-1,6-dihydropyridine-3-carboxylate (68, R^1 = Ph; R^2 = CO_2Me; R^3 = 4-$MeOC_6H_4$); Typical Procedure:[104]

A soln of propargyl vinyl ether **67** (R^1 = Ph; R^2 = CO_2Me; 1.0 mmol) and 4-methoxyaniline (1.1 mmol) in toluene (5 mL) was placed in a microwave-specific closed vial and the soln was irradiated for 30 min in a single-mode microwave oven (150 W, 120 °C). The mixture was dried (Na_2SO_4) and filtered using CH_2Cl_2 as solvent. After removing the solvent under reduced pressure, the products were purified by flash column chromatography (silica gel, hexane/EtOAc 4:1) to give the product; yield: 95%.

References

[1] Grossmann, A.; Enders, D., *Angew. Chem.*, (2012) **124**, 320; *Angew. Chem. Int. Ed.*, (2012) **51**, 314.
[2] Claisen, L., *Chem. Ber.*, (1912) **45**, 3157.
[3] Cope, A. C.; Hardy, E. M., *J. Am. Chem. Soc.*, (1940) **62**, 441.
[4] Rhoads, S. J.; Raulins, N. R., *Org. React. (N. Y.)*, (1975) **22**, 1.
[5] Lutz, R. P., *Chem. Rev.*, (1984) **84**, 205.
[6] Wilson, S. R., *Org. React. (N. Y.)*, (1993) **43**, 93.
[7] Zeh, J.; Hiersemann, M., In *Science of Synthesis: Stereoselective Synthesis*, Evans, P. A., Ed.; Thieme: Stuttgart, (2011); Vol. 3, 347.
[8] Davies, H. M. L.; Lian, Y., *Acc. Chem. Res.*, (2012) **45**, 923.
[9] Alnajjar, M.; Quigley, D.; Kuntamukkula, M.; Simmons, F.; Freshwater, D.; Bigger, S., *J. Chem. Health Saf.*, (2011) **18**, 5.
[10] Berson, J. A.; Jones, M., Jr., *J. Am. Chem. Soc.*, (1964) **86**, 5019.
[11] Evans, D. A.; Golob, A. M., *J. Am. Chem. Soc.*, (1975) **97**, 4765.
[12] Paquette, L. A., *Synlett*, (1990), 67.
[13] Paquette, L. A., *Tetrahedron*, (1997) **53**, 13 971.
[14] Paquette, L. A.; Combrink, K. D.; Elmore, S. W.; Rogers, R. D., *J. Am. Chem. Soc.*, (1991) **113**, 1335.
[15] Paquette, L. A.; Elmore, S. W.; Combrink, K. D.; Hickey, E. R.; Rogers, R. D., *Helv. Chim. Acta*, (1992) **75**, 1755.
[16] Paquette, L. A.; Su, Z.; Bailey, S.; Montgomery, F. J., *J. Org. Chem.*, (1995) **60**, 897.
[17] Paquette, L. A.; Bailey, S., *J. Org. Chem.*, (1995) **60**, 7849.
[18] Paquette, L. A.; Montgomery, F. J.; Wang, T.-Z., *J. Org. Chem.*, (1995) **60**, 7857.
[19] Sworin, M.; Lin, K.-C., *J. Am. Chem. Soc.*, (1989) **111**, 1815.
[20] Paquette, L. A.; Colapret, J. A.; Andrews, D. R., *J. Org. Chem.*, (1985) **50**, 201.
[21] Paquette, L. A.; Reagan, J.; Schreiber, S. L.; Teleha, C. A., *J. Am. Chem. Soc.*, (1989) **111**, 2331.
[22] Paquette, L. A.; Shi, Y.-J., *J. Am. Chem. Soc.*, (1990) **112**, 8478.
[23] Chen, C.; Layton, M. E.; Shair, M. D., *J. Am. Chem. Soc.*, (1998) **120**, 10 784.
[24] Sheehan, S. M.; Lalic, G.; Chen, J. S.; Shair, M. D., *Angew. Chem.*, (2000) **112**, 2826; *Angew. Chem. Int. Ed.*, (2000) **39**, 2714.
[25] Chen, C.; Layton, M. E.; Sheehan, S. M.; Shair, M. D., *J. Am. Chem. Soc.*, (2000) **122**, 7424.
[26] Lolkema, L. D. M.; Semeyn, C.; Ashek, L.; Hiemstra, H.; Speckamp, W. N., *Tetrahedron*, (1994) **50**, 7129.
[27] Dalgard, J. E.; Rychnovsky, S. D., *J. Am. Chem. Soc.*, (2004) **126**, 15 662.
[28] Bolla, M. L.; Patterson, B.; Rychnovsky, S. D., *J. Am. Chem. Soc.*, (2005) **127**, 16 044.
[29] Ent, H.; de Koning, H.; Speckamp, W. N., *J. Org. Chem.*, (1986) **51**, 1687; and references cited therein.
[30] Overman, L. E.; Humphreys, P. G.; Welmaker, G. S., *Org. React. (N. Y.)*, (2011) **75**, 747.
[31] Overman, L. E.; Kakimoto, M., *J. Am. Chem. Soc.*, (1979) **101**, 1310.
[32] Overman, L. E., *Tetrahedron*, (2009) **65**, 6432.
[33] Overman, L. E.; Kakimoto, M.; Okazaki, M. E.; Meier, G. P., *J. Am. Chem. Soc.*, (1983) **105**, 6622.
[34] Overman, L. E.; Fukaya, C., *J. Am. Chem. Soc.*, (1980) **102**, 1454.
[35] Overman, L. E.; Freerks, R. L., *J. Org. Chem.*, (1981) **46**, 2833.
[36] Overman, L. E.; Mendelson, L. T., *J. Am. Chem. Soc.*, (1981) **103**, 5579.
[37] Overman, L. E.; Shim, J., *J. Org. Chem.*, (1991) **56**, 5005.
[38] Overman, L. E.; Shim, J., *J. Org. Chem.*, (1993) **58**, 4662.
[39] Overman, L. E.; Sworin, M.; Burk, R. M., *J. Org. Chem.*, (1983) **48**, 2685.
[40] Knight, S. D.; Overman, L. E.; Pairaudeau, G., *J. Am. Chem. Soc.*, (1993) **115**, 9293.
[41] Knight, S. D.; Overman, L. E.; Pairaudeau, G., *J. Am. Chem. Soc.*, (1995) **117**, 5776.
[42] Angle, S. R.; Fevig, J. M.; Knight, S. D.; Marquis, R. W., Jr.; Overman, L. E., *J. Am. Chem. Soc.*, (1993) **115**, 3966.
[43] Martin, C. L.; Overman, L. E.; Rohde, J. M., *J. Am. Chem. Soc.*, (2008) **130**, 7568.
[44] Martin, C. L.; Overman, L. E.; Rohde, J. M., *J. Am. Chem. Soc.*, (2010) **132**, 4894.
[45] Agami, C.; Cases, M.; Couty, F., *J. Org. Chem.*, (1994) **59**, 7937.
[46] Armstrong, A.; Bhonoah, Y.; Shanahan, S. E., *J. Org. Chem.*, (2007) **72**, 8019.
[47] Overman, L. E., *J. Am. Chem. Soc.*, (1974) **96**, 597.
[48] Overman, L. E., *J. Am. Chem. Soc.*, (1976) **98**, 2901.
[49] Momose, T.; Hama, N.; Higashino, C.; Sato, H.; Chida, N., *Tetrahedron Lett.*, (2008) **49**, 1376.

[50] Ichiki, M.; Tanimoto, H.; Miwa, S.; Saito, R.; Sato, T.; Chida, N., *Chem.–Eur. J.*, (2013) **19**, 264.
[51] Nakayama, Y.; Sekiya, R.; Oishi, H.; Hama, N.; Yamazaki, M.; Sato, T.; Chida, N., *Chem.–Eur. J.*, (2013) **19**, 12 052.
[52] Chorlton, A. P.; Morris, G. A.; Sutherland, J. K., *J. Chem. Soc., Perkin Trans. 1*, (1991), 1205.
[53] Janardhanam, S.; Rajagopalan, K., *J. Chem. Soc., Perkin Trans. 1*, (1992), 2727.
[54] Janardhanam, S.; Devan, B.; Rajagopalan, K., *Tetrahedron Lett.*, (1993) **34**, 6761.
[55] Shanmugam, P.; Rajagopalan, K., *Tetrahedron*, (1996) **52**, 7737.
[56] Warrington, J. M.; Yap, G. P. A.; Barriault, L., *Org. Lett.*, (2000) **2**, 663.
[57] Arns, S.; Barriault, L., *Chem. Commun. (Cambridge)*, (2007), 2211.
[58] Poulin, J.; Grisé-Bard, C. M.; Barriault, L., *Chem. Soc. Rev.*, (2009) **38**, 3092.
[59] Barriault, L.; Deon, D. H., *Org. Lett.*, (2001) **3**, 1925.
[60] Barriault, L.; Denissova, I., *Org. Lett.*, (2002) **4**, 1371.
[61] Warrington, J. M.; Barriault, L., *Org. Lett.*, (2005) **7**, 4589.
[62] Barriault, L.; Denissova, I.; Goulet, N., *Synthesis*, (2012) **43**, 1833.
[63] Sauer, E. L. O.; Barriault, L., *J. Am. Chem. Soc.*, (2004) **126**, 8569.
[64] Arns, S.; Barriault, L., *J. Org. Chem.*, (2006) **71**, 1809.
[65] Clément, R.; Grisé, C. M.; Barriault, L., *Chem. Commun. (Cambridge)*, (2008), 3004.
[66] Malherbe, R.; Bellus, D., *Helv. Chim. Acta*, (1978) **61**, 3096.
[67] Malherbe, R.; Rist, G.; Bellus, D., *J. Org. Chem.*, (1983) **48**, 860.
[68] Yoon, T. P.; Dong, V. M.; MacMillan, D. W. C., *J. Am. Chem. Soc.*, (1999) **121**, 9726.
[69] Dong, V. M.; MacMillan, D. W. C., *J. Am. Chem. Soc.*, (2001) **123**, 2448.
[70] Uemura, D.; Chou, T.; Haino, T.; Nagatsu, A.; Fukuzawa, S.; Zheng, S.; Chen, H., *J. Am. Chem. Soc.*, (1995) **117**, 1155.
[71] Chou, T.; Kamo, O.; Uemura, D., *Tetrahedron Lett.*, (1996) **37**, 4023.
[72] Evans, D. A.; Andrews, G. C., *Acc. Chem. Res.*, (1974) **7**, 147; and references cited therein.
[73] Pelc, M. J.; Zakarian, A., *Org. Lett.*, (2005) **7**, 1629.
[74] Pelc, M. J.; Zakarian, A., *Tetrahedron Lett.*, (2006) **47**, 7519.
[75] Lu, C.-D.; Zakarian, A., *Org. Lett.*, (2007) **9**, 3161.
[76] Stivala, C. E.; Zakarian, A., *J. Am. Chem. Soc.*, (2008) **130**, 3774.
[77] Chantarasriwong, O.; Batova, A.; Chavasiri, W.; Theodorakis, E. A., *Chem.–Eur. J.*, (2010) **16**, 9944; and references cited therein.
[78] Saouf, A.; Guerra, F. M.; Rubal, J. J.; Moreno-Dorado, F. J.; Akssira, M.; Mellouki, F.; López, M.; Pujadas, A. J.; Jorge, Z. D.; Massanet, G. M., *Org. Lett.*, (2005) **7**, 881.
[79] Appendino, G.; Prosperini, S.; Valdivia, C.; Ballero, M.; Colombano, G.; Billington, R. A.; Genazzani, A. A.; Sterner, O., *J. Nat. Prod.*, (2005) **68**, 1213.
[80] Nicolaou, K. C.; Li, J., *Angew. Chem.*, (2001) **113**, 4394; *Angew. Chem. Int. Ed.*, (2001) **40**, 4264.
[81] Nicolaou, K. C.; Sasmal, P. K.; Xu, H., *J. Am. Chem. Soc.*, (2004) **126**, 5493.
[82] Nicolaou, K. C.; Xu, H.; Wartmann, M., *Angew. Chem.*, (2005) **117**, 766; *Angew. Chem. Int. Ed.*, (2005) **44**, 756.
[83] Quillinan, A. J.; Scheinmann, F., *J. Chem. Soc. D*, (1971), 966.
[84] Tisdale, E. J.; Chowdhury, C.; Vong, B. G.; Li, H.; Theodorakis, E. A., *Org. Lett.*, (2002) **4**, 909.
[85] Tisdale, E. J.; Vong, B. G.; Li, H.; Kim, S. H.; Chowdhury, C.; Theodorakis, E. A., *Tetrahedron*, (2003) **59**, 6873.
[86] Tisdale, E. J.; Slobodov, I.; Theodorakis, E. A., *Org. Biomol. Chem.*, (2003) **1**, 4418.
[87] Tisdale, E. J.; Slobodov, I.; Theodorakis, E. A., *Proc. Natl. Acad. Sci. U. S. A.*, (2004) **101**, 12 030.
[88] Li, X.; Zhang, X.; Wang, X.; Li, N.; Lin, C.; Gao, Y.; Yu, Z.; Guo, Q.; You, Q., *Chin. J. Chem.*, (2012) **30**, 35.
[89] Zhang, X.; Li, X.; Sun, H.; Jiang, Z.; Tao, L.; Gao, Y.; Guo, Q.; You, Q., *Org. Biomol. Chem.*, (2012) **10**, 3288.
[90] Zhang, X.; Li, X.; Sun, H.; Wang, X.; Zhao, L.; Gao, Y.; Liu, X.; Zhang, S.; Wang, Y.; Yang, Y.; Zeng, S.; Guo, Q.; You, Q., *J. Med. Chem.*, (2013) **56**, 276.
[91] Zhou, X.; Wu, W.; Liu, X.; Lee, C.-S., *Org. Lett.*, (2008) **10**, 5525.
[92] Kozytska, M. V.; Dudley, G. B., *Tetrahedron Lett.*, (2008) **49**, 2899.
[93] Larsson, R.; Sterner, O.; Johansson, M., *Org. Lett.*, (2009) **11**, 657.
[94] Larsson, R.; Scheeren, H. W.; Aben, R. W. M.; Johansson, M.; Sterner, O., *Eur. J. Org. Chem.*, (2013), 6955.
[95] Nelson, H. M.; Stoltz, B. M., *Org. Lett.*, (2008) **10**, 25.
[96] Nelson, H. M.; Stoltz, B. M., *Tetrahedron Lett.*, (2009) **50**, 1699.

[97] Nelson, H. M.; Murakami, K.; Virgil, S. C.; Stoltz, B. M., *Angew. Chem.*, (2011) **123**, 3772; *Angew. Chem. Int. Ed.*, (2011) **50**, 3688.
[98] Nelson, H. M.; Gordon, J. R.; Virgil, S. C.; Stoltz, B. M., *Angew. Chem.*, (2013) **125**, 6831; *Angew. Chem. Int. Ed.*, (2013) **52**, 6699.
[99] Tejedor, D.; Méndez-Abt, G.; Cotos, L.; García-Tellado, F., *Chem. Soc. Rev.*, (2013) **42**, 458.
[100] Herzon, S. B.; Myers, A. G., *J. Am. Chem. Soc.*, (2005) **127**, 5342.
[101] Tejedor, D.; Méndez-Abt, G.; Cotos, L.; Ramirez, M. A.; García-Tellado, F., *Chem.–Eur. J.*, (2011) **17**, 3318.
[102] Reber, K. P.; Tilley, S. D.; Sorensen, E. J., *Chem. Soc. Rev.*, (2009) **38**, 3022.
[103] Danheiser, R. L.; Brisbois, R. G.; Kowalczyk, J. J.; Miller, R. F., *J. Am. Chem. Soc.*, (1990) **112**, 3093.
[104] Tejedor, D.; Méndez-Abt, G.; García-Tellado, F., *Chem.–Eur. J.*, (2010) **16**, 428.

2.2 Intermolecular Alkylative Dearomatizations of Phenolic Derivatives in Organic Synthesis

J. A. Porco, Jr., and J. Boyce

General Introduction

In the last ten years, the synthetic community has shown increasing interest in domino transformations involving alkylative dearomatization reactions capable of accessing highly complex architectures in fewer steps and in higher yields than would otherwise be possible.[1–23] Strategically, these reactions allow the synthetic chemist to begin with a stable aromatic molecule that can undergo tandem dearomatization/annulation processes to access enantioenriched spiro, fused, and bridging polycyclic ring systems. The utility of dearomatizations to access a wide range of structural diversity is best exemplified by several chemical syntheses and also the proposed biosyntheses of polyprenylated acylphloroglucinol (PPAP) natural products from dearomatized adduct **1** (Scheme 1).[24–33]

The representative set of polyprenylated acylphloroglucinol natural products shown in Scheme 1 is thought to derive biosynthetically from dearomatized adduct **1** via a series of prenylation reactions, proton transfers, and double-bond isomerizations initiated by prenyl pyrophosphate (**2**, Scheme 2). Alkylative dearomatization reactions play a fundamental role in the biogenesis of polyprenylated acylphloroglucinols by serving as the first step of a domino transformation leading to these highly bioactive natural products.[31–33] Interestingly, this reaction type is highly unusual in that its efficiency in the laboratory is able to parallel that in nature.[1–13,34]

The use of intermolecular alkylative dearomatization as part of a domino transformation was first demonstrated in an efficient four-step synthesis of (±)-seychellene utilizing a tandem alkylative dearomatization/[4+2]-cycloaddition sequence (Section 2.2.3.1).[34] More examples came to light in the ensuing 30 years with the development of novel alkylative dearomatization/cationic and radical annulation sequences.[1–12] These works describe several key accomplishments including a notably efficient substitution strategy for the dearomatization using unactivated trifluoromethanesulfonate electrophiles and an asymmetric alkylative dearomatization/annulation process providing up to 90% enantiomeric excess (Section 2.2.3.4).[7,8]

Scheme 1 Biosynthesis of Polyprenylated Acylphloroglucinol Natural Products from a Dearomatized Adduct[24–33]

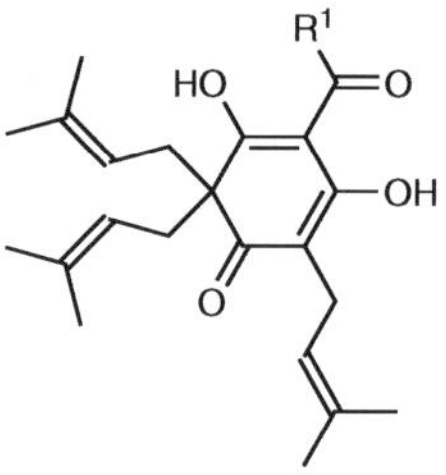

1 R^1 = Ph, iPr, 3,4-$(HO)_2C_6H_3$

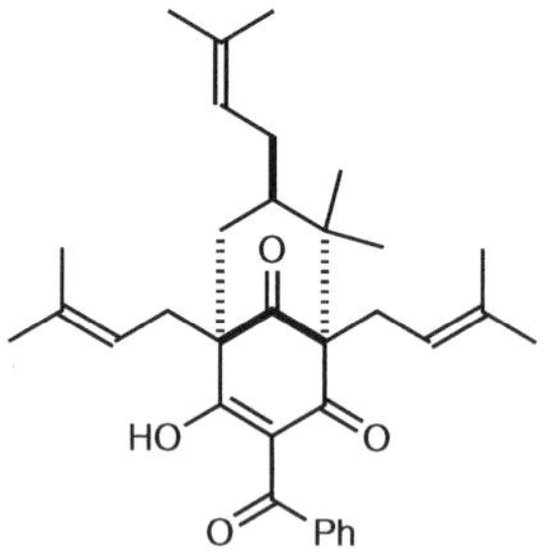

(+)-clusianone

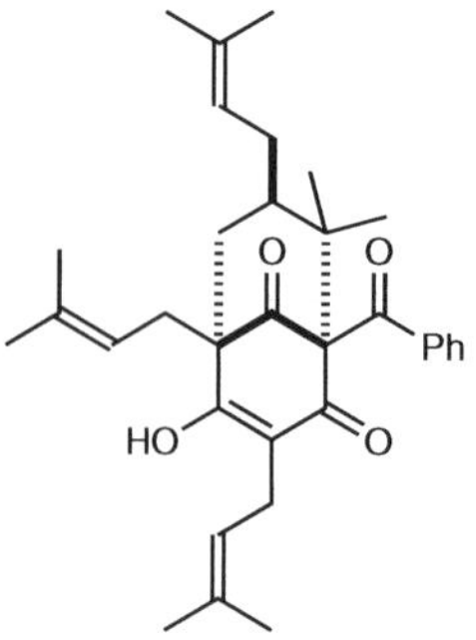

(+)-nemorosone

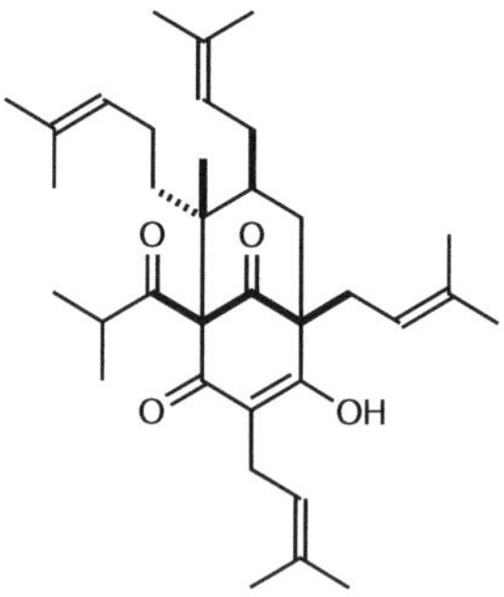

(+)-hyperforin

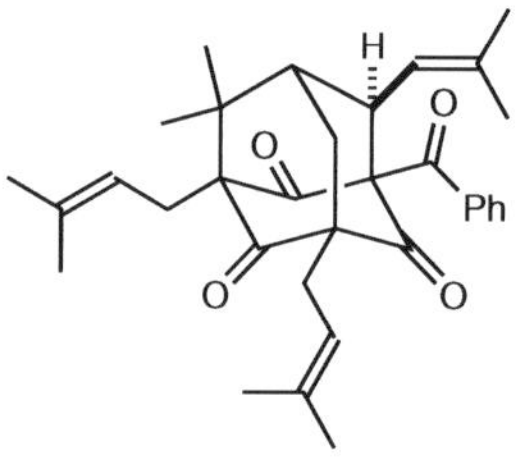

(−)-hyperibone K

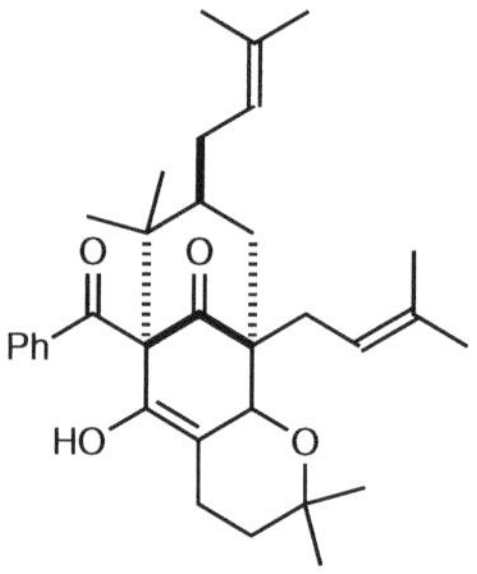

propolone A

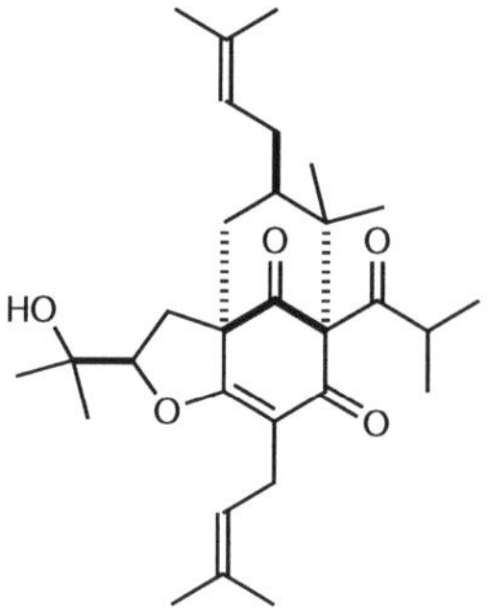

garsubellin A

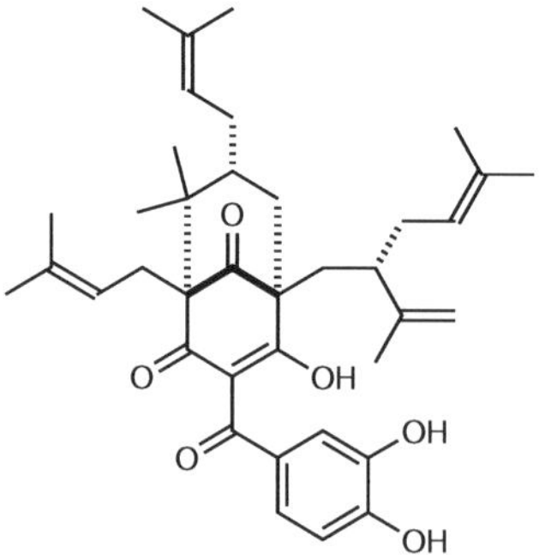

garcinol (camboginol)

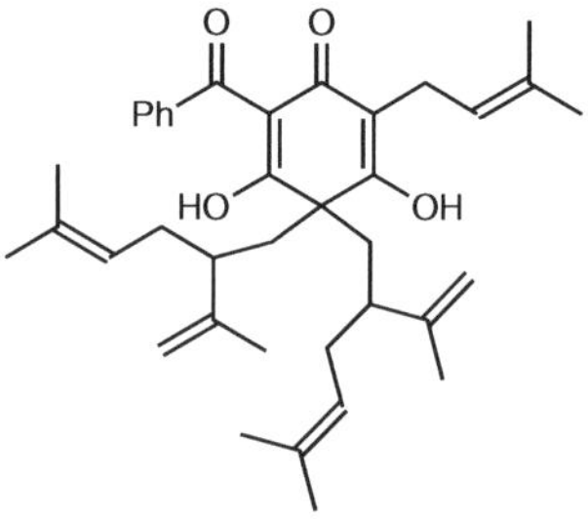

weddellianone B

Scheme 2 Proposed Biosynthesis of Common Intermediates[31–33]

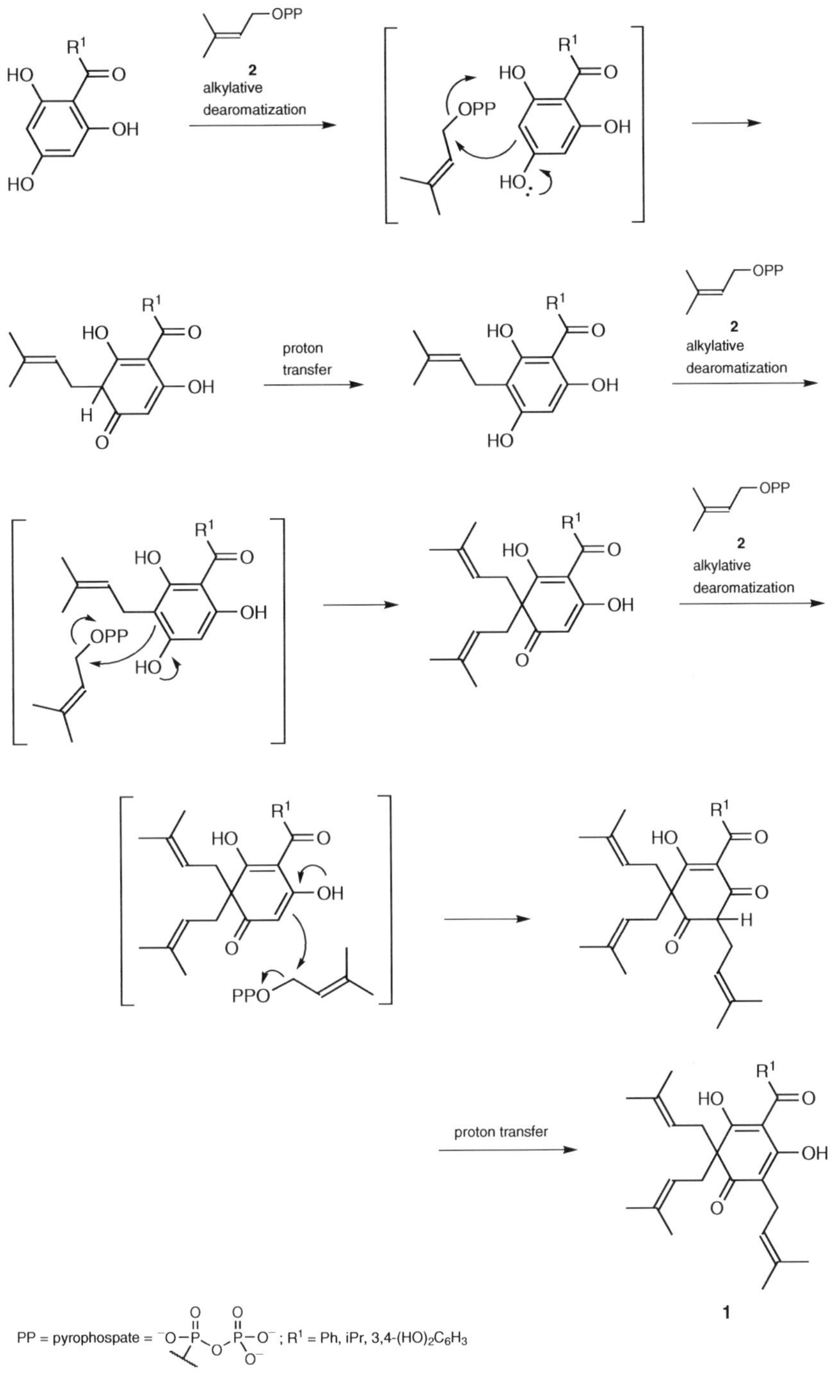

2
alkylative dearomatization

proton transfer

2
alkylative dearomatization

2
alkylative dearomatization

proton transfer

1

PP = pyrophospate = $^{-}O-P(O)(O^{-})-O-P(O)(O^{-})-O^{-}$; R^1 = Ph, iPr, 3,4-$(HO)_2C_6H_3$

for references see p 290

Despite the great synthetic potential that alkylative dearomatizations bring, much of the reaction scope and factors governing its reactivity and selectivity remain largely undocumented; advances in this area have been recently reviewed.[29,35–37] In this chapter, we will focus on intermolecular alkylative dearomatizations and recent methods for executing these reactions in the laboratory, placing special emphasis on reaction efficiency, yield, selectivity, scalability, and robustness as a means to distinguish examples. Also shown are examples of alkylative dearomatizations used as part of domino transformations to access highly complex architectures. Ultimately, we hope to encourage further improvements in this area and assist in making the outcomes of these reactions more predictable and reliable.

2.2.1 Metal-Mediated Intermolecular Alkylative Dearomatization

2.2.1.1 Osmium(II)-Mediated Intermolecular Alkylative Dearomatization

In the early 1990s, it was discovered that alkylative dearomatizations can be achieved under mild conditions utilizing η^2-coordinated osmium(II) complexes of substituted phenols to generate *ortho*- and *para*-substituted adducts with high regio- and stereocontrol.[38–41]

The binding of osmium(II) to a single π-bond of the aromatic phenol causes the unbound electrons to localize because of partial dearomatization; overall, this feature has the effect of creating a more electron-rich nucleophile.[38,39] In the case of β-estradiol (**3**), the η^2-pentaammineosmium(II) complex is formed after one-electron reduction of pentaammine(trifluoromethanesulfonato)osmium(III) trifluoromethanesulfonate with magnesium metal (Scheme 3). Addition of 1.0 equivalent of ethyl(diisopropyl)amine followed by 1.0 equivalent of methyl vinyl ketone selectively provides the *para*-substituted cyclohexadienone complex **5** at −40 °C in 97% yield. At room temperature, the reaction affords a mixture of both *ortho*- and *para*-substituted products **4** and **5**. Decomplexation of **5** is achieved in good yield to provide steroid **6** as a single diastereomer.[38]

Scheme 3 Osmium(II)–η²-Arene Complex Mediated Alkylative Dearomatization of Substituted Phenols[38]

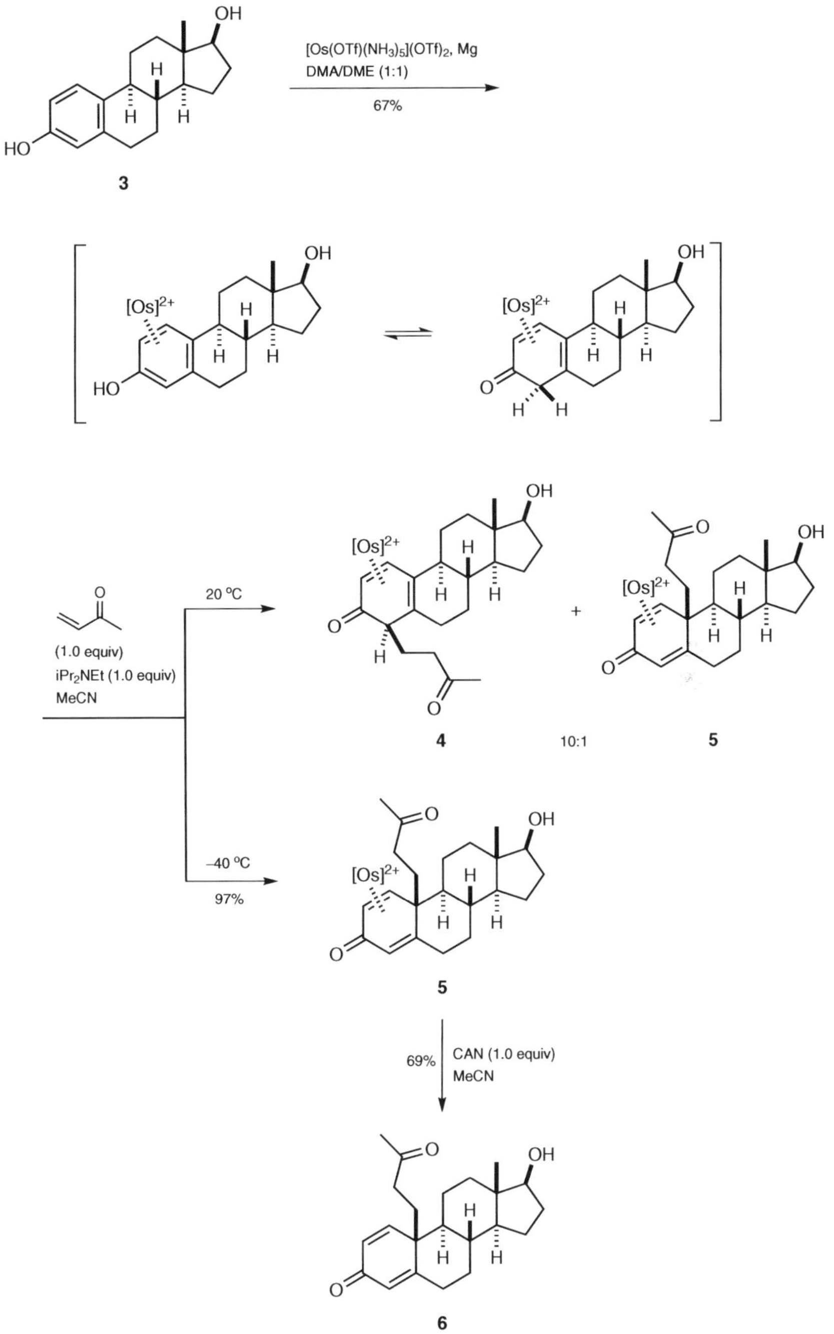

HO OH H H H

3

$[Os(OTf)(NH_3)_5](OTf)_2$, Mg
DMA/DME (1:1)
67%

$[Os]^{2+}$ OH H H H HO

$[Os]^{2+}$ OH H H H O H H

O
(1.0 equiv)
iPr_2NEt (1.0 equiv)
MeCN

20 °C

$[Os]^{2+}$ OH H H H O H O

\+

O OH $[Os]^{2+}$ H H H O

4 10:1 5

–40 °C
97%

O OH $[Os]^{2+}$ H H H O

5

69% CAN (1.0 equiv) MeCN

O OH H H H O

6

The excellent diastereoselectivity is best explained by analysis of osmium complex **7** (Scheme 4) in which the osmium metal center is blocking one face of the phenol substrate such that conjugate addition to a Michael acceptor occurs from the opposite face.

Scheme 4 Stereocontrol Arising from the Osmium(II)–η^2-Arene Complex[39]

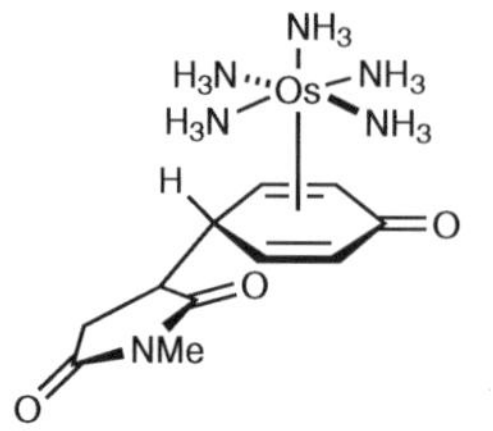

Small-scale reactions are carried out under an inert atmosphere with rigorous exclusion of moisture. However, the method can be applied on gram-scale conveniently on the benchtop using reagents and solvents directly from their commercial sources without further purification. The reactions proceed overnight and in good yields between –40 °C and room temperature, providing air-stable products. The gram-scale synthesis only requires 60 mol% pentaammine(trifluoromethanesulfonato)osmium(III) trifluoromethanesulfonate, which undergoes a one-electron reduction by amalgamated zinc leading to osmium(II) complexation with *p*-cresol to provide osmium complex **8** (Scheme 5). Upon addition of 1.0 equivalent of pyridine, *N*,*N*-dimethylaniline, or 2,6-dimethylpyridine, the solution changes to a deep red color suggestive of the formation of the phenolate anion. Although the use of base dramatically enhances the rate of reaction, phenol complexes are in many cases sufficiently electron rich to react with the electrophile in the absence of base, albeit over longer reaction times.[40] Lower temperatures or shorter reaction times are required for stronger bases, such as ethyl(diisopropyl)amine. Michael acceptors such as methyl acrylate, methyl vinyl ketone, 3-methylbut-3-en-2-one, acrylonitrile, *N*-methylmaleimide, and but-2-enal are employed to successfully provide the corresponding Michael addition adducts with high *para* selectivity.[40]

Interestingly, *ortho* addition is not observed for phenol, but complex **8** and other *para*-substituted phenols give *ortho* alkylation products depending on the temperature or the presence of Lewis acid (Scheme 5). When less reactive electrophiles such as methyl acrylate are used, a Lewis acid (>0.5 equiv) is required for the reaction to proceed. Zinc(II) trifluoromethanesulfonate causes a complete reversal in selectivity to favor *ortho* addition, which may be a result of zinc chelation to both the phenolic oxygen and Michael acceptor (e.g., methyl vinyl ketone). When the reactions are carried out in acetonitrile, a mixture of *para* and *ortho* substitution products **9** and **10** are observed at room temperature whereas *para* substitution is only observed at –40 °C. Despite much discussion in the literature,[42–49] it is not well understood why nucleophilic attack preferentially takes place at the more substituted position of the phenol; one possibility may result, in part, from the relative instability of the cross-conjugated anion of complex **8** compared to the more stable conjugated α-anion and phenoxide anion formed upon deprotonation. The less stable anion should react with the electrophile at a faster rate, leading to the exclusive formation of *para*-substitution product **9** at lower temperatures. The *para*-disubstituted cyclohexadienone **11** is directly provided after decomplexation with a strong oxidant, such as ammonium cerium(IV) nitrate or 2,3-dichloro-5,6-dicyanobenzo-1,4-quinone. This oxidant poses a limitation for the use of osmium-mediated alkylative dearomatizations in total synthesis by requiring the exclusion of any easily oxidizable functionalities. Although the method is operationally simple and has promise for enantioselective variants, the high toxicity of pentaammine(trifluoromethanesulfonato)osmium(III) trifluoromethanesulfonate and amalgamated zinc renders the methodology less attractive for execution on a multigram scale.

Scheme 5 Osmium(II)-Mediated Alkylative Dearomatization: Effect of Lewis Acid and Solvent on *ortho* versus *para* Addition[40]

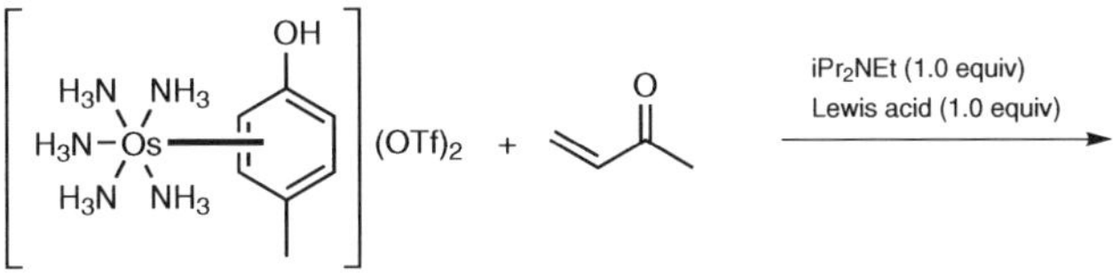

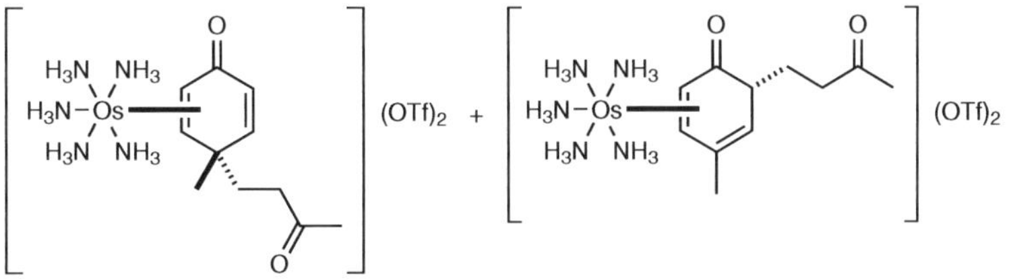

Lewis Acid	Solvent	Temp (°C)	Ratio (**9**/**10**)	Ref
–	MeCN	−40	100:0	[40]
$Zn(OTf)_2$	MeCN	25	5:85	[40]
BF_3	MeCN	−40	100:0	[40]
$Zn(OTf)_2$	MeOH	25	100:0	[40]

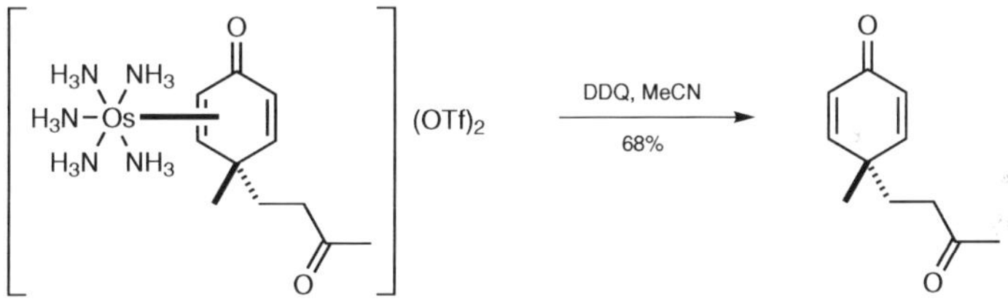

4-Methyl-4-(3-oxobutyl)cyclohexa-2,5-dien-1-one (11); Typical Procedure:[40]
A 50-mL Erlenmeyer flask was charged with a stirrer bar, $[Os(OTf)(NH_3)_5](OTf)_2$ (1.00 g, 1.39 mol), and MeOH (15 g) and flushed with argon (no attempts were made to rigorously dry solvents or glassware prior to their use). A soln of *p*-cresol (250 mg, 2.31 mmol) in MeOH was added followed immediately by Zn/Hg (3 g), at which point the flow of argon was discontinued and the flask was sealed with a septum. After the mixture had been stirred for 1 h, the dark soln was treated with methyl vinyl ketone (160 mg, 2.28 mmol) and iPr_2NEt (80 mg, 0.62 mmol), and the mixture was stirred for an additional 6 h and filtered. The filtrate was added to Et_2O (150 mL), and the tan precipitate was collected in air to give osmium(II) complex **9**; yield: 1.02 g (98%).

Complex **9** (1.0 g, 1.33 mmol) was dissolved in MeCN (3 g) under air and was treated with DDQ (303 mg, 1.33 mmol, 1.0 equiv). The mixture immediately turned deep black and was stirred for 1 min. The mixture was then dissolved in 10% aq $NaHCO_3$ (40 mL) and

for references see p 290

the aqueous phase was extracted with Et_2O (3 × 40 mL) and CH_2Cl_2 (40 mL). The combined pale yellow extracts were dried (Na_2SO_4). After evaporation of the solvent, a significant amount of H_2O remained. The residue was dissolved in CH_2Cl_2 and redried (Na_2SO_4). After evaporation of the solvent, the remaining yellow oil [186 mg (84%)] was purified by preparative TLC to give the product as an analytically pure, clear oil; yield: 150 mg (68%; 65% from *p*-cresol).

2.2.1.2 Palladium-Catalyzed Intermolecular Alkylative Dearomatization

Palladium-catalyzed methods for alkylative dearomatizations are extremely important processes with high potential for further development. This reaction requires a catalytic amount of palladium and may be carried out under neutral conditions, rendering it a promising candidate for future use in industrial processes, complex total synthesis, and enantioselective methods. The method was first introduced in 1996 for the C-allylation of naphthol derivatives.[50] Catalytic amounts of palladium(II) acetate and 1,1′-bis(diphenylphosphino)ferrocene ligand and 4-Å molecular sieves were used without the need for a Lewis acid activator. However, long reaction times (7 d) and high temperatures (100 °C) are required, limiting the scope to thermally stable substrates.

While investigating the synthesis of allyl phenyl ethers,[51] it was discovered that 3,5-dimethoxyphenol undergoes bisallylation at C4 to provide product **12** in 56% yield (Scheme 6). The reaction is catalytic with respect to palladium, triphenylphosphine, and titanium(IV) isopropoxide, and allows for lower temperatures (50 °C) and shorter reaction times (4–20 h). Although these conditions are used for the O-allylation of phenols, only C-allylation products are observed when using the electron-rich benzene-1,3,5-triol (phloroglucinol).

Scheme 6 Palladium-Catalyzed Intermolecular Alkylative Dearomatization[51]

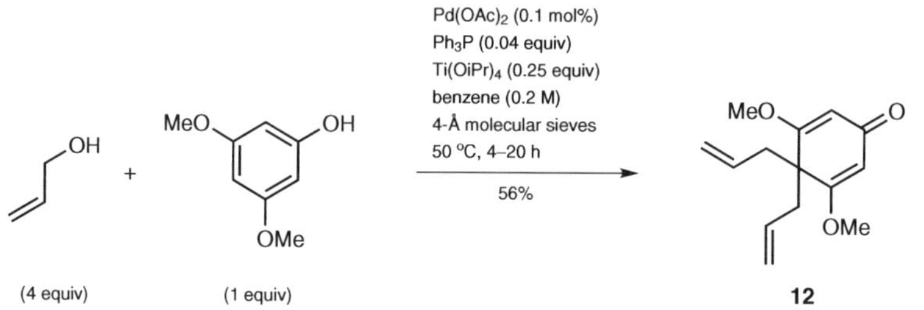

OH (4 equiv) + MeO, OH, OMe (1 equiv)

$Pd(OAc)_2$ (0.1 mol%)
Ph_3P (0.04 equiv)
$Ti(OiPr)_4$ (0.25 equiv)
benzene (0.2 M)
4-Å molecular sieves
50 °C, 4–20 h
56%

MeO, O, OMe **12**

The method was modified ten years later wherein excess triethylborane was added as a Lewis acid activator to promote oxidative addition, thereby allowing the reaction to be carried out at lower temperatures, in shorter times, and with fewer equivalents of allyl alcohol. For substrates that are not sufficiently electron rich for alkylative dearomatization in tetrahydrofuran such as resorcinol, it is often possible to achieve high yields of dearomatized products in toluene. When naphthols are used as substrates in toluene, the reactions are generally complete in 2 hours. Although phloroglucinol is not soluble in toluene, it is sufficiently electron rich to produce hexasubstituted cyclohexenetrione products **13** in tetrahydrofuran (Scheme 7).[52] The main disadvantage of this method is a 5.0 equivalent excess of the highly toxic triethylborane to activate the allyl alcohol for oxidative addition to the palladium–π-allyl complex.

Scheme 7 Palladium-Catalyzed Alkylative Dearomatization Promoted by Triethylborane[52]

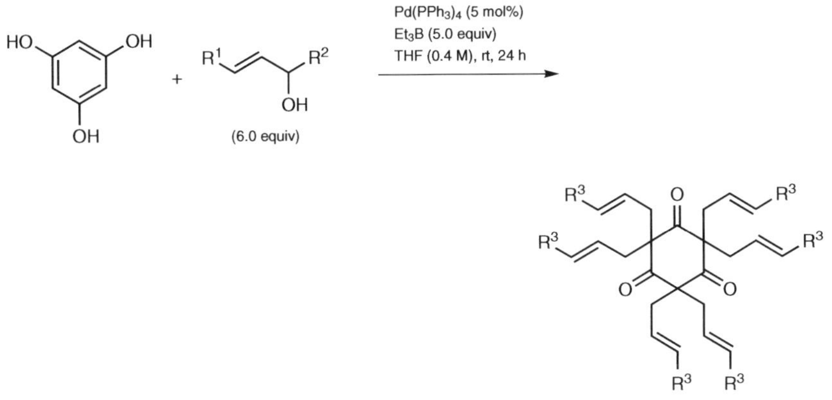

HO OH OH + R^1 R^2 OH (6.0 equiv)

$Pd(PPh_3)_4$ (5 mol%)
Et_3B (5.0 equiv)
THF (0.4 M), rt, 24 h

13

R^1	R^2	R^3	Yield (%)	Ref
H	H	H	83	[52]
Ph	H	Ph	57	[52]
H	Ph	Ph	64	[52]

In general, palladium-catalyzed allylative dearomatizations require the use of electron-rich aromatic substrates, such as naphthol, resorcinol, and phloroglucinol derivatives. The reactions are not compatible with benzene-1,2- or benzene-1,4-diol substrates or phenols, which tend to give O-allylation products;[52] the electrophile is generally limited to allyl, isoprenyl, and cinnamyl units. Therefore, further studies to enhance the scope of substrates and electrophiles are needed.

2.2.1.3 Tandem Palladium-Catalyzed Intermolecular Alkylative Dearomatization/Annulation

To the best of our knowledge, the use of palladium-catalyzed intermolecular alkylative dearomatizations in synthesis is limited to two examples,[1,2] each involving derivatives of 3,5-dimethoxyphenol. In 2006, an efficient decarboxylative variant of the palladium-catalyzed allylation method was utilized en route to garsubellin A (Scheme 8).[2] The observed regioselectivity favoring alkylation of **14** at C4 may be attributed to relief in $A_{1,3}$-strain and to the enhanced reactivity of the more substituted, cross-conjugated anion at C4 relative to the less substituted, conjugated anion at C2. The anion at C4 should be less stable and more reactive, resulting in the production of dearomatized adduct **15** in good yield. This reaction sets the stage for an efficient deketalization/Michael addition/β-elimination/demethylation sequence to construct fused furanocyclohexenedione **16** (Scheme 8).

Scheme 8 Domino Reaction Sequence Leading to the Fused Furanocyclohexenedione Core of Garsubellin A[2]

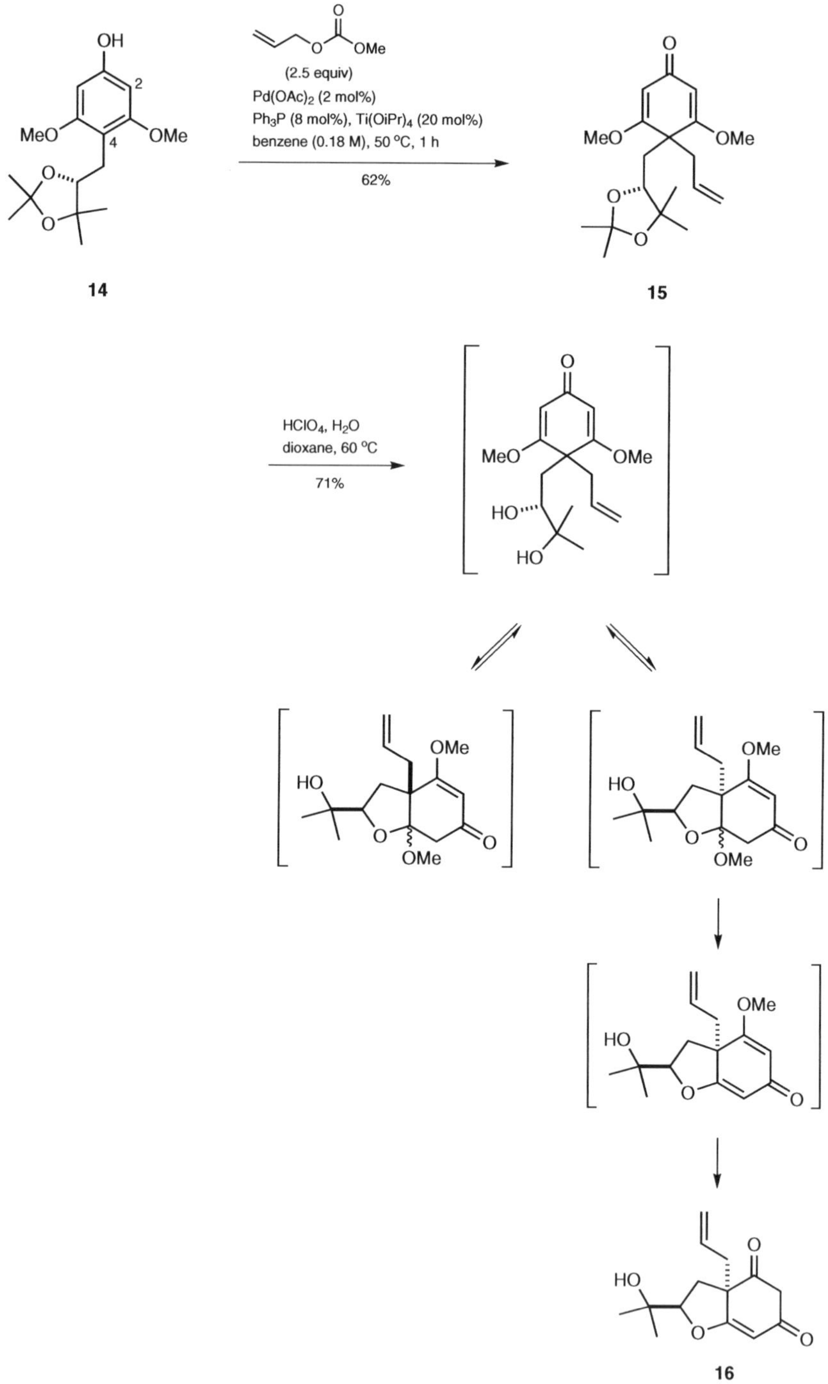

(2.5 equiv)
$Pd(OAc)_2$ (2 mol%)
Ph_3P (8 mol%), $Ti(OiPr)_4$ (20 mol%)
benzene (0.18 M), 50 °C, 1 h
62%

14 **15**

$HClO_4$, H_2O
dioxane, 60 °C
71%

16

The aforementioned alkylative dearomatization is highly efficient, with catalyst loadings of 2 mol%, reaction temperatures not exceeding 50 °C, and short reaction times (ca. 1 h). The only disadvantage is the use of toxic titanium(IV) isopropoxide, which may be avoided because 1 equivalent of carbonylate (base) is generated upon oxidative addition of allyl methyl carbonate by palladium(0) (Scheme 9). The resulting carbonylate anion can deprotonate pronucleophile **14**, resulting in an "outer sphere" S_N2-type attack of phenolate **17** on the palladium–π-allyl electrophile to provide product **15**.[53] An investigation of different ligands for palladium may render this reaction asymmetric.[54–56]

Scheme 9 Proposed "Outer Sphere" Tsuji–Trost-Type Mechanism for Palladium-Catalyzed Decarboxylative Allylation of Phloroglucinols[1,53]

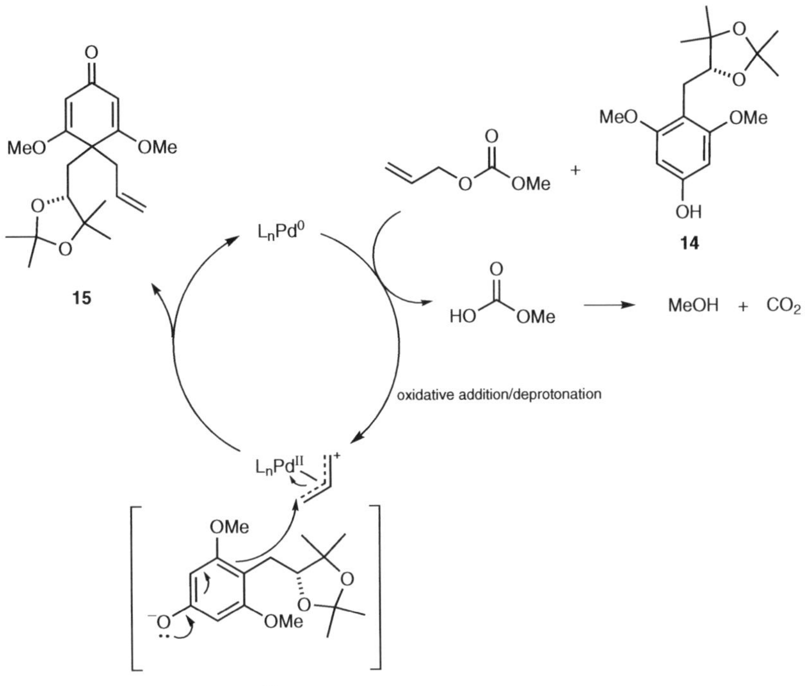

(R)-4-Allyl-3,5-dimethoxy-4-[(2,2,5,5-tetramethyl-1,3-dioxolan-4-yl)methyl]cyclohexa-2,5-dien-1-one (15); Typical Procedure:[2]

Solid $Pd(OAc)_2$ (200 mg, 0.892 mmol, 0.02 equiv) and Ph_3P (936 mg, 3.57 mmol, 0.08 equiv) were added to a suspension of phenol derivative **14** (13.2 g, 45 mmol, 1.0 equiv) in benzene (250 mL) (**CAUTION:** *carcinogen*) at 23 °C. While stirring, allyl methyl carbonate (12.7 mL, 0.112 mol, 2.5 equiv) was added followed by $Ti(OiPr)_4$ (2.6 mL, 8.9 mmol, 0.2 equiv) whereupon the soln turned a deep red. The mixture was placed in a 50 °C oil bath and stirred, and after 5 min the solids had dissolved. After the mixture had been stirred for 40 min, extensive gas evolution occurred and continued for 10 min. The soln was removed from the heating bath. Sat. aq NH_4Cl (50 mL) was added and the soln was stirred at 23 °C for 5 min. 1 M aq HCl (100 mL) was added and the organic phase was collected. The aqueous phase was extracted with EtOAc (2 × 250 mL) and the combined organic extracts were dried (Na_2SO_4), filtered, and concentrated to provide a red sludge. The crude ketone was purified by flash column chromatography (EtOAc/hexanes 1:2 to 1:0) to provide the product as an off-white solid; yield: 9.7 g (62%); mp 107 °C.

2.2.2 Non-Metal-Mediated Intermolecular Alkylative Dearomatization

2.2.2.1 Alkylative Dearomatizations of Phenolic Derivatives with Activated Electrophiles

Although many discussions of alkylative dearomatizations with simple phenol derivatives ensued throughout the 1950s and 1960s,[42,43,57–63] the reproducibility of the results was often a challenge due to product instability[42,46,57] and the factors governing the outcome of these reactions were not fully understood.[42–44,57–63] For example, inconsistencies[57,58] with previous observations[44] were reported. Interestingly, alkylative dearomatizations using lithium phenolates were deemed inefficient as they were shown to provide mixtures of products.[57]

Increasing steric bulk at *ortho* substituents causes simple unactivated electrophiles to prefer either *para* or O-alkylation.[60] This was later confirmed, and it was found that the efficiency of *para*-alkylative dearomatization nearly doubles for both activated and unactivated electrophiles in the presence of a *para*-methyl substituent as shown in Scheme 10 for the formation of products **18–20** from 2,6-di-*tert*-butyl-4-methylphenol.[59] In the event that the electrophile approach is not prevented by the steric bulk of a substituent at C2, C4, or C6, the substituent appears to enhance the nucleophilicity of the site to which it is attached on the aromatic ring in many cases.[59,60] A proposed transition state **21** also illustrates the tendency for the electrophile to react with the more-substituted phenolic position (Scheme 11).[61]

Scheme 10 Effect of Bulky Substituents at *ortho* Positions[59]

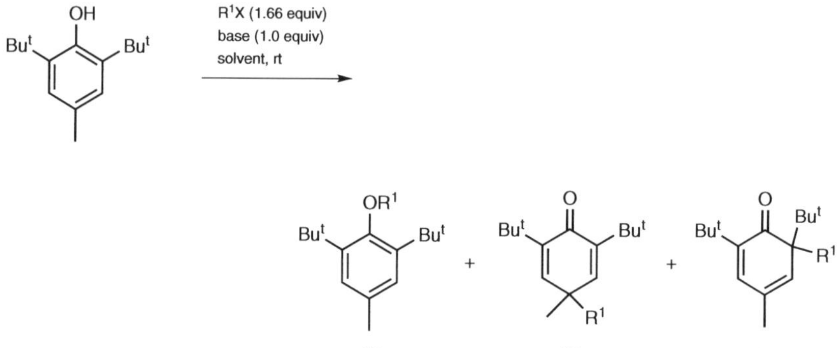

OH
Bu^t Bu^t
R^1X (1.66 equiv)
base (1.0 equiv)
solvent, rt

OR^1 Bu^t Bu^t **18** + O Bu^t Bu^t R^1 **19** + O Bu^t Bu^t R^1 **20**

R^1	X	Base	Solvent	Time	Yield (%)			Ref
					18	**19**	**20**	
Et	I	*t*-BuOK	*t*-BuOH	65 h	19	73	8	[59]
iPr	Br	*t*-BuOK	*t*-BuOH	65 h	20	76	4[a]	[59]
(*E*/*Z*)-$CH_2CH{=}CHMe$	Br	NaH	DMSO[b]	10 min	9	70	22	[59]
$CH_2CH{=}CH_2$	Br	NaH	DMSO[b]	10 min	11	65	23	[59]

[a] Estimated by UV spectroscopy, assuming ε_{320} 4000.
[b] 1.0 equiv of R^1X was used.

Scheme 11 Proposed Transition State for Alkylative Dearomatization[61]

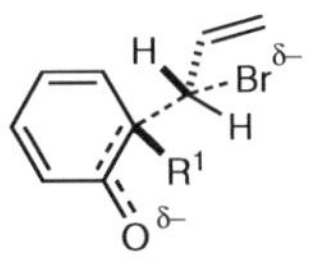

Chelation is important in understanding the outcome of *ortho*-selective alkylative dearomatizations.[45] Chloromethyl methyl ether, benzyl chloromethyl ether, and chloromethyl methyl sulfide are highly activated electrophiles because the oxygen or sulfur lone pair can participate in anchimeric assistance to improve leaving group ability. A series of these electrophiles provide high yields for the resulting *ortho*-products of dearomatized phenols (Scheme 12).

Scheme 12 High *ortho* Selectivity Observed for Activated Electrophiles Capable of Anchimeric Assistance[45]

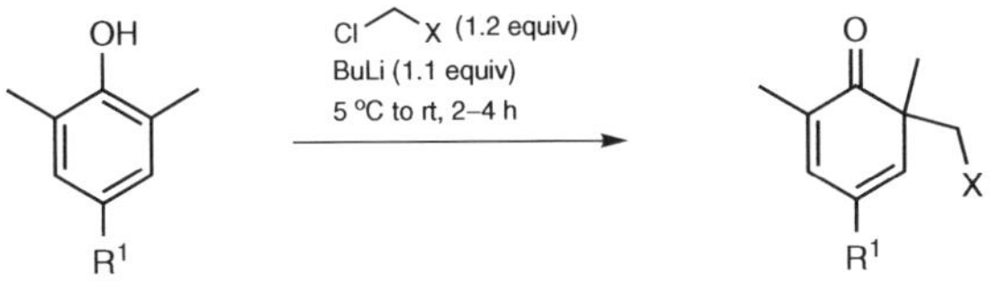

R^1	X	Yield[a] (%)	Ref
H	OMe	65	[45]
H	$O(CH_2)_2TMS$	85	[45]
H	SMe	95	[45]
Me	$O(CH_2)_2OMe$	85	[45]
Me	$O(CH_2)_2TMS$	85	[45]
Me	SMe	95	[45]

[a] Yields based on conversion of the starting phenol by ^1H NMR analysis (CCl_4); acceptable reaction solvents include pentane, hexanes, benzene, and toluene.

The choice of lithium as a base counterion is strategic in achieving *ortho* selectivity in these reactions; the lithium chelates to both the phenolate oxygen and the α-oxygen lone pair or the α-thioether lone pair leading to an optimal six-membered transition state (Scheme 13).[45,59]

Scheme 13 Transition State Rationale for the Observed *ortho* Selectivity with α-Sulfanyl and α-Oxy Halides[45,59]

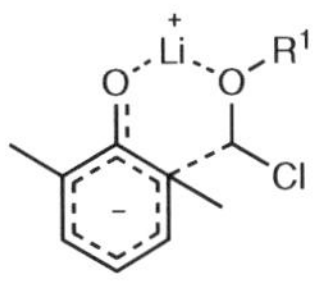

for references see p 290

A synthetically valuable study has explored a variety of activated chlorides for alkylative dearomatization of 2,6-dimethylphenol on a multigram scale (Scheme 14).[46] Decent scope with regard to activated chlorides is demonstrated, showing the effect of electron-withdrawing groups, variation of steric bulk, and the low activity of nonbulky substrates (not shown).[46] The substrates and reagents are commercially available and isolated product yields are reported, a rare phenomenon due to the high instability of the products.

Scheme 14 Alkylative Dearomatization with Activated Electrophiles Followed by In Situ Hydrogenation[46]

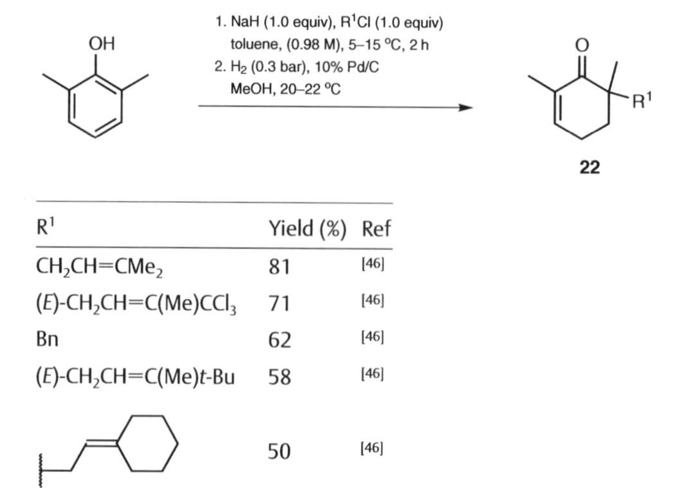

R^1	Yield (%)	Ref
$CH_2CH{=}CMe_2$	81	[46]
(*E*)-$CH_2CH{=}C(Me)CCl_3$	71	[46]
Bn	62	[46]
(*E*)-$CH_2CH{=}C(Me)$*t*-Bu	58	[46]
(structure)	50	[46]

However, to avoid undesired retro-Claisen rearrangement of the cyclohexadienones to allyl phenyl ethers, the temperatures require careful monitoring at 5–15 °C and the products require in situ hydrogenation prior to isolation. The reactions are carried out by preforming the phenolate anion of 2,6-dimethylphenol using 1 equivalent of sodium hydride in a 0.98 M solution of toluene with subsequent addition of the chloride electrophile. After hydrogenation, products **22** are provided in good yield following distillation. Interestingly, the reactions are highly regioselective, favoring the formation of *ortho*-substituted dearomatized adducts. The observed *ortho* selectivity supports the possible involvement of steric influences in the transition state as the nucleophile tends to react at the more-substituted carbon of the aromatic ring (see discussion of Scheme 11 above). Another major contributor to the high regioselectivity is the choice of sodium as the counterion, which is optimal for *ortho*-selective alkylative dearomatization of 2,6-dimethylphenol. The ability of sodium to attract the chloride leaving group of the electrophile in toluene while coordinating to the phenolate oxygen could boost the *ortho* selectivity.[46]

Enantioselective alkylative dearomatization of 2,6-dimethylphenol utilizes chiral amine ligands (Scheme 15).[46] Optimal results are achieved with (−)-sparteine (**23A**) or α-isosparteine (**23B**) using a lithium counterion; reactions likely proceed through chelated assembly **24**. The low enantioselectivity may be explained by poor facial selectivity imparted by the catalyst.[46]

Scheme 15 Asymmetric Alkylative Dearomatization of 2,6-Dimethylphenol[46]

23A

23B

1. chiral amine
BuLi, benzene
2. Cl
rt, 1 h

OH

24

Chiral Amine (Equiv)	ee (%)	Yield (%)	Ref
23A (1)	4	28	[46]
23A (2)	8	14	[46]
23B (1)	9	17	[46]

When the more electron-rich naphthols **25** and **27** are reacted under these conditions, the enantioselectivity for the formation of the corresponding products **26** and **28** quadruples, indicating a high degree of substrate dependence (Scheme 16).[46]

Scheme 16 Enantioselective Alkylative Dearomatization of Naphthols[46]

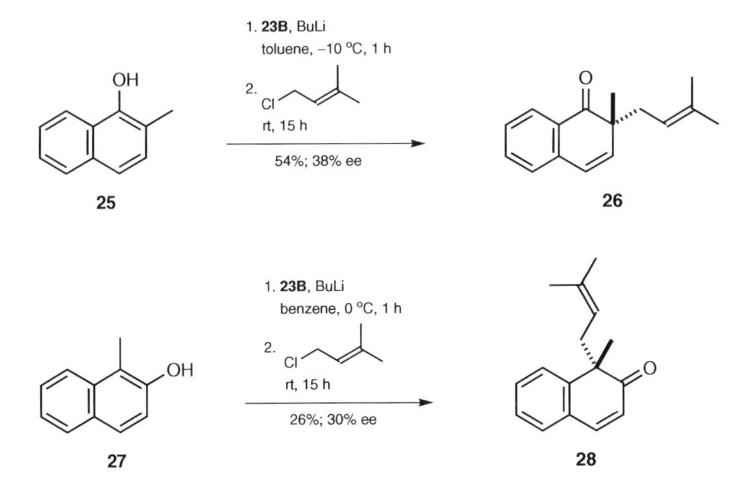

for references see p 290

Alternative methods for alkylative dearomatization of electron-rich phloroglucinols include a diverse set of activated electrophiles for both aqueous and organic solvents (Scheme 17).[3] Hydrogen-bonding is likely responsible for the C-selectivity observed in aqueous media, as the oxygen lone pairs are highly solvated to the extent that they are unavailable for bonding.[47,62] Allyl bromide is much less reactive than prenyl bromide, requiring higher temperatures for adequate yields. Phenyl lupone (**30**) is generated in 40% yield after treating a 0.2 M aqueous solution of clusiaphenone B (**29**) at 0 °C with 4.0 equivalents of potassium hydroxide and 4.0 equivalents of prenyl bromide.[3] Aqueous media provide lower yields of dearomatized products presumably due to partial hydrolysis of prenyl bromide. However, yields of **30** can potentially be improved through the use of additional prenyl bromide (~6.0 equiv) and potassium hydroxide (~6.0 equiv), raising temperatures to 60 °C, and/or increasing the concentration to 1.0 M.

Scheme 17 Alkylative Dearomatization of Acylphloroglucinols with Activated Electrophiles[3,5,8]

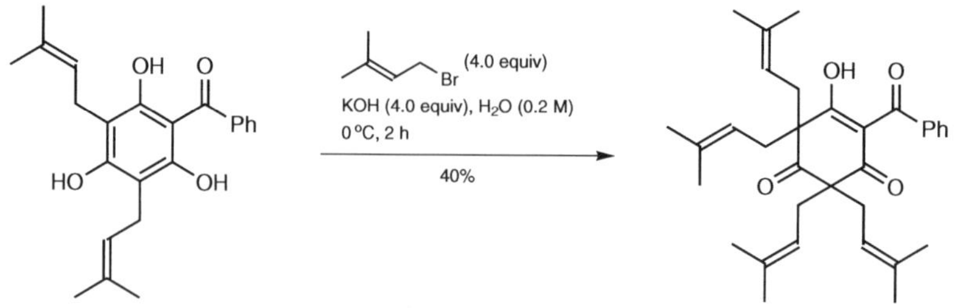

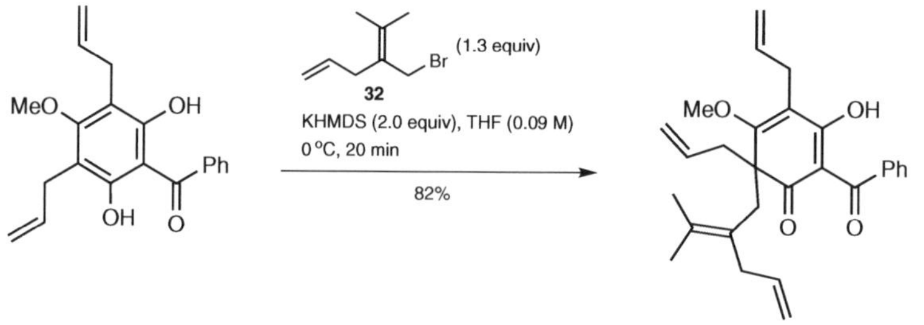

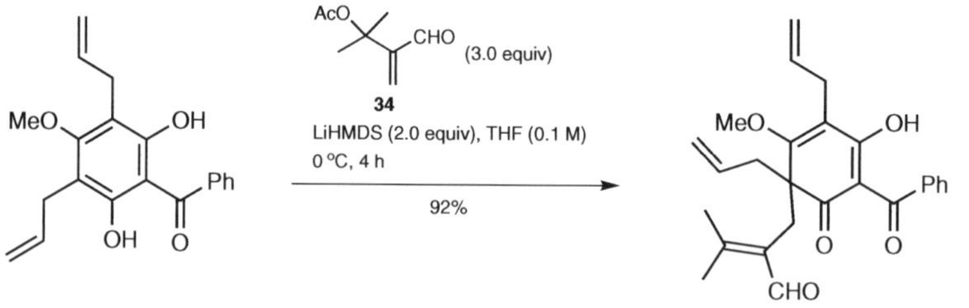

High yields of alkylative dearomatization product **33** can be achieved with activated bromide **32** in a 0.09 M solution of tetrahydrofuran.[8] The dark red phenolate anion of 4-methoxy-substituted acylphloroglucinol **31** forms instantaneously at 0 °C following the dropwise addition of 2.0 equivalents of potassium hexamethyldisilazanide (0.5 M solution in toluene). The activated bromide electrophile **32** is added as a 2.4 M solution in benzene. This method is general for nine different electrophile variations (see also Section 2.2.3.5, Scheme 32). The highly reactive bromide electrophile should be stored under argon as a solution in benzene in a freezer at −20 °C.

Dearomatized Michael adduct **35** is isolated in 92% yield by reaction of deprotonated 4-methoxy-substituted acylphloroglucinol **31** with activated Michael acceptor **34** in a 0.1 M solution of tetrahydrofuran.[5] The addition of 2.0 equivalents of lithium hexamethyldisilazanide (1.0 M in THF) to **31** provides a light red phenolate anion at 0 °C. It is important to use a lithium counterion because chelation to both the phenolate oxygen of **31** and the carbonyl of enal **34** places them within optimal proximity and activates Michael acceptor **34** for electrophilic attack. The elimination of the α-acetoxy group then proceeds via an $E1_{CB}$ mechanism to provide Michael adduct **35** in high yield.

For the best results in organic solvents, the reaction concentration usually should not exceed 0.1 M. A combination of tetrahydrofuran and benzene often provides optimal results. It is critical to note that electrophile **31** and potassium hexamethyldisilazanide are both stored as solutions in benzene giving an overall ratio of tetrahydrofuran/benzene of 2.4:1 after the addition of **31** is complete. For the case of Michael adduct **35**, the α-acetoxy enal **34** is added as a 0.55 M solution in benzene giving an overall ratio of tetrahydrofuran/benzene of 1.7:1 after the addition of the electrophile is complete. The reactions in organic solvents are less robust than in aqueous media because they are more sensitive to heat and moisture. Degassing of the solvent prior to the addition of base decreases the sensitivity to higher temperatures for each method. Products **33** and **35** are difficult to purify, as dearomatized enols often bind to silica gel. However, these compounds can also be purified by precipitation via basic extraction without the need for column chromatography. Substrates **29** and **31** are unstable in air and should be stored under argon in a freezer at −20 °C. One of the greatest disadvantages of each method is a required excess of highly toxic and reactive electrophile. Overall, the methods are general for most analogues of benzoylphloroglucinols, and may be compatible with phenols, resorcinols, and phloroglucinols.

Three alternative procedures to synthesize alkylated dearomatized acylphloroglucinols from activated γ-hydroxy halide electrophiles have been described.[4] The most efficient of these involves a biphasic mixture of chlorobenzene, a 3 M aqueous solution of potassium hydroxide, and Aliquat 336 (8 mol%) as a phase-transfer catalyst; the desired quantities of base, electrophile, and solvent vary for each reaction (Scheme 18). In general, it is effective to use more base than halide to prevent formation of hydrochloric or hydrobromic acid, which may react undesirably with the substrate [deoxycohumulone (**36**)], the halide **37**, or the colupulone analogue product **38**. Using a slight excess of halide **37** relative to substrate **36** is often favored. The concentration of the halide in the mixture is varied depending on its reactivity; larger concentrations appear optimal for aqueous media compared to other methods described earlier (see Scheme 17) using organic solvents. The reactions are amenable to large-scale modification as many reactions in Scheme 18 are carried out on a multigram scale; equivalencies and molarities are clearly outlined in Scheme 18. The alternative methods involve use of sodium methoxide in methanol and potassium carbonate in dimethylformamide, but the yields are often significantly lower than the biphasic method with activated electrophiles.[4]

Scheme 18 Alkylative Dearomatization of Deoxycohumulone with Activated γ-Hydroxy Halides[4]

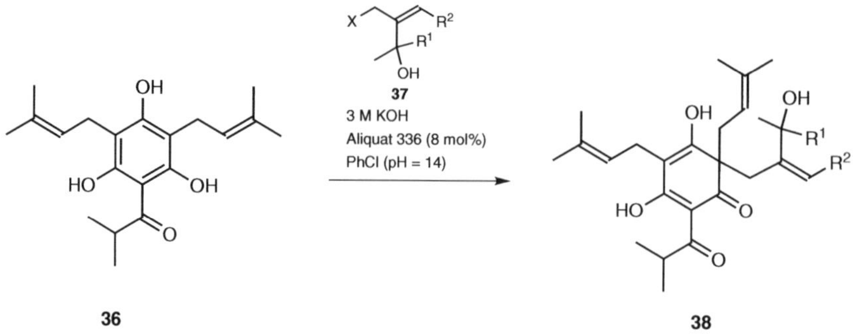

37: 3 M KOH, Aliquat 336 (8 mol%), PhCl (pH = 14); 36 → 38

R^1	R^2	X	Equiv of **37**	KOH (Equiv)	Concentration[a]	Yield (%)	Ref
Me	$CH_2OTBDMS$	Br	0.8	3.9	0.38 M	79	[4]
Me	$CH_2OTBDMS$	Cl	1.15	3.0	1.5 M	97	[4]
$(CH_2)_2CH{=}CMe_2$	H	Br	2.0	1.8	0.5 M	62[b]	[4]
$(CH_2)_2CH{=}CMe_2$	$CH_2OTBDMS$	Cl	1.2	2.5	2.2 M	99[b]	[4]
$(CH_2)_2CH{=}CMe_2$	$CH_2OTBDMS$	Br	0.5	2.0	0.74 M	44[b]	[4]

[a] Concentration of substrate **36** in PhCl.
[b] Product was formed as a ~1:1 mixture of diastereomers (^{1}H NMR).

2,6-Dimethyl-6-(3-methylbut-2-enyl)cyclohex-2-en-1-one (22, $R^1 = CH_2CH{=}CMe_2$); Typical Procedure:[46]
NaH (60% in mineral oil; 85 g, 2.13 mol, 1.04 equiv) was added in portions to a soln of 2,6-dimethylphenol (250 g, 2.05 mol, 1.0 equiv) in toluene (2 L) at 10–15 °C. The resulting suspension was stirred for 45 min. The mixture was cooled to 5 °C, and 1-chloro-3-methylbut-2-ene (2.13 mol, 1.04 equiv) was added over 1.5 h keeping the temperature at 5 °C. The mixture was then stirred for a further 2 h at 10–15 °C. 10% Pd/C (2.5 g) was added followed by MeOH (1 L), and the gray suspension was hydrogenated (0.3 bar H_2) keeping the temperature at 20–22 °C (ice bath). The suspension was then filtered through a pad of Celite. The yellow filtrate was washed with H_2O (0.5 L), aq NaOH (0.5 L), and brine (0.5 L), dried ($MgSO_4$), and concentrated under reduced pressure. The residue was distilled over a 5-cm Vigreux column to yield the product as a colorless oil; yield: 318 g (81%); bp 78–82 °C/0.05 Torr.

(*R*)-2-Methyl-2-(3-methylbut-2-enyl)naphthalen-1(2*H*)-one (26); Typical Procedure:[46]
A soln of α-isosparteine (**23B**; 440 mg, 1.80 mmol) in toluene (5.0 mL) was cooled to −10 °C. A 1.6 M soln of BuLi in hexane (1.1 mL, 1.80 mmol) was added dropwise. A bright yellow soln was obtained, which was stirred for 30 min at −0 °C. A soln of 2-methylnaphthalen-1-ol (**25**; 284 mg, 1.80 mmol) in toluene (2.0 mL) was added dropwise. After the mixture had been stirred at −10 °C for 30 min, the solids precipitated and the mixture turned gray/green. Additional toluene (2.0 mL) was added to maintain a movable suspension. 1-Chloro-3-methylbut-2-ene (234 mg, 1.80 mmol) was added causing the mixture to turn darker. The reaction was left stirring at rt for 15 h. More solids formed and analysis of the mixture by TLC showed that all the starting materials had been consumed. Silica gel and hexane were added to the mixture. The solids were removed by filtration and washed with hexane/*t*-BuOMe (1:1). The filtrate was concentrated and the residue was purified by chromatography (alumina, hexane/*t*-BuOMe 95:5). Short-path distillation (120 °C, 0.05 mbar) afforded the product as a light yellow oil; yield: 220 mg (54%); 38% ee; $[\alpha]_D^{25}$ +118.8 (*c* 1.0,

EtOH). The absolute configuration was determined by vibrational circular dichroism (VCD) spectroscopy. Due to the high conformational mobility of the side chain, only VCD signals resulting from stretching vibrations of the C=O group and the C=C double bonds (the 1750–1550 cm^{-1} region) can be interpreted. The CD spectrum of naphthalenone **26** in MeCN contains only positive bands with fine structures in the 250–400 nm region. Below 250 nm, the spectrum is governed by the positive ^1L_a band at ~230 nm and a negative one below 200 nm. These bands mostly result from the phenyl chromophore. Applying the sector rule of the saturated ketones to the positive long-wavelength band at 333 nm resulted in an *R* configuration of the stereogenic center, assuming the axial position of the methyl group.

Phenyl Lupone (30)**; Typical Procedure:**[3]

To a soln of clusiaphenone B (**29**; 183 mg, 0.5 mmol) and KOH (112 mg, 2 mmol) in H_2O (2.5 mL) at 0 °C under argon was added prenyl bromide (243 μL, 2 mmol) dropwise over 10 min. The resulting mixture was stirred at 0 °C for 2 h. The mixture was quenched with aq HCl (pH 1) and extracted with EtOAc (3 ×). The combined organic layers were washed with brine, dried ($MgSO_4$), filtered, and concentrated under reduced pressure. Purification by chromatography (silica gel, hexane/EtOAc 20:1) provided the product as a colorless oil; yield: 100 mg (40%).

4,6-Diallyl-2-benzoyl-3-hydroxy-5-methoxy-6-[2-(propan-2-ylidene)pent-4-enyl]cyclohexa-2,4-dien-1-one (33)**; Typical Procedure:**[8]

To a soln of (3,5-diallyl-2,6-dihydroxy-4-methoxyphenyl)(phenyl)methanone (**31**; 150 mg, 0.462 mmol) in THF (5 mL) was added a 0.5 M soln of KHMDS in toluene (1.850 mL, 0.925 mmol) at 0 °C under argon. After 2 min, bromide **32** (114 mg, 0.601 mmol) was added dropwise as a soln in benzene (450 mg/mL) (**CAUTION:** *carcinogen*). After 20 min, the mixture was poured into sat. aq NH_4Cl and then extracted with EtOAc (3 ×). The combined organic layers were washed with brine, dried (Na_2SO_4), and concentrated under reduced pressure. The crude residue was purified by chromatography (silica gel, hexane/EtOAc 20:1 to 3:1) to afford the product as a colorless to light brown oil; yield: 164 mg (82%).

2-[(1,3-Diallyl-5-benzoyl-4-hydroxy-2-methoxy-6-oxocyclohexa-2,4-dienyl)methyl]-3-methylbut-2-enal (35)**; Typical Procedure:**[5]

(3,5-Diallyl-2,6-dihydroxy-4-methoxyphenyl)(phenyl)methanone (**31**; 360 mg, 1.11 mmol) was dissolved in THF (10 mL). The soln was stirred at rt for 3 min, and then cooled to 0 °C. A 1 M soln of LiHMDS in THF (2.22 mL, 2.22 mmol, 2.0 equiv) was added dropwise forming a dark orange soln. After 5 min, 3-formyl-2-methylbut-3-en-2-yl acetate (**34**; 519 mg, 3.33 mmol, 3.0 equiv) in benzene (6 mL) (**CAUTION:** *carcinogen*) was added dropwise. The resulting mixture was stirred at 0 °C for 4 h. The mixture was finally quenched with 1 M aq HCl and extracted with EtOAc (3 ×). The combined organic layers were washed with brine, dried (Na_2SO_4), decanted, and concentrated under reduced pressure. The residue was purified by chromatography (silica gel, hexane/EtOAc 7:3) to afford the product as an orange oil; yield: 430 mg (92%); 1.6:1 ratio of enol tautomers as determined by ^1H NMR analysis in $CDCl_3$.

Colupulone Analogues 38**; General Procedure:**[4]

3 M aq KOH, methyltrioctylammonium chloride (Aliquat 336; 8 mol%), and a halide **37** were added at 25 °C to a slurry of deoxycohumulone (**36**) in chlorobenzene. The mixture was degassed and stirred vigorously at 25 °C under argon. After consumption of **36** (TLC), the reaction was quenched with sat. aq NH_4Cl, and the mixture was extracted with EtOAc

for references see p 290

(3 ×). The combined organic extracts were washed successively with H_2O and brine, dried (Na_2SO_4), and concentrated under reduced pressure. The residue was purified by flash column chromatography (silica gel, hexanes/EtOAc/MeOH).

2.2.2.2 Alkylative Dearomatizations of Phenolic Derivatives with Unactivated Electrophiles

The synthetic community has witnessed relatively few developments of alkylative dearomatization methods utilizing unactivated alkyl electrophiles, especially with respect to acylphloroglucinol substrates. Prior to a breakthrough utilizing primary alkyl trifluoromethanesulfonates (Scheme 19),[8] unactivated alkyl electrophiles were reported to give solely O-alkylation products when reacted with acylphloroglucinols.[47,48] Pent-4-enyl trifluoromethanesulfonate was isolated and added immediately in a dropwise fashion to a dilute solution of the sodium anion of (2,6-dihydroxy-4-methoxy-3,5-dipropylphenyl)-(phenyl)methanone (**39**) in tetrahydrofuran/benzene (3:1; 0.057 M). After 15 minutes at 0 °C, the reaction is complete, providing product **40** in good yield. The ratio of base to trifluoromethanesulfonate should always be greater than or equal to one, as the formation of trifluoromethanesulfonic acid can lead to undesired byproducts. Since the introduction of this method in 2009, it has been further optimized by the use of 2.5 equivalents of lithium hexamethyldisilazanide, which reduces the formation of O-alkylation byproducts. As one possibility for this outcome, it is proposed that lithium coordinates tightly to the phenolate oxygen and the S=O oxygen of the sulfonate ester to form either a six- or eight-membered transition state assembly as shown in Scheme 20.[49,64–70] To avoid undesired oxidation of substrate **39**, the base is added at −20 °C to form the phenolate anion and then warmed to 5 °C during the addition of trifluoromethanesulfonate. To avoid isolation of the highly unstable trifluoromethanesulfonate, it is pulled into a syringe with a thick disposable needle, which is then fitted with a cotton plug, and the solution of trifluoromethanesulfonate is filtered directly into the mixture leaving white ammonium salts of ethyl(diisopropyl)amine behind in the syringe. The successful formation of the trifluoromethanesulfonate largely depends on the quality of trifluoromethanesulfonic anhydride, which should be tightly sealed under an inert atmosphere and stored in a refrigerator at 5–8 °C; storing it cold significantly reduces the rate of decomposition. The time required for reaction completion is highly substrate and temperature dependent and may range from 15 minutes to 2 hours. The reaction may be carried out on a multigram scale without a change in yield, but the trifluoromethanesulfonate should be added in portions to avoid decomposition during the addition process.

Scheme 19 Alkylative Dearomatization of Acylphloroglucinols Utilizing Unactivated Electrophiles[4,8–11]

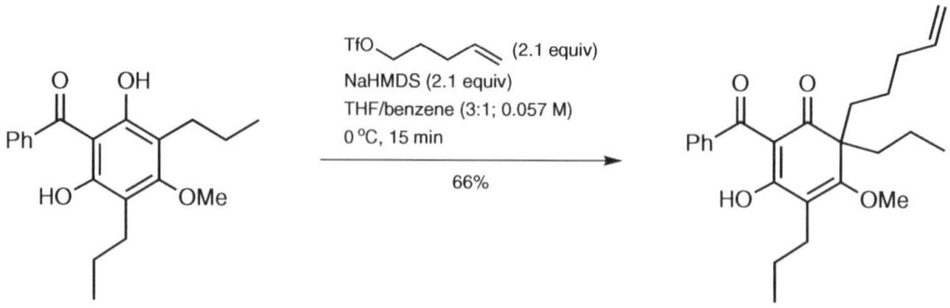

TfO(CH2)3CH=CH2 (2.1 equiv)
NaHMDS (2.1 equiv)
THF/benzene (3:1; 0.057 M)
0 °C, 15 min
66%

39 → **40**

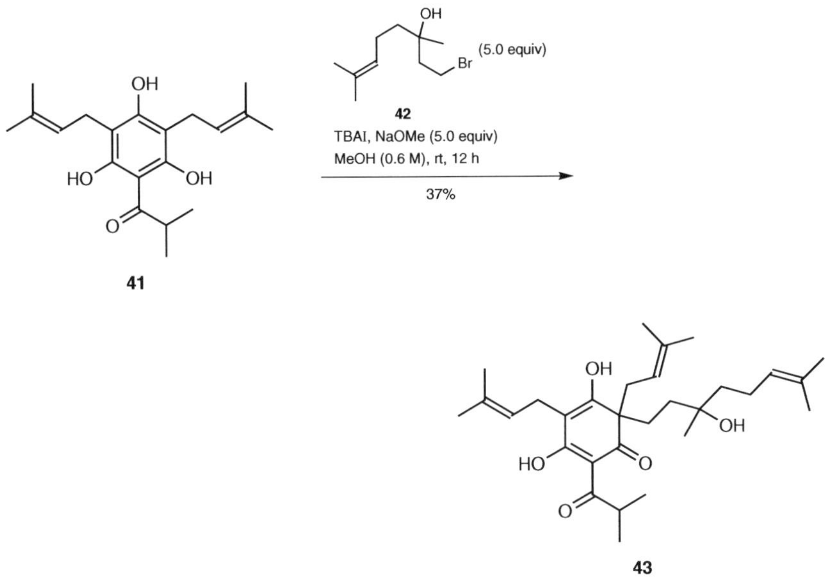

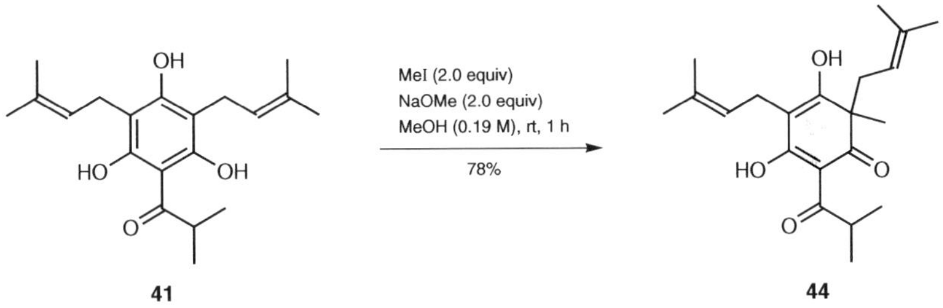

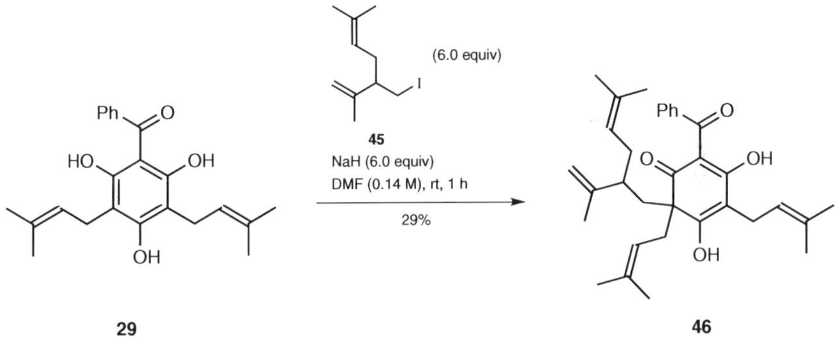

42 (5.0 equiv)

TBAI, NaOMe (5.0 equiv)

MeOH (0.6 M), rt, 12 h

37%

41 → 43

MeI (2.0 equiv)

NaOMe (2.0 equiv)

MeOH (0.19 M), rt, 1 h

78%

41 → 44

45 (6.0 equiv)

NaH (6.0 equiv)

DMF (0.14 M), rt, 1 h

29%

29 → 46

for references see p 290

Scheme 20 Lithium Coordination May Facilitate C-Alkylative Dearomatization[49,64–70]

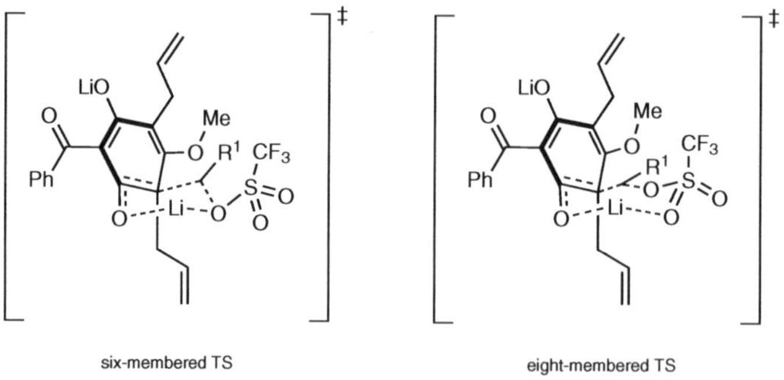

six-membered TS

eight-membered TS

A novel method for the alkylative dearomatization of acylphloroglucinol **41** applies unactivated γ-hydroxy bromide **42** (Scheme 19).[4] The reaction uses an excess of sodium methoxide in a solution of dry methanol along with tetrabutylammonium iodide to enhance the reactivity of the phenoxide anion, providing product **43** in 37% yield after 12 hours at room temperature. The large ammonium counterion is non-coordinating, generating a more nucleophilic phenoxide anion in methanol. Hydrogen-bonding of methanol to the resulting phenoxide anion presumably reduces the extent of O-alkylation.[47,62] The following factors may improve the yield: (1) higher temperatures with reduced reaction times, as bromide **42** is significantly less reactive than an activated variant (e.g., **37**) (Scheme 18); (2) reducing the equivalents of sodium methoxide, as reaction of γ-hydroxy bromide **42** with sodium methoxide may occur faster than reaction of bromide **42** with the phenolate of **41**; and (3) increasing the reaction concentration to at least 1.0 M.

Although methylation of phloroglucinols is known,[71–73] there are not many reports for C-methylation of acylphloroglucinols.[9,12] An efficient procedure gives access to C-monomethylated acylphloroglucinol **44** in 78% yield (Scheme 19).[9] The sodium phenolate of **41** is formed in dry methanol using solid sodium methoxide; because sodium methoxide is very hygroscopic, caution should be used to minimize its exposure to air. After the addition of iodomethane, the reaction is complete after 1 hour at room temperature; the reaction is also carried out on a multigram scale. In contrast, using a larger excess of iodomethane (5.0 equiv) and sodium methoxide (5.0 equiv) at lower temperatures (–20 to 0 °C) in a 0.1 M solution of methanol gives **44** in 70% yield.[12]

Alkylative dearomatization of the sodium phenolate ion of clusiaphenone B (**29**) with 6.0 equivalents of unactivated alkyl iodide **45** in a 0.14 M solution of dimethylformamide provides the product **46** in 29% yield (Scheme 19).[10,11] The relatively low yield may be a result of using 6.0 equivalents of sodium hydride, because iodide **45** is added to the solution of excess sodium hydride causing possible room for elimination prior to reaction with substrate. Elimination of the iodide is a problematic side reaction,[10,11] which may be avoided if 2.5–3.0 equivalents of base are used. The alkyl iodide **45** is also relatively unreactive toward S_N2 displacement in comparison to its trifluoromethanesulfonate analogue, and higher temperatures may be required assuming fewer equivalents of base are used.

2-Benzoyl-3-hydroxy-5-methoxy-6-(pent-4-enyl)-4,6-dipropylcyclohexa-2,4-dien-1-one (40); Typical Procedure:[8]

CAUTION: *It is important to take necessary precautions when handling trifluoromethanesulfonic anhydride (Tf_2O). Avoid inhalation and contact of this reagent with eyes, skin, or clothing as this could cause serious harm.*

To a soln of pent-4-en-1-ol (0.069 mL, 0.685 mmol) in hexane/benzene (**CAUTION:** *carcinogen*) (3:1; 4.0 mL) was added iPr_2NEt (0.159 mL, 0.914 mmol) at 0 °C under argon. Tf_2O (0.115 mL, 0.685 mmol) was added dropwise over 1 min. After 5 min, the white suspension was collected into a syringe and filtered through a plug of cotton. The solvents were removed under reduced pressure to an approximate volume of 0.5 mL, and this soln of pent-4-enyl trifluoromethanesulfonate was used in the next step without further purification.

To a soln of (2,6-dihydroxy-4-methoxy-3,5-dipropylphenyl)(phenyl)methanone (**39**; 75 mg, 0.228 mmol) in THF/benzene (3:1; 4.0 mL) was added a 1.0 M soln of NaHMDS in THF (0.457 mL, 0.457 mmol) at 0 °C under argon. After 2 min, the previously prepared soln of pent-4-enyl trifluoromethanesulfonate was added dropwise over 1 min. After 15 min, the soln was poured into sat. aq NH_4Cl and the mixture was extracted with EtOAc (3 ×). The combined organic layers were washed with brine, dried (Na_2SO_4), and concentrated under reduced pressure. The crude residue was purified by chromatography (silica gel, hexane/EtOAc 20:1 to 3:1) to afford the product as a colorless oil; yield: 60 mg (66%); 1.2:1 mixture of tautomers.

3,5-Dihydroxy-6-(3-hydroxy-3,7-dimethyloct-6-enyl)-4,6-bis(3-methylbut-2-enyl)-2-(2-methylpropanoyl)cyclohexa-2,4-dien-1-one (43); Typical Procedure:[4]

To a degassed soln of deoxycohumulone (**41**; 100 mg, 0.30 mmol) in anhyd MeOH (0.5 mL) was added NaOMe (81 mg, 1.50 mmol) under argon at 0 °C. After the mixture had been stirred for 5 min at this temperature, 1-bromo-3,7-dimethyloct-6-en-3-ol (**42**; 396 mg, 1.68 mmol) was added followed by a catalytic amount of TBAI. The mixture was stirred at 25 °C for 12 h. The reaction was quenched with sat. aq NH_4Cl and the mixture was extracted with EtOAc (3 ×). The combined organic extracts were washed successively with H_2O and brine, dried (Na_2SO_4), and concentrated under reduced pressure. The residue was purified by flash column chromatography to give the product as a light yellow oil; yield: 55 mg (37%).

3,5-Dihydroxy-6-methyl-4,6-bis(3-methylbut-2-enyl)-2-(2-methylpropanoyl)cyclohexa-2,4-dien-1-one (44); Typical Procedure:[9]

CAUTION: *Inhalation, ingestion, or skin absorption of iodomethane can be fatal.*

To a soln of deoxycohumulone (**41**; 3.75 g, 11.3 mmol) in anhyd MeOH (60 mL) at rt was added NaOMe (1.22 g, 22.6 mmol). The mixture was stirred at rt for 5 min, then MeI (1.41 mL, 22.6 mmol) was added dropwise. The mixture was stirred at rt for 1 h, and then acidified with 1 M aq HCl and extracted with EtOAc (3 × 100 mL). The combined organic layers were washed with H_2O (100 mL) and brine (100 mL), dried ($MgSO_4$), filtered, and concentrated under reduced pressure. The residue was purified by flash chromatography (silica gel, petroleum ether/EtOAc 5:1 to 1:1) to give the product as a yellow oil; yield: 3.05 g (78%). Crystallization (neat hexane) at 0 °C gave colorless crystals with significant loss of material due to aerobic oxidation; yield: 1.95 g (50%).

2-Benzoyl-3,5-dihydroxy-6-[5-methyl-2-(prop-1-en-2-yl)hex-4-enyl]-4,6-bis(3-methylbut-2-enyl)cyclohexa-2,4-dien-1-one (46); Typical Procedure:[11]

To a soln of clusiaphenone B (**29**; 616 mg, 1.68 mmol) in anhyd DMF (12 mL) was added NaH (242 mg, 10.1 mmol, 6 equiv) at rt. The mixture was stirred at rt for 10 min. 3-(Iodomethyl)-2,6-dimethylhepta-1,5-diene (**45**; 2.66 g, 10.1 mmol, 6 equiv) in anhyd DMF (2 mL)

for references see p 290

was then added at rt. The mixture was stirred at rt for 1 h. The reaction was quenched with 1 M aq HCl (15 mL) and the mixture was extracted with EtOAc (3 × 20 mL). The combined organics were washed sequentially with H_2O (2 × 30 mL) and brine (30 mL), dried ($MgSO_4$), filtered, and concentrated under reduced pressure. The residue was purified by flash chromatography (silica gel, petroleum/EtOAc 50:1 to 15:1) to give the product as a viscous yellow oil; yield: 242 mg (29%, 50% based on recovered starting material).

2.2.3 Tandem Intermolecular Alkylative Dearomatization/Annulation

2.2.3.1 Tandem Alkylative Dearomatization/[4+2] Cycloaddition

One of the first and most notable examples of an efficient dearomative domino transformation was reported in 1974 (Scheme 21).[34] The reaction is highly efficient despite the instability of alkylative dearomatization intermediate **47**, which immediately undergoes [4+2] cycloaddition upon heating to 80 °C, providing the desired bridged tricyclic core **48**. Subsequent alkene hydrogenation is followed by a stereodivergent methyllithium addition allowing for the final acid-catalyzed rearrangement to access the natural product seychellene (**49**) in only four steps from commercial starting materials.

Scheme 21 Tandem Alkylative Dearomatization/[4+2] Cycloaddition To Initiate a Tandem Four-Step Synthesis of Seychellene[34]

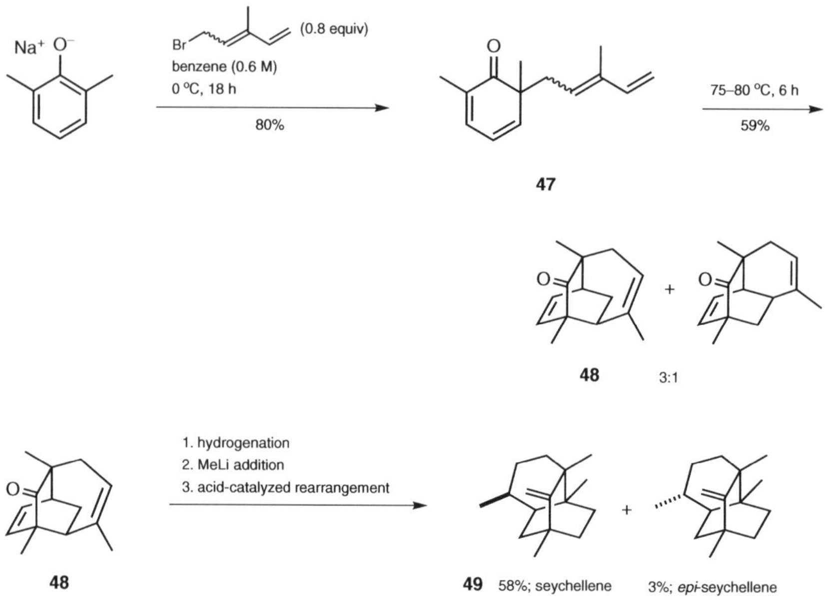

Na+ O−
Br (0.8 equiv)
benzene (0.6 M)
0 °C, 18 h
80%
47
75–80 °C, 6 h
59%
48 3:1

48
1. hydrogenation
2. MeLi addition
3. acid-catalyzed rearrangement
49 58%; seychellene
3%; *epi*-seychellene

2.2.3.2 Tandem Alkylative Dearomatization/Hydrogenation Followed by Lewis Acid Catalyzed Cyclization

An efficient domino sequence gives access to [3.2.1]-bicyclic ring systems in addition to fused cyclopropane and cyclobutane tricyclic ketones. Numerous dearomatized precursors (e.g., **50**) for this sequence with differing substitution patterns can be prepared by C-alkylation of 2,6-dimethylphenol on a 250-g scale (Scheme 22).[34,74] When precursors **51** are treated with 1.5 equivalents of ethylaluminum dichloride in toluene, a tandem cationic 1,4-addition followed by two consecutive Wagner–Meerwein 1,2-rearrangements provide the [3.2.1]-bicyclic compounds **52** in high yield (Scheme 23).[13]

Scheme 22 One-Pot Alkylative Dearomatization/Hydrogenation on 250-g Scale[34,74]

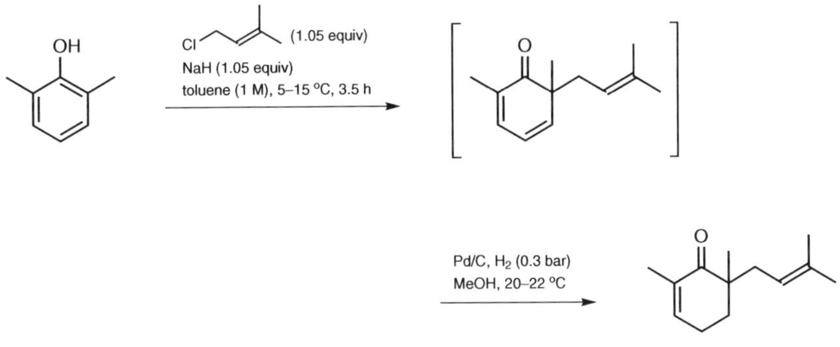

OH

Cl (1.05 equiv)
NaH (1.05 equiv)
toluene (1 M), 5–15 °C, 3.5 h

O

Pd/C, H_2 (0.3 bar)
MeOH, 20–22 °C

O

50 81%

Scheme 23 Formation of [3.2.1]-Bicyclic Systems from Dearomatized Intermediates[13]

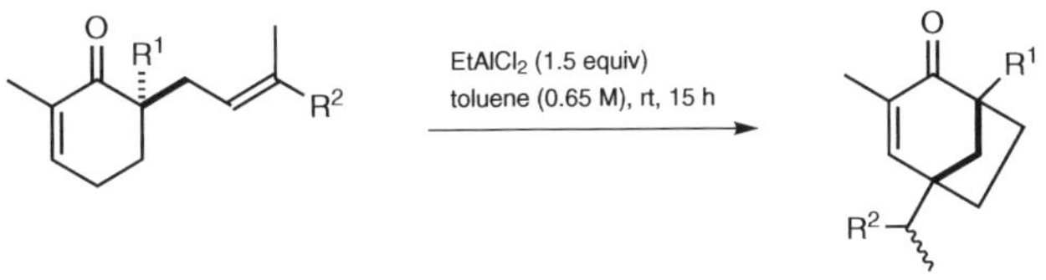

O R1 R2

$EtAlCl_2$ (1.5 equiv)
toluene (0.65 M), rt, 15 h

O R1 R2

51 **52**

R^1	R^2	dr	Yield[a] (%)	Ref
Me	Me	–	95	[13]
H	Me	–	86	[13]
H	iPr	1.6:1	90	[13]
H	$(CH_2)_2CO_2Me$	1.5:1	86[b]	[13]

[a] Products isolated by chromatography.
[b] Conditions: $EtAlCl_2$ (3.0 equiv), 8 h, 80 °C.

Under the same conditions, it is possible to access tricyclic compounds **54** and **55** by simply changing the substitution pattern of the dearomatized substrate (Scheme 24).[13] This cationic cyclization strategy allows for efficient access to 17 diverse architectures in only two steps from dearomatized substrates **51** (Scheme 23) and **53** (Scheme 24), each provided by a one-pot alkylative dearomatization/hydrogenation sequence.

Scheme 24 Formation of Fused Cyclobutane and Cyclopropane Tricyclic Ketones from Dearomatized Intermediates[13]

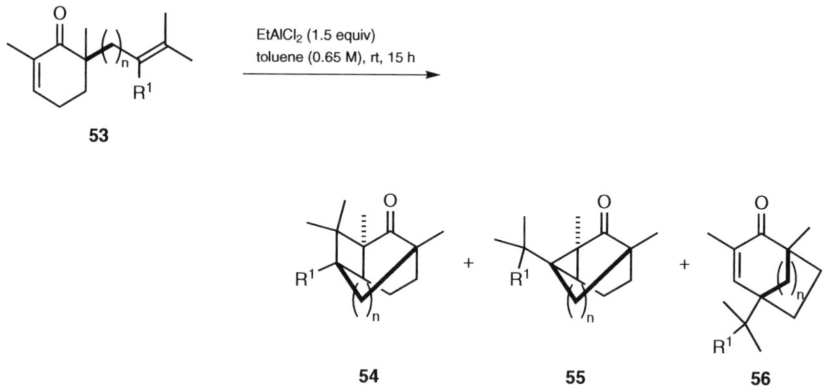

EtAlCl$_2$ (1.5 equiv)
toluene (0.65 M), rt, 15 h

53

54 + 55 + 56

R^1	n	Ratio (**54**/**55**/**56**)	Yield[a] (%)	Ref
Me	1	1:0:1[b]	61	[13]
H	2	–[c]	60	[13]

[a] Products isolated by chromatography.
[b] **55** was not detected.
[c] **55** only.

2.2.3.3 Tandem Alkylative Dearomatization/Annulation To Access Type A and B Polyprenylated Acylphloroglucinol Derivatives

Porco and co-workers have described an efficient and highly diastereoselective domino transformation to access type B polyprenylated acylphloroglucinol derivatives from α-acetoxy acrylates such as **57** (Scheme 25).[3] In this case, clusiaphenone B (**29**) is treated with 3.0 equivalents of lithium hexamethyldisilazanide in a 0.07 M solution of tetrahydrofuran at 0 °C providing the phenolate, which reacts with 2.0 equivalents of α-acetoxymethyl acrylate **57**[75] as a solution in tetrahydrofuran causing a tandem intermolecular Michael addition/E1$_{CB}$ elimination/intramolecular Michael addition domino sequence. The resulting ester enolate **58** quickly reacts with the second equivalent of **57** leading to another intermolecular Michael addition/E1$_{CB}$ elimination to provide product **59** in 70% yield.

Scheme 25 Domino Construction of a Type B Polyprenylated Acylphloroglucinol Derivative from Clusiaphenone B[3]

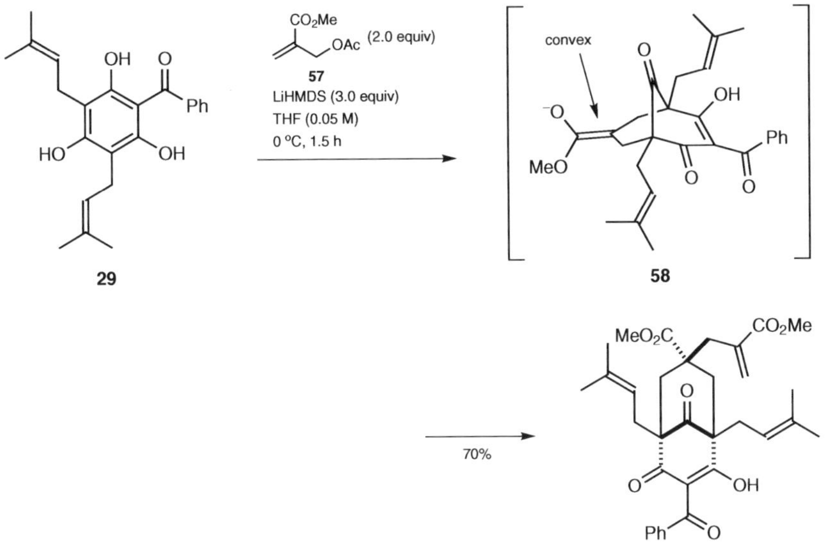

Tandem alkylative dearomatization/annulation of 29 to 59 (via 58): reagents and conditions

Reagents and conditions	Yield
57 (2.0 equiv), LiHMDS (3.0 equiv), THF (0.05 M), 0 °C, 1.5 h	70%

Tandem alkylative dearomatization/annulation of acylphloroglucinols appears to be general for a variety of substituted α-acetoxymethyl acrylates (Scheme 26).[3] It is notable that acylphloroglucinol **60** provides product **61** in high yield despite the presence of an α-carbonyl proton. All bicyclo[3.3.1]nonane derivatives **62** derived from clusiaphenone (**29**), except **62** where $R^1 = R^2 = Me$ and $R^3 = CO_2Me$, are provided as single diastereomers with functional handles at C7 for further manipulation (Scheme 26). This reaction demonstrates a high degree of atom economy since two substrates combine to make a united product.

Scheme 26 Bicyclo[3.3.1]nonane Derivatives via Alkylative Dearomatization/Annulation[3]

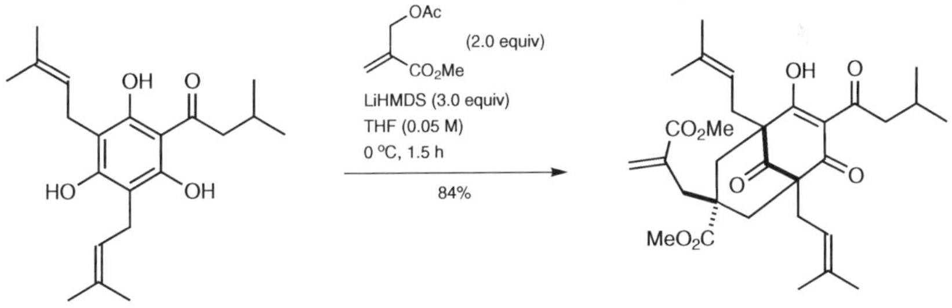

Conversion of 60 to 61: reagents and conditions

Reagents and conditions	Yield
methyl 2-(acetoxymethyl)acrylate (2.0 equiv), LiHMDS (3.0 equiv), THF (0.05 M), 0 °C, 1.5 h	84%

for references see p 290

1. AcO, R^1, R^2 (1.0 equiv), R^3
KHMDS (2.0 equiv)
THF (0.11 M), 65 °C, 4 h
2. TMSCHN$_2$ (2 equiv)
iPr$_2$NEt (1.5 equiv)
MeCN, MeOH

29 → **62**

R^1	R^2	R^3	Yield[a] (%)	Ref
H	H	CN	41	[3]
Me	Me	CO_2Me	63[a,b]	[3]
Me	Me	$CO_2CH_2CF_3$	55[a,c]	[3]
Me	Me	SO_2Ph	58[a]	[3]
Me	Me	CHO	54[a]	[3]

[a] Mixture of enol ether isomers produced.
[b] dr (7β/7α) 4:1.
[c] Reaction performed at rt.

A novel tandem alkylative dearomatization/intramolecular Michael addition/aldol cyclization process gives access to a type B adamantane derivative of type **64** (Scheme 27). A solution of clusiaphenone B (**29**) in tetrahydrofuran is treated with 3.0 equivalents of potassium hexamethyldisilazanide followed by 1.5 equivalents of 3-formyl-2-methylbut-3-en-2-yl acetate as a solution in tetrahydrofuran to provide transient Michael/E1$_{CB}$ adduct **63**. In the same reaction, product **63** undergoes a tandem intramolecular Michael addition/aldol cyclization to provide adamantane **64** in 65% yield and with good stereoselectivity.[3]

Scheme 27 Domino Transformation To Access a Type B Polyprenylated Acylphloroglucinol Adamantane Core[3]

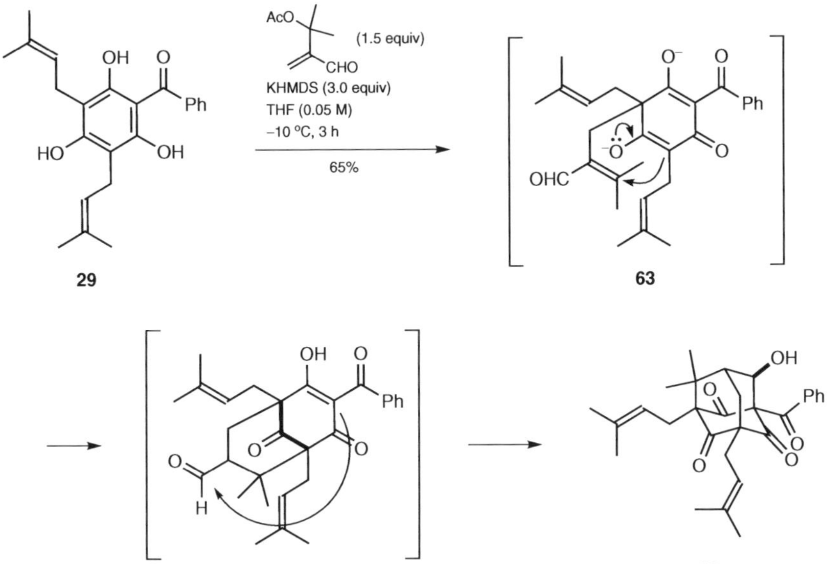

AcO; (1.5 equiv); CHO; KHMDS (3.0 equiv); THF (0.05 M); −10 °C, 3 h; 65%

29 → [**63**] → [ ] → **64**

A domino sequence to access the type A polyprenylated acylphloroglucinol adamantane core **65** utilizes 4-methoxy-substituted acylphloroglucinol (**31**) (Scheme 28).[5] As shown earlier in Scheme 17 (Section 2.2.2.1), alkylative dearomatization by intermolecular 1,4-addition provides product **35** in 92% yield. Upon adding 4.0 equivalents of concentrated hydrochloric acid to a 0.1 M solution of **35** in tetrahydrofuran, a 1,4-conjugate addition is followed by a demethylation/aldol cyclization sequence to provide type A polyprenylated acylphloroglucinol adamantane core **65** in 75% yield. Product **65** serves as a common intermediate to type A bicyclo[3.3.1]nonane derivative **66**[6] and type B polyprenylated acylphloroglucinol derivative **68**.[5] The organocerium reagent, generated from vinylmagnesium bromide and cerium(III) chloride, avoids deprotonation of the α-proton in intermediate **67** and preferentially reacts with the aldehyde at C7 leading to type A polyprenylated acylphloroglucinol core **66**.[6] However, the Grignard reagent favors deprotonation causing fragmentation of **67** followed by intramolecular Michael addition to provide a type B bicyclo[3.3.1]nonane derivative (not shown) after quenching with an aqueous 1 M hydrochloric acid solution.[5] Methylation with diazo(trimethylsilyl)methane provides type B polyprenylated acylphloroglucinol core **68**. This example shows the dynamic reactivity profile of alkylative dearomatization products which allow access to different classes of natural product cores from one dearomatization/cyclization intermediate **65**.

Scheme 28 Domino Sequence To Access a Type A Adamantane Core, a Common Intermediate En Route to both Type A and B Polyprenylated Acylphloroglucinols[5,6]

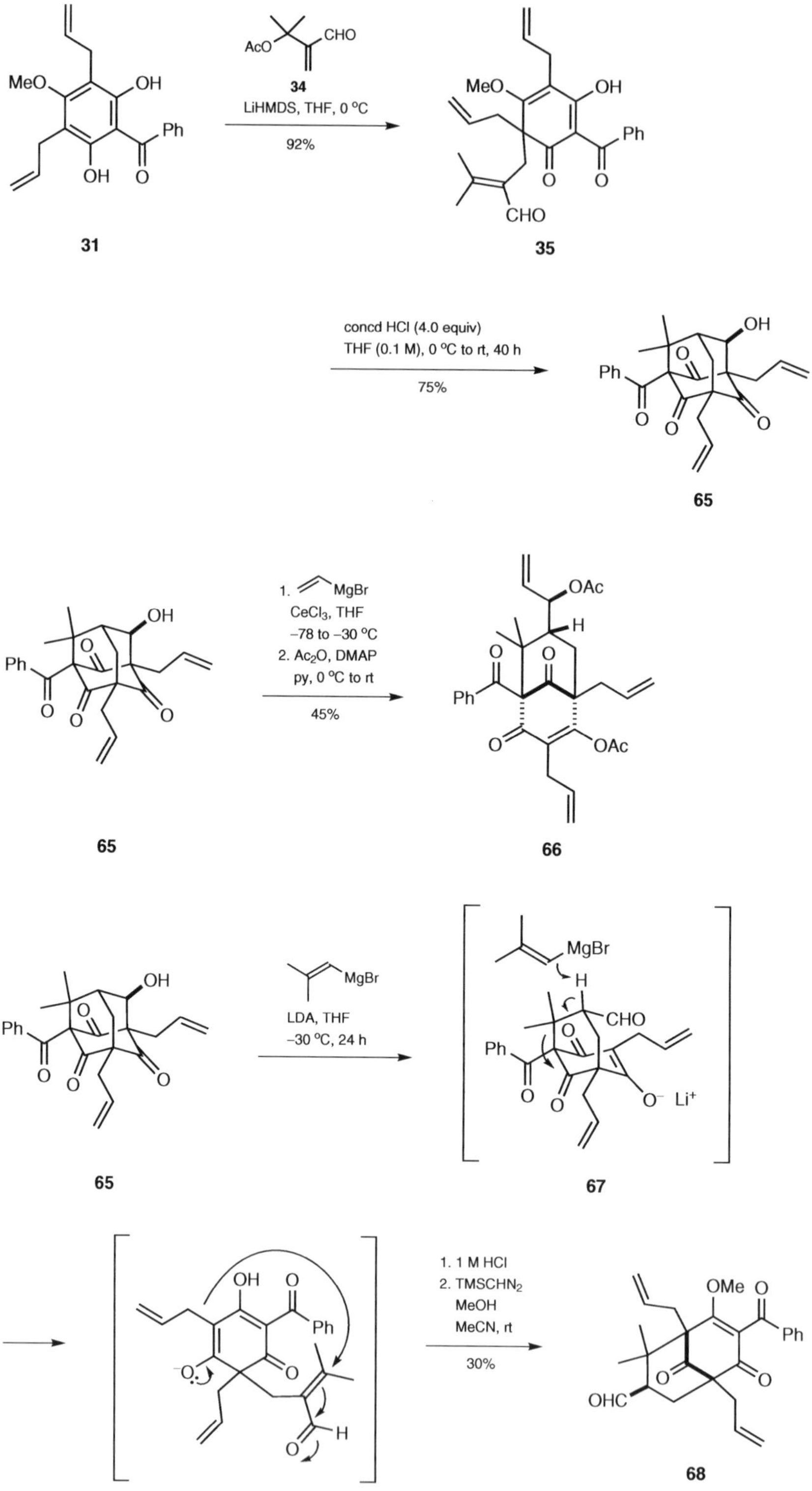

31 → 35: **34**, LiHMDS, THF, 0 °C, 92%

35 → 65: concd HCl (4.0 equiv), THF (0.1 M), 0 °C to rt, 40 h, 75%

65 → 66: 1. vinyl-MgBr, $CeCl_3$, THF, –78 to –30 °C; 2. Ac_2O, DMAP, py, 0 °C to rt; 45%

65 → [67] → [intermediate] → 68: 2-methylprop-1-enyl-MgBr, LDA, THF, –30 °C, 24 h; then 1. 1 M HCl; 2. $TMSCHN_2$, MeOH, MeCN, rt; 30%

A two-step domino sequence involving alkylative dearomatization of acylphloroglucinols followed by cationic cyclization of an allylic cation provides type A polyprenylated acylphloroglucinol derivatives (Scheme 29).[4] As described in Section 2.2.2.1, intermediate **69** is prepared in 81% yield by alkylative dearomatization of acylphloroglucinol **36** (Scheme 18) utilizing a biphasic method employing phase-transfer catalyst Aliquat 336; this method is compatible with activated electrophiles bearing a γ-hydroxide function. Dearomatized product **69** is acylated in high yield to deactivate the C3 enol of compound **70**, which then renders the C1 enol more electron-rich. The combination of triethylamine and methanesulfonyl chloride generates a strong electrophile capable of mesylating the tertiary alcohol in compound **70**.[4,16] This presumably leads to the formation of a resonance stabilized carbocation **71** with possible ion-pair stabilization. Intramolecular cyclization provides a 1:1 mixture of C-cyclization product **72** (via path a) and O-cyclization product **73** (via path b).[4]

Scheme 29 Efficient Construction of Type A Polyprenylated Acylphloroglucinols via Cationic Cyclization of Dearomatized Intermediates[4]

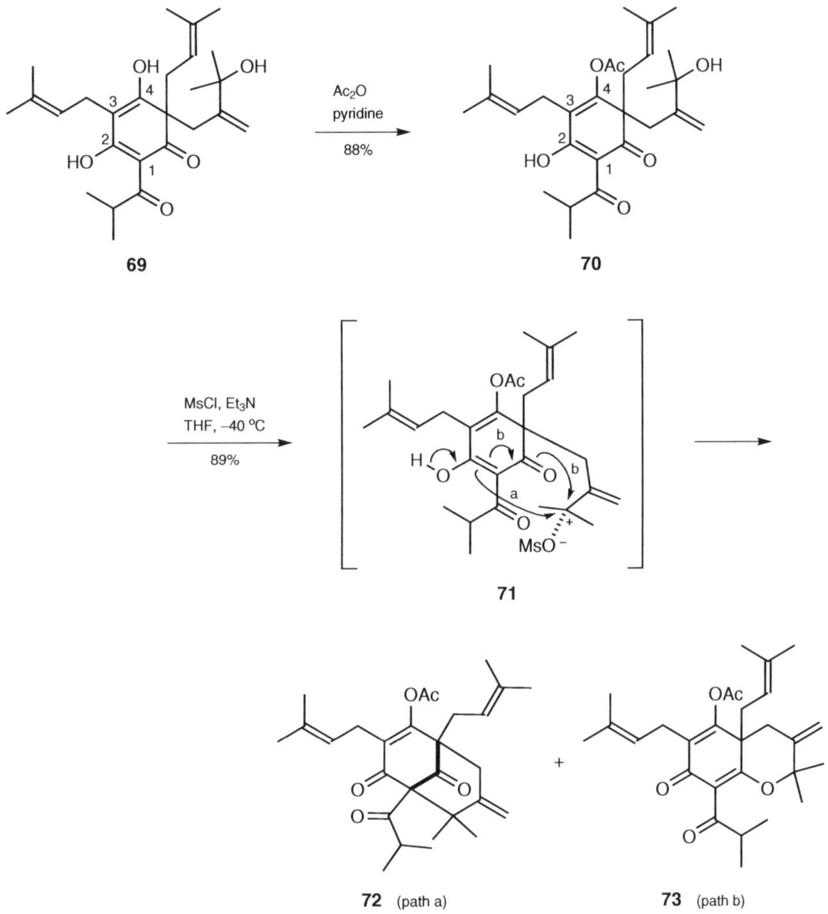

3-Benzoyl-4-methoxy-1,5-bis(3-methylbut-2-enyl)bicyclo[3.3.1]non-3-ene-2,9-diones 62; **General Procedure:**[3]

To a soln of clusiaphenone B (**29**; 201 mg, 0.55 mmol) in THF (3.0 mL) at 0 °C under argon was added a 0.5 M soln of KHMDS in toluene (2.2 mL, 1.1 mmol, 2.0 equiv) dropwise. The resulting mixture was stirred at 0 °C for 5 min. A soln of an α-acetoxymethyl acrylate

(0.55 mmol) in THF (2.0 mL) was added slowly to the mixture. The resulting mixture was stirred at 65 °C for 4 h, and then quenched with aq HCl (pH 1) and extracted with EtOAc (3 ×). The combined organic layers were washed with brine, dried ($MgSO_4$), filtered, and concentrated under reduced pressure. To a soln of the crude mixture in MeCN (9.0 mL) and MeOH (1.0 mL) under argon was added iPr_2NEt (139 μL, 0.8 mmol) and a 2 M soln of $TMSCHN_2$ in hexanes (0.6 mL, 1.2 mmol). The resulting mixture was stirred at rt for 5 h and finally quenched with aq HCl (pH 1) and extracted with EtOAc (3 ×). The combined organic layers were washed with brine, dried ($MgSO_4$), filtered, and concentrated under reduced pressure.

(1*R,3*S**,5*R**,7*R**,8*R**)-1-Benzoyl-8-hydroxy-6,6-dimethyl-3,5-bis(3-methylbut-2-enyl)-adamantane-2,4,9-trione (64); Typical Procedure:**[3]

To a soln of clusiaphenone B (**29**; 37 mg, 0.1 mmol) in THF (1.5 mL) at −10 °C under argon was added a 0.5 M soln of KHMDS in THF (0.6 mL, 0.3 mmol) dropwise. The resulting mixture was stirred for 5 min. A soln of 3-formyl-2-methylbut-3-en-2-yl acetate (24 mg, 0.15 mmol) in THF (0.5 mL) was added slowly to the mixture. The resulting mixture was stirred at −10 °C for 3 h. The reaction was quenched with aq HCl (pH 1), and the mixture was extracted with EtOAc (3 ×). The combined organic layers were washed with brine, dried ($MgSO_4$), filtered, and concentrated under reduced pressure. The residue was purified by chromatography (silica gel, hexane/EtOAc 12:1) to provide the product as a colorless oil; yield: 30 mg (65%).

2.2.3.4 Enantioselective, Tandem Alkylative Dearomatization/Annulation

A highly efficient catalytic enantioselective alkylative dearomatization/annulation procedure utilizes a chiral phase-transfer catalyst (Scheme 30).[7] Optimized reaction conditions are achieved by the addition of 5.0 equivalents of cesium hydroxide monohydrate to a 0.03 M solution of clusiaphenone B (**29**) and *Cinchona* alkaloid derived catalyst **78** (25 mol%) in dichloromethane over 4-Å molecular sieves at room temperature followed by cooling to −50 °C. Improved results are achieved when cesium hydroxide monohydrate is stored in a glovebox because it is extremely hygroscopic; indeed, the reaction is highly sensitive to moisture. The reaction rate depends on the reactivity of the electrophile, noting that enal **80** ($R^1 = Me$) is less reactive and requires up to 22 hours of reaction time whereas enal **80** [$R^1 = (CH_2)_5Me$] requires only 10 hours. Better yields and higher enantioselectivities for the adamantane-2,4,9-trione product **81** are also obtained when using the more reactive electrophile **80** [$R^1 = (CH_2)_5Me$].

In searching for the optimum phase-transfer catalyst, it was found that acceptable yields (68%) and enantiomeric excesses (75%) can be achieved using stoichiometric amounts of monomeric catalyst **74** (Scheme 30).[7] Dimeric catalysts **76** and **77** provide low enantiomeric excesses and yields, but better results (68% ee) are achieved with catalyst **75** utilizing a *meta*-substituted linker. Further manipulation of the catalyst structure by switching from O-allylated catalyst **75** to O-benzylated catalyst **78** allows for further improvement (48% yield, 84% ee).

Scheme 30 Enantioselective Alkylative Dearomatization Using *Cinchona* Alkaloid Derived Phase-Transfer Catalysts[7]

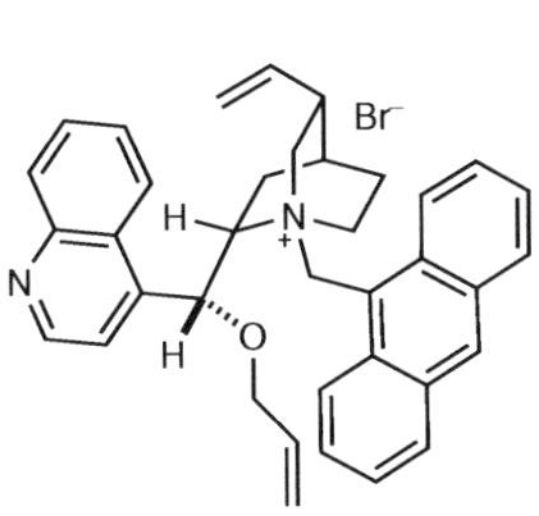

Br⁻ H N O H 74

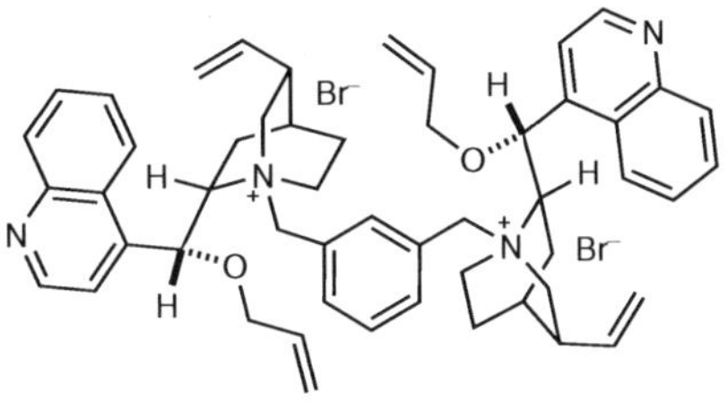

75 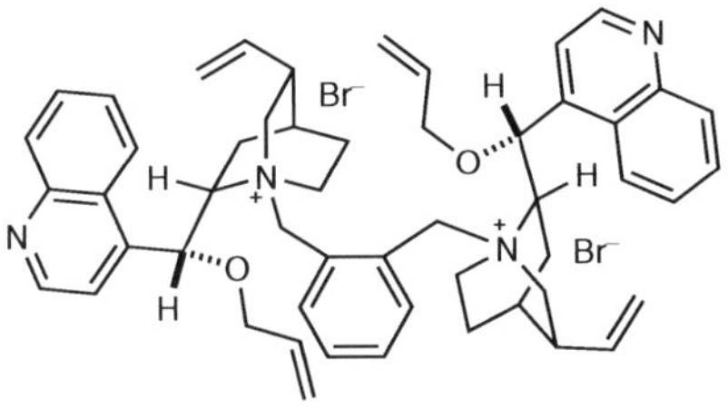 76 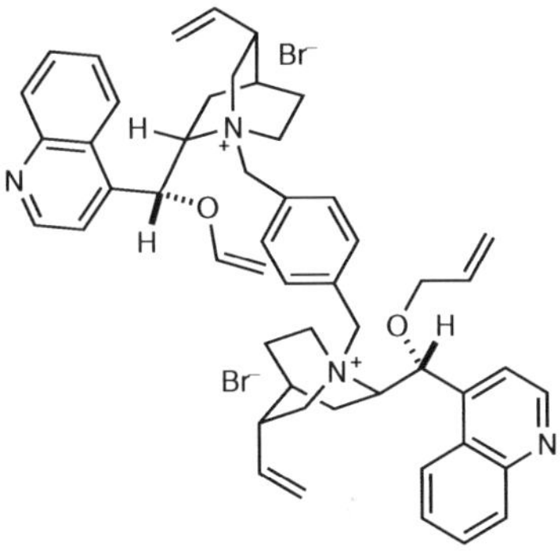 77 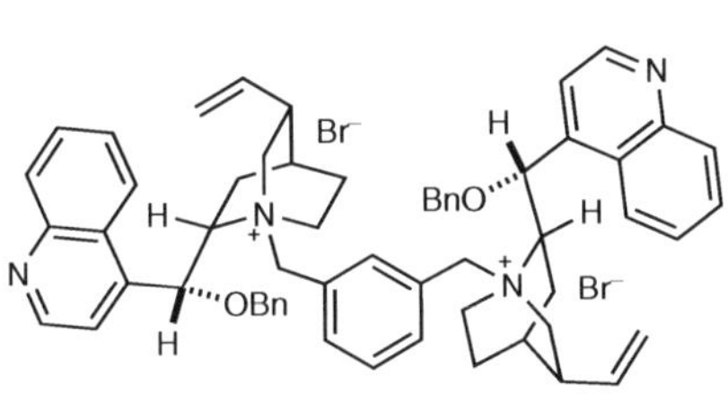 78 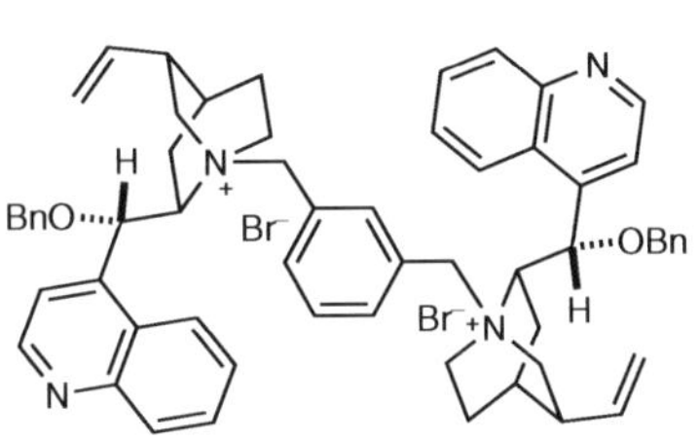 *ent*-78 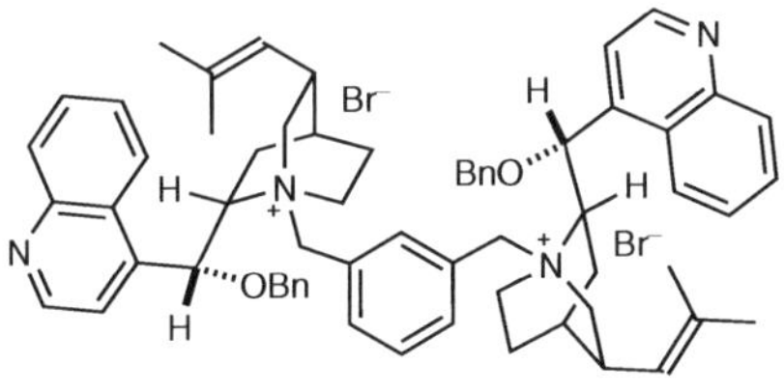 79

(1.05 equiv)
80
chiral phase-transfer catalyst
4-Å molecular sieves
$CsOH\cdot H_2O$ (5.0 equiv)
CH_2Cl_2, –50 °C, 10–22 h
29
81

R^1	Catalyst	Time (h)	ee (%)	Yield (%)	Ref
Me	**74** (1 equiv)	15	75 (8*R*)	68	[7]
Me	**75** (25 mol%)	22	68 (8*R*)	41	[7]
Me	**76** (25 mol%)	22	11 (8*R*)	22	[7]
Me	**77** (25 mol%)	22	20 (8*R*)	~10	[7]
Me	**78** (25 mol%)	22	84 (8*R*)	48	[7]
$(CH_2)_5Me$	**75** (25 mol%)	10	76 (8*R*)	65	[7]
$(CH_2)_5Me$	**78** (25 mol%)	10	86 (8*R*)	61	[7]
$(CH_2)_5Me$	**79** (25 mol%)	10	90 (8*R*)	71	[7]
$(CH_2)_5Me$	*ent*-**78** (25 mol%)	10	60 (8*S*)	53	[7]

As determined by mechanistic studies, there are two important factors relating to the interaction between the catalyst **78** and active substrate **82** (Scheme 31). First, ion-pairing interactions are optimized by placing the *para* oxygen phenolate ion of **82** in close proximity to the more accessible quaternary ammonium ion. Presumably, the presence of the relatively non-coordinative cesium counterion could maximize this interaction. Second, the enantioselectivity likely depends on hydrophobic binding interactions arising between a prenyl group on dearomatized substrate **82** and the *O*-benzyl and allyl substituents on phase-transfer catalyst **78**.[7] Interestingly, conversion of the allyl group in **78** into a prenyl group generates catalyst **79**, which results in a 4% increase in enantiomeric excess (71% yield, 90% ee), further confirming the importance of this key hydrophobic interaction.

This methodology is currently the only example of an asymmetric, tandem alkylative dearomatization/annulation of acylphloroglucinols. The method provides higher enantiomeric excesses in comparison to the enantioselective alkylative dearomatization of 2,6-dimethylphenol employing stoichiometric quantities of chiral sparteine and α-isosparteine ligands (see Section 2.2.2.1).[46] The reaction is operationally simple and does not require pyrophoric reagents. Pleasingly, the opposite enantiomer of the product can be afforded using the opposite enantiomer of catalyst **78** (*ent*-**78**), albeit with 60% enantiomeric excess.

Despite these advantages, there is tremendous room for improvement with several notable limitations. *Cinchona* alkaloid derived catalysts **78** and **79** are not commercially available and require multistep synthesis for their preparation. Most of the reagents used are highly toxic, but the overall operational simplicity of the reaction makes it feasible on a multigram scale. The instability of the substrates is somewhat problematic in that oxidation of clusiaphenone B (**29**) can occur with prolonged exposure to air and the enal electrophile **80** [$R^1 = (CH_2)_5Me$] is moisture-sensitive and should be freshly prepared.

Clusiaphenone B (**29**) should be stored under argon in a freezer at −20 °C, and the enal **80** [$R^1 = (CH_2)_5Me$] should be sealed under argon and stored in a refrigerator (5–8 °C).

Scheme 31 Proposed Reactive Enolate for Enantioselective Dearomatization/Annulation[7]

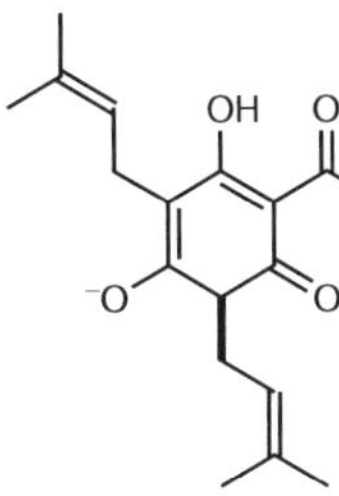

82

(1*R*,3*S*,5*R*,7*R*,8*R*)-1-Benzoyl-8-hydroxy-6,6-dimethyl-3,5-bis(3-methylbut-2-enyl)-adamantane-2,4,9-trione (81); General Procedure:[7]

To a mixture of clusiaphenone B (**29**; 20 mg, 0.055 mmol), a chiral phase-transfer catalyst (25 mol%), 4-Å molecular sieves (500 mg, 8–12 mesh, beads activated in 130 °C vacuum oven), and $CsOH•H_2O$ (46 mg, 0.27 mmol) was added CH_2Cl_2 (2.0 mL). The mixture was stirred at rt for 5 min and cooled to −50 °C. A soln of an aldehyde **80** (0.057 mmol) in benzene (300 μL) (**CAUTION:** *carcinogen*) was added dropwise. The resulting mixture was stirred at −50 °C for 22 h in the case of aldehyde **80** (R^1 = Me) or 10 h in the case of aldehyde **80** [$R^1 = (CH_2)_5Me$]. The reaction was quenched with aq HCl and the mixture was extracted with EtOAc (3 ×). The combined organic layers were washed with brine, dried (Na_2SO_4), filtered, and concentrated under reduced pressure. The residue was purified by chromatography (silica gel, hexane/EtOAc 13:1).

2.2.3.5 Tandem Alkylative Dearomatization/Radical Cyclization

Since the first example of cascade oxidative radical cyclizations of acylphloroglucinols was demonstrated in 2009,[8] this approach has been utilized to access a number of diverse scaffolds including several natural products.[9–12] In the first dearomatization/radical cyclization domino sequence, nine dearomatized acylphloroglucinol variants were synthesized by alkylative dearomatization methods with a variety of activated and unactivated electrophiles (see Scheme 17, Section 2.2.2.1 and Scheme 19, Section 2.2.2.2). Upon treatment with 2.1 equivalents of manganese(III) acetate dihydrate and 1.0 equivalent of copper(II) acetate monohydrate in a 0.033 M solution of glacial acetic acid, substrates **83**, **84**, **86**, **89**, **91**, and **92** undergo a formal oxidative [4+2] cycloaddition either in a stepwise sequence of radical cyclizations or in a concerted fashion (Scheme 32).[8] For substrates **91** and **92**, the formal [4+2] cycloaddition is followed by two consecutive radical 5-*exo* cyclizations due to the presence of prenyl or allyl moieties at C3 and C5. Products **85**, **87**, and **90** are each formed as the Z-isomer, and bicyclo[3.3.1]nonane derivative **88** is formed as a major product in 60% yield resulting from 6-*exo* cyclization of a radical at C1 onto the terminal double bond in substrate **86**. Interestingly, O-cyclization is observed for substrates **93** and **94**, e.g. to give **95**.[8]

Under the same conditions in a 0.05 M solution of glacial acetic acid, the natural products ialibinone A (**97A**) and ialibinone B (**97B**) are obtained in 14% and 21% yield, respectively, from the dearomatized methylation adduct **96** (Scheme 32),[12] which can be synthesized according to the method shown in Scheme 19.[9] A more efficient synthesis of natural products **97A** and **97B** utilizes (diacetoxyiodo)benzene (see below) and gives the products in 58% yield as a 1:1 mixture (Scheme 33).[9] For both methods, substrate **96** likely undergoes two consecutive 5-*exo-trig* cyclizations prior to radical termination.[9]

In a degassed 0.06 M solution of acetic acid containing 2.1 equivalents of manganese(III) acetate dihydrate and 1.0 equivalent of copper(II) acetate monohydrate, chiral, racemic substrate **98** undergoes a formal [4+2] cycloaddition followed by two additional radical-mediated ring-closing events to furnish a 14% yield of the natural product (+)-garcibracteatone (**99**) and 8% of (−)-garcibracteatone (*ent*-**99**) (Scheme 32).[10,11] The natural products (±)-doitunggarcinone A (**101A**) and (±)-5-*epi*-doitunggarcinone A (**101B**) are produced in the same manner from substrate **100** in yields of 13 and 12%, respectively (Scheme 32).[10] For all examples listed in Scheme 32, the radical intermediate is thought to terminate upon reduction by copper(II) with subsequent elimination of copper(I) acetate and acetic acid.

Scheme 32 Oxidative Radical Cyclizations of Dearomatized Acylphloroglucinols with Manganese(III)[8,10–12]

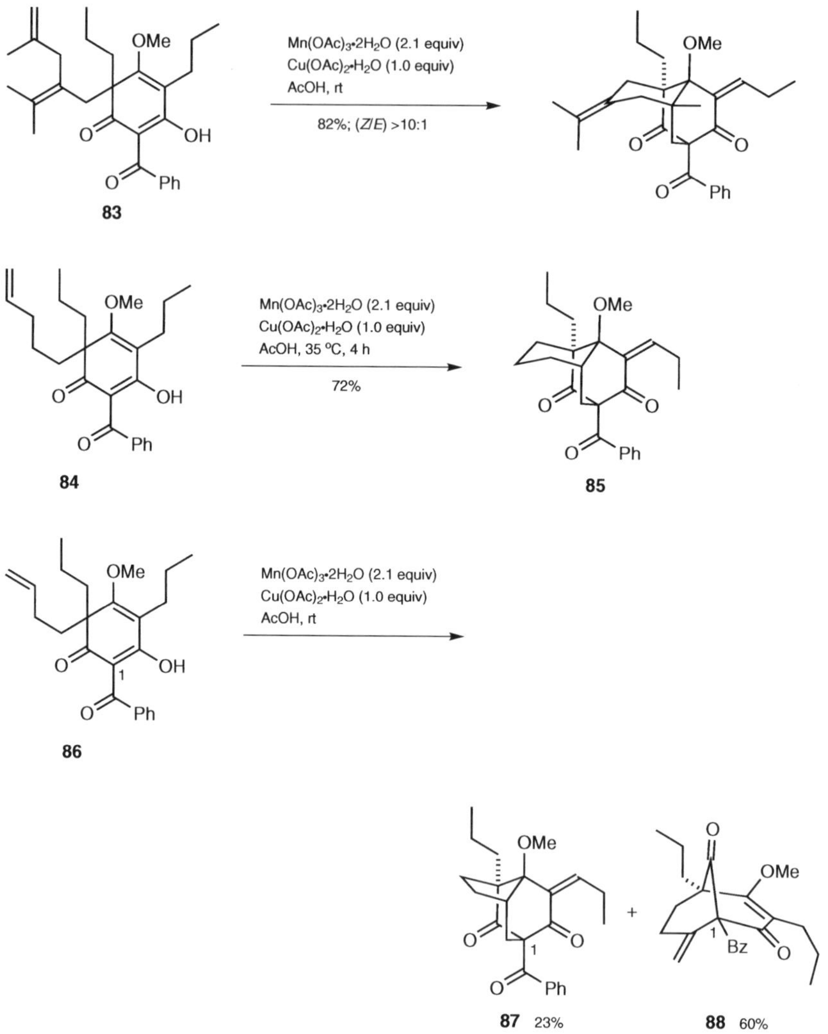

OMe, OH, O, O, Ph

89

$Mn(OAc)_3{\cdot}2H_2O$ (2.1 equiv)
$Cu(OAc)_2{\cdot}H_2O$ (1.0 equiv)
AcOH, rt

69%

OMe, O, O, O, Ph

90

OMe, OH, O, O, Ph

91

$Mn(OAc)_3{\cdot}2H_2O$ (2.1 equiv)
$Cu(OAc)_2{\cdot}H_2O$ (1.0 equiv)
AcOH, rt

74%; dr 4:1

OMe, H, O, O, O, Ph

OMe, OH, O, O, Ph

92

$Mn(OAc)_3{\cdot}2H_2O$ (2.1 equiv)
$Cu(OAc)_2{\cdot}H_2O$ (1.0 equiv)
AcOH, rt

76%

OMe, H, O, O, O, Ph

OMe, OH, O, O, Ph

93

$Mn(OAc)_3{\cdot}2H_2O$ (2.1 equiv)
$Cu(OAc)_2{\cdot}H_2O$ (1.0 equiv)
AcOH, rt

70%

OMe, O, O, O, Ph

OMe, OH, O, O, Ph

94

$Mn(OAc)_3{\cdot}2H_2O$ (2.1 equiv)
$Cu(OAc)_2{\cdot}H_2O$ (1.0 equiv)
AcOH, 65 °C, 15 min

76%

OMe, O, O, O, Ph

95

for references see p 290

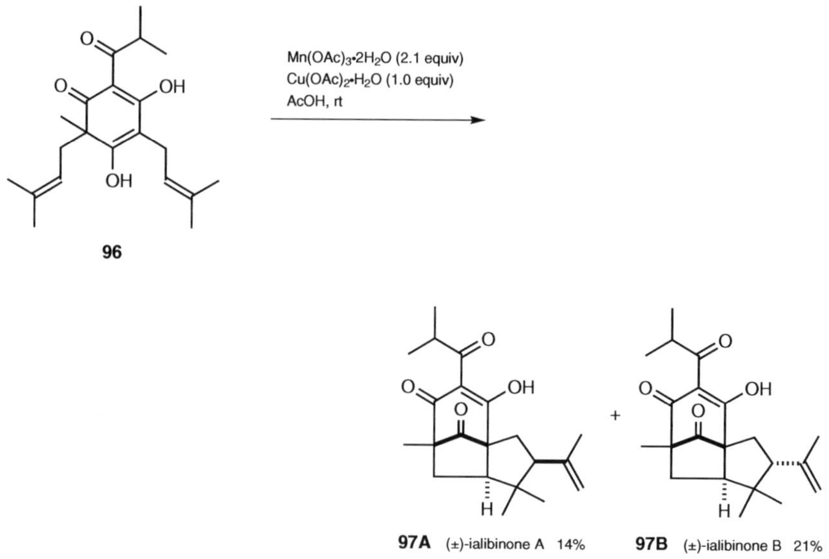

$Mn(OAc)_3 \cdot 2H_2O$ (2.1 equiv)
$Cu(OAc)_2 \cdot H_2O$ (1.0 equiv)
AcOH, rt

96

97A (±)-ialibinone A 14%

97B (±)-ialibinone B 21%

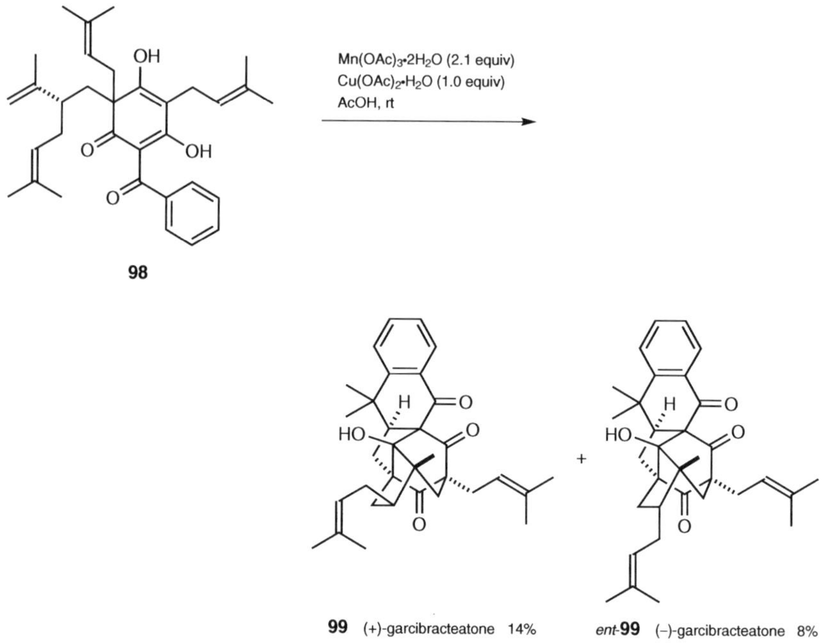

$Mn(OAc)_3 \cdot 2H_2O$ (2.1 equiv)
$Cu(OAc)_2 \cdot H_2O$ (1.0 equiv)
AcOH, rt

98

99 (+)-garcibracteatone 14%

ent-99 (−)-garcibracteatone 8%

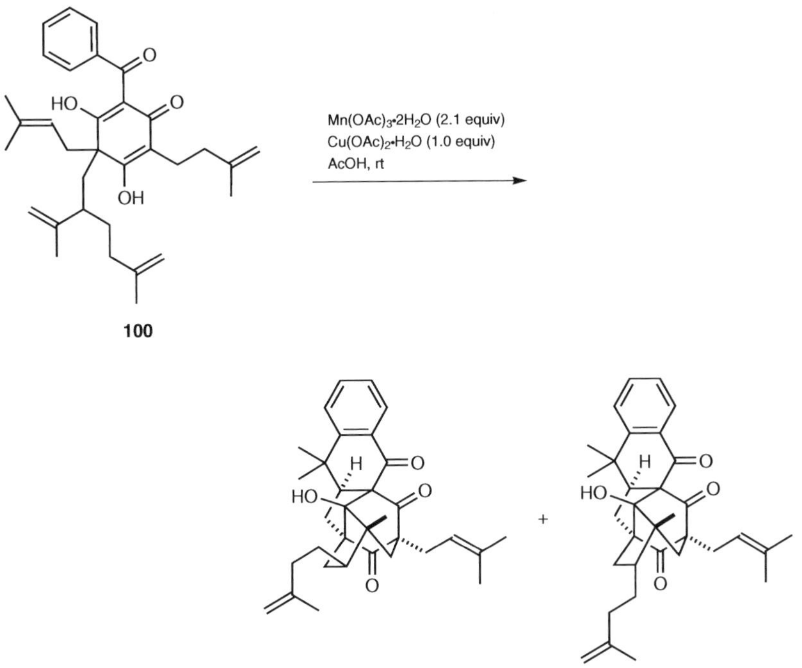

$Mn(OAc)_3 \cdot 2H_2O$ (2.1 equiv)
$Cu(OAc)_2 \cdot H_2O$ (1.0 equiv)
AcOH, rt

100

101A (±)-doitunggarcinone A 13% **101B** (±)-5-*epi*-doitunggarcinone A 12%

Substrate **96** is oxidized in the presence of (diacetoxyiodo)benzene and 2,2,6,6-tetramethylpiperidin-1-oxyl radical (TEMPO) to form a quinomethane, which undergoes a 6π-electrocyclization to generate (±)-hyperguinone B (**102**) in 73% yield (Scheme 33). The same result is achieved with dearomatized triprenylated substrate **103** to give product **104** in high yield (Scheme 33).[9]

Scheme 33 Oxidative Radical Cyclizations of Dearomatized Acylphloroglucinols Utilizing (Diacetoxyiodo)benzene[9]

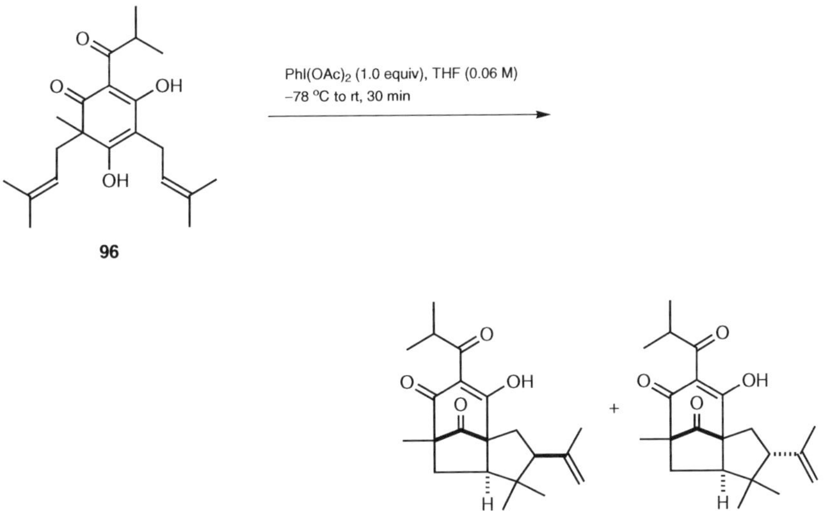

$PhI(OAc)_2$ (1.0 equiv), THF (0.06 M)
−78 °C to rt, 30 min

96

97A (±)-ialibinone A 29% **97B** (±)-ialibinone B 29%

for references see p 290

$PhI(OAc)_2$ (1.1 equiv)
TEMPO (2.0 equiv)
THF (0.06 M)
–78 °C to rt, 30 min
73%

96 102

TEMPO =

$PhI(OAc)_2$ (1.1 equiv)
TEMPO (2.0 equiv)
THF (0.62 M)
–78 °C to rt, 30 min
64%

103 104

2.2.4 Recent Methods for Alkylative Dearomatization of Phenolic Derivatives

In this final section, an update describing the latest developments for intermolecular alkylative dearomatization methods is provided, which may provide insight for current and future developments.

2.2.4.1 Recent Applications to Intermolecular Alkylative Dearomatization of Naphthols

In 2013, the first asymmetric palladium-catalyzed allylic dearomatization of 2-naphthol derivatives was described, with good yields, enantioselectivity, and chemoselectivity.[76] Although this reaction has not been used as part of a domino transformation in the following examples and discussion, its wide substrate scope and high enantioselectivities and yields bode well for its use in future domino applications. A 0.0025 M solution of 5 mol% allylpalladium(II) chloride dimer and 11 mol% of ligand **105** in 1,4-dioxane forms a cationic complex upon stirring for 30 minutes. The dearomatized 2-naphthols **113** are provided in high yields and enantioselectivities upon adding the prepared catalyst solution to 2.0 equivalents of a 2-naphthol **111** followed by the addition of 2.0 equivalents of 1,8-diazabicyclo[5.4.0]undec-7-ene, 1.0 equivalent of an allyl carbonate **112**, and 1,4-dioxane (0.05 M with respect to allyl carbonate **112**). The workup is simple and only requires removal of solvent and direct placement of the resulting residue on a silica gel column. The regioisomeric O-alkylation product **114** is avoided using 1,8-diazabicyclo[5.4.0]undec-7-ene as a base in combination with ligand **105**; O-alkylation product **114** is also avoided with ligand **109** using lithium carbonate as a base in 1,4-dioxane (0.1 M), albeit without asymmetric induction. Poor yields are obtained utilizing ligands **108** and **110**, whereas ligands **106**, **107**, and **109** provide low enantioselectivities. Fortunately, ligand **105** proves highly efficient for this asymmetric alkylative dearomatization reaction at room temperature with catalyst loadings of 5 mol% and relatively short reaction times of approximately 4–5 hours (Scheme 34). The reaction times are significantly reduced in the presence of excess 1,8-diazabicyclo[5.4.0]undec-7-ene.[76]

Scheme 34 Palladium-Catalyzed Intermolecular Asymmetric Allylic Dearomatization of 2-Naphthol Derivatives[76]

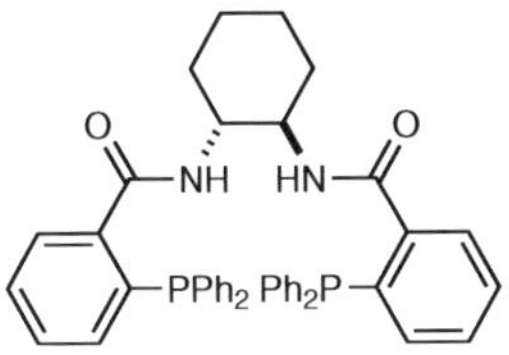

(R,R)-**105** 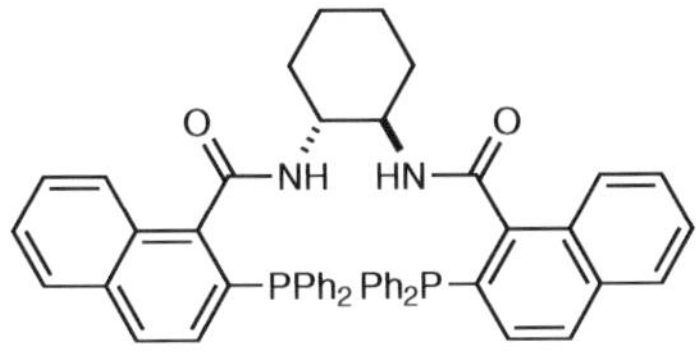 (R,R)-**106**

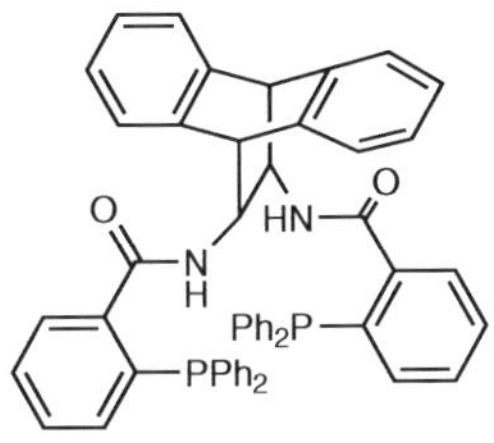

(R,R)-**107** 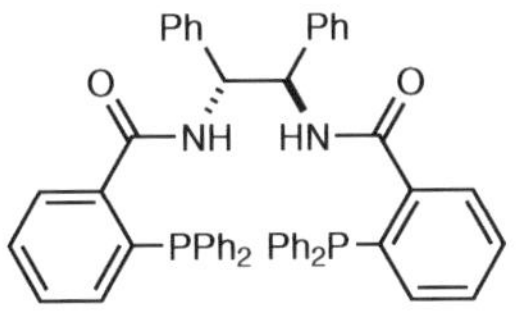 (R,R)-**108**

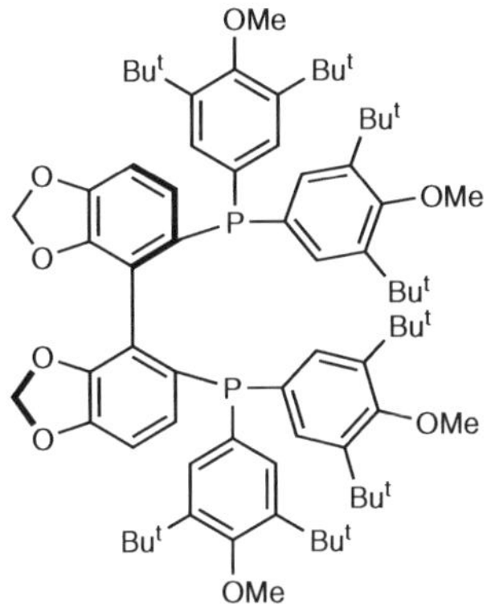

(R_a)-**109** 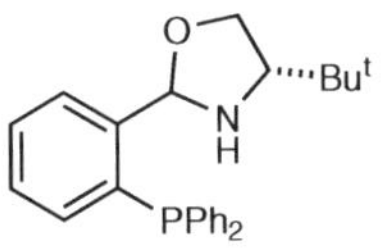 (S)-**110**

$\{Pd(CH_2CH{=}CH_2)Cl\}_2$ (5 mol%)
105 (11 mol%), DBU
1,4-dioxane (0.025 M), rt

111 + **112** → **113** + **114**

R^1	R^2	R^3	Equiv		Time (h)	Ratio[a] (**113**/**114**)	ee[b] (%)	Yield[c] (%) of **113**	Ref
			111	DBU					
Me	Me	Ph	2.0	2.0	4	>95:5	90	84	[76]
Et	Me	Ph	2.0	2.0	4	>95:5	90	71	[76]
$CH_2CH{=}CH_2$	Me	Ph	1.1	1.1	24	>95:5	94	67	[76]
Me	Et	Ph	2.0	2.0	5	>95:5	84	67	[76]
Me	$CH_2CH{=}CH_2$	Ph	1.5	0.9	11	>95:5	83	78	[76]
Me	Cl	Ph	2.0	2.0	4	>95:5	81	66	[76]
Me	Br	Ph	2.0	2.0	5	>95:5	96	91	[76]
Me	C≡CPh	Ph	2.0	2.0	4	>95:5	97	74	[76]
Me	Br	Ph	1.0	1.0	4.5	>95:5	97	86	[76]
Me	Br	4-Tol	2.0	2.0	5	>95:5	96	72	[76]
Me	Br	4-MeOC_6H_4	2.0	2.0	5	>95:5	96	71	[76]
Me	Br	4-FC_6H_4	1.0	1.0	15	>95:5	97	76	[76]
Me	Br	4-ClC_6H_4	2.0	2.0	5	>95:5	93	68	[76]
Me	Br	4-BrC_6H_4	2.0	2.0	5	>95:5	92	82	[76]
Me	Br	2-thienyl	2.0	2.0	5	>95:5	97	82	[76]
Me	Br	Me	2.0	2.0	4.5	>95:5	96[d]	50	[76]
Me	Me	Ph	1.0	1.0[e]	28	77:23	83	>95[a]	[76]

[a] Determined by ^1H NMR analysis of the crude reaction mixture.
[b] Determined by HPLC analysis.
[c] Yield of isolated product.
[d] Determined by Agilent SFC analysis.
[e] Li_2CO_3 was used as base.

In the case of naphthols **111**, dioxane is an optimal solvent for achieving high yields and enantioselectivity when compared with dichloromethane, 1,2-dichloroethane, tetrahydrofuran, 1,2-dimethoxyethane, and diethyl ether. Lithium carbonate, potassium carbonate, sodium carbonate, cesium carbonate, potassium acetate, tripotassium phosphate, and 1,8-diazabicyclo[5.4.0]undec-7-ene are effective bases for dearomatization to provide the dearomatized products **113** with a slight improvement in chemoselectivity for bases with less-coordinative counterions (e.g., potassium, cesium, and ammonium ions) in 1,4-dioxane. There is a slight increase in enantioselectivity with the use of 1,8-diazabicyclo-

[5.4.0]undec-7-ene in place of lithium carbonate as a base; a transition state is shown that provides insight into the origin of asymmetric induction and the mechanism of the reaction (Scheme 35).[76]

Scheme 35 Origin of Asymmetric Induction for Palladium-Catalyzed Alkylative Dearomatization of Naphthol Derivatives[76]

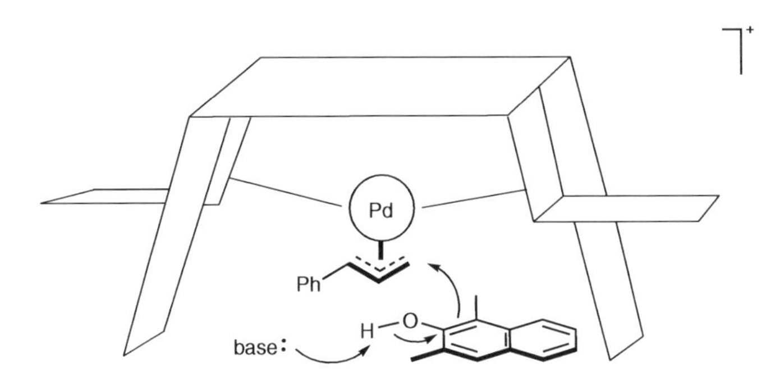

Based on comparison of the reaction data presented for palladium-catalyzed alkylative dearomatization of 2-naphthols **111**,[76,77] the biggest factors affecting the chemoselectivity and enantioselectivity for the reaction are the ligand, solvent, and the 2-naphthol substrate. For reactions of naphthols **111** conducted in dioxane, variations in the allyl carbonate and base do not dramatically influence the chemoselectivity or enantioselectivity of the reaction but do largely affect the yield in some cases. The base has a much larger influence on the chemoselectivity when the substrate is changed to a naphthol possessing an α-ester group (*vide infra*). The enantioselectivity and yield are often improved with the use of 2.0 equivalents of base and 2.0 equivalents of 2-naphthol.

Naphthol **115** lacking a substituent at C3 provides product **116** in poor yield (17%) and low chemoselectivity [ratio (**116/117**) 64:36] using 1 equivalent of 1,8-diazabicyclo[5.4.0]undec-7-ene in 1,4-dioxane at room temperature (Scheme 36).[76] However, a fourfold improvement in yield (65%) is achieved upon changing to the non-coordinating and less-polar degassed solvent system benzene/cyclohexane (1:4; 0.04 M), increasing the amount of substrate **115** and base to 2.0 equivalents, using potassium carbonate as a base, and reducing the temperature to 0 °C.

Scheme 36 Optimized Conditions in the Absence of a Naphthol Substituent at C3[76]

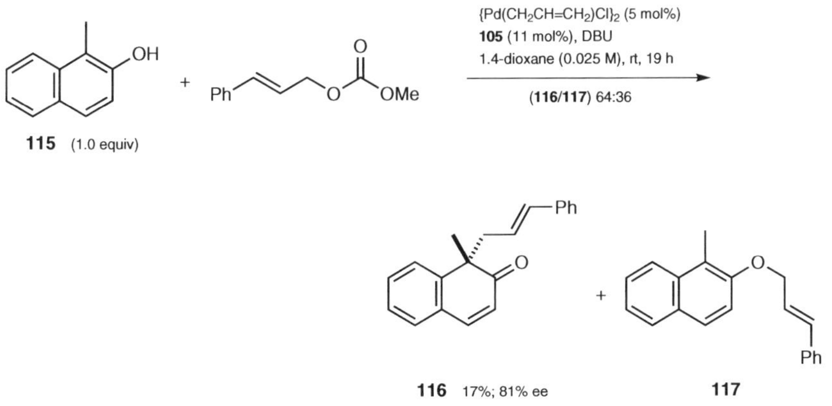

OH + Ph O OMe O

115 (1.0 equiv)

$\{Pd(CH_2CH{=}CH_2)Cl\}_2$ (5 mol%)
105 (11 mol%), DBU
1,4-dioxane (0.025 M), rt, 19 h

(**116/117**) 64:36

Ph O + O Ph

116 17%; 81% ee **117**

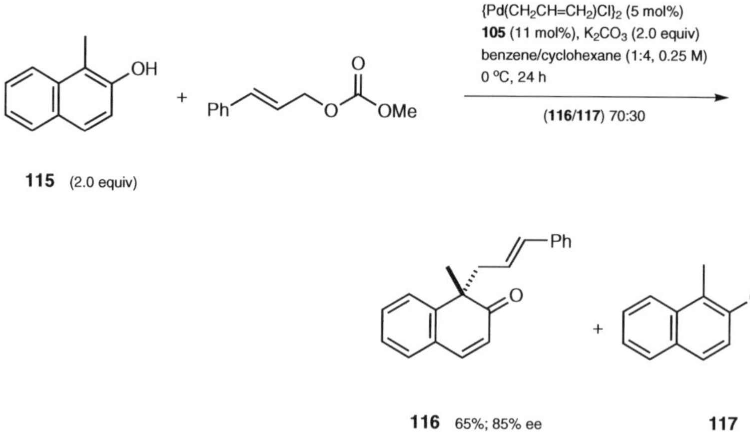

{Pd(CH$_2$CH=CH$_2$)Cl}$_2$ (5 mol%)
105 (11 mol%), K$_2$CO$_3$ (2.0 equiv)
benzene/cyclohexane (1:4, 0.25 M)
0 °C, 24 h
(**116**/**117**) 70:30

115 (2.0 equiv)

116 65%; 85% ee

117

To better understand how to further improve the reaction yield the relative thermodynamic stability of the naphtholate anion of **115** as compared to that of **111** ($R^1 = R^2 = Me$) in 1,4-dioxane needs to be considered. Also, a potassium counterion is more coordinating relative to a 1,8-diazabicycloundec-7-enium counterion and should also have a more pronounced effect in benzene/cyclohexane than in dioxane because it is less solvated in the former and could provide higher quantities of dearomatized C-alkylation adduct **117**.[67]

The mechanism of the palladium-catalyzed alkylative dearomatization of naphthols likely proceeds via oxidative addition of the allyl carbonate to generate a cationic palladium(II) π-allyl complex followed by nucleophilic attack of the naphtholate to generate dearomatized naphthol product (Scheme 37; also see Scheme 9, Section 2.2.1.3 for a related mechanism).[76]

Scheme 37 Proposed Catalytic Cycle of Palladium-Catalyzed Alkylative Dearomatization of 2-Naphthol Derivatives[76]

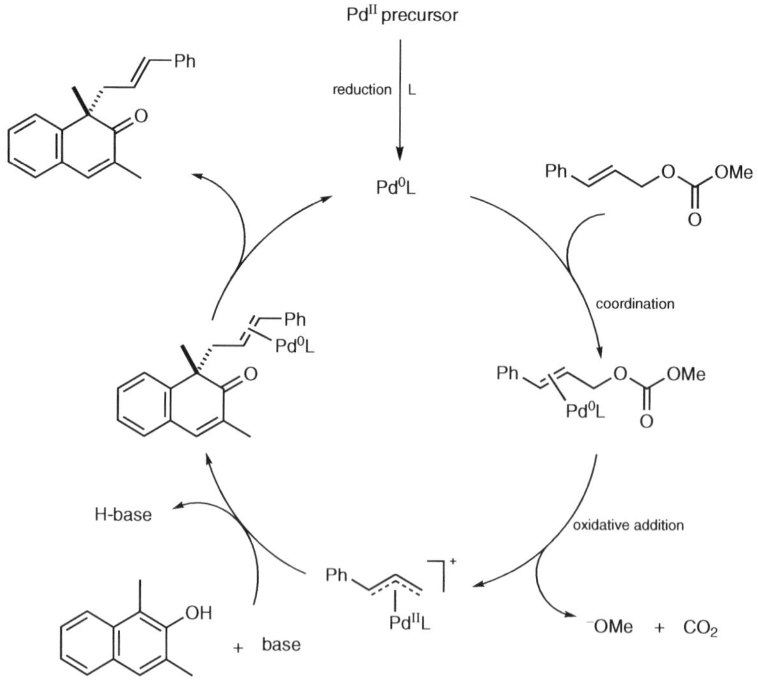

Pd^{II} precursor
reduction | L
Pd^{0}L
coordination
oxidative addition
$^-$OMe + CO$_2$
Pd^{II}L
H-base
+ base

In an effort to overcome challenges with respect to yield and chemoselectivity on substrates analogous to **115** lacking a substituent at C3, substrate **118** along with other 2-naphthols possessing an ester or ketone functionality at C1 were investigated as possible candidates for palladium-catalyzed alkylative dearomatization (Scheme 38 and Scheme 39).[77]

When the reaction is carried out under modified conditions using 5 mol% tetrakis(triphenylphosphine)palladium(0) [instead of allylpalladium(II) chloride dimer], 2.0 equivalents of 1,8-diazabicyclo[5.4.0]undec-7-ene, 1.0 equivalents of 2-naphthol **118**, and 1.2 equivalents of allyl methyl carbonate as a 0.1 M solution in 1,4-dioxane, the reaction provides dearomatized naphthol **119** and O-alkylated naphthol **120** in >95% conversion as a 17:83 ratio favoring the undesired product **120** after 41 hours of reaction time (Scheme 38).[77]

The yield of **119** can be improved to 85% under modified conditions by changing the base from 2.0 equivalents of 1,8-diazabicyclo[5.4.0]undec-7-ene to 1.0 equivalent of lithium carbonate, increasing the concentration from 0.1 M to 0.2 M, and reducing the amount of allyl methyl carbonate from 1.2 equivalents to 0.5 equivalents. These modifications allow for reaction completion in 27 hours at room temperature providing a reversed selectivity with a product ratio of >95:5 (**119**/**120**) (Scheme 38). In contrast to the case of naphthols **111** in 1,4-dioxane (Scheme 34), the base has a dramatic impact on the chemoselectivity in the case of naphthol **118** in 1,4-dioxane.[77]

Scheme 38 Optimization of the Reaction Conditions for Palladium-Catalyzed Alkylative Dearomatization of Methyl 2-Hydroxy-1-naphthoate[77]

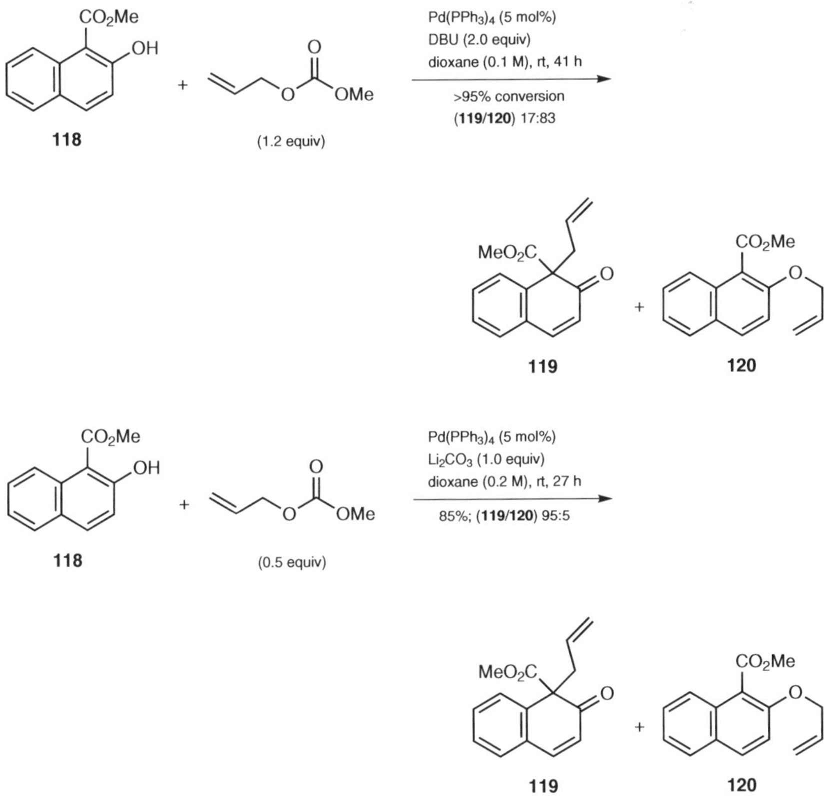

CO_2Me OH **118** + O OMe O (1.2 equiv)

$Pd(PPh_3)_4$ (5 mol%)
DBU (2.0 equiv)
dioxane (0.1 M), rt, 41 h
>95% conversion
(**119**/**120**) 17:83

MeO_2C O **119** + CO_2Me O **120**

CO_2Me OH **118** + O OMe O (0.5 equiv)

$Pd(PPh_3)_4$ (5 mol%)
Li_2CO_3 (1.0 equiv)
dioxane (0.2 M), rt, 27 h
85%; (**119**/**120**) 95:5

MeO_2C O **119** + CO_2Me O **120**

for references see p 290

With 2.0 equivalents of a 2-naphthol **121** and 1.0 equivalent of an allyl methyl carbonate **122** in the presence of 5 mol% tetrakis(triphenylphosphine)palladium(0) and 2.0 equivalents of lithium carbonate in a 0.05 M solution of 1,4-dioxane, a large substrate scope is demonstrated with good chemoselectivity favoring products **123** over **124** in moderate to excellent yield (Scheme 39).[77]

Scheme 39 Palladium-Catalyzed Alkylative Dearomatization of 1-(Alkoxycarbonyl)- and 1-Acyl-2-naphthols[77]

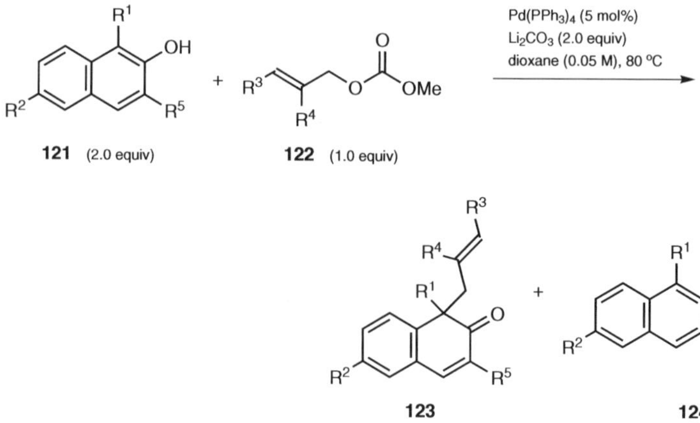

$Pd(PPh_3)_4$ (5 mol%)
Li_2CO_3 (2.0 equiv)
dioxane (0.05 M), 80 °C

121 (2.0 equiv) + **122** (1.0 equiv) → **123** + **124**

R^1	R^2	R^3	R^4	R^5	Time (h)	Ratio[a] (**123**/**124**)	Yield[b] (%) of **123**	Ref
CO_2Me	H	H	H	H	27	>95:5	85[c]	[77]
CO_2Me	H	Ph	H	H	28	75:25	72	[77]
CO_2Me	H	Ph	H	H	34	95:5	94	[77]
CO_2Me	H	Ph	H	H	24	>95:5	84[d]	[77]
CO_2Me	H	H	Ph	H	18	78:22	60	[77]
CO_2Me	H	$4-MeOC_6H_4$	H	H	11	>95:5	98	[77]
CO_2Me	H	$3-MeOC_6H_4$	H	H	7	>95:5	87	[77]
CO_2Me	H	$2-ClC_6H_4$	H	H	11	>95:5	90	[77]
CO_2Me	H	$4-ClC_6H_4$	H	H	18	>95:5	83	[77]
CO_2Me	H	$4-F_3CC_6H_4$	H	H	7	>95:5	69	[77]
CO_2Me	H	2-thienyl	H	H	11	>95:5	85	[77]
Ac	H	H	H	H	21	>95:5	85[c]	[77]
Ac	H	Ph	H	H	5	>95:5	99	[77]
Ac	H	H	H	CO_2Me	5	>95:5	71	[77]
Ac	H	Ph	H	CO_2Me	24	>95:5	66	[77]
Ac	OMe	H	H	H	7	>95:5	77	[77]
Ac	OMe	Ph	H	H	7	>95:5	95	[77]
Me	H	H	H	H	6	89:11	75[c]	[77]
Me	H	Ph	H	H	11	88:12	83	[77]

[a] Determined by ^1H NMR analysis of the crude reaction mixture.
[b] Yield of isolated product.
[c] Reaction was conducted at room temperature.
[d] Reaction was conducted with 1 mol% $Pd(PPh_3)_4$ and at 0.1 M.

Ligands **105**, **125**, **126**, and **127** provide poor enantioselectivity for the alkylative dearomatization of methyl 2-hydroxy-1-naphthoate (**118**; 2.0 equiv) with allyl methyl carbonate (1.0 equiv) in the presence of lithium carbonate (2.0 equiv) in dichloromethane at room temperature. Further solvent and base screening along with a more extensive ligand investigation may lead to improved results (Scheme 40).[77]

Scheme 40 Attempts To Improve Enantioselectivity[77]

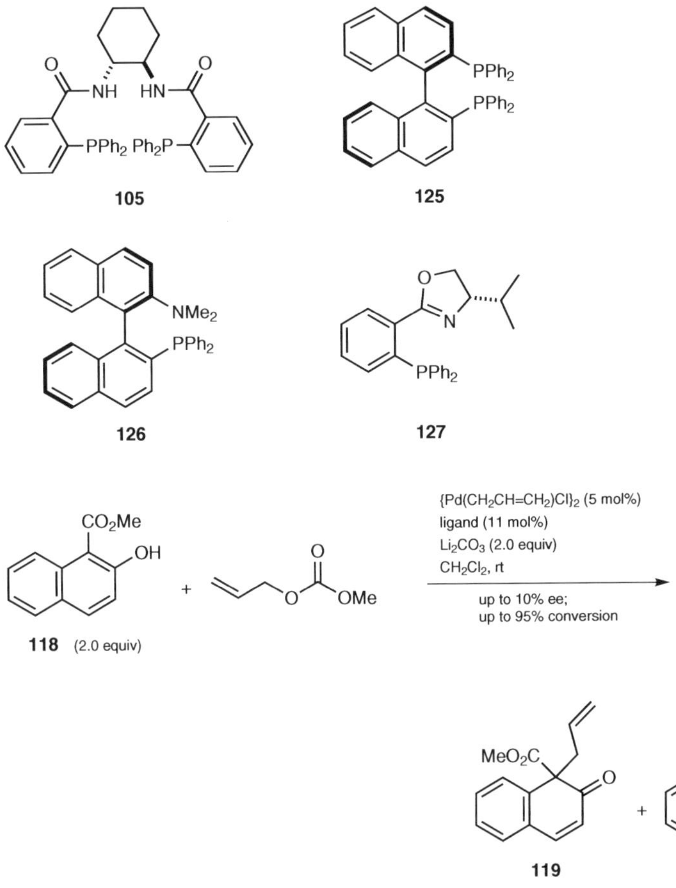

105 125 126 127

118 (2.0 equiv)

$\{Pd(CH_2CH{=}CH_2)Cl\}_2$ (5 mol%)
ligand (11 mol%)
Li_2CO_3 (2.0 equiv)
CH_2Cl_2, rt

up to 10% ee;
up to 95% conversion

119 + 120

In 2015, an asymmetric intermolecular alkylative dearomatization of 2-naphthols was demonstrated using *meso*-aziridine electrophiles activated by a chiral bifunctional magnesium catalyst.[78] After investigating 16 different chiral ligands (**128–132** and **136–138**, Scheme 41 and Scheme 42), successful application of novel 2-[(4,5-dihydrooxazol-2-yl)methyl]phenol **132** provided excellent enantioselectivity (>99% ee in most cases), diastereoselectivity (dr >20:1 in most cases), and yields (99% in most cases) for a diverse group of dearomatized products containing a functional handle that allows the method to serve as part of a domino transformation (Scheme 41). The reaction is catalytic in magnesium where 20 mol% of a dibutylmagnesium solution (1.0 M in heptane) is added to a 0.01 M solution of ligand **132** (5 mol%) in toluene with 13x molecular sieves; a solution of 2.0 equivalents of a naphthol **133** and 1.0 equivalent of an aziridine **134** is transferred to the catalyst solution providing product **135** in high yield, diastereoselectivity, and enantioselectivity after workup and column chromatography.

for references see p 290

Scheme 41 Ligand Investigation and Naphthol Substrate Scope for Enantioselective Magnesium-Catalyzed Alkylative Dearomatization[78]

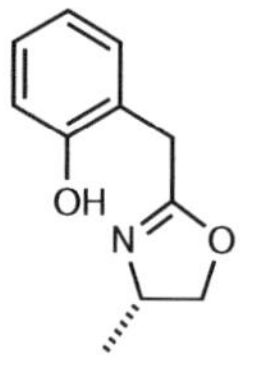

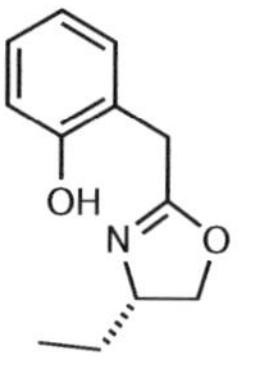

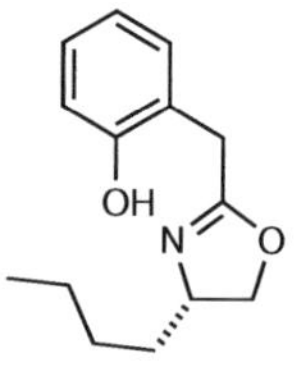

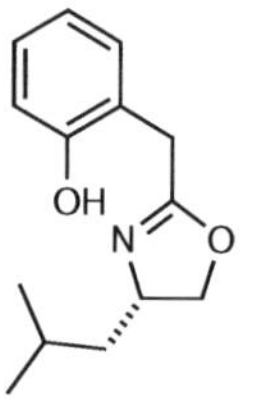

R1	R2	Ligand	dr	ee (%)	Yield[a] (%)	Ref
Me	Me	**128**	>20:1	94	93	[78]
Me	Me	**129**	>20:1	86	82	[78]
Me	Me	**130**	>20:1	91	87	[78]
Me	Me	**131**	>20:1	90	99	[78]
Me	Me	(*S*)-**132** (20 mol%)	>20:1	>99	98	[78]
Me	Me	(*S*)-**132** (10 mol%)	>20:1	>99	97	[78]
Me	Me	(*S*)-**132** (5 mol%)	>20:1	>99	97	[78]
Me	Me	(*S*)-**132** (2 mol%)	>20:1	>92	41	[78]
Me	Me	(*S*)-**132** (1 mol%)	>20:1	–	trace	[78]
Me	Me	–	>20:1	–	85[b]	[78]
Me	Me	(*S*)-**132** (20 mol%)	>20:1	–	n.r.[c]	[78]
Me	Et	(*S*)-**132**	>20:1	>99	99	[78]
Me	TMS	(*S*)-**132**	>20:1	>99	98	[78]
Me	Cl	(*S*)-**132**	>20:1	>99	72	[78]
Me	Br	(*S*)-**132**	>20:1	>99	80	[78]
Me	Br	(*R*)-**132**	>20:1	>99	83	[78]
Me	I	(*S*)-**132**	>20:1	>99	99	[78]

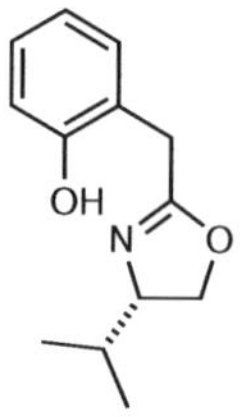

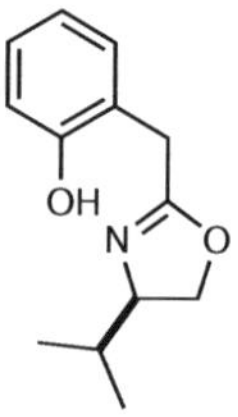

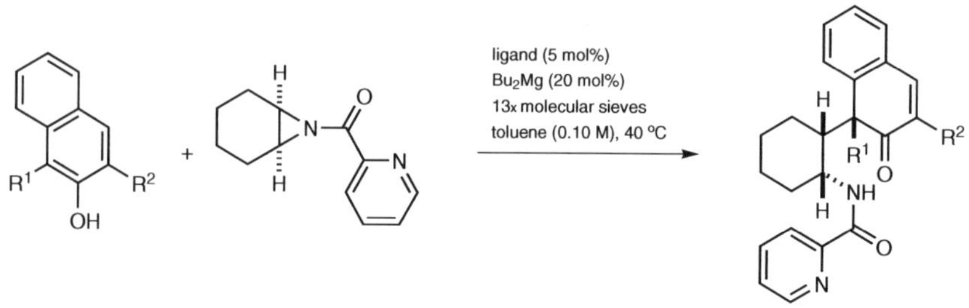

R^1	R^2	Ligand	dr	ee (%)	Yield[a] (%)	Ref
Me	Ph	(*S*)-**132**	>20:1	>99	99	[78]
Me	C≡CPh	(*S*)-**132**	4.6:1	>99	94	[78]
Et	Me	(*S*)-**132**	4.6:1	>99	97	[78]
Pr	Me	(*S*)-**132**	14:1	98	99	[78]
$(CH_2)_6Me$	Me	(*S*)-**132**	11:1	97	99	[78]
Me	H	(*S*)-**132**	14:1	91	59[d]	[78]
$(CH_2)_6Me$	H	(*S*)-**132**	11:1	>99	74[d]	[78]

[a] Yield of isolated product; determined by ^1H NMR analysis of the crude reaction mixture.
[b] No chiral ligand was used.
[c] n.r. = not reported; reaction performed without Bu_2Mg.
[d] 20 mol% Bu_2Mg was used.

Some naphthols **133** [R^2 = H or C≡CPh, or R^1 = $(CH_2)_6Me$] provide lower but still synthetically advantageous yields and diastereoselectivities. An increase in enantioselectivity is observed when the length of R^1 increases in the absence of a substituent at R^2. Either enantiomer of product **135** can be accessed depending on the specific enantiomer used for ligand **132**. The magnesium is thought to activate aziridine **134**, rendering it more electrophilic in addition to coordinating the naphthol substrate and chiral ligand **132**. Butane gas is released upon deprotonation and complexation of magnesium to a naphthol **133**. Quinine (**136**), six variations of **137**, and four variations of **138** provide the dearomatized product in low yields, enantioselectivities, and diastereoselectivities (Scheme 42).[78,79]

Scheme 42 Ligands Providing Low Enantio- and Diastereoselectivity[78,79]

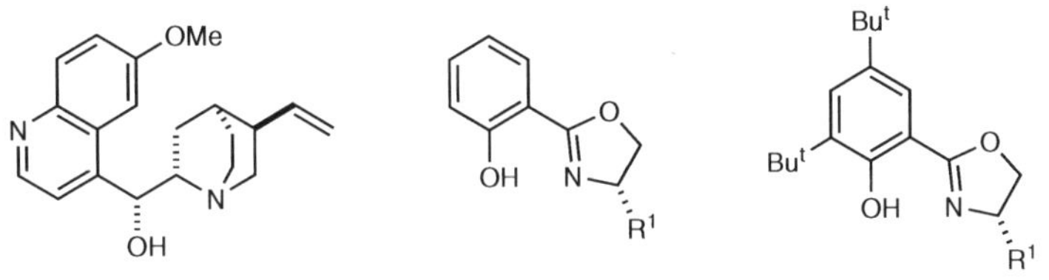

Acyclic and cyclic *meso*-aziridines serve as suitable electrophiles for dearomatization of 1,3-dimethylnaphthalen-2-ol (Table 1). A moderate yield of 48% is obtained with a cycloheptane-fused aziridine, possibly due to ring strain (entry 5).[78]

for references see p 290

Table 1 Examination of Aziridine Electrophiles for Enantioselective Magnesium-Catalyzed Alkylative Dearomatization of 1,3-Dimethylnaphthalen-2-ol[78]

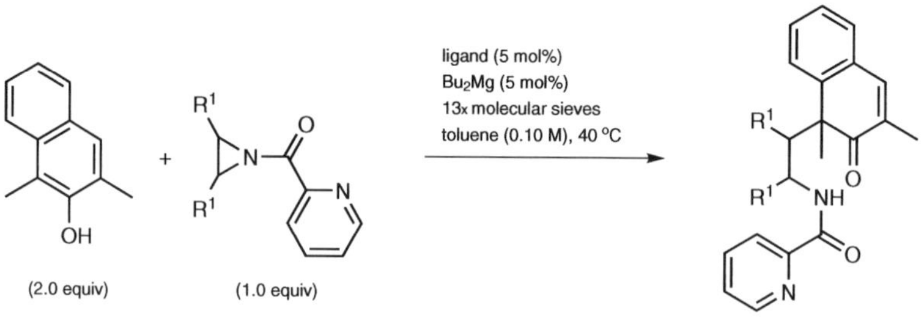

ligand (5 mol%)
Bu_2Mg (5 mol%)
13x molecular sieves
toluene (0.10 M), 40 °C

(2.0 equiv) (1.0 equiv)

Entry	Aziridine	Ligand	Product	ee (%)	Yield[a] (%)	Ref
1	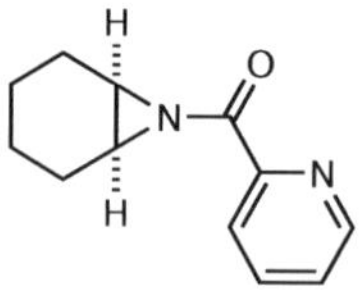	(*R*)-**132**	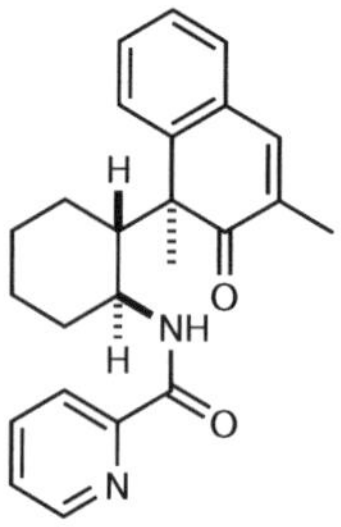	97	97	[78]
2	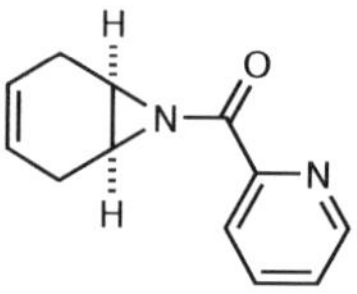	(*S*)-**132**	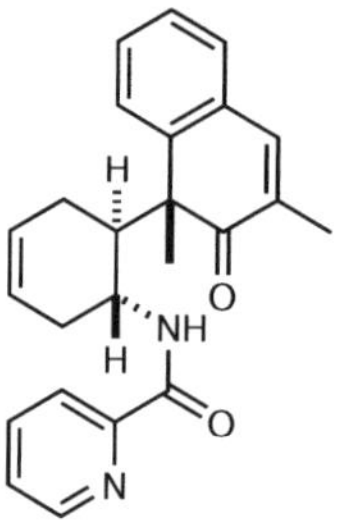	>99	93	[78]
3	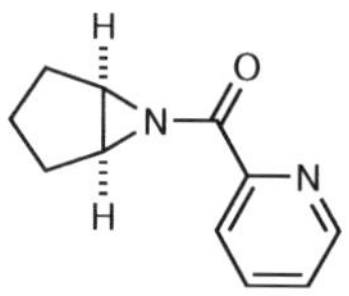	(*S*)-**132**	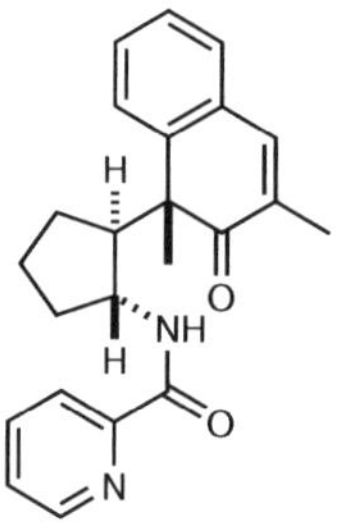	>99	98	[78]
4	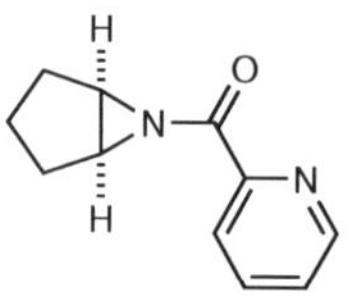	(*R*)-**132**	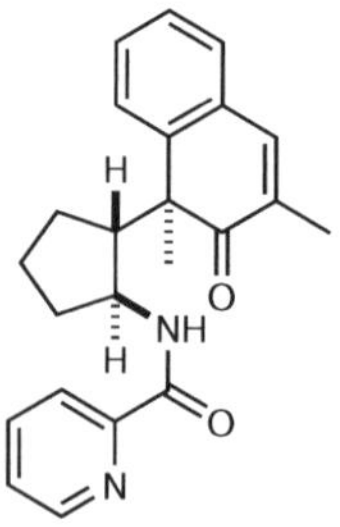	>99	96	[78]

Table 1 (cont.)

Entry	Aziridine	Ligand	Product	ee (%)	Yield[a] (%)	Ref
5	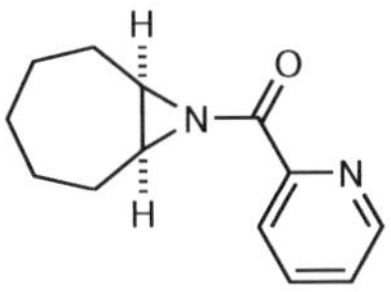	(*S*)-**132**	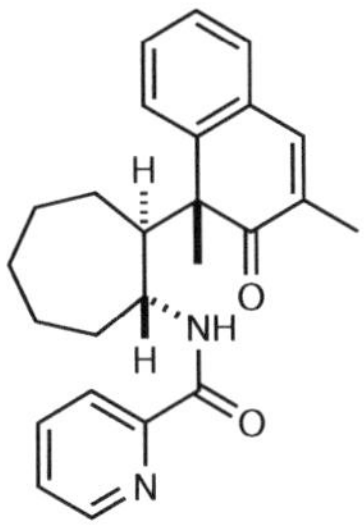	>99	48	[78]
6	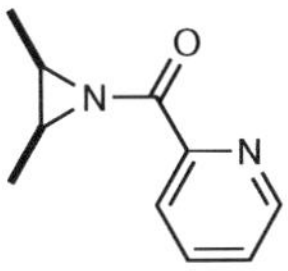	(*S*)-**132**	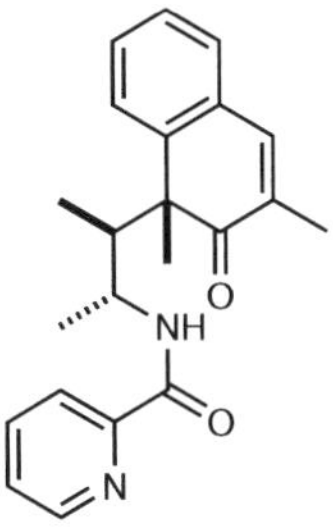	>99	99	[78]
7	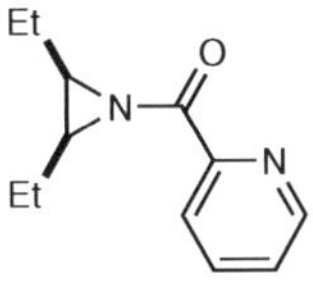	(*S*)-**132**	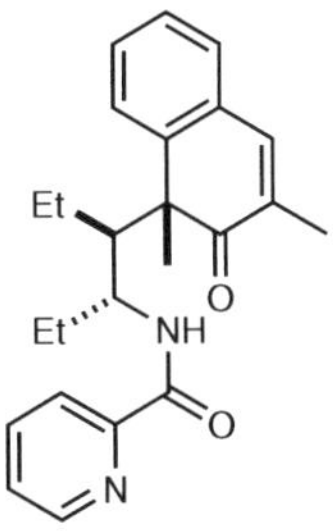	>99	99	[78]
8	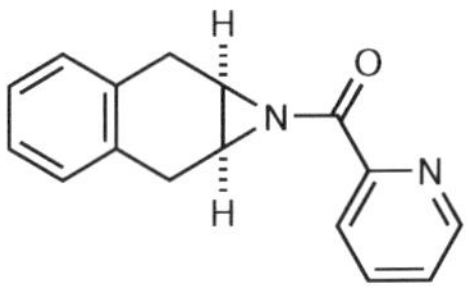	(*S*)-**132**	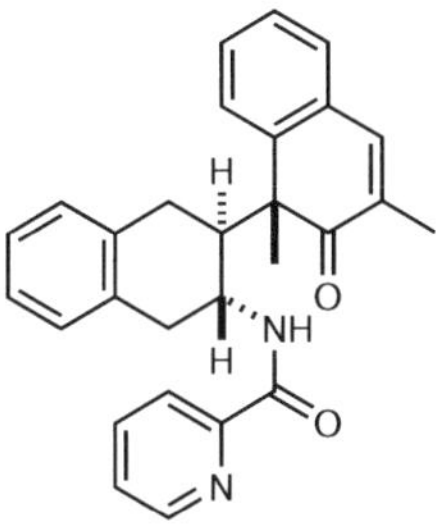	>99	99	[78]

for references see p 290

Table 1 (cont.)

Entry	Aziridine	Ligand	Product	ee (%)	Yield[a] (%)	Ref
9	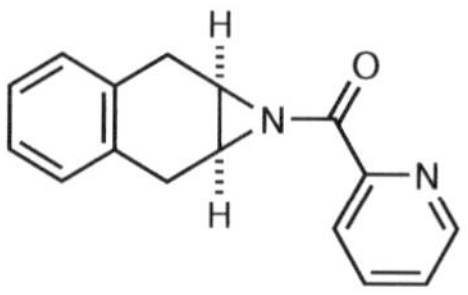	(*R*)-**132**	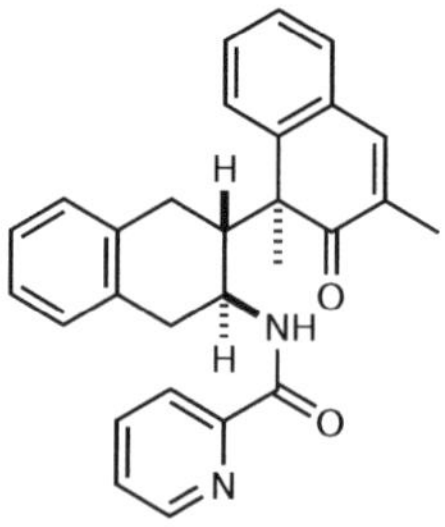	>99	95	[78]
10	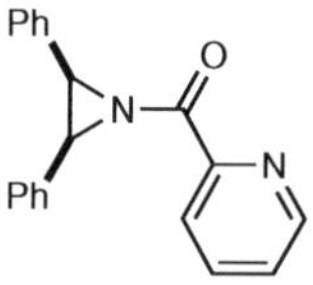	(*S*)-**132**	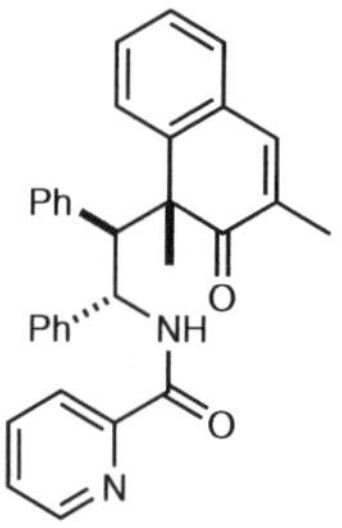	>99	99	[78]
11	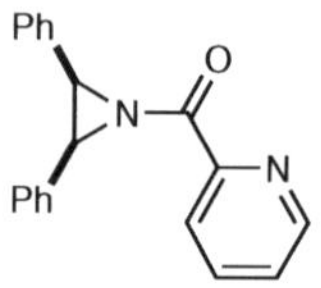	(*R*)-**132**	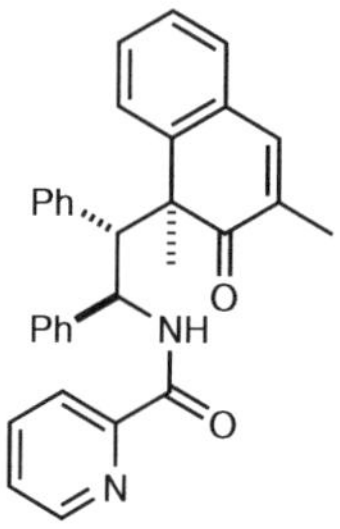	>99	94	[78]

[a] Each product was obtained with dr >20:1.

(*S*,*E*)-Naphthalen-2(1*H*)-ones 113; **General Procedure:**[76]
Two flame-dried Schlenk tubes were cooled to rt and filled with argon. To one flask were added $\{Pd(CH_2CH{=}CH_2)Cl\}_2$ (0.010 mmol, 5 mol%) and chiral ligand **105** (0.022 mmol, 11 mol%). The flask was evacuated and refilled with argon. To the flask was added freshly distilled 1,4-dioxane (4.0 mL), and the resultant soln was stirred for 30 min. A 2-naphthol derivative **111** (0.40 mmol, 2 equiv) was added to another dry Schlenk tube. The flask was evacuated and refilled with argon. To this flask were added the catalyst soln, DBU (0.40 mmol, 2.0 equiv), an allyl carbonate **112** (0.20 mmol, 1 equiv), and freshly distilled 1,4-dioxane (4.0 mL). The mixture was stirred at 25 °C. When the reaction was complete (monitored by TLC), the solvents were removed under reduced pressure. The ratio of **113**/**114** was determined by ^1H NMR analysis of the crude residue, which was purified by column chromatography (silica gel, petroleum ether/EtOAc 50:1).

Naphthalen-2(1*H*)-ones 123; **General Procedure:**[77]
A flame-dried Schlenk tube was cooled to rt and filled with argon. To this flask were added $Pd(PPh_3)_4$ (11.5 mg, 0.010 mmol, 5 mol%), Li_2CO_3 (29.6 mg, 0.40 mmol, 2.0 equiv), a 2-naphthol **121** (0.40 mmol, 2.0 equiv), an allyl carbonate **122** (0.20 mmol, 1.0 equiv), and 1,4-dioxane (2.0 mL). The mixture was stirred at 80 °C. When the reaction was complete (monitored by TLC), the crude mixture was filtered through Celite (CH_2Cl_2). The solvents were

removed under reduced pressure. The ratio of **123/124** was determined by ^{1}H NMR analysis of the crude residue, which was purified by column chromatography (silica gel, petroleum ether/EtOAc 23:1).

***N*-[2-(2-Oxo-1,2-dihydronaphthalen-1-yl)cyclohexyl]pyridine-2-carboxamides** 135; **General Procedure:**[78]

To a stirred soln of ligand **132** (0.01 mmol, 5 mol%) and 13x molecular sieves (200 mg) in toluene (1.0 mL) was added a 1.0 M soln of Bu_2Mg in heptane (10.0 μL, 0.01 mmol) under argon and the mixture was stirred at rt for 30 min to generate the catalyst. A soln of a 2-naphthol **133** (0.40 mmol, 2.0 equiv) and aziridine **134** (0.20 mmol, 1.0 equiv) in toluene (1.0 mL) was transferred into the flask containing the in situ generated catalyst. After the addition, the mixture was stirred at 40 °C for 24 h. It was then cooled to rt, quenched with sat. aq NH_4Cl, and extracted with CH_2Cl_2. The combined organic layers were dried (Na_2SO_4) and concentrated under reduced pressure. The residue was purified by column chromatography (silica gel).

2.2.4.2 Dearomatization Reactions as Domino Transformations To Access Type A and B Polyprenylated Acylphloroglucinol Analogues

A highly efficient palladium-catalyzed dearomative conjunctive allylic annulation provides access to both type A and B polyprenylated acylphloroglucinol analogues possessing a bicyclo[3.3.1]nonane framework in a single step.[80] The methylenepropane-1,3-diol derivative **139** mediates the conjunctive bond-forming process by undergoing two sequential oxidative addition steps in a single chemical reaction to first generate a dearomatized intermediate after nucleophilic attack by phenol **31**, followed by further cyclization to the [3.3.1]-bicyclic core. Dearomative conjunctive allylic annulation provides type A polyprenylated acylphloroglucinol analogues via two alternative procedures (Scheme 43).[80] The first one involves adding 2.5 mol% of tetrakis(triphenylphosphine)palladium(0) as a 0.0125 M solution in cyclohexane to a 0.2 M cyclohexane solution containing conjunctive reagent **139** (1.0 equivalent) and monomethoxyphloroglucinol **31** (1.0 equivalent) and is heated for 30 minutes under argon to provide bicyclic product **140** in 87% yield.

Scheme 43 Conditions To Access Type A Polyprenylated Acylphloroglucinol Analogues in a Dearomative Domino Transformation[80]

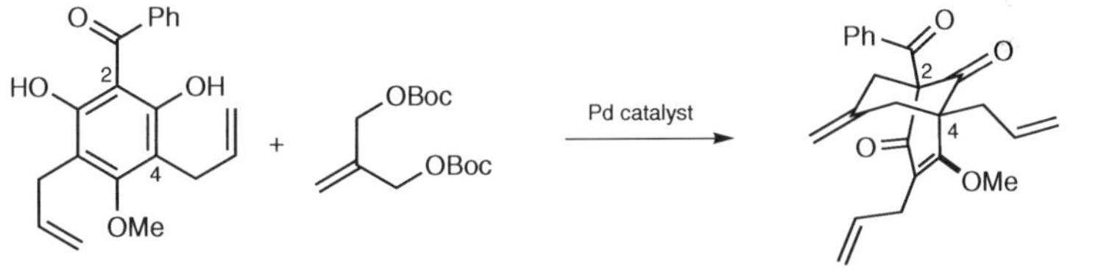

31 (1.0 equiv) + **139** (1.0 equiv) → (Pd catalyst) **140**

Conditions	Yield (%)	Ref
$Pd(PPh_3)_4$ (2.5 mol%), cyclohexane (0.2 M), 75 °C, 0.5 h	87	[80]
$Pd_2(dba)_3$ (2 mol%), *rac*-BINAP (4 mol%), toluene (0.2 M), 100 °C, 1 h	81	[80]

When other solvents such as dichloroethane, toluene, and tetrahydrofuran are used in the reaction under these conditions, product **140** is not produced. Alternatively, the second conditions prove to be slightly more reproducible due to the more robust 2,2′-bis(diphenylphosphino)-1,1′-binaphthylpalladium(0) complex formed from heating a solution of tris(dibenzylideneacetone)dipalladium(0) and racemic 2,2′-bis(diphenylphosphino)-

for references see p 290

1,1′-binaphthyl for 2–3 minutes at 100 °C in toluene (0.01 M with respect to palladium catalyst) or until the solution becomes clear (red in color). The palladium complex is transferred to a solution of 4-methoxy substituted acylphloroglucinol **31** (1.0 equivalent) and conjunctive reagent **139** (1.0 equivalent) in toluene. Heating to 100 °C for 1 hour provides product **140** in 81% yield. Two equivalents of the base *tert*-butoxide are generated in situ, one equivalent for each oxidative addition.

This method is effective on a gram-scale and is general for six different substrate variations, four of which are shown in Scheme 44. The conjunctive reagent **139** is stable and can be stored at room temperature in air. Both substrates **141** and type A product analogues **142** were stored under argon in a freezer at −20 °C.

Scheme 44 Dearomative Conjunctive Allylic Annulation To Access Type A Polyprenylated Acylphloroglucinol Analogues[80]

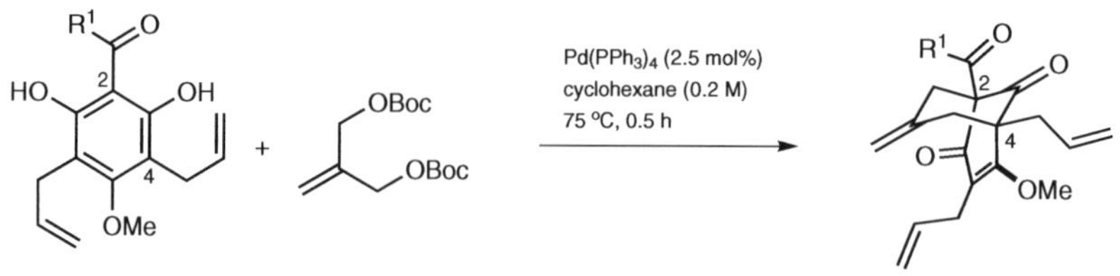

R^1	Yield (%)	Ref
Ph	87	[80]
Me	82	[80]
iPr	93[a]	[80]
iBu	92	[80]

[a] The reaction was carried out on gram-scale with 0.5 mol% $Pd(PPh_3)_4$.

Neither set of conditions is suitable for the production of type B analogues **144** that result from reaction at C4 and C6 of acylphloroglucinols **143** with a palladium(II)–π-allyl species generated in situ from **139** in the presence of 5 mol% racemic 2,2′-bis(diphenylphosphino)-1,1′-binaphthylpalladium(0) (Scheme 45). The reaction is carried out in a similar manner to that described in Scheme 43; however, the method requires longer reaction times (24 hours), higher temperatures (90 °C), higher catalyst loadings (5 mol%), and a sealed system to avoid changes in concentration. The slower reaction rate observed for substrates **143** compared with substrate **31** may result from the enhanced nucleophilicity of the phenol at C5 causing substrates **143** to bind with the palladium metal center.[80,81]

Scheme 45 Dearomative Conjunctive Allylic Annulation To Access Type B Polyprenylated Acylphloroglucinol Analogues[80]

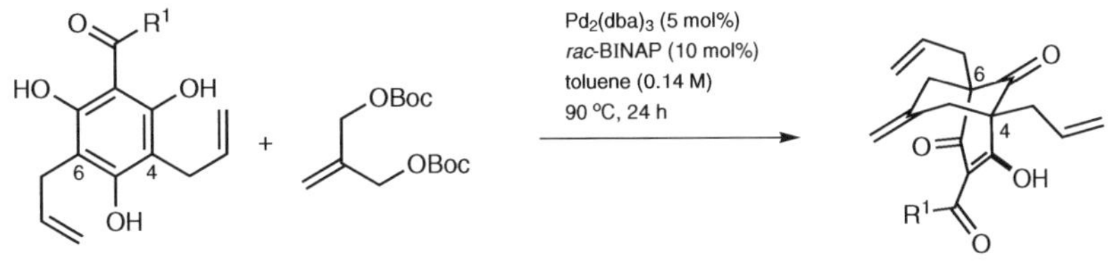

$Pd_2(dba)_3$ (5 mol%)
rac-BINAP (10 mol%)
toluene (0.14 M)
90 °C, 24 h

143 (1.0 equiv) + **139** (0.86 equiv) → **144**

R^1	Yield (%)	Ref
Ph	69	[80]
Me	80	[80]
iBu	80	[80]

The following factors may contribute to a loss of catalytic activity in the reaction of phenols **143** to provide type B polyprenylated acylphloroglucinol analogues **144**: (1) continuous fluctuation in temperature over time; hence, it is important to ensure that the oil bath is preset to the desired temperature so as to prevent slow heating to the desired temperature; (2) solvent evaporation; hence, it is important to ensure that the reaction vessel is completely sealed, especially for small-scale reactions; (3) changes in concentration; when transferring reagents in toluene, try to accomplish this in a single transfer and avoid transferring additional toluene rinses to the reaction mixture; (4) insufficient stirring; (5) bubbling argon through the reaction mixture; (6) temperatures above 100 °C; and (7) using excess (±)-2,2′-bis(diphenylphosphino)-1,1′-binaphthyl ligand (>2:1 ligand/metal).

It is a precautionary measure to bubble a stream of argon through the reaction solvent for 5–10 minutes prior to setup. The presence of oxygen does correlate with a loss in catalytic activity causing precipitation of palladium black, especially at high temperatures; this may result from ligand oxidation. However, it is not always necessary to degas the solvent depending on the reaction time, the stability of the catalyst, and temperature of the reaction.

Dearomative conjunctive allylic annulation is successful in further application of the described conditions to a variety of different phenol substrates and one dearomatized enol substrate (Scheme 46).

Scheme 46 Examination of Substrate Scope for Dearomative Conjunctive Allylic Annulation[80]

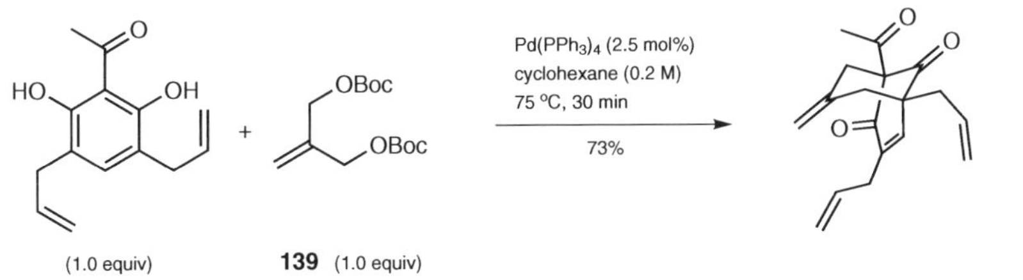

$Pd(PPh_3)_4$ (2.5 mol%)
cyclohexane (0.2 M)
75 °C, 30 min
73%

(1.0 equiv) + **139** (1.0 equiv)

$Pd_2(dba)_3$ (2 mol%)
BINAP (4 mol%)
toluene (0.2 M)
100 °C, 1 h

(1.0 equiv) **139** (1.0 equiv)

R^1	Yield (%)	Ref
Ph	81	[80]
Me	61	[80]
iBu	83	[80]

$Pd_2(dba)_3$ (2 mol%)
BINAP (4 mol%)
toluene (0.2 M)
100 °C, 1 h
72%

(1.0 equiv) **139** (1.0 equiv)

$Pd_2(dba)_3$ (2 mol%)
BINAP (4 mol%)
toluene (0.2 M)
100 °C, 1 h

(1.0 equiv) **139** (1.0 equiv)

63% 27%

$Pd(PPh_3)_4$ (2.5 mol%)
cyclohexane (0.2 M)
75 °C, 30 min
55%

(1.0 equiv) **139** (1.0 equiv)

A modified protocol that serves as a complement to that for alkylative dearomatization[8] uses the highly reactive unactivated trifluoromethanesulfonate electrophile **145** in a 3:1 mixture of tetrahydrofuran/toluene [Scheme 47; also refer to Scheme 19 (Section 2.2.2.2) for comparison of data].[82] The method can be applied on a multigram scale with high yields. The relatively non-coordinating potassium base counterion allows for exclusive formation of O-alkylation product (not shown) in 53% yield, whereas the lithium base counterion strongly coordinates to the phenolate oxygens in (3,5-diallyl-2,6-dihydroxy-4-methoxyphenyl)(phenyl)methanone (**31**) leading to dearomatized product **146** in 73% yield.[82] Product **146** exists as a combination of two diastereomers (Scheme 47).

Scheme 47 Effect of Base Counterion for Alkylative Dearomatization and In Situ Preparation of a Trifluoromethanesulfonate Electrophile[82]

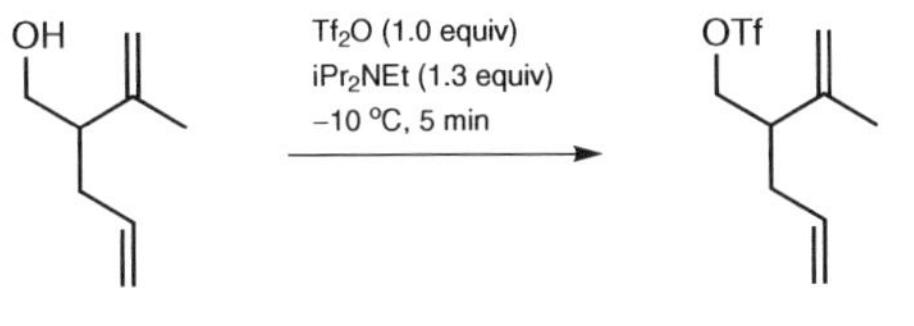

Tf_2O (1.0 equiv)
iPr_2NEt (1.3 equiv)
−10 °C, 5 min

145

0 °C, 15 min

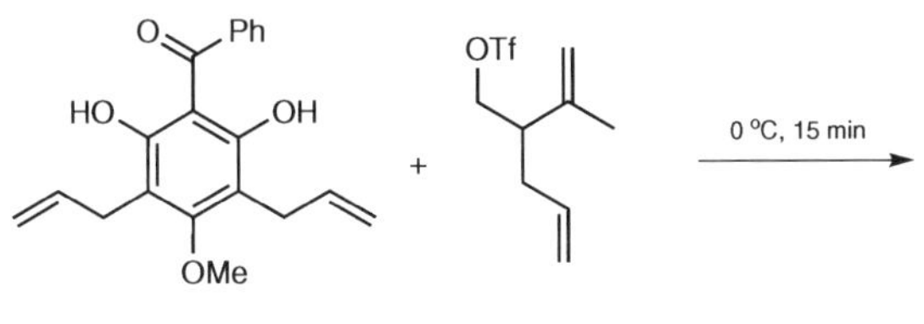

31

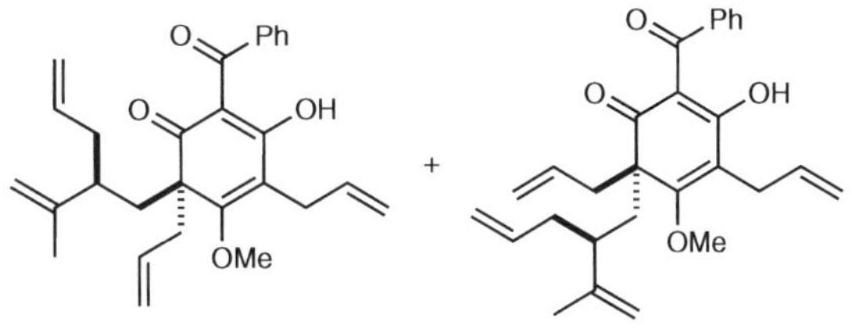

146A

146B

Conditions	Yield (%)		Ref
	C-Alkylation	O-Alkylation	
KHMDS, THF/toluene (3:1)	0	53	[82]
NaHMDS, THF/benzene (3:1)	47	31	[82]
LiHMDS, THF/toluene (3:1)	73[a]	6	[82]

[a] 83% yield was obtained under optimized conditions.

Dearomatized product **146** can undergo further cyclization under acidic conditions to provide five novel architectures selectively depending on the specific Brønsted acid applied. Using formic acid, the bicyclo[3.3.1]nonane type B framework of clusianone (**147**) is provided in 30% yield in addition to an O-cyclized Cope rearrangement product **148** in 34% yield (Scheme 48).[82]

Scheme 48 Cationic Biomimetic Cyclization of a Dearomatized Phenol Leads to Five Novel Architectures and the Functionalized Core of Clusianone[82]

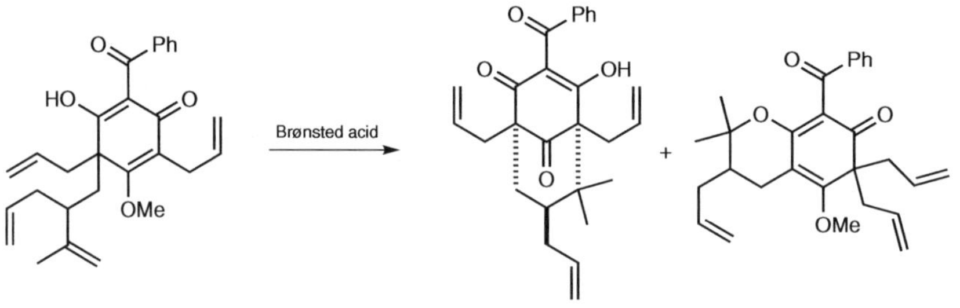

Brønsted acid

146 **147** **148**

Brønsted Acid	Conditions	Ratio (**147**/**148**)	Yield[a] (%)	Ref
HCO_2H	neat, 10 °C to rt, 72 h	1:1.13	64	[82]
TsOH (10 equiv), LiOTf (10 equiv)	$(CF_3)_2CHOH$ (0.0115 M), 50 °C, 12 h	0:100	89	[82]
TFA (0.14 M)	neat, rt, 24 h	0:100	97	[82]
TsOH (10.0 equiv), LiBr (10 equiv)	$MeNO_2$ (0.015 M), rt, 10 min	0:100	55	[82]

[a] Yield of isolated product.

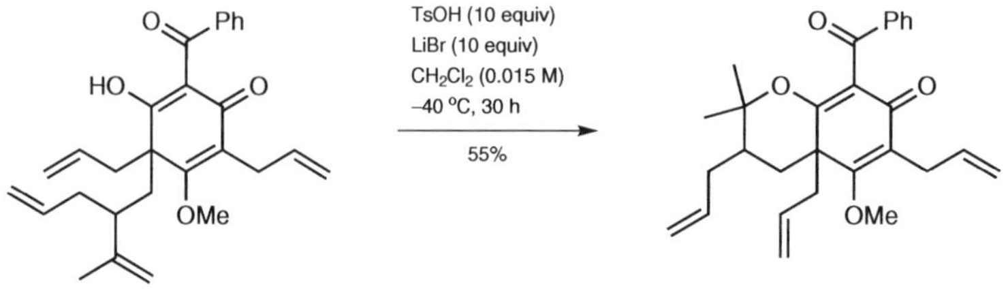

146

TsOH (10 equiv)
LiBr (10 equiv)
CH_2Cl_2 (0.015 M)
–40 °C, 30 h
55%

146

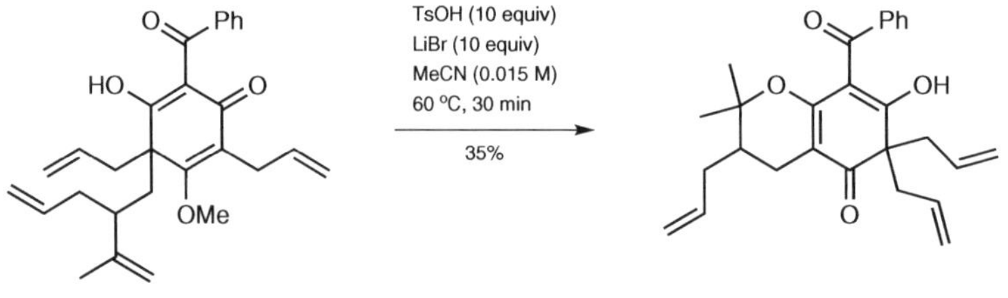

TsOH (10 equiv)
LiBr (10 equiv)
MeCN (0.015 M)
60 °C, 30 min
35%

146

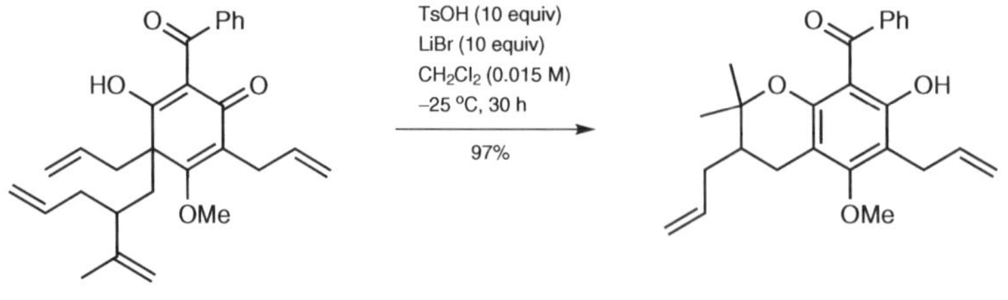

TsOH (10 equiv)
LiBr (10 equiv)
CH_2Cl_2 (0.015 M)
–25 °C, 30 h
97%

(1*R,5*R**)-3,5-Diallyl-1-benzoyl-4-methoxy-7-methylenebicyclo[3.3.1]non-3-ene-2,9-dione (140); Typical Procedure Using Tetrakis(triphenylphosphine)palladium(0):**[80]
A 10-mL flame-dried Schlenk flask equipped with a stirrer bar was charged with (3,5-diallyl-2,6-dihydroxy-4-methoxyphenyl)(phenyl)methanone (**31**; 162 mg, 0.5 mmol, 1.0 equiv) and conjunctive reagent **139** (144 mg, 0.5 mmol, 1.0 equiv). Cyclohexane (2.5 mL) was added and the flask was heated to 75 °C and sparged with argon for 2 min. $Pd(PPh_3)_4$ (14 mg, 12.5 μmol, 2.5 mol%) was added to a 1-dram vial followed by cyclohexane (1.0 mL). The catalyst slurry was then added via Pasteur pipet to the mixture. The flask was sealed with a rubber septum equipped with an argon balloon and heated at 75 °C until complete conversion of the starting material was observed (<1 h, commonly 30 min.). After the allotted reaction time, the vessel was removed from the oil bath, and the mixture was diluted with hexanes (5 mL) and directly subjected to column chromatography (silica gel, EtOAc/hexanes 1:9 to 1:4) to yield the pure product; yield: 87%.

(1*R,5*R**)-3,5-Diallyl-1-benzoyl-4-methoxy-7-methylenebicyclo[3.3.1]non-3-ene-2,9-dione (140); Typical Procedure Using 2,2′-Bis(diphenylphosphino)-1,1′-binaphthylpalladium(0):**[80]
A 10-mL flame-dried Schlenk flask equipped with a stirrer bar was charged with (3,5-diallyl-2,6-dihydroxy-4-methoxyphenyl)(phenyl)methanone (**31**; 162 mg, 0.5 mmol, 1.0 equiv) and conjunctive reagent **139** (144 mg, 0.5 mmol, 1.0 equiv). Toluene (2.5 mL) was added and the mixture was sparged with argon for 2 min. In a 1-dram vial, $Pd_2(dba)_3$ (10 mg, 0.010 mmol, 2 mol%) and *rac*-BINAP (13 mg, 0.020 mmol, 4 mol%) were dissolved in toluene (1.0 mL) and were heated with a heat gun until a bright red, clear soln was formed. The activated catalyst was then transferred to the reaction flask via Pasteur pipet. The reaction vessel was sealed with a rubber septum equipped with an argon balloon and submerged in an oil bath at 100 °C until the reaction was complete as monitored by TLC analysis (typically ~1 h). After the allotted reaction time, the vessel was removed from the oil bath, and the mixture was diluted with hexanes (5.0 mL) and directly subjected to column chromatography (silica gel, EtOAc/hexanes 1:9 to 1:4) to yield the pure product; yield: 81%.

(1*R,5*R**)-1,5-Diallyl-3-benzoyl-4-hydroxy-7-methylenebicyclo[3.3.1]non-3-ene-2,9-dione (144, R^1 = Ph); Typical Procedure:**[80]
To a flame-dried round-bottomed flask fitted with a stirrer bar was added $Pd_2(dba)_3$ (2.2 mg, 2.45 μmol, 5 mol%) and *rac*-BINAP (3.0 mg, 4.90 μmol, 10 mol%). The flask was purged with argon, and degassed toluene (0.1 mL) was added to provide a purple suspension, which was then heated to 110 °C for 5 min or until the formation of a clear red soln. After removal of the flask from the oil bath, the clear red soln was pulled into a 1-mL syringe and transferred to a soln of (3,5-diallyl-2,4,6-trihydroxyphenyl)(phenyl)methanone (**143**, R^1 = Ph; 15.2 mg, 0.0490 mmol, 1.0 equiv) and **139** (11.7 μL, 0.0421 mmol, 0.86 equiv) in degassed toluene (0.26 mL) at rt. The argon balloon and septum were removed and the system was quickly sealed with a polyethylene yellow cap, Teflon tape, and Keck clamp; the mixture was then placed in an oil bath at 90 °C and stirred for 24 h. The crude mixture was cooled to rt, diluted with hexanes (3.0 mL), and filtered through a cotton plug into a round-bottomed flask to remove BINAP and palladium (pentane was used to rinse the flask). The pentane/hexane soln was capped and placed in a refrigerator at −20 °C for 1 h; the soln was then decanted to a separatory funnel. The pentane/hexane soln was mixed vigorously with sat. aq K_2CO_3 in a separatory funnel. H_2O was added and the mixture was again shaken vigorously. A precipitate formed and the layers were separated, leaving the precipitate in the separatory funnel. The precipitate was then dissolved in acetone and concentrated to provide the product as the potassium salt (>97% purity as indicated by ^1H NMR analysis); yield: 8.3 mg (49%). The pentane/hexane layer was re-extracted in this manner (2×) and then concentrated under reduced pressure. The residue was washed with pentane to remove impurities, providing additional product as the potassium salt

for references see p 290

(>97% purity as indicated by ^{1}H NMR analysis). The combined aqueous extracts were washed with pentane (3×) [enol **144** (R^1 = Ph) was isolated from the pentane extracts] and EtOAc (3×). The organic layers containing product as indicated by TLC were separately washed with 1 M aq HCl and brine, dried (Na_2SO_4), and concentrated under reduced pressure to provide the product as the vinylogous acid (>97% purity as indicated by ^{1}H NMR analysis); yield: 3.1 mg (20%); total yield: 11.4 mg (69%).

2-(Prop-1-en-2-yl)pent-4-enyl Trifluoromethanesulfonate (145):[82]
To a flame-dried pear-shaped flask under argon containing 2-(prop-1-en-2-yl)pent-4-en-1-ol (187 mg, 1.48 mmol) was added toluene (1.3 mL), hexane (3.0 mL), and iPr_2NEt (0.34 mL, 1.97 mmol). The soln was cooled to −10 °C and Tf_2O (0.26 mL, 1.5 mmol) was added dropwise over the course of 1 min. The mixture was stirred for 5 min and then taken forward without further purification as a soln in the synthesis of cyclohexadienones **146**.

(*R)-2,4-Diallyl-6-benzoyl-5-hydroxy-3-methoxy-4-[(*S**)-2-(prop-1-en-2-yl)pent-4-enyl]cyclohexa-2,5-dien-1-one (146A) and (*S**)-2,4-Diallyl-6-benzoyl-5-hydroxy-3-methoxy-4-[(*S**)-2-(prop-1-en-2-yl)pent-4-enyl]cyclohexa-2,5-dien-1-one (146B); Typical Procedure on Small Scale:**[82]
To a flame-dried 25-mL round-bottomed flask fitted with a stirrer bar was added (3,5-diallyl-2,6-dihydroxy-4-methoxyphenyl)(phenyl)methanone (**31**; 160 mg, 0.49 mmol). The flask was charged with argon, and toluene (2.16 mL) and THF (6.49 mL) were added, and the reaction was then cooled to −20 °C. LiHMDS (1.47 mL, 1.0 M in THF) was added over the course of 1 min giving a dark red soln. The reaction was stirred for 3 min, and the soln of trifluoromethanesulfonate **145** (379 mg, 1.48 mmol) was then pulled into a Luer Lock syringe; the needle was quickly replaced with a wide needle fitted with a cotton plug, and the soln was filtered directly into the mixture over 30 min. The reaction was allowed to warm to 5 °C and the mixture was stirred for 2 h. If not complete, the reaction was warmed to rt until (3,5-diallyl-2,6-dihydroxy-4-methoxyphenyl)(phenyl)methanone (**31**) had been consumed (TLC). The reaction was quenched with sat. aq NH_4Cl, and the mixture was poured into H_2O and extracted with CH_2Cl_2 (3×). The combined organic layers were washed with sat. aq $NaHCO_3$, H_2O, and brine. The organic soln was then dried (Na_2SO_4), filtered, and concentrated under reduced pressure. The residue was purified by chromatography (silica gel, EtOAc/hexanes 0:1 to 15:85); yield: 176 mg (83%); dr 1.3:1 as determined by ^{1}H NMR analysis. The diastereomers could be separated by preparative TLC.

(*R)-2,4-Diallyl-6-benzoyl-5-hydroxy-3-methoxy-4-[(*S**)-2-(prop-1-en-2-yl)pent-4-enyl]cyclohexa-2,5-dien-1-one (146A) and (*S**)-2,4-Diallyl-6-benzoyl-5-hydroxy-3-methoxy-4-[(*S**)-2-(prop-1-en-2-yl)pent-4-enyl]cyclohexa-2,5-dien-1-one (146B); Typical Procedure on Large Scale:**[82]

CAUTION: *For safety measures, it is important to use a Luer Lock syringe for this reaction. For multigram scale reactions, consider adding the trifluoromethanesulfonate in portions to prevent decomposition during the addition.*

To an oven-dried 1-L round-bottomed flask fitted with a stirrer bar was added (3,5-diallyl-2,6-dihydroxy-4-methoxyphenyl)(phenyl)methanone (**31**; 5.0 g, 15.4 mmol). The flask was charged with argon, THF (203 mL), and toluene (68 mL). The soln was cooled to −20 °C and a 1.0 M soln of LiHMDS in THF (56.2 mL) was added over the course of 3 min. Removal of the protonated amine salts was crucial for success of the reaction. The soln of trifluoromethanesulfonate **145** (10.0 g, 38.5 mmol) was filtered dropwise into the mixture as described above for the small-scale preparation. After the addition had been complete, the mixture was warmed between 5 and 10 °C for 2 h. The reaction was then stirred at rt for an additional 5.5 h and quenched with sat. aq NH_4Cl. The mixture was poured into H_2O and extracted with CH_2Cl_2 (3×). The combined organic layers were washed with sat. aq $NaHCO_3$, H_2O, and brine. The aqueous bicarbonate layer was re-extracted with CH_2Cl_2. The combined organic extracts were dried (Na_2SO_4), filtered, and concentrated under reduced pressure to yield a dark brown oil, which was purified by chromatography (silica gel, EtOAc/hexanes 0:1 to 15:85); yield: 4.5 g (67%); dr 1.3:1 as determined by ^1H NMR analysis. An additional fraction of diastereomer **146B** was isolated as a single diastereomer from the column (silica gel, EtOAc/hexanes 15:85 to 1:1); yield: 701 mg (10%, 59% purity by NMR analysis).

Alternatively, the diastereomers **146** could be purified without column chromatography using a general purification procedure for dearomatized phloroglucinols and type B polyprenylated acylphloroglucinol derivatives, which has been applied to the purification of diverse dearomatized architectures with much better efficiency than column chromatography: the crude product residue (protonated enol form) was dissolved in pentane (a few drops of AcOH was sometimes found to be necessary to ensure that the product was in the protonated form) and mixed vigorously with sat. aq K_2CO_3 in a separatory funnel. H_2O was added and the mixture was again shaken vigorously. A precipitate formed and the layers were separated, leaving the precipitate in the separatory funnel. The precipitate was then dissolved in acetone and the soln was concentrated to provide the product as the potassium salt (dr 1.3:1). Re-extraction of the pentane layer in this manner was repeated (2×). The combined aqueous extracts were washed with pentane (3×) (pentane layers were concentrated to confirm the presence of impurities or product), hexanes, Et_2O (3×), and EtOAc (3×). The combined Et_2O and EtOAc layers were washed with 1 M aq HCl and brine, dried (Na_2SO_4), and concentrated under reduced pressure to provide the product as a 1.3:1 mixture of diastereomers.

for references see p 290

References

[1] Tsukano, C.; Siegel, D. R.; Danishefsky, S. J., *Angew. Chem. Int. Ed.*, (2007) **46**, 8840.
[2] Siegel, D. R.; Danishefsky, S. J., *J. Am. Chem. Soc.*, (2006) **128**, 1048.
[3] Qi, J.; Porco, J. A., Jr., *J. Am. Chem. Soc.*, (2007) **129**, 12682.
[4] Couladouros, E. A.; Dakanali, M.; Demadis, K. D.; Vidali, V. P., *Org. Lett.*, (2009) **11**, 4430.
[5] Zhang, Q.; Mitasev, B.; Porco, J. A., Jr., *J. Am. Chem. Soc.*, (2010) **132**, 14212.
[6] Zhang, Q.; Porco, J. A., Jr., *Org. Lett.*, (2012) **14**, 1796.
[7] Qi, J.; Beeler, A. B.; Zhang, Q.; Porco, J. A., Jr., *J. Am. Chem. Soc.*, (2010) **132**, 13642.
[8] Mitasev, B.; Porco, J. A., Jr., *Org. Lett.*, (2009) **11**, 2285.
[9] George, J. H.; Hesse, M. D.; Baldwin, J. E.; Adlington, R. M., *Org. Lett.*, (2010) **12**, 3532.
[10] Pepper, H. P.; Lam, H. C.; Bloch, W. M.; George, J. H., *Org. Lett.*, (2012) **14**, 5162.
[11] Pepper, H. P.; Tulip, S. J.; Nakano, Y.; George, J. H., *J. Org. Chem.*, (2014) **79**, 2564.
[12] Simpkins, N. S.; Weller, M. D., *Tetrahedron Lett.*, (2010) **51**, 4823.
[13] Goeke, A.; Mertl, D.; Brunner, G., *Angew. Chem. Int. Ed.*, (2005) **44**, 99.
[14] Dai, M.; Danishefsky, S. J., *Tetrahedron Lett.*, (2008) **49**, 6610.
[15] Wang, Z.; Dai, M.; Park, P. K.; Danishefsky, S. J., *Tetrahedron*, (2011) **67**, 10249.
[16] Surendra, K.; Corey, E. J., *J. Am. Chem. Soc.*, (2009) **131**, 13928.
[17] McGrath, N. A.; Bartlett, E. S.; Sittihan, S.; Njardarson, J. T., *Angew. Chem. Int. Ed.*, (2009) **48**, 8543.
[18] Wu, Q.-F.; Liu, W.-B.; Zhuo, C.-X.; Rong, Z.-R.; Ye, K.-Y.; You, S.-L., *Angew. Chem. Int. Ed.*, (2011) **50**, 4455.
[19] Nemoto, T.; Ishige, Y.; Yoshida, M.; Kohno, Y.; Kanematsu, M.; Hamada, Y., *Org. Lett.*, (2010) **12**, 5020.
[20] Yoshida, M.; Nemoto, T.; Zhao, Z.; Ishige, Y.; Hamada, Y., *Tetrahedron: Asymmetry*, (2012) **23**, 859.
[21] Leon, R.; Jawalekar, A.; Redert, T.; Gaunt, M. J., *Chem. Sci.*, (2011) **2**, 1487.
[22] Jia, M.-Q.; You, S.-L., *Chem. Commun. (Cambridge)*, (2012) **48**, 6363.
[23] Xiao, Q.; Jackson, J. J.; Basak, A.; Bowler, J. M.; Miller, B. G.; Zakarian, A., *Nat. Chem.*, (2013) **5**, 410.
[24] Ciochina, R.; Grossman, R. B., *Chem. Rev.*, (2006) **106**, 3963.
[25] Singh, I. P.; Sidana, J.; Bharate, S. B.; Foley, W. J., *Nat. Prod. Rep.*, (2010) **27**, 393.
[26] Tsukano, C.; Siegel, D. R.; Danishefsky, S. J., *J. Synth. Org. Chem., Jpn.*, (2010) **68**, 592.
[27] Dakanali, M.; Theodorakis, E. A., In *Biomimetic Organic Synthesis*, Poupon, E.; Nay, B., Eds.; Wiley-VCH: Weinheim, Germany, (2011); pp 433–467.
[28] Njardarson, J. T., *Tetrahedron*, (2011) **67**, 7631.
[29] Richard, J.-A.; Pouwer, R. H.; Chen, D. Y.-K., *Angew. Chem. Int. Ed.*, (2012) **51**, 4536.
[30] Simpkins, N. S., *Chem. Commun. (Cambridge)*, (2013) **49**, 1042.
[31] Adam, P.; Arigoni, D.; Bacher, A.; Eisenreich, W., *J. Med. Chem.*, (2002) **45**, 4786.
[32] Richard, J.-A., *Eur. J. Org. Chem.*, (2014), 273.
[33] Cuesta-Rubio, O.; Velez-Castro, H.; Frontana-Uribe, B. A.; Cárdenas, J., *Phytochemistry.*, (2001) **57**, 279.
[34] Fráter, G., *Helv. Chim. Acta*, (1974) **57**, 172.
[35] Roche, S. P.; Porco, J. A., Jr., *Angew. Chem. Int. Ed.*, (2011) **50**, 4068.
[36] Ding, Q.; Ye, Y.; Fan, R., *Synthesis*, (2013) **45**, 1.
[37] Zhuo, C.-X.; Zhang, W.; You, S.-L., *Angew. Chem. Int. Ed.*, (2012) **51**, 12.
[38] Kopach, M. E.; Kelsh, L. P.; Stork, K. C.; Harman, W. D., *J. Am. Chem. Soc.*, (1993) **115**, 5322.
[39] Kopach, M. E.; Hipple, W. G.; Harman, W. D., *J. Am. Chem. Soc.*, (1992) **114**, 1736.
[40] Kopach, M. E.; Harman, W. D., *J. Am. Chem. Soc.*, (1994) **116**, 6581.
[41] Kopach, M. E.; Gonzalez, J.; Harman, W. D., *J. Am. Chem. Soc.*, (1991) **113**, 8972.
[42] Curtin, D. Y.; Fraser, R. R., *Chem. Ind. (London)*, (1957), 1358.
[43] Brown, T. L.; Curtin, D. Y.; Fraser, R. R., *J. Am. Chem. Soc.*, (1958) **80**, 4339.
[44] Claisen, L.; Kremers, F.; Roth, F.; Tietze, E., *Justus Liebigs Ann. Chem.*, (1925) **442**, 210.
[45] Topgi, R. S., *J. Org. Chem.*, (1989) **54**, 6125.
[46] Lovchik, M. A.; Goeke, A.; Fráter, G., *Tetrahedron: Asymmetry*, (2006) **17**, 1693.
[47] Raikar, S. B.; Nuhant, P.; Delpech, B.; Marazano, C., *Eur. J. Org. Chem.*, (2008), 1358.
[48] Brajeul, S.; Delpech, B.; Marazano, C., *Tetrahedron Lett.*, (2007) **48**, 5597.
[49] Olsher, U.; Izatt, R. M.; Bradshaw, J. S.; Dalley, N. K., *Chem. Rev.*, (1991) **91**, 137.
[50] Tada, Y.; Satake, A.; Shimizu, I.; Yamamoto, A., *Chem. Lett.*, (1996), 1021.
[51] Satoh, T.; Ikeda, M.; Miura, M.; Nomura, M., *J. Org. Chem.*, (1997) **62**, 4877.
[52] Kimura, M.; Fukasaka, M.; Tamaru, Y., *Synthesis*, (2006), 3611.

[53] Trost, B. M.; Xu, J.; Schmidt, T., *J. Am. Chem. Soc.*, (2009) **131**, 18343.
[54] Mohr, J. T.; Nishimata, T.; Behenna, D. C.; Stoltz, B. M., *J. Am. Chem. Soc.*, (2006) **128**, 11348.
[55] Behenna, D. C.; Stoltz, B. M., *J. Am. Chem. Soc.*, (2004) **126**, 15044.
[56] Burger, E. C.; Tunge, J. A., *Org. Lett.*, (2004) **6**, 4113.
[57] Curtin, D. Y.; Fraser, R. R., *J. Am. Chem. Soc.*, (1958) **80**, 6016.
[58] Curtin, D. Y.; Crawford, R. J.; Wilhelm, M., *J. Am. Chem. Soc.*, (1958) **80**, 1391.
[59] Miller, B.; Margulies, H., *J. Org. Chem.*, (1965) **30**, 3895.
[60] Kornblum, N.; Seltzer, R., *J. Am. Chem. Soc.*, (1961) **83**, 3668.
[61] Kornblum, N.; Lurie, A. P., *J. Am. Chem. Soc.*, (1959) **81**, 2705.
[62] Kornblum, N.; Berrigan, P. J.; le Noble, W. J., *J. Am. Chem. Soc.*, (1960) **82**, 1257.
[63] Kornblum, N.; Seltzer, R.; Haberfield, P., *J. Am. Chem. Soc.*, (1963) **85**, 1148.
[64] Pearson, R. G., *J. Am. Chem. Soc.*, (1963) **85**, 3533.
[65] Ho, T.-L., *Tetrahedron*, (1985) **41**, 3.
[66] Shing, T. K. M.; Li, L.-H.; Narkunan, K., *J. Org. Chem.*, (1997) **62**, 1617.
[67] Mayr, H.; Breugst, M.; Ofial, A. R., *Angew. Chem. Int. Ed.*, (2011) **50**, 6470.
[68] Tanikaga, R.; Hamamura, K.; Hosoya, K.; Kaji, A., *J. Chem. Soc., Chem. Commun.*, (1988), 817.
[69] Godenschwager, P. F.; Collum, D. B., *J. Am. Chem. Soc.*, (2008) **130**, 8726.
[70] Reich, H. J., *Chem. Rev.*, (2013) **113**, 7130.
[71] Stein, A. R., *Can. J. Chem.*, (1965) **43**, 1508.
[72] Herzig, J.; Zeisel, S., *Monatsh. Chem.*, (1888) **9**, 217.
[73] Herzig, J.; Zeisel, S., *Monatsh. Chem.*, (1890) **11**, 2119.
[74] Goeke, A., EP 1213276, (2002); *Chem. Abstr.*, (2002) **137**, 19737.
[75] Takagi, R.; Nerio, T.; Miwa, Y.; Matsumura, S.; Ohkata, K., *Tetrahedron Lett.*, (2004) **45**, 7401.
[76] Zhuo, C.-X.; You, S.-L., *Angew. Chem. Int. Ed.*, (2013) **52**, 10056.
[77] Zhuo, C.-X.; You, S.-L., *Adv. Synth. Catal.*, (2014) **356**, 2020.
[78] Yang, D.; Wang, L.; Han, F.; Li, D.; Zhao, D.; Wang, R., *Angew. Chem. Int. Ed.*, (2015) **54**, 2185.
[79] Yang, D.; Wang, L.; Han, F.; Li, D.; Zhao, D.; Cao, Y.; Ma, Y.; Kong, W.; Sun, Q.; Wang, R., *Chem.–Eur. J.*, (2014) **20**, 16478.
[80] Grenning, A. J.; Boyce, J. H.; Porco, J. A., Jr., *J. Am. Chem. Soc.*, (2014) **136**, 11799.
[81] Zalesskiy, S. S.; Ananikov, V. P., *Organometallics*, (2012) **31**, 2302.
[82] Boyce, J. H.; Porco, J. A., Jr., *Angew. Chem. Int. Ed.*, (2014) **53**, 7832.

2.3 Additions to Alkenes and C=O and C=N Bonds

2.3.1 Additions to Nonactivated C=C Bonds

Z. W. Yu and Y.-Y. Yeung

General Introduction

The addition of electrophilic reagents to nonactivated C=C bonds is one of the most useful organic transformations.[1–12] These "nonactivated" C=C bonds include unfunctionalized cyclic and acyclic alkenes, conformationally strained alkenes, and conjugated alkenes (Scheme 1).[6] The addition of an electrophile, such as a proton, halonium, or chalconium, to an alkene followed by an intra- or intermolecular attack by a nucleophile can give vicinally functionalized compounds. Clearly, these products are valuable precursors to many useful complex organic compounds. Furthermore, if these electrophile-initiated events can be combined with other chemical transformations in a "domino" manner without isolation of the intermediate, in which three or more bonds are created, they can serve as a rapid and efficient way to construct natural and non-natural products such as those shown in Scheme 1.

Scheme 1 Schematic Illustration of Domino Reactions Involving Electrophilic Addition to Nonactivated Alkenes, Types of Alkenes Involved, and Some Natural and Non-natural Products Obtained by This Strategy

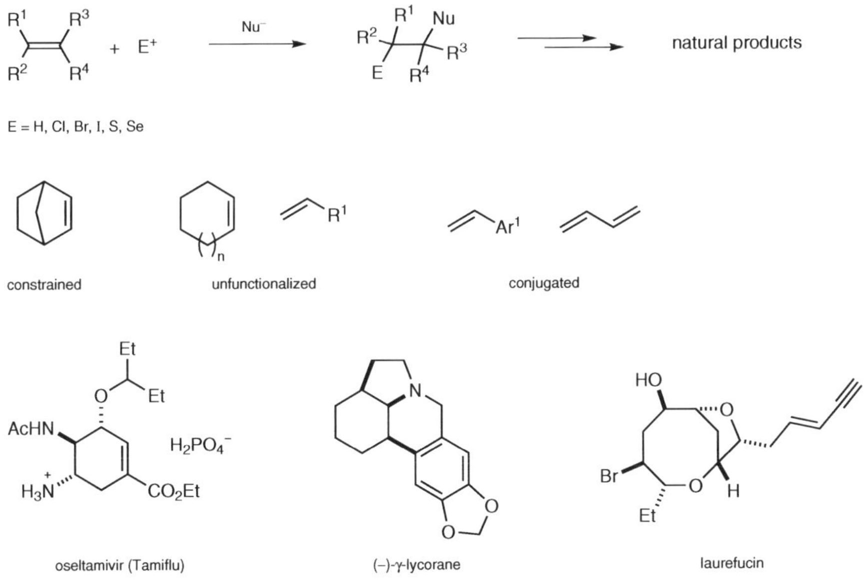

The purpose of this chapter is to provide an overview on domino transformations triggered by electrophilic addition to nonactivated C=C bonds. Four different classes of reac-

for references see p 332

tions will be covered in Sections 2.3.1.1 to 2.3.1.4. Each reaction class is subdivided into subsections based on the specific type of electrophilic initiator, with the examples focused on the domino sequence in cases that often apply to natural product total synthesis.

2.3.1.1 Domino Amination

The addition of an electrophile and a nitrogen-containing nucleophile across a C=C bond creates new carbon—electrophile and C—N bonds, which is useful for the preparation of nitrogen-containing products. Amines, amides, ureas, and thioureas are the most commonly incorporated nitrogen moieties. One common domino approach is the use of dienamines incorporating primary amines to achieve consecutive hydroamination reactions.

2.3.1.1.1 Proton-Initiated Events

Nitrogen-containing heterocycles such as pyrrole, pyrazine, and dihydroindole (indoline) exist as the core structures in many natural products. In recent years, there has been a surge in interest by synthetic chemists to assemble indoline alkaloids.[13–15]

Beller and co-workers reported a one-pot, base-catalyzed intermolecular domino reaction of commercially available 1-chloro-2-vinylbenzene with aniline in the presence of potassium *tert*-butoxide to give 1-phenyl-2,3-dihydroindole (**2**) in moderate yield of up to 53% (Scheme 2).[16,17] Although this method shows improved yields and high atom economy over previous methods that use 2-(2-chlorophenyl)ethylamine as the starting material, the generation of side products from dehydrogenation of 1-phenyl-2,3-dihydroindole (**2**) and nucleophilic substitution of 1-chloro-2-vinylbenzene is unavoidable.[18–21] The problem of nucleophilic substitution is reduced by decreasing the equivalents of aniline and potassium *tert*-butoxide. Nonetheless, this domino transformation represents a method for direct construction of the dihydroindole core. It is believed that an aryne intermediate **1** is part of the mechanistic sequence. A base-catalyzed domino isomerization/hydroamination reaction has also been developed.[22]

Scheme 2 Base-Catalyzed Domino Hydroamination/Aryne Cyclization Reaction[16,17]

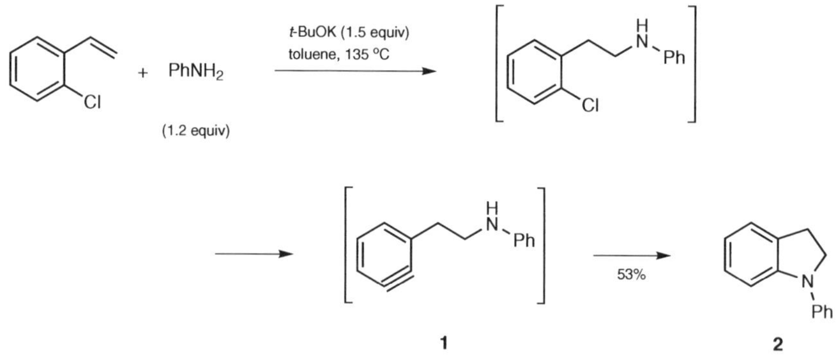

Cl + $PhNH_2$ (1.2 equiv) → t-BuOK (1.5 equiv), toluene, 135 °C → [HN-Ph, Cl] → [**1**] → 53% → **2** (N-Ph)

Hultzsch and co-workers reported that dienamine **3** undergoes an asymmetric hydroamination reaction with 5 mol% of the dimeric dilithium salt of chiral diaminobinaphthyl **4** as catalyst (Scheme 3).[23] Allyl-substituted pyrrolidines **5** are obtained after the first hydroamination process with moderate enantioselectivities (64% ee for **5A** and 72% ee for **5B**) and low diastereoselectivity (**5A**/**5B** = 1.2:1) in 2 hours with 79% yield. Interestingly, only pyrrolidine **5B** undergoes a second hydroamination/cyclization sequence under the same reaction conditions, ultimately affording 2,4,6-trimethyl-1-azabicyclo[2.2.1]heptanes **6A**/**6B** with moderate yield and diastereoselectivity in 6 days. Critically, this transformation represents a rapid construction of a specific bicyclic amine scaffold that may have potential pharmacological or chemotherapeutic activity.[24,25] Subsequently, Markó and co-workers showed that dienamine **3** also undergoes a cascade sequence of base-catalyzed hydroamination reactions to give **6** in 80% yield when butyllithium is employed.[26]

Scheme 3 Asymmetric Double Hydroamination of an Aminoalkene Catalyzed by the Dilithium Salt of a Chiral Diaminobinaphthyl[23]

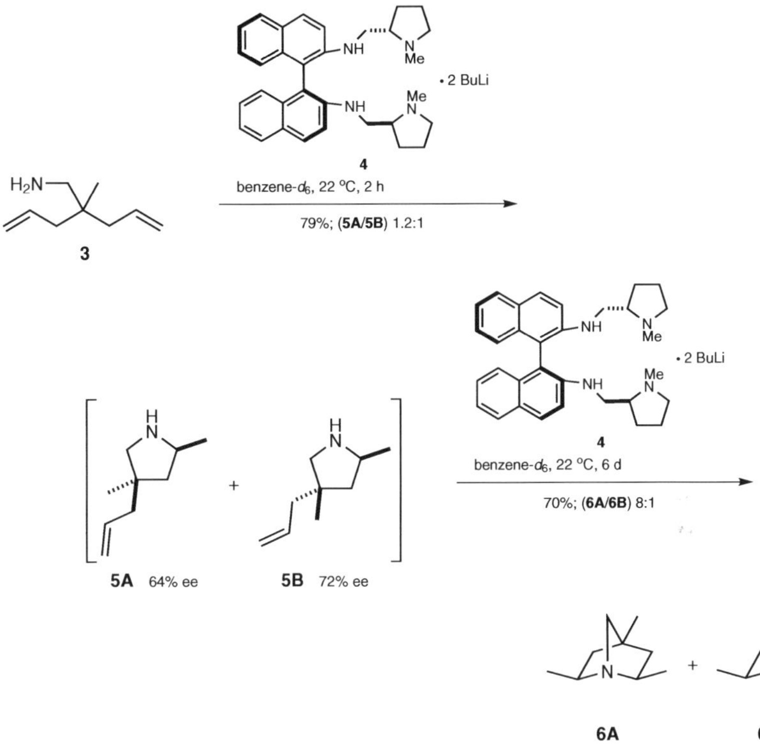

2.3.1.1.2 Transition-Metal-Initiated Events

Nitrogen-containing saturated heterocycles such as pyrrolidine, piperidine, azepane, and indolizidine have been the subject of extensive synthetic interest as they occur widely in many natural products.[27–30] In addition, they are common building blocks for drug molecules.[31]

An intramolecular hydroamination/double cyclization of dialkenylamine **7**, promoted by an organosamarium reagent, to give fused heterocycles such as pyrrolizidines and indolizidines was reported by Marks and co-workers (Scheme 4).[32] Mechanistic studies found that the reaction sequence involves an initial protonolytic reaction between the organolanthanide catalyst and substrate **7** to form intermediate **8**, which then undergoes alkene insertion to form intermediate **9**. Finally, alkene insertion into the Sm—C bond in **9** furnishes the desired bicyclic product **10**.[33,34] The formation of monocyclic products through competing intermolecular protonolysis of **9** is dependent on the steric effect of the organolanthanide, concentration of the dialkenylamine, and the reaction temperature. This catalytic process is indicated by distinct color changes as the reaction proceeds. Additionally, the synthesis of the bicyclic products **10** can be carried out on both NMR- and preparative scale. In a separate report, the scope of the reaction was expanded using a similar catalyst to effect intermolecular hydroaminations of polyvinylarenes (Scheme 4). For the case of trivinylarene **11**, racemic hexahydrocyclopent[*ij*]isoquinoline **12** is ob-

for references see p 332

tained with high yield and excellent chemo-, regio-, and diastereoselectivities in a consecutive formation of C—N and C—C bonds.[35] This methodology has since been applied by Molander and co-workers in the diastereoselective synthesis of quinolizine and isoindolizine ring systems.[36]

Scheme 4 Organolanthanide-Catalyzed Intra- and Intermolecular Domino Hydroamination/Cyclization Reaction[32–35]

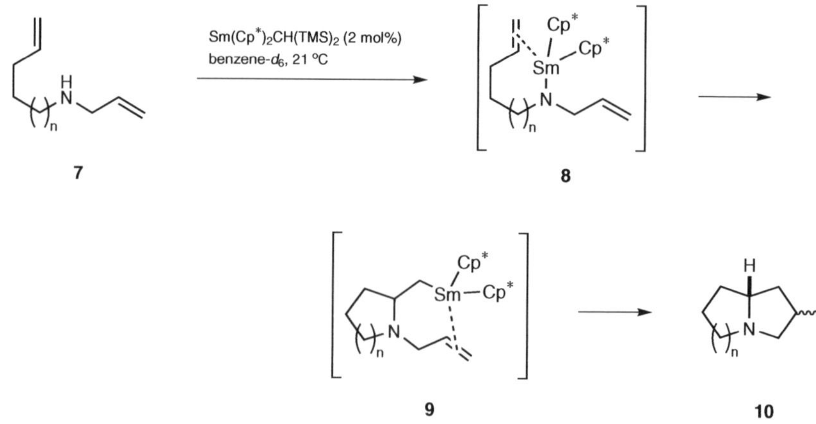

Sm(Cp*)$_2$CH(TMS)$_2$ (2 mol%)
benzene-d_6, 21 °C

7 → 8 → 9 → 10

n	Ratio (*cis*/*trans*)	Yield (%)	Ref
1	45:55	93	[32]
2	85:15	88	[32]

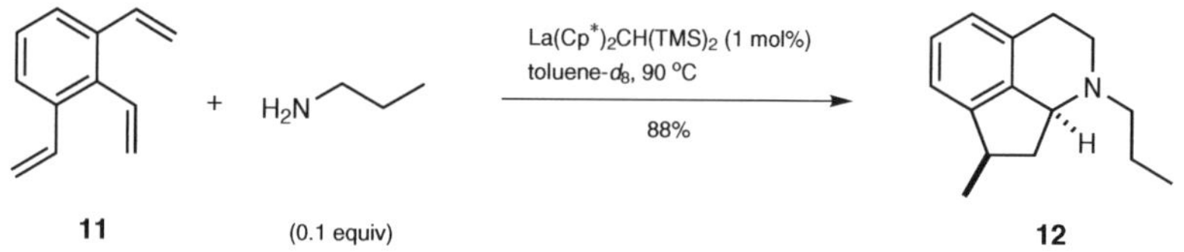

11 + H_2N (0.1 equiv) → 12

La(Cp*)$_2$CH(TMS)$_2$ (1 mol%)
toluene-d_8, 90 °C
88%

Another simple domino approach to construct pyrrolidine derivatives, exemplified by Shi and co-workers, is a catalytic domino ring-opening/ring-closing hydroamination of methylenecyclopropanes **13** with sulfonamides (Scheme 5).[37] A catalyst system consisting of equimolar amounts of chloro(triphenylphosphine)gold(I) and silver(I) trifluoromethanesulfonate is essential for both the ring-opening hydroamination to give intermediate **14** and subsequent ring-closing hydroamination to furnish the corresponding pyrrolidine derivatives **15**.

Scheme 5 Gold(I)-Catalyzed Domino Ring-Opening/Ring-Closing Hydroamination Reaction of Methylenecyclopropanes with 4-Toluenesulfonamide[37]

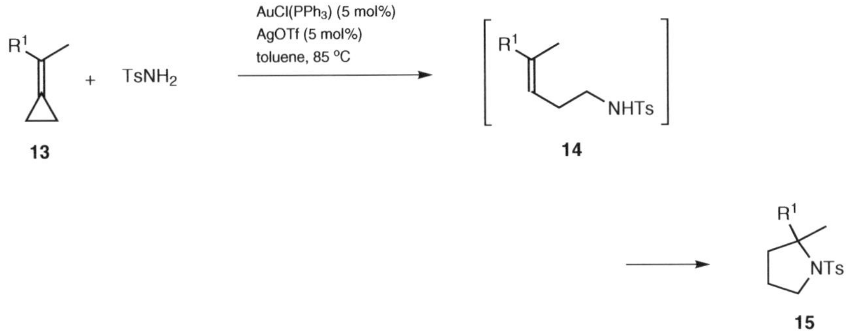

AuCl(PPh3) (5 mol%)
AgOTf (5 mol%)
toluene, 85 °C

13 + TsNH2 → [14] → 15

R^1	Yield (%)	Ref
4-Tol	72	[37]
$(CH_2)_6Me$	64	[37]

Simple linear aliphatic amines are broadly utilized by the bulk and fine chemical industries for the preparation of pharmaceutical intermediates, polymerizations, and the synthesis of emulsifiers and fabric softeners.[38] Therefore, straightforward synthetic pathways that use inexpensive and widely available materials such as alkenes as chemical feedstock for conversion into linear aliphatic amines are highly valuable.

Beller and co-workers have reported an elegant catalytic hydroaminomethylation of internal aliphatic alkenes to give linear amine products with high regio- and chemoselectivity (Scheme 6).[39] Although internal alkenes are much more widely available as feedstock than terminal alkenes, hydroaminomethylations of internal alkenes frequently encounter challenges such as lower reactivity and a mixture of linear and branched regioisomeric products. Alkene isomerization of (*E*)-but-2-ene is catalyzed by a rhodium catalyst to give the terminal alkene but-1-ene. In the presence of carbon monoxide and molecular hydrogen, but-1-ene is then converted into pentanal. Next, amination gives enamine **17**, which is subsequently hydrogenated to give the saturated product **18**. The steric and electronic properties of 1,1′-binaphthyl phosphine ligand **16** are essential for both the selectivity and reaction rate. This domino transformation has wide applicability and can accommodate a variety of 2,2′,6,6′-tetrakis[(diphenylphosphino)methyl]-1,1′-biphenyl (Tetrabi) and phenoxaphosphino- and dibenzophosphole-modified 4,5-bis(diphenylphosphino)-9,9-dimethylxanthene (Xantphos-type) ligands, which improve the reactivity and selectivity in the case of longer-chain internal alkenes.[40–42] A hydrogen-free ruthenium-catalyzed strategy using either syngas or carbon monoxide in water improves the practicability and reduces the economic cost of this process.[43]

for references see p 332

Scheme 6 Hydroaminomethylation of an Internal Alkene for Selective Synthesis of a Linear Amine[39]

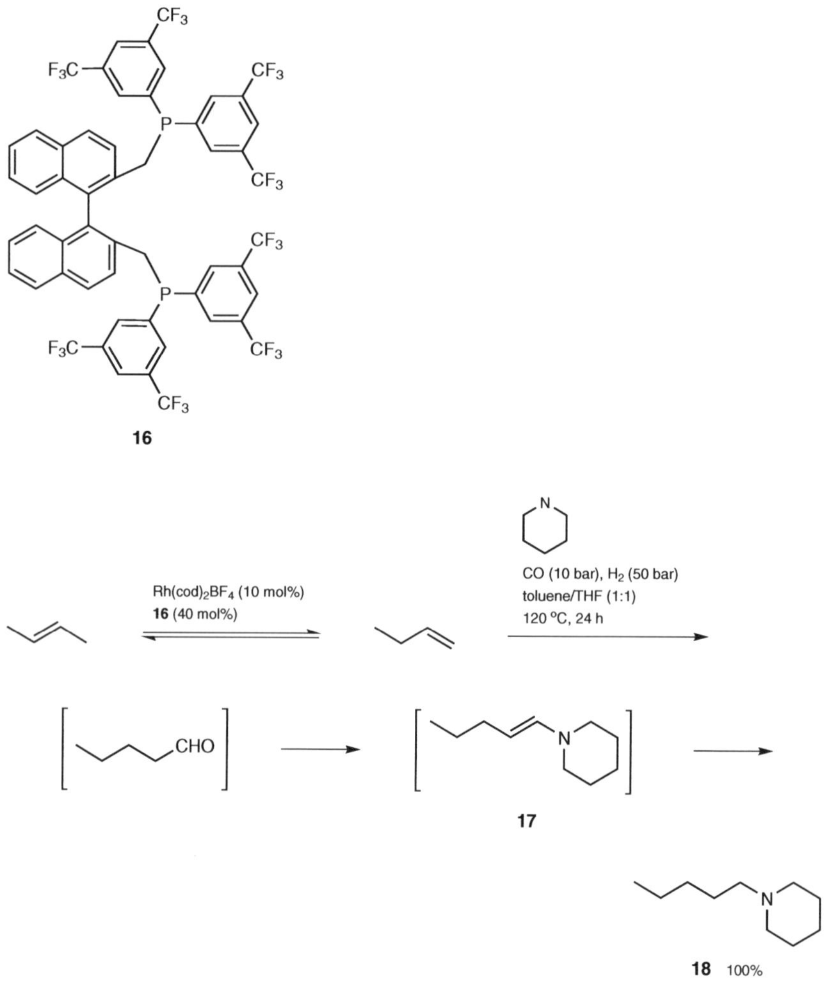

CF_3 F_3C CF_3 P CF_3 CF_3 P F_3C CF_3 CF_3

16

$Rh(cod)_2BF_4$ (10 mol%)
16 (40 mol%)

N
CO (10 bar), H_2 (50 bar)
toluene/THF (1:1)
120 °C, 24 h

CHO

N

17

N

18 100%

2-Methylhexahydro-1*H*-pyrrolizine (10, n = 1); Typical Procedure:[33]
In a glovebox, $Sm(Cp^*)_2CH(TMS)_2$ (14.5 mg, 25 μmol) was loaded into a 25-mL reaction vessel equipped with a stirrer bar. Benzene-d_6 (2.0 mL) (**CAUTION:** *carcinogen*) was vacuum-transferred onto the precatalyst, followed by addition of a mixture of *N*-allylpent-4-en-1-amine (**7**, n = 1; 0.21 g, 1.68 mmol) and benzene (~1.5 mL) via syringe onto the frozen benzene and precatalyst mixture. The mixture was then freeze–pump–thaw degassed and warmed to rt. The clear yellow soln was stirred under argon for 5 d. Filtration followed by vacuum transfer afforded a mixture of benzene and the product. Benzene was removed by distillation at atmospheric pressure to give the product as a colorless oil; yield: 194 mg (93%); ratio (*cis/trans*) 45:55; >95% pure by ^{1}H NMR and GC/MS analyses.

2.3.1.1.3 Halogen-Initiated Events

Vicinal diamines constitute an important functional group in organic compounds and occur frequently in natural products, pharmaceutical agents, ligands, and catalysts.[44–48] The direct diamination of alkenes is thus an attractive transformation as it provides ready access to these vicinal diamines.

Recently, Muñiz and co-workers reported that an iodonium ion promotes intramolecular diamination of alkenes with tethered ureas as the nucleophilic nitrogen source (Scheme 7).[49] In this reaction sequence, iodonium ion **20** is presumed to be formed when the substrate **19** is treated with bis(pyridine)iodonium(I) tetrafluoroborate, effecting a subsequent 5-*exo* N-cyclization to give intermediate **21** and lastly a nucleophilic substitution of the iodide by the other nitrogen to give pyrroloimidazole **22**. Performing the reaction at higher temperature is required to suppress competing 7-*exo* O-cyclization of intermediate **20**. The alkene scope is broad and includes various mono- or disubstituted terminal and internal C=C bonds. Additionally, this domino transformation is suitable for the construction of bicyclic ureas, sulfamides, and guanidines. Newer catalytic protocols use *N*-iodosuccinimide and bromine, and furnish bicyclic imidazolidinium salts and thioureas successfully.[50–53] The alkaloid absouline (**23**) was obtained in a concise synthesis from an enantiopure analogue of a pyrroloimidazole of type **22**.[54]

Scheme 7 Iodonium Ion Mediated Intramolecular Diamination of Alkenes with Bis(pyridine)iodonium(I) Tetrafluoroborate[49]

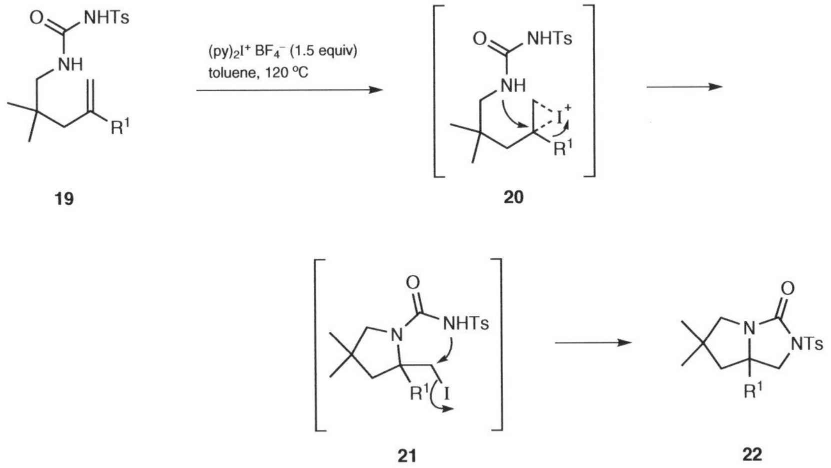

$(py)_2I^+ BF_4^-$ (1.5 equiv)
toluene, 120 °C

19 20 21 22

R^1	Yield (%)	Ref
H	92	[49]
Me	75	[49]

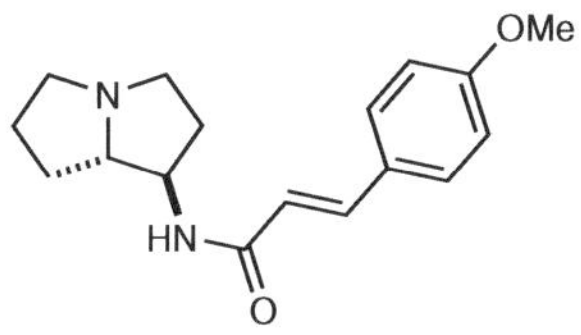

23 absouline

In addition to the metal-free intramolecular diamination of alkenes, the first intermolecular enantioselective diamination of alkenes used chiral iodine(III) reagent **24** and *N*-(methylsulfonyl)methanesulfonamide as the nitrogen source (Scheme 8).[55] The reaction sequence is believed to be initiated by an electrophilic addition of reagent **24** to styrene, forming a hypervalent iodonium intermediate **25**, followed by a nucleophilic displacement with inversion of configuration to provide 1,2-vicinal λ^3-iodoamine **26**. On the basis of *anti* selectivity in diamine product **29**, it is therefore proposed that intermedi-

for references see p 332

ate **26** undergoes protonolysis to generate the ion pair **27**, which leads to the formation of aziridinium intermediate **28**, which then undergoes a second nucleophilic displacement with inversion of configuration to furnish the product **29**. The products of diamination are useful synthetic intermediates that are amenable to further modifications, as exemplified in the synthesis of the known anthelmintic (*S*)-levamisole.

Scheme 8 Enantioselective Intermolecular Diamination of Styrene with a Chiral Iodine(III) Reagent[55]

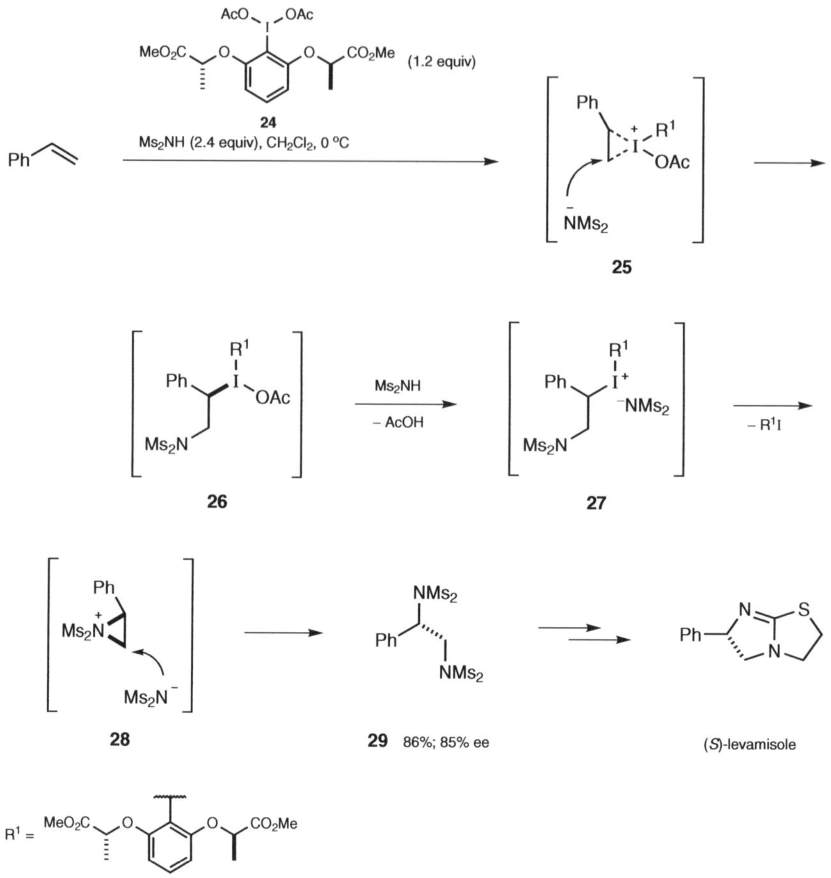

24 (1.2 equiv)
Ms_2NH (2.4 equiv), CH_2Cl_2, 0 °C

25

Ms_2NH
– AcOH

26

27

– R^1I

28

29 86%; 85% ee

(*S*)-levamisole

R^1 =

Oseltamivir (Tamiflu), an orally effective anti-influenza drug, is a clinically approved drug that contains a 1,2-diamine moiety.[56] In 2006, Corey and co-workers successfully shortened the previous methods for the total synthesis of oseltamivir.[57] During the same period, a new influenza pandemic in the form of the H5N1 virus emerged.[58–62] Crucially, the key in the concise synthesis of oseltamivir involves a Lewis acid catalyzed bromoamidation domino transformation (Scheme 9). The reaction sequence involves Lewis acid ($SnCl_4$) activation of *N*-bromoacetamide and regioselective reaction with dienyl ester **30** to form bromonium ion **31**, which is then captured by acetonitrile to furnish intermediate **32** stereoselectively. Finally, hydration gives the *trans*-haloacetamide product **33** stereoselectively. The isolation of the Markovnikov product is thus consistent with the postulated bromonium ion **31**. Under optimized conditions with boron trifluoride–diethyl ether complex as the Lewis acid catalyst the substrate range is expanded to various alkenes and nitriles.[63] Recently, Yeung and co-workers reported that catalytic diphenyl diselenide serves as an efficient Lewis base activator for *N*-chlorosuccinimide in the chloroamidation of a number of alkenes.[64]

Scheme 9 Lewis Acid Mediated Haloamidation of Alkenes[57]

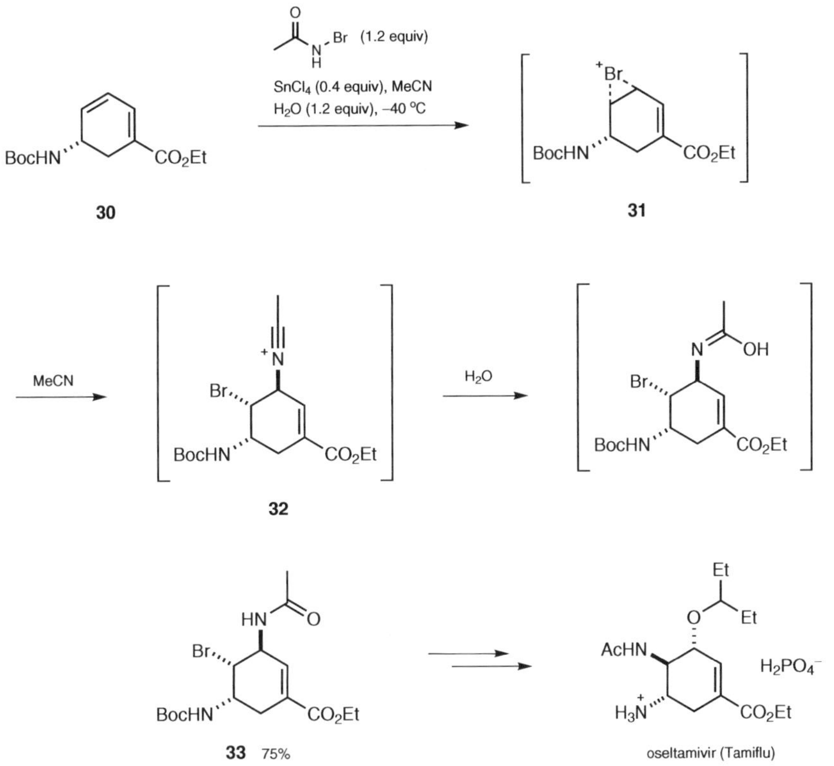

Yeung and co-workers have developed a catalyst-free multicomponent reaction for the haloamidination of alkenes (Scheme 10).[65,66] In this electrophile-initiated cascade, bromonium ion **34** is formed when cyclohexene is treated with *N*-bromosuccinimide. The bromonium ion **34** is first attacked by the nitrile and subsequently by 4-toluenesulfonamide to give bromoamidine product **35** ($R^1 = Me$). Alternatively, guanidine **35** ($R^1 = NMe_2$) is obtained using the same sequence by replacing the nitrile with *N*,*N*-dimethylcyanamide. Based on mechanistic studies, 4-toluenesulfonamide is proposed to activate *N*-bromosuccinimide either by in situ generation of *N*-bromo-4-toluenesulfonamide (**37**) or coordination of the sulfonyl or carbonyl oxygen of *N*-bromo-4-toluenesulfonamide (**37**) or *N*-bromosuccinimide via adducts **38** or **39**, respectively (Scheme 10). This methodology has been applied to the synthesis of dihydroimidazole (e.g., **36**) and guanidine derivatives.

Scheme 10 Multicomponent Synthesis of Guanidines and Dihydroimidazoles[65,66]

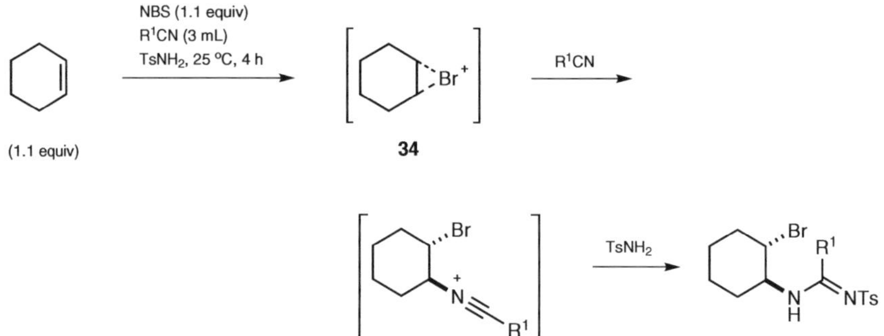

NBS (1.1 equiv)
R^1CN (3 mL)
$TsNH_2$, 25 °C, 4 h

(1.1 equiv)

34

R^1CN

$TsNH_2$

35

R^1	Yield (%) of **35**	Ref
Me	91	[65]
NMe_2	72	[66]

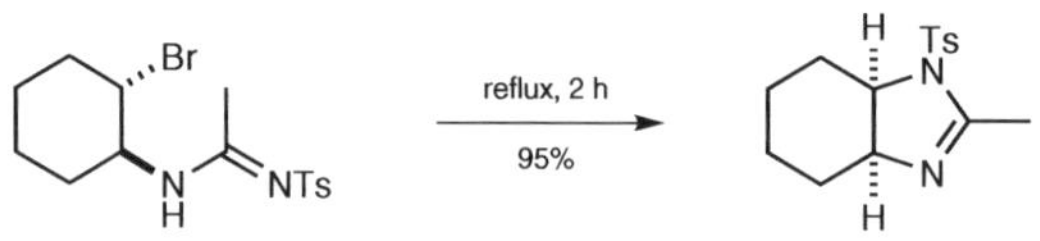

reflux, 2 h
95%

36

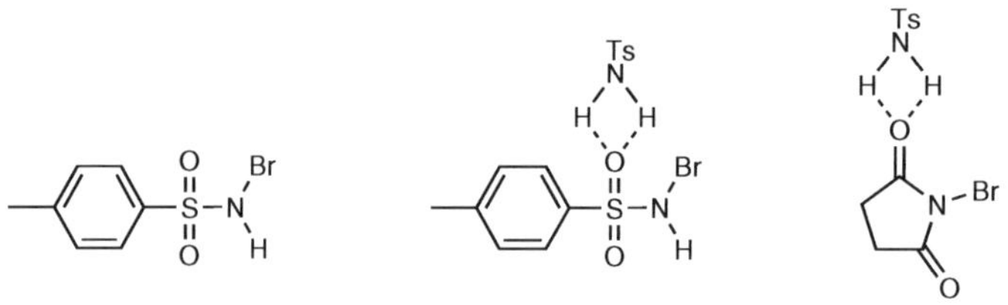

37 38 39

A similar catalyst-free multicomponent strategy was also adopted by Yeung and co-workers to give nitrogen-containing saturated heterocycles including pyrrolidines, piperidines, and azepanes. The highly stereoselective bromoaminocyclization of the unsaturated aziridine **40** results in the formation of an enantiopure azepane derivative **43** with three stereogenic centers (Scheme 11).[67] Similar to the previously proposed activation modes of *N*-bromosuccinimide, a highly electrophilic bromonium ion is generated which leads to the formation of intermediate **41**. The aziridine moiety in **41** reacts with the bromonium ion intramolecularly to form aziridinium intermediate **42**, followed by a second stereoselective nucleophilic attack by 4-nitrobenzenesulfonamide to give the product **43** in high yield. Notably, racemization of the chiral center in the azepane product **43** is not observed. A diversity of products, such as piperidine and urea derivatives, and the bicyclic amine **44**, found in the cinchona alkaloid scaffold, can be prepared from azepanes of type **43**. Interestingly, a mixture of pyrrolidine **46** and piperidine **47** is obtained when cinnamylaziridine **45** is subjected to similar conditions to those used in the cyclization of aziridine **40** (Scheme 11).[68]

Scheme 11 Bromonium-Initiated Aminocyclization/Ring Expansion Reaction Cascade[67,68]

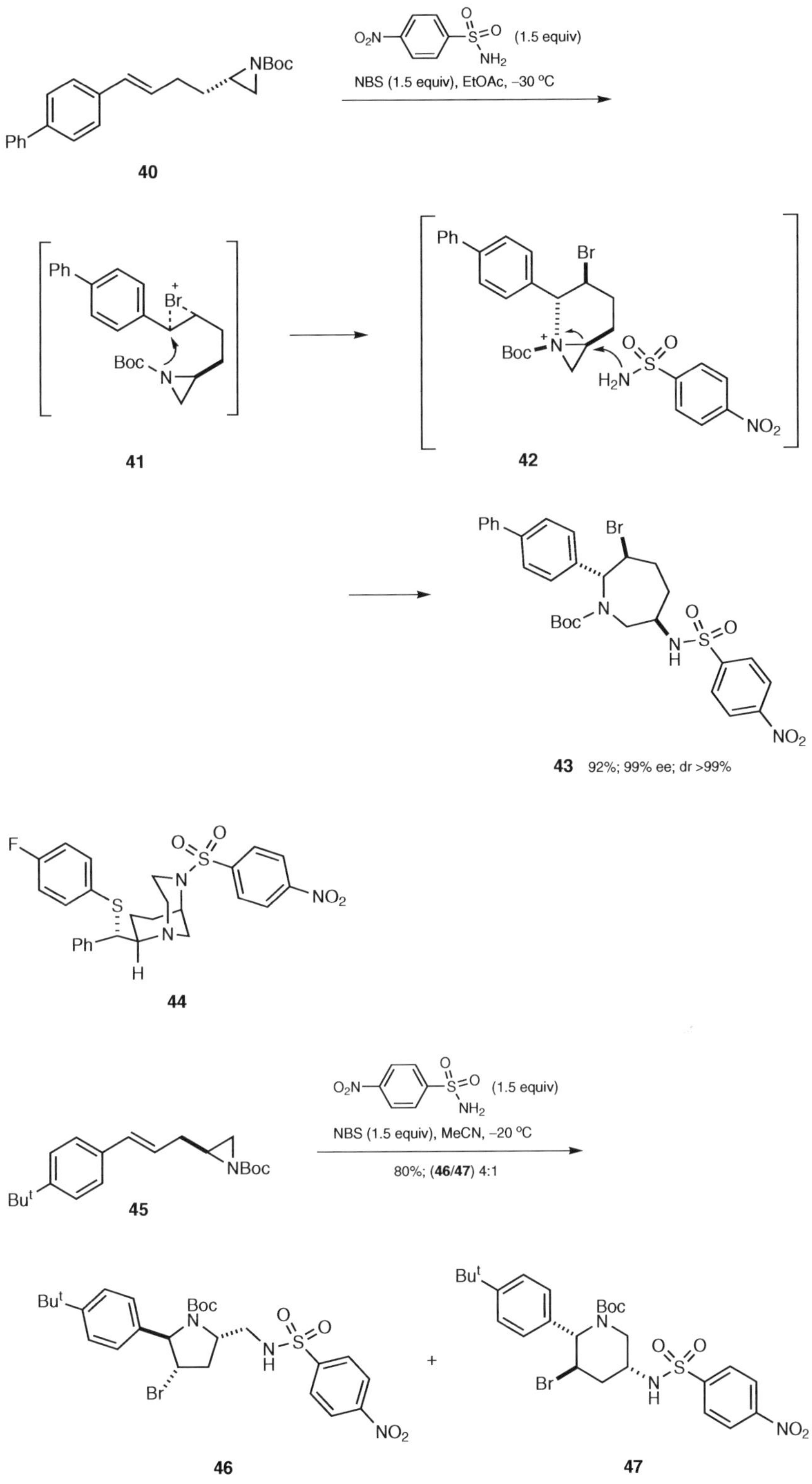

O_2N–C_6H_4–SO_2NH_2 (1.5 equiv)
NBS (1.5 equiv), EtOAc, –30 °C

40 → [**41**] → [**42**] → **43** 92%; 99% ee; dr >99%

44

O_2N–C_6H_4–SO_2NH_2 (1.5 equiv)
NBS (1.5 equiv), MeCN, –20 °C
80%; (**46**/**47**) 4:1

45 → **46** + **47**

for references see p 332

In an approach by Davies and co-workers toward the synthesis of pyrrolidine derivatives, iodine is used as the initiator for a stereoselective iodoamination and deprotection reaction sequence of unsaturated β-amino ester **48** for the asymmetric synthesis of β-amino acids (e.g., **52**; Scheme 12).[69] The sequence involves the formation of iodonium ion intermediate **49** which undergoes a stereoselective nucleophilic attack by the tethered tertiary amine to afford quaternary ammonium intermediate **50**. Finally, a chemoselective deprotection of the *N*-1-phenylethyl group furnishes a diastereomeric mixture of pyrrolidines **51A** and **51B**. This methodology was successfully applied toward the synthesis of polyhydroxylated pyrrolidine β-amino acid **52**.

Scheme 12 Iodoamination/Deprotection Sequence toward the Synthesis of β-Amino Acids[69]

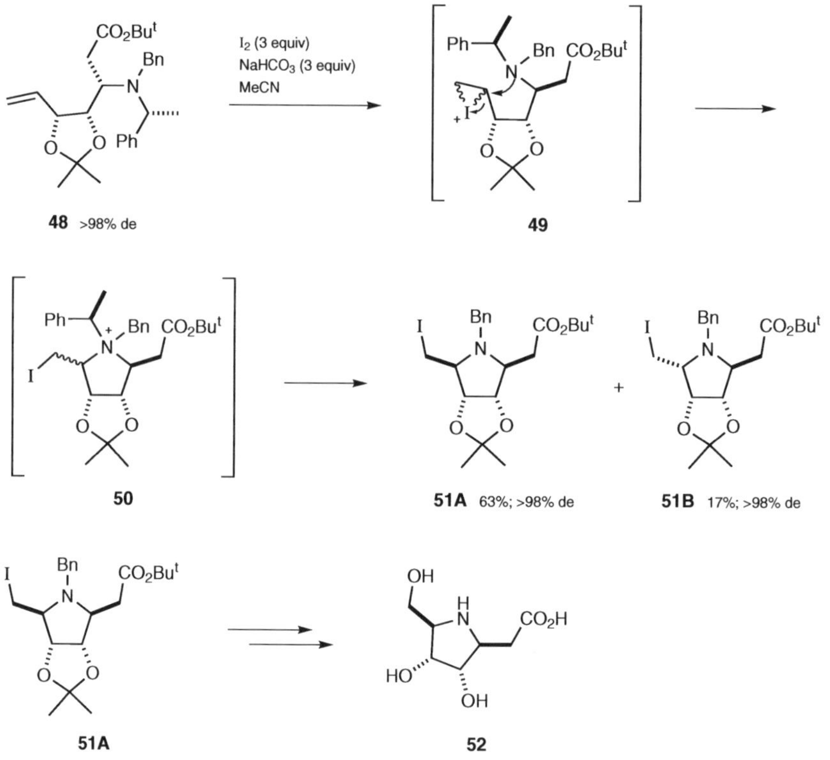

I_2 (3 equiv)
$NaHCO_3$ (3 equiv)
MeCN

48 >98% de

49

50

51A 63%; >98% de

51B 17%; >98% de

51A

52

Amaryllidaceae alkaloids are regarded as a special class of isoquinoline alkaloids.[70,71] Recent efforts in the isolation, biosynthesis, and evaluation of biological activities of amaryllidaceae alkaloids have revealed potential therapeutic applications.[72] (−)-γ-Lycorane (**58**) belongs to this family of amaryllidaceae alkaloids and was obtained by Fujioka and co-workers in a concise asymmetric synthesis using an intramolecular bromoamination/oxidation reaction sequence (Scheme 13).[73] This sequence requires the use of enantiopure cyclohexa-1,4-diene aminal **53**. Initially, bromonium intermediate **54** is generated using *N*-bromosuccinimide as the electrophilic halogen source. Next, intramolecular trapping of **54** by the nitrogen atom in the dihydroimidazole unit gives intermediate **55**. Subsequently, with *N*-bromosuccinimide as oxidant, the aminal unit in **55** is converted into the N-brominated compound **56**. Finally, the sequence is completed with elimination of hydrogen bromide to furnish the imidazo[1,2-*a*]indole product **57** as a single stereoisomer.

Scheme 13 Bromoamination/Oxidation Sequence toward the Synthesis of (−)-γ-Lycorane[73]

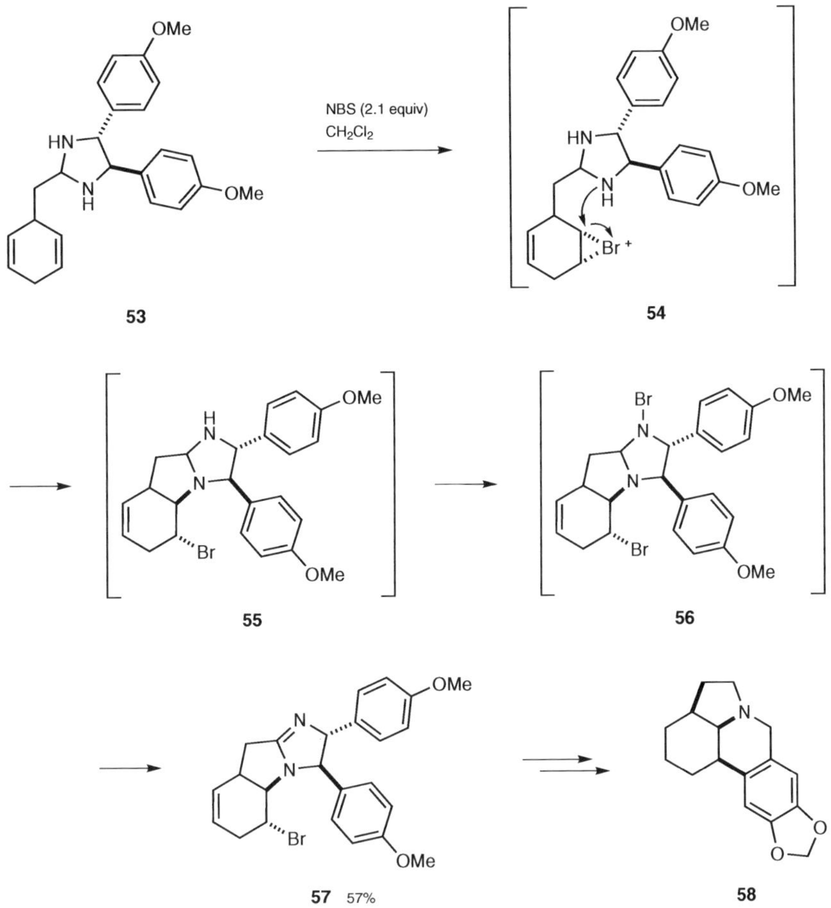

NBS (2.1 equiv)
CH_2Cl_2

53 54 55 56 57 57% 58

(2*R*,3*R*,4a*R*,5*R*)-5-Bromo-2,3-bis(4-methoxylphenyl)-2,4a,5,6,8a,9-hexahydro-3*H*-imidazo-[1,2-*a*]indole (57); Typical Procedure:[73]

(1*R*,2*R*)-1,2-Bis(4-methoxyphenyl)ethane-1,2-diamine (4.39 g, 16.1 mmol) was added to a soln of 2-(cyclohexa-2,5-dien-1-yl)acetaldehyde (1.97 g, 16.1 mmol) in CH_2Cl_2 (322 mL) at 0 °C under N_2. The mixture was stirred for 1 h. To the obtained soln of (4*R*,5*R*)-2-(cyclohexa-2,5-dien-1-ylmethyl)-4,5-bis(4-methoxyphenyl)imidazolidine (**53**) in CH_2Cl_2 was added NBS (6.02 g, 33.8 mmol) and the resulting soln was stirred for 15 min at 0 °C. The reaction was quenched by addition of sat. aq $Na_2S_2O_3$ and sat. aq $NaHCO_3$. The resulting soln was extracted with CH_2Cl_2. The organic layer was dried (Na_2SO_4) and concentrated under reduced pressure. The residue was purified by column chromatography (silica gel, EtOAc/Et_3N 20:1 to EtOAc/MeOH/Et_3N 20:1:1); yield: 4.15 g (57%).

2.3.1.2 Domino Etherification

The addition of an electrophile and an oxygen-containing nucleophile across a C=C bond creates new carbon—electrophile and C—O bonds for the assembly of ether-containing products. The key strategic component for these domino sequences is the formation of an oxonium ion intermediate.

2.3.1.2.1 Halogen-Initiated Events

Haufe demonstrated the halonium ion mediated transannular O-heterocyclization of 9-oxabicyclo[6.1.0]non-4-ene (**59**) employing an *N*-halosuccinimide **60** as the halogen source and excess triethylamine tris(hydrofluoride) (Et_3N•3HF) as nucleophile to give ethers **63** and **64** via intermediates **61** and **62** (Scheme 14).[74] This reaction is compatible with various halogens (Cl^+, Br^+, and I^+) as electrophile. By replacing triethylamine tris(hydrofluoride) with nucleophilic solvents such as alcohol and water in the presence of a catalytic amount of tetramethylguanidine (1 mol%), solvent-incorporated products similar to **63** and **64** can be isolated.[75] Likewise, two other different transannular cyclization reactions using cycloocta-1,5-diene have also been reported.[76,77]

Scheme 14 Halonium Ion Initiated Transannular Cyclization Reaction[74]

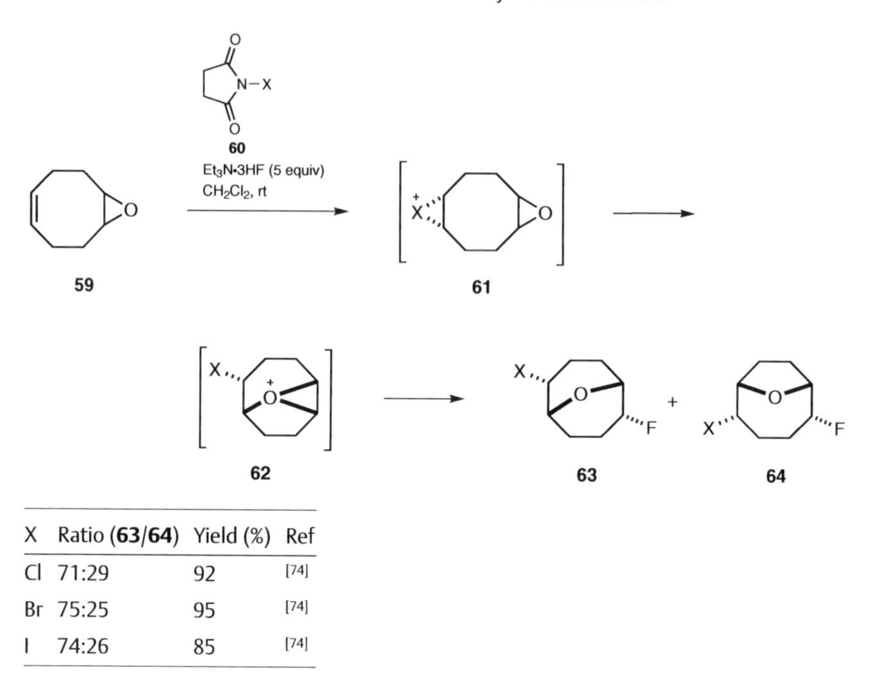

X	Ratio (**63**/**64**)	Yield (%)	Ref
Cl	71:29	92	[74]
Br	75:25	95	[74]
I	74:26	85	[74]

At around the same time, Thomas and co-workers also demonstrated that treatment of epoxyalkene **65** with molecular bromine leads to a mixture of ring-expanded ethereal dibromide regio- and diastereoisomers **69–71** via bromonium **66**, then oxonium ion intermediates **67** and **68** (Scheme 15).[78] The intermolecular reaction of cyclohexene and oxirane under the same conditions gives *trans*-1-bromo-2-(2-bromoethoxy)cyclohexane (**74**) via intermediates **72** and **73** (Scheme 15).[79] In both cases, the domino sequence starts with an electrophilic attack, followed by an attack from the epoxide oxygen to form an oxonium ion intermediate (e.g., **62**, **67**, **68**, and **73**), and is completed by a nucleophilic ring opening to give the respective products. These pioneering works by Haufe and Thomas were the basis for the later advancement in the biosynthesis of halogenated marine natural products and the design of multicomponent reactions covered in this section.

Scheme 15 Bromine-Assisted Epoxide Ring Expansion[78,79]

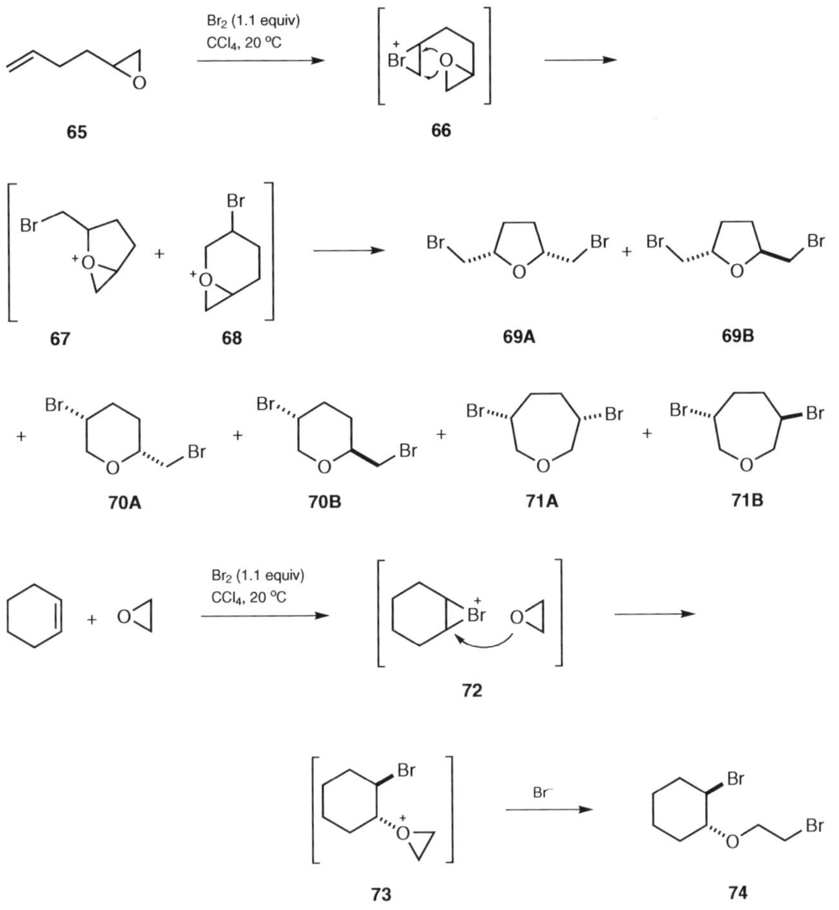

Br2 (1.1 equiv)
CCl4, 20 °C
65
66
67
68
69A
69B
70A
70B
71A
71B
Br2 (1.1 equiv)
CCl4, 20 °C
72
Br−
73
74

Marine red algae of *Laurencia* sp. have a general characteristic of producing C_{15}-acetogenic medium to macrocyclic ring ethers containing two bromine atoms and up to eight stereogenic centers. An enzymatic approach by Murai and co-workers affords laureatin (**77**) in 0.07% yield from straight-chain laurediol (**75**) via prelaureatin (**76**) by two successive electrophilic bromoetherification reactions (Scheme 16).[80,81] The reaction is carried out with sodium bromide, hydrogen peroxide, and bromoperoxidase as catalyst in a 1:100 mixture of dimethyl sulfoxide/phosphate buffer (pH 5.5) under argon at 5 °C for 24 hours. Although the yield of **77** is very low, laurediol (**75**) is widely regarded as the biosynthetic precursor to the *Laurencia* family of natural products.

for references see p 332

Scheme 16 Enzymatic Synthesis of Laureatin Using Bromoperoxidase[80,81]

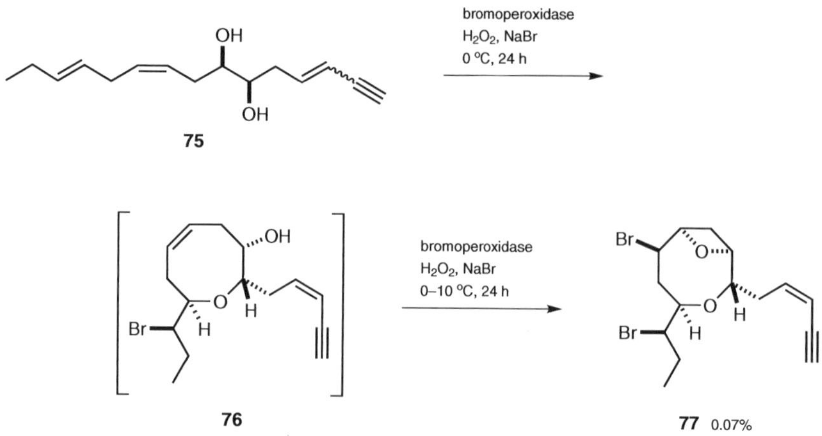

bromoperoxidase
H_2O_2, NaBr
0 °C, 24 h

75

bromoperoxidase
H_2O_2, NaBr
0–10 °C, 24 h

76

77 0.07%

Braddock and co-workers reasoned that macrocyclic obtusallenes, a subset of a family of natural products from *Laurencia* sp., could be assembled through a bromonium ion initiated transannular oxonium ion formation (Scheme 17).[82] Indeed, the obtusallene II model compound **78** is transformed to **79**, which contains the macrocyclic carbon framework of obtusallene VII, by employing *N*-bromosuccinimide as bromine electrophile, acetic acid as nucleophile, and tetramethylguanidine as catalyst.[83] The isolation of dibromobicyclo[5.5.1]tridecanyl acetate **80** and the relative stereochemistry assignment of the acetate group by X-ray crystallography suggests the presence of bromonium and oxonium ion events. By applying similar reaction conditions, linear, unsaturated epoxide C_{15}-precursor **81** can undergo direct cyclization in two successive bromonium ion initiated transannular oxonium ion formation events to give hexahydrolaureoxanyne (**82**; Scheme 17).[84] In this case, water functions both as a solvent and as an external nucleophile to capture the first oxonium ion, and subsequently participates as the second oxonium ion. The use of *N*-bromosuccinimide/water conditions is, however, not compatible with analogous C_{15}-precursors with a well-positioned intramolecular nucleophile such as a hydroxy group at the terminus.[85]

Scheme 17 Synthesis of Obtusallenes Using *N*-Bromosuccinimide[82–84]

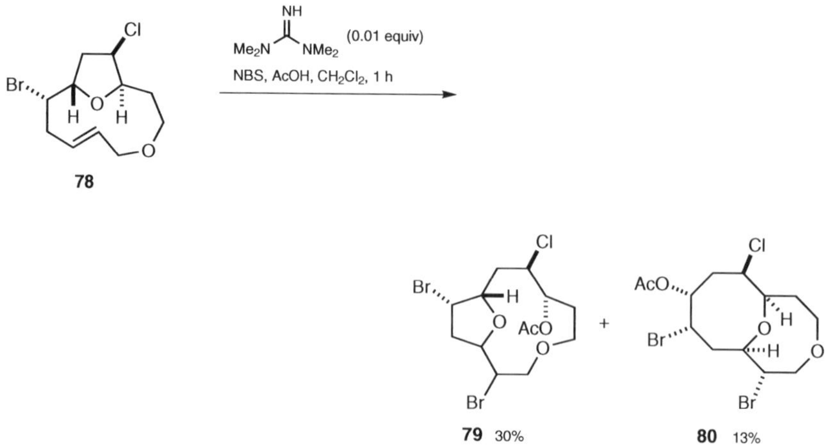

$Me_2N(C=NH)NMe_2$ (0.01 equiv)
NBS, AcOH, CH_2Cl_2, 1 h

78

79 30% + **80** 13%

NBS (2 equiv), H_2O
rt, dark, 18 h
2.5%

81

82

Epoxyalkene derivative **83**, with a well-positioned intramolecular nucleophile, undergoes bromonium ion initiated epoxide opening with different halogen sources and/or solvents. In the total synthesis of *ent*-dioxepandehydrothyrsiferol (**85**) and armatol A, Jamison and co-workers used a combination of the highly polar non-nucleophilic solvent 1,1,1,3,3,3-hexafluoropropan-2-ol, *N*-bromosuccinimide, and 4-Å molecular sieves to effect the bromonium ion initiated epoxide opening cascade of carbonate functionalized epoxyalkene **83** to give bromooxepanes **84** as a 1:1 mixture of epimers (Scheme 18).[86,87] When these reaction conditions were applied to a C_{15}-precursor of type **81** "armed" with a similar carbonate group, the reaction was not successful.[85] On the other hand, McDonald and co-workers successfully employed bis(2,4,6-collidine)bromonium perchlorate as the electrophilic bromine source in acetonitrile at −40 °C for the stereoselective synthesis of bromooxepanes from an analogous carbonate-functionalized epoxyalkene in its simplest form, containing an epoxide and alkene separated by two methylene groups.[88]

for references see p 332

Scheme 18 Synthesis of *ent*-Dioxepandehydrothyrsiferol Using *N*-Bromosuccinimide/1,1,1,3,3,3-Hexafluoropropan-2-ol/4-Å Molecular Sieves[86,87]

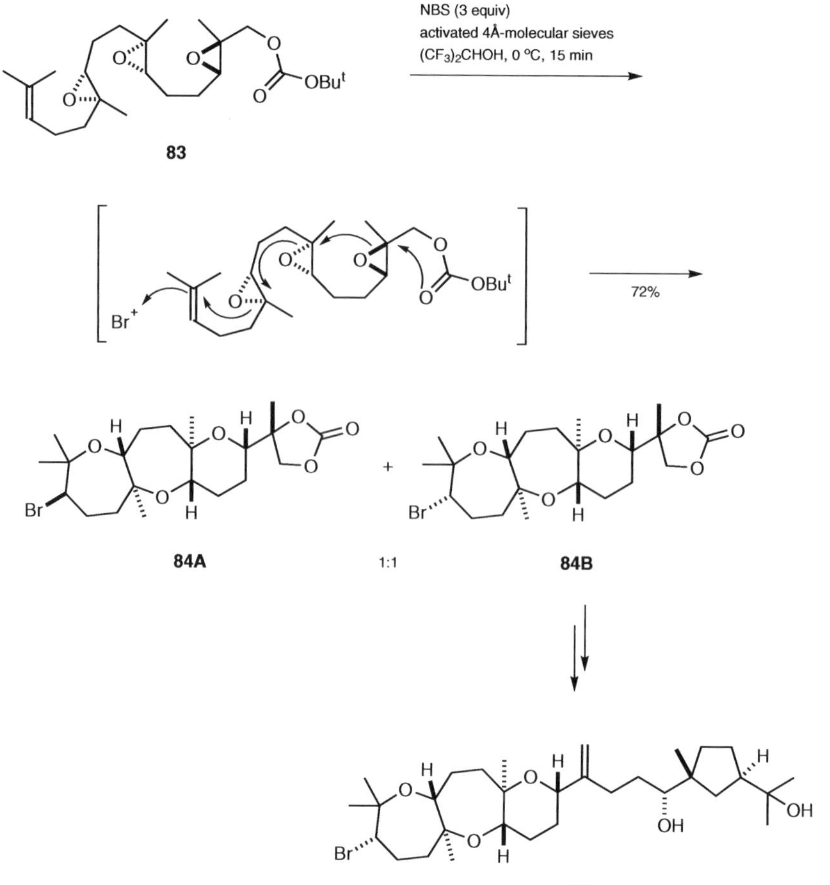

NBS (3 equiv)
activated 4Å-molecular sieves
$(CF_3)_2CHOH$, 0 °C, 15 min

83

72%

84A 1:1 84B

85

Similarly, Snyder and co-workers devised an elegant approach for the synthesis of the monocyclic core **89** of the bromo ethers produced by *Laurencia* sp., starting from linear, unsaturated precursor **86** (Scheme 19).[89,90] One of the key reasons for this successful bromonium ion **87** initiated event is the use of the highly reactive bromodiethylsulfonium bromopentachloroantimonate(V) (**164**) as brominating agent (see also Scheme 36, Section 2.3.1.4.2). In fact, bromodiethylsulfonium bromopentachloroantimonate(V) also effectively induces the intramolecular bromonium ion initiated diastereoselective epoxide-ring opening of a C_{15}-precursor of type **81** "armed" with a carbonate group to give the medium-ring ether.[85] Another key feature is the favorable formation of the five-membered oxonium ion **88** using tetrahydrofuran alkene **86** instead of epoxyalkene **81** as precursor. Bromodiethylsulfonium bromopentachloroantimonate(V) was effectively used by Snyder and co-workers to access the natural products laurefucin (Scheme 19) and (*E*)- and (*Z*)-pinnatifidenyne within the *Laurencia* family from tetrahydrofuran alkene **86** and derivatives thereof, respectively. The syntheses of these three natural products was achieved through the shortest sequences to date.[90]

Scheme 19 Concise Synthesis of Laurefucin Using Bromodiethylsulfonium Bromopentachloroantimonate(V)[89,90]

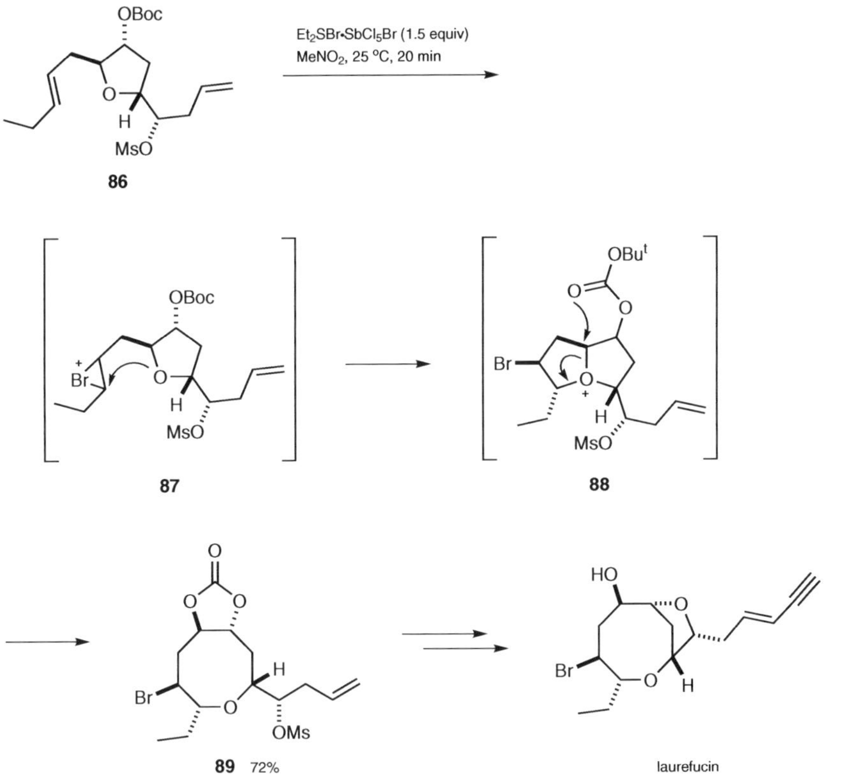

$Et_2SBr{\cdot}SbCl_5Br$ (1.5 equiv)
$MeNO_2$, 25 °C, 20 min

86

87

88

89 72%

laurefucin

Other approaches in the synthesis of natural products of the *Laurencia* family via electrophilic bromoetherification were less successful but still revealing.[91,92] A recent example by Howell and co-workers uses oxetane alcohol **90**, a structural motif previously proposed by Suzuki and co-workers as a suitable biosynthesis precursor,[93] in the synthesis of laureatin (**77**).[94] Although the medium-ring ether of the laureatin core was not formed, an unexpected *N*-bromosuccinimide-initiated rearrangement of **90** furnished epoxytetrahydrofuran **93**. The involvement of an initial bromination event is proposed to give **91**, followed by a highly strained four/five-membered oxonium ion **92**, which is then attacked by the alcohol group to furnish **93** (Scheme 20).

for references see p 332

Scheme 20 *N*-Bromosuccinimide-Initiated Oxetane Alcohol Rearrangement[94]

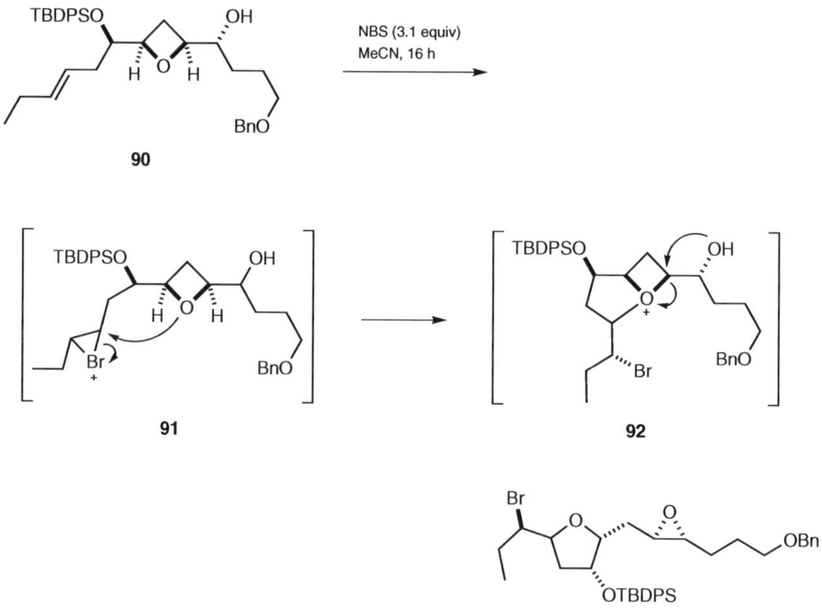

TBDPSO OH NBS (3.1 equiv) MeCN, 16 h BnO **90**

TBDPSO OH Br BnO **91**

TBDPSO OH Br BnO **92**

Br OTBDPS OBn **93** 51%

The catalyst-free multicomponent approach for the haloamidation reaction of alkenes (Section 2.3.1.1.3, Schemes 10 and 11) was further extended by Yeung and co-workers for the efficient synthesis of oxygenated compounds **98** by replacing the nitrile as solvent and nucleophilic partner with cyclic ethers **94** of varying size (Scheme 21).[95–98] In this electrophilic cascade, bromonium ion **96** is initially formed when styrene is treated with *N*-bromosuccinimide. Attack by a cyclic ether **94** forms oxonium ion intermediate **97**. Subsequent ring opening of **97** with a suitable nucleophilic partner **95**, such as a sulfonamide, carboxylic acid, or phenol, gives the product **98**. Bromo ether **98** (R^1 = 4-$O_2NC_6H_4SO_2$; X = N) can be further transformed into morpholine derivative **99**. Interestingly, when the nucleophilic solvent is replaced with a relatively nonpolar and non-nucleophilic solvent such as dichloromethane, a direct haloamination takes place efficiently without the need for any external *N*-bromosuccinimide activator.[99] The reaction proceeds smoothly with both cyclic and acyclic alkenes, furnishing a range of 1,2-bromoamine products.

Scheme 21 Multicomponent Aminoalkoxylation, Acyloxyetherification, and Phenoxyetherification[95–98]

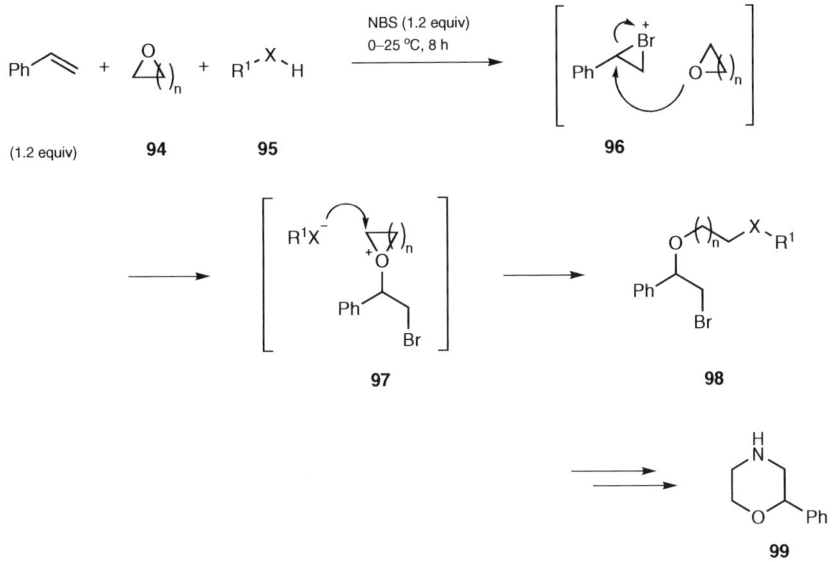

NBS (1.2 equiv)
0–25 °C, 8 h

(1.2 equiv) **94** **95** **96** **97** **98** **99**

R^1	X	n	Yield (%) of **98**	Ref
$4\text{-}O_2NC_6H_4SO_2$	NH	1	78	[95]
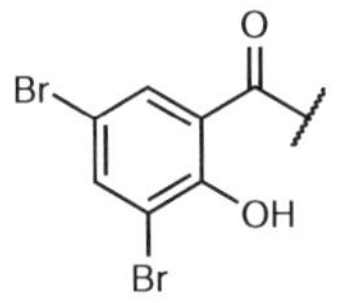	O	2	91	[96]
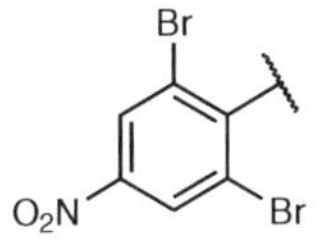	O	2	89	[98]

The 2-substituted morpholine products from Yeung's *N*-bromosuccinimide-initiated multicomponent reactions are useful precursors in the synthesis of bioactive molecules, as exemplified by the formal synthesis of the 2-substituted morpholine reboxetine (**100**) and the enantioselective synthesis of carnitine acetyltransferase inhibitor **101** (Scheme 22). The 2-substituted morpholine products were obtained on a gram-scale.[100]

for references see p 332

Scheme 22 Synthesis of a Reboxetine Intermediate and a Carnitine Acetyltransferase Inhibitor via *N*-Bromosuccinimide-Initiated Multicomponent Reaction[100]

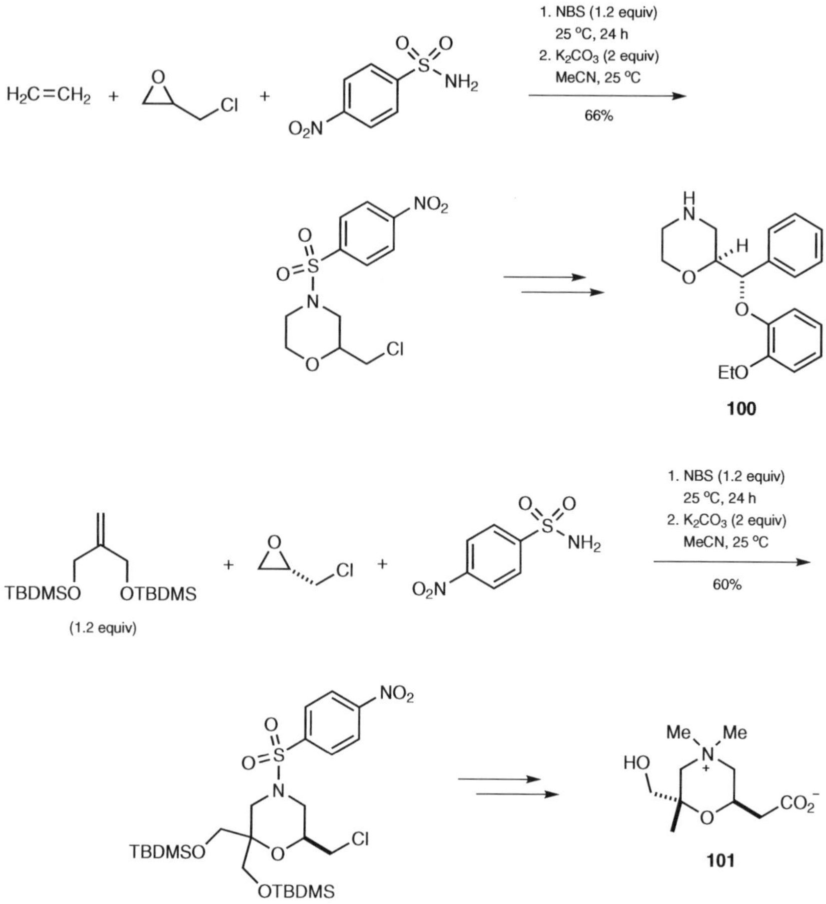

$H_2C{=}CH_2$ + (epichlorohydrin) + 4-nitrobenzenesulfonamide → 1. NBS (1.2 equiv), 25 °C, 24 h; 2. K_2CO_3 (2 equiv), MeCN, 25 °C; 66% → 4-(4-nitrophenylsulfonyl)-2-(chloromethyl)morpholine → → **100**

(1.2 equiv) TBDMSO/OTBDMS alkene + (epoxide) + 4-nitrobenzenesulfonamide → 1. NBS (1.2 equiv), 25 °C, 24 h; 2. K_2CO_3 (2 equiv), MeCN, 25 °C; 60% → sulfonyl morpholine (TBDMSO, OTBDMS, Cl) → → **101**

In correlation to the earlier discussion on Fujioka's domino bromoamination/oxidation approach to the synthesis of (−)-γ-lycorane (**58**; Scheme 13, Section 2.3.1.1.3), the same group has also made significant contributions to the domino intramolecular haloetherification with a variety of chiral ene acetals[101–103] and cyclohexadiene acetals[104,105] as substrates. In the former reaction, chiral ene acetals derived from diols and ene aldehydes are applicable for the asymmetric desymmetrization of *meso*-1,2-diols. The latter reaction was applied to the total synthesis of (−)-stenine[106] and the cyclohexenone core structure of scyphostatin.[107] The double domino intramolecular iodoetherification of chiral diene acetals of type **102** gives bicyclic diiodide **105** with four newly created stereocenters in a one-pot reaction (Scheme 23).[108] The domino sequence involves an iodoetherification reaction to form the oxonium ion **103**, which is then followed by a nucleophilic attack by water to furnish hemiacetal intermediate **104**. The second iodoetherification reaction on the other double bond gives octahydrofuro[2,3-*e*][1,4]dioxocin **105**. This methodology has been employed in the asymmetric total synthesis of (+)-rubrenolide.[109]

Scheme 23 Double Domino Iodoetherification Sequence toward the Synthesis of (+)-Rubrenolide[108]

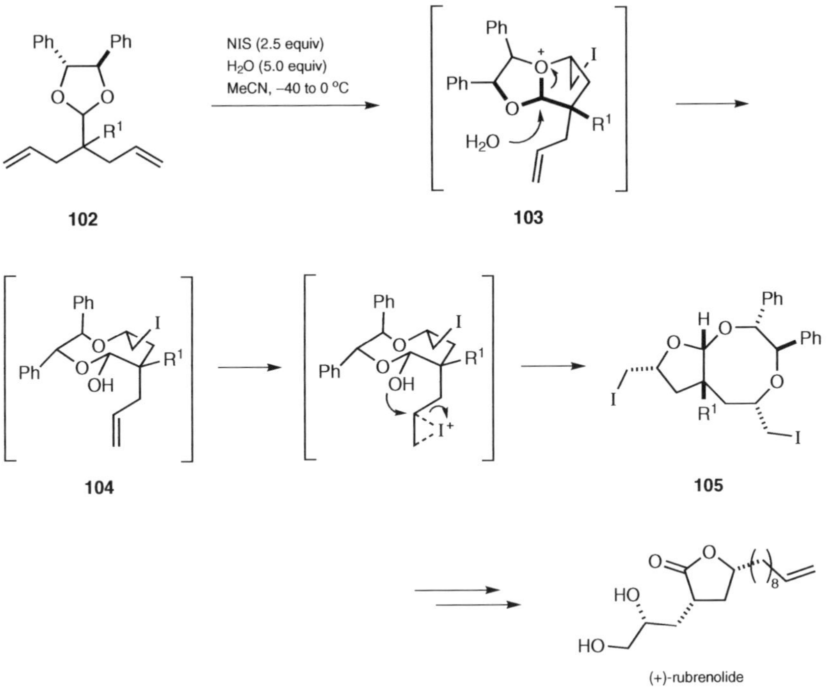

NIS (2.5 equiv)
H_2O (5.0 equiv)
MeCN, −40 to 0 °C

102 103 104 105 (+)-rubrenolide

R^1	dr	Yield (%) of **105**	Ref
H	3.5:1	80	[108]
OTBDMS	11:1	84	[108]

Mootoo and co-workers reported a domino iodoetherification reaction for the desymmetrization of alkene **106** as an efficient method for the synthesis of bis(tetrahydrofuran)-containing acetogenins (Scheme 24).[110] The domino sequence begins with the formation of oxonium cation **107** via iodoetherification on treatment with bis(2,4,6-collidine)iodonium perchlorate. Subsequent acetal cleavage of **107** gives an oxocarbenium intermediate, which is further cyclized to yield **108**.

Scheme 24 Domino Iodoetherification Sequence toward the Synthesis of the Asimicin Core[110]

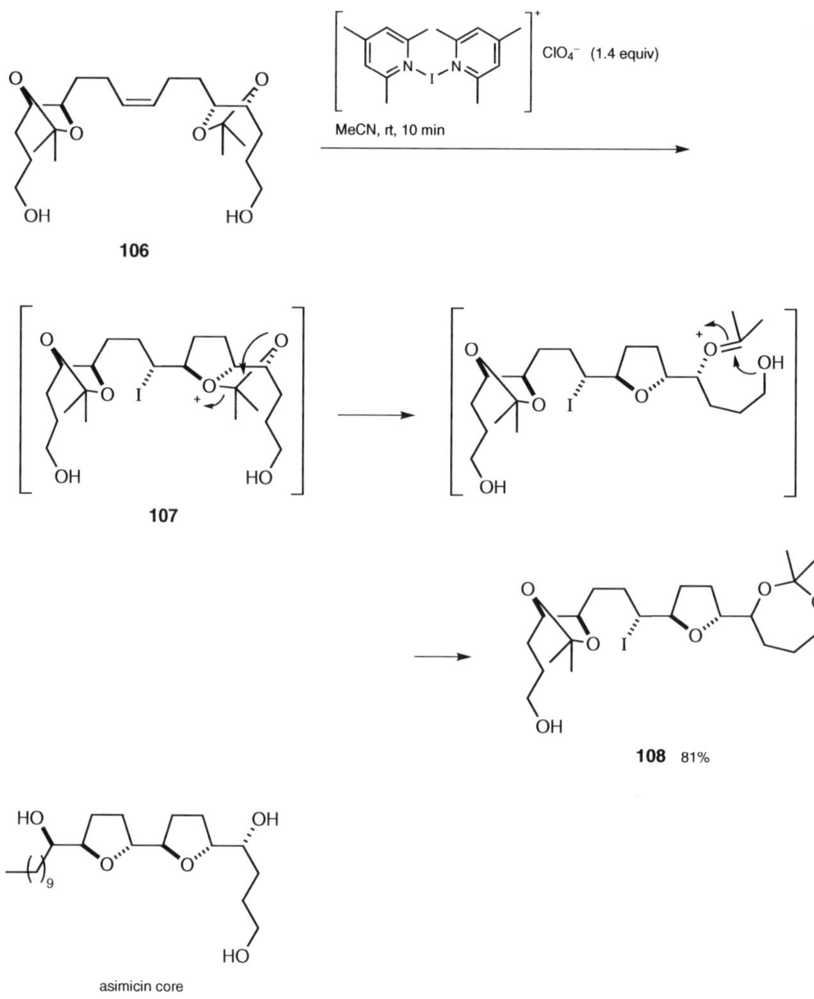

ClO_4^- (1.4 equiv)

MeCN, rt, 10 min

106

107

108 81%

asimicin core

N-[2-(2-Bromo-1-phenylethoxy)ethyl]-4-nitrobenzenesulfonamide (98, $R^1 = 4\text{-}O_2NC_6H_4SO_2$; X = N; n = 1); Typical Procedure:[95]

CAUTION: *Oxirane is extremely flammable, an eye, skin, and respiratory tract irritant, and a probable human carcinogen.*

A mixture of styrene (45 μL, 0.39 mmol), 4-nitrobenzenesulfonamide (**95**, $R^1 = 4\text{-}O_2NC_6H_4SO_2$; X = N; 65 mg, 0.32 mmol), NBS (69 mg, 0.39 mmol), and oxirane (**94**, n = 1; 100 mg, 2.3 mmol) in CH_2Cl_2 (2 mL) was stirred at 0 °C. The resultant soln was shielded from light and stirred at 25 °C for 8 h. The solvent was then removed under reduced pressure and the residue was purified by chromatography (silica gel, hexanes/EtOAc 3:1) to give the product as a clear oil; yield: 107 mg (78%). The reaction time and yield were not affected when the addition sequence of NBS, sulfonamide, cyclic ether, and alkene was changed.

2.3.1.3 Domino Carbonylation

The addition of an electrophile and a carbon-containing nucleophile across a C=C bond creates new carbon—electrophile and C—C bonds which can serve as a basis for addition of carbon atoms onto the molecule or an intramolecular cyclization reaction triggered by an electrophile. Critically, such domino sequences allow for the production of heterocycles in a single operation.

2.3.1.3.1 Transition-Metal-Initiated Events

Hydroxycarbonylation, hydroesterification, and hydroformylation reactions involve the addition of carbon monoxide and either water, an alcohol, or molecular hydrogen across C=C bonds. The resultant products include saturated carboxylic acids, esters, or aldehydes with an additional carbon atom incorporated (Scheme 25).[111–114] These atom-economical processes typically require transition-metal catalysts such as palladium-, ruthenium-, and rhodium-based catalysts for good chemo-, regio-, and stereoselectivities.[115–120] Two general mechanisms have been proposed which differ in the insertion of the alkene either into a metal—hydride bond or an (alkoxycarbonyl)—metal bond.[118,121] For this section, the presentation of the domino transformations will start from the product isolated from the hydroformylation reaction.

Scheme 25 General Representation of Hydroxycarbonylation, Hydroesterification, and Hydroformylation[111–114]

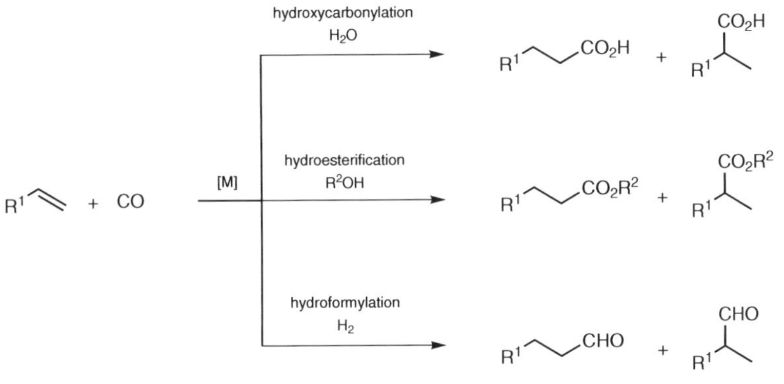

R^1CH=CH$_2$ + CO, [M]:
- hydroxycarbonylation, H_2O → R^1CH$_2$CH$_2$$CO_2H$ + R^1CH(CO_2H)CH$_3$
- hydroesterification, R^2OH → R^1CH$_2$CH$_2$$CO_2R^2$ + R^1CH(CO_2R^2)CH$_3$
- hydroformylation, H_2 → R^1CH$_2$CH$_2$CHO + R^1CH(CHO)CH$_3$

Breit and co-workers reported that a rhodium-catalyzed domino hydroformylation/Wittig reaction with methallyl alcohol derivative **109** as substrate in the presence of a mono- or disubstituted stabilized Wittig ylide (**110** or **113**, respectively) gives saturated or α,β-unsaturated carbonyl derivatives **112** and **114**, respectively (Scheme 26).[122,123] The introduction of the catalyst-directing *ortho*-(diphenylphosphinyl)benzoyl group in **109** is crucial for stereocontrol in the hydroformylation reaction.[124] The sequence begins with hydroformylation to give an aldehyde, which then undergoes a Wittig reaction to give the α,β-unsaturated carbonyl intermediate **111**. Under these reaction conditions, only the product derived from the monosubstituted ylide undergoes further rhodium-catalyzed hydrogenation to give the saturated carbonyl derivative **112**, whereas in the case of the disubstituted ylide, the reaction stops at α,β-unsaturated carbonyl derivative **114**. Helmchen and co-workers showed that the isolation of α,β-unsaturated carbonyls from monosubstituted ylides is achieved at a lower temperature and with shorter reaction times by employing the ligand **115** (BIPHEPHOS; Scheme 27).[125] Breit also showed that by replacing the ylide with dimethyl malonate and including an addition of base, a new domino hydroformylation/Knoevenagel/hydrogenation reaction sequence gives malonate derivatives in high yields with good regio- and stereocontrol.[126]

Scheme 26 Rhodium-Catalyzed Domino Hydroformylation/Wittig Alkenation/Hydrogenation Reaction[122,123]

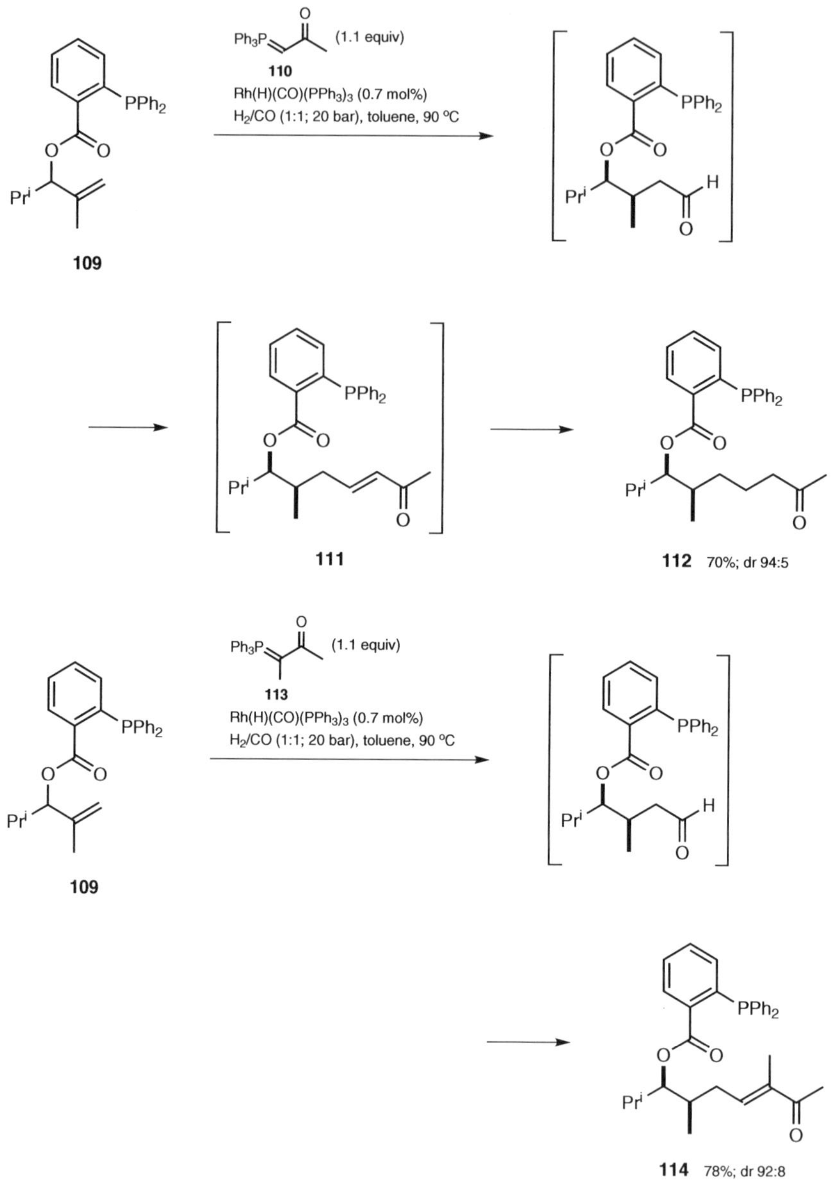

Hoffmann and co-workers reported a rhodium-catalyzed domino hydroformylation/allylboration/hydroformylation reaction of amidoallylboronate **116** in the presence of ligand **115** to give the *trans*-fused perhydropyrano[3,2-*b*]pyridine derivative **119** (Scheme 27).[127] The reaction sequence begins with a regioselective hydroformylation to give the linear aldehyde. Subsequent allylboration gives piperidine **117** with three stereogenic centers and a terminal C=C bond. Piperidine **117** then undergoes a second hydroformylation to give intermediate **118**, with a final ring-closure reaction yielding lactol **119** as a 1:1 mixture of anomers.

Scheme 27 Rhodium-Catalyzed Domino Hydroformylation/Allylboration/Hydroformylation Reaction[127]

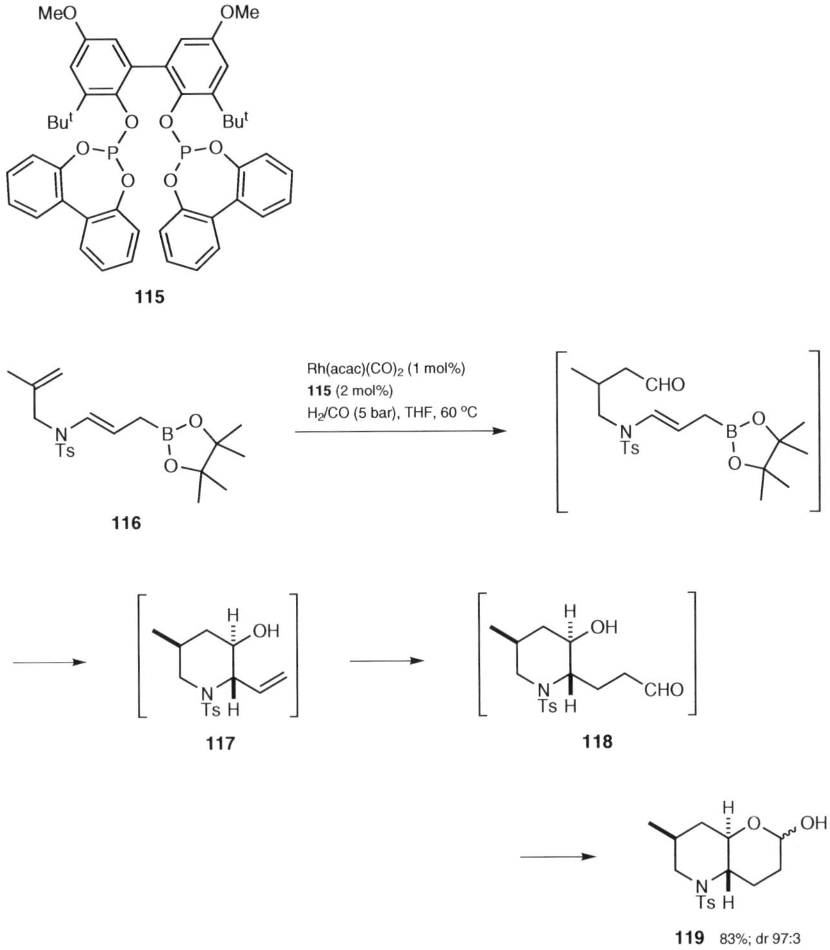

MeO OMe Bu^t O O Bu^t O–P P–O O O

115

Rh(acac)(CO)$_2$ (1 mol%)
115 (2 mol%)
H_2/CO (5 bar), THF, 60 °C

N Ts B O O **116**

CHO N Ts B O O

H OH N Ts H **117**

H OH N Ts H CHO **118**

H O OH N Ts H

119 83%; dr 97:3

A similar reaction using the same rhodium/**115** catalyst system was employed by Mann and co-workers in the domino hydroformylation/nucleophilic addition reaction of but-3-enamide derivatives (e.g., **120**) with a tethered nucleophile (Scheme 28). Critically, a variety of aza-heterocycles could be synthesized. The domino sequence begins with the formation of aldehyde **121** from a regioselective hydroformylation of the alkenic amide substrate, followed by a nucleophilic attack by the amide to form *N*-acyliminium ion **122**. The reactive *N*-acyliminium ion is captured by the tethered nucleophile to give nitrogen heterocycle **123**.[128] Taddei and co-workers showed that nitrogen heterocycle **123** can also be synthesized in 79% yield and with high diastereoselectivity under microwave irradiation for 30 minutes.[129] The rhodium/**115** catalyst system was also successfully applied by Helmchen and co-workers to the domino hydroformylation/reductive amination reaction of a homoallylamine to synthesize enantiopure (*S*)-nicotine.[130]

for references see p 332

Scheme 28 Rhodium-Catalyzed Domino Hydroformylation/Reductive Amination Reaction[128]

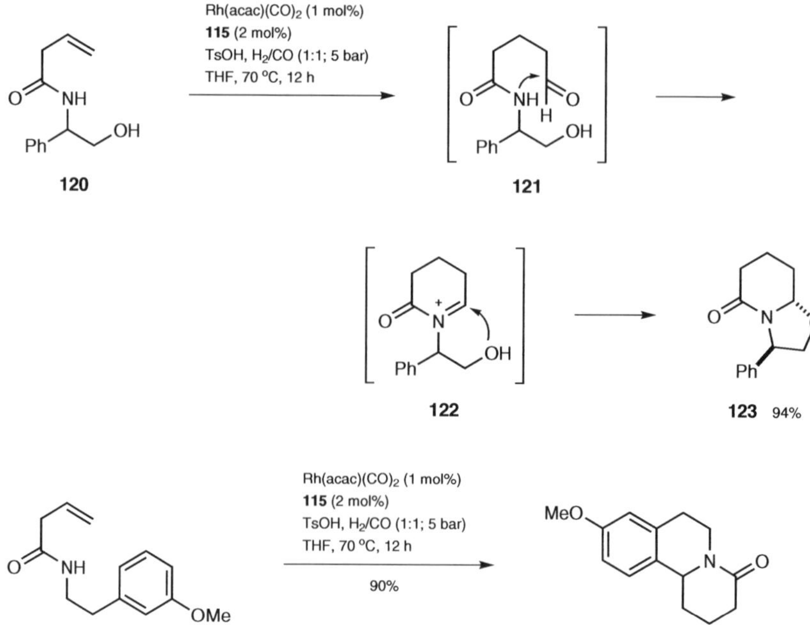

Rh(acac)(CO)$_2$ (1 mol%)
115 (2 mol%)
TsOH, H_2/CO (1:1; 5 bar)
THF, 70 °C, 12 h

120 → [**121**] → [**122**] → **123** 94%

Rh(acac)(CO)$_2$ (1 mol%)
115 (2 mol%)
TsOH, H_2/CO (1:1; 5 bar)
THF, 70 °C, 12 h
90%

2.3.1.3.2 Halogen-Initiated Events

A diastereoselective iodine-initiated carbocyclization of polyenes employing bis(pyridine)iodonium(I) tetrafluoroborate as the iodinating agent together with tetrafluoroboric acid was reported by Barluenga and co-workers.[131] The scope of this reaction also includes acetonide terpene derivatives[132] and an interesting iodonium-initiated polyene carbocyclization cascade (Scheme 29).[133] In this domino sequence, polyene **124** reacts with bis(pyridine)iodonium(I) tetrafluoroborate to form an iodonium intermediate **125**, which subsequently rearranges to form oxonium intermediate **126**. A cascade of C—C bond formations via ene and arylation events gives tricyclic product **127**. A faster rate in the formation of **127** is observed when the polyene isomer **128** is used as the substrate under otherwise identical conditions.

Scheme 29 Iodonium-Initiated Rearrangement/Carbocyclization Reaction Using Bis(pyridine)iodonium(I) Tetrafluoroborate/Tetrafluoroboric Acid[133]

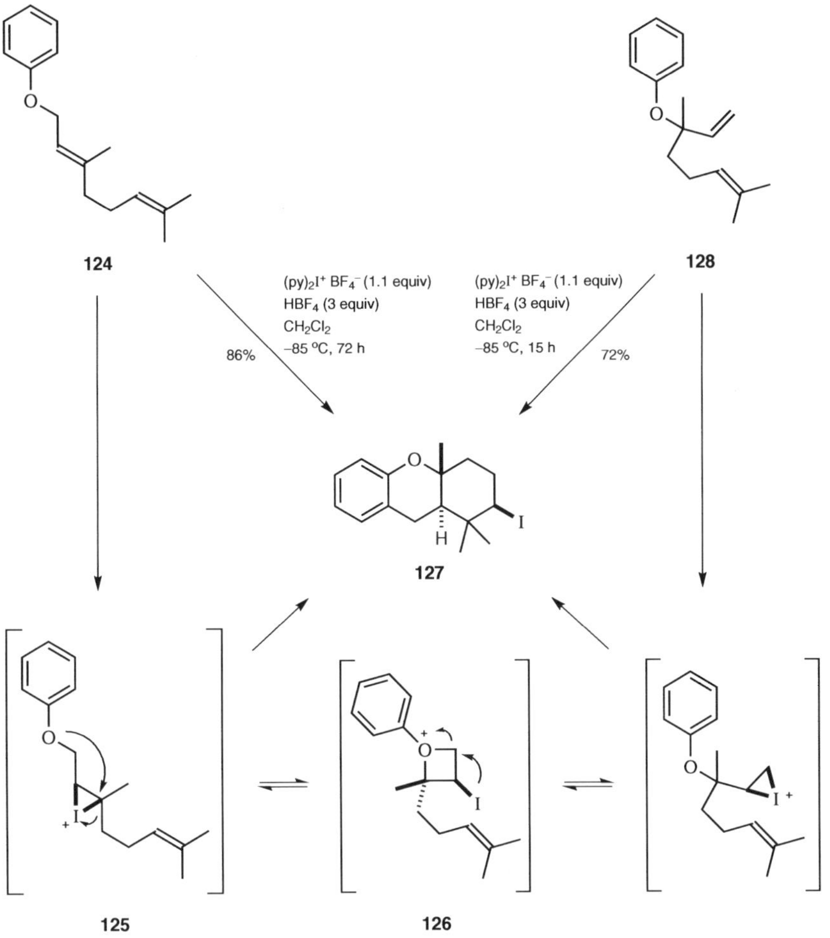

124 → 127: $(py)_2I^+ BF_4^-$ (1.1 equiv), HBF_4 (3 equiv), CH_2Cl_2, −85 °C, 72 h; 86%

128 → 127: $(py)_2I^+ BF_4^-$ (1.1 equiv), HBF_4 (3 equiv), CH_2Cl_2, −85 °C, 15 h; 72%

124 → 125 ⇌ 126 ⇌ [128-derived intermediate] → 127

(2*R,4a*R**,9a*R**)-2-Iodo-4a-methyl-2,3,4,4a,9,9a-hexahydro-1*H*-xanthene (127); Typical Procedure:**[133]

CAUTION: *Tetrafluoroboric acid is extremely destructive to the skin, eyes, and respiratory tract.*

Bis(pyridine)iodonium(I) tetrafluoroborate (372 mg, 1 mmol, 1 equiv) was stirred in CH_2Cl_2 (10 mL) at rt under N_2 for 5 min until a homogeneous soln was obtained. The soln was cooled at −85 °C and a 54% soln of HBF_4 (3 equiv) in Et_2O was added. The mixture was stirred for 10 min, and the allyl ether **124** or **128** (1 mmol, 1 equiv) was added. The reaction was monitored by TLC and, after completion, the mixture was poured onto crushed ice and H_2O and 5% aq $Na_2S_2O_3$ (20 mL) was added. The mixture was extracted with CH_2Cl_2 (3 × 20 mL). The combined organic layers were washed with H_2O and dried (Na_2SO_4). The solvents were removed under reduced pressure, and the residue was purified by column chromatography [silica gel, hexanes (to remove traces of starting material) then hexanes/EtOAc 100:1]; yield: 72–86%.

for references see p 332

2.3.1.4 Domino Polyene Cyclization

Polyene cyclization is defined as a cascade of C—C bond formations in substrates that contain strategically positioned C=C bonds. Electrophilic addition to C=C bonds, commonly occurring at the terminal end, is one of the methods to initiate this process. For example, the biogenetic conversion of the open-chain polyene squalene (**129**) involves an initial protonation step to form a chair-like transition state **130** which then undergoes a cascade of C—C bond formations to give carbocation **131** with *trans*-fused rings (Scheme 30).[134–139] Hop-22(29)-ene (**132**) and hopan-22-ol (**133**), which both contain multiple newly created stereogenic centers, are formed by proton elimination (path a) and hydrolysis (path b) of **131**.[138,140]

Scheme 30 Hypothetical Representation of Biogenetic Conversion of Squalene into Hopene[134–140]

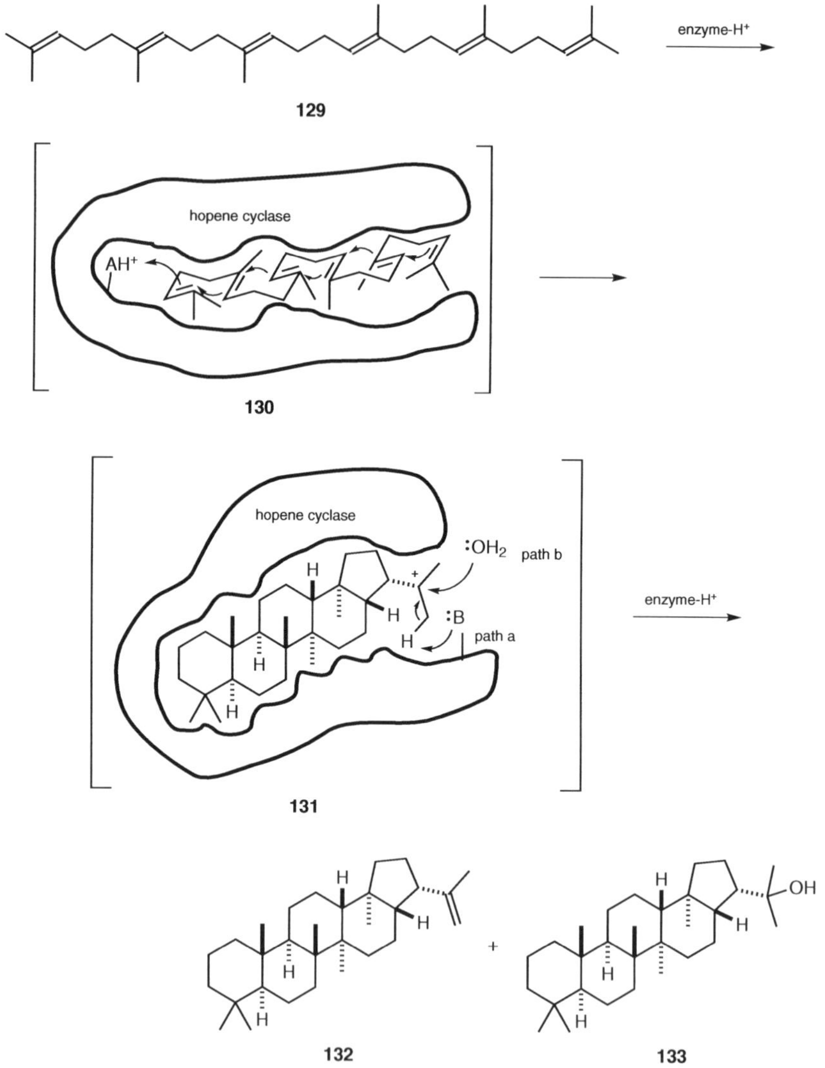

enzyme-H⁺

129

hopene cyclase

AH⁺

130

hopene cyclase

:OH₂ path b

:B path a

enzyme-H⁺

131

132 + 133

Early reports by Stork and Eschenmoser on the stereochemistry of polyene cyclization[141,142] triggered interest in research into the nonenzymatic biomimetic approach for the rapid assembly of polycyclic fused-ring systems from linear polyenes. This section begins with transition-metal-initiated events (Section 2.3.1.4.1), followed by recent efforts on halogen-initiated events (Section 2.3.1.4.2) and comparatively less-explored chalcogen-initiated events (Section 2.3.1.4.3). Interested readers on the topic of Brønsted acid (H^+) initiated events can refer to Section 1.1. These domino sequences, although different in their respective initiators, are commonly terminated with heteroatoms.

2.3.1.4.1 Transition-Metal-Initiated Events

An early report by Nishizawa and co-workers suggested that a mercury(II) reagent can serve as an effective electrophilic initiator in the polyene cyclization of various isoprenoid polyenes to give mercury-functionalized polycyclic rings.[143] The substrate scope includes dienol **134** and the product distribution of the cascade reaction of this substrate can be changed by additives (Scheme 31).[144] It is proposed that the cascade sequence first involves an electrophilic addition of mercury(II) cation followed by an equilibration of isomers where intermediate **135** leads to the formation of mercury bicycle **137** and mercury monocycle **138** whereas intermediate **136** leads to the formation of the monocyclic mercury product **139**. This methodology has been applied to the synthesis of polycyclic terpenoids such as racemic drimenol,[143] racemic driman-8,11-diol,[143] (+)-karatavic acid,[145] (+)-α,γ-onoceradienedione,[146] and racemic baiyunol.[147] Gopalan and co-workers employed these mercury(II) reagents for the cyclization of polyenes with carbonyl oxygen as the terminating nucleophile to give fused tricyclic products.[148]

Scheme 31 Mercury(II) Reagents for Achiral Polyene Cyclization[144]

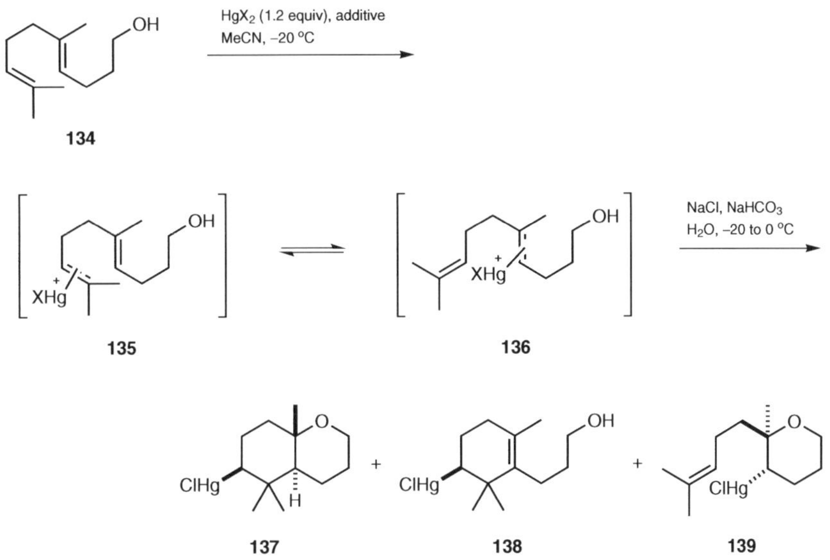

HgX_2 (1.2 equiv), additive
MeCN, –20 °C

134

135

136

NaCl, $NaHCO_3$
H_2O, –20 to 0 °C

137 + 138 + 139

X	Additive (Equiv)	Yield (%)			Ref
		137	**138**	**139**	
OAc	–	–	–	95	[144]
OAc	TfOH (1.2)	42	30	27	[144]
OTf	–	20	60	–	[144]
OTf	1,1,3,3-tetramethylurea (1.2)	47	40	–	[144]

for references see p 332

In 2004, Gagné and co-workers reported a new electrophilic palladium(II)- and platinum(II)-initiated cyclization of diene **143** using a phenol moiety as the terminator. For example, with platinum catalyst **140**, the reaction gives the platinum-containing tricyclic fused ring system **144** in high yield and diastereoselectivity (Scheme 32).[149,150] The redesigned platinum(II) catalyst **141** with an open *cis* site achieves reaction turnover by protodemetalation through β-hydrogen elimination and regeneration of platinum(II) by trityl-assisted oxidation (Scheme 32).[151] Asymmetric induction is obtained by using platinum catalyst **142** with the chiral bisphosphine ligand (*S*)-(+)-4,12-bis[bis(3,5-dimethylphenyl)phosphino][2.2]-paracyclophane, leading to good yields and enantioselectivities of the product **145** (Scheme 32).[152] The substrate scope also includes polyenes with a less-reactive terminating group such as an alkene.[153] The use of different reagents such as xenon difluoride,[154] molecular oxygen,[155] and proton[156,157] to functionalize the Pd—C bond gives fluorinated, oxygenated, and saturated polycyclic ring systems, respectively.

Scheme 32 Asymmetric Platinum(II)-Catalyzed Polyene Cyclization[149–152]

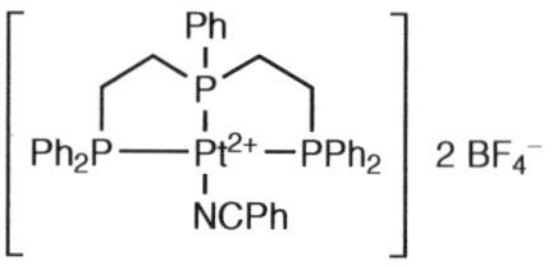

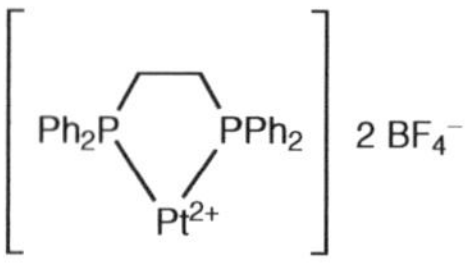

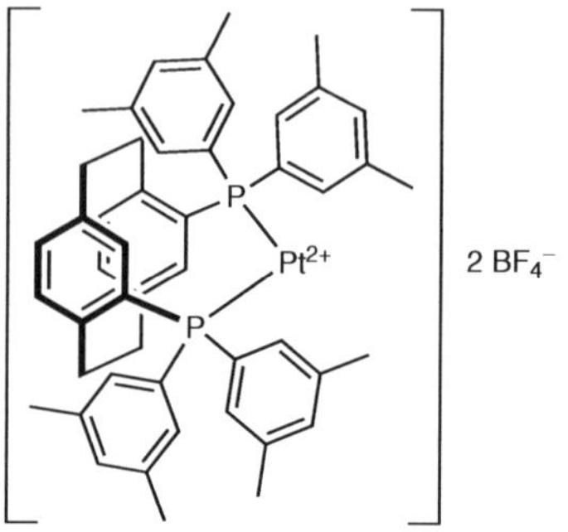

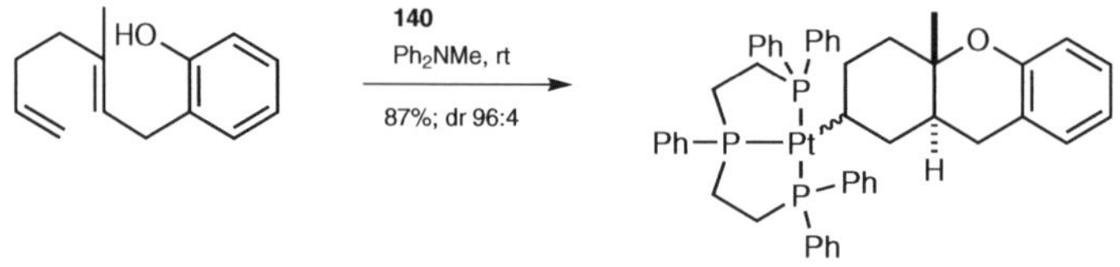

Catalyst	Yield (%)	Ref
141	73	[151]
142	73[a]	[152]

[a] 75% ee.

2.3.1.4.2 Halogen-Initiated Events

Nonenzymatic halonium-initiated polyene cyclization reactions for the synthesis of halogen-containing polycyclic natural products are challenging in several respects. For example, in the case of (*E*)-4,8-dimethylnona-3,7-dienoic acid (**146**), the challenges faced are (1) competitive halo-initiated (Br^+) and undesired proton-initiated (H^+) cyclization to give the desired bromobicycle **149** and the cyclized product **150**, respectively;[158] (2) regioselective halogenation of only the "terminal" and not the internal C=C bond in **146**;[159] (3) competitive intramolecular C—C bond formation and hydrolysis of bromonium **147** (Scheme 33, path a vs b) to give intermediate **148** and bromohydrin **151**, respectively;[160,161] and (4) competitive intramolecular ring-closing reaction to give the desired product **149** (Scheme 33, path c) and α-, β-, or γ-deprotonation (Scheme 33, path d, e, or f) to give the bromomonocycles **152–154**.[159,162–170] In fact, the desired halobicycle **149** is obtained in low yield with no diastereoselectivity.[159–172] Therefore, it is important to suppress the formation of **150–154**. Kato and co-workers reported that the use of 2,4,4,6-tetrabromocyclohexa-2,5-dienone in nitromethane instead of *N*-bromosuccinimide or bromine in aqueous acetone as an alternative brominating agent suppresses the formation of bromohydrin **151**. However, the desired product **149** is obtained in a relatively low yield of 15%.[171]

Scheme 33 General Challenges Faced in Halo-Initiated Polyene Cyclization Reactions[158–172]

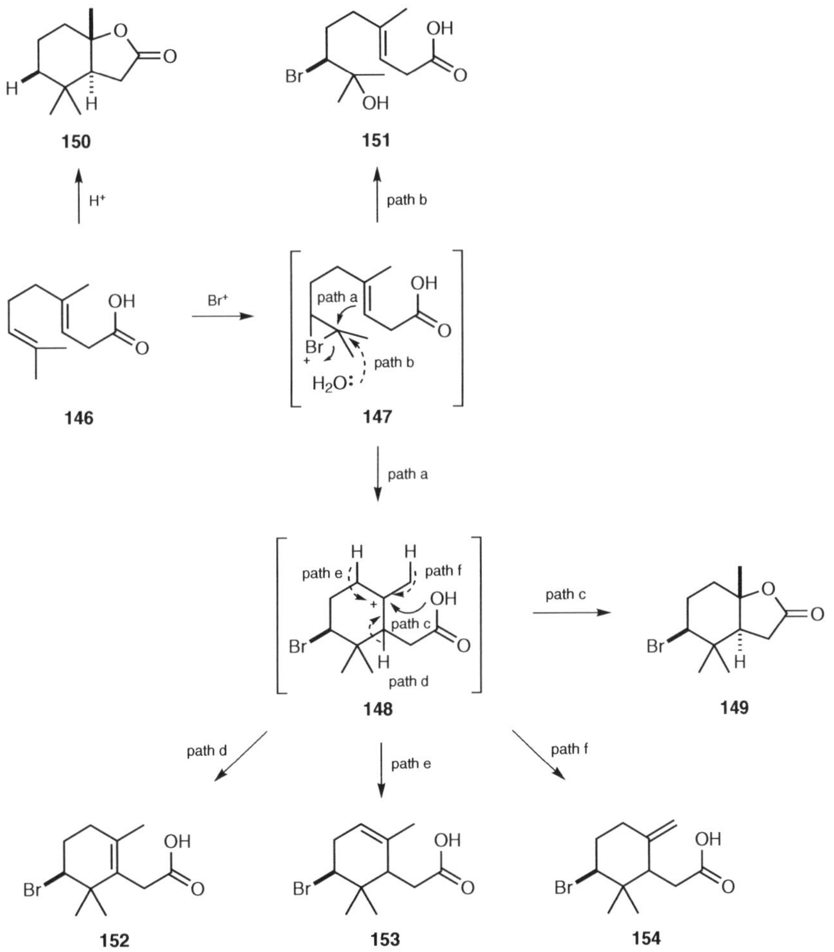

Butler and co-workers reported the first vanadium bromoperoxidase catalyzed, bromonium ion initiated cyclization of terpene **155** in aqueous solution to give snyderols **156** and **157** (Scheme 34).[173] The hydrophobic nature of the active site in vanadium bromoperoxidase that binds to the terpene suppresses the formation of the corresponding bromohydrin in the presence of water. The results are consistent with Kato's biomimetic cyclization in nitromethane as a non-nucleophilic solvent.[171]

Scheme 34 Vanadium Bromoperoxidase Catalyzed Cyclization of a Terpene[173]

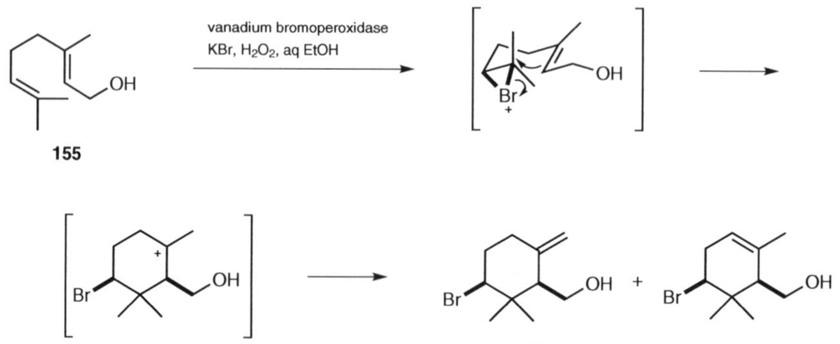

vanadium bromoperoxidase
KBr, H_2O_2, aq EtOH

155

156

157

Ishihara and co-workers reported the first example of a highly enantioselective iodocyclization of (*E*)-1-(4,8-dimethylnona-3,7-dien-1-yl)-4-methylbenzene (**158**) mediated by a stoichiometric amount of chiral phosphoramidite **159** (Scheme 35).[174] The iodonium ion initiated polyene cyclization sequence begins by activation of *N*-iodosuccinimide with phosphoramidite **159** through an ion-pair interaction followed by a regioselective delivery of the electrophilic iodine to the terminal alkene to form intermediate **160**. This triggers a cascade of C—C bond formations to give the single diastereomer **161** and the partially cyclized products **162** and **163** [**161**/(**162** + **163**) = 3:2], which are transformed to tricycle **161** on treatment with chlorosulfuric acid. However, when *N*-iodosuccinimide is replaced with *N*-bromosuccinimide, the brominated polyene products are generated in lower yield with significant side-product formation.

Scheme 35 Asymmetric Iodonium-Initiated Polyene Cyclization of Polyprenoids[174]

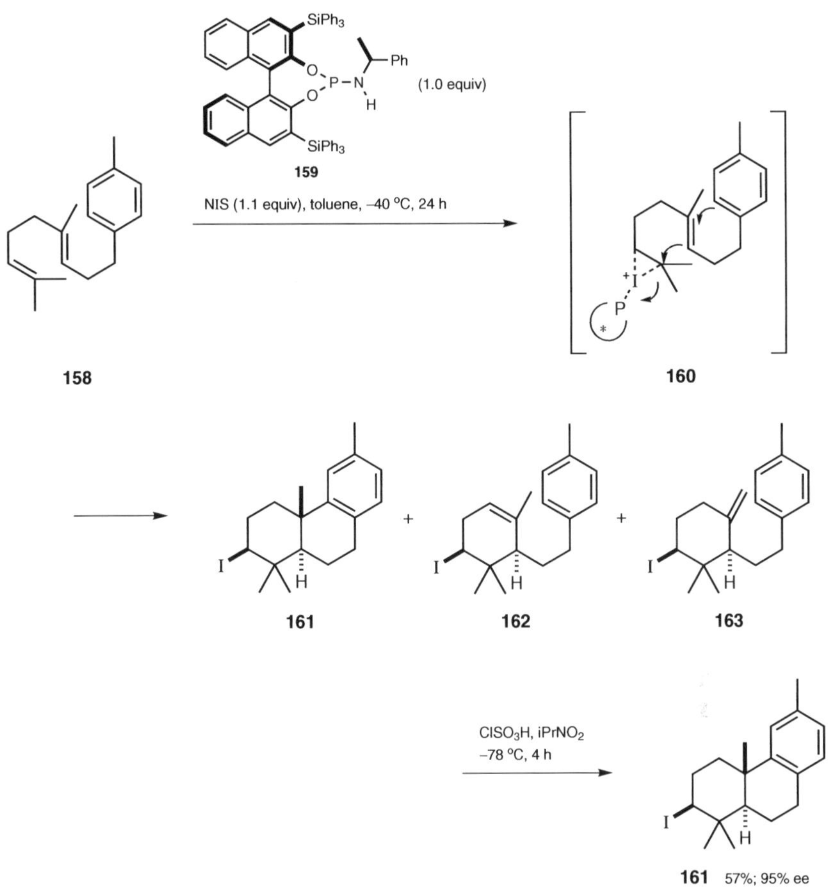

SiPh3
Ph
(1.0 equiv)
SiPh3
159
NIS (1.1 equiv), toluene, −40 °C, 24 h
158
160
161
162
163
ClSO3H, iPrNO2
−78 °C, 4 h
161 57%; 95% ee

A highly selective and catalytic bromonium ion initiated polyene cyclization of homogeranylarenes employs a nucleophilic phosphite–urea cooperative catalyst.[175] The urea moiety and the bulky aryl groups in the catalyst are important to suppress byproduct formation and catalyst decomposition. This method was successfully applied to the bromonium ion initiated polyene cyclization of 2,4-dibromo-6-geranylphenol. In a related phosphoramidite-catalyzed bromonium-initiated polyene cyclization by McErlean and coworkers, the synthesis of the furan-containing snyderane (+)-luzofuran uses a catalytic amount of a trichlorophenyltriazole phosphoramidite with *N*-bromosuccinimide and gives a mixture of (+)-luzofuran and *epi*-luzofuran.[176]

A related polyene cyclization by Snyder and co-workers uses bromodiethylsulfonium bromopentachloroantimonate(V) (**164**; see also Scheme 19, Section 2.3.1.2.1) as the electrophilic bromine source (Scheme 36).[177,178] This reagent initiates bromonium ion cation-π cyclization in various electron-rich and electron-deficient systems (e.g., **167**), shows far superior performance to other common brominating agents such as *N*-bromosuccinimide and 2,4,4,6-tetrabromocyclohexa-2,5-dienone, and has been applied in the total synthesis of the bromine-containing polycyclic natural products peyssonol A and peyssonoic acid A, and a formal synthesis of aplysin-20 (**171**) via cyanides **168–170** (Scheme 36).[178] Iodonium- and chloronium-initiated cation-π cyclizations proceed smoothly with various substrates using the iodine and chlorine variants of bromodiethylsulfonium bromopentachloroantimonate(V) (**164**), **165** and chlorodiethylsulfonium hexachloroantimonate(V) (**166**), respectively (Scheme 36). Reagent **165** has been applied to the total syntheses of io-

for references see p 332

dine-containing loliolide and stemodin as well as non-halogenated targets such as K-76. An asymmetric polyene cyclization uses stoichiometric amounts of mercury(II) trifluoromethanesulfonate with a chiral bis(dihydrooxazole) ligand as the promoter, and generally gives good enantioselectivities.[179]

Scheme 36 Halonium-Initiated Polyene Cyclization Using Bromodiethylsulfonium Bromopentachloroantimonate(V) and Other Hexachloroantimonate(V) Reagents[177,178]

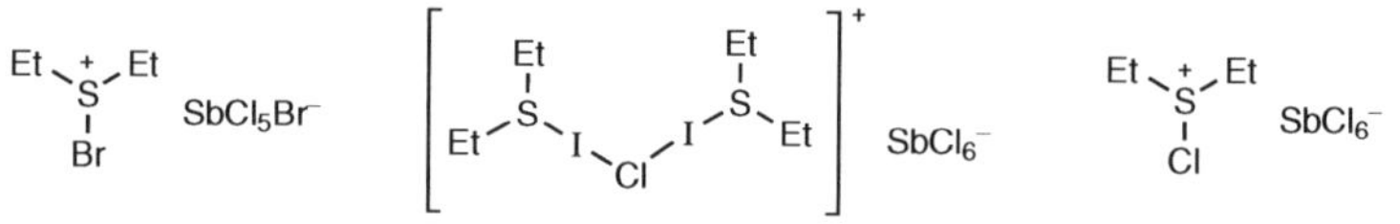

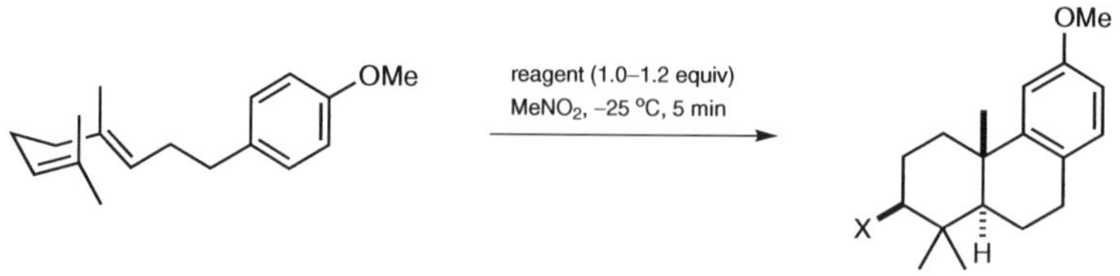

Reagent	X	Yield (%)	Ref
164	Br	76	[177]
165	I	90	[177]

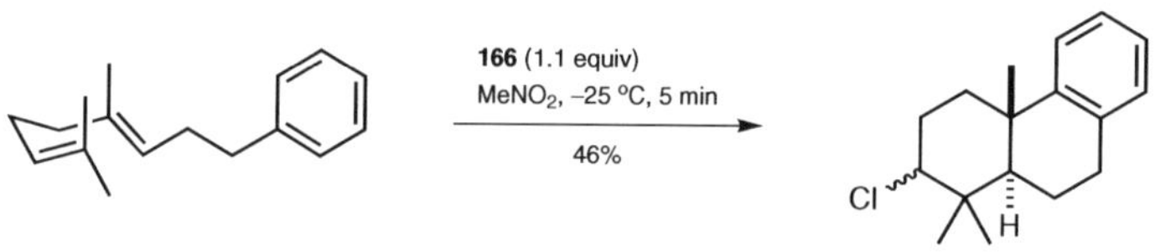

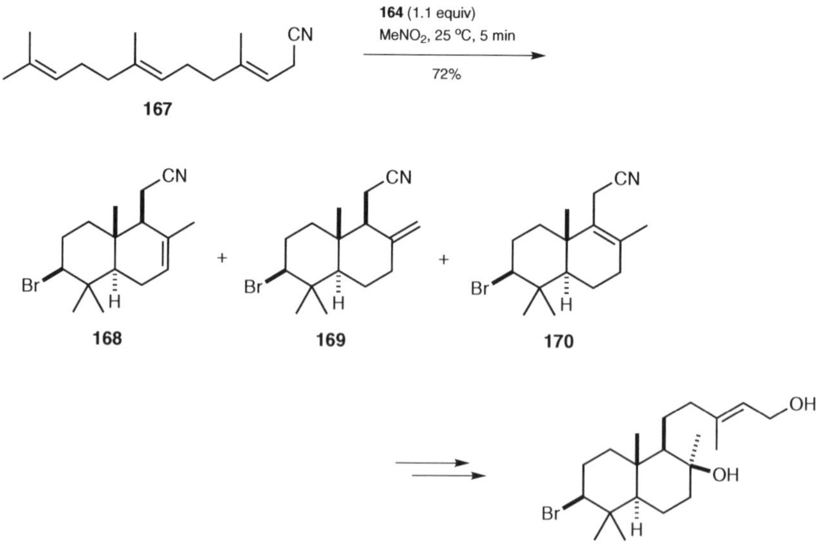

Whereas nonenzymatic bromonium-initiated polyene cyclization reactions (Scheme 33) suffered from low yield in the first reported work and were not asymmetric, seminal works by Ishihara with the chiral phosphoramidite **159**/*N*-iodosuccinimide combination (Scheme 35) and Snyder with bromodiethylsulfonium bromopentachloroantimonate(V) (**164**) as the reactive bromonium reagent (Scheme 36) demonstrated improved yield. However, despite the continued isolation of naturally occurring brominated polycyclic compounds in enantiopure form, no asymmetric electrophilic addition of bromonium ion to unactivated alkenes has been developed to date to initiate polyene cyclization. In a different approach, Braddock and co-workers demonstrated than an enantiopure bromonium ion generated from an enantiopure bromohydrin undergoes polyene cyclization with high yield and enantioselectivity.[180–182]

Bromodiethylsulfonium Bromopentachloroantimonate(V) (164); **Typical Procedure:**[178]

CAUTION: *Bromine is a severe irritant of the eyes, mucous membranes, lungs, and skin. Liquid bromine causes severe and painful burns on contact with eyes and skin.*

Et_2S (2.97 mL, 27.5 mmol, 1.1 equiv) and a 1.0 M soln of $SbCl_5$ in CH_2Cl_2 (30.0 mL, 30.0 mmol, 1.2 equiv) were added slowly and sequentially to a soln of Br_2 (1.28 mL, 25.0 mmol, 1.0 equiv) in 1,2-dichloroethane (60 mL) at −30 °C. The dark-red heterogeneous mixture was stirred for 20 min at −30 °C, and then warmed slowly using a water bath until the soln became homogeneous (~30 °C). At this time, the reaction flask was allowed to cool slowly to 0 °C (4 h) and then −20 °C (12 h), and large orange plates crystallized from the soln. The solvent was decanted and the crystals were rinsed with cold CH_2Cl_2 (2 × 5 mL) and dried under reduced pressure; yield: 11.9 g (87%).

Acetonitriles 168–170; **Typical Procedure:**[178]

CAUTION: *Nitromethane is flammable, a shock- and heat-sensitive explosive, and an eye, skin, and respiratory tract irritant.*

A soln of bromodiethylsulfonium bromopentachloroantimonate(V) (**164**; 60 mg, 0.11 mmol, 1.1 equiv) in $MeNO_2$ (0.5 mL) was added quickly via syringe into a soln of (3*E*,7*E*)-4,8,12-trimethyltrideca-3,7,11-trienenitrile (**167**; 23 mg, 0.10 mmol, 1.0 equiv) in $MeNO_2$ (1.5 mL) at 0 °C. After the mixture had been stirred at 25 °C for 5 min, the reaction was quenched by sequential addition of 5% aq Na_2SO_3 (5 mL) and sat. aq $NaHCO_3$ (5 mL). The heterogeneous mixture was stirred vigorously for 15 min, then poured into brine (5 mL), and extracted with CH_2Cl_2 (3 × 5 mL). The combined organic layers were washed with brine (100 mL), dried ($MgSO_4$), filtered, and concentrated. The residue was purified by flash column chromatography (silica gel, hexanes/EtOAc 1:0 to 3:1); yield: 22 mg (72%); ratio (**168**/**169**/**170**) 5.3:1.3:1.0.

2.3.1.4.3 Chalcogen-Initiated Events

A selenium-initiated polyene cyclization of trienol **172** was first demonstrated by White and co-workers using benzeneselenenyl hexafluorophosphate to give selenium-incorporated oxepane **173** and spiroindane **174** in 10 and 15% yield, respectively (Scheme 37).[183]

for references see p 332

Scheme 37 Selenium-Initiated Polyene Cyclization Using Benzeneselenenyl Chloride/Silver(I) Hexafluorophosphate[183]

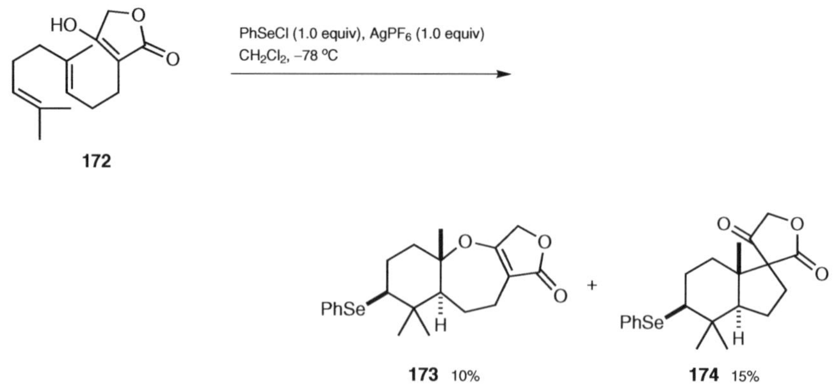

PhSeCl (1.0 equiv), $AgPF_6$ (1.0 equiv)
CH_2Cl_2, −78 °C

172

173 10%

174 15%

A related sulfenium-initiated polyene cyclization using methyl phenyl sulfoxide as an electrophilic sulfenium reagent was first reported by Livinghouse and co-workers.[184] Polyene cyclization of **175** is initiated using a Lewis acid (BF_3) to activate the sulfenium species and gives octahydrophenanthrene **176** in 85% yield without diastereoselectivity (Scheme 38).[184] Additionally, *N*-(phenylselanyl)succinimide can also be used as the electrophilic reagent. The sulfenium methodology has been employed to synthesize racemic nimbidiol (**177**).[185] The scope of sulfenium-initiated polyene cyclization also includes various electrophilic sulfenium reagents and the most effective is 2-methoxyethyl 4-chlorobenzenesulfenate.[186]

Scheme 38 Sulfenium-Initiated Polyene Cyclization Using Methyl Phenyl Sulfoxide[184]

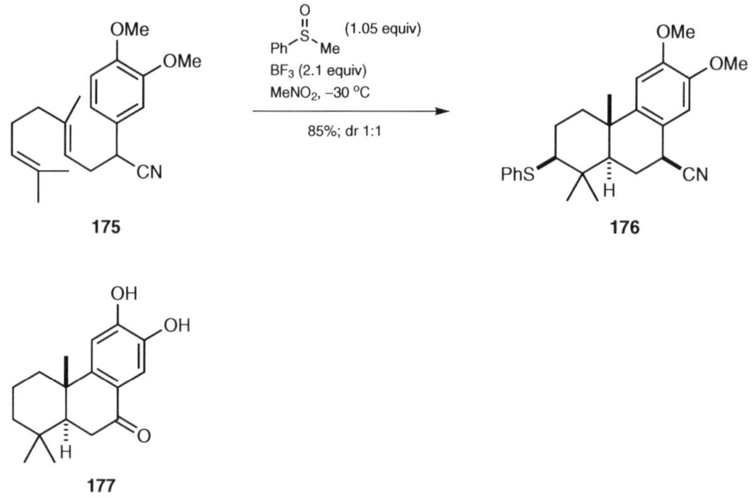

PhS(O)Me (1.05 equiv)
BF_3 (2.1 equiv)
$MeNO_2$, −30 °C

85%; dr 1:1

175

176

177

Shaw and co-workers recently detailed a catalytic approach for chalconium (S^+ and Se^+) initiated polyene cyclization of geranylated anilines (e.g., **178**) using a benzenesulfenyl or benzeneselenenyl chloride as the electrophilic reagent and scandium(III) trifluoromethanesulfonate as catalyst (Scheme 39).[187] Various electron-rich and electron-deficient systems can be used, but benzeneselenenyl chloride affords higher yields than benzenesulfenyl chloride. Bicyclization gives almost no regioselectivity when compared to monocyclization under the same reaction conditions.

Scheme 39 Scandium(III) Trifluoromethanesulfonate Catalyzed Chalconium-Initiated Polyene Cyclization[187]

PhXCl (1.2 equiv)
$Sc(OTf)_3$ (10 mol%)
$MeNO_2$, –20 °C

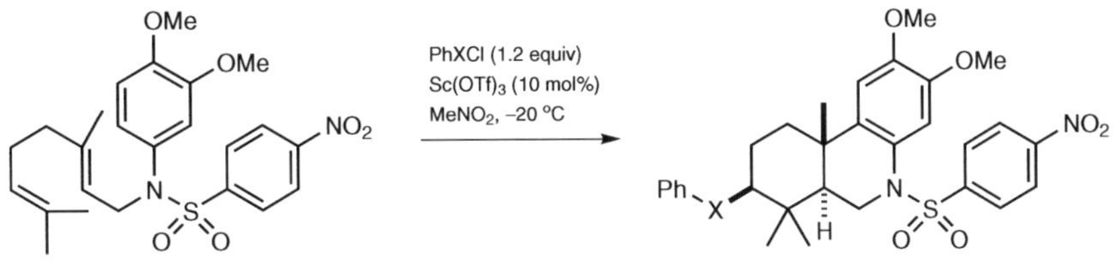

178

X	dr	Yield (%)	Ref
S	52:48	61	[187]
Se	55:45	72	[187]

At present, there are no reports on asymmetric, chalcogenium ion initiated polyene cyclization, although there are sporadic reports on catalytic and enantioselective carbosulfenylation to form monocyclized compounds.[188,189] Another challenge is the development of a regioselective chalcogenium-ion-initiated polyene cyclization.[187]

for references see p 332

References

[1] Müller, T. E.; Beller, M., *Chem. Rev.*, (1998) **98**, 675.
[2] Hultzsch, K. C., *Org. Biomol. Chem.*, (2005) **3**, 1819.
[3] Delacroix, O.; Gaumont, A. C., *Curr. Org. Chem.*, (2005) **9**, 1851.
[4] Severin, R.; Doye, S., *Chem. Soc. Rev.*, (2007) **36**, 1407.
[5] Liu, C.; Bender, C. F.; Han, X.; Widenhoefer, R. A., *Chem. Commun. (Cambridge)*, (2007), 3607.
[6] Taylor, J. G.; Adrio, L. A.; Hii, K. K., *Dalton Trans.*, (2010) **39**, 1171.
[7] Antoniotti, S.; Poulain-Martini, S.; Duñach, E., *Synlett*, (2010), 2973.
[8] Hesp, K. D.; Stradiotto, M., *ChemCatChem*, (2010) **2**, 1192.
[9] Yadav, J. S.; Antony, A.; Rao, T. S.; Reddy, B. V. S., *J. Organomet. Chem.*, (2011) **696**, 16.
[10] Andreatta, J. R.; McKeown, B. A.; Gunnoe, T. B., *J. Organomet. Chem.*, (2011) **696**, 305.
[11] Hannedouche, J.; Schulz, E., *Chem.–Eur. J.*, (2013) **19**, 4972.
[12] Chiarucci, M.; Bandini, M., *Beilstein J. Org. Chem.*, (1991) **9**, 2586.
[13] Liu, D.; Zhao, G.; Xiang, L., *Eur. J. Org. Chem.*, (2010), 3975.
[14] Zhang, D.; Song, H.; Qin, A. Y., *Acc. Chem. Res.*, (2011) **44**, 447.
[15] Zi, W.; Zuo, Z.; Ma, D., *Acc. Chem. Res.*, (2015) **48**, 702.
[16] Beller, M.; Breindl, C.; Riermeier, T. H.; Eichberger, M.; Trauthwein, H., *Angew. Chem. Int. Ed.*, (1998) **37**, 3389.
[17] Beller, M.; Breindl, C.; Riermeier, T. H.; Tillack, A., *J. Org. Chem.*, (2001) **66**, 1403.
[18] Huisgen, R.; König, H., *Chem. Ber.*, (1959) **92**, 203.
[19] Huisgen, R.; König, H.; Bleeker, N., *Chem. Ber.*, (1959) **92**, 424.
[20] König, H.; Huisgen, R., *Chem. Ber.*, (1959) **92**, 429.
[21] Iida, H.; Aoyagi, S.; Kibayashi, C., *J. Chem. Soc., Perkin Trans. 1*, (1975), 2502.
[22] Beller, M.; Breindl, C., *Tetrahedron*, (1998) **54**, 6359.
[23] Martínez, P. H.; Hultzsch, K. C.; Hampel, F., *Chem. Commun. (Cambridge)*, (2006), 2221.
[24] Rubtsov, M. V.; Mikhkina, E. E.; Yakhontov, L. N., *Russ. Chem. Rev. (Engl. Transl.)*, (1960) **29**, 37.
[25] Mashkovsky, M. D.; Yakhontov, L. N.; Churyukanov, V. V., In *Quinuclidinium Compounds*, Kharkevich, D. A., Ed.; Springer: Berlin, (1986); Vol. 79, pp 371–382.
[26] Quinet, C.; Jourdain, P.; Hermans, C.; Ates, A.; Lucas, I.; Markó, I. E., *Tetrahedron*, (2008) **64**, 1077.
[27] Kemp, J. E. G., In *Comprehensive Organic Synthesis*, Trost, B. M.; Fleming, I., Eds.; Pergamon: Oxford, (1991): Vol. 7, pp 469–513.
[28] Pearson, W. H.; Lian, B. W.; Bergmeier, S. C., In *Comprehensive Heterocyclic Chemistry II*, Katritzky, A. R.; Rees, C. W.; Scriven, E. F. V., Eds.; Pergamon: Oxford, (1996); Vol. 1A, pp 1–60.
[29] Rai, K. M. L.; Hassner, A., In *Comprehensive Heterocyclic Chemistry II*, Katritzky, A. R.; Rees, C. W.; Scriven, E. F. V., Eds.; Pergamon: Oxford, (1996); Vol. 1A, pp 61–96.
[30] Grimmett, M. R., In *Comprehensive Heterocyclic Chemistry II*, Katritzky, A. R.; Rees, C. W.; Scriven, E. F. V., Eds.; Pergamon: Oxford, (1996); Vol. 3, pp 77–22.
[31] Bolleddula, J.; DeMent, K.; Driscoll, J. P.; Worboys, P.; Brassil, P. J.; Bourdet, D. L., *Drug Metab. Rev.*, (2014) **46**, 379.
[32] Li, Y.; Marks, T. J., *J. Am. Chem. Soc.*, (1996) **118**, 707.
[33] Li, Y.; Marks, T. J., *J. Am. Chem. Soc.*, (1998) **120**, 1757.
[34] Ryu, J.-S.; Marks, T. J.; McDonald, F. E., *J. Org. Chem.*, (2004) **69**, 1038.
[35] Ryu, J.-S.; Li, G. Y.; Marks, T. J., *J. Am. Chem. Soc.*, (2003) **125**, 12 584.
[36] Molander, G. A.; Pack, S. K., *Tetrahedron*, (2003) **59**, 10 581.
[37] Shi, M.; Liu, L.-P.; Tang, J., *Org. Lett.*, (2006) **8**, 4043.
[38] Beller, M., *Eur. J. Lipid Sci. Technol.*, (2008) **110**, 789.
[39] Seayad, A.; Ahmed, M.; Klein, H.; Jackstell, R.; Gross, T.; Beller, M., *Science (Washington, D. C.)* (2002) **297**, 1676.
[40] Ahmed, M.; Seayad, A. M.; Jackstell, R.; Beller, M., *J. Am. Chem. Soc.*, (2003) **125**, 10 311.
[41] Ahmed, M.; Bronger, R. P. J.; Jackstell, R.; Kamer, P. C. J.; van Leeuwen, P. W. N. M.; Beller, M., *Chem.–Eur. J.*, (2006) **12**, 8979.
[42] Liu, G.; Huang, K.; Cao, B.; Chang, M.; Li, S.; Yu, S.; Zhou, L.; Wu, W.; Zhang, X., *Org. Lett.*, (2012) **14**, 102.
[43] Gülak, S.; Wu, L.; Liu, Q.; Franke, R.; Jackstell, R.; Beller, M., *Angew. Chem. Int. Ed.*, (2014) **53**, 7320.
[44] de Figueiredo, R. M., *Angew. Chem. Int. Ed.*, (2009) **48**, 1190.
[45] Cardona, F.; Goti, A., *Nat. Chem.*, (2009) **1**, 269.
[46] De Jong, S.; Nosal, D. G.; Wardrop, D. J., *Tetrahedron*, (2012) **68**, 4067.

[47] Muñiz, K.; Martínez, C., *J. Org. Chem.*, (2013) **78**, 2168.
[48] Muñiz, K., *Pure Appl. Chem.*, (2013) **85**, 755.
[49] Muñiz, K.; Hövelmann, C. H.; Campos-Gómez, E.; Barluenga, J.; González, J. M.; Streuff, J.; Nieger, M., *Chem.–Asian J.*, (2008) **3**, 776.
[50] Li, H.; Widenhoefer, R. A., *Tetrahedron*, (2010) **66**, 4827.
[51] Chávez, P.; Kirsch, J.; Hövelmann, C. H.; Streuff, J.; Martínez-Belmonte, M.; Escudero-Adán, E. C.; Martin, E.; Muñiz, K., *Chem. Sci.*, (2012) **3**, 2375.
[52] Zhang, J.; Zhang, G.; Wu, W.; Zhang, X.; Shi, M., *Chem. Commun. (Cambridge)*, (2014) **50**, 15 052.
[53] Zhang, J.; Zhang, X.; Wu, W.; Zhang, G.; Xu, S.; Shi, M., *Tetrahedron Lett.*, (2015) **56**, 1505.
[54] Muñiz, K.; Streuff, J.; Chávez, P.; Hövelmann, C. H., *Chem.–Asian J.*, (2008) **3**, 1248.
[55] Röben, C.; Souto, J. A.; González, Y.; Lishchynskyi, A.; Muñiz, K., *Angew. Chem. Int. Ed.*, (2011) **50**, 9478.
[56] Trost, B. M.; Zhang, T., *Chem.–Eur. J.*, (2011) **17**, 3630.
[57] Yeung, Y.-Y.; Hong, S.; Corey, E. J., *J. Am. Chem. Soc.*, (2006) **128**, 6310.
[58] Kim, C. U.; Lew, W.; Williams, M. A.; Liu, H.; Zhang, L.; Swaminathan, S.; Bischofberger, N.; Chen, M. S.; Mendel, D. B.; Tai, C. Y.; Laver, W. G.; Stevens, R. C., *J. Am. Chem. Soc.*, (1997) **119**, 681.
[59] Rohloff, J. C.; Kent, K. M.; Postich, M. J.; Becker, M. W.; Chapman, H. H.; Kelly, D. E.; Lew, W.; Louie, M. S.; McGee, L. R.; Prisbe, E. J.; Schultze, L. M.; Yu, R. H.; Zhang, L., *J. Org. Chem.*, (1998) **63**, 4545.
[60] Moscona, A., *N. Engl. J. Med.*, (2005) **353**, 1363.
[61] Moscona, A., *N. Engl. J. Med.*, (2005) **353**, 2633.
[62] Shibasaki, M.; Kani, M., *Eur. J. Org. Chem.*, (2008), 1839.
[63] Yeung, Y.-Y.; Gao, X.; Corey, E. J., *J. Am. Chem. Soc.*, (2006) **128**, 9644.
[64] Tay, D. W.; Tsoi, I. T.; Er, J. C.; Leung, G. Y. C.; Yeung, Y.-Y., *Org. Lett.*, (2013) **15**, 1310.
[65] Zhou, L.; Zhou, J.; Tan, C. K.; Chen, J.; Yeung, Y.-Y., *Org. Lett.*, (2011) **13**, 2448.
[66] Zhou, L.; Chen, J.; Zhou, J.; Yeung, Y.-Y., *Org. Lett.*, (2011) **13**, 5804.
[67] Zhou, J.; Yeung, Y.-Y., *Org. Lett.*, (2014) **16**, 2134.
[68] Zhou, J.; Yeung, Y.-Y., *Org. Biomol. Chem.*, (2014) **12**, 7482.
[69] Davies, S. G.; Nicholson, R. L.; Price, P. D.; Roberts, P. M.; Smith, A. D., *Synlett*, (2004), 901.
[70] Jin, Z., *Nat. Prod. Rep.*, (2005) **22**, 111.
[71] Jin, Z., *Nat. Prod. Rep.*, (2007) **24**, 886.
[72] He, M.; Qu, C.; Gao, O.; Hu, X.; Hong, X., *RSC Adv.*, (2015) **5**, 16 562.
[73] Fujioka, H.; Murai, K.; Ohba, Y.; Hirose, H.; Kita, Y., *Chem. Commun. (Cambridge)*, (2006), 832.
[74] Haufe, G., *J. Fluorine Chem.*, (1990) **46**, 83.
[75] Bonney, K. J.; Braddock, D. C.; White, A. J. P.; Yaqoob, M., *J. Org. Chem.*, (2011) **76**, 97.
[76] Haufe, G.; Alvernhe, G.; Laurent, A., *Tetrahedron Lett.*, (1986) **27**, 4449.
[77] Haufe, G.; Kleinpeter, E., *Tetrahedron Lett.*, (1982) **23**, 3555.
[78] Davies, S. G.; Polywka, M. E. C.; Thomas, S. E., *Tetrahedron Lett.*, (1985) **26**, 1461.
[79] Davies, S. G.; Polywka, M. E. C.; Thomas, S. E., *J. Chem. Soc., Perkin Trans. 1*, (1986), 1277.
[80] Fukuzawa, A.; Aye, M.; Takasugi, Y.; Nakamura, M.; Tamura, M.; Murai, A., *Chem. Lett.*, (1994), 2307.
[81] Ishihara, J.; Shimada, Y.; Kanoh, N.; Takasugi, Y.; Fukuzawa, A.; Murai, A., *Tetrahedron*, (1997) **53**, 8371.
[82] Braddock, D. C., *Org. Lett.*, (2006) **8**, 6055.
[83] Braddock, D. C.; Millan, D. S.; Pérez-Fuertes, Y.; Pouwer, R. H.; Sheppard, R. N.; Solanki, S.; White, A. J. P., *J. Org. Chem.*, (2009) **74**, 1835.
[84] Braddock, D. C.; Sbircea, D.-T., *Chem. Commun. (Cambridge)*, (2014) **50**, 12 691.
[85] Bonney, K. J.; Braddock, D. C., *J. Org. Chem.*, (2012) **77**, 9574.
[86] Tanuwidjaja, J.; Ng, S.-S.; Jamison, T. F., *J. Am. Chem. Soc.*, (2009) **131**, 12 084.
[87] Underwood, B. S.; Tanuwidjaja, J.; Ng, S.-S.; Jamison, T. F., *Tetrahedron*, (2013) **69**, 5205.
[88] Bravo, F.; McDonald, F. E.; Neiwert, W. A.; Hardcastle, K. I., *Org. Lett.*, (2004) **6**, 4487.
[89] Snyder, S. A.; Treitler, D. S.; Brucks, A. P.; Sattler, W., *J. Am. Chem. Soc.*, (2011) **133**, 15 898.
[90] Snyder, S. A.; Brucks, A. P.; Treitler, D. S.; Moga, I., *J. Am. Chem. Soc.*, (2012) **134**, 17 714.
[91] Sugimoto, M.; Suzuki, T.; Hagiwara, H.; Hoshi, T., *Tetrahedron Lett.*, (2007) **48**, 1109.
[92] Taylor, M. T.; Fox, J. M., *Tetrahedron Lett.*, (2015) **56**, 3560.
[93] Kikuchi, H.; Suzuki, T.; Kurosawa, E.; Suzuki, M., *Bull. Chem. Soc. Jpn.*, (1991) **64**, 1763.
[94] Keshipeddy, S.; Martínez, I.; Castillo, B. F., II; Morton, A. D.; Howell, A. R., *J. Org. Chem.*, (2012) **77**, 7883.

[95] Zhou, L.; Tan, C. K.; Zhou, J.; Yeung, Y.-Y., *J. Am. Chem. Soc.*, (2010) **132**, 10245.
[96] Chen, J.; Chng, S.; Zhou, L.; Yeung, Y.-Y., *Org. Lett.*, (2011) **13**, 6456.
[97] Zhou, J.; Zhou, L.; Yeung, Y.-Y., *Org. Lett.*, (2012) **14**, 5250.
[98] Ke, Z.; Yeung, Y.-Y., *Org. Lett.*, (2013) **15**, 1906.
[99] Yu, W. Z.; Chen, F.; Cheng, Y. A.; Yeung, Y.-Y., *J. Org. Chem.*, (2015) **80**, 2815.
[100] Zhou, J.; Yeung, Y.-Y., *J. Org. Chem.*, (2014) **79**, 4644.
[101] Fujioka, H.; Nagatomi, Y.; Kotoku, N.; Kitagawa, H.; Kita, Y., *Tetrahedron*, (2000) **56**, 10141.
[102] Fujioka, H.; Kitagawa, H.; Nagatomi, Y.; Kita, Y., *J. Org. Chem.*, (1996) **61**, 7309.
[103] Fujioka, H.; Nakahara, K.; Hirose, H.; Hirano, K.; Oki, T.; Kita, Y., *Chem. Commun. (Cambridge)*, (2011) **47**, 1060.
[104] Fujioka, H.; Nagatomi, Y.; Kotoku, N.; Kitagawa, H.; Kita, Y., *Tetrahedron Lett.*, (1998) **39**, 7309.
[105] Fujioka, H.; Fujita, T.; Kotoku, N.; Ohba, Y.; Nagatomi, Y.; Hiramatsu, A.; Kita, Y., *Chem.–Eur. J.*, (2004) **10**, 5386.
[106] Fujioka, H.; Nakahara, K.; Kotoku, N.; Ohba, Y.; Nagatomi, Y.; Wang, T.-L.; Sawama, Y.; Murai, K.; Hirano, K.; Oki, T.; Wakamatsu, S.; Kita, Y., *Chem.–Eur. J.*, (2012) **18**, 13861.
[107] Fujioka, H.; Kotoku, N.; Sawama, Y.; Nagatomi, Y.; Kita, Y., *Tetrahedron Lett.*, (2002) **43**, 4825.
[108] Fujioka, H.; Ohba, Y.; Hirose, H.; Nakahara, K.; Murai, K.; Kita, Y., *Tetrahedron*, (2008) **64**, 4233.
[109] Fujioka, H.; Ohba, Y.; Hirose, H.; Murai, K.; Kita, Y., *Angew. Chem. Int. Ed.*, (2005) **44**, 734.
[110] Ruan, Z.; Mootoo, D. R., *Tetrahedron Lett.*, (1999) **40**, 49.
[111] Breit, B., *Acc. Chem. Res.*, (2003) **36**, 264.
[112] Pordea, A.; Ward, T. R., *Synlett*, (2009), 3225.
[113] Airiau, E.; Chemin, C.; Girard, N.; Lonzi, G.; Mann, A.; Petricci, E.; Salvadori, J.; Taddei, M., *Synthesis*, (2010), 2901.
[114] Gusevskaya, E. V.; Jiménez-Pinto, J.; Börner, A., *ChemCatChem*, (2014) **6**, 382.
[115] Breit, B., *Chem.–Eur. J.*, (2000) **6**, 1519.
[116] Breit, B.; Seiche, W., *Synthesis*, (2001), 1.
[117] Kiss, G., *Chem. Rev.*, (2001) **101**, 3435.
[118] Brennführer, A.; Neumann, H.; Beller, M., *ChemCatChem*, (2009) **1**, 28.
[119] Wu, X.-F.; Neumann, H., *ChemCatChem*, (2012) **4**, 447.
[120] Wu, L.; Liu, Q.; Fleischer, I.; Jackstell, R.; Beller, M., *Nat. Commun.*, (2014) **5**, 3091.
[121] del Río, I.; Claver, C.; van Leeuwen, P. W. N. M., *Eur. J. Inorg. Chem.*, (2001) **11**, 2719.
[122] Breit, B.; Zahn, S. K., *Angew. Chem. Int. Ed.*, (1999) **38**, 969.
[123] Breit, B.; Zahn, S. K., *Tetrahedron*, (2005) **61**, 6171.
[124] Breit, B.; Zahn, S. K., *Polyhedron*, (2000) **19**, 513.
[125] Dübon, P.; Farwick, A.; Helmchen, G., *Synlett*, (2009), 1413.
[126] Breit, B.; Zahn, S. K., *Angew. Chem. Int. Ed.*, (2001) **40**, 1910.
[127] Hoffmann, R. W.; Brückner, D., *New J. Chem.*, (2001) **25**, 369.
[128] Airiau, E.; Spangenberg, T.; Girard, N.; Schoenfelder, A.; Salvadori, J.; Taddei, M.; Mann, A., *Chem.–Eur. J.*, (2008) **14**, 10938.
[129] Cini, E.; Airiau, E.; Girard, N.; Mann, A.; Salvadori, J.; Taddei, M., *Synlett*, (2011), 199.
[130] Farwick, A.; Helmchen, G., *Adv. Synth. Catal.*, (2010) **352**, 1023.
[131] Barluenga, J.; González, J. M.; Campos, P. J.; Asensio, G., *Angew. Chem. Int. Ed. Engl.*, (1988) **27**, 1546.
[132] Barluenga, J.; Alvarez-Pérez, M.; Rodríguez, F.; Fañanás, F. J.; Cuesta, J. A.; García-Granda, S., *J. Org. Chem.*, (2003) **68**, 6583.
[133] Barluenga, J.; Trincado, M.; Rubio, E.; González, J. M., *J. Am. Chem. Soc.*, (2004) **126**, 3416.
[134] Woodward, R. B.; Bloch, K., *J. Am. Chem. Soc.*, (1953) **75**, 2023.
[135] Johnson, W. S., *Bioorg. Chem.*, (1976) **5**, 51.
[136] Johnson, W. S., *Angew. Chem. Int. Ed. Engl.*, (1976) **15**, 9.
[137] Yoder, R. A.; Johnston, J. N., *Chem. Rev.*, (2005) **105**, 4730.
[138] Abe, I., *Nat. Prod. Rep.*, (2007) **24**, 1311.
[139] Peters, R. J., *Nat. Prod. Rep.*, (2010) **27**, 1521.
[140] Wendt, K. U.; Schulz, G. E.; Corey, E. J.; Liu, D. R., *Angew. Chem. Int. Ed.*, (2000) **39**, 2812.
[141] Stork, G.; Burgstahler, A. W., *J. Am. Chem. Soc.*, (1955) **77**, 5068.
[142] Eschenmoser, A.; Arigoni, D., *Helv. Chim. Acta*, (2005) **88**, 3011.
[143] Nishizawa, M.; Takenaka, H.; Nishide, H.; Hayashi, Y., *Tetrahedron Lett.*, (1983) **24**, 2581.
[144] Takao, H.; Wakabayashi, A.; Takahashi, K.; Imagawa, H.; Sugihara, T.; Nishizawa, M., *Tetrahedron Lett.*, (2004) **45**, 1079.

[145] Nishizawa, M.; Takenaka, H.; Hayashi, Y., *Tetrahedron Lett.*, (1984) **25**, 437.
[146] Nishizawa, M.; Nishide, H.; Hayashi, Y., *Tetrahedron Lett.*, (1984) **25**, 5071.
[147] Nishizawa, M.; Yamada, H.; Hayashi, Y., *J. Org. Chem.*, (1987) **52**, 4878.
[148] Gopalan, A. S.; Prieto, R.; Mueller, B.; Peters, D., *Tetrahedron Lett.*, (1992) **33**, 1679.
[149] Koh, J. H.; Mascarenhas, C.; Gagné, M. R., *Tetrahedron*, (2004) **60**, 7405.
[150] Koh, J. H.; Gagné, M. R., *Angew. Chem. Int. Ed.*, (2004) **43**, 3459.
[151] Mullen, C. A.; Gagné, M. R., *J. Am. Chem. Soc.*, (2007) **129**, 11880.
[152] Mullen, C. A.; Campbell, A. N.; Gagné, M. R., *Angew. Chem. Int. Ed.*, (2008) **47**, 6011.
[153] Sokol, J. G.; Cochrane, N. A.; Becker, J. J.; Gagné, M. R., *Chem. Commun. (Cambridge)*, (2013) **49**, 5046.
[154] Cochrane, N. A.; Nguyen, H.; Gagné, M. R., *J. Am. Chem. Soc.*, (2013) **135**, 628.
[155] Geier, M. J.; Gagné, M. R., *Organometallics*, (2013) **32**, 380.
[156] Nguyen, H.; Gagné, M. R., *ACS Catal.*, (2014) **4**, 855.
[157] Geier, M. J.; Gagné, M. R., *J. Am. Chem. Soc.*, (2014) **136**, 3032.
[158] Hoye, T. R.; Kurth, M. J., *J. Org. Chem.*, (1978) **43**, 3693.
[159] Kato, T.; Ishii, K.; Ichinose, I.; Nakai, Y.; Kumagai, T., *J. Chem. Soc., Chem. Commun.*, (1980), 1106.
[160] Wolinsky, L. E.; Faulkner, D. J., *J. Org. Chem.*, (1976) **41**, 597.
[161] Yamamura, S.; Terada, Y., *Tetrahedron Lett.*, (1977), 2171.
[162] Greenwood, J. M.; Sutherland, J. K.; Torre, A., *Chem. Commun.*, (1965), 410.
[163] van Tamelen, E. E.; Hessler, E. J., *Chem. Commun.*, (1966), 411.
[164] Kato, T.; Ichinose, I.; Kamoshida, A.; Kitahara, Y., *J. Chem. Soc., Chem. Commun.*, (1976), 518.
[165] van Tamelen, E. E.; Storni, A.; Hessler, E. J.; Schwartz, M. A., *Bioorg. Chem.*, (1982) **11**, 133.
[166] Kato, T.; Mochizuki, M.; Hirano, T.; Fujiwara, S.; Uyehara, T., *J. Chem. Soc., Chem. Commun.*, (1984), 1077.
[167] Yamaguchi, Y.; Uyehara, T.; Kato, T., *Tetrahedron Lett.*, (1985) **26**, 343.
[168] Fujiwara, S.; Takeda, K.; Uyehara, T.; Kato, T., *Chem. Lett.*, (1986), 1763.
[169] Tanaka, A.; Sato, M.; Yamashita, K., *Agric. Biol. Chem.*, (1990) **54**, 121.
[170] Tanaka, A.; Oritani, T., *Biosci., Biotechnol., Biochem.*, (1995) **59**, 516.
[171] Kato, T.; Ichinose, I.; Kumazawa, S.; Kitahara, Y., *Bioorg. Chem.*, (1975) **4**, 188.
[172] Shieh, H.-M.; Prestwich, G. D., *Tetrahedron Lett.*, (1982) **23**, 4643.
[173] Carter-Franklin, J. N.; Parrish, J. D.; Tschirret-Guth, R. A.; Little, R. D.; Butler, A., *J. Am. Chem. Soc.*, (2003) **125**, 3688.
[174] Sakakura, A.; Ukai, A.; Ishihara, K., *Nature (London)*, (2007) **445**, 900.
[175] Sawamura, Y.; Nakatsuji, H.; Sakakura, A.; Ishihara, K., *Chem. Sci.*, (2013) **4**, 4181.
[176] Recsei, C.; Chan, B.; McErlean, C. S. P., *J. Org. Chem.*, (2014) **79**, 880.
[177] Snyder, S. A.; Treitler, D. S., *Angew. Chem. Int. Ed.*, (2009) **48**, 7899.
[178] Snyder, S. A.; Treitler, D. S.; Brucks, A. P., *J. Am. Chem. Soc.*, (2010) **132**, 14303.
[179] Snyder, S. A.; Treitler, D. S.; Schall, A., *Tetrahedron*, (2010) **66**, 4796.
[180] Braddock, D. C.; Hermitage, S. A.; Kwok, L.; Pouwer, R.; Redmond, J. M.; White, A. J. P., *Chem. Commun. (Cambridge)*, (2009), 1082.
[181] Braddock, D. C.; Marklew, J. S.; Thomas, A. J. F., *Chem. Commun. (Cambridge)*, (2011) **47**, 9051.
[182] Braddock, D. C.; Marklew, J. S.; Foote, K. M.; White, A. J. P., *Chirality*, (2013) **25**, 692.
[183] White, J. D.; Nishiguchi, T.; Skeean, R. W., *J. Am. Chem. Soc.*, (1982) **104**, 3923.
[184] Edstrom, E. D.; Livinghouse, T., *J. Org. Chem.*, (1987) **52**, 949.
[185] Harring, S. R.; Livinghouse, T., *Tetrahedron Lett.*, (1989) **30**, 1499.
[186] Harring, S. R.; Livinghouse, T., *J. Org. Chem.*, (1997) **62**, 6388.
[187] Moore, J. T.; Soldi, C.; Fettinger, J. C.; Shaw, J. T., *Chem. Sci.*, (2013) **4**, 292.
[188] Denmark, S. E.; Jaunet, A., *J. Am. Chem. Soc.*, (2013) **135**, 6419.
[189] Denmark, S. E.; Jaunet, A., *J. Org. Chem.*, (2014) **79**, 140.

2.3.2 Organocatalyzed Addition to Activated C=C Bonds

P. Renzi, M. Moliterno, R. Salvio, and M. Bella

General Introduction

Organocatalytic reactions are especially attractive because they do not require anhydrous conditions and expensive catalysts.[1–9] Most of all, waste management is significantly easier with respect to processes employing transition metals, which present issues with the disposal of residues and the phenomenon of "metal leaching" into the products. The most well-known organocatalytic process, the Hajos–Parrish–Eder–Sauer–Wiechert reaction,[10–12] is actually a domino transformation which has found significant industrial application. Despite this key result, this field has surprisingly been neglected for nearly thirty years before its renaissance at the beginning of the millennium. In particular, organocatalytic domino reactions are appealing transformations. The mild operational conditions required allow that in the same reaction vessel several processes can operate consecutively. This situation would be more complex if the several transformations composing that domino reaction had to be run under significantly different conditions.

The goal of this chapter is to highlight and critically discuss some selected examples of organocatalytic domino reactions, not to write a comprehensive treatise on the subject. It will begin by reviewing the first ground-breaking examples of organocatalytic domino reactions (Section 2.3.2.1). Subsequently, more specific applications from the chemistry of oxindoles (Section 2.3.2.2), and finally large-scale transformations, specifically the synthesis of (−)-oseltamivir (Section 2.3.2.3), ABT-341 (Section 2.3.2.4), and industrial-scale processes for the synthesis of chiral diene ligands (Section 2.3.2.5) are described.

2.3.2.1 Organocatalyzed Domino Reactions with Activated Alkenes: The First Examples

Among the organocatalysts developed so far, secondary amines stand out for their peculiar mode of carbonyl compound activation by forming enamines or iminium ions. More specifically, they can form enamines with aldehydes and ketones, thus raising the HOMO and allowing their functionalization with an electrophile at the α-position (Scheme 1).[5] In the case of α,β-unsaturated carbonyl compounds, iminium ions with lowered LUMO energy are formed, thus facilitating functionalization with a nucleophile at the β-position (Scheme 1).[13] This double activation mode makes secondary amines privileged catalysts in cascade reactions because one can easily design a transformation in which an α,β-unsaturated carbonyl compound is first subjected to a conjugate addition and then to a functionalization on the α-position (Scheme 1).

Scheme 1 Activation of Carbonyl Compounds via an Enamine or Iminium Ion[5,13]

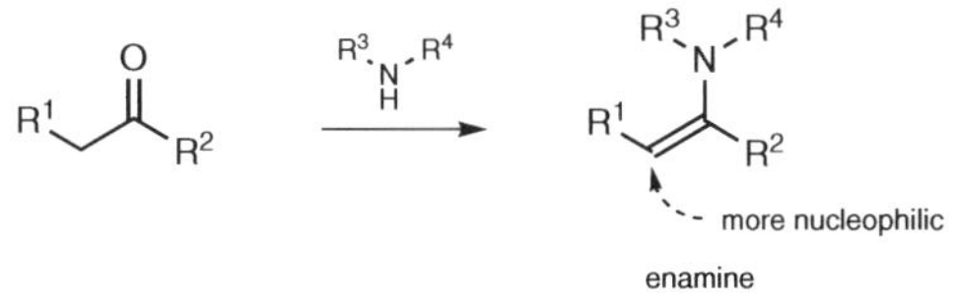

R^3NHR^4

more nucleophilic

enamine

for references see p 384

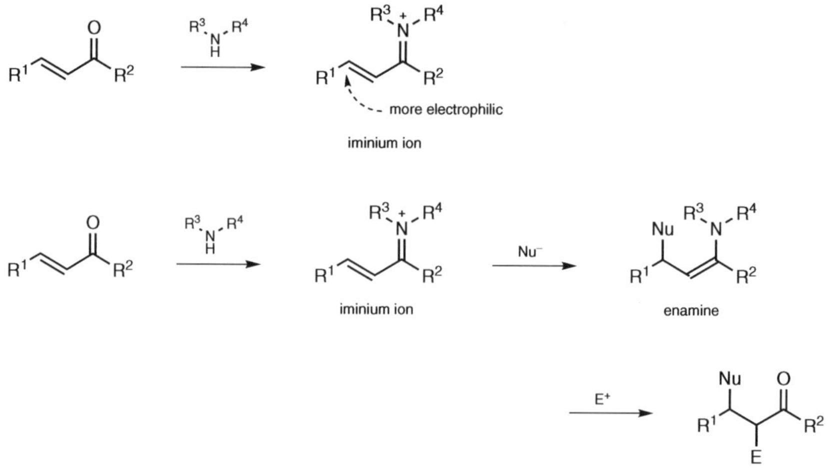

O
R1 R2
R3 N R4
H
R3 + R4
N
R1 R2
more electrophilic
iminium ion

O
R1 R2
R3 N R4
H
R3 + R4
N
R1 R2
iminium ion
Nu−
Nu R3 N R4
R1 R2
enamine

E+
Nu O
R1 R2
E

Many synthetic efforts have therefore been made to design asymmetric domino processes which employ chiral secondary amines as catalysts, conveniently combining the separate elements of more conventional enamine and iminium ion catalysis. Scheme 2 and Scheme 3 summarize some of the first results reported in this field, findings that illustrate the versatility of these methods, which are discussed later in detail.[14–19]

Scheme 2 Some of the First Organocatalytic Domino Reactions for the Asymmetric Functionalization of α,β-Unsaturated Aldehydes by Iminium–Enamine Catalysis[14,15,17,18]

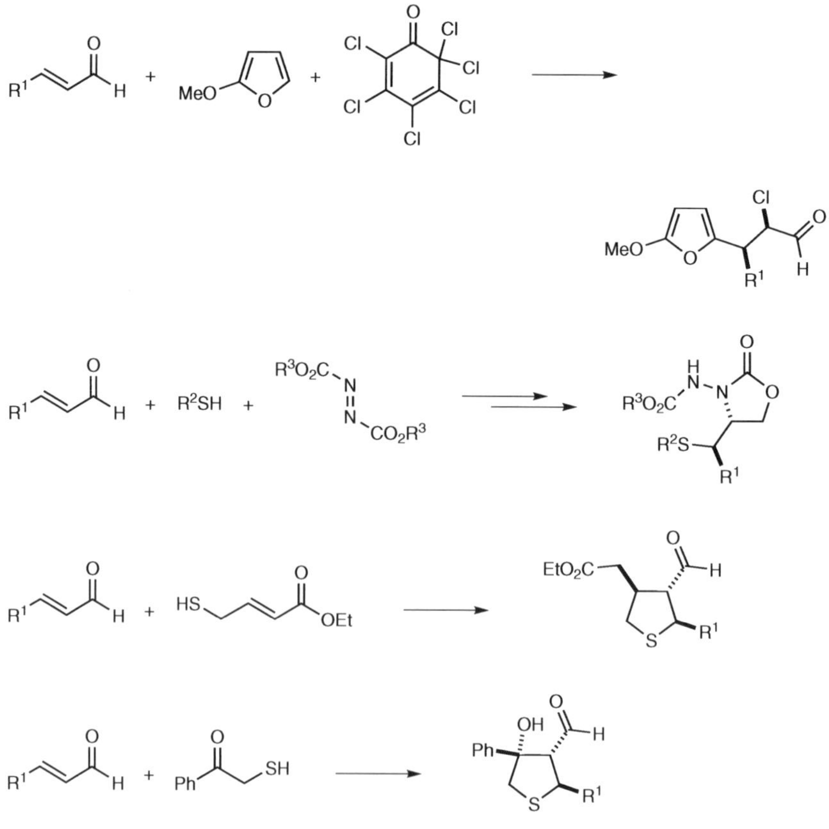

O
R1 H + MeO O + Cl O Cl Cl Cl Cl Cl Cl

Cl
MeO O O
R1 H

O
R1 H + R^2SH + R^3O_2C N=N CO_2R^3

H O
R^3O_2C N N O
R^2S
R1

O O
R1 H + HS OEt

O
EtO_2C H
S R1

O O
R1 H + Ph SH

O
OH H
Ph
S R1

Scheme 3 Some of the First Iminium-Based Organocatalytic Domino Reactions for the Asymmetric Functionalization of α,β-Unsaturated Aldehydes[16,17,19]

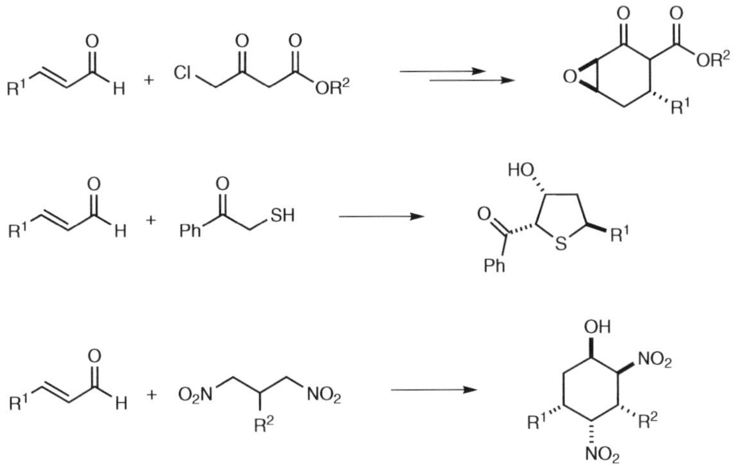

One of the first examples of this iminium–enamine approach to organocatalytic domino reactions was reported in 2004 by Jørgensen and co-workers. Here, an enantioselective Michael/aldol reaction of α,β-unsaturated ketones **2** with β-diketones, β-oxo esters, and β-oxo sulfones **3** yields cyclohexanones **4** with up to four stereocenters in high diastereo- and enantioselectivity (Scheme 4).[20,21]

Scheme 4 Michael/Aldol Domino Reaction of α,β-Unsaturated Ketones with β-Diketones, β-Oxo Esters, and β-Oxo Sulfones[20,21]

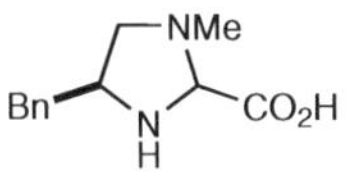

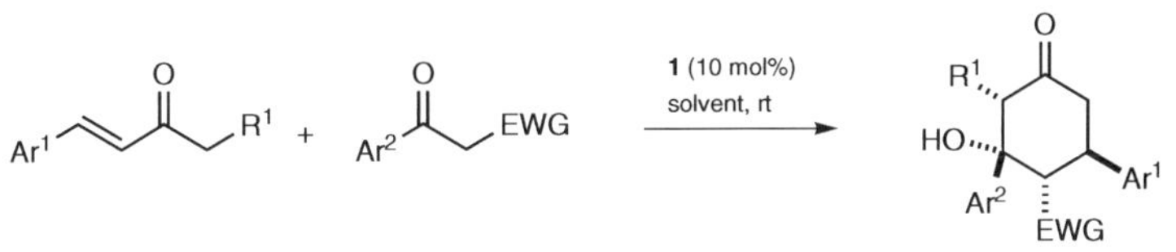

R^1	Ar^1	Ar^2	EWG	Solvent	Time (h)	dr	ee (%)	Yield (%)	Ref
H	Ph	Ph	CO_2Bn	EtOH	192	>97:3	95	80	[20]
H	2-naphthyl	Ph	CO_2Bn	EtOH	95	>97:3	91	85	[20]
Me	Ph	Ph	CO_2Bn	EtOH	95	>97:3	95	50	[20]
H	4-ClC_6H_4	Ph	Bz	$CHCl_3$	135	>97:3	80	87	[21]
H	Ph	Ph	Bz	$CHCl_3$	160	>97:3	91	56	[21]
H	1-naphthyl	Ph	SO_2Ph	CH_2Cl_2	140	>97:3	97	52	[21]
H	Ph	Ph	SO_2Ph	CH_2Cl_2	140	>97:3	96	93	[21]

The broad variety of substituents tolerated by this protocol allows the preparation of highly functionalized cyclohexanones **4**, materials that can be easily transformed into valu-

for references see p 384

able chiral molecules such as ε-lactones, other cyclohexanones, and cyclohexanediols. The reaction is catalyzed by the phenylalanine-derived imidazolidine **1**, a species which facilitates the conjugate addition by forming an iminium ion with the substrate **2** and by deprotonating the nucleophile **3**. The authors propose that the subsequent intramolecular aldol reaction is catalyzed by imidazolidine **1** acting as a base toward the Michael intermediate. The observed stereoselection, with the formation of essentially just a single diastereomer (dr >97:3), can be explained by the control exerted by the stereocenter on the Michael adduct. Nevertheless, the authors do not exclude that the aldol reaction portion of the domino process could proceed through an enamine mechanism.

The first direct evidence of a domino reaction proceeding through an iminium–enamine mechanism was reported by MacMillan and co-workers in 2005.[14] They described the conjugate addition of aromatic π-nucleophiles to α,β-unsaturated aldehydes, with subsequent halogenation of the resulting adduct terminating the sequence (Scheme 5).

Scheme 5 Domino Nucleophilic/Electrophilic Addition to α,β-Unsaturated Aldehydes[14]

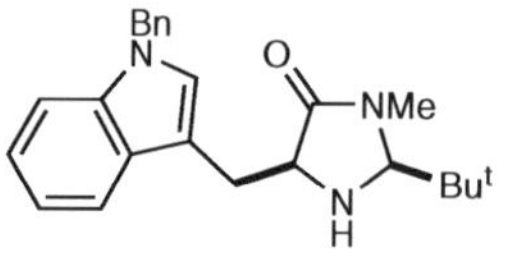

5

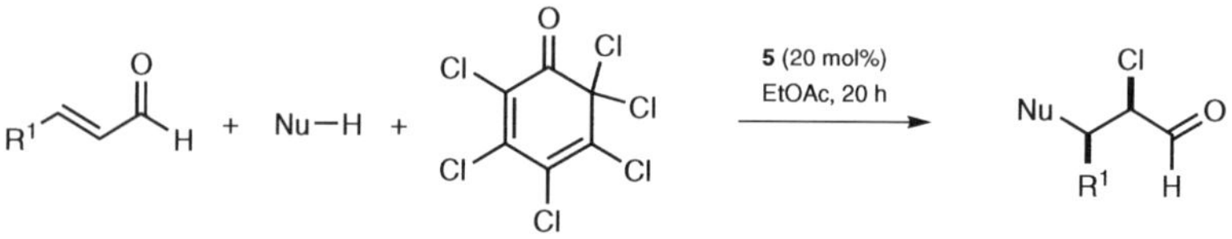

+ Nu—H +

5 (20 mol%)
EtOAc, 20 h

R^1	Nu	Temp (°C)	dr	ee (%)	Yield (%)	Ref
Me	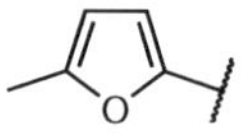	−50	14:1	99	86	[14]
CH_2OAc	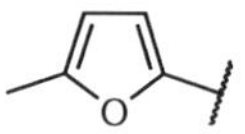	−40	11:1	>99	82	[14]
Me	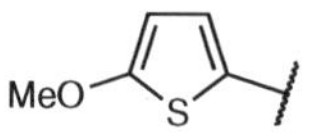 MeO	−50	11:1	99	77	[14]
Me	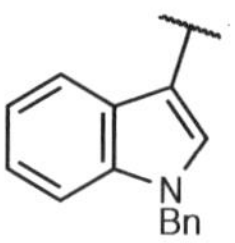 N Bn	−60	12:1	>99	75	[14]

5 (20 mol%), EtOAc
−40 °C, 20 h
97%; dr 9:1; >99% ee

5 (20 mol%), EtOAc
−55 °C, 20 h
71%; dr >25:1; >99% ee

This reaction, catalyzed by imidazolidinone **5**, is postulated to proceed through two different, yet related, catalytic cycles. As depicted in Scheme 6, in the first cycle the unsaturated aldehyde **6** and the catalyst **5** combine into iminium ion **7**, a species which can then undergo an enantioselective nucleophilic addition in the presence of a π-nucleophile to give intermediate **8**. After hydrolysis of the iminium ion **9**, the resultant Michael adduct **10** enters the second cycle in which enamine catalysis now enables a highly diastereoselective addition of the electrophile (E), affording cascade product **11**. This mechanistic pathway is in agreement with the observed stereochemistry of the products. In fact, the electrophilic addition is controlled by the catalyst architecture rather than by the stereochemistry of the intermediate **9** as in the Jørgensen hypothesis. The result is a predominantly *syn* addition in all cases studied to date.

It is noteworthy that the excellent result in terms of enantioselectivity (≥99% ee) is due to the presence of two catalytic asymmetric events in sequence. Although each of these would provide more modest enantioselection individually, the statistical amplification, according to the Horeau principle, leads to a much higher level of enantioselectivity in the final major diastereomer produced.[22]

for references see p 384

Scheme 6 Mechanism of Iminium–Enamine Catalysis[14]

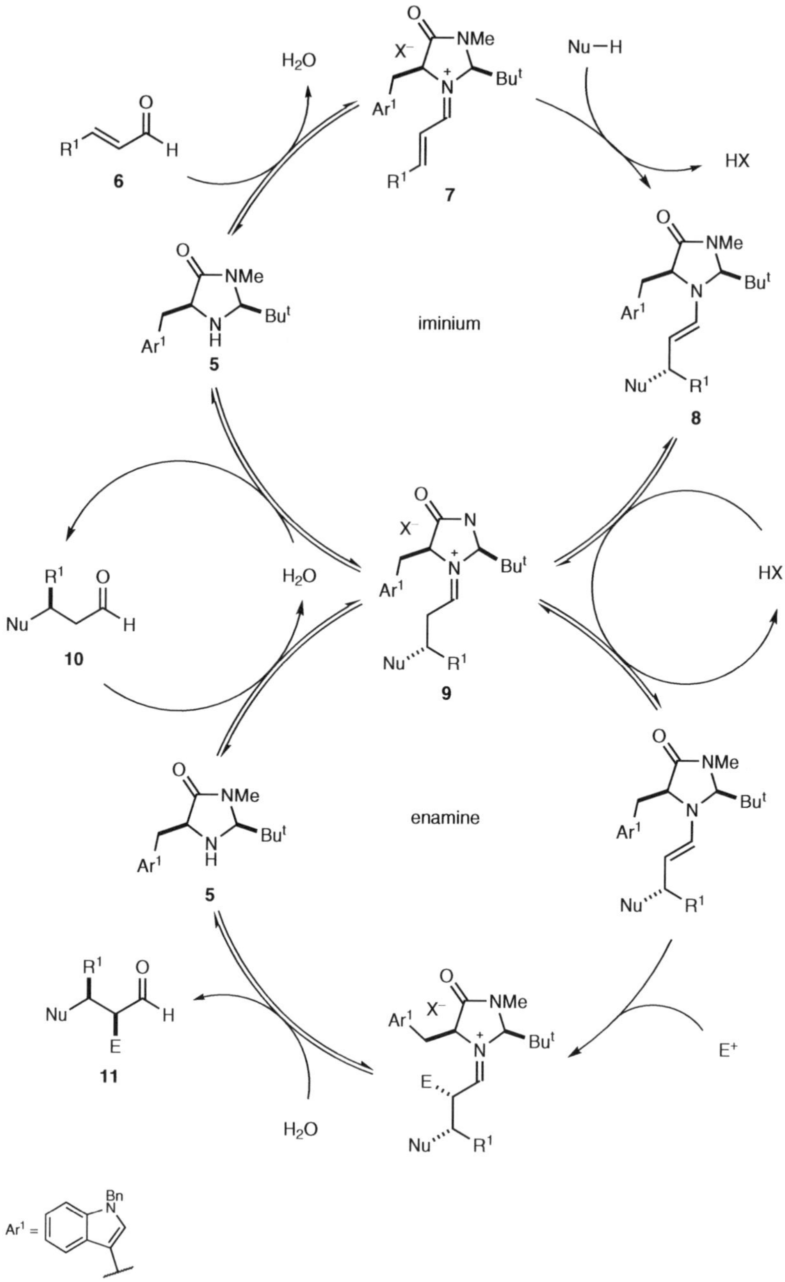

Since the publication of the MacMillan studies, iminium–enamine catalysis has been recognized as a powerful method to develop domino reactions involving activated double bonds. Jørgensen and co-workers have used this strategy to achieve the first enantioselective, organocatalytic conjugate thiol addition/amination reaction (Scheme 7).[15] In the presence of trimethylsilyl prolinol ether **12** as the catalyst, α,β-unsaturated aldehydes undergo iminium activation toward the conjugate addition of thiol nucleophiles **13**, and the resulting enamines are intercepted by the electrophilic azodicarboxylates **14**, thus affording the sulfanyl- and hydrazino-substituted aldehydes **15**. The products of the domino transformation are reduced in situ and treated with sodium hydroxide, triggering the intramolecular transesterification which leads to oxazolidinones **16** in high diastereoselectivity and excellent enantioselectivity.

Scheme 7 Conjugate Thiol Addition/Amination Reaction[15]

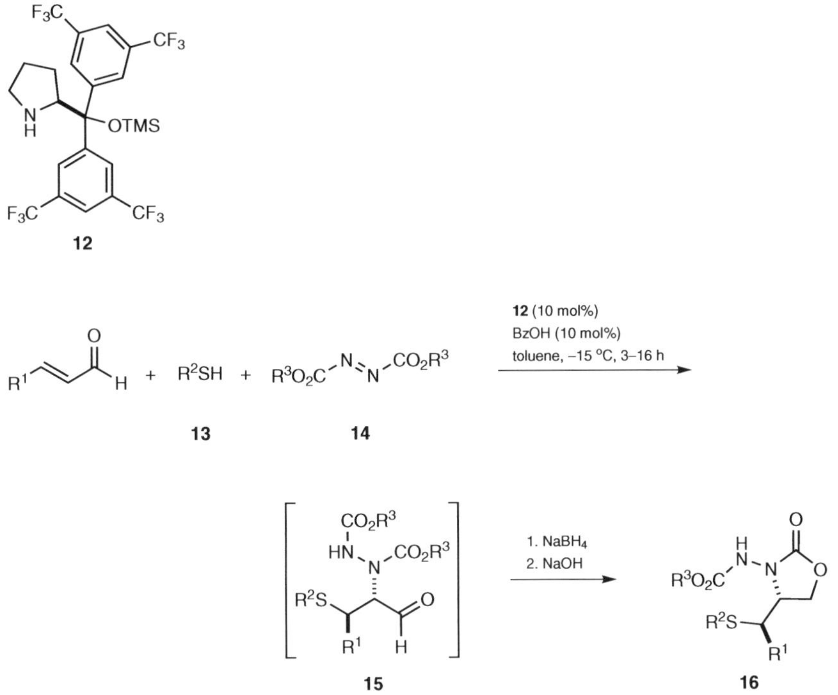

R^1	R^2	R^3	dr	ee (%)	Yield (%)	Ref
Me	Bn	Et	93:7	>99	72	[15]
Et	Et	Et	95:5	>99	63	[15]
Me	Bn	Bn	88:12	>99	51	[15]

To avoid racemization of the sulfa-Michael adduct, it is necessary to perform the reaction at −15 °C and to employ benzoic acid as co-catalyst to accelerate the reaction, which would otherwise be prohibitively slow. This work paved the way for the use of diarylprolinol silyl ethers such as **12** as "privileged" catalysts in the field of asymmetric domino transformations proceeding through iminium and enamine intermediates.

3-[(Ethoxycarbonyl)amino]-4-(1-sulfanylalkyl)oxazolidin-2-ones 16 **(R^3 = Et); General Procedure:**[15]

The aldehyde (0.38 mmol), catalyst **12** (0.025 mmol), and BzOH (0.025 mmol) were stirred in toluene and cooled to −15 °C. The thiol **13** (0.25 mmol) was added and the mixture was stirred for 30 min. DEAD (**14**, R^3 = Et; 0.33 mmol) was then added and the mixture was stirred until consumption of the substrate (3 to 16 h). The crude mixture was diluted with MeOH (2 mL) and cooled to 0 °C, followed by the addition of $NaBH_4$ (0.5 mmol). After 10 min, 2 M aq NaOH (2 mL) and THF (2 mL) were added and the crude mixture was stirred for 2 h. After standard aqueous workup, the product was purified by flash chromatography (silica gel). The dr and ee of the product were determined by chiral HPLC.

for references see p 384

2.3.2.1.1 Prolinol Trimethylsilyl Ethers as Privileged Catalysts for Enamine and Iminium Ion Activation

Diarylprolinol silyl ethers such as **12** (Scheme 7) catalyze a number of reactions involving the α-functionalization of carbonyl compounds, including their amination, halogenation, and Mannich reaction.[23–25] These molecules, derived from the natural amino acid proline, have been developed to extend the potential of proline itself, noting that their stereocontrol is complementary, yet also quite different in some cases. Taking as an example the addition of an electrophile to an enamine intermediate, in the case of proline the attack is directed by a hydrogen bond between the electrophile and the acidic proton of the catalyst, whereas diarylprolinol ethers exert a steric control due to their bulky substituents. Therefore, the stereochemical outcome is opposite in these two cases, though the catalyst scaffold is largely the same (Scheme 8).[26]

Scheme 8 Stereochemical Control Exerted by Proline and Diarylprolinol Ethers in Enamine Catalysis[26]

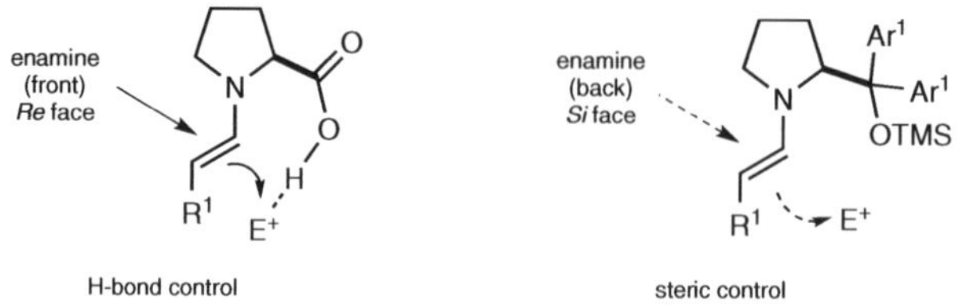

Two reports by Jørgensen and co-workers published in 2006 expanded the employment of trimethylsilyl prolinol ether **12** as a catalyst for asymmetric domino transformations. The first one deals with the Michael/aldol reaction of enals **17** and γ-chloro β-oxo esters **18** (Scheme 9).[16] In this case, the conjugate addition follows the iminium ion pathway, whereas the intramolecular aldol reaction is catalyzed by sodium acetate. Treatment of intermediate **19** with potassium carbonate leads to an S_N2-type intramolecular ring closure, which affords the bicyclic products **20** in excellent diastereo- and enantioselectivity.

Scheme 9 Domino Michael/Aldol and Intramolecular S_N2 Reaction Leading to Epoxy Cyclohexanones[16]

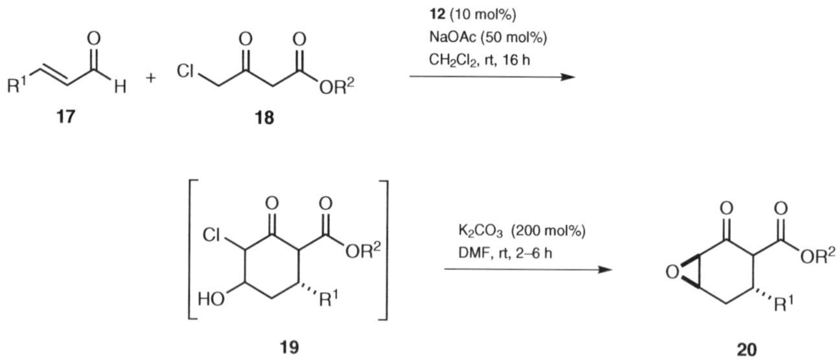

R^1	R^2	dr	ee (%)	Yield (%)	Ref
Me	$CH_2CH=CH_2$	99.5:0.5	85	56	[16]
Et	Et	>99.5:0.5	92	53	[16]
iPr	Me	>99.5:0.5	97	50	[16]

R^1	R^2	dr	ee (%)	Yield (%)	Ref
CH_2OTIPS	$CH_2CH=CH_2$	>99.5:0.5	86	51	[16]
$(CH_2)_2CH=CHEt$	$CH_2CH=CH_2$	>99.5:0.5	88	42	[16]
Ph	Et	>99.5:0.5	90	47	[16]

Although the aldol intermediates **19** may exist as eight different diastereomers, only two of them bear the chlorine and the hydroxy group antiperiplanar to each other (i.e., both in axial position) and can undergo the S_N2 reaction. The reversibility of the processes, and, ultimately, the energy differences between these two diastereomers that can go forward, account for the formation of a single diastereomer of epoxides **20** in the reaction.

The second report concerns the asymmetric synthesis of highly functionalized tetrahydrothiophenes.[17] The reaction of α,β-unsaturated aldehydes with thiols (e.g., **21**) in the presence of Jørgensen catalyst **12** affords either optically active tetrahydrothiophenes **23** or **24**, depending on the final additive employed (Scheme 10). The first event of the domino process is actually the same in both cases, namely the Michael addition of the thiol to the iminium ion formed between the α,β-unsaturated aldehyde and catalyst **12**, a process which leads to enamine intermediate **22**. Under basic conditions, the latter is hydrolyzed and the resulting thioether undergoes fast enolization on the most acidic position, affording tetrahydrothiophene **24** by intramolecular aldol addition. The asymmetric induction in the last step is therefore due only to the stereogenic center formed in the conjugate addition. On the other hand, when using benzoic acid as additive, enamine intermediate **22** is the reactive species in a second catalytic cycle and tetrahydrothiophene **23** results instead. In this iminium–enamine mode of activation, the stereochemical outcome is controlled by the catalyst itself in both steps, accounting for the much higher enantioenrichment of tetrahydrothiophenes of type **23** compared to that of **24**. Nevertheless, in both cases, the formation of a single diastereomer of product is observed.

Scheme 10 Regiodivergent Synthesis of Chiral Tetrahydrothiophenes via Michael/Aldol Domino Reaction[17]

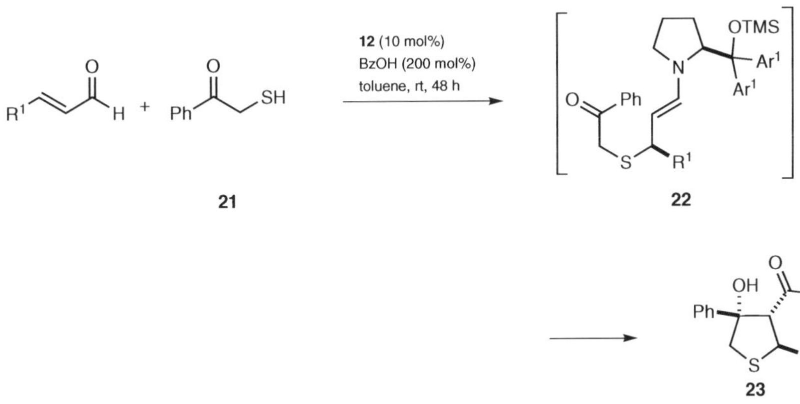

R^1	ee (%)	Yield (%)	Ref
Me	90	59	[17]
Pr	95	74	[17]
(*Z*)-$(CH_2)_2CH=CHEt$	93	61	[17]

for references see p 384

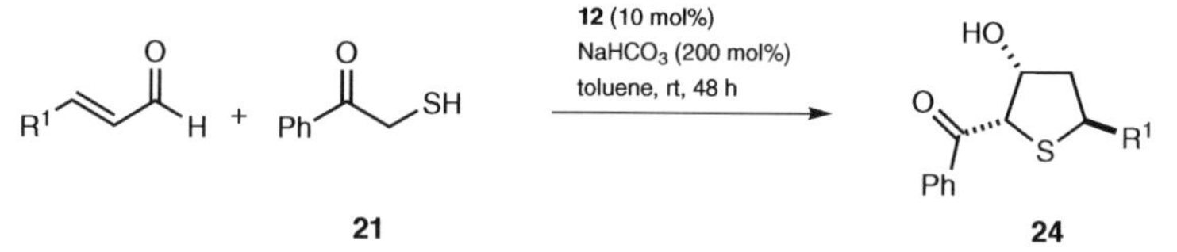

R^1	ee (%)	Yield (%)	Ref
iPr	80	61	[17]
Pr	82	43	[17]
$(CH_2)_2OTBDMS$	70	61	[17]

Chiral tetrahydrothiophenes can also be prepared via a Michael/Michael domino sequence.[18] The reaction is catalyzed by the diphenylprolinol trimethylsilyl ether **25** through an iminium–enamine mechanism (Scheme 11). In this transformation, compound **26** acts both as the nucleophile and the electrophile. Indeed, after the first conjugate addition of the thiol of **26** to the iminium ion of the enal, the resulting enamine undergoes an intramolecular conjugate addition onto the double bond of α,β-unsaturated ester **26**. The stereocontrol of the chiral catalyst on both steps results in excellent enantioselectivity (94 to >99% ee) and good to high diastereoselectivity (dr 6:1 to 18:1).

Scheme 11 Double Michael Domino Reaction Leading to Tetrahydrothiophenes[18]

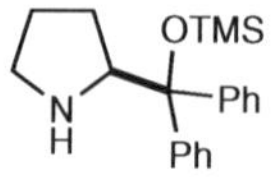

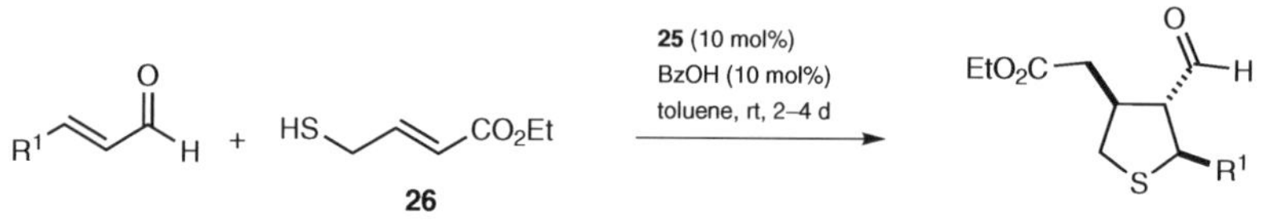

R^1	dr	ee (%)	Yield (%)	Ref
Ph	15:1	>99	76	[18]
$4-O_2NC_6H_4$	7:1	99	84	[18]
$4-F_3CC_6H_4$	11:1	98	69	[18]
$3-MeO-4-AcOC_6H_3$	17:1	99	89	[18]
$2-MeOC_6H_4$	18:1	98	96	[18]
CH_2OBn	7:1	94	62	[18]

4-Hydroxy-4-phenyltetrahydrothiophene-3-carbaldehydes 23 **and 2-Benzoyl-3-hydroxytetrahydrothiophenes** 24**; General Procedure:**[17]

Catalyst **12** (15.0 mg, 0.025 mmol) was added to a stirred soln of the α,β-unsaturated aldehyde (0.75 mmol) in toluene (2.5 mL), followed by the addition of 2-sulfanyl-1-phenylethanone (**21**; 38.1 mg, 0.25 mmol) and BzOH (61.0 mg, 0.50 mmol) to obtain tetrahydrothiophene-3-carbaldehydes **23** or $NaHCO_3$ (42.0 mg, 0.50 mmol) to obtain 2-benzoyl-3-hydroxytetrahydrothiophenes **24**. The mixture was stirred for 48 h at rt and then filtered on a small pad of silica gel (1–2 cm), washing through with Et_2O. The solvents were evaporated under reduced pressure. The crude product was purified by flash chromatography (silica gel, Et_2O/pentane). The ee of the product was determined by chiral HPLC after reduction to the corresponding alcohol.

2.3.2.1.2 Increasing Complexity in Organocatalyzed Domino Reactions

Employing diarylprolinol trimethylsilyl ether catalysts in the design of new domino transformations has rapidly led to the development of synthetic strategies that can be at the same time easy to manage and mechanistically complex, like all cascade reactions. A striking example is the triple cascade process developed by Enders and co-workers which affords cyclohexenecarbaldehydes **29** bearing four stereogenic centers through an enamine–iminium–enamine activation mode from α,β-unsaturated aldehydes, nitroalkenes **27**, and linear aldehydes **28** (Scheme 12).[27]

Scheme 12 Multicomponent Triple Domino Reaction: Control of Four Stereogenic Centers[27]

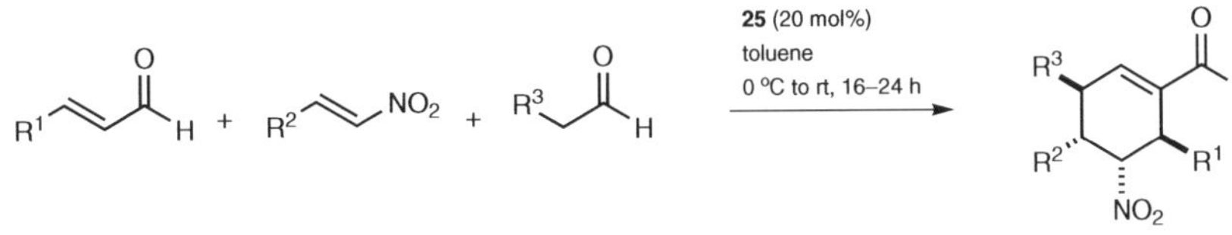
25 (20 mol%)
toluene
0 °C to rt, 16–24 h
R1
H
R2
NO2
R3
O
H
NO2
R1
R2
R3
O
H

27 28 29

R^1	R^2	R^3	dr	ee (%)	Yield (%)	Ref
Ph	Ph	Me	7.8:2.2	>99	40	[27]
Ph	2-ClC_6H_4	Me	8.4:1.6	>99	51	[27]
Ph	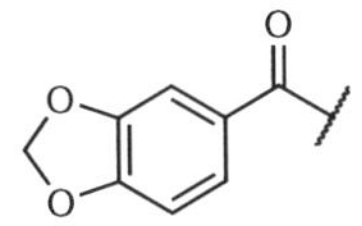	Me	8.7:1.3	>99	39	[27]
Ph	Ph	Et	8.0:2.0	>99	58	[27]
Ph	Ph	$CH_2OTBDMS$	9.9:0.1	99	54	[27]
Me	Ph	Me	6.8:3.2	>99	25	[27]

Scheme 13 provides the reported mechanism for the catalytic cycle of this triple domino reaction. In the first step, catalyst **25** activates the aldehyde by formation of an enamine, which then selectively adds to the nitroalkene in a Michael-type reaction. After hydrolysis, the catalyst **25** forms an iminium ion with the α,β-unsaturated aldehyde, a species which undergoes conjugate addition with the nitroalkane intermediate **30**. In the last step, the resulting enamine intermediate **31** undergoes an intramolecular aldol condensation to give intermediate **32**. Hydrolysis liberates the catalyst and delivers the product in exquisite enantioselectivity (≥99% in all cases) after elimination of water.[27]

for references see p 384

Scheme 13 Enamine–Iminium–Enamine Catalytic Cycle[27]

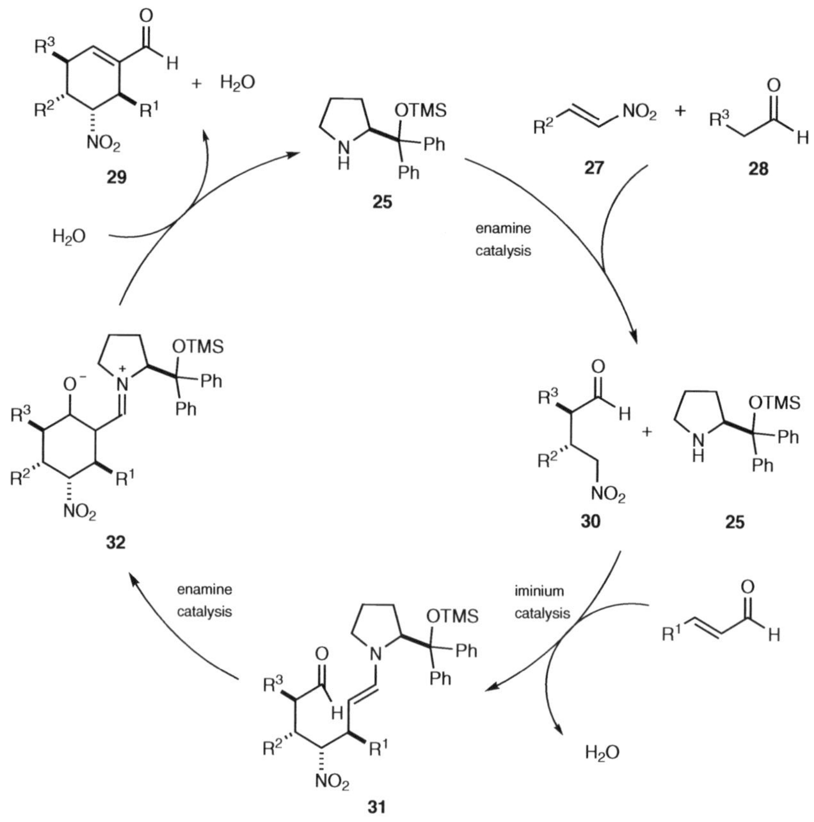

In a one-pot procedure, cyclohexenecarbaldehydes **29** can be directly converted into tricyclic systems (with eight stereogenic centers) upon treatment with a Lewis acid, a species which facilitates an asymmetric intramolecular Diels–Alder reaction.[28]

Cyclohexenecarbaldehydes can also be prepared by another three-component domino reaction (Scheme 14).[29] The products **34** are less structurally complex as compared to those produced from the Enders approach, but, nevertheless, the mode of activation and the reaction pathway is intriguing. Once again, the domino transformation is catalyzed by diarylprolinol trimethylsilyl ether **12**, which participates in all three reactive events with an iminium–iminium–enamine activation mode. The α,β-unsaturated aldehyde is first activated by catalyst **12** toward the conjugate nucleophilic attack of the active methylene compound **33** via iminium ion formation. The resulting adduct is again a nucleophilic species and can perform another Michael addition on the iminium ion of the α,β-unsaturated aldehyde, affording an enamine capable of intramolecular aldol condensation. The final products **34** bear up to three stereogenic centers (if EWG ≠ CN).

Scheme 14 Iminium–Iminium–Enamine Domino Process[29]

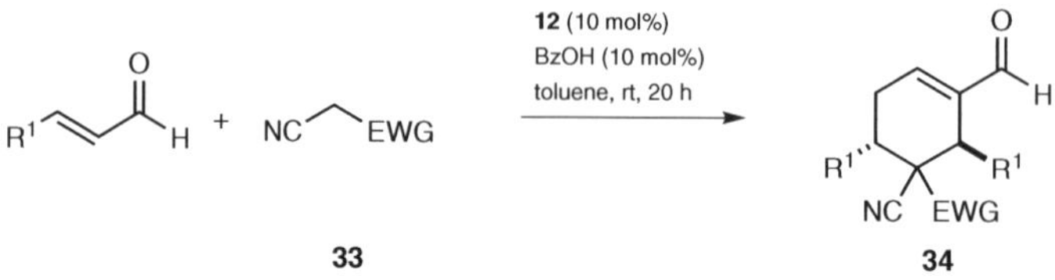

R^1	EWG	dr	ee (%)	Yield (%)	Ref
iPr	CN	>95:5	>99	68	[29]
Me	CO_2Me	86:14	98	47	[29]
Ph	CO_2iPr	>98:2	>99	53	[29]

Finally, diarylprolinol trimethylsilyl ether **12** has been successfully applied to the synthesis of cyclohexanols **36** via a nitro-Michael/Henry domino reaction, with the selective formation of up to five stereocenters (Scheme 15).[19] Once again, an iminium ion activates the α,β-unsaturated aldehyde toward the conjugate addition of the nitroalkane **35**, affording an intermediate which then undergoes an intramolecular aldol addition. The configuration of the stereocenters formed in this second step is thought to be controlled exclusively by those already present on the initial Michael adduct, as the catalyst does not participate in the final reactive events of the domino process.

Scheme 15 Formation of Five Stereocenters in One Domino Process[19]

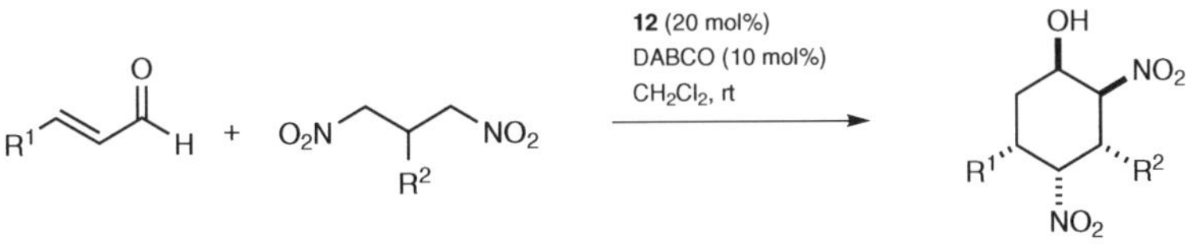

12 (20 mol%), DABCO (10 mol%), CH_2Cl_2, rt

35 → **36**

R^1	R^2	dr	ee (%)	Yield (%)	Ref
Et	Ph	4:2:1	90	45	[19]
iPr	4-MeOC_6H_4	5:1:1	89	53	[19]
Et	2-furyl	4:2:1	86	60	[19]
Et	2-thienyl	4:0:1	80	42	[19]
CH_2OTIPS	Ph	3:1:0	94	56	[19]

5-Nitrocyclohexene-1-carbaldehydes 29**; General Procedure:**[27]
To a soln of catalyst **25** (65 mg, 0.20 mmol) and nitroalkene **27** (1.00 mmol) in toluene (0.8 mL) was added subsequently under stirring aldehyde **28** (1.20 mmol) and the α,β-unsaturated aldehyde (1.05 mmol) at 0 °C. After 1 h, the soln was allowed to reach rt, and stirred until complete conversion of the starting materials (16–24 h, monitored by GC). The mixture was directly purified by flash column chromatography (silica gel, EtOAc/pentane 1:8 to 1:6). The dr and ee of the product were determined by chiral HPLC.

2.3.2.2 Domino Organocatalyzed Reactions of Oxindole Derivatives

Oxindole is an important heteroaromatic compound. It features a bicyclic structure, consisting of a six-membered benzene ring fused to a five-membered nitrogen-containing ring. Spiro compounds have always been prevalent in organic synthesis investigations due to their pronounced biological activities.[30] In particular, the spirocyclic oxindoles have emerged as attractive synthetic targets because of their occurrence in numerous natural products and biologically active molecules.[31–34] The key structural characteristic of these compounds is the presence of a spiro-fused ring attached to the C3 position of the oxindole core possessing varied heterocyclic motifs. These spirooxindoles seem to be promising candidates for drug discovery, because they concurrently incorporate both ox-

for references see p 384

indole and other heterocyclic moieties. Recently, oxindoles have drawn increasing attention as anticancer agents.[35] The spirooxindole ring system is present in a number of naturally-occurring alkaloids such as compounds **37–39** (Scheme 16). The non-spirocyclic quaternary stereocenter at the C3 position of oxindole can be found also in designed compounds with an important pharmaceutical activity such as the anticancer agent **40**, developed by Hoffmann-La Roche (Scheme 16).[36]

Scheme 16 Examples of Alkaloids Bearing the Spirooxindole Moiety[35–37]

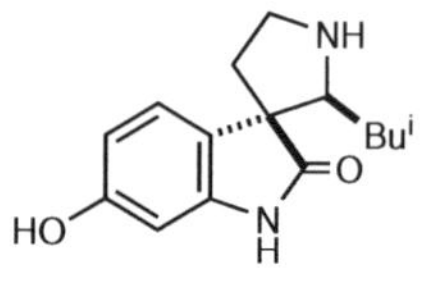

37 (+)-elacomine

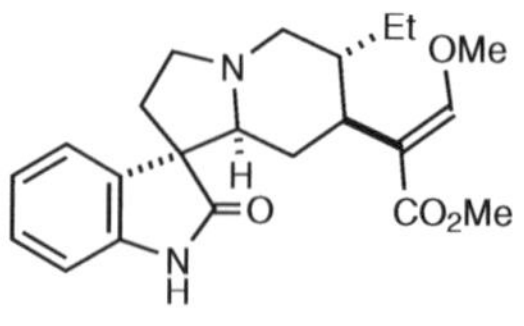

38 rhynchophylline

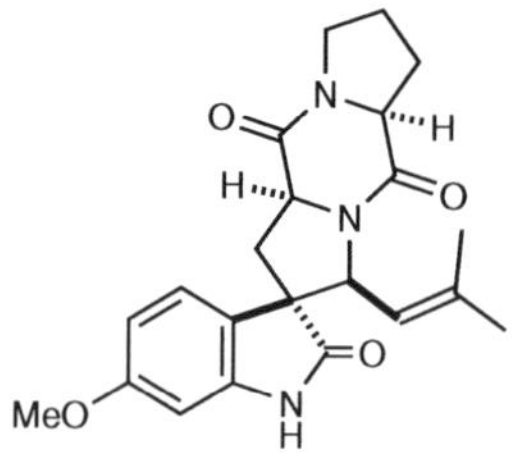

39 spirotyprostatin A

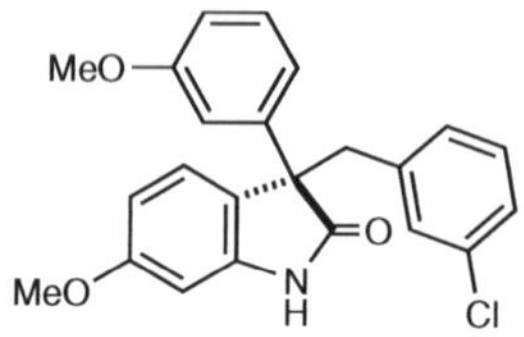

40

A recent interesting finding is that steroid oxindole hybrids show excellent cytotoxicity against human cancer cell lines and they can inhibit cell cycle and induce cell apoptosis.[37–39]

For the above reasons, the oxindole scaffold has been a recurring target for domino strategies. Here we list and comment upon a series of representative and meaningful examples.

2.3.2.2.1 From Enders' Domino Reactions to Melchiorre's Methylene Oxindole

In 2006, Enders and co-workers reported the development of an asymmetric organocatalytic triple cascade reaction for the synthesis of tetrasubstituted cyclohexenecarbaldehydes **29** (see Section 2.3.2.1.2).[27] This three-component domino reaction proceeds via a catalyzed Michael/Michael/aldol condensation sequence that affords the final products with good yields. During this sequence, four stereogenic centers are generated with complete enantioselectivity and very high diastereoselectivity. Furthermore, the starting materials can be varied to obtain diverse polyfunctional cyclohexene derivatives that can be used as versatile building blocks in useful synthetic processes.

This strategy inspired Melchiorre and his co-workers to develop analogous procedures that replace the nitro compounds with oxindole derivatives and related compounds. In the Melchiorre approach for the construction of spirocyclic oxindoles having multiple stereocenters,[39] complementary organocatalytic multicomponent domino reactions based on two distinct organocatalysts (**25** and **41**) were developed. These molecules efficiently activate carbonyl compounds (ketones and aldehydes, respectively) toward multiple asymmetric transformations in a well-defined cascade sequence. Both strategies provide straightforward access to a variety of natural-product-inspired compounds which would be difficult to obtain using other enantioselective methods in terms of either step count or enantioselection.

According to the authors, the recent advances achieved in this field have set the conditions for the development of many asymmetric cascade reactions based on the efficient activation of aldehydes and ketones by chiral secondary amines.[40,41] Catalyst **41**, a primary amine derived in a single step from the *Cinchona* alkaloid dihydroquinine, is a general and selective catalyst for ketone activation.[42] The formation of spirooxindoles **45** involves double Michael additions via an enamine–iminium activation sequence (Scheme 17).[39] The reaction of catalyst **41** with the α,β-unsaturated ketone produces enamine intermediate **43**, which reacts via Michael addition with oxindole **42** to give iminium intermediate **44**. Intramolecular Michael addition of this intermediate then furnishes product **45**.

Scheme 17 Double Michael Additions via an Enamine–Iminium Activation Sequence for the Synthesis of Spirocyclic Oxindole Cyclohexanones[39]

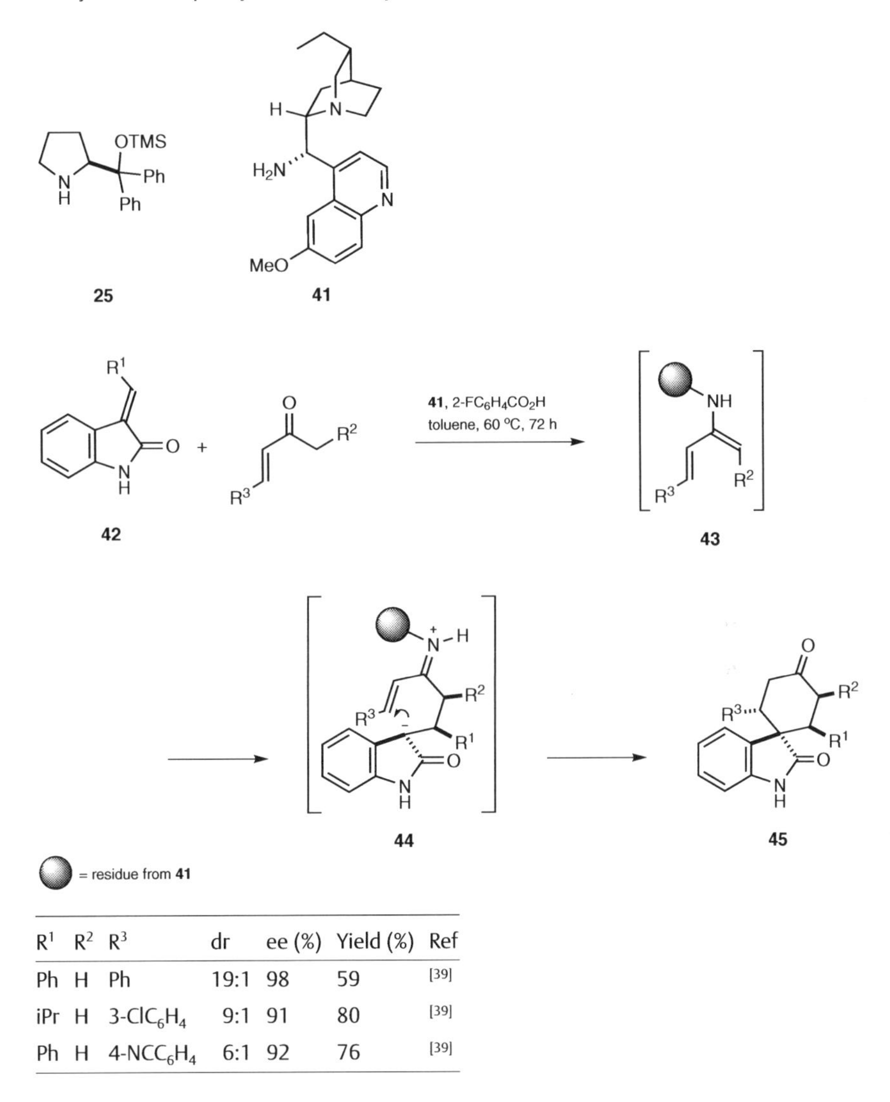

R^1	R^2	R^3	dr	ee (%)	Yield (%)	Ref
Ph	H	Ph	19:1	98	59	[39]
iPr	H	3-ClC_6H_4	9:1	91	80	[39]
Ph	H	4-NCC_6H_4	6:1	92	76	[39]

In the same paper, a complementary organocascade strategy based on the activation of aldehydes is reported in which the spirooxindole cyclohexene products **46** are built up with the simultaneous creation of three bonds and four stereogenic centers in a single chemical step catalyzed by chiral secondary amine **25** (Scheme 18).[39]

Scheme 18 Triple Cascade Mediated by a Chiral Secondary Amine by Enamine–Iminium–Enamine Activation of Aldehydes[39]

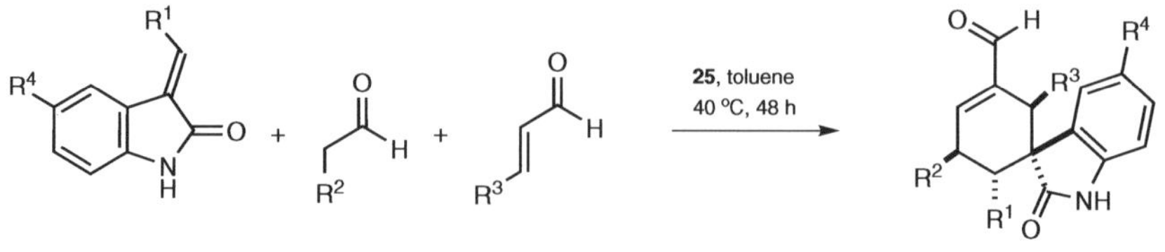

R^1	R^2	R^3	R^4	dr	ee (%)	Yield (%)	Ref
Ph	Me	Ph	H	12:1	>99	74	[39]
CO_2Et	Bn	Ph	H	19:1	>99	65	[39]
Ph	Me	$4\text{-}FC_6H_4$	H	19:1	>99	50	[39]

In 2011, the same authors reported the synthesis of a range of nearly enantiopure spirocyclic benzofuranones **50** in higher than 90% enantiomeric excess, high diastereoselectivity, and good yields (Scheme 19).[43] These products are generated by a three-component domino Michael/Michael/aldol/dehydration sequence involving aldehydes, α,β-unsaturated aldehydes, and benzofuranone-based compounds **47**. The catalytic action is performed by amine **25** in combination with 2-fluorobenzoic acid. The reaction sequence starts with the addition of the aldehydes onto the Michael acceptors **47**, evolving through enamine activation to give intermediate aldehydes **48**. Then, a Michael/aldol sequence occurs between aldehydes **48** and α,β-unsaturated aldehydes through iminium–enamine activation, which leads to intermediates **49**. These intermediates subsequently dehydrate to afford the desired spirocyclic benzofuranones **50**.

Scheme 19 Triple Cascade Mediated by a Chiral Secondary Amine by Enamine–Iminium–Enamine Activation of Aldehydes[43]

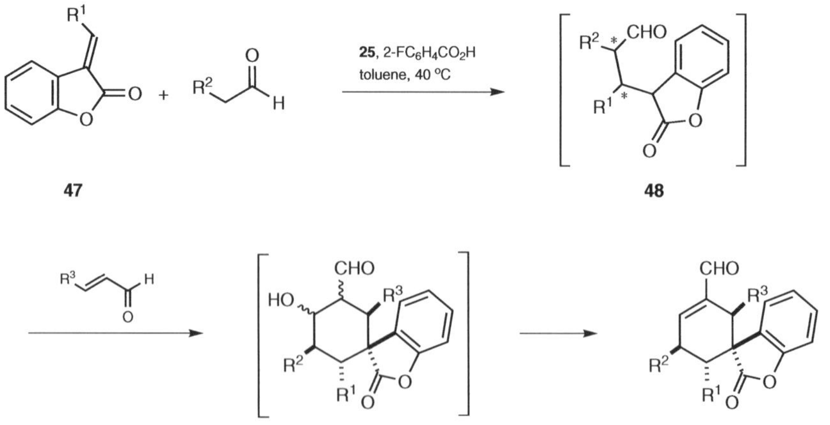

R^1	R^2	R^3	dr	ee (%)	Yield (%)	Ref
Ph	Me	Ph	>19:1	>99	56	[43]
4-ClC_6H_4	Me	Ph	>19:1	99	57	[43]
CO_2Et	Me	Ph	>19:1	>99	54	[43]

The remote control of the stereochemistry in the synthesis of spirocyclic oxindoles (e.g., **54**) is also achieved by vinylogous organocascade reactions of α,β,γ,δ-unsaturated dienones **52** and oxindoles **53** catalyzed by *Cinchona*-alkaloid-based primary amines (e.g., **51**; Scheme 20).[44] These compounds propagate the electronic effects inherent in iminium ion and enamine reactivity modes (i.e., the LUMO lowering and the HOMO raising activating effects) through the conjugated π-system of cyclic α,β,γ,δ-unsaturated dienones while transmitting the stereochemical information to distal positions.

Scheme 20 Vinylogous Cascade Catalysis Promoted by a Chiral Primary Amine[44]

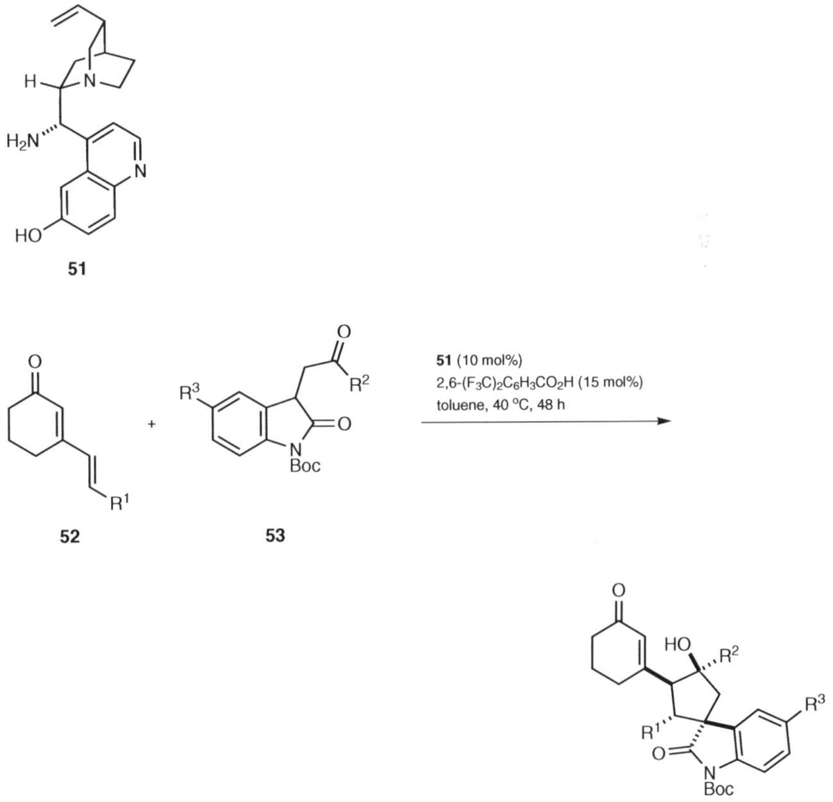

51

52 + **53**

51 (10 mol%)
2,6-$(F_3C)_2C_6H_3CO_2H$ (15 mol%)
toluene, 40 °C, 48 h

54

R^1	R^2	R^3	dr	ee (%)	Yield (%)	Ref
Ph	Me	H	12:1	97	83	[44]
3-ClC_6H_4	Me	H	7:2:1	97	80	[44]
Me	Me	H	3:2:1	94	62	[44]

for references see p 384

The Knoevenagel reaction consists of the condensation of aldehydes or ketones with active methylene compounds in the presence of a base.[45] The first organocatalytic enantioselective domino multicomponent reaction using this process as a component was developed by Barbas and co-workers in 2001 and involved an enantioselective Knoevenagel/Michael-type reaction between benzaldehyde, acetone, and diethyl malonate catalyzed by L-proline.[46] Ever since, the Knoevenagel reaction has been used to initiate other enantioselective domino multicomponent reactions. A recent example was reported by Yuan and co-workers who developed the first enantioselective organocatalytic three-component domino Knoevenagel/Michael addition/cyclization reaction, giving access to a range of chiral spiro[4*H*-pyran-3,3′-oxindoles] **57** (Scheme 21).[47]

Scheme 21 Three-Component Domino Knoevenagel/Michael Addition/Cyclization Reaction[47]

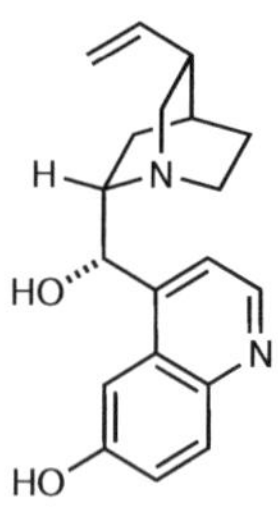

H N HO HO N

55

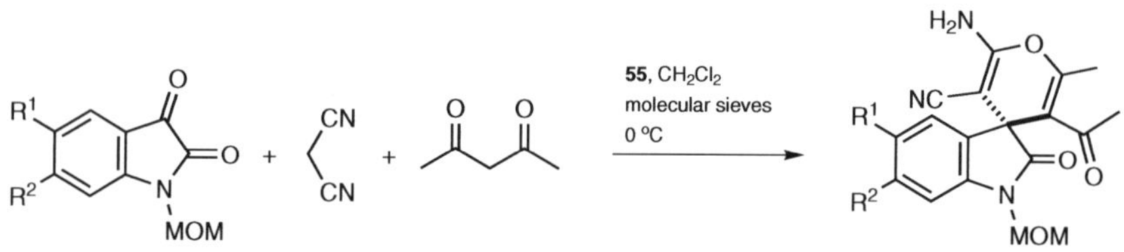

R1 R2 O O N MOM **56** + CN CN + O O

55, CH_2Cl_2
molecular sieves
0 °C

H_2N O NC R1 R2 O N MOM O **57**

R^1	R^2	ee (%)	Yield (%)	Ref
H	H	95	93	[47]
F	H	95	93	[47]
Cl	H	72	95	[47]

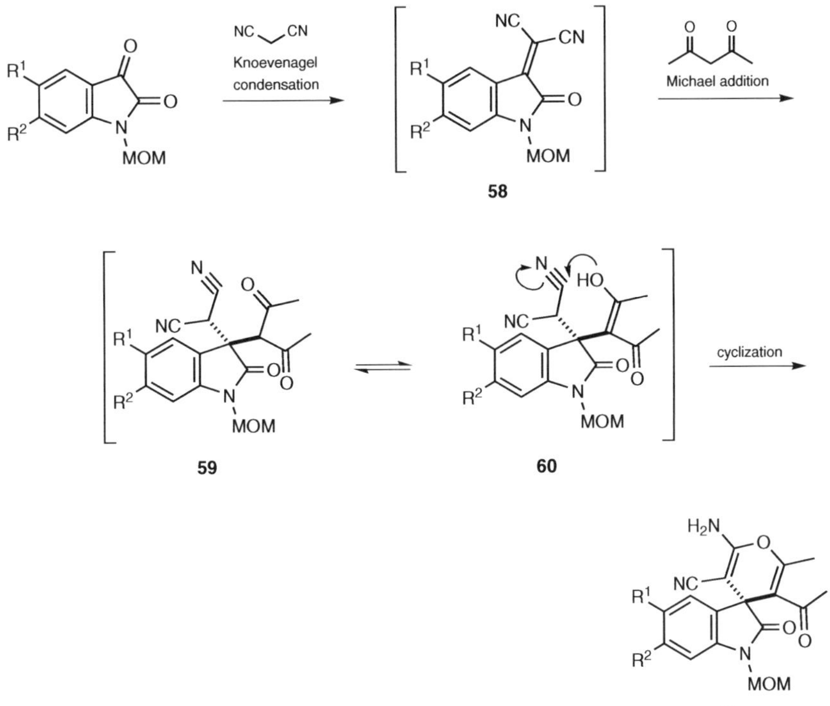

This novel and highly efficient sequence is catalyzed by cupreine (**55**) and involves malononitrile, a 1,3-diketone, and isatin derivatives **56**. The spiro heterocyclic products **57** are generally obtained in both high yields and enantioselectivities of up to 97% enantiomeric excess. The authors proposed the mechanism depicted in Scheme 21, in which the sequence starts with the Knoevenagel condensation of malononitrile onto the isatin derivatives to give intermediates **58**; then, Michael addition of a 1,3-diketone leads to intermediates **59**, in equilibrium with compounds **60**, which cyclize to afford the products.

More recently, Macaev and co-workers investigated the reaction of isatin with malononitrile and acetylacetone catalyzed by brevicolline (10 mol% in CH_2Cl_2 at 0 °C).[48] The corresponding spiro[4*H*-pyran-3,3′-oxindole] was formed in 62% yield and 94% enantiomeric excess.

The efficient assembly of hydroindane derivatives incorporating a spirooxindole motif is achieved through three-component enantioselective domino reactions of two molecules of an α,β-unsaturated aldehyde with (*E*)-4-(1-methyl-2-oxoindolin-3-ylidene)-3-hydroxybut-2-enoates **61** catalyzed by chiral amine **25** in combination with benzoic acid as co-catalyst.[49] According to a Michael/Michael/Michael/aldol sequence, the process provides enantio- and diastereopure fused tetracyclic products **62** (Scheme 22). The yields range from moderate to excellent. The products bear six contiguous stereogenic centers. The authors propose a quadruple iminium–enamine–iminium–enamine catalysis as outlined in Scheme 22. After the first domino Michael/Michael reaction of the butanoates **61** and 1 equivalent of an α,β-unsaturated aldehyde, compounds **64** are generated through intermediates **63**. The second domino Michael/aldol reaction occurs between adducts **64** and a second equivalent of an α,β-unsaturated aldehyde, leading to the final products via intermediates **65**.

for references see p 384

Scheme 22 Synthesis of Fused Tetracyclic Products and Postulated Mechanism[49]

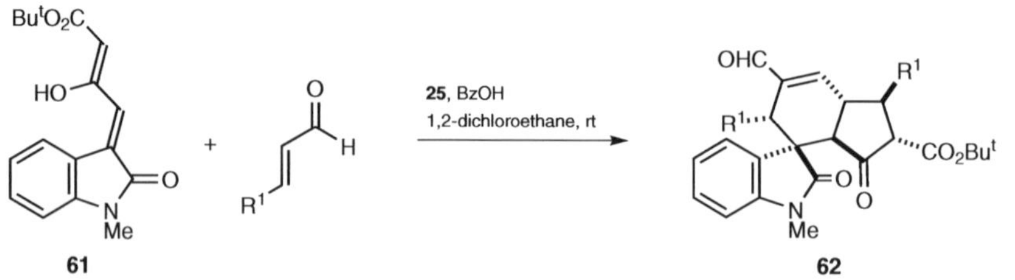

R^1	ee (%)	Yield (%)	Ref
Me	>99	66	[49]
2-BrC_6H_4	>99	84	[49]
Ph	>99	97	[49]

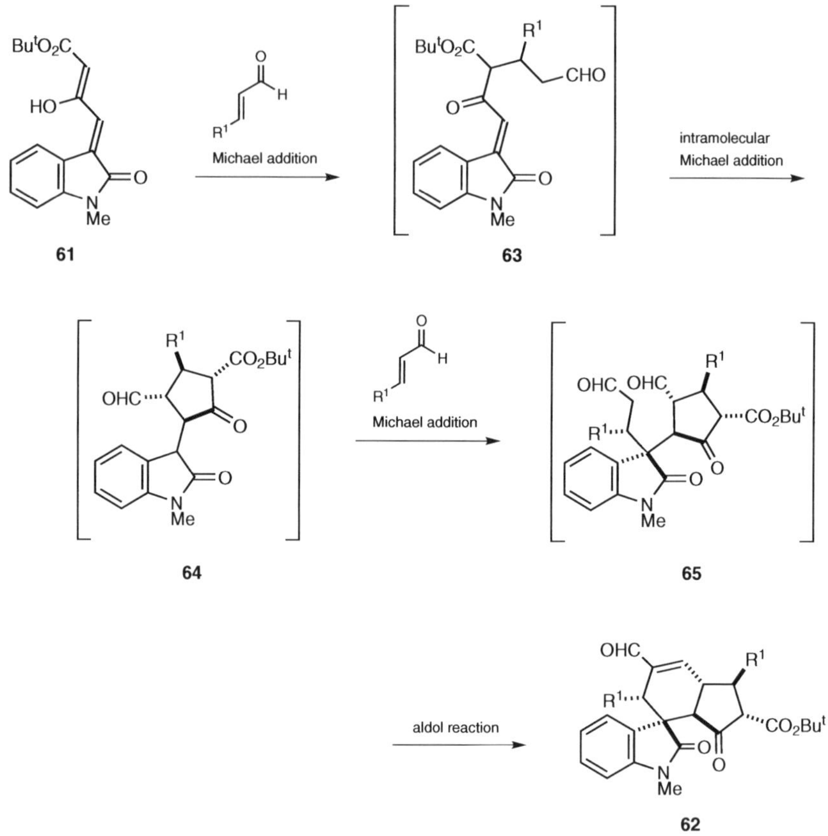

Spirocyclic Oxindoles 54; **General Procedure:**[44]

An ordinary vial equipped with a Teflon-coated stirrer bar and a plastic screw cap was charged with amine catalyst **51** (3.1 mg, 0.01 mmol, 10 mol%). 2,6-Bis(trifluoromethyl)benzoic acid (3.9 mg, 0.015 mmol, 15 mol%) and toluene (0.1 mL) were sequentially added and the resulting soln was stirred at ambient temperature for 10 min to allow catalyst salt formation. The reaction was started by the sequential addition of the cyclic α,β,γ,δ-unsaturated dienone **52** (0.1 mmol) and the oxindole derivative **53** (0.15 mmol,

1.5 equiv). The vial was sealed and immerged in a silicone-oil bath (thermostated at 40 °C) and stirring was continued over 48 h. The crude mixture was flushed through a short plug of silica gel [CH_2Cl_2/Et_2O 1:1 (5 mL)]. The solvent was removed under reduced pressure and the products were isolated by flash column chromatography (silica gel).

2.3.2.2.2 Michael Addition to Oxindoles

Asymmetric organocatalytic domino reactions are not limited to amine catalysis. Indeed, significant contributions have also been made in the field of organocatalysis operating through hydrogen bonding and Brønsted acid catalysis. These catalysts activate the substrates by formation of a LUMO-lowering hydrogen bond and promote the construction of C—C and C—heteroatom bonds. The interaction between the catalyst and the substrate is noncovalent and the chiral ion pair is the intrinsically activated species. In particular, chiral thiourea-based derivatives and phosphoric acid derivatives are well-known for their application as effective Brønsted acid organocatalysts. A recent example of this type of activation was reported by Gong and Wei who investigated a series of Takemoto-type catalysts to promote the domino Michael/Michael reaction of Nazarov reagents **67** with methyleneindolones **68** to provide the corresponding spiro[4-cyclohexanone-1,3′-oxindoline] derivatives **69**.[50] Among a series of bifunctional (thio)urea catalysts studied, urea **66** was selected as the optimal catalyst, allowing the formal [4+2] cycloadducts to be obtained in excellent yields and diastereo- and enantioselectivities (Scheme 23). The activation of the substrates through hydrogen bond formation with the catalyst possessing Brønsted acidic (HB) and Lewis basic (LB) functionalities is depicted in the same scheme.

Scheme 23 Domino Michael/Michael Reaction of Nazarov Reagents with Methyleneindolines through the Hydrogen-Bonding Activation Mode[50]

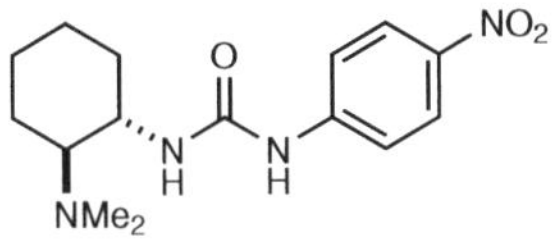

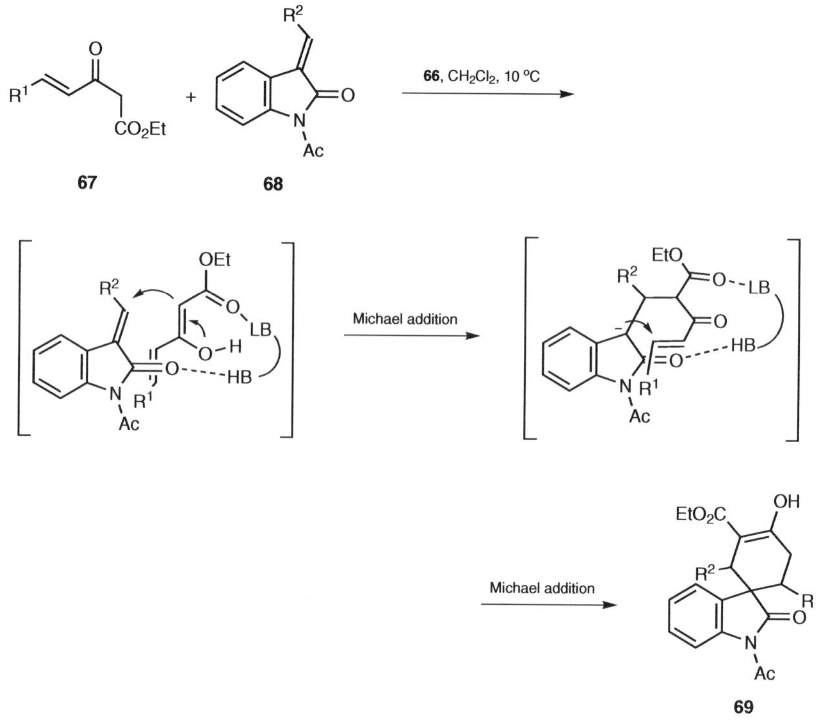

R^1	R^2	dr	ee (%)	Yield (%)	Ref
Ph	4-$MeOC_6H_4$	98:2	93	80	[50]
Ph	4-ClC_6H_4	92:8	91	91	[50]
Pr	Ph	97:3	90	89	[50]

The hydrogen-bonding activation mode was also recently implicated by Barbas and coworkers to explain the formation of chiral bispirooxindoles **71** on the basis of a novel asymmetric domino Michael/aldol reaction occurring between 3-substituted oxindoles and a range of methyleneindolinones.[51] This process is catalyzed by a novel multifunctional organocatalyst **70** containing tertiary and primary amines as well as a thiourea moiety to activate the substrates simultaneously, providing ultimately for extraordinary levels of stereocontrol in the generation of four stereocenters, two of which are quaternary carbon centers.

As shown in Scheme 24, this novel methodology provides a facile access to a number of potent, biologically active, multisubstituted bispirooxindoles **71** with high yields and excellent diastereo- and enantioselectivities of up to >98% diastereomeric excess and 96% enantiomeric excess. Even more interestingly, the authors have proposed the possibility to prepare the opposite enantiomers of these novel multifunctionalized products **71**. Indeed, they demonstrated that by performing the reaction of oxindole with methyleneindolinone in the presence of a reconfigured catalyst, one in which the tertiary amine and the thiourea configurations are changed in comparison to catalyst **70**, can give rise to the opposite enantiomer in good yield combined with high diastereoselectivity (88% de) and high enantioselectivity (90% ee). The possibility to obtain both enantiomers of these novel products will be highly useful in the investigation of their biological activity. Furthermore, these studies highlighted the growing potential of reaction and catalyst design in executing full control over these powerful processes.

Scheme 24 Domino Michael/Aldol Reactions of 3-Substituted Oxindoles with Methyleneindolinones through the Hydrogen-Bonding Activation Mode[51]

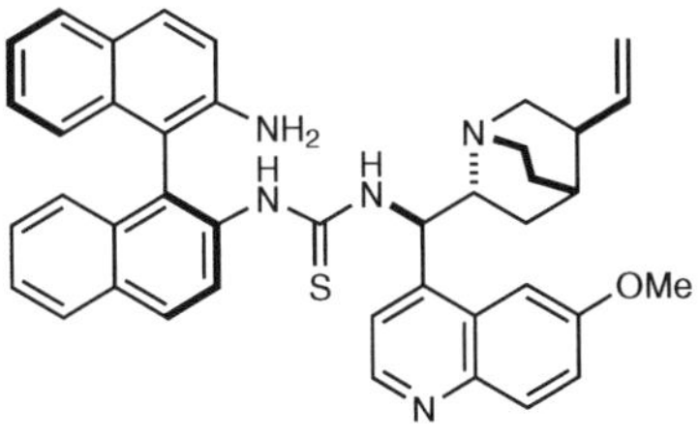

70

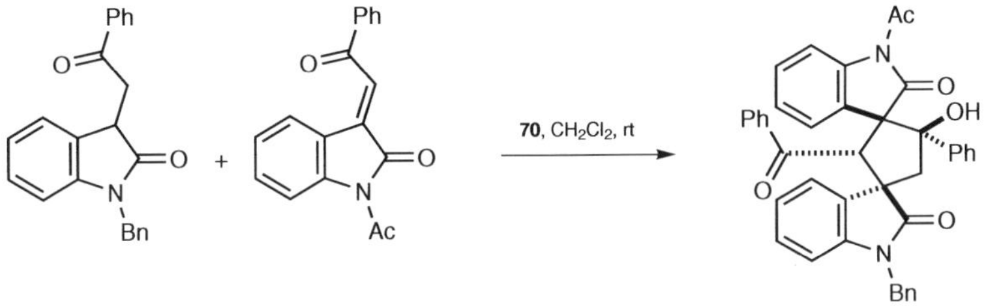

71

Wang and co-workers developed an efficient Michael/ketone aldol/dehydration domino reaction with indole **73** catalyzed by chiral alkaloid **72** to afford spiro[cyclohex-2-enone–oxindole] motifs (e.g., **74**) with high yields (up to 99%), excellent diastereoselectivities (dr >20:1), and enantioselectivities (up to 96% ee; Scheme 25).[52]

Scheme 25 Domino Reaction To Afford a Spiro[cyclohex-2-enone–oxindole] Catalyzed by a Chiral Amine[52]

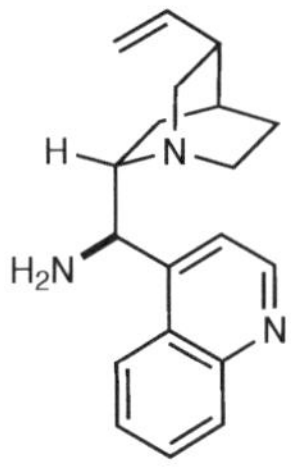

72

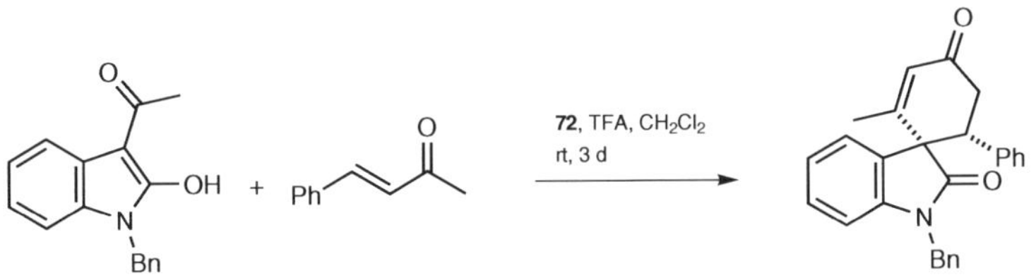

73 74

Rios and co-workers also reported the synthesis of spiro compounds **75** via a cascade Michael/Michael/aldol reaction.[53] The reaction process, the proposed mechanism, and the substrate scope are outlined in Scheme 26.

for references see p 384

Scheme 26 Synthesis of Spiro Compounds via a Cascade Michael/Michael/Aldol Reaction Catalyzed by a Chiral Amine[53]

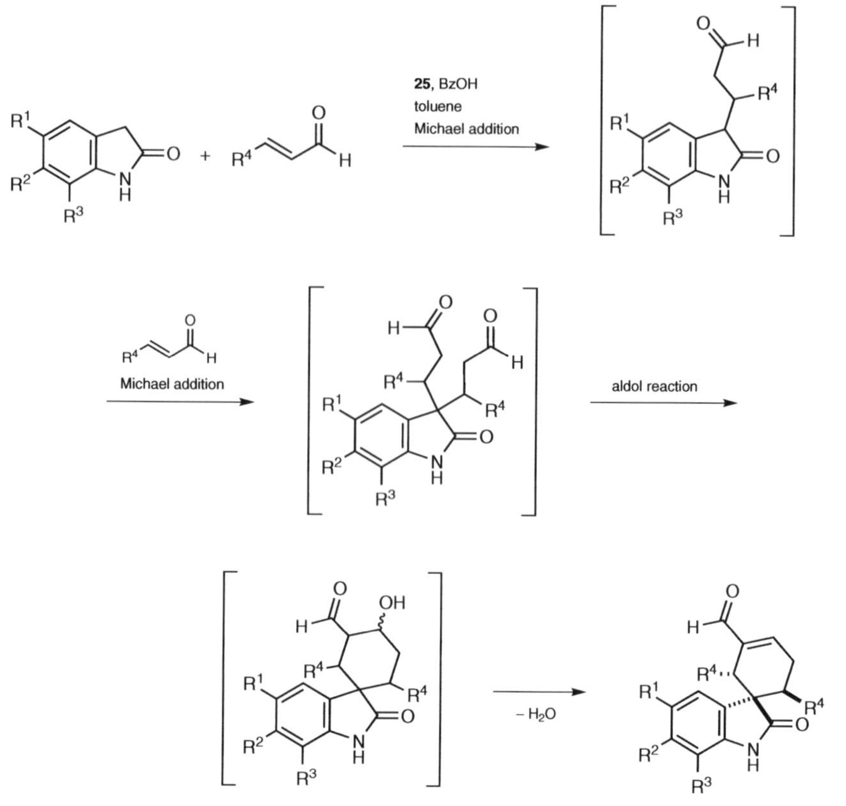

25, BzOH
toluene
Michael addition

Michael addition

aldol reaction

$- H_2O$

75

R^1	R^2	R^3	R^4	dr	ee (%)	Yield (%)	Ref
H	H	H	Ph	>25:1	>99	71	[53]
NO_2	H	H	Ph	>25:1	>99	62	[53]
H	H	Cl	Ph	>25:1	>99	90	[53]

More recently, Bartoli and co-workers reported an enantioselective nitrocyclopropanation of oxindoles **77** with bromo(nitro)methane which is induced through hydrogen-bonding activation by 9-*epi*-9-thiourea-9-deoxydihydroquinidine (**76**) in the presence of sodium hydrogen carbonate.[54] As shown in Scheme 27, the corresponding spiro nitrocyclopropyl oxindoles **78** are obtained in good to high yields, moderate to high diastereoselectivities, and excellent enantioselectivities (up to 98% ee).

Scheme 27 Synthesis of Spiro Nitrocyclopropyl Oxindoles through the Hydrogen-Bonding Activation Mode[54]

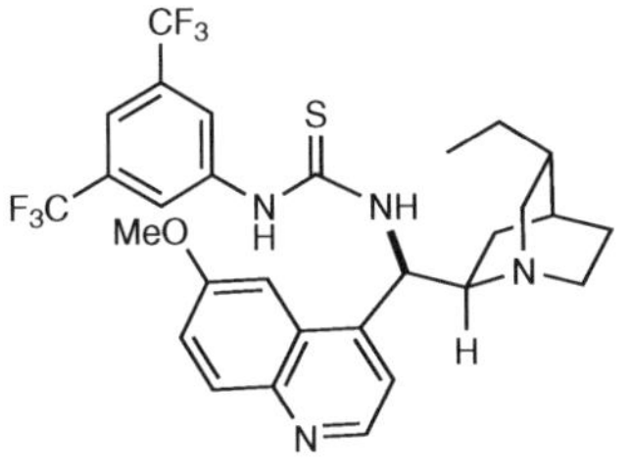

76

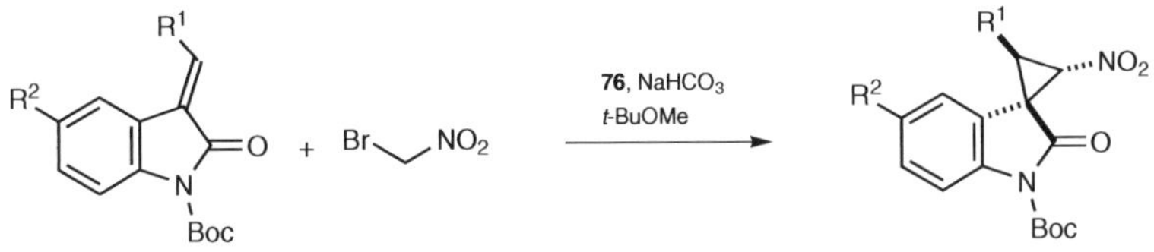

77 + Br⌄NO$_2$ → (76, NaHCO$_3$, t-BuOMe) → 78

R^1	R^2	dr	ee (%)	Yield (%)	Ref
CO_2Et	H	19:1	98	77	[54]
Bz	H	19:1	98	82	[54]
Ph	Cl	8:1	90	84	[54]

In another noteworthy paper, Barbas and co-workers presented an organocatalytic strategy for the synthesis of highly substituted spirocyclopentaneoxindoles **81** employing nitroalkenes and 3-substituted oxindoles **80** as starting materials.[55] Michael–Henry cascade reactions, facilitated by *Cinchona* alkaloids (e.g., **79**), provide the products in a single step with high yield and excellent enantioselectivity (Scheme 28).

Scheme 28 Synthesis of Spirocyclopentaneoxindoles Catalyzed by a *Cinchona* Alkaloid[55]

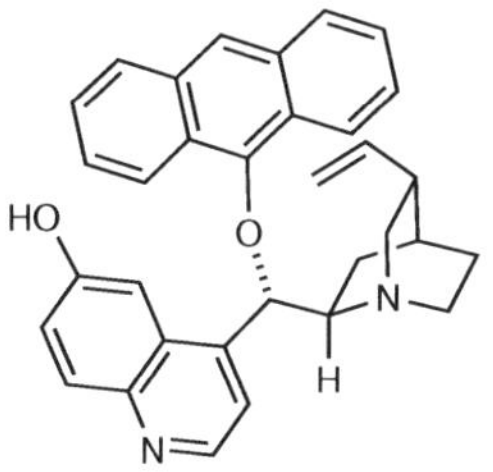

79

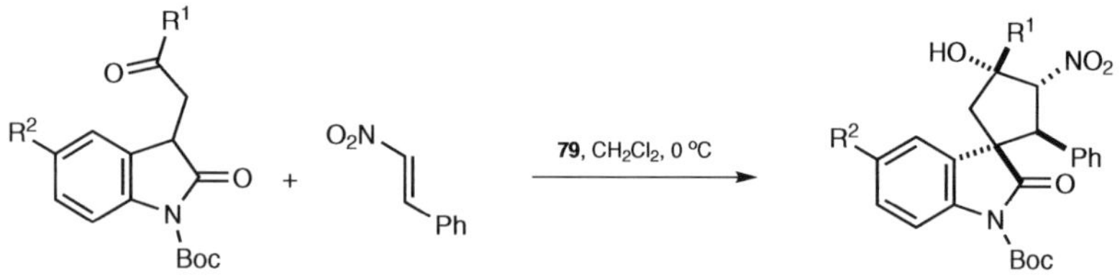

R^1	R^2	dr	ee (%)	Yield (%)	Ref
Me	H	11:1	94	39	[55]
Me	Cl	8:1	90	97	[55]
Et	H	11:1	91	96	[55]

Wang and co-workers have reported a highly efficient procedure for the synthesis of bispirooxindole derivatives **84** containing three stereocenters, including two spiro quaternary centers (Scheme 29).[56] The products are obtained through a stereocontrolled cascade Michael/cyclization reaction between methyleneindolinones **82** and (isothiocyanato)oxindole **83** catalyzed by multifunctional thiourea organocatalyst **70**. Mild conditions give the bispirooxindoles **84** with excellent enantio- and diastereoselectivity. Amazingly, these reactions take place in less than 1 minute; furthermore, employing a quasi-enantiomeric catalyst offers access to the opposite enantiomer.

Scheme 29 Synthesis of Spirooxindoles via Organocatalytic Cascade Michael/Cyclization Reactions[56]

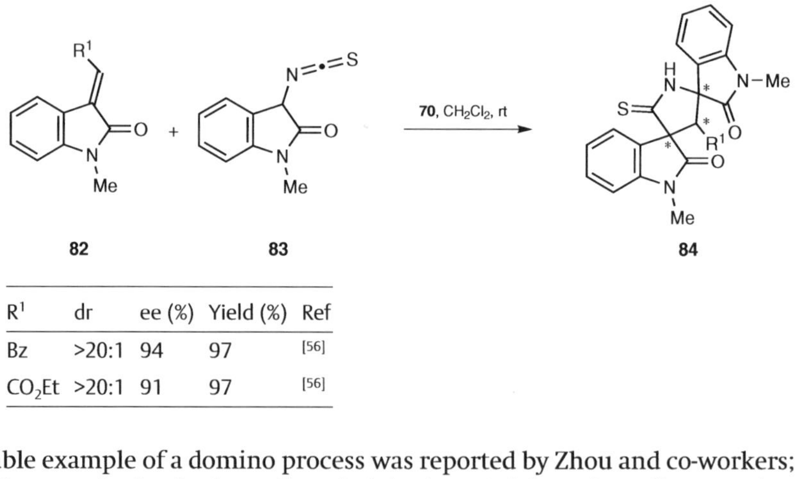

R^1	dr	ee (%)	Yield (%)	Ref
Bz	>20:1	94	97	[56]
CO_2Et	>20:1	91	97	[56]

A recent notable example of a domino process was reported by Zhou and co-workers; it is a one-pot tandem synthesis of spirocyclic oxindoles (e.g., **87**) featuring adjacent spiro stereocenters from isatin, propenal, and 1-methyl-1*H*-pyrrolo[2,3-*b*]pyridine-2,3-dione (**86**).[57] This approach consists of a Morita–Baylis–Hillman reaction/bromination/[3+2] annulation sequence and involves two distinct nucleophilic catalytic steps using the same tertiary amine catalyst **85** (Scheme 30).

Scheme 30 One-Pot Organocatalyzed Morita–Baylis–Hillman/Bromination/[3+2] Annulation Reaction Sequence[57]

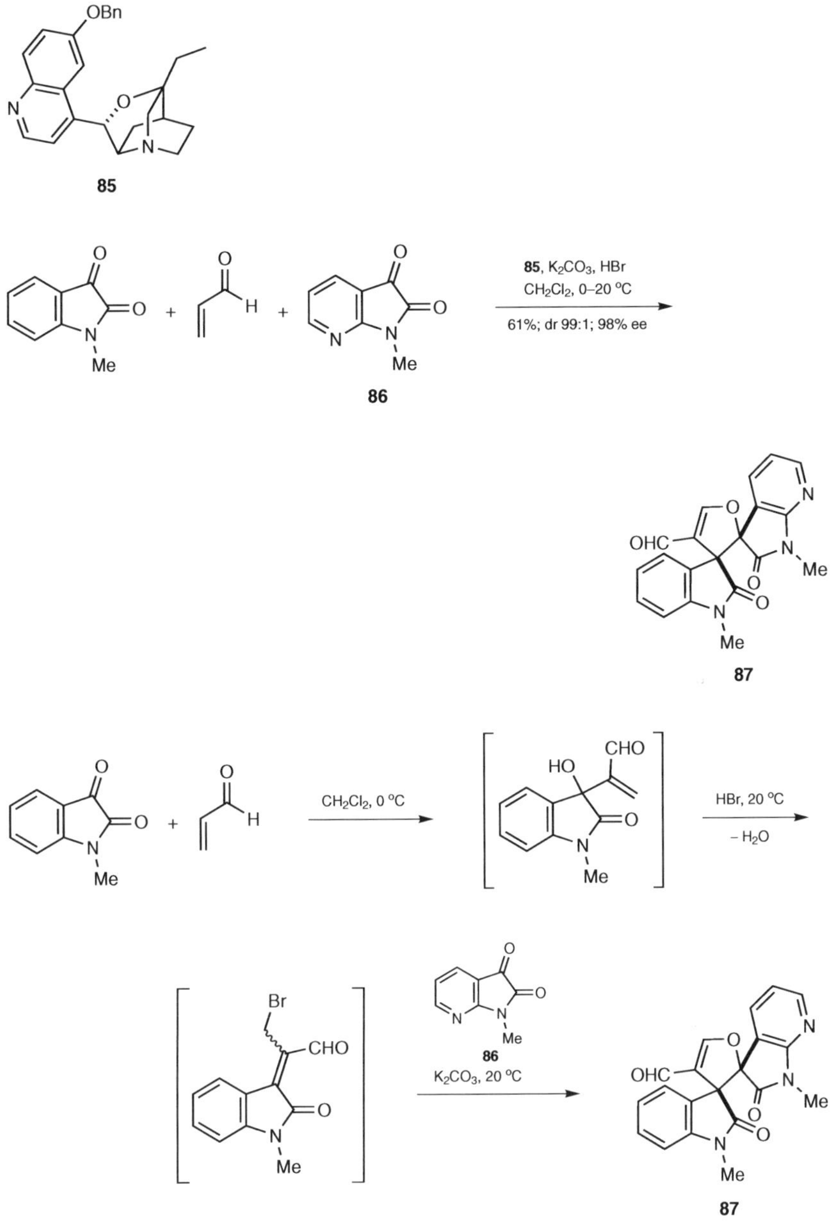

85, K_2CO_3, HBr
CH_2Cl_2, 0–20 °C
61%; dr 99:1; 98% ee

CH_2Cl_2, 0 °C

HBr, 20 °C
– H_2O

86
K_2CO_3, 20 °C

As already pointed out, methyleneindolinones can serve as essentially perfect electron-deficient alkenes because of their high reactivity as Michael acceptors, as well as their unique structural characteristic for the construction of 2′-thioxo-3,3′-pyrrolidinyl spirooxindoles. Wang and co-workers reported the enantioselective synthesis of compounds **91** through a Michael addition/cyclization sequence of α-isothiocyanato imide **90** and methyleneindolinone **89** catalyzed by chiral thiourea **88** (Scheme 31).[58]

Scheme 31 Strategy for the Synthesis of the 2′-Thioxo-3,3′-pyrrolidinyl Spirooxindole Scaffold Using a Bifunctional Chiral Catalyst[58]

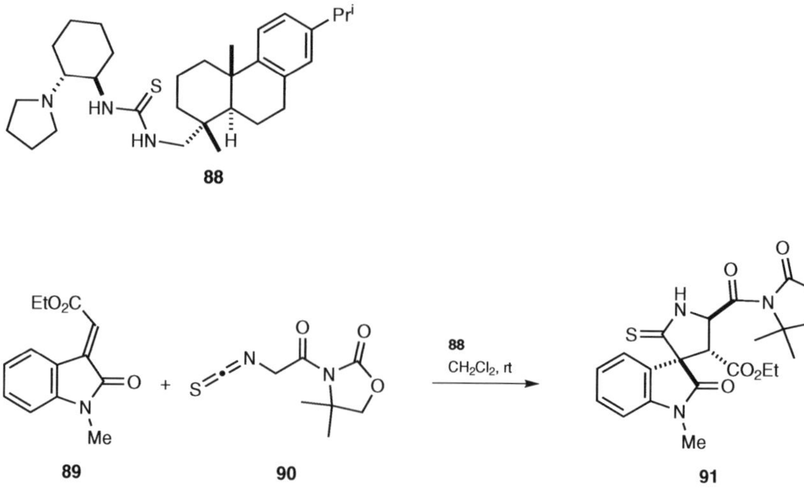

An asymmetric organocatalytic one-pot synthesis of six-membered spirocyclic oxindoles was developed by Li and co-workers through a relay Michael/Michael/aldol addition reaction between oxindoles **93**, nitroalkenes **94**, and α,β-unsaturated aldehydes **95** catalyzed by the combination of diarylprolinol trimethylsilyl ether **25** and bifunctional quinine thiourea **92**.[59] This protocol affords substituted spirocyclic oxindoles **96** in high yields and excellent stereoselectivity (Scheme 32).

Scheme 32 Construction of Spirocyclic Oxindoles Catalyzed by a Combination of Amines[59]

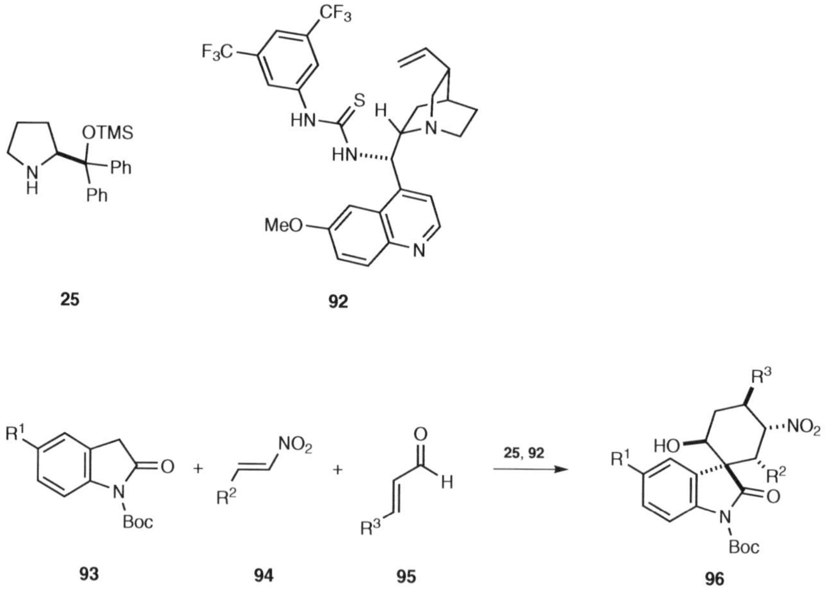

R^1	R^2	R^3	dr	ee (%)	Yield (%)	Ref
H	Ph	Ph	6:2.5:1	>99	85	[59]
H	2-ClC_6H_4	Ph	4.5:3.1:1	>99	94	[59]
OMe	Ph	Ph	6:7.5:1	>99	94	[59]

Zhang and co-workers reported the enantioselective synthesis of 3,3′-disubstituted oxindoles by the organocatalytic Michael addition of indoles to 2-(2-oxoindol-3-ylidene)acetaldehydes with trimethylsilyl prolinol ether **12** as catalyst, a process which can be used for the total synthesis of (−)-chimonanthine and the core structure of (+)-gliocladin C.[60]

Ethyl 1′-Acetyl-4-hydroxy-2′-oxospiro[cyclohexane-1,3′-indolin]-3-ene-3-carboxylates 69; **General Procedure:**[50]

A mixture of a Nazarov reagent **67** (0.4 mmol), urea catalyst **66** (6.2 mg, 0.02 mmol), a methyleneindolinone **68** (0.2 mmol), and 4-Å molecular sieves (200 mg) in CH_2Cl_2 (1.0 mL) was stirred at 10 °C (some of the reactions under other temperatures) until the reaction was complete (1–7 d, monitored by TLC). The resultant soln was purified by flash column chromatography (silica gel, petroleum ether/EtOAc 30:1 to 15:1).

1′-(*tert*-Butoxycarbonyl)-2-nitro-2′-oxospiro[cyclopropane-1,3′-indolines] 78; **General Procedure:**[54]

All the reactions were carried out in undistilled *t*-BuOMe. Bromo(nitro)methane (42 mg, 0.3 mmol, 1.5 equiv) was added to thiourea catalyst **76** (6.0 mg, 0.01 mmol, 5 mol%) in *t*-BuOMe (1 mL) in a vial equipped with a Teflon-coated stirrer bar. After 5 min, *N-tert*-butoxycarbonyl-protected oxindole **77** (0.2 mmol, 1.0 equiv) and Na_2CO_3 (0.2 mmol, 1 equiv) were added and the resulting soln was stirred at 0 °C for 48 h. The crude mixture was diluted with CH_2Cl_2 and flushed through a short plug of silica gel [CH_2Cl_2/EtOAc 1:1 (100 mL)]. The solvent was removed under reduced pressure and the product was isolated by flash column chromatography.

2.3.2.3 Synthesis of Tamiflu: The Hayashi Approach

(−)-Oseltamivir (**98**), marketed as the corresponding phosphate under the name Tamiflu, is a neuroamidase inhibitor used in the treatment of type A and B human influenza.[61] In the last few years, the fear of a possible influenza outbreak prompted many nations to plan the storage of a significant amount of this compound, so the scientific community was asked to develop efficient preparations of this drug and its derivatives in an effort to keep costs low and to enable ready preparation if needed. Two fully functionalized key intermediates **97** and **99** for the synthesis of (−)-oseltamivir (**98**) are shown in Scheme 33.[62–64]

for references see p 384

Scheme 33 Fully Functionalized Intermediates for (−)-Oseltamivir Synthesis Obtained through Different One-Pot Domino Sequences[62–64]

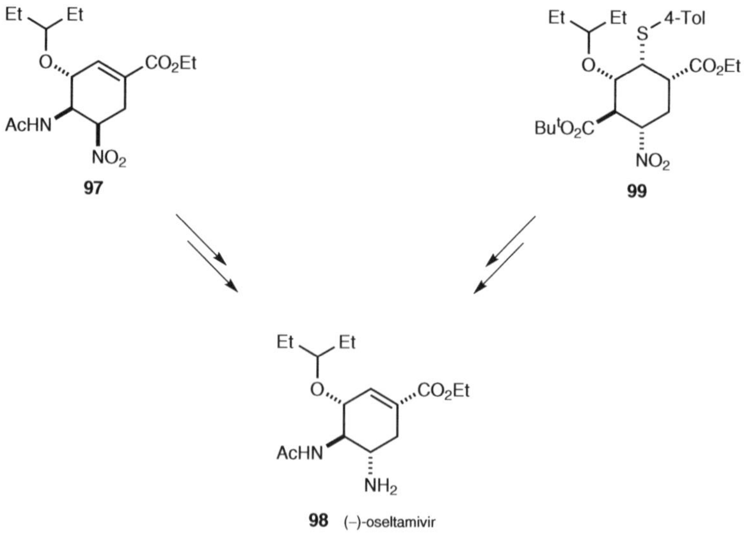

Et Et O CO2Et AcHN NO2 **97**

Et Et S 4-Tol O CO2Et Bu^tO_2C NO2 **99**

Et Et O CO2Et AcHN NH2 **98** (−)-oseltamivir

With this framework in mind, with the key objective being the preparation of large amounts of (−)-oseltamivir (**98**) in a short time and at low cost, the Hayashi group developed three elegant strategies for its synthesis based on one-pot operations and domino reactions (Scheme 34).[62]

The first synthesis of (−)-oseltamivir (**98**) from Hayashi's group was reported in 2009 and relied on the construction of the key functionalized cyclohexane carboxylate intermediate **99** in a single-pot operation, leaving only functional-group manipulation to complete the synthesis (Scheme 34).[62]

The construction of compound **99** is a premier example of the power of domino transformations and one-pot reactions in organic synthesis. Indeed, it results from the Michael addition of aldehyde **100** to nitroalkene **101** followed by a domino Michael–Horner–Wadsworth–Emmons reaction combined with retro-aldol and retro-Michael reactions and finally a thiol-Michael addition followed by a base-catalyzed isomerization.

Scheme 34 Total Synthesis of (−)-Oseltamivir via Three One-Pot Operations[62]

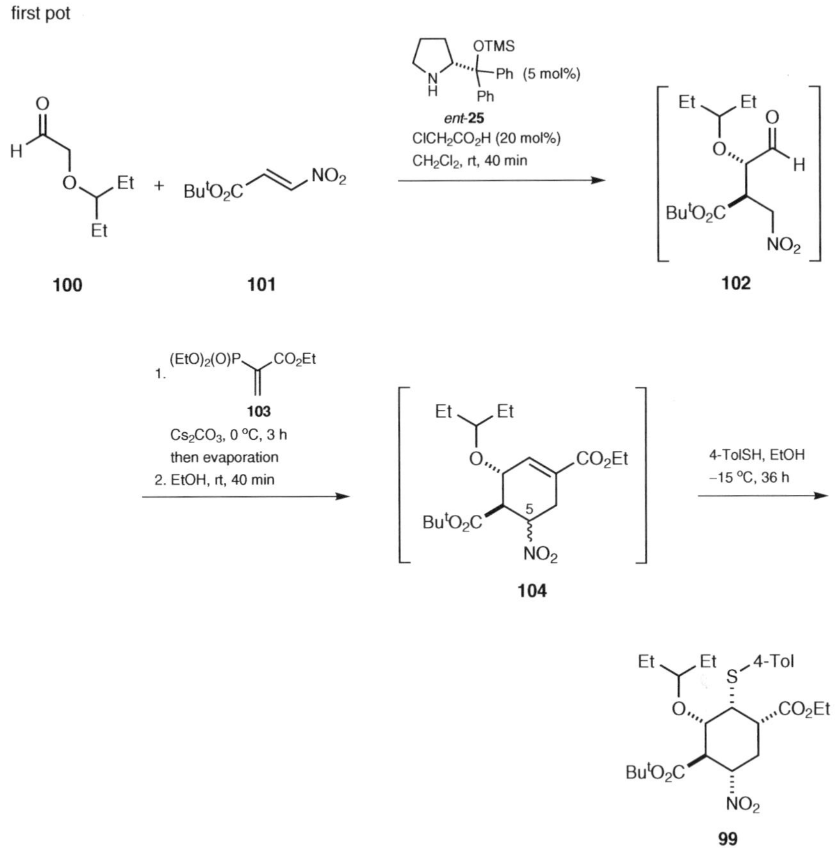

first pot

100 + 101 → 102

ent-25 (5 mol%)
$ClCH_2CO_2H$ (20 mol%)
CH_2Cl_2, rt, 40 min

1. $(EtO)_2(O)P$–C(=CH_2)–CO_2Et 103
Cs_2CO_3, 0 °C, 3 h
then evaporation
2. EtOH, rt, 40 min

→ 104

4-TolSH, EtOH
−15 °C, 36 h

→ 99

for references see p 384

second pot

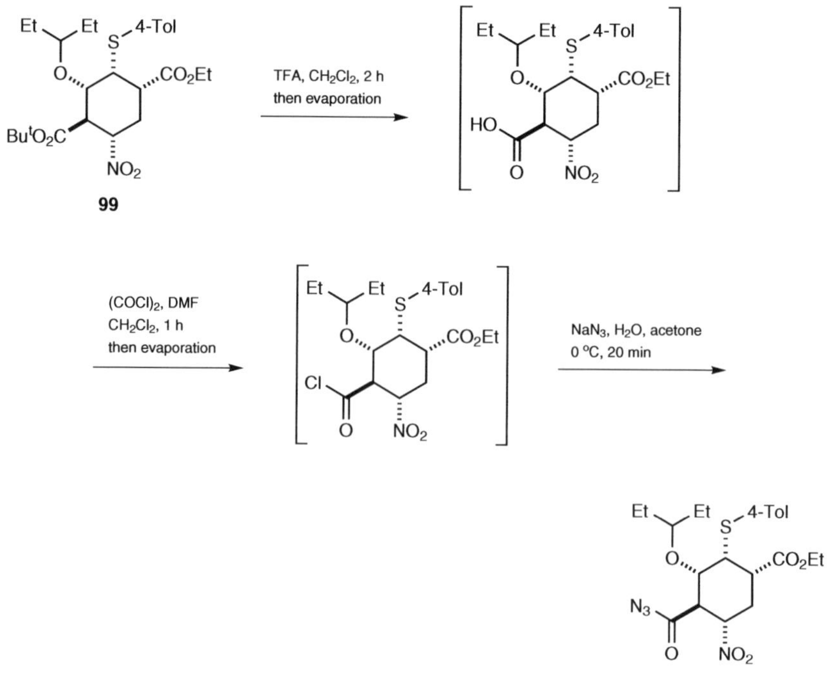

99 → TFA, CH_2Cl_2, 2 h then evaporation → (COCl)$_2$, DMF CH_2Cl_2, 1 h then evaporation → NaN_3, H_2O, acetone 0 °C, 20 min → 105

third pot

105 → AcOH, Ac_2O, rt, 49 h then evaporation → 106 → Zn, TMSCl EtOH, 70 °C, 2 h → 107 → NH_3, then K_2CO_3 EtOH, 6 h → 98 57%

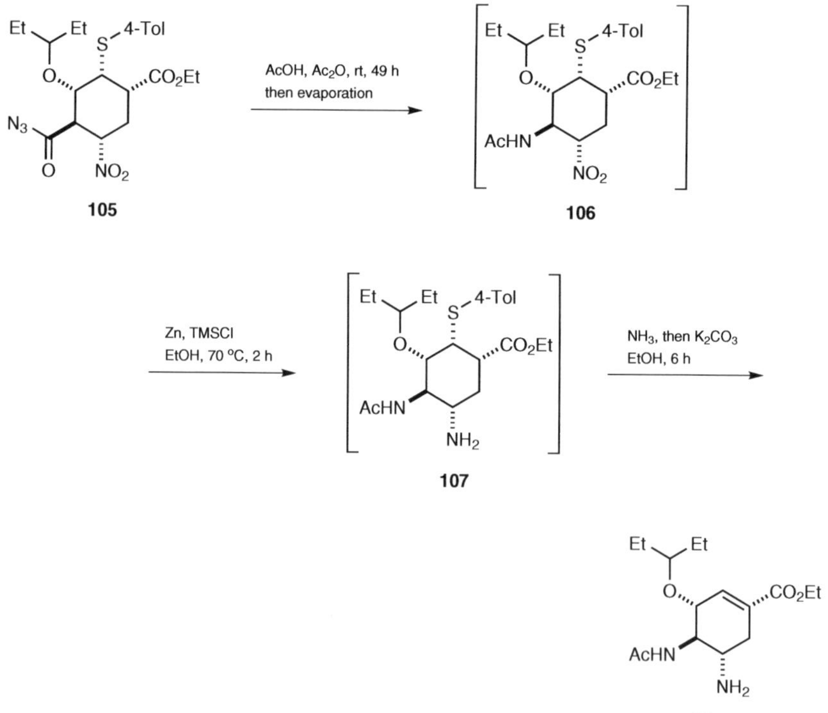

This process, and its success, stems from the effective combination of many past studies reported from both the group of Hayashi and of Ma concerning the asymmetric Michael addition of aldehydes to nitroalkenes and β-nitro acrylates.[65,66] The first key reaction is an enantioselective Michael addition of alkoxyaldehyde **100** to β-nitro acrylate **101** catalyzed by diarylprolinol trimethylsilyl ether *ent*-**25** and chloroacetic acid as additive in dichloromethane. Under optimized conditions, the reaction proceeds in quantitative yield, 96% enantiomeric excess for the *syn*-enantiomer **102** (87% ee for the minor *anti*-isomer), and 6:1 diastereomeric ratio (*syn*/*anti*) in 1 hour if the reaction is quenched at this stage. In the hope of developing a one-pot process, compound **102** was not purified but activated by deprotonation as a nitronate ion and directly reacted with vinylphosphonate **103** in a Michael reaction. The resultant phosphonate is a highly reactive compound that undergoes a domino intramolecular Horner–Wadsworth–Emmons reaction to generate cyclohexene **104**. This reaction is also characterized by the formation of compounds **108** and **109** as byproducts (Scheme 35).[62–64]

Scheme 35 Byproducts Formed in the Domino Intramolecular Horner–Wadsworth–Emmons Reaction[62–64]

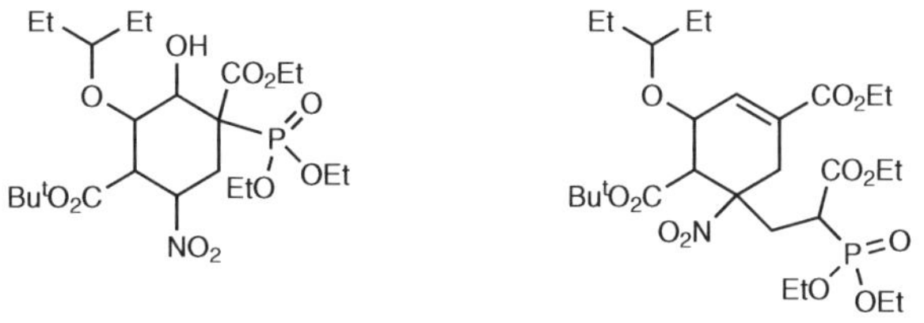

108 **109**

Byproduct **108** arises from an *anti* arrangement of the hydroxy and diethoxyphosphoryl groups, whereas byproduct **109** is formed by a second Michael addition of the target compound **104** with vinylphosphonate **103**. Pleasingly, the byproducts can be converted directly into the target compound **104** by treatment of the mixture with cesium carbonate in ethanol, because the base catalyzes the retro-aldol reaction of **108** followed by a new Horner–Wadsworth–Emmons reaction and **109** undergoes a retro-Michael reaction. The target compound **104** is obtained as a 5*R*/5*S* epimeric mixture with the undesired 5*R*-isomer being the most abundant compound. Isomerization into the desired diastereomer is quite complex. Treatment of this inseparable mixture of isomers with different bases in ethanol led to the same epimeric ratio of products. As shown by DFT calculations, there is a small energy difference between the two epimers, but if the cyclohexene framework is converted into a cyclohexane one, the epimerization would be possible because the epimerizable substituents would lead to an equatorial orientation. Therefore, on treatment of the epimeric mixture of cyclohexene **104** with 4-toluenethiol and cesium carbonate already present in the reaction flask, a sulfa-Michael reaction proceeds along with the needed epimerization. In fact, the two isomers equilibrate. The major isomer (5*R*)-**104** reacts predominantly with 4-toluenethiol, but the Michael adduct formed is easily isomerized into compound (5*R*)-**99**, which is more stable under basic conditions. The target compound **99** is then purified by column chromatography at the end of the first pot.

The second pot is realized by deprotection of the *tert*-butyl ester with trifluoroacetic acid. The excess of trifluoroacetic acid is then removed by evaporation, and the carboxylic acid functionality is converted into an acyl azide by addition of oxalyl chloride and a catalytic amount of dimethylformamide followed by sodium azide in aqueous acetone. The resultant acyl azide **105** is used without any further purification as starting material for the third pot.

for references see p 384

The third pot starts with a domino reaction initiated by treating acyl azide **105** with acetic acid in acetic anhydride at room temperature. The Curtius rearrangement, which proceeds at room temperature thus decreasing potential hazards, is followed by amide formation to afford compound **106**, which can be purified by crystallization. The last two steps require the reduction of the nitro function to afford the primary amine **107** and the retro-Michael reaction of the thiol to restore the double bond. (−)-Oseltamivir (**98**) is obtained in 57% overall yield after acid/base extraction. All reagents employed are inexpensive and the reactions are performed without exclusion of water and air.

Modification of the previously reported three one-pot sequences allowed the Hayashi group to carry out the complete total synthesis of (−)-oseltamivir (**98**) in just two pots, also increasing the overall yield from 57 to 60%.[63] The second and the third pot were combined by replacement of the aqueous conditions for the formation of acyl azide **105** by nonaqueous ones using azidotrimethylsilane and pyridine in toluene. Through the use of azidotrimethylsilane, the extraction and concentration of the acyl azide **105**, a potential hazard, could be omitted, which makes the synthesis safer. Additionally, no halogenated solvents were used in the modified sequence. In particular, dichloromethane was replaced with toluene, which is more environmentally friendly.[63]

The authors also established a column-free synthesis of (−)-oseltamivir (**98**). In the two one-pot sequences, *tert*-butyl ester **99** was separated from phosphoric acid diethyl ester, the side product of the Horner–Wadsworth–Emmons reaction, by column chromatography, but it could also be removed by washing the organic phase containing *tert*-butyl ester **99** with aqueous ammonia. Crude *tert*-butyl ester **99** was then treated with trifluoroacetic acid to afford the carboxylic acid analogue, which was purified by acid–base extraction and filtration over a short pad of silica gel.[63]

The synthesis of (−)-oseltamivir (**98**) through two one-pot sequences has also been implemented with a microreactor flow system for the Curtius rearrangement. The use of flow reactions reduces the risk of explosion from shock-sensitive compounds such as acyl azide **105**. This third-generation synthesis is more efficient, more practical, and safer. It has been scaled up to 10 grams. It is also column free, with one single recrystallization and three acid–base extractions.[67]

Both syntheses of (−)-oseltamivir (**98**) employ *tert*-butyl (*E*)-3-nitroacrylate (**101**) as Michael acceptor and the ester moiety is converted into the acetylamino group using sodium azide or azidotrimethylsilane late in the synthesis. As recently described by the groups of Ma, Šebesta, and Lu, the acetylamino group can be introduced in the first step by an organocatalytic Michael reaction of (*Z*)-*N*-(2-nitrovinyl)acetamide (**111**) with aldehyde **100** catalyzed by trimethylsilyl prolinol ethers **25** or **110**, avoiding the formation of the potentially explosive acyl azide intermediate as a result (Scheme 36).[68–70] The previously reported synthesis by two one-pot sequences by the Hayashi group has been modified to realize a one-pot sequence without any evaporation or solvent exchanges.[64] Because the procedure reported by the groups of Ma, Šebesta, and Lu employed halogenated solvents, they were modified to avoid this class of solvents and to obtain a clean enough reaction to be the first process of a multistep synthesis, in light of possible scale up.

Scheme 36 Use of (*Z*)-*N*-(2-Nitrovinyl)acetamide in the First Step En Route to Oseltamivir[68–70]

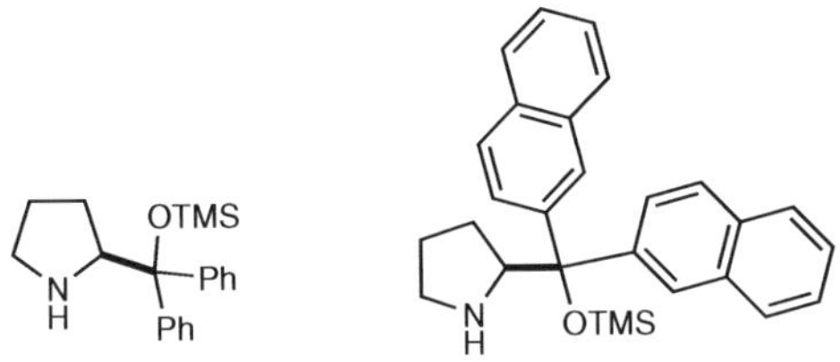

Conditions	Ratio (*syn*/*anti*)	Yield (%)	Ref
110 (10 mol%), BzOH (30 mol%), CH_2Cl_2, −5 °C	5:1	80	[68]
25 (20 mol%), $ClCH_2CO_2H$ (20 mol%), $CHCl_3/H_2O$, 0 °C	4.3:1	88	[69]
25 (20 mol%), $ClCH_2CO_2H$ (40 mol%), CH_2Cl_2, 0 °C	4:1	70	[70]

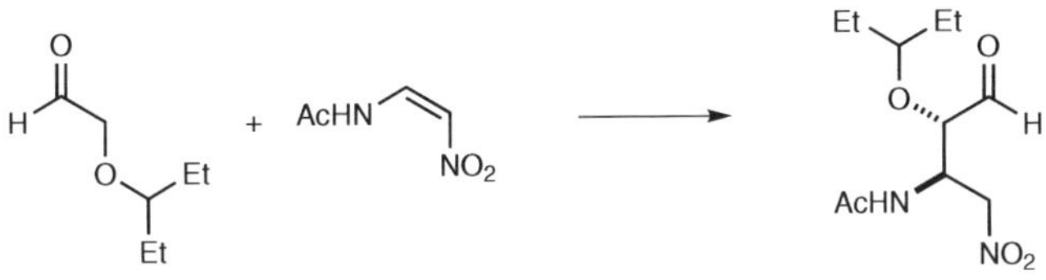

Diphenylprolinol trimethylsilyl ether catalyst **25** is not suitable for the Michael addition of aldehyde **100** and (*Z*)-*N*-(2-nitrovinyl)acetamide (**111**) because it produces several byproducts. Replacing this compound with the bulky diphenylmethylsilyl ether **113** developed by Seebach and co-workers leads to a cleaner reaction.[71] In terms of the solvent, toluene, which in the two-pot sequence replaced dichloromethane, is substituted in the one-pot sequence with chlorobenzene or acetonitrile, solvents which are also suitable for large-scale production. To identify the best reaction conditions, a study on the solvent/additive interactions showed that with the solvent of choice, chlorobenzene, the addition of benzoic acid leads to a fast reaction characterized by low selectivity, whereas chloroacetic acid makes the reaction slow but with excellent diastereoselectivity. The right compromise between reaction rate and selectivity was found with formic acid (Scheme 37).[64]

Scheme 37 Synthesis of the Key Functionalized Intermediate[64]

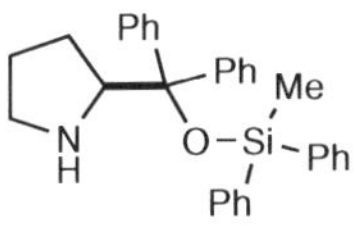

for references see p 384

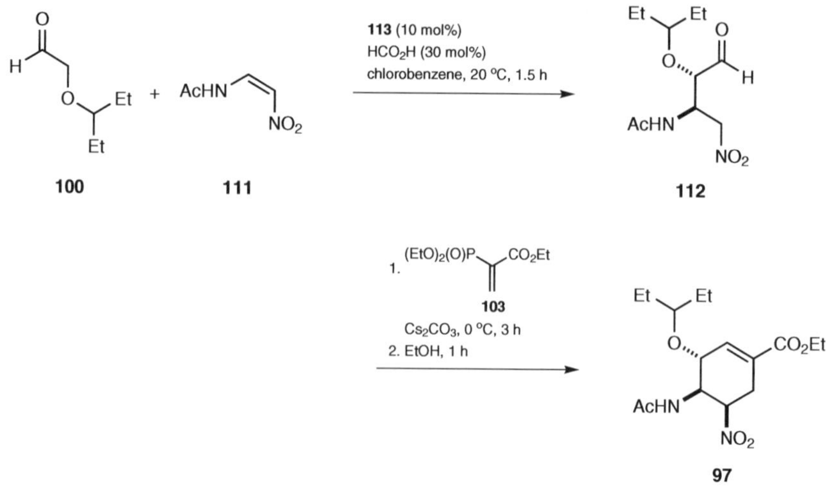

In order to scale up the first step and run the reaction without evaporation or chromatographic separation, it is important to control the temperature to keep the diastereoselectivity high and to add the aldehyde **100** slowly to suppress self-condensation. The cascade reaction is then performed by addition of vinylphosphonate **103** and cesium carbonate after 1.5 hours to nitroaldehyde **112** in the same pot. The different products formed are converted into the target compound **97** by addition of ethanol. (−)-Oseltamivir (**98**) is then obtained in 28% yield on gram-scale from compound **97** after Michael addition of 4-toluenethiol, reduction of the nitro group, and retro-Michael reaction (Scheme 38).[64] No evaporation or solvent exchange is needed, even if chlorobenzene is not the ideal solvent for all the reactions. No interference with the desired reaction course is detected.

Scheme 38 Final Steps for the Synthesis of (−)-Oseltamivir via the One-Pot Sequence[64]

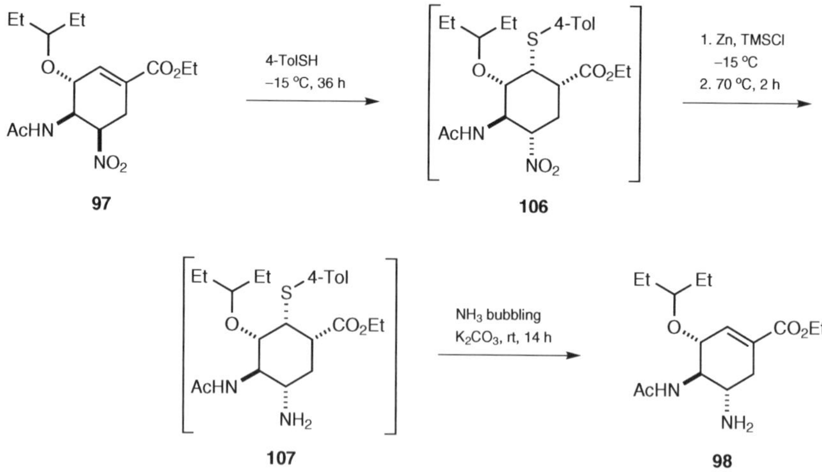

2.3.2.4 One-Pot Synthesis of ABT-341, a DPP4-Selective Inhibitor

[(4*R*,5*S*)-5-Amino-4-(2,4,5-trifluorophenyl)cyclohex-1-enyl]-[3-(trifluoromethyl)-5,6-dihydro-[1,2,4]triazolo[4,3-*a*]pyrazin-7(8*H*)-yl]methanone (ABT-341; **120**), developed by Abbott Laboratories,[72] is a drug applied in the therapy of type-2 diabetes owing to its inhibitory activity toward dipeptidyl peptidase IV (DPP4), a serine protease which deactivates glucose-regulating hormones such as GLP-1 and GIP.[73,74] The Abbott Laboratories synthesis consists of 11 steps with one separation of enantiomers by preparative HPLC on a chiral stationary phase.[72] More recently, the group of Hayashi, transferring the knowhow acquired during the preparation of (−)-oseltamivir (**98**), developed an efficient strategy by which ABT-341 (**120**) can be obtained through a one-pot, high-yielding synthesis consisting of 6 steps.[75]

The synthesis of ABT-341 (**120**) reported by Hayashi and his team starts with a Michael reaction between acetaldehyde and nitroalkene **114** under the catalysis of diphenylprolinol trimethylsilyl ether *ent*-**25** (Scheme 39).[75] This reaction, previously reported by the groups of Hayashi[76] and List,[77] was further optimized to obtain a quantitative yield and to reduce the amount of byproducts, such as self-aldolization compounds, whose accumulation could interrupt subsequent reactions in the designed one-pot sequence. Under the optimized reaction conditions, the use of only 2 equivalents of acetaldehyde with 10 mol% of catalyst *ent*-**25** in 1,4-dioxane gives the nitroaldehyde product **115** with 93% yield and 97% enantiomeric excess following quenching of the reaction after 5 hours. The small excess of acetaldehyde can be removed by evaporation under reduced pressure. Compound **115** is not isolated, but reacted directly with *tert*-butyl 2-(diethoxyphosphoryl)acrylate in the presence of cesium carbonate as a base in dichloromethane at 0 °C to promote a Michael reaction followed in a domino fashion by a Horner–Wadsworth–Emmons reaction to produce the desired *cis*-substituted cyclohexene **116** with the phosphonate ester derivative **121** as byproduct. The byproduct **121** can be converted into cyclohexene **116** by addition of ethanol, which promotes retro-aldol and intramolecular Horner–Wadsworth–Emmons reaction (Scheme 40).[75]

for references see p 384

Scheme 39 One-Pot Synthesis of ABT-341[75]

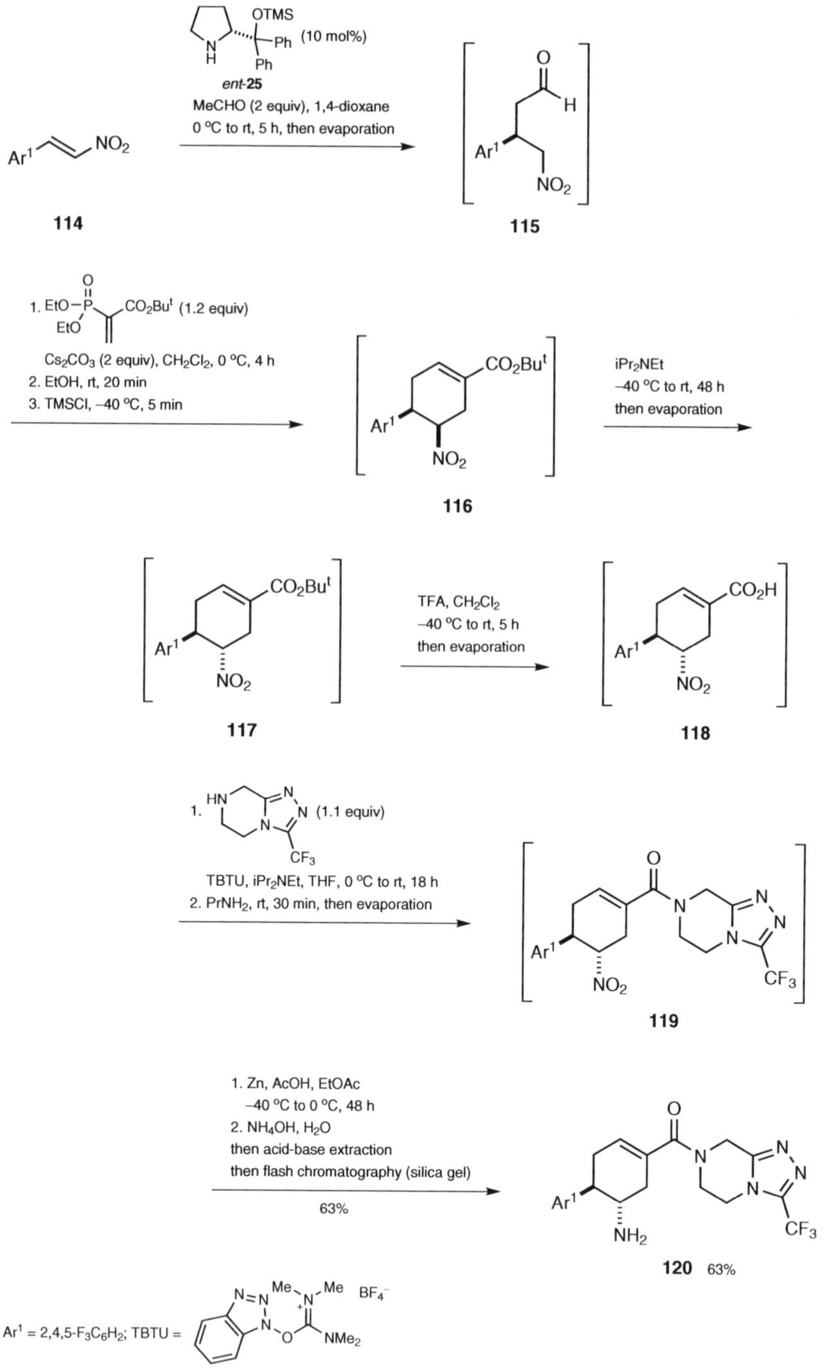

ent-25 (10 mol%)

MeCHO (2 equiv), 1,4-dioxane

0 °C to rt, 5 h, then evaporation

114 → 115

1. $(EtO)_2P(O)C(=CH_2)CO_2Bu^t$ (1.2 equiv)

Cs_2CO_3 (2 equiv), CH_2Cl_2, 0 °C, 4 h

2. EtOH, rt, 20 min

3. TMSCl, –40 °C, 5 min

→ 116

iPr_2NEt

–40 °C to rt, 48 h

then evaporation

→ 117

TFA, CH_2Cl_2

–40 °C to rt, 5 h

then evaporation

→ 118

1. 3-(trifluoromethyl)-5,6,7,8-tetrahydro-[1,2,4]triazolo[4,3-a]pyrazine (1.1 equiv)

TBTU, iPr_2NEt, THF, 0 °C to rt, 18 h

2. $PrNH_2$, rt, 30 min, then evaporation

→ 119

1. Zn, AcOH, EtOAc

–40 °C to 0 °C, 48 h

2. NH_4OH, H_2O

then acid-base extraction

then flash chromatography (silica gel)

63%

→ 120 63%

Ar^1 = 2,4,5-$F_3C_6H_2$; TBTU = O-(benzotriazol-1-yl)-N,N,N',N'-tetramethyluronium BF_4^-

To deactivate cesium carbonate, which causes problems for the subsequent isomerization, the mixture is cooled to −40 °C and chlorotrimethylsilane is added to form the insoluble and neutral cesium chloride. Subsequent addition of *N*,*N*-diisopropylethylamine at room temperature promotes the quantitative isomerization of **116** to give *trans*-isomer **117**. When the reaction is performed employing the isolated nitroaldehyde **115**, the yield of isolated *trans*-cyclohexene **117** is 92%. After removal of all volatile materials under reduced pressure, deprotection of the *tert*-butyl ester with trifluoroacetic acid in dichloromethane gives cyclohexenecarboxylic acid **118**. Amide bond formation to give **119** and reduction of the nitro group affords the target compound ABT-341 (**120**) in 63% overall yield from nitroalkene **114** after purification by acid–base extraction and column chromatography.

Scheme 40 Conversion of the Side Product[75]

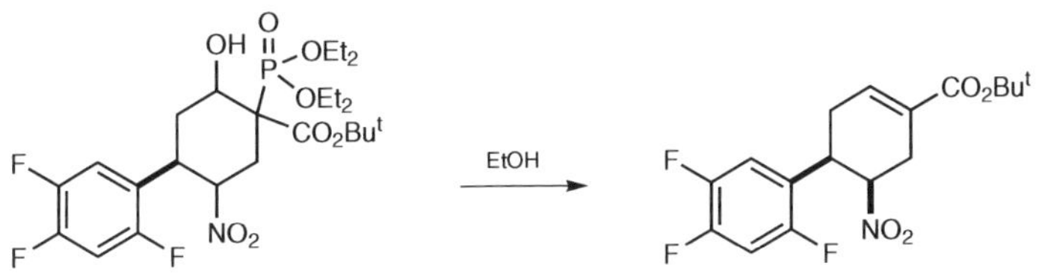

[(4*R*,5*S*)-5-Amino-4-(2,4,5-trifluorophenyl)cyclohex-1-enyl]-[3-(trifluoromethyl)-5,6-dihydro-[1,2,4]triazolo[4,3-*a*]pyrazin-7(8*H*)-yl]methanone (ABT-341; 120); Typical Procedure:[75]
(*R*)-Diphenylprolinol trimethylsilyl ether (*ent*-**25**; 13.0 mg, 0.039 mmol) was added to a soln of (*E*)-1,2,4-trifluoro-5-(2-nitrovinyl)benzene (**114**; 80 mg, 0.393 mmol) and acetaldehyde (44 μL, 0.786 mmol) in 1,4-dioxane (0.4 mL) at 0 °C under argon. The mixture was stirred for 5 h at 23 °C before removing the excess of acetaldehyde and solvent under reduced pressure. *tert*-Butyl 2-(diethoxyphosphoryl)acrylate (124.1 mg, 0.472 mmol) and Cs_2CO_3 (257.0 mg, 0.786 mmol) were added to a soln of the crude material in CH_2Cl_2 (2 mL) at 0 °C under argon. The mixture was stirred for 4 h at 0 °C before removal of the solvent under reduced pressure. EtOH (2 mL) was then added to the crude material at 23 °C and the resulting mixture was stirred for 20 min before being cooled to −40 °C. TMSCl (200.0 μL, 1.57 mmol) was slowly added to the resulting mixture at −40 °C. The mixture was stirred for 5 min at the same temperature followed by slow addition of iPr_2NEt (685.0 μL, 3.93 mmol). The mixture was stirred for additional 48 h at 23 °C. Excess of reagents and solvent were removed under reduced pressure. TFA (582.0 μL, 7.86 mmol) was slowly added to the crude mixture in CH_2Cl_2 (5 mL) at −40 °C under argon. After the resulting mixture had been warmed up to 23 °C, it was stirred for 5 h at 23 °C. The excess reagent and solvent were removed under reduced pressure and the resulting mixture was dried to give (4*R*,5*S*)-5-nitro-4-(2,4,5-trifluorophenyl)cyclohex-1-ene-1-carboxylic acid (**118**).

O-(Benzotriazol-1-yl)-*N*,*N*,*N*′,*N*′-tetramethyluronium tetrafluoroborate (TBTU; 252.3 mg, 0.786 mmol) was added to a soln of crude **118**, 3-(trifluoromethyl)-5,6,7,8-tetrahydro-[1,2,4]triazolo[4,3-*a*]pyrazine (83.2 mg, 0.432 mmol), and iPr_2NEt (685.0 μL, 3.93 mmol) in THF (5 mL) at 0 °C under argon. The mixture was stirred for 18 h at 23 °C. $PrNH_2$ (646 μL, 7.86 mmol) was added and the resulting mixture was stirred for additional 30 min. Excess reagent and solvent were removed under reduced pressure to give **119**.

AcOH (2 mL) was slowly added to a soln of the crude mixture in EtOAc (5 mL) at −40 °C under argon, followed by the addition of activated Zn powder (2.57 g, 39.3 mmol; washed with 2 M aq HCl, H_2O, EtOH, and Et_2O before use). The mixture was stirred for 48 h at 0 °C before filtration though a Celite pad. Excess 28% aq NH_4OH was added to the filtrate. The aqueous layer was extracted with EtOAc (3 ×). The combined organic layer was washed with sat. aq NaCl, dried ($MgSO_4$), and concentrated under reduced pressure. The crude material was dissolved with 6 M aq HCl and EtOAc. The resulting mixture was stirred for 20 min to become a clear soln. The organic layer was removed and the aqueous layer was adjusted to pH 11 with 28% aq NH_4OH, followed by extraction with EtOAc (3 ×). The

for references see p 384

combined organic layer was washed with sat. aq NaCl, dried ($MgSO_4$), and concentrated under reduced pressure. Flash chromatography (silica gel, MeOH/EtOAc 10:90) provided ABT-341 (**120**) as a pale yellow amorphous powder; yield: 110.5 mg (63%).

2.3.2.5 Large-Scale Industrial Application of Organocatalytic Domino Reactions: A Case Study

2.3.2.5.1 Transferring Organocatalytic Reactions from Academia to Industry: Not Straightforward

For the reasons pointed out earlier (see the General Introduction),[1–10] organocatalytic domino reactions should be highly appealing transformations for large-scale industrial application because they combine the advantages of domino processes (especially the so-called "pot-economy")[78,79] with the mild operational conditions of organocatalysis. Surprisingly, although these transformations in general appear to be extremely promising as candidate processes for industrial scale-up, only a handful of reactions actually meet these requirements, despite the vast number of publications in this field. In particular, beside the above mentioned Hajos–Parrish–Eder–Sauer–Wiechert reaction,[5,11,12] the other well-known examples of large-scale organocatalytic reactions (Julia–Colonna epoxidation,[80–85] Jacobsen–Rhodia hydrocyanation of imines,[86] and the alkylation of cyclic ketones and imines[87–92]) are not domino processes. The lack of organocatalytic domino processes in industry might be due to the fact that a significant time is often required between a novel transformation being discovered and its application in industry, or to some less evident but more substantial issues. As an example, the addition of carbonyl compounds to nitroalkenes is one of the most widely published transformations in organocatalysis,[93] and the domino version of this key reaction has had a strong impact on the scientific community.[94] High yields and stereoselectivities, together with minimal catalyst loading, render this transformation the apparent ideal nominee for industrial applications. However, the technical issues (safety) associated with the large-scale production of potentially explosive nitro compounds are a significant hurdle to overcome. Even more serious is the impurity profile of the slightly unstable aliphatic nitro compounds, with potential genotoxic impurities present. For these and other reasons, the transfer of organocatalytic reactions from academia into the real world of industrial production is less straightforward than anticipated and requires the development of specific expertise. These general observations on organocatalytic reactions are especially true for organocatalytic domino processes, where the complexity of the process might generate a variety of impurities which are difficult to detect, analyze, quantify, and readily separate. Lastly, it can be argued that intellectual property issues might also be responsible for the delay which spans from the discovery of a novel reaction and the publication of its large-scale version. Although patents are the preferred method to protect the intellectual rights regarding active pharmaceutical ingredients, this might not be the most cost-effective strategy to defend the processes leading to these compounds. In the latter case, an approach defined "publish to grant freedom to operate" might represent a better option. If a novel process becomes public knowledge through a scientific paper, anyone can exploit it but nobody can patent it. Therefore, anyone can manufacture the final compound but without any significant commercial gain, because only those possessing the intellectual right for the final compound might be interested in producing it. The drawback of such a disclosure is sharing knowhow with potential generics, which could use this knowledge to hit the market earlier with the drug post-patent-life. The purpose of this section is to describe a successful example of an organocatalytic process which made its way from academic research into large-scale production.

2.3.2.5.2 The Reaction Developed in the Academic Environment

In 2009, the group in our laboratory developed a new organocatalytic domino reaction which, starting from readily available phenylacetaldehyde and cyclohex-2-en-1-one, gives access to the bicyclo[2.2.2]octan-2-one skeleton of diastereomers **126** in a single operation (Scheme 41).[95] Employing L-proline (**122**) alone, the reaction affords only traces of the desired product, whereas quinine (**123**) alone is ineffective. Instead, the combination of L-proline (**122**) and quinine (**123**) promotes the reaction, which proceeds in even higher yield and stereoselectivity when the *Cinchona* alkaloid derivative dihydrocupreine (**124**) is used instead of quinine (**123**). Finally, employing the catalyst combination of L-thiazolidine **125** and quinine (**123**) the reaction proceeds in good enantioselectivity (ee up to 87%) and diastereoselectivity (dr >10:1), but with moderate yield (20–41%) and prolonged reaction times.

Scheme 41 Domino Conjugate Addition/Aldol Reaction between Phenylacetaldehyde and Cyclohex-2-en-1-one for the Synthesis of a Bicyclic Adduct[95]

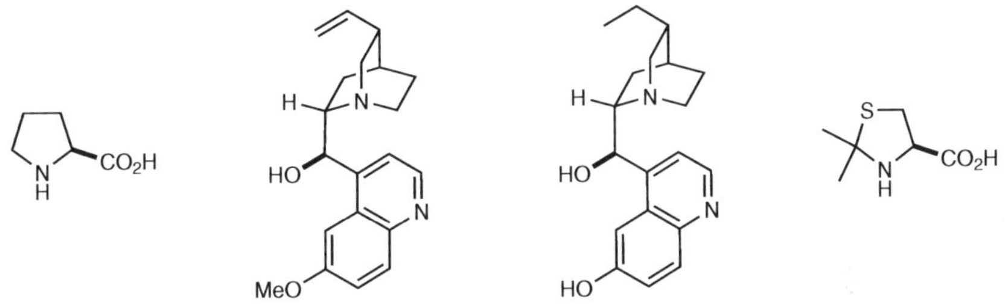

122 L-proline　**123** quinine　**124** dihydrocupreine　**125**

for references see p 384

catalyst, toluene, rt

126A **126B**

Catalyst (mol%)	Time (d)	dr (**126A**/**126B**)	ee[a] (%) **126A**	ee[a] (%) **126B**	Yield[b] (%)	Ref
122 (25)	3	–	–	–	trace	[95]
123 (25)	3	–	–	–	–	[95]
122 (25) + **123** (25)	3	2.5:1	62	41	74	[95]
122 (25) + **124** (25)	3	4:1	66	48	83	[95]
125 (25) + **123** (25)	3	6:1	−82	n.d.	22	[95]
125 (25) + **123** (25)	7	>10:1	−87[c]	n.d.	41	[95]

[a] Negative ee indicates the prevalent formation of the opposite enantiomer; n.d. = not determined.

[b] Reactions performed with 0.5 mmol of phenylacetaldehyde and 0.55 mmol of cyclohex-2-en-1-one in toluene (1 mL).

[c] Veratrol (1,2-dimethoxybenzene) was employed as solvent at 4 °C.

Generally, in enamine activation the absolute configuration of the major enantiomer can be predicted by hypothesizing that the attack of the electrophile occurs on the most hindered (*Re* face) of the enamine intermediate thanks to hydrogen bonding between the carboxylic group and the electrophile (Scheme 42; see also Scheme 8).[2] Using a tertiary amine [e.g., quinine (**123**) or dihydrocupreine (**124**)] as an additional base, deprotonation enhances the nucleophilicity of the enamine intermediate and therefore its reactivity (Scheme 42). Notably, in this specific case, the stereochemistry of the products is not the one usually expected resulting from an attack on the enamine intermediate *Re* face, but rather on the less hindered *Si* face. The enantiomer prevalently obtained from the catalyst combination L-thiazolidine **125** and quinine (**123**) has the opposite absolute configuration, probably due to the steric hindrance of the two methyl groups which favors the enamine intermediate with the opposite double-bond conformation (Scheme 42). These findings (enhanced reactivity of deprotonated enamine and reverse stereoselectivity) are in agreement with the studies of the Armstrong/Blackmond[96–98] and Gschwind[99] groups, who described a similar phenomenon in the amination of aldehydes.

Scheme 42 Proposed Mode of Activation of Enamines in the Domino Organocatalytic Synthesis of a Key Bicyclic Intermediate[96–99]

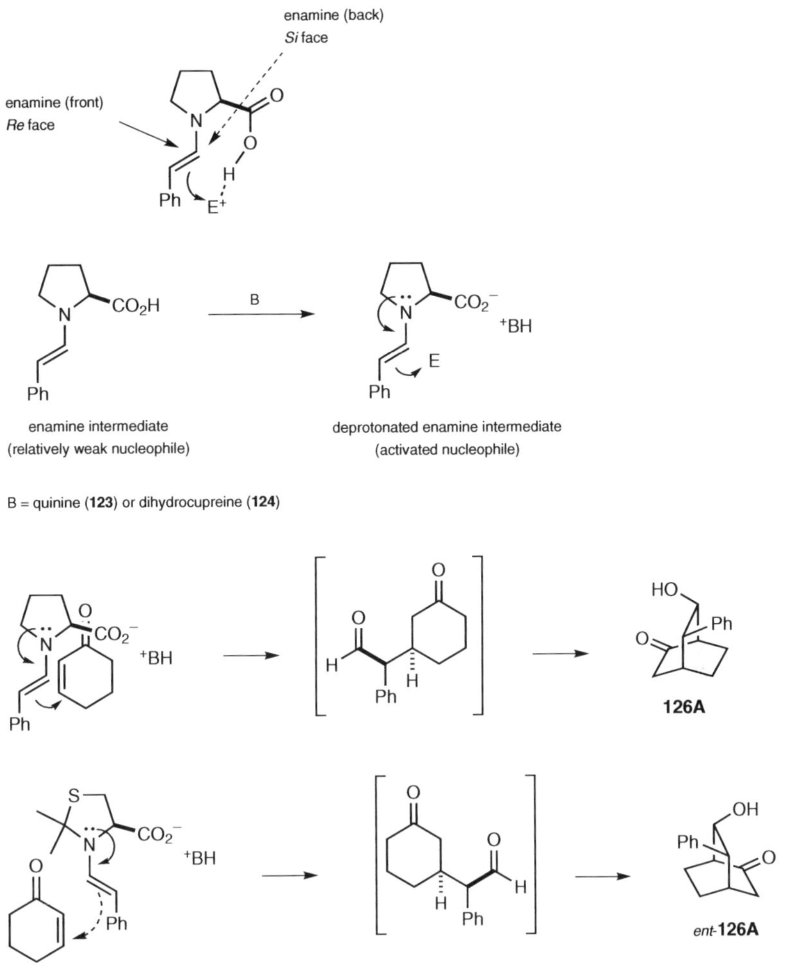

enamine (back)
Si face

enamine (front)
Re face

enamine intermediate
(relatively weak nucleophile)

deprotonated enamine intermediate
(activated nucleophile)

B = quinine (**123**) or dihydrocupreine (**124**)

126A

ent-**126A**

This reaction is a key example of how two separate organocatalysts can act synergistically to promote a reaction which does not afford any product employing only one of them. Furthermore, different combinations of the two enantiomers of the secondary amine with the quasi-enantiomeric catalysts derived from quinine or quinidine can modulate the stereoselectivity of the reaction. The effect of two distinct chiral catalysts in asymmetric reactions has been the subject of a review.[100]

In conclusion, no ideal solution was found, because the yield was higher employing the combination of L-proline (**122**) and dihydrocupreine (**124**), but the stereoselectivity was only moderate in this case.

2.3.2.5.3 The Reaction Developed in the Industrial Environment

Despite these limitations, this reaction was scaled up to kilogram scale, becoming an industrial process in just three years after its publication, thanks to the efforts of the Abele group at the Swiss-based chemical company Actelion.

The private company was interested in the synthesis of the bicyclic ketone **128**, an intermediate both for the manufacturing of L/T calcium channel blocker ACT-280778 (**129**)[101] and the preparation of enantiomerically pure bicyclo[2.2.2]octadiene ligands

for references see p 384

127,[101] which were commercially available but very expensive because they were produced by resolution of chiral racemic compound **130** (Scheme 43).[102–105] Furthermore, the preparation of the latter compound is not straightforward, because it requires the Diels–Alder reaction between trimethylsilyl-protected cyclohexenone **131** and 1-cyanovinyl acetate. The most significant issues identified with the industrial scale-up of this route were the low yield of the Diels–Alder step, associated with heavy tar formation (chromatography) that thwarted scale-up of the synthesis of this symmetrical diketone, and, most of all, the impractical resolution of the enantiomers of diketone **130** on a large scale. Still, to supply material for imminent clinical studies, Actelion developed a Diels–Alder approach to access the ketone **128** on a 90-kg scale[105] solving all intrinsic safety issues.[106–108]

Scheme 43 Retrosynthetic Scheme for the Synthesis of ACT-280778 and Bicyclo[2.2.2]octadiene Ligands from a Common Intermediate[102–105]

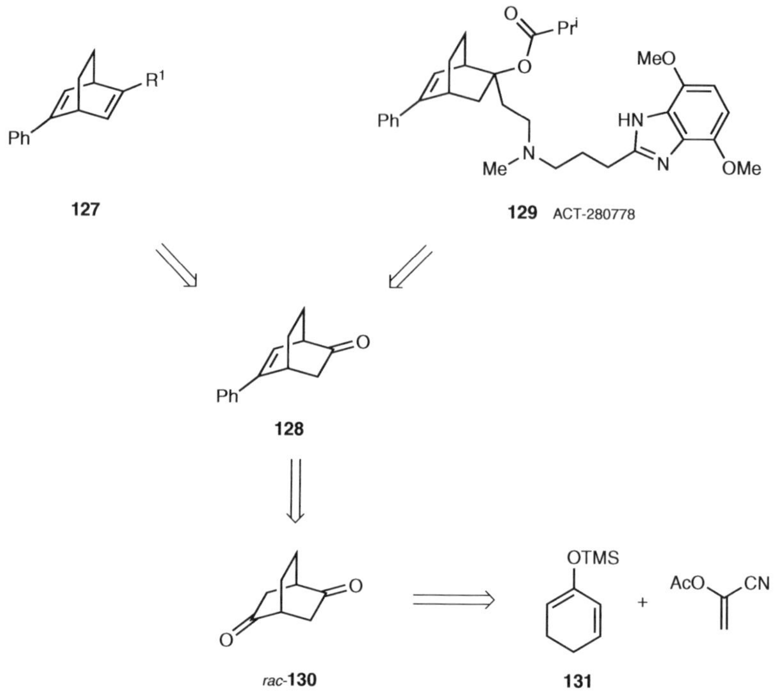

In the first-generation synthesis, dimethyl malonate is added to cyclohex-2-en-1-one,[109–111] and ketone protection gives intermediate **132**. One ester moiety of this compound is hydrolyzed and removed by decarboxylation to afford compound **133**. Deprotonation of this compound and coupling with an aryl moiety gives compound **134** as a mixture of diastereomers. The mixture is carried directly to the next step ("telescoped") and reduction with lithium aluminum hydride gives alcohol **135**, which is then reoxidized to afford aldehyde **136**. Finally, treatment with a diluted hydrochloric acid gives intermediate **126A** as a single stereoisomer thanks to an epimerization process favoring the most stable diastereomer (Scheme 44).

Scheme 44 First-Generation Large-Scale Synthesis of a Key Bicyclic Intermediate for the Preparation of Chiral Diene Ligands[109–111]

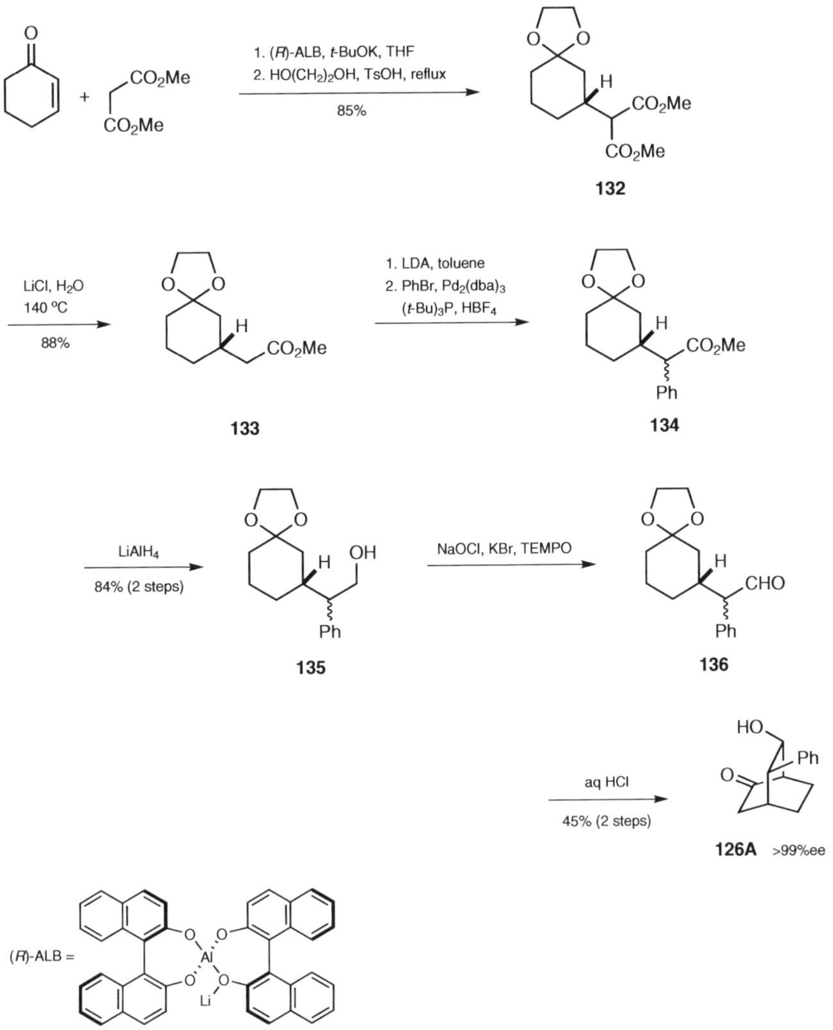

When the paper describing the organocatalytic synthesis of bicyclic adduct **126A** was published, it was obvious that this route could be appealing. The major issues identified with this transformation were the relatively high cost of the *Cinchona* alkaloid and the low yield of the process when the thiazolidine was employed as the secondary catalyst. An in-depth screening was performed by process chemists at Actelion Pharmaceuticals Ltd varying the organocatalysts, chiral and nonchiral additives, and solvents.[112] The best compromise between high diastereomeric ratio, enantiomeric excess, reaction time, and cost of the organocatalyst was L-proline (**122**). Therefore, the industrial process was developed employing L-proline (**122**) and *N*,*N*-diisopropylethylamine as catalysts (Scheme 45). The scale-up was conducted with the same solvent (toluene) originally employed.[95] Notably, similar diastereo- and enantioselectivities were obtained when L-proline (**122**) was used in toluene without any base.

Scheme 45 Second-Generation Large-Scale Domino Organocatalytic Synthesis of the Key Bicyclic Intermediate[112]

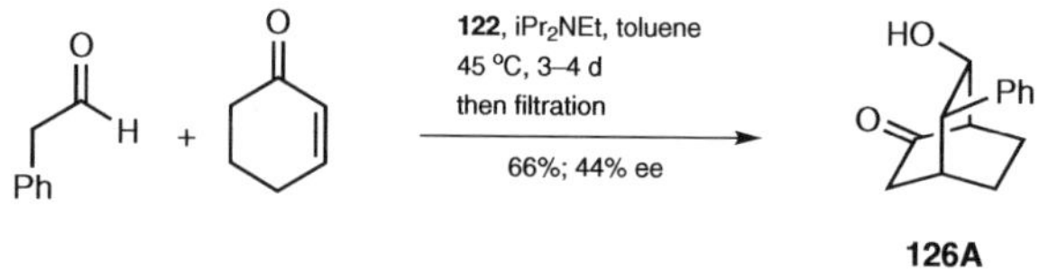

122, iPr_2NEt, toluene
45 °C, 3–4 d
then filtration
66%; 44% ee

126A

The synthesis of the desired bicyclic ligands is then completed by transformation of the hydroxy functional group into a double bond via mesylation. The issue of low enantiomeric excess of bicyclic adduct **126A** is addressed later in the synthesis by recrystallization of intermediate **128**, which increases the enantiopurity of the product (from initially 79% yield and 44% ee to 33% yield and >99% ee after two recrystallizations). Finally, the second aryl functionality is introduced by the addition of an aryl Grignard reagent and subsequent elimination of water (mesylation/triethylamine) to give C2-symmetric or pseudosymmetric ligands **137** (e.g., in 65% yield for $Ar^1 = Ph$).[112] Alternatively, an alkyl group can be added to the skeleton of the molecule instead. First, deprotonation and addition of Comins' reagent affords enol ether **138**. The addition of alkyl Grignard reagents in the presence of tris(acetylacetonato)iron(III) gives ligands **139** (Scheme 46).[112,113]

Scheme 46 Completion of Second-Generation Large-Scale Synthesis of Chiral Diene Ligands[112,113]

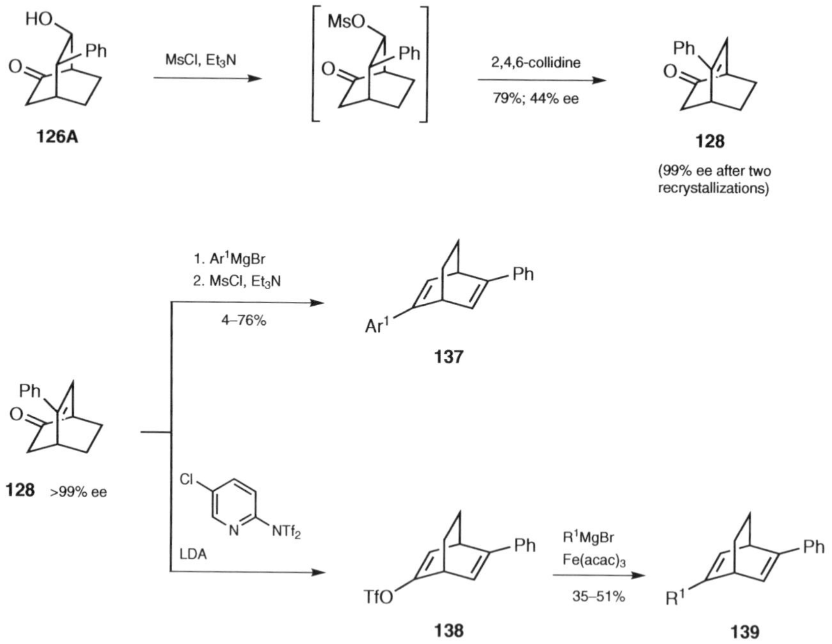

126A
MsCl, Et_3N
2,4,6-collidine
79%; 44% ee
128
(99% ee after two recrystallizations)

128 >99% ee
1. Ar^1MgBr
2. MsCl, Et_3N
4–76%
137
LDA
138
R^1MgBr
$Fe(acac)_3$
35–51%
139

It is noteworthy to compare the original route[104] to prepare the chiral bicyclic ligands [e.g., **137** ($Ar^1 = Ph$)] with the organocatalytic one with parameters well-known within the industrial environment but which are not so popular in the academic world. The organocatalytic route affords the final product **137** ($Ar^1 = Ph$) in higher yield with respect to the first-generation synthesis (11 vs 1%) and reduces the number of chemical steps (5 vs 10). The number of isolated intermediates is reduced from 7 to 2. This outcome is significant because all intermediates and their impurities in industrial processes need to be fully characterized with a toxicological and safety profile (e.g., testing that they are safe to handle and that the risk of explosions is negligible). Most of all, the total mass of materials required to produce 1 kilogram of final compound (PMI) and the quantity of solvent used to prepare 1 kilogram of the desired molecule are crucial for an industrial process. In this case, the PMI is reduced from over 1300 kilograms to just 36 kilograms and the solvent use is diminished nearly 50 times (from over 10 000 liters to just 207, a typical value for an industrial process is around 200 liters of solvent per kilogram of compound). The development of "green chemistry" is a priority for novel industrial production, but above all, "green" means "efficient". Surely, a process which allows savings of nearly 10 000 liters of solvent per kilogram of product, solvent which is derived from non-renewable fossil sources, is intrinsically a green process, beside the well-known advantage of not having to use transition metals.

(1*R*,4*R*,5*S*,6*S*)-6-Hydroxy-5-phenylbicyclo[2.2.2]octan-2-one (126A); **Typical Procedure:**[112]
In a 4-L double-jacketed reactor, L-proline (**122**; 146.0 g, 0.25 equiv) and iPr_2NEt (217 mL, 0.25 equiv) were added to a mixture of cyclohex-2-en-1-one (98% purity; 496.5 g, 5.06 mol) and phenylacetaldehyde (743.3 g, 90% purity, 1.10 equiv) in toluene (3 L) at 23 °C (**CAUTION:** *the addition of* iPr_2NEt *is exothermic*). The mixture was stirred at 45 °C for 4 d. The initial cloudy yellow suspension turned into a thick, well-stirred suspension. The suspension was cooled to 20 °C, stirred at 20 °C for 1 h, and filtered. The filter cake was washed with H_2O (pH 9–10; 3 × 500 mL) and toluene (3 × 400 mL). The filter cake was dried under air for 2 h to yield the product as a colorless solid; yield: 717 g (66%); 100% purity (LC/MS); 44% ee (chiral HPLC). The product (5 g) was dissolved in THF (50 mL) at 66 °C. After the soln had cooled to 23 °C, the suspension was filtered, washed with THF (2 × 1 mL), and dried under reduced pressure to afford the product as a white solid; yield: 1.17 g (23%); 92% ee (chiral HPLC).

for references see p 384

References

[1] Dalko, P. I.; Moisan, L., *Angew. Chem. Int. Ed.*, (2004) **43**, 5138.
[2] Berkessel, A.; Gröger, H., *Asymmetric Organocatalysis*, Wiley-VCH: Weinheim, Germany, (2004).
[3] *Acc. Chem. Res.*, (2004) **37**, 487–631.
[4] Seayad, J.; List, B., *Org. Biomol. Chem.*, (2005) **3**, 719.
[5] List, B., *Chem. Commun. (Cambridge)*, (2006), 819.
[6] Dalko, P. I., *Enantioselective Organocatalysis*, Wiley-VCH: Weinheim, Germany, (2007).
[7] *Chem. Rev.*, (2007) **107**, 5413–5883.
[8] Dondoni, A.; Massi, A., *Angew. Chem. Int. Ed.*, (2008) **47**, 4638.
[9] Bella, M.; Gasperi, T., *Synthesis*, (2009), 1583.
[10] Hajos, Z. G.; Parrish, D. R.; DE 2 102 623, (1971); *Chem. Abstr.*, (1972) **76**, 59 072.
[11] Hajos, Z. G.; Parrish, D. R., *J. Org. Chem.*, (1974) **39**, 1615.
[12] Eder, U.; Sauer, G.; Wiechert, R., *Angew. Chem. Int. Ed. Engl.*, (1971) **10**, 496.
[13] Ahrendt, K. A.; Borths, C. J.; MacMillan, D. W. C., *J. Am. Chem. Soc.*, (2000) **122**, 4243.
[14] Huang, Y.; Walji, A. M.; Larsen, C. H.; MacMillan, D. W. C., *J. Am. Chem. Soc.*, (2005) **127**, 15 051.
[15] Marigo, M.; Schulte, T.; Franzén, J.; Jørgensen, K. A., *J. Am. Chem. Soc.*, (2005) **127**, 15 710.
[16] Marigo, M.; Bertelsen, S.; Landa, A.; Jørgensen, K. A., *J. Am. Chem. Soc.*, (2006) **128**, 5475.
[17] Brandau, S.; Maerten, E.; Jørgensen, K. A., *J. Am. Chem. Soc.*, (2006) **128**, 14 986.
[18] Li, H.; Zu, L.; Xie, H.; Wang, J.; Jiang, W.; Wang, W., *Org. Lett.*, (2007) **9**, 1833.
[19] Reyes, E.; Jiang, H.; Milelli, A.; Elsner, P.; Hazell, R. G.; Jørgensen, K. A., *Angew. Chem. Int. Ed.*, (2007) **46**, 9202.
[20] Halland, N.; Aburel, P. S.; Jørgensen, K. A., *Angew. Chem. Int. Ed.*, (2004) **43**, 1272.
[21] Pulkkinen, J.; Aburel, P. S.; Halland, N.; Jørgensen, K. A., *Adv. Synth. Catal.*, (2004) **346**, 1077.
[22] Vigneron, J. P.; Dhaenens, M.; Horeau, A., *Tetrahedron*, (1977) **33**, 497.
[23] Marigo, M.; Jørgensen, K. A., *Chem. Commun. (Cambridge)*, (2006), 2001.
[24] Guillena, G.; Ramón, D. J., *Tetrahedron: Asymmetry*, (2006) **17**, 1465.
[25] Franzén, J.; Marigo, M.; Fielenbach, D.; Wabnitz, T. C.; Kjaersgard, A.; Jørgensen, K. A., *J. Am. Chem. Soc.*, (2005) **127**, 18 296.
[26] Palomo, C.; Mielgo, A., *Angew. Chem. Int. Ed.*, (2006) **45**, 7876.
[27] Enders, D.; Hüttl, M. R. M.; Grondal, C.; Raabe, G., *Nature (London)*, (2006) **441**, 861.
[28] Enders, D.; Hüttl, M. R. M.; Runsink, J.; Raabe, G.; Wendt, B., *Angew. Chem. Int. Ed.*, (2007) **46**, 467.
[29] Carlone, A.; Cabrera, S.; Marigo, M.; Jørgensen, K. A., *Angew. Chem. Int. Ed.*, (2007) **46**, 1101.
[30] Rios, R., *Chem. Soc. Rev.*, (2012) **41**, 1060.
[31] Ball-Jones, N. R.; Badillo, J. J.; Franz, A. K., *Org. Biomol. Chem.*, (2012) **10**, 5165.
[32] Galliford, C. V.; Scheidt, K. A., *Angew. Chem. Int. Ed.*, (2007) **46**, 8748.
[33] Honga, L.; Wang, R., *Adv. Synth. Catal.*, (2013) **355**, 1023.
[34] Lin, H.; Danishefsky, S. J., *Angew. Chem. Int. Ed.*, (2003) **42**, 36.
[35] Yu, B.; Yu, Z.; Qi, P.-P.; Yu, D.-Q.; Liu, H.-M., *Eur. J. Med. Chem.*, (2015) **95**, 35.
[36] Trost, B. M.; Brennan, M. K., *Synthesis*, (2009), 3003.
[37] Yu, B.; Shi, X.-J.; Qi, P.-P.; Yu, D.-Q.; Liu, H.-M., *J. Steroid Biochem. Mol. Biol.*, (2014) **141**, 121.
[38] Yu, B.; Qi, P.-P.; Shi, X.-J.; Shan, L.-H.; Yu, D.-Q.; Liu, H.-M., *Steroids*, (2014) **88**, 44.
[39] Bencivenni, G.; Wu, L.-Y.; Mazzanti, A.; Giannichi, B.; Pesciaioli, F.; Song, M.-P.; Bartoli, G.; Melchiorre, P., *Angew. Chem. Int. Ed.*, (2009) **48**, 7200.
[40] Melchiorre, P.; Marigo, M.; Carlone, A.; Bartoli, G., *Angew. Chem. Int. Ed.*, (2008) **47**, 6138.
[41] Barbas, C. F., III, *Angew. Chem. Int. Ed.*, (2008) **47**, 42.
[42] Bartoli, G.; Melchiorre, P., *Synlett*, (2008), 1759.
[43] Cassani, C.; Tian, X.; Escudero-Adán, E. C.; Melchiorre, P., *Chem. Commun. (Cambridge)*, (2011) **47**, 233.
[44] Tian, X.; Melchiorre, P., *Angew. Chem. Int. Ed.*, (2013) **52**, 5360.
[45] List, B., *Angew. Chem. Int. Ed.*, (2010) **49**, 1730.
[46] Betancort, J. M.; Sakthivel, K.; Thayumanavan, R.; Barbas, C. F., III, *Tetrahedron Lett.*, (2001) **42**, 4441.
[47] Chen, W.-B.; Wu, Z.-J.; Pei, Q.-L.; Cun, L.-F.; Zhang, X.-M.; Yuan, W.-C., *Org. Lett.*, (2010) **12**, 3132.
[48] Macaev, F.; Sucman, N.; Shepeli, F.; Zveaghintseva, M.; Pogrebnoi, V., *Symmetry*, (2011) **3**, 165.
[49] Jiang, K.; Jia, Z.-J.; Yin, X.; Wu, L.; Chen, Y.-C., *Org. Lett.*, (2010) **12**, 2766.
[50] Wei, Q.; Gong, L.-Z., *Org. Lett.*, (2010) **12**, 1008.
[51] Tan, B.; Candeias, N. R.; Barbas, C. F., III, *Nat. Chem.*, (2011) **3**, 473.

[52] Wang, L.-L.; Peng, L.; Bai, J.-F.; Huang, Q.-C.; Xu, X.-Y.; Wang, L.-X., *Chem. Commun. (Cambridge)*, (2010) **46**, 8064.
[53] Companyo, X.; Zea, A.; Alba, A. N.; Mazzanti, A.; Moyano, A.; Rios, R., *Chem. Commun. (Cambridge)*, (2010) **46**, 6953.
[54] Pesciaioli, F.; Righi, P.; Mazzanti, A.; Bartoli, G.; Bencivenni, G., *Chem.–Eur. J.*, (2011) **17**, 2842.
[55] Albertshofer, K.; Tan, B.; Barbas, C. F., III, *Org. Lett.*, (2012) **14**, 1834.
[56] Wu, H.; Zhang, L.-L.; Tian, Z.-Q.; Huang, Y.-D.; Wang, Y.-M., *Chem.–Eur. J.*, (2013) **19**, 1747.
[57] Liu, Y.-L.; Wang, X.; Zhao, Y.-L.; Zhu, F.; Zeng, X.-P.; Chen, L.; Wang, C.-H.; Zhao, X.-L.; Zhou, J., *Angew. Chem. Int. Ed.*, (2013) **52**, 13 735.
[58] Cao, Y.; Jiang, X.; Liu, L.; Shen, F.; Zhang, F.; Wang, R., *Angew. Chem. Int. Ed.*, (2011) **50**, 9124.
[59] Zhou, B.; Yang, Y.; Shi, J.; Luo, Z.; Li, Y., *J. Org. Chem.*, (2013) **78**, 2897.
[60] Liu, R.; Zhang, J., *Org. Lett.*, (2013) **15**, 2266.
[61] Kim, C. U.; Lew, W.; Williams, M. A.; Liu, H.; Zhang, L.; Swaminathan, S.; Bischofberger, N.; Chen, M. S.; Mendel, D. B.; Tai, C. Y.; Laver, W. G.; Stevens, R. C., *J. Am. Chem. Soc.*, (1997) **119**, 681.
[62] Ishikawa, H.; Suzuki, T.; Hayashi, Y., *Angew. Chem. Int. Ed.*, (2009) **48**, 1304.
[63] Ishikawa, H.; Suzuki, T.; Orita, H.; Uchimaru, T.; Hayashi, Y., *Chem.–Eur. J.*, (2010) **16**, 12 616.
[64] Mukaiyama, T.; Ishikawa, H.; Koshino, H.; Hayashi, Y., *Chem.–Eur. J.*, (2013) **19**, 17 789.
[65] Hayashi, Y.; Gotoh, H.; Hayashi, T.; Shoji, M., *Angew. Chem. Int. Ed.*, (2005) **44**, 4212.
[66] Zhu, S.; Yu, S.; Ma, D., *Angew. Chem. Int. Ed.*, (2008) **47**, 545.
[67] Ishikawa, H.; Bondzic, B. P.; Hayashi, Y., *Eur. J. Org. Chem.*, (2011), 6020.
[68] Zhu, S.; Yu, S.; Wang, Y.; Ma, D., *Angew. Chem. Int. Ed.*, (2010) **49**, 4656.
[69] Rehák, J.; Hut'ka, M.; Latika, A.; Brath, H.; Almássy, A.; Hajzer, V.; Durmis, J.; Toma, Š.; Šebesta, R., *Synthesis*, (2012) **44**, 2424.
[70] Wenig, J.; Li, Y.-B.; Wang, R.-B.; Lu, G., *ChemCatChem*, (2012) **4**, 1007.
[71] Grošelj, U.; Seebach, D.; Badine, D. M.; Schweizer, W. B.; Beck, A. K.; Krossing, I.; Klose, P.; Hayashi, Y.; Uchimaru, T., *Helv. Chim. Acta*, (2009) **92**, 1225.
[72] Pei, Z.; Li, X.; von Golden, T. W.; Madar, D. J.; Longenecker, K.; Yong, H.; Lubben, T. H.; Steward, K. D.; Zinker, B. A.; Backes, B. J.; Judd, A. S.; Mulhern, M.; Ballaron, S. J.; Stashko, M. A.; Mika, A. M.; Beno, D. W. A.; Reinhart, G. A.; Fryer, R. M.; Preusser, L. C.; Kempf-Grote, A. J.; Sham, H. L.; Trevillyan, J. M., *J. Med. Chem.*, (2006) **49**, 6439.
[73] Weber, A. E., *J. Med. Chem.*, (2004) **47**, 4135.
[74] Gwaltney, S. L., II; Stafford, J. A., *Annu. Rep. Med. Chem.*, (2005) **40**, 149.
[75] Ishikawa, H.; Honma, M.; Hayashi, Y., *Angew. Chem. Int. Ed.*, (2011) **50**, 2824.
[76] Hayashi, Y.; Itoh, T.; Ohkubo, M.; Ishikawa, H., *Angew. Chem. Int. Ed.*, (2008) **47**, 4722.
[77] García-García, P.; Ladépéche, A.; Halder, R.; List, B., *Angew. Chem. Int. Ed.*, (2008) **47**, 4719.
[78] Hayashi, Y.; Umemiya, S., *Angew. Chem. Int. Ed.*, (2013) **52**, 3450.
[79] Bradshaw, B.; Luque-Corredera, C.; Bonjoch, J., *Chem. Commun. (Cambridge)*, (2014) **50**, 7099.
[80] Julia, S.; Masana, J.; Vega, J., *Angew. Chem. Int. Ed. Engl.*, (1980) **19**, 929.
[81] Julia, S.; Guixer, J.; Masana, J.; Rocas, J.; Colonna, S.; Annunziata, R.; Molinari, H., *J. Chem. Soc., Perkin Trans. 1*, (1982), 1317.
[82] Geller, T.; Gerlach, A.; Krüger, C. M.; Militzer, H.-C., *Tetrahedron Lett.*, (2004) **45**, 5065.
[83] Geller, T.; Krüger, C. M.; Militzer, H.-C., EP 1 279 670, (2003); *Chem. Abstr.*, (2003) **138**, 137 150.
[84] Geller, T.; Krüger, C. M.; Militzer, H.-C., EP 1 279 671, (2003); *Chem. Abstr.*, (2003) **138**, 137 151.
[85] Geller, T.; Krüger, C. M.; Militzer, H.-C., EP 1 279 672, (2003); *Chem. Abstr.*, (2003) **138**, 137 152.
[86] http://www.rhodiachirex.com/techpages/amino_acid_technology.htm.
[87] Grabowski, E. J. J., In *Chemical Process Research*, Abdel-Magid, A. F.; Ragan, J. A., Eds.; ACS Symposium Series 870; American Chemical Society: Washington, DC, (2003); p 1.
[88] Dolling, U.-H.; Davis, P.; Grabowski, E. J. J., *J. Am. Chem. Soc.*, (1984) **106**, 446.
[89] Hughes, D. L.; Dolling, U.-H.; Ryan, E. F.; Schoenewaldt, E. F.; Grabowski, E. J. J., *J. Org. Chem.*, (1987) **52**, 4745.
[90] O'Donnell, M. J., *Aldrichimica Acta*, (2001) **34**, 3.
[91] O'Donnell, M. J., *Acc. Chem. Res.*, (2004) **37**, 506.
[92] O'Donnell, M. J.; Bennett, W. D.; Wu, S., *J. Am. Chem. Soc.*, (1989) **111**, 2353.
[93] Aitken, L.; Arezki, N.; Dell'Isola, A.; Cobb, A. A., *Synthesis*, (2013) **45**, 2627.
[94] Grondal, C.; Jeanty, M.; Enders, D., *Nat. Chem.*, (2010) **2**, 167.
[95] Bella, M.; Scarpino Schietroma, D. M.; Cusella, P. P.; Gasperi, T.; Visca, V., *Chem. Commun. (Cambridge)*, (2009), 597.

[96] Hein, J. E.; Burés, J.; Lam, Y.-H.; Hughes, M.; Houk, K. N.; Armstrong, A.; Blackmond, D. G., *Org. Lett.*, (2011) **13**, 5644.
[97] Hein, J. E.; Armstrong, A.; Blackmond, D. G., *Org. Lett.*, (2011) **13**, 4300.
[98] Blackmond, D. G.; Moran, A.; Hughes, M.; Armstrong, A., *J. Am. Chem. Soc.*, (2010) **132**, 7598.
[99] Schmidt, M. B.; Zeitler, K.; Gschwind, R. M., *Chem.–Eur. J.*, (2012) **18**, 3362.
[100] Piovesana, S.; Scarpino Schietroma, D. M.; Bella, M., *Angew. Chem. Int. Ed.*, (2011) **50**, 6216.
[101] Funel, J.-A.; Brodbeck, S.; Guggisberg, Y.; Litjens, R.; Seidel, T.; Struijk, M.; Abele, S., *Org. Process Res. Dev.*, (2014) **18**, 1674.
[102] Hayashi, T.; Ueyama, K.; Tokunaga, N.; Yoshida, K., *J. Am. Chem. Soc.*, (2003) **125**, 11 508.
[103] Tokunaga, N.; Otomaru, Y.; Okamoto, K.; Ueyama, K.; Shintani, R.; Hayashi, T., *J. Am. Chem. Soc.*, (2004) **126**, 13 584.
[104] Otomaru, Y.; Okamoto, K.; Shintani, R.; Hayashi, T., *J. Org. Chem.*, (2005) **70**, 2503.
[105] Funel, J.-A.; Schmidt, G.; Abele, S., *Org. Process Res. Dev.*, (2011) **15**, 1420.
[106] Abele, S.; Höck, S.; Schmidt, G.; Funel, J.-A.; Marti, R., *Org. Process Res. Dev.*, (2012) **16**, 1114.
[107] Abele, S.; Schwaninger, M.; Fierz, H.; Schmidt, G.; Funel, J.-A.; Stoessel, F., *Org. Process Res. Dev.*, (2012) **16**, 2015.
[108] Funel, J.-A.; Abele, S., *Angew. Chem. Int. Ed.*, (2013) **52**, 3822.
[109] Abele, S.; Inauen, R.; Funel, J.-A.; Weller, T., *Org. Process Res. Dev.*, (2012) **16**, 129.
[110] Ohshima, T.; Xu, Y.; Takita, R.; Shibasaki, M., *Tetrahedron*, (2004) **60**, 9569.
[111] Jiricek, J.; Blechert, S., *J. Am. Chem. Soc.*, (2004) **126**, 3534.
[112] Abele, S.; Inauen, R.; Spielvogel, D.; Moessner, C., *J. Org. Chem.*, (2012) **77**, 4765.
[113] Brönnimann, R.; Chun, S.; Marti, R.; Abele, S., *Helv. Chim. Acta*, (2012) **95**, 1809.

2.3.3 Addition to Monofunctional C=O Bonds

A. Song and W. Wang

General Introduction

Addition to widely distributed monofunctional C=O bonds has been explored extensively since the introduction of asymmetric catalysis to both aldol and Henry reactions. Incorporation of these reactions into catalytic, enantioselective domino addition processes for monofunctional C=O bonds provides valuable methods for the highly efficient construction of complex structures. In this chapter, representative examples are used to illustrate the most important discoveries in this area. In particular, three catalytic systems used in promoting these domino processes, namely transition metals, organocatalysts, and Lewis acids, will be discussed. Transition metals activate the carbonyl functionalities to initiate nucleophilic addition to the monofunctional C=O bonds in the domino processes. Asymmetric addition to monofunctional C=O bonds has been significantly expanded by organocatalysis through enamine-, iminium-, and hydrogen-bond-mediated catalytic systems with aldehydes, ketones, and α,β-unsaturated carbonyls. Furthermore, Lewis acid catalyzed enantioselective domino addition to monofunctional C=O bonds has been developed which allows access to unique optically pure structures.

2.3.3.1 Transition-Metal-Catalyzed Domino Addition to C=O Bonds

2.3.3.1.1 Domino Reactions Involving Carbonyl Ylides

Carbonyl ylides, generated from metal-catalyzed decomposition of carbenes or carbenoids, can be subjected to 1,3-dipolar cycloaddition with various dipolarophiles.[1] For example, carbonyl ylide **2**, formed by treatment of α-diazo carbonyl precursor **1** with bis(acetylacetonato)copper(II), can be trapped by 2,6-dichlorobenzaldehyde to produce cycloadduct **3** with 68% yield and good stereoselectivity (Scheme 1).[2]

Scheme 1 Cycloaddition of a Carbonyl Ylide with an Aldehyde[2]

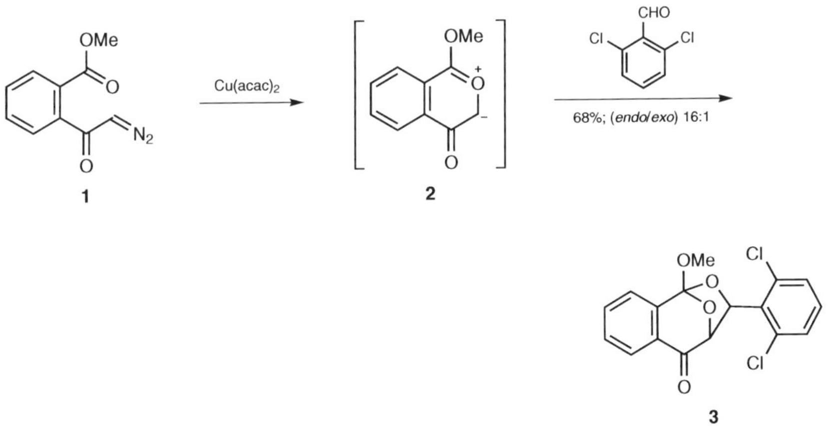

However, the general development of catalytic, enantioselective 1,3-dipolar cycloadditions with carbonyl ylides is problematic due to the fact that the free carbonyl ylide (e.g., **2**) primarily engages in the cycloaddition reaction rather than a metal-associated ylide.[3] To achieve an enantioselective version, a binary catalytic system is used (Scheme 2).[4] Here, the carbonyl ylide **2**, derived from dirhodium(II) acetate catalyzed decomposition of diazo compound **1**, participates in the subsequent scandium(III)–2,6-bis[(4*S*)-4-isopropyl-4,5-dihydrooxazol-2-yl]pyridine complex mediated enantioselective 1,3-dipolar cycloaddition reaction to afford the adduct **4** in 93% enantiomeric excess and 5:1 *endo*/*exo* ratio.

Scheme 2 Enantioselective 1,3-Dipolar Cycloaddition[4]

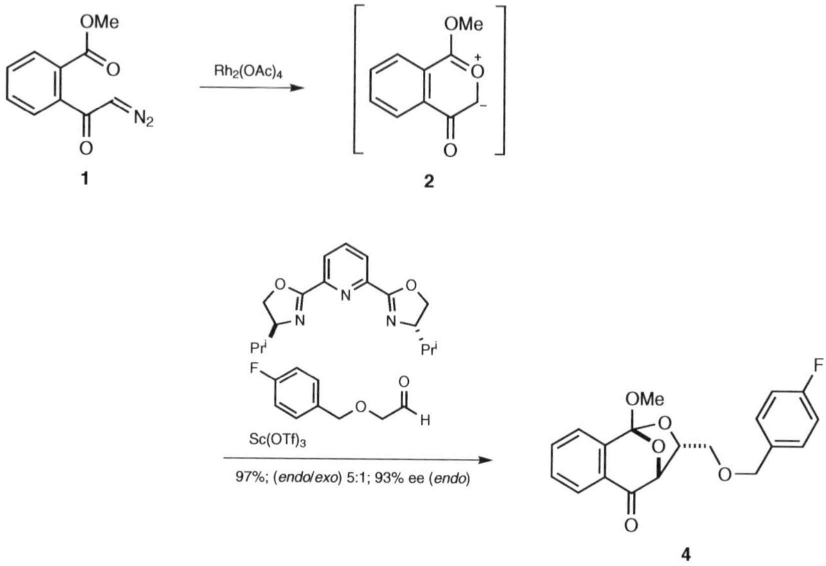

$Rh_2(OAc)_4$

$Sc(OTf)_3$

97%; (*endo*/*exo*) 5:1; 93% ee (*endo*)

An alternative mechanism, invoking a rhodium(II)-associated carbonyl ylide as the primary species involved in the transition state of the cycloaddition reaction, has also been proposed.[5] In addition, it is believed that the dirhodium(II) complex catalyzed formation of the carbonyl ylide under mild conditions is beneficial to the asymmetric induction observed.

Enantioselective intramolecular cycloaddition reactions of α-diazo-β-oxo esters (e.g., **5**) involving carbonyl ylides, catalyzed by dirhodium(II) prolinate catalyst **6**, afford cycloadducts such as **7** at 28 °C (Scheme 3).[6]

Scheme 3 Intramolecular Cycloaddition of a Carbonyl Ylide[6]

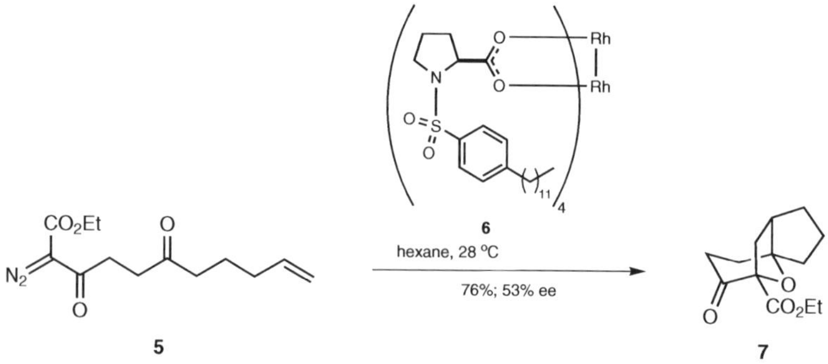

6
hexane, 28 °C

76%; 53% ee

Intermolecular enantioselective cycloaddition of diazo diones **8** with dimethyl acetylenedicarboxylate catalyzed by dirhodium(II) complex **9** to generate adducts **10** also involves carbonyl ylides (Scheme 4).[7] Similar carbonyl ylides are also trapped by aromatic aldehydes,[8] alkenes,[9] alkynes,[9,10] and indoles,[11] revealing that the process as a whole is fairly general.

Scheme 4 Intermolecular Enantioselective Cycloaddition[7]

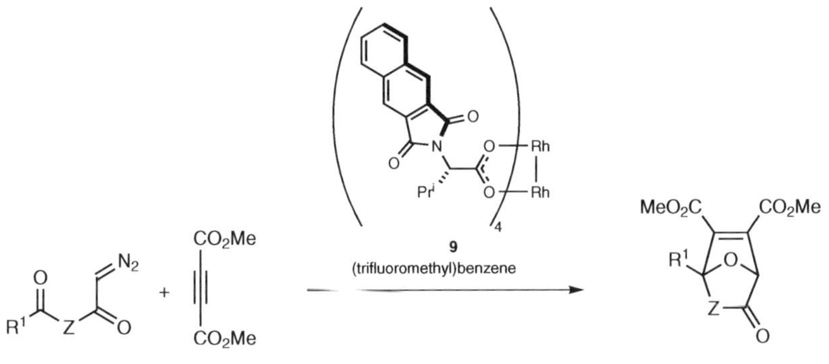

R^1	Z	Yield (%)	ee (%)	Ref
Ph	$(CH_2)_2$	77	90	[7]
4-$MeOC_6H_4$	$(CH_2)_2$	65	90	[7]
Et	$(CH_2)_2$	53	84	[7]
Ph	CMe_2	75	68	[7]

Dimethyl 4-Oxo-1-phenyl-8-oxabicyclo[3.2.1]oct-6-ene-6,7-dicarboxylate [10, R^1 = Ph; Z = $(CH_2)_2$]; Typical Procedure:[7]

To a vigorously stirred soln of 5-diazo-1-phenylpentane-1,4-dione [**8**, R^1 = Ph; Z = $(CH_2)_2$; 50.0 mg, 0.248 mmol] and DMAD (71.0 mg, 0.50 mmol) in $PhCF_3$ (3 mL) was added the bis(tetrahydrofuran) adduct of dirhodium catalyst **9** (3.5 mg, 2.5 μmol) in one portion at 0 °C under argon. The mixture was stirred for 5 min and the solvent was removed under reduced pressure. The greenish residue was purified by column chromatography to give the product as a colorless oil; yield: 60.3 mg (77%); 90% ee.

2.3.3.1.2 Reductive Aldol Reactions

The reductive aldol reaction is a domino sequence involving initial reduction of α,β-unsaturated esters or ketones followed by subsequent coupling with aldehydes or ketones. Although valuable aldol products can be obtained directly from simple silanes, acrylates, and aldehydes in a one-pot fashion via the reductive aldol process as designed, challenges are also obvious due to the side reactions that are possible based on the reaction conditions and reactants. For instance, hydrosilylation of aldehydes[12] or acrylates[13] can lead to undesired reduction products. Nevertheless, there are powerful variants.

One example, the asymmetric catalytic reductive aldol process shown in Scheme 5, is mediated by chloro(cycloocta-1,5-diene)rhodium(I) dimer and (*R*)-2,2′-bis(diphenylphosphino)-1,1′-binaphthyl and gives *syn*-aldol products **11**.[14]

Scheme 5 Asymmetric Catalytic Reductive Aldol Reaction[14]

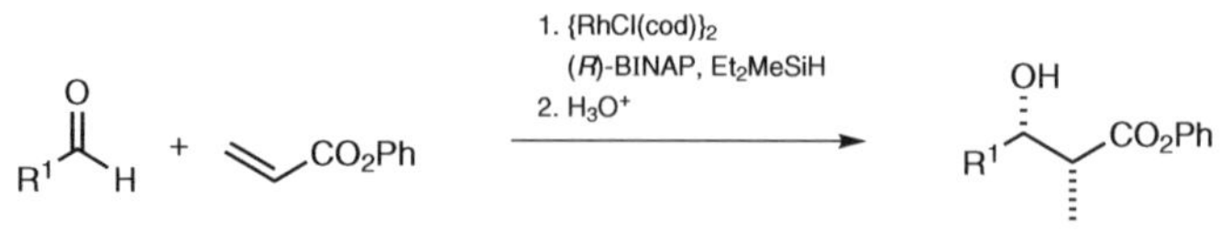

1. $\{RhCl(cod)\}_2$ (*R*)-BINAP, Et_2MeSiH
2. H_3O^+

R^1CHO + $CH_2{=}CHCO_2Ph$ → **11**

R^1	Yield (%)	dr	ee (%)	Ref
Ph	72	3.4:1	87	[14]
Et	59	5.1:1	88	[14]
t-Bu	48	1.8:1	45	[14]

anti-Reductive aldol products can be formed selectively with excellent diastereoselectivities and enantioselectivities by utilizing chiral rhodium complex **12**. For example, *anti*-aldol product **13** is obtained by the reductive aldol reaction between benzaldehyde and *tert*-butyl acrylate in 93% yield, 94:6 *anti*/*syn* ratio, and 96% enantiomeric excess (Scheme 6).[15] Highly diastereoselective and enantioselective reduction of vinyl ketones and coupling with aldehydes can also be achieved.[16]

Scheme 6 *anti*-Selective Reductive Aldol Reaction[15]

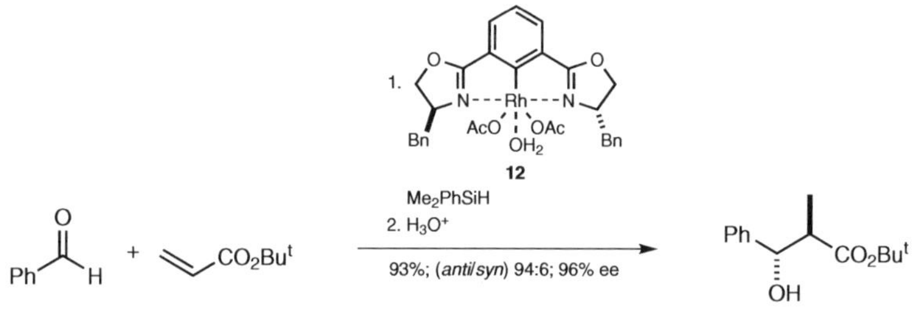

1. **12** (Rh complex with AcO, OAc, OH_2, Bn), Me_2PhSiH
2. H_3O^+

PhCHO + $CH_2{=}CHCO_2Bu^t$ → **13**

93%; (*anti*/*syn*) 94:6; 96% ee

Coupling with ketones in the second step of a reductive aldol reaction is typically difficult due to the weak activity and poor differentiation of the prochiral faces of ketones. By using a chiral copper complex as catalyst, as derived from fluorotris(triphenylphosphine)-copper(I) and ferrocene ligand **14**, the reductive aldol reaction of benzaldehyde and methyl acrylate produces hydroxy ester **15** in 98% yield, 92:8 diastereomeric ratio, and 95% enantiomeric excess (Scheme 7).[17]

Intramolecular reductive aldol reactions involving reduction of α,β-unsaturated ketones and subsequent coupling with ketones afford six-membered rings in high yields and good stereoselectivities.[18] In addition, catalytic asymmetric reductive[19] and alkylative[20] aldol reactions of allenic esters with ketones afford an array of highly valuable materials.

Scheme 7 Reductive Aldol Reaction of a Ketone[17]

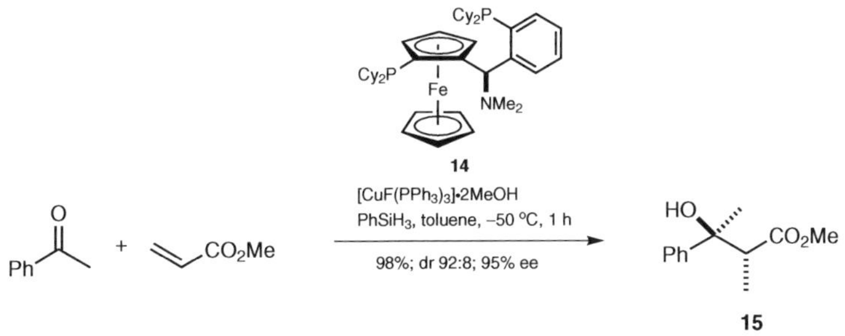

Catalytic asymmetric reductive aldol reactions between alkynes and aldehydes are even more useful because they generate chiral allylic alcohols which can be used for an array of further reactions. For example, nickel-catalyzed asymmetric reductive coupling of 1-phenylprop-1-yne with isobutyraldehyde affords allylic alcohol **17** in high yield and with high stereoselectivity using (*S*)-(+)-neomenthyldiphenylphosphine (**16**) as the ligand (Scheme 8).[21]

Scheme 8 Formation of a Chiral Allylic Alcohol via Reductive Coupling[21]

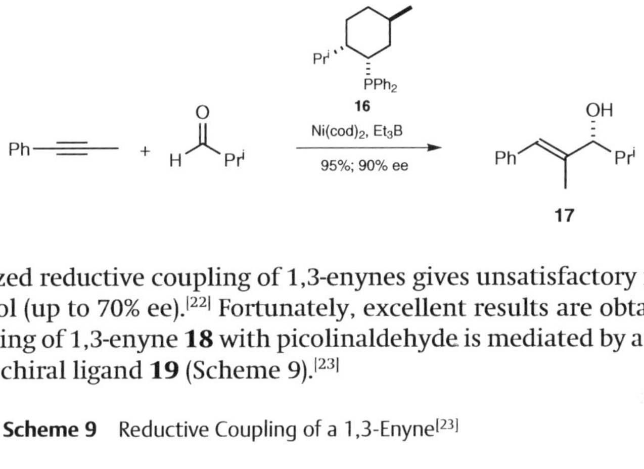

Nickel-catalyzed reductive coupling of 1,3-enynes gives unsatisfactory results in terms of enantiocontrol (up to 70% ee).[22] Fortunately, excellent results are obtained when the reductive coupling of 1,3-enyne **18** with picolinaldehyde is mediated by a rhodium catalytic system using chiral ligand **19** (Scheme 9).[23]

Scheme 9 Reductive Coupling of a 1,3-Enyne[23]

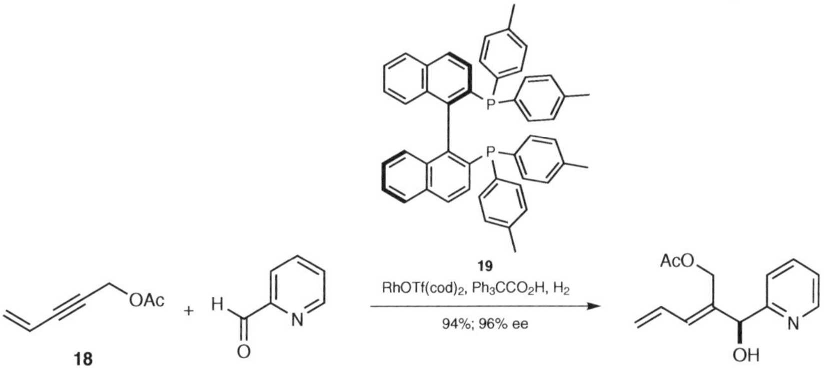

Compared to the reductive coupling of 1,3-enynes, unsatisfactory results are obtained for 1,3-dienes due to the difficulty in controlling the stereoselectivity for the final products.[24] The application of spiro phosphoramidite ligand **21** in the nickel-catalyzed reductive coupling of 1,3-diene **20** and benzaldehyde affords chiral bishomoallylic alcohol **22** with excellent yield and enantiomeric excess (only the *anti*-isomer is obtained, although the absolute configuration is not determined) (Scheme 10).[25]

Scheme 10 Reductive Coupling of a 1,3-Diene[25]

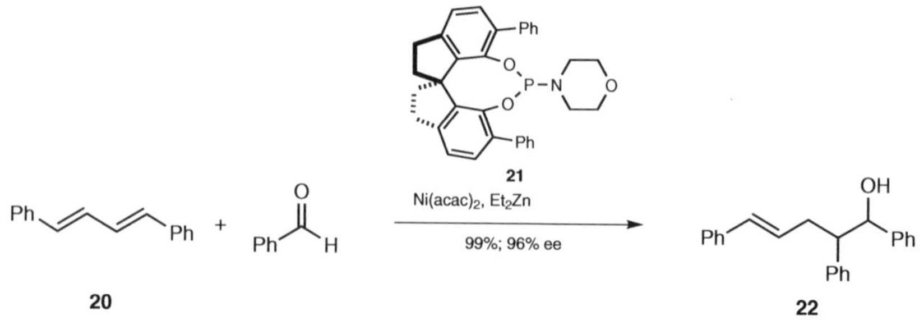

Ni(acac)$_2$, Et$_2$Zn, **21**; 99%; 96% ee

Interestingly, replacement of the reducing reagent in the reductive coupling of 1,3-dienes with a silylborane leads to the formation of α-chiral allylsilanes. One example of this reaction type is the nickel-catalyzed asymmetric coupling of silylborane **25**, 1,3-diene **23**, and an aldehyde to give products **26** as single diastereomers using chiral phosphoramidite ligand **24** (Scheme 11).[26]

Scheme 11 Formation of Chiral Allylsilanes[26]

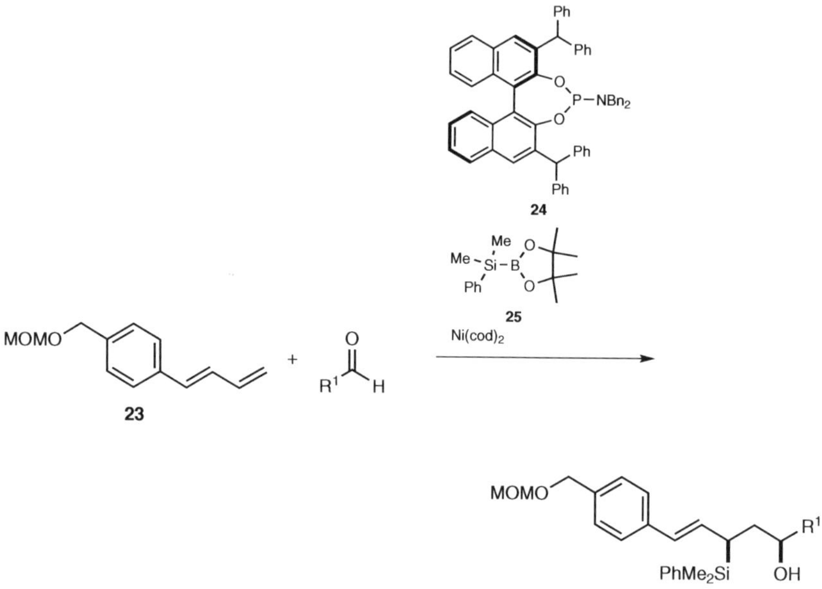

Ni(cod)$_2$, **24**, **25**

R^1	Yield (%)	ee (%)	Ref
Ph	71	97	[26]
iPr	68	96	[26]
4-MeOC$_6$H$_4$	92	92	[26]
4-F$_3$COC$_6$H$_4$	29	85	[26]

Methyl (2*R*,3*S*)-3-Hydroxy-2-methyl-3-phenylbutanoate (15); Typical Procedure:[17]
A flame-dried, 10-mL round-bottomed flask equipped with a magnetic stirrer was charged with $CuF(PPh_3)_3$•2MeOH (9.0 mg, 0.01 mmol), ligand **14** (0.01 mmol), and toluene (4.8 mL). This catalyst soln was stirred for 30 min at rt and then $PhSiH_3$ (180 μL, 1.5 mmol) was added at the same temperature. After cooling the soln at −50 °C, methyl acrylate (110 μL, 1.2 mmol) and acetophenone (117 μL, 1.0 mmol) were added simultaneously. The resulting mixture was stirred for 1 h at −50 °C under argon and then the reaction was quenched by adding aq NH_4F soln (5 mL). The aqueous layer was extracted with Et_2O (3 × 5 mL) and the combined organic layers were washed with brine (20 mL), dried ($MgSO_4$), filtered, and concentrated under reduced pressure. The crude residue was purified by flash chromatography to give the product; yield: 98%.

2.3.3.1.3 Michael/Aldol Reactions

Michael/aldol domino sequences can be designed by replacing the reducing reagent in the reductive aldol reaction with carbon nucleophiles to produce structurally diverse frameworks. In this context, a highly enantioselective catalytic tandem Michael/aldol reaction of cyclopent-4-ene-1,3-dione monoacetals **27** and aldehydes in the presence of dialkylzinc reagents, under copper catalysis with chiral phosphoramidite ligand **28**, gives the products **29** after oxidation (Scheme 12).[27,28]

Scheme 12 Copper-Catalyzed Michael/Aldol Reaction[27,28]

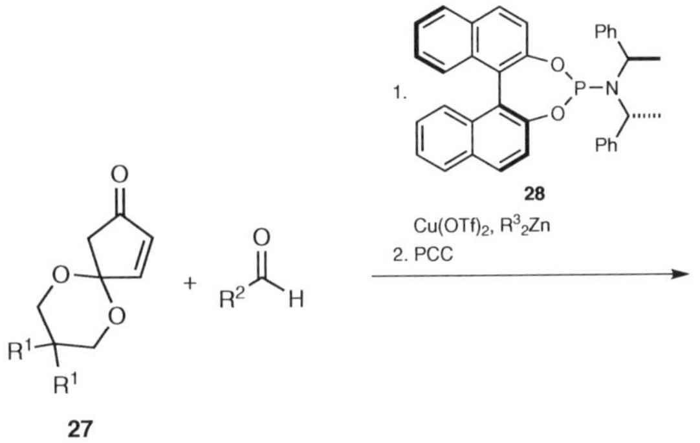

1. **28**, $Cu(OTf)_2$, $R^3{}_2Zn$
2. PCC

R^1	R^2	R^3	Yield (%)	ee (%)	Ref
Me	Ph	Et	67	87	[27]
Ph	Ph	Bu	69	94	[27]
Ph	4-BrC$_6$H$_4$	Et	69	96	[27]

Modest results are obtained in a preliminary study of a 1,4-addition/aldol sequence employing 9-aryl-9-borabicyclo[3.3.1]nonanes.[29] However, soon after this, excellent outcomes are achieved in a rhodium–2,2′-bis(diphenylphosphino)-1,1′-binaphthyl catalyzed intramolecular Michael/aldol reaction of ene trione **30** to produce bicycle **31** in 83% yield, >99:1 diastereomeric ratio, and 90% enantiomeric excess (Scheme 13).[30,31]

Scheme 13 Intramolecular 1,4-Addition/Aldol Reaction[30,31]

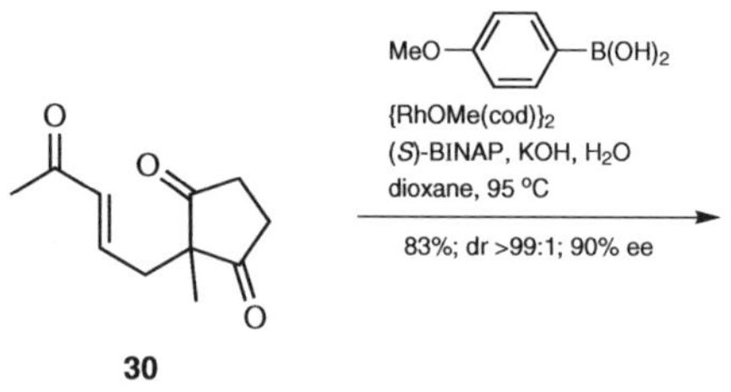
MeO
B(OH)2
{RhOMe(cod)}2
(S)-BINAP, KOH, H2O
dioxane, 95 °C
83%; dr >99:1; 90% ee
30
31

(3a*R*,4*R*,5*S*,6a*R*)-4-Acetyl-3a-hydroxy-5-(4-methoxyphenyl)-6a-methylhexahydropentalen-1(2*H*)-one (31); Typical Procedure:[31]

Anhyd dioxane (5.0 mL) was added to a flame-dried Schlenk tube charged with 4-methoxyphenylboronic acid (122.0 mg, 1 mmol), (*S*)-BINAP (23.3 mg, 37.5 μmol), and $\{RhOMe(cod)\}_2$ (6.1 mg, 12.5 μmol) under an atmosphere of argon. The resulting mixture was allowed to stir at ambient temperature for 5 min, at which point aq KOH [KOH (2.8 mg, 50 μmol) in H_2O (45 μL, 2.5 mmol)] was added and the whole was allowed to stir for 10 min at ambient temperature. Finally, ene trione **30** (97.1 mg, 0.50 mmol) was added and the reaction vessel was placed in an oil bath preheated to 95 °C. The mixture was allowed to stir at 95 °C until complete consumption of the starting material, as monitored by TLC, at which point the mixture was adsorbed onto silica gel and purified by flash chromatography to give the product; yield: 125.5 mg (83%).

2.3.3.1.4 Other Domino Addition Reactions

Similar to the 1,4-addition/aldol sequence, a domino Friedel–Crafts/Henry reaction between 1*H*-indole, (*E*)-(2-nitrovinyl)benzene, and benzaldehyde gives the product **33** in 79% yield and 99% enantiomeric excess under copper catalysis with chiral ligand **32** (Scheme 14).[32] Poor diastereoselectivities are obtained for other substrates, although the yields are generally good.

Scheme 14 Friedel–Crafts/Henry Domino Reaction[32]

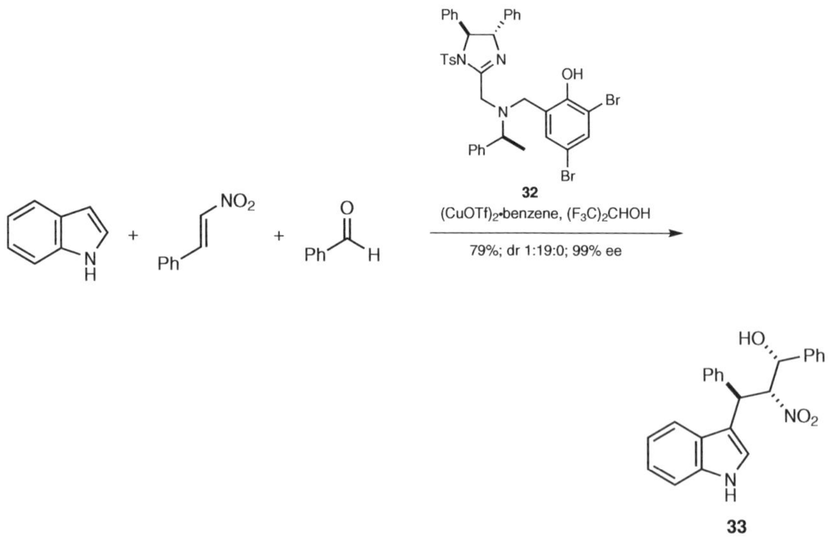
Ph
Ph
TsN
N
OH
Br
N
Ph
Br
32
(CuOTf)2•benzene, (F3C)2CHOH
79%; dr 1:19:0; 99% ee
NO2
Ph
O
Ph
H
N
H
HO
Ph
Ph
NO2
N
H
33

In addition, transition-metal-catalyzed Michael/Michael/Henry sequences,[33] domino α-addition of isocyanides to aldehydes,[34] and domino sequences for the synthesis of chiral acetals[35] lead to even greater degrees of molecular complexity in a single pot.

2.3.3.2 Organocatalytic Domino Addition to C=O Bonds

Although a large number of domino sequences have been realized using catalytic amounts of transition-metal systems, the scope of reaction models and substrates remains limited. Small-molecule organic catalysts (often called organocatalysts) have established a completely new platform for domino addition to C=O bonds since 2000. In particular, novel combinations of various kinds of reaction sequences are possible due to the discovery of interconvertible enamine and iminium catalytic models in chiral amine catalysis. In addition, chiral thiourea catalytic systems provide an important alternative for the expansion of cascade sequences. Furthermore, valuable chiral sulfur-containing structures, materials which are often poisonous to transition-metal catalysts, can be obtained from thiourea-catalyzed domino sequences.

2.3.3.2.1 Amine-Catalyzed Domino Addition to C=O Bonds

Proline has been applied successfully as a class I aldolase mimic for the asymmetric aldol reaction between unmodified acetone and aldehydes via an enamine catalytic model.[36] The reactive enamines formed from the amine catalyst and aldehydes or ketones are suitable nucleophiles to attack other carbonyl groups in a domino addition. The carbonyl groups released from the enamines are also suitable electrophiles to be attacked by other nucleophiles in the same pot in further extension of the sequence. At the same time, the chiral amine catalyzed asymmetric Diels–Alder reaction of enals via an iminium catalytic model has been developed.[37] Several sites of the reactive iminium formed from enals and the amine catalyst can participate in domino addition reactions to C=O bonds in general terms, examples of which are provided in the following sections.

2.3.3.2.1.1 Enamine-Catalyzed Aldol/Aldol Reactions

Propanal trimerizes by enamine catalysis through an aldol/aldol/hemiacetalization sequence to provide the lactol product **34** in 53% yield and 8:1 diastereomeric ratio (Scheme 15). The enantiomeric excess of the product was determined to be 47% after oxidation to lactone **35**.[38]

Scheme 15 Proline-Catalyzed Aldol/Aldol/Hemiacetalization[38]

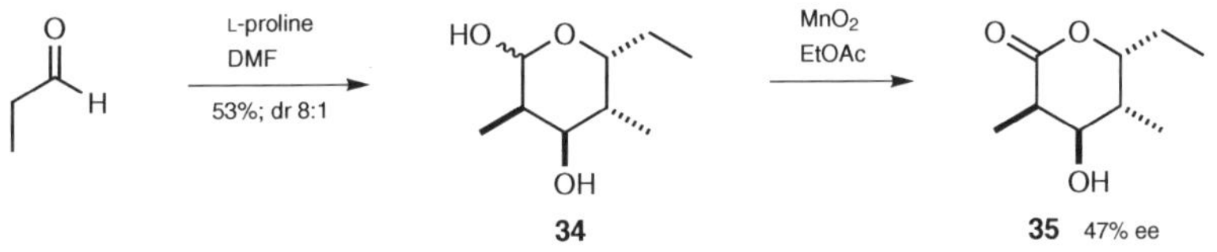

L-proline, DMF, 53%; dr 8:1 → **34**; MnO_2, EtOAc → **35** 47% ee

Significant improvement in the enantioselectivity is achieved by a two-step aldol/Mukaiyama aldol sequence starting from α-siloxy-substituted aldehyde **36** (Scheme 16) and proceeding via hydroxybis(siloxy)-substituted aldehyde **37**, which reacts with silyl enol ether **38**.[39] The ingenious design provides simple routes to optically pure, protected D-glucose **39A**, D-mannose **39B**, and D-allose **39C** derivatives by judicious choice of Lewis acid and solvent.

Scheme 16 Aldol/Mukaiyama Aldol/Hemiacetalization[39]

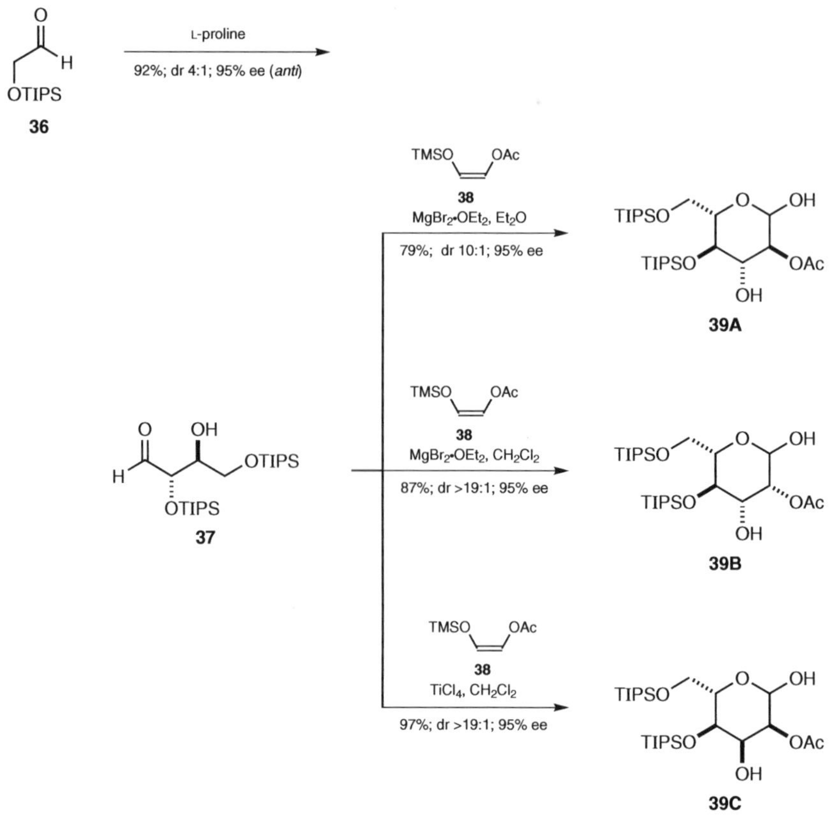

36 → L-proline, 92%; dr 4:1; 95% ee (*anti*) → 37

37 + TMSO/OAc alkene 38, $MgBr_2•OEt_2$, Et_2O; 79%; dr 10:1; 95% ee → 39A

37 + 38, $MgBr_2•OEt_2$, CH_2Cl_2; 87%; dr >19:1; 95% ee → 39B

37 + 38, $TiCl_4$, CH_2Cl_2; 97%; dr >19:1; 95% ee → 39C

A two-step process for the proline-catalyzed aldol/aldol/hemiacetalization sequence also improves the enantioselectivity (Scheme 17).[40] The initial product **40**, formed by an L-proline-catalyzed aldol reaction, is isolated and subjected to a D-proline-catalyzed aldol reaction with propanal to furnish optically pure trimer **41** in 42% yield as a single diastereomer.

Scheme 17 Proline-Catalyzed Aldol/Aldol/Hemiacetalization in Two Steps[40]

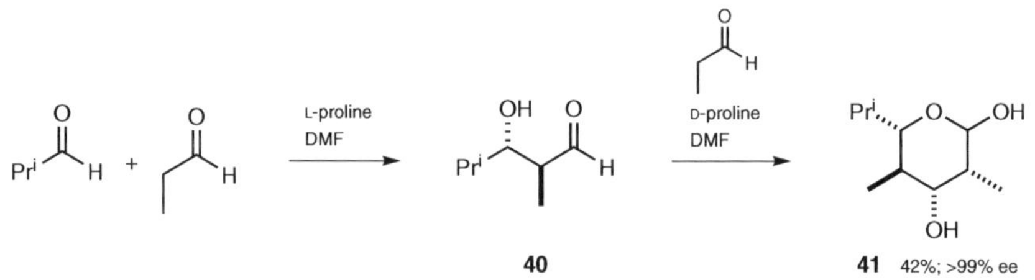

Pr^iCHO + propanal → L-proline, DMF → 40 → propanal, D-proline, DMF → 41 42%; >99% ee

2.3.3.2.1.2 **Enamine-Catalyzed Aldol/Michael Reactions**

An interesting domino aldol/Michael/hemiacetalization sequence is used for the synthesis of α-tocopherol via dienamine catalysis.[41] A similar reaction between salicylaldehyde and 3-methylbut-2-enal, catalyzed by chiral proline derivative **42** to furnish product **43** in 66% yield and 98% enantiomeric excess, is also applied to the synthesis of 4-dehydroxydiversonol (Scheme 18).[42]

Scheme 18 Synthesis of a Bicyclic Structure via Aldol/Michael/Hemiacetalization[42]

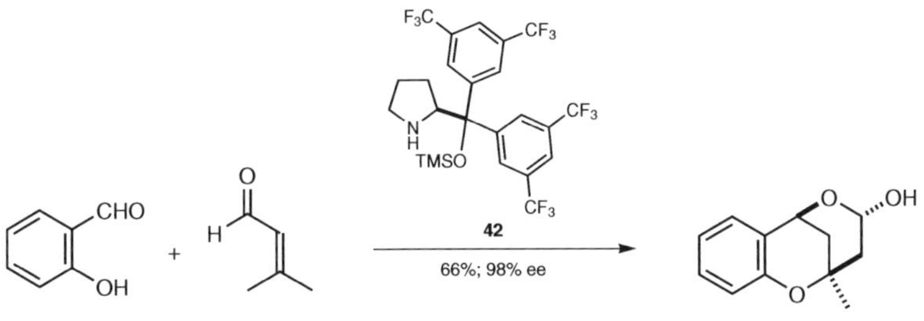

2.3.3.2.1.3 Enamine-Catalyzed Diels–Alder Reactions

Enamines can also serve as electron-rich dienophiles for inverse-electron-demand Diels–Alder reactions to yield valuable six-membered rings. The major challenge of the process is catalyst regeneration if a catalytic version of the reaction is desired. Pleasingly, catalyst **44** can be released from the catalyst-bound intermediate by the simple addition of silica gel. In the example presented in Scheme 19, the Diels–Alder products are oxidized to lactones **45** for the determination of enantioselectivity, which is found to be excellent.[43]

Scheme 19 Organocatalytic Oxy-Diels–Alder Reactions[43]

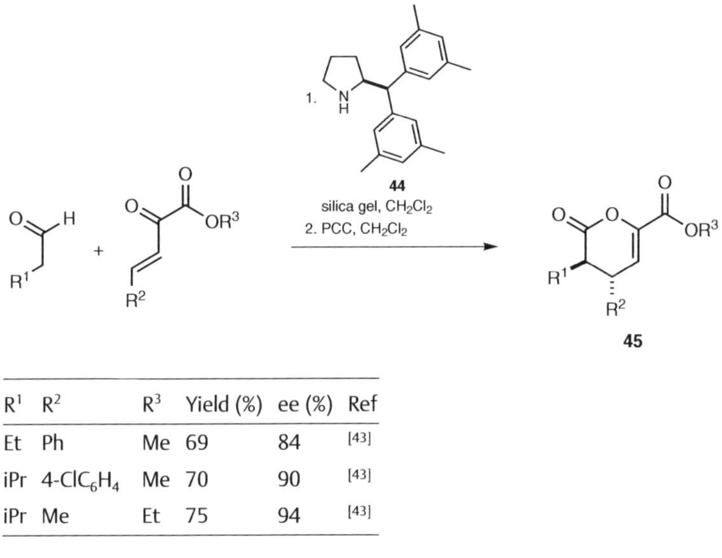

R^1	R^2	R^3	Yield (%)	ee (%)	Ref
Et	Ph	Me	69	84	[43]
iPr	4-ClC_6H_4	Me	70	90	[43]
iPr	Me	Et	75	94	[43]

Application of the strategy to other oxygen-containing electron-deficient dienes was realized soon afterwards in follow-up studies shown in general terms in Scheme 20.[44–47] For those examples containing undefined stereochemistry, the enantiomeric excess is determined by HPLC analysis of the oxidized derivatives. One exception is the final example, where HPLC analysis is conducted on the reduced derivatives.

for references see p 415

Scheme 20 Extension of Oxy-Diels–Alder Reactions[44–47]

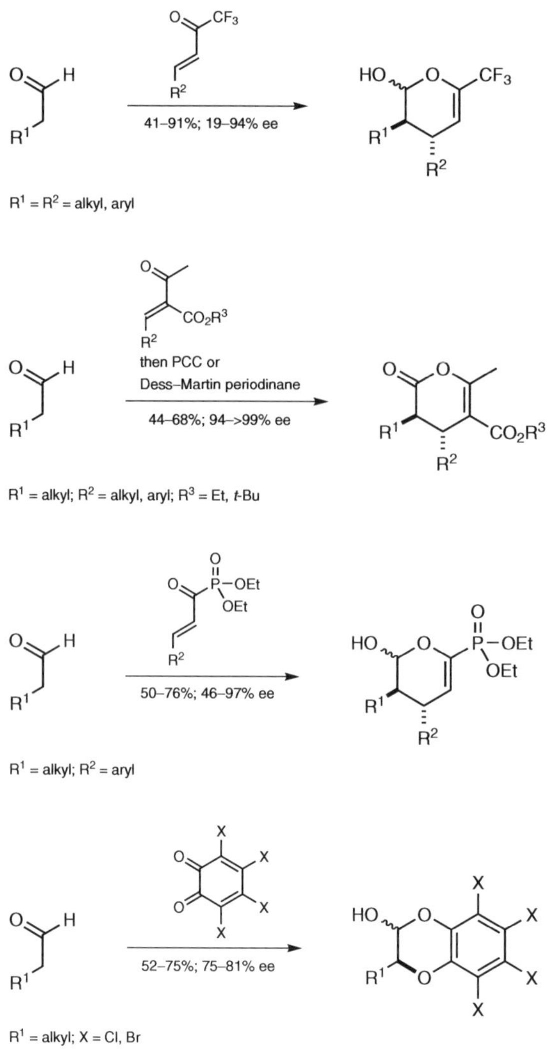

$R^1 = R^2$ = alkyl, aryl

then PCC or Dess–Martin periodinane

44–68%; 94–>99% ee

R^1 = alkyl; R^2 = alkyl, aryl; R^3 = Et, *t*-Bu

50–76%; 46–97% ee

R^1 = alkyl; R^2 = aryl

52–75%; 75–81% ee

R^1 = alkyl; X = Cl, Br

Inverse-electron-demand Diels–Alder reactions of *N*-sulfonylazabuta-1,3-diene **46** with enamines[48] and dienamines[49] using proline derivative **47** as the organocatalyst produce highly enantioenriched materials, such as **48** and **49**, in excellent yields (Scheme 21).[48–51]

Scheme 21 Inverse-Electron-Demand Aza-Diels–Alder Reactions[48–51]

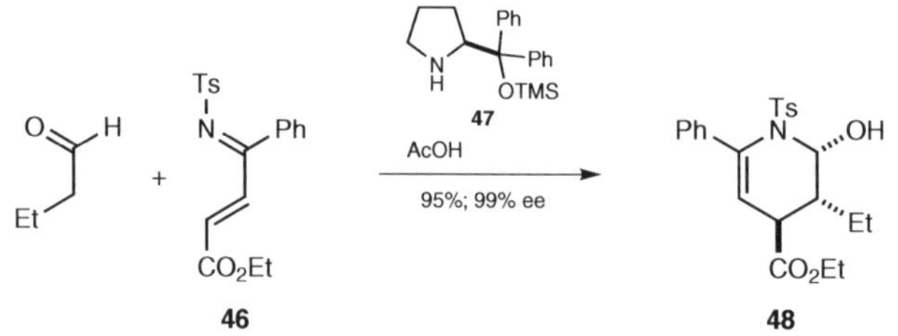

47, AcOH; 95%; 99% ee

46 **48**

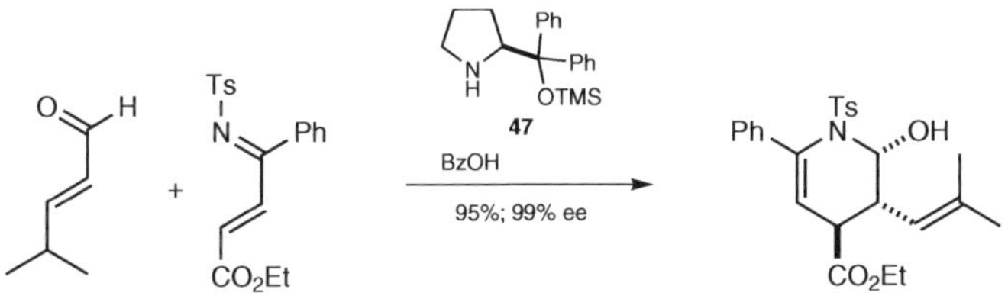

Alkyl (3*R*,4*S*)-3-Alkyl-4-aryl-2-oxo-3,4-dihydro-2*H*-pyran-6-carboxylates 45; General Procedure:[43]

The aldehyde (0.50 mmol) and the enone (1.00 mmol) were dissolved in CH_2Cl_2 (0.5 mL) and cooled to −15 °C. Catalyst **44** (0.05 mmol) was then added followed by silica gel (50 mg), and the resulting mixture was allowed to warm to rt overnight with stirring. The mixture of two anomers of cyclized hemiaminal (and uncyclized linear aldehyde) was isolated by flash chromatography. Oxidation of this mixture was in turn performed in CH_2Cl_2 on addition of PCC (1 equiv) at rt. After 1 h, further PCC (1 equiv) was added and, after 2 h, the lactone product was isolated by flash chromatography (CH_2Cl_2).

2.3.3.2.1.4 Enamine-Catalyzed Michael/Henry Reactions

A Michael/intramolecular Henry sequence between pentane-1,5-dial and (*E*)-(2-nitrovinyl)-benzene leads to valuable chiral cyclohexane derivative **50** in 66% yield and 99% enantiomeric excess using organocatalyst **47** (Scheme 22).[52] It is proposed that the intermediate enamine attacks the β-position of the nitroalkene, an event that is followed by an intramolecular Henry reaction to furnish the observed product.

Scheme 22 Organocatalytic Michael/Intramolecular Henry Reaction[52]

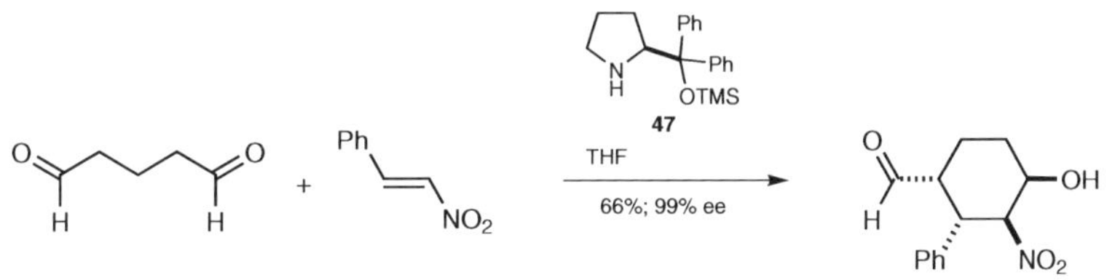

The Michael/intermolecular Henry sequence can be extended to prepare other highly substituted six-membered rings by hemiacetalization (Scheme 23).[53] Here, intermediate **51**, formed from Michael addition of propanal and (*E*)-(2-nitrovinyl)benzene, undergoes Henry reaction with the reactive aldehyde ethyl glyoxylate to give product **52** in 76% yield, 11:1 diastereomeric ratio, and >99% enantiomeric excess. A similar reaction mediated by primary thiourea catalysts gives the diastereomer of **52**.[54]

for references see p 415

Scheme 23 Michael/Intermolecular Henry/Hemiacetalization Sequence[53]

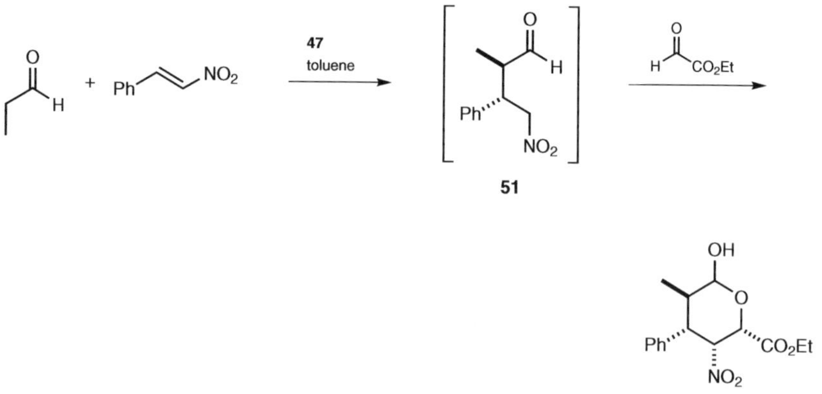

2.3.3.2.1.5 Enamine-Catalyzed Michael/Aldol Reactions

Similar to the Michael/intramolecular Henry sequence, a Michael/intramolecular aldol sequence of pentane-1,5-dial has also been developed.[55] In addition, an interesting Michael/intramolecular aldol reaction between cyclohexanone and β,γ-unsaturated α-oxo ester **53** produces bicycle **55** in 80% yield and 90% enantiomeric excess using proline derivative **54** as the organocatalyst (Scheme 24).[56] One α-face of cyclohexanone attacks the γ-position of β,γ-unsaturated α-oxo ester **53** through the first enamine-catalyzed process; then, the other side of cyclohexanone undergoes an intramolecular aldol reaction as the second enamine-catalyzed process.

Scheme 24 Michael/Intramolecular Aldol Reaction[56]

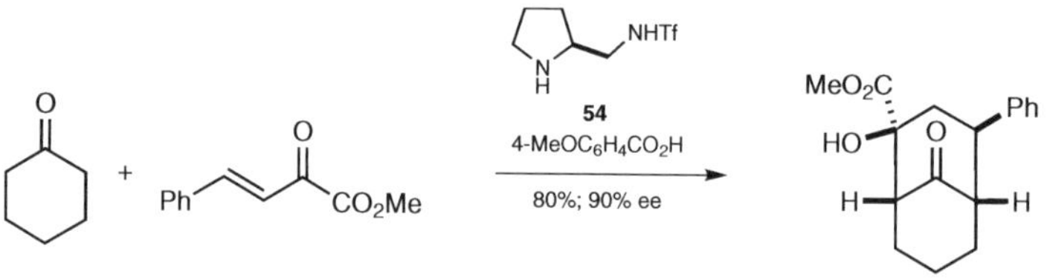

Methyl (1*R*,2*S*,4*R*,5*S*)-2-Hydroxy-9-oxo-4-phenylbicyclo[3.3.1]nonane-2-carboxylate (55); Typical Procedure:[56]

A mixture of catalyst **54** (0.04 mmol, 0.2 equiv) and 4-methoxybenzoic acid (0.04 mmol, 0.2 equiv) in cyclohexanone (1 mL) was stirred for 10 min at rt and then enone **53** (0.2 mmol, 1 equiv) was added. Once the reaction was complete, as monitored by TLC, the resulting mixture was concentrated under reduced pressure and the residue was purified by flash chromatography to give the product; yield: 80%.

2.3.3.2.1.6 Enamine-Catalyzed Michael/Hemiacetalization Reactions

Nucleophilic hydroxy groups at a suitable position within the product of an enamine-activated Michael addition can then attack the carbonyl group released from the enamine to furnish chiral cyclic hemiacetals. In 2007, the first Michael/hemiacetalization reaction sequence was reported via enamine catalysis (Scheme 25).[57] The aromatized Michael product **56** is in equilibrium with the hemiacetal **57**. However, hemiacetal **57** can be obtained as a single diastereomer in 90% yield and 99% enantiomeric excess after separation by flash chromatography using Iatrobeads.

Scheme 25 Organocatalytic α-Arylation of an Aldehyde[57]

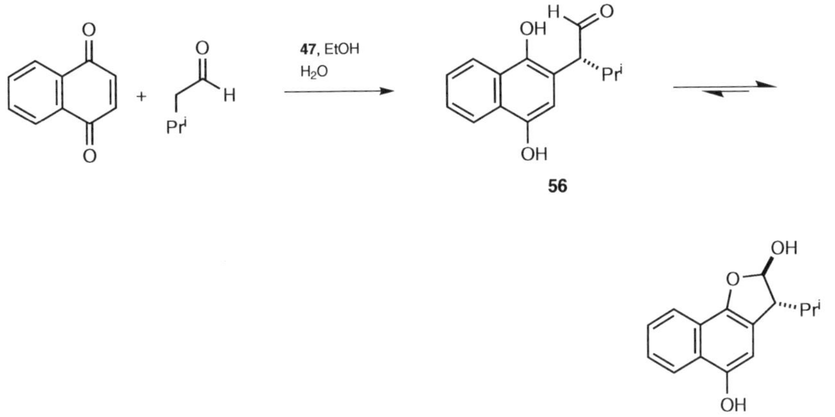

Critically, the formation of hemiacetals not only provides an efficient way to produce cyclic structures, but also becomes a driving force for the formation of novel and potentially valuable building blocks. Coupling between dienamines and nitroalkenes has been conducted to produce the Michael addition products.[58,59] However, when the dienamine derived from α,β-unsaturated aldehyde **58** and organocatalyst **47** is exposed to an α-(hydroxymethyl)-substituted nitroalkene **59**, a [2+2] cycloaddition reaction is followed by hemiacetal formation to afford adduct **61** (Scheme 26).[60] In this case, thiourea catalyst **60** activates the nitroalkene, and the formation of the hemiacetal is believed to be the crucial driving force for the [2+2] cycloaddition reaction. Other hemiacetalization reactions for the synthesis of cyclic structures have also been reported.[61–68]

Scheme 26 [2+2] Cycloaddition Reaction Followed by Hemiacetal Formation[60]

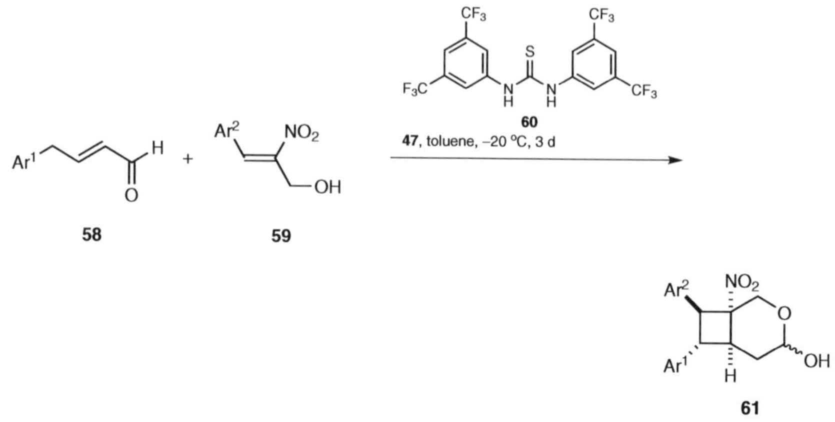

60
47, toluene, −20 °C, 3 d

58 59 61

Ar^1	Ar^2	Yield (%)	ee[a] (%)	Ref
Ph	Ph	86	91	[60]
4-Tol	4-MeOC$_6$H$_4$	91	94	[60]
Ph	2-thienyl	52	94	[60]

[a] The ee was determined after oxidation to the corresponding lactone.

1-Nitro-3-oxabicyclo[4.2.0]octan-4-ols 61; **General Procedure:**[60]
An ordinary vial equipped with a magnetic stirrer bar was charged with proline derivative **47** (0.07 mmol), thiourea **60** (0.07 mmol), a 3-aryl-2-nitroprop-2-en-1-ol **59** (0.52 mmol), and toluene (1.5 mL). The mixture was cooled to −20 °C and a soln of an α,β-unsaturated aldehyde **58** (0.35 mmol) in toluene (1 mL) was added. The mixture was stirred at −20 °C for 3 d and then concentrated. The residue was charged directly onto silica gel and subjected to flash chromatography.

2.3.3.2.1.7 Iminium-Catalyzed Michael/Aldol Reactions

The first organocatalytic domino Michael/aldol reaction, of β-oxo ester **62** and α,β-unsaturated ketone **63**, was reported in 2004 (Scheme 27).[69] The process is mediated by imidazolidine catalyst **64** via iminium catalysis to give adduct **65** in 80% yield, >97:3 diastereomeric ratio, and 95% enantiomeric excess.

Scheme 27 Michael/Aldol Reaction of a β-Oxo Ester[69]

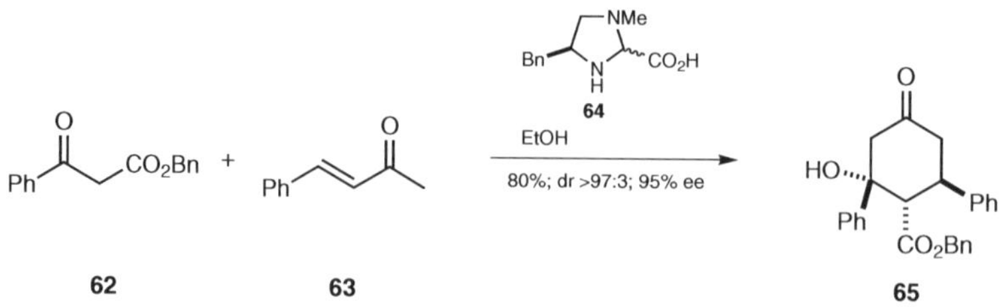

64
EtOH
80%; dr >97:3; 95% ee

62 63 65

Similar Michael/aldol reactions of 1-phenyl-2-sulfanylethanone (**66**) with enal **67** give highly functionalized, optically active tetrahydrothiophenes **68** and **69** (Scheme 28).[70] Interestingly, the formation of either **68** or **69** can be controlled entirely by the choice of simple additives (either BzOH or $NaHCO_3$), with both reactions performed in toluene.

Scheme 28 Michael/Aldol Reactions of 1-Phenyl-2-sulfanylethanone[70]

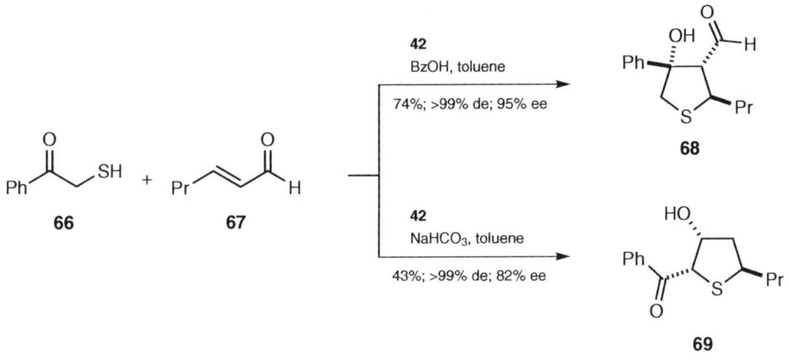

42, BzOH, toluene; 74%; >99% de; 95% ee → 68

42, $NaHCO_3$, toluene; 43%; >99% de; 82% ee → 69

66 + 67

The Michael/aldol sequence has also been extended to γ-chloro-β-oxo esters **70** (Scheme 29).[71] The advantage of this particular system is that a base-catalyzed S_N2 reaction can occur after the standard iminium-activated Michael/aldol domino sequence to furnish epoxides **71**.

Scheme 29 Domino Michael/Aldol/S_N2 Reactions[71]

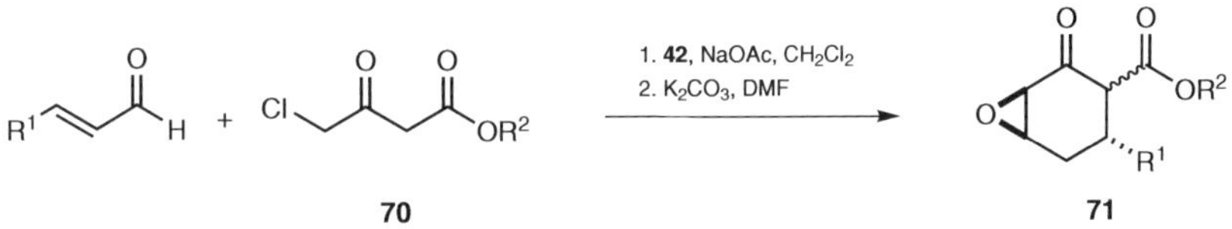

1. **42**, NaOAc, CH_2Cl_2
2. K_2CO_3, DMF

70 → 71

R^1	R^2	Yield (%)	ee[a] (%)	Ref
Ph	Et	47	90	[71]
Et	Bn	47	84	[71]
Bn	$CH_2CH=CH_2$	40	92	[71]

[a] The ee was determined by GC after decarboxylation or α-chlorination of **71**.

Cyclohexane-1,2-dione (**72**) is an excellent substrate for the Michael/aldol reaction with cinnamaldehyde (**73**) to produce bicyclic adduct **74** in 77% yield, >99:1 diastereomeric ratio, and 96% enantiomeric excess (Scheme 30).[72]

Scheme 30 Domino Michael/Aldol Reaction of Cyclohexane-1,2-dione[72]

47, EtOH, rt, 24 h; 77%; dr >99:1; 96% ee

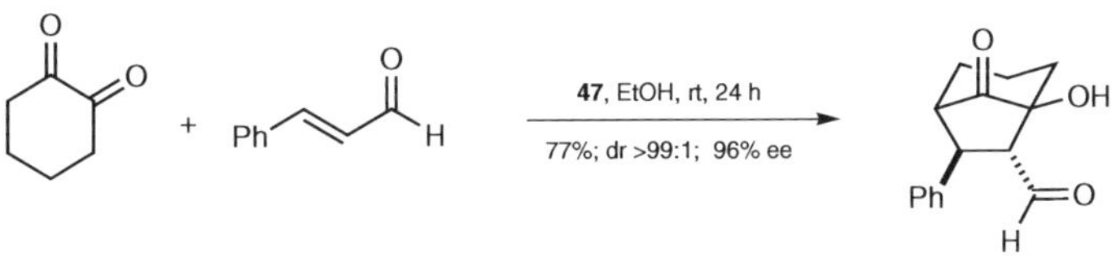

72 + 73 → 74

for references see p 415

Although various Michael-initiated domino processes involving carbon or sulfur nucleophiles have been disclosed, oxo-Michael-initiated tandem sequences are particularly difficult to achieve because of the reversibility of the conjugate addition process. Therefore, domino sequences in the form of oxo-Michael/aldol/hemiacetalization processes were designed for iminium activation.[73] For instance, coupling of dihydroxyacetone dimer (**75**) with enal **67** affords adduct **76** in 96% yield, >10:1 diastereomeric ratio, and 99% enantiomeric excess (Scheme 31).

Scheme 31 Organocatalytic Oxo-Michael/Aldol/Hemiacetalization[73]

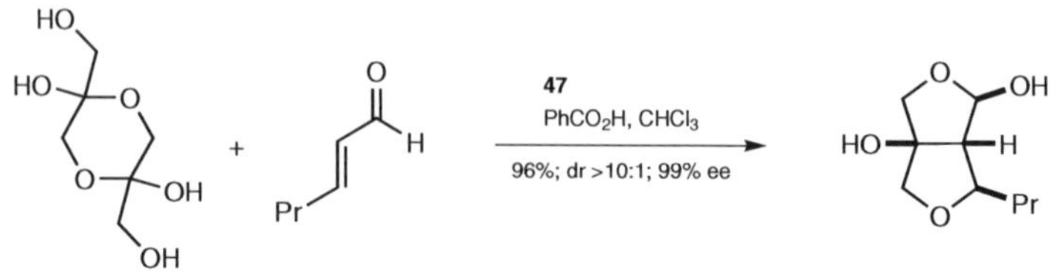

(1*S*,5*S*,6*R*,7*R*)-5-Hydroxy-8-oxo-7-phenylbicyclo[3.2.1]octane-6-carbaldehyde (74); **Typical Procedure:**[72]

A soln of catalyst **47** (10 mol%) and cyclohexane-1,2-dione (**72**; 0.24 mmol) in EtOH (1 mL) was placed in a screw-capped test tube equipped with stirrer bar. The aldehyde **73** (0.2 mmol) was added and the resulting mixture was stirred at rt for 24 h. The solvent was then removed under vacuum and the mixture was purified by column chromatography (silica gel) to afford the product; yield: 77%.

2.3.3.2.1.8 Iminium-Catalyzed Michael/Henry Reactions

Similar to the Michael/aldol reactions already described, it is not surprising that Michael/Henry reactions leading to six-membered rings such as **77** with five stereocenters have also been developed (Scheme 32).[74] In this case, only 4:2:1 diastereomeric ratio, 86% enantiomeric excess, and 44% yield are obtained for the desired product **77** using proline derivative **42** as the organocatalyst in the presence of 1,4-diazabicyclo[2.2.2]octane. However, given the array of processes occurring here, the result is still impressive and further investigations are certainly warranted.

Scheme 32 Enantioselective Michael/Henry Reaction[74]

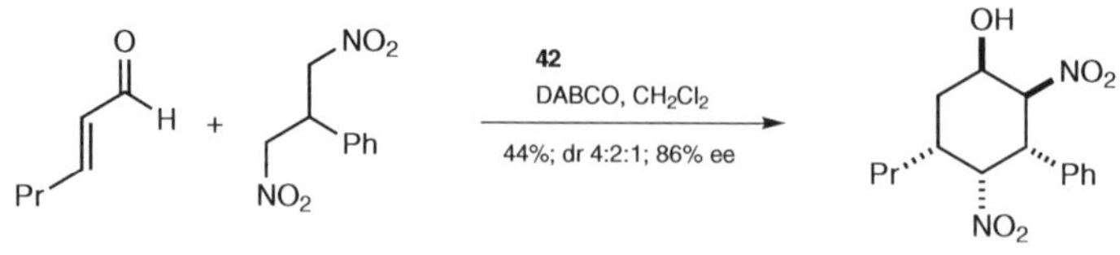

2.3.3.2.1.9 Iminium-Catalyzed Michael/Morita–Baylis–Hillman Reactions

An interesting and unexpected domino Michael/Morita–Baylis–Hillman reaction has been developed.[75] Here, a domino reaction between cinnamaldehyde (**73**) and Nazarov reagent **78** was originally designed in hope of producing a sequential [4+2] product via a Michael/Michael sequence. However, in practice, the Michael/Morita–Baylis–Hillman product **79** is obtained in lieu of the originally desired Michael/Michael adduct, a result of value in terms of the reasonable scope of this alternate process (Scheme 33).

Scheme 33 Michael/Morita–Baylis–Hillman Reaction[75]

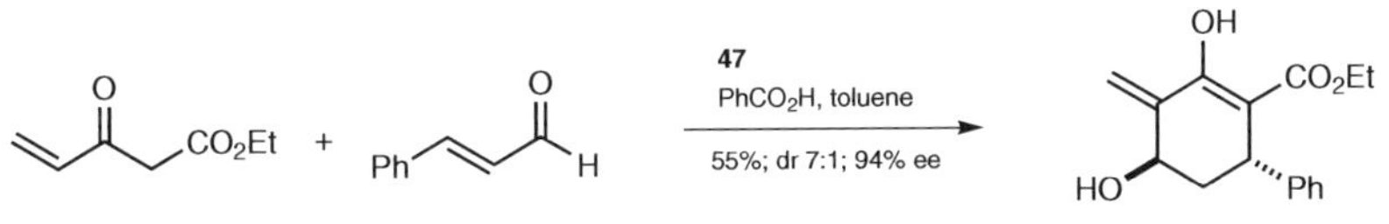

2.3.3.2.1.10 Iminium-Catalyzed Michael/Hemiacetalization Reactions

Enols and enamides are also suitable reaction partners for Michael/hemiacetalization reactions of enals through a [3+3] mode. Employment of enols[76,77] (e.g., **80**) and enamides[78,79] (e.g., **82**) in [3+3] reactions with enals produces hemiacetals (e.g., **81**) and hemiaminals (e.g., **83A** and **83B**), respectively (Scheme 34).

Scheme 34 Michael/Hemiacetalization Reactions of Enols and Enamides[76,78]

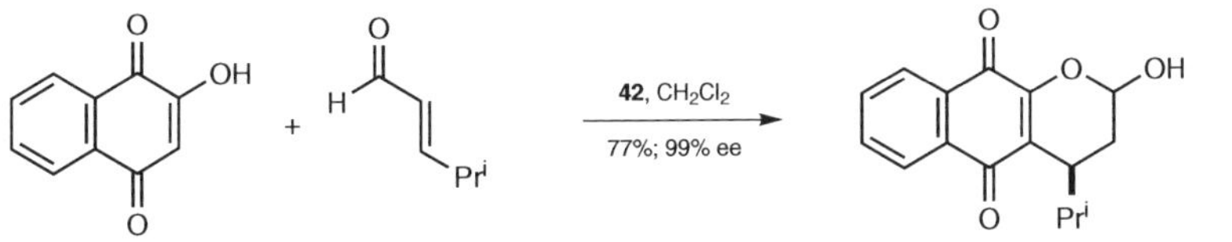

2.3.3.2.2 Thiourea-Catalyzed Domino Addition to C=O Bonds

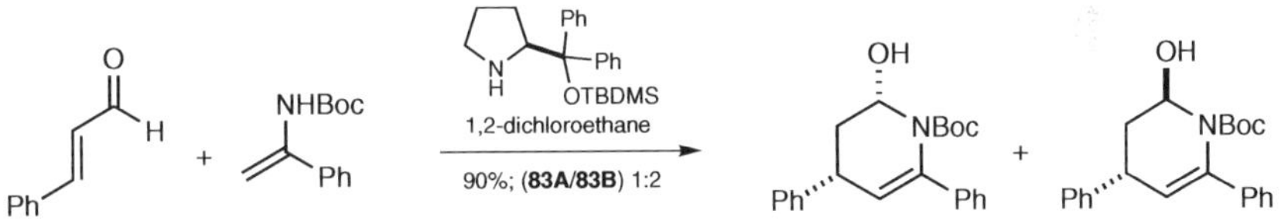

Bifunctional thiourea catalysts bearing both Brønsted basic sites and hydrogen-bonding units can be employed to activate nucleophiles and carbonyl groups simultaneously for asymmetric nucleophilic 1,2- or 1,4-addition reactions.[80–82] The oxygen anion or carbanion produced from the nucleophilic 1,2- or 1,4-addition in this dual activation model can then act as an efficient nucleophile for additional attack. Thus, domino reactions mediated by thiourea catalysts can be designed in aldol/cyclization, Michael/aldol, Michael/Henry, and Michael/hemiacetalization sequences.

2.3.3.2.2.1 Aldol/Cyclization Reactions

The oxygen anion produced from aldol addition to an aldehyde can attack isothiocyanate groups to form cyclic structures.[83] Indeed, the desired products **86** are obtained from a domino aldol/cyclization between α-isothiocyanato imide **84** and aromatic aldehydes promoted by thiourea catalyst **85** (Scheme 35). α-Isothiocyanato imide **84** reacts in aldol/cyclization reactions with isatins[84] and α-oxo esters[85,86] through similar reactive modes.

for references see p 415

Scheme 35 Chiral Thiourea Catalyzed Aldol/Cyclization Reactions[83]

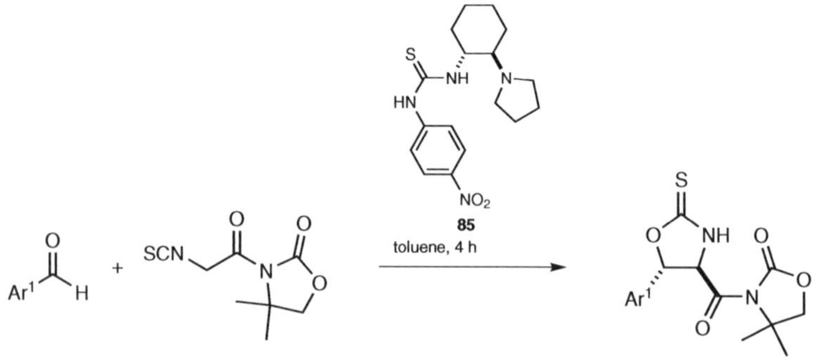

Ar^1	Yield (%)	dr	ee (%)	Ref
Ph	99	93:7	94	[83]
4-$O_2NC_6H_4$	97	95:5	93	[83]
2-naphthyl	97	94:6	91	[83]

2.3.3.2.2.2 Michael/Aldol Reactions

A cascade Michael/aldol reaction between 2-sulfanylbenzaldehyde and α,β-unsaturated *N*-acyloxazolidinone **87**, catalyzed by chiral thiourea **88**, gives benzothiopyran **89** in high yield and 99% enantiomeric excess (Scheme 36).[87] Extensions of this strategy to maleimides,[88] benzylidenemalonates,[89] and α,β-unsaturated *N*-acylpyrazoles[90] were also reported soon afterwards.

Scheme 36 Enantioselective Michael/Aldol Reaction[87]

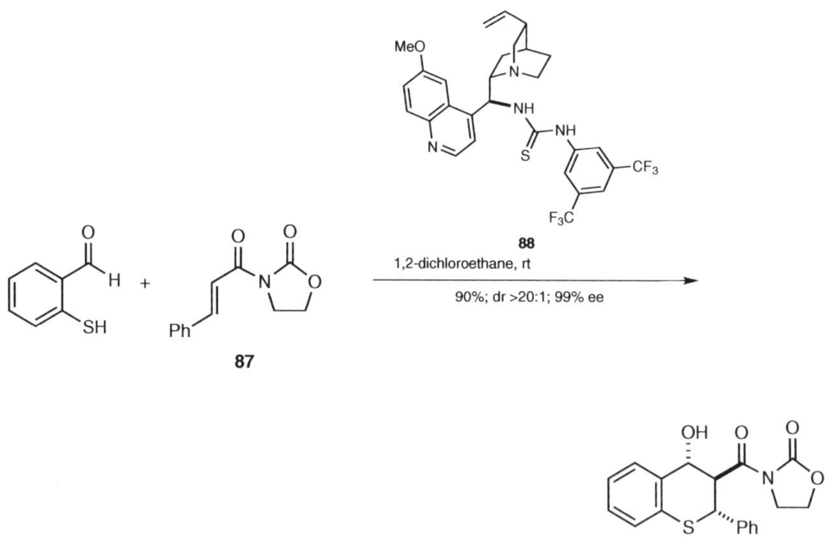

A novel multifunctional cinchona alkaloid catalyst **92** containing a primary amine and an axially chiral moiety promotes asymmetric domino Michael/aldol reactions between

3-substituted 2-oxoindoles **90** and **91** (Scheme 37).[91] Bispirooxindoles, such as **93**, containing three chiral quaternary carbon centers are obtained.

Scheme 37 Domino Michael/Aldol Reactions[91]

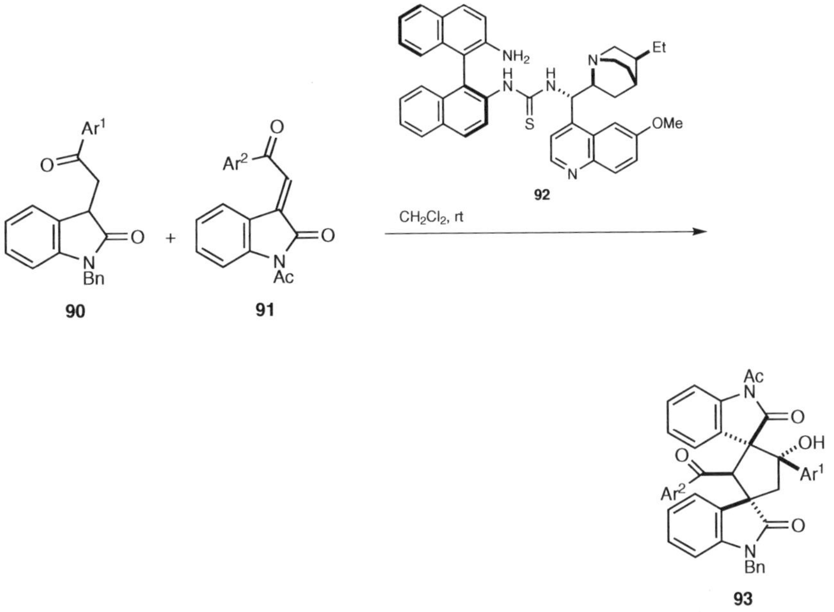

Ar^1	Ar^2	Yield (%)	dr	ee (%)	Ref
Ph	Ph	84	96:4	94	[91]
2-furyl	Ph	94	>99:1	94	[91]
Ph	2-Tol	69	95:5	82	[91]

3-[(2*R*,3*S*,4*R*)-4-Hydroxy-2-phenylthiochromane-3-carbonyl]oxazolidin-2-one (89); **Typical Procedure:**[87]

A mixture of 2-sulfanylbenzaldehyde (0.15 mmol), α,β-unsaturated oxazolidinone **87** (0.1 mmol), and catalyst **88** (0.001 mmol) was stirred at rt in 1,2-dichloroethane (0.2 mL). The crude product was purified by column chromatography to give the desired product; yield: 90%

2.3.3.2.2.3 Michael/Henry Reactions

Similar to the iminium-mediated Michael/aldol reaction between cyclohexane-1,2-dione (**72**) and cinnamaldehyde (**73**) (see Scheme 30), a Michael/Henry reaction between cyclohexane-1,2-dione (**72**) and (*E*)-(2-nitrovinyl)benzene catalyzed by chiral thiourea **88** produces [3+2] adduct **94** in 75% yield and 97% enantiomeric excess but only 7:1 diastereomeric ratio (Scheme 38).[92,93]

Scheme 38 Thiourea-Catalyzed Michael/Henry Reaction[92,93]

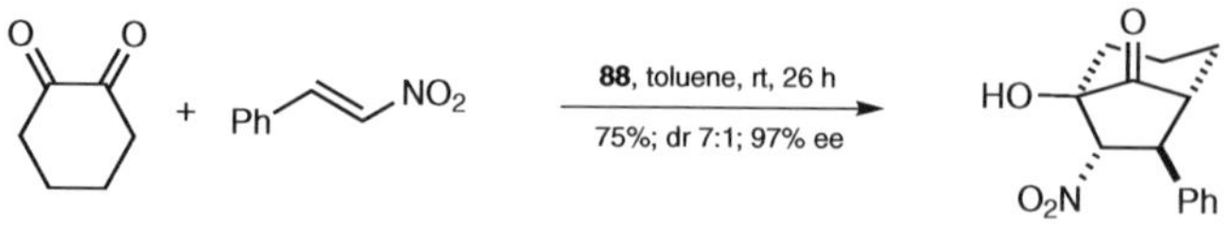

72 + Ph–CH=CH–NO_2 → (**88**, toluene, rt, 26 h; 75%; dr 7:1; 97% ee) **94**

However, excellent diastereoselectivity is achieved when cyclohexane-1,2-dione (**72**) is replaced with 1,4-dione **95** in the [3+2] reaction with (*E*)-(2-nitrovinyl)benzene catalyzed by thiourea **96** in benzonitrile (not toluene) as the solvent (Scheme 39).[94]

Scheme 39 Enantioselective Michael/Henry Reaction[94]

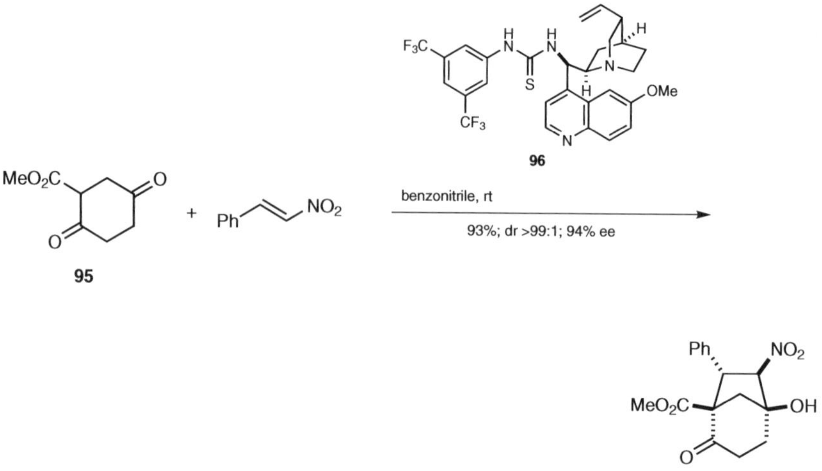

95 + Ph–CH=CH–NO_2 → (**96**, benzonitrile, rt; 93%; dr >99:1; 94% ee)

(1*R*,5*R*,6*S*,7*S*)-1-Hydroxy-7-nitro-6-phenylbicyclo[3.2.1]octan-8-one (94); **Typical Procedure:**[92]

The nitroalkene (0.25 mmol) was added to a solution of cyclohexane-1,2-dione (**72**; 0.30 mmol) and catalyst **88** (0.038 mmol) in toluene (1.5 mL). The resulting mixture was stirred at rt for 26 h and then the solvent was removed under reduced pressure. The residue obtained was purified by chromatography (silica gel) to afford the product; yield: 75%.

2.3.3.2.2.4 Michael/Hemiacetalization Reactions

Chiral thioureas **97** and **99** catalyze [3+3] reactions of β,γ-unsaturated α-oxo ester **53** with 4-hydroxycoumarin or 2-hydroxynaphtho-1,4-quinone via a Michael/hemiacetalization sequence to give the products **98** and **100**, respectively (Scheme 40).[95,96]

Scheme 40 Chiral Thiourea-Catalyzed [3+3] Reactions[95,96]

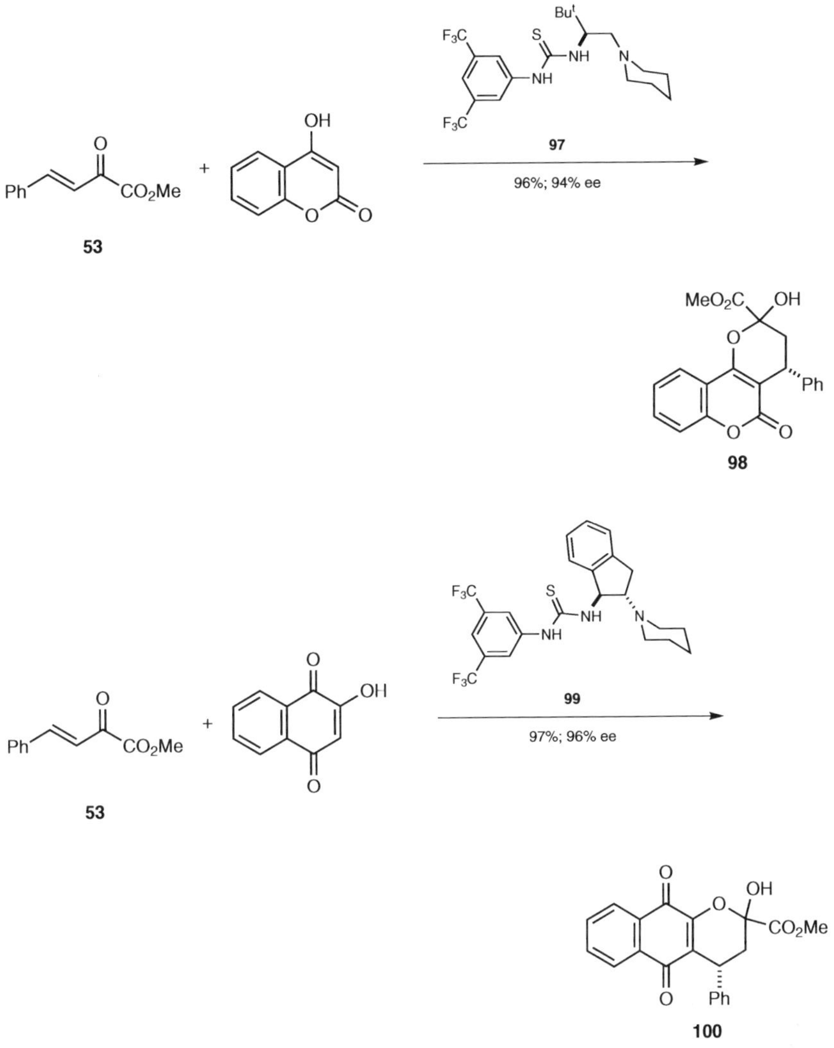

Ph–CH=CH–C(O)–CO2Me (53) + 4-hydroxycoumarin → 98 (cat. 97; 96%; 94% ee)

Ph–CH=CH–C(O)–CO2Me (53) + 2-hydroxynaphtho-1,4-quinone → 100 (cat. 99; 97%; 96% ee)

Replacement of 4-hydroxycoumarin or 2-hydroxynaphtho-1,4-quinone with ethyl 4,4,4-trifluoro-3-oxobutanoate in the Michael/hemiacetalization reaction of β,γ-unsaturated α-oxo ester **53** catalyzed by thiourea **88** leads to [4+2] product **101** in 93% yield, >30:1 diastereomeric ratio, and 96% enantiomeric excess (Scheme 41).[97]

Scheme 41 Thiourea-Catalyzed Michael/Hemiacetalization Reaction[97]

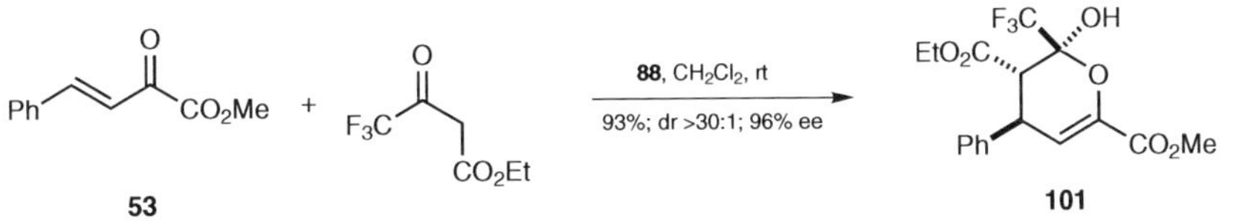

53 + F_3C–C(O)–CH_2–CO_2Et → 101 (**88**, CH_2Cl_2, rt; 93%; dr >30:1; 96% ee)

In addition, chiral thiourea catalyzed Michael/hemiacetalization reactions of β,γ-unsaturated α-oxo esters with β-oxoaldehydes[98] or α-cyano ketones with either α,β-unsaturated trifluoromethyl ketones[99] or α,β-unsaturated trichloromethyl ketones[100] have also been disclosed subsequently.

2.3.3.2.3 Phosphoric Acid Catalyzed Domino Addition to C=O Bonds

Desymmetrization of 4-hydroperoxy-4-methylcyclohexa-2,5-dien-1-one (**102**) via an acetalization/oxa-Michael sequence is mediated by chiral phosphoric acid **103** together with thiourea **104** at 50 °C and furnishes 1,2,4-trioxanes **105** as single diastereomers (Scheme 42). The presence of 4-Å molecular sieves is crucial to the high enantioselectivity observed in the optimized system.[101]

Scheme 42 Organocatalytic Desymmetrization of 4-Hydroperoxy-4-methylcyclohexa-2,5-dien-1-one[101]

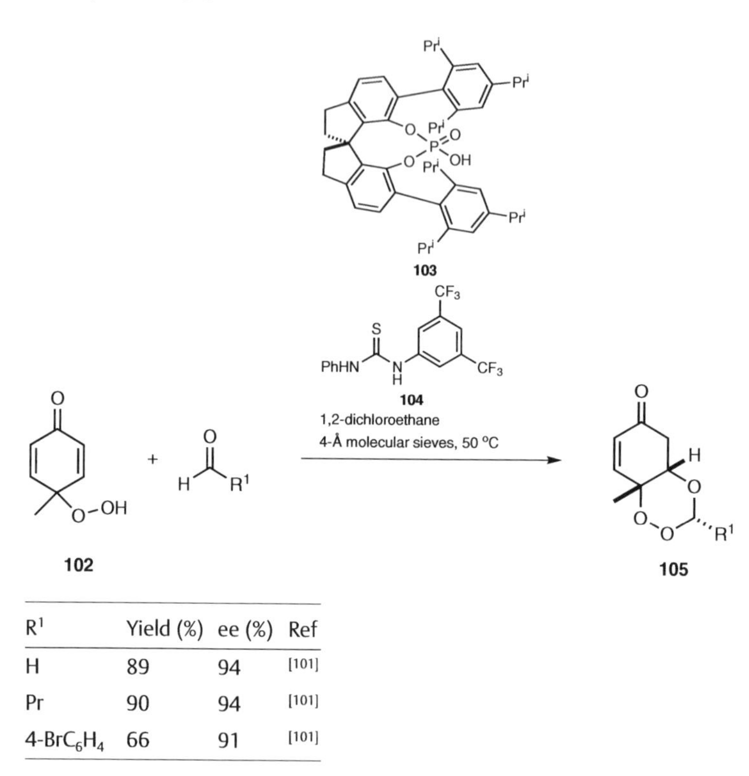

R1	Yield (%)	ee (%)	Ref
H	89	94	[101]
Pr	90	94	[101]
4-BrC_6H_4	66	91	[101]

(4a*S*,8a*R*)-8a-Methyl-4a,8a-dihydrobenzo[*e*][1,2,4]trioxin-6(5*H*)-one (105, **R1=H**)**; Typical Procedure:**[101]

Peroxide **102** (0.1 mmol), phosphoric acid **103** (0.005 mmol), thiourea **104** (0.005 mmol), formaldehyde (R1 = H; 0.15 mmol), activated 4-Å molecular sieves (25 mg), and 1,2-dichloroethane (0.4 mL) were added to an oven-dried, 1-mL vial containing a magnetic stirrer bar. The vial was then sealed and heated to 45 °C, and the mixture was stirred until the starting material was consumed, as monitored by TLC, and then concentrated in vacuo. The resulting yellow residue was subjected to column chromatography to give the analytically pure product; yield: 89%.

2.3.3.3 Lewis Acid Catalyzed Domino Addition to C=O Bonds

It is well-known that chiral Lewis acid complexes can also be applied as efficient catalysts in asymmetric reactions. Activation of carbonyl compounds with these complexes is similar to the models involving transition-metal complexes. However, totally different products can be obtained from similar substrates in reactions mediated by the two kinds of catalytic systems. For instance, transition-metal-catalyzed alkylation reactions of α,β-unsaturated ketones prefer to undergo 1,4-addition to produce Michael addition products. Conversely, in Lewis acid catalyzed alkylation reactions of α,β-unsaturated ketones, aldol products are always obtained via 1,2-addition. Therefore, chiral Lewis acid catalysis provides a viable method for the enantioselective synthesis of chiral allylic alcohols from α,β-unsaturated ketones or aldehydes.[102,103] Furthermore, tandem reaction systems can be designed if the allylic alcohols are further functionalized.

For instance, asymmetric alkylation of enone **106** catalyzed by titanium–ligand **107** complex, followed by epoxidation of the in situ prepared optically active allylic alcohol with molecular oxygen, furnishes chiral epoxy alcohol **108** in 82% yield and 99% enantiomeric excess as a single diastereomer (Scheme 43).[103] This ingenious design strategy can be extended to an asymmetric allylation/epoxidation sequence to produce epoxy alcohol **109** in 84% yield and 96% enantiomeric excess as a single diastereomer (Scheme 43).[104]

Scheme 43 Enantioselective Synthesis of Epoxy Alcohols from an Enone[102–104]

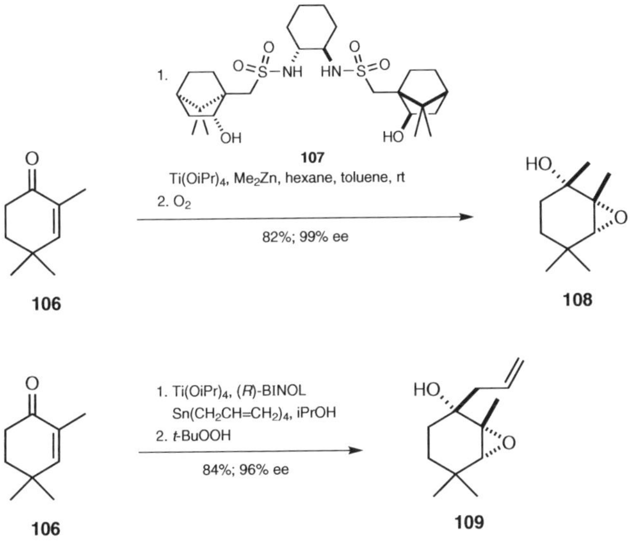

1. **107**, Ti(OiPr)$_4$, Me$_2$Zn, hexane, toluene, rt
2. O$_2$

82%; 99% ee

106 → **108**

1. Ti(OiPr)$_4$, (*R*)-BINOL, Sn(CH$_2$CH=CH$_2$)$_4$, iPrOH
2. *t*-BuOOH

84%; 96% ee

106 → **109**

Similarly, acyclic optically active epoxy alcohols **112** are produced from α,β-unsaturated aldehydes **110** using chiral ligand **111** (Scheme 44).[105,106] An alternative approach for the synthesis of epoxy alcohols **114** from vinylzinc reagents **113** and aldehydes gives comparable results (Scheme 44).[106] In this latter case, the *E*- or *Z*-vinylzinc reagents can be prepared in situ from alkynes.[107,108]

for references see p 415

Scheme 44 Enantioselective Synthesis of Acyclic Epoxy Alcohols[105,106]

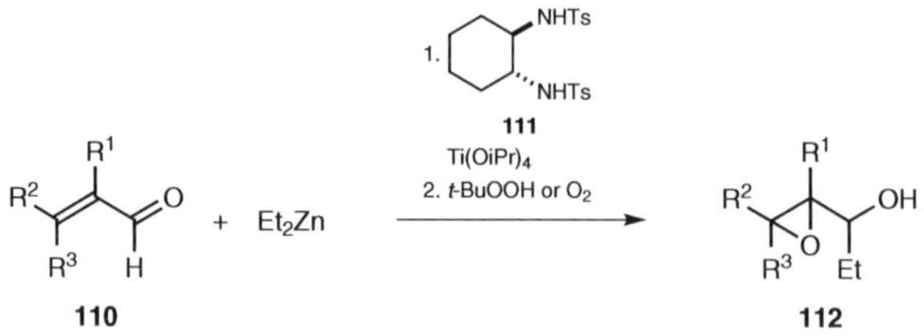

R^1	R^2	R^3	Yield (%)	dr	ee (%)	Ref
Me	Ph	H	90	20:1	92	[106]
H	H	Ph	60	10:1	96	[106]
H	Me	Me	93	20:1	95	[106]

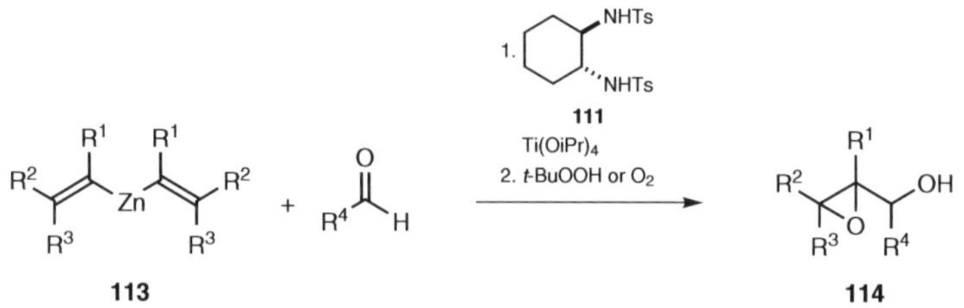

R^1	R^2	R^3	R^4	Yield (%)	dr	ee (%)	Ref
Me	H	H	4-Tol	89	16:1	94	[106]
H	Me	Me	Cy	90	20:1	97	[106]

Another interesting approach for the synthesis of epoxy alcohols involves enals **115** and divinylzinc reagents **116**. The reaction is challenging due to the complexity of controlling chemoselectivity in the epoxidation of unsymmetrical bis(allylic alkoxide) intermediates **117** produced from enal vinylation (Scheme 45).[106]

Scheme 45 Challenging Chemoselectivity in the Epoxidation of Bis(allylic alkoxides)[106]

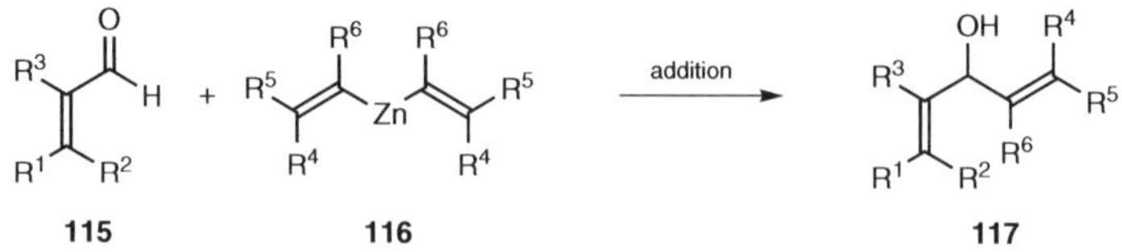

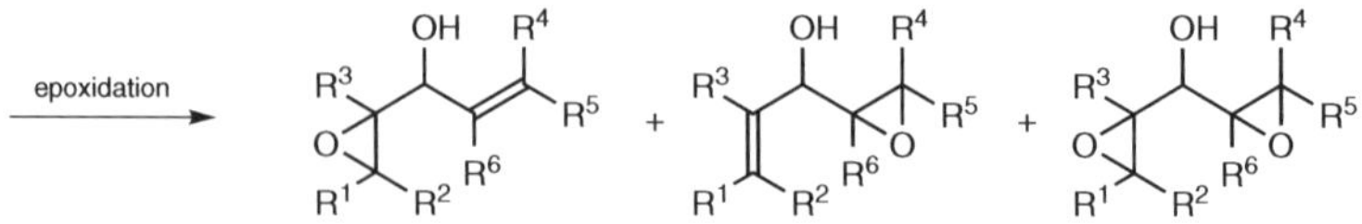

The key to success is the electrophilic nature of the proximal titanium species, which reacts faster with more electron-rich alkenes.[109] Thus, with correct design, allylic epoxy alcohol **119** is obtained in 90% yield, 95% enantiomeric excess, and 20:1 diastereomeric ratio from cyclohex-1-ene-1-carbaldehyde and di(prop-1-en-2-yl)zinc in the presence of chiral ligand **118** (Scheme 46). The vinylzinc reagents can be prepared in situ from alkynes to furnish, for example, allylic epoxy alcohol **120** in 78% yield and 99% enantiomeric excess as a single diastereomer (Scheme 46).[106]

Scheme 46 Enantioselective Synthesis of Allylic Epoxy Alcohols[106,109]

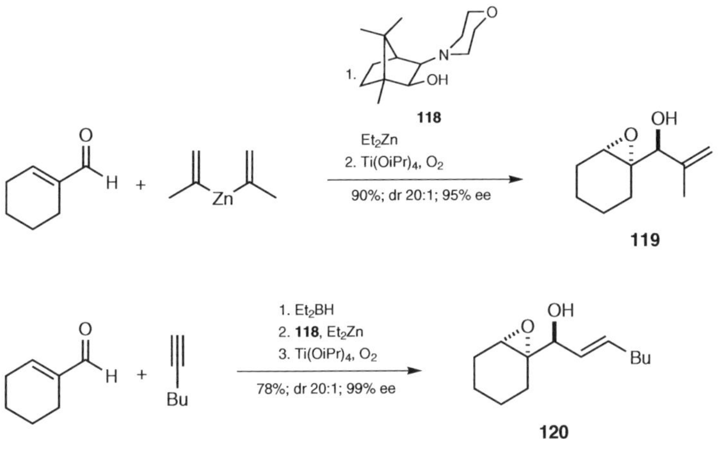

1. **118**, Et_2Zn
2. $Ti(OiPr)_4$, O_2
90%; dr 20:1; 95% ee
119

1. Et_2BH
2. **118**, Et_2Zn
3. $Ti(OiPr)_4$, O_2
78%; dr 20:1; 99% ee
120

Chiral Lewis acid catalyzed aldol/cyclization reactions[110] and Passerini-type reactions[111,112] have also been developed to furnish valuable optically active building blocks through additional domino processes. These molecular architectures are valuable in natural product synthesis.

(*S*,*E*)-1-[(1*S*,6*S*)-7-Oxabicyclo[4.1.0]heptanyl]hept-2-en-1-ol (120); **Typical Procedure:**[106]
Hex-1-yne (1.2 mmol) and a 1.0 M soln of Et_2BH in hexanes (1.2 mL, 1.2 mmol) were stirred at rt for 1 h. The solvent was removed under reduced pressure, and hexanes (1 mL) were added. In a separate flask, 1.0 M Et_2Zn in hexane (2.2 mL, 2.2 mmol), chiral ligand **118** (9.6 mg, 0.04 mmol), and cyclohex-1-ene-1-carbaldehyde (114 μL, 1.0 mmol) were combined and stirred at −78 °C. The hydroboration product was cannulated into the reaction flask and, after 30 min, the temperature was increased to −10 °C. The reaction was monitored by TLC until the addition was complete (6 h). The temperature was then decreased to −20 °C, the apparatus headspace was purged, and the mixture was exposed to O_2. After 30 min, a 1.0 M soln of $Ti(OiPr)_4$ in hexanes (200 μL, 0.20 mmol) was added and the reaction proceeded under 1 atm of O_2. After the reaction was complete (24 h), it was quenched with sat. aq NH_4Cl and the mixture was extracted with hexanes (3 × 10 mL). The combined organic phases were dried ($MgSO_4$) and concentrated under reduced pressure, and the residue was purified by column chromatography (silica gel, EtOAc/hexanes 9:1) to afford the product as a colorless oil; yield: 78%; dr 20:1; 99% ee.

for references see p 415

2.3.3.4 Conclusions

In conclusion, considerable efforts have been devoted to the development of catalytic asymmetric domino addition processes for monofunctional C=O bonds in terms of organometallic catalysis, organocatalysis, and Lewis acid catalysis to provide efficient methods for the rapid construction of valuable chiral building blocks. However, application of these methods is still limited due to the high catalyst loading in organocatalysis or limited substrate scope in organometallic and Lewis acid catalysis. The development of organometallic and Lewis acid catalyzed domino reactions with broader substrate scope and the corresponding more efficient organocatalytic systems would thus present a fascinating opportunity for the promotion of these methods and are worthy of further investigation.

References

[1] Padwa, A.; Weingarten, M. D., *Chem. Rev.*, (1996) **96**, 223.

[2] Ibata, T.; Toyoda, J., *Bull. Chem. Soc. Jpn.*, (1985) **58**, 1787.

[3] Clark, J. S.; Fretwell, M.; Whitlock, G. A.; Burns, C. J.; Fox, D. N. A., *Tetrahedron Lett.*, (1998) **39**, 97.

[4] Suga, H.; Inoue, K.; Inoue, S.; Kakehi, A., *J. Am. Chem. Soc.*, (2002) **124**, 14 836.

[5] Doyle, M. P.; Forbes, D. C.; Vasbinder, M. M.; Peterson, C. S., *J. Am. Chem. Soc.*, (1998) **120**, 7653.

[6] Hodgson, D. M.; Stupple, P. A.; Johnstone, C., *Tetrahedron Lett.*, (1997) **38**, 6471.

[7] Kitagaki, S.; Anada, M.; Kataoka, O.; Matsuno, K.; Umeda, C.; Watanabe, N.; Hashimoto, S., *J. Am. Chem. Soc.*, (1999) **121**, 1417.

[8] Tsutsui, H.; Shimada, N.; Abe, T.; Anada, M.; Nakajima, M.; Nakamura, S.; Nambu, H.; Hashimoto, S., *Adv. Synth. Catal.*, (2007) **349**, 521.

[9] Shimada, N.; Anada, M.; Nakamura, S.; Nambu, H.; Tsutsui, H.; Hashimoto, S., *Org. Lett.*, (2008) **10**, 3603.

[10] Shimada, N.; Hanari, T.; Kurosaki, Y.; Takeda, K.; Anada, M.; Nambu, H.; Shiro, M.; Hashimoto, S., *J. Org. Chem.*, (2010) **75**, 6039.

[11] Shimada, N.; Oohara, T.; Krishnamurthi, J.; Nambu, H.; Hashimoto, S., *Org. Lett.*, (2011) **13**, 6284.

[12] Chakraborty, S.; Krause, J. A.; Guan, H., *Organometallics*, (2009) **28**, 582.

[13] Appella, D. H.; Moritani, Y.; Shintani, R.; Ferreira, E. M.; Buchwald, S. L., *J. Am. Chem. Soc.*, (1999) **121**, 9473.

[14] Taylor, S. T.; Duffey, M. O.; Morken, J. P., *J. Am. Chem. Soc.*, (2000) **122**, 4528.

[15] Nishiyama, H.; Shiomi, T.; Tsuchiya, Y.; Matsuda, I., *J. Am. Chem. Soc.*, (2005) **127**, 6972.

[16] Bee, C.; Han, S. B.; Hassan, A.; Iida, H.; Krische, M. J., *J. Am. Chem. Soc.*, (2008) **130**, 2746.

[17] Deschamp, J.; Chuzel, O.; Hannedouche, J.; Riant, O., *Angew. Chem.*, (2006) **118**, 1314; *Angew. Chem. Int. Ed.*, (2006) **45**, 1292.

[18] Lipshutz, B. H.; Amorelli, B.; Unger, J. B., *J. Am. Chem. Soc.*, (2008) **130**, 14 378.

[19] Zhao, D.; Oisaki, K.; Kanai, M.; Shibasaki, M., *J. Am. Chem. Soc.*, (2006) **128**, 14 440.

[20] Oisaki, K.; Zhao, D.; Kanai, M.; Shibasaki, M., *J. Am. Chem. Soc.*, (2007) **129**, 7439.

[21] Miller, K. M.; Huang, W.-S.; Jamison, T. F., *J. Am. Chem. Soc.*, (2003) **125**, 3442.

[22] Miller, K. M.; Jamison, T. F., *Org. Lett.*, (2005) **7**, 3077.

[23] Komanduri, V.; Krische, M. J., *J. Am. Chem. Soc.*, (2006) **128**, 16 448.

[24] Sato, Y.; Saito, N.; Mori, M., *J. Am. Chem. Soc.*, (2000) **122**, 2371.

[25] Yang, Y.; Zhu, S.-F.; Duan, H.-F.; Zhou, C.-Y.; Wang, L.-X.; Zhou, Q.-L., *J. Am. Chem. Soc.*, (2007) **129**, 2248.

[26] Saito, N.; Kobayashi, A.; Sato, Y., *Angew. Chem.*, (2012) **124**, 1254; *Angew. Chem. Int. Ed.*, (2012) **51**, 1228.

[27] Arnold, L. A.; Naasz, R.; Minnaard, A. J.; Feringa, B. L., *J. Am. Chem. Soc.*, (2001) **123**, 5841.

[28] Arnold, L. A.; Naasz, R.; Minnaard, A. J.; Feringa, B. L., *J. Org. Chem.*, (2002) **67**, 7244.

[29] Yoshida, K.; Ogasawara, M.; Hayashi, T., *J. Am. Chem. Soc.*, (2002) **124**, 10 984.

[30] Cauble, D. F.; Gipson, J. D.; Krische, M. J., *J. Am. Chem. Soc.*, (2003) **125**, 1110.

[31] Bocknack, B. M.; Wang, L.-C.; Krische, M. J., *Proc. Natl. Acad. Sci. U. S. A.*, (2004) **101**, 5421.

[32] Arai, T.; Yokoyama, N., *Angew. Chem.*, (2008) **120**, 5067; *Angew. Chem. Int. Ed.*, (2008) **47**, 4989; corrigendum: *Angew. Chem.*, (2010) **122**, 9748; *Angew. Chem. Int. Ed.*, (2010) **49**, 9555.

[33] Shi, D.; Xie, Y.; Zhou, H.; Xia, C.; Huang, H., *Angew. Chem.*, (2012) **124**, 1274; *Angew. Chem. Int. Ed.*, (2012) **51**, 1248.

[34] Mihara, H.; Xu, Y.; Shepherd, N. E.; Matsunaga, S.; Shibasaki, M., *J. Am. Chem. Soc.*, (2009) **131**, 8384.

[35] Handa, S.; Slaughter, L. M., *Angew. Chem.*, (2012) **124**, 2966; *Angew. Chem. Int. Ed.*, (2012) **51**, 2912.

[36] List, B.; Lerner, R. A.; Barbas, C. F., III, *J. Am. Chem. Soc.*, (2000) **122**, 2395.

[37] Ahrendt, K. A.; Borths, C. J.; MacMillan, D. W. C., *J. Am. Chem. Soc.*, (2000) **122**, 4243.

[38] Chowdari, N. S.; Ramachary, D. B.; Córdova, A.; Barbas, C. F., III, *Tetrahedron Lett.*, (2002) **43**, 9591.

[39] Northrup, A. B.; MacMillan, D. W. C., *Science (Washington, D. C.)*, (2004) **305**, 1752.

[40] Casas, J.; Engqvist, M.; Ibrahem, I.; Kaynak, B.; Córdova, A., *Angew. Chem.*, (2005) **117**, 1367; *Angew. Chem. Int. Ed.*, (2005) **44**, 1343.

[41] Liu, K.; Chougnet, A.; Woggon, W.-D., *Angew. Chem.*, (2008) **120**, 5911; *Angew. Chem. Int. Ed.*, (2008) **47**, 5827.

[42] Volz, N.; Bröhmer, M. C.; Nieger, M.; Bräse, S., *Synlett*, (2009), 550.

[43] Juhl, K.; Jørgensen, K. A., *Angew. Chem.*, (2003) **115**, 1536; *Angew. Chem. Int. Ed.*, (2003) **42**, 1498.

[44] Samanta, S.; Krause, J.; Mandal, T.; Zhao, C.-G., *Org. Lett.*, (2007) **9**, 2745.
[45] Hernandez-Juan, F. A.; Cockfield, D. M.; Dixon, D. J., *Tetrahedron Lett.*, (2007) **48**, 1605.
[46] Wang, J.; Yu, F.; Zhang, X.; Ma, D., *Org. Lett.*, (2008) **10**, 2561.
[47] Zhao, Y.; Wang, X.-J.; Liu, J.-T., *Synlett*, (2008), 1017.
[48] Han, B.; Li, J.-L.; Ma, C.; Zhang, S.-J.; Chen, Y.-C., *Angew. Chem.*, (2008) **120**, 10119; *Angew. Chem. Int. Ed.*, (2008) **47**, 9971.
[49] Han, B.; He, Z.-Q.; Li, J.-L.; Li, R.; Jiang, K.; Liu, T.-Y.; Chen, Y.-C., *Angew. Chem.*, (2009) **121**, 5582; *Angew. Chem. Int. Ed.*, (2009) **48**, 5474.
[50] He, Z.-Q.; Han, B.; Li, R.; Wu, L.; Chen, Y.-C., *Org. Biomol. Chem.*, (2010) **8**, 755.
[51] Li, J.-L.; Han, B.; Jiang, K.; Du, W.; Chen, Y.-C., *Bioorg. Med. Chem. Lett.*, (2009) **19**, 3952.
[52] Hayashi, Y.; Okano, T.; Aratake, S.; Hazelard, D., *Angew. Chem.*, (2007) **119**, 5010; *Angew. Chem. Int. Ed.*, (2007) **46**, 4922.
[53] Ishikawa, H.; Sawano, S.; Yasui, Y.; Shibata, Y.; Hayashi, Y., *Angew. Chem.*, (2011) **123**, 3858; *Angew. Chem. Int. Ed.*, (2011) **50**, 3774.
[54] Uehara, H.; Imashiro, R.; Hernández-Torres, G.; Barbas, C. F., III, *Proc. Natl. Acad. Sci. U. S. A.*, (2010) **107**, 20672.
[55] Zhao, G.; Dziedzic, P.; Ullah, F.; Eriksson, L.; Córdova, A., *Tetrahedron Lett.*, (2009) **50**, 3458.
[56] Cao, C.-L.; Sun, X.-L.; Kang, Y.-B.; Tang, Y., *Org. Lett.*, (2007) **9**, 4151.
[57] Alemán, J.; Cabrera, S.; Maerten, E.; Overgaard, J.; Jørgensen, K. A., *Angew. Chem.*, (2007) **119**, 5616; *Angew. Chem. Int. Ed.*, (2007) **46**, 5520.
[58] Han, B.; Xiao, Y.-C.; He, Z.-Q.; Chen, Y.-C., *Org. Lett.*, (2009) **11**, 4660.
[59] Bencivenni, G.; Galzerano, P.; Mazzanti, A.; Bartoli, G.; Melchiorre, P., *Proc. Natl. Acad. Sci. U. S. A.*, (2010) **107**, 20642.
[60] Talavera, G.; Reyes, E.; Vicario, J. L.; Carrillo, L., *Angew. Chem.*, (2012) **124**, 4180; *Angew. Chem. Int. Ed.*, (2012) **51**, 4104.
[61] Chandrasekhar, S.; Mallikarjun, K.; Pavankumarreddy, G.; Rao, K. V.; Jagadeesh, B., *Chem. Commun. (Cambridge)*, (2009), 4985.
[62] Wang, Y.; Zhu, S.; Ma, D., *Org. Lett.*, (2011) **13**, 1602.
[63] Lu, D.; Li, Y.; Gong, Y., *J. Org. Chem.*, (2010) **75**, 6900.
[64] Enders, D.; Wang, C.; Yang, X.; Raabe, G., *Synlett*, (2011), 469.
[65] Hong, B.-C.; Kotame, P.; Liao, J.-H., *Org. Biomol. Chem.*, (2011) **9**, 382.
[66] Ramachary, D. B.; Prasad, M. S.; Madhavachary, R., *Org. Biomol. Chem.*, (2011) **9**, 2715.
[67] Wang, Y.; Yu, D.-F.; Liu, Y.-Z.; Wei, H.; Luo, Y.-C.; Dixon, D. J.; Xu, P.-F., *Chem.–Eur. J.*, (2010) **16**, 3922.
[68] Belot, S.; Vogt, K. A.; Besnard, C.; Krause, N.; Alexakis, A., *Angew. Chem.*, (2009) **121**, 9085; *Angew. Chem. Int. Ed.*, (2009) **48**, 8923.
[69] Halland, N.; Aburel, P. S.; Jørgensen, K. A., *Angew. Chem.*, (2004) **116**, 1292; *Angew. Chem. Int. Ed.*, (2004) **43**, 1272.
[70] Brandau, S.; Maerten, E.; Jørgensen, K. A., *J. Am. Chem. Soc.*, (2006) **128**, 14986.
[71] Marigo, M.; Bertelsen, S.; Landa, A.; Jørgensen, K. A., *J. Am. Chem. Soc.*, (2006) **128**, 5475.
[72] Rueping, M.; Kuenkel, A.; Tato, F.; Bats, J. W., *Angew. Chem.*, (2009) **121**, 3754; *Angew. Chem. Int. Ed.*, (2009) **48**, 3699.
[73] Reyes, E.; Talavera, G.; Vicario, J. L.; Badía, D.; Carrillo, L., *Angew. Chem.*, (2009) **121**, 5811; *Angew. Chem. Int. Ed.*, (2009) **48**, 5701.
[74] Reyes, E.; Jiang, H.; Milelli, A.; Elsner, P.; Hazell, R. G.; Jørgensen, K. A., *Angew. Chem.*, (2007) **119**, 9362; *Angew. Chem. Int. Ed.*, (2007) **46**, 9202.
[75] Cabrera, S.; Alemán, J.; Bolze, P.; Bertelsen, S.; Jørgensen, K. A., *Angew. Chem.*, (2008) **120**, 127; *Angew. Chem. Int. Ed.*, (2008) **47**, 121.
[76] Rueping, M.; Sugiono, E.; Merino, E., *Angew. Chem.*, (2008) **120**, 3089; *Angew. Chem. Int. Ed.*, (2008) **47**, 3046.
[77] Rueping, M.; Merino, E.; Sugiono, E., *Adv. Synth. Catal.*, (2008) **350**, 2127.
[78] Hayashi, Y.; Gotoh, H.; Masui, R.; Ishikawa, H., *Angew. Chem.*, (2008) **120**, 4076; *Angew. Chem. Int. Ed.*, (2008) **47**, 4012.
[79] Zu, L.; Xie, H.; Li, H.; Wang, J.; Yu, X.; Wang, W., *Chem.–Eur. J.*, (2008) **14**, 6333.
[80] Okino, T.; Hoashi, Y.; Furukawa, T.; Xu, X.; Takemoto, Y., *J. Am. Chem. Soc.*, (2005) **127**, 119.
[81] Li, H.; Wang, Y.; Tang, L.; Wu, F.; Liu, X.; Guo, C.; Foxman, B. M.; Deng, L., *Angew. Chem.*, (2005) **117**, 107; *Angew. Chem. Int. Ed.*, (2005) **44**, 105.
[82] Hamza, A.; Schubert, G.; Soos, T.; Papai, I., *J. Am. Chem. Soc.*, (2006) **128**, 13151.

[83] Li, L.; Klauber, E. G.; Seidel, D., *J. Am. Chem. Soc.*, (2008) **130**, 12 248.
[84] Jiang, X.; Cao, Y.; Wang, Y.; Liu, L.; Shen, F.; Wang, R., *J. Am. Chem. Soc.*, (2010) **132**, 15 328.
[85] Vecchione, M. K.; Li, L.; Seidel, D., *Chem. Commun. (Cambridge)*, (2010) **46**, 4604.
[86] Jiang, X.; Zhang, G.; Fu, D.; Cao, Y.; Shen, F.; Wang, R., *Org. Lett.*, (2010) **12**, 1544.
[87] Zu, L.; Wang, J.; Li, H.; Xie, H.; Jiang, W.; Wang, W., *J. Am. Chem. Soc.*, (2007) **129**, 1036.
[88] Zu, L.; Xie, H.; Li, H.; Jiang, W.; Wang, W., *Adv. Synth. Catal.*, (2007) **349**, 1882.
[89] Dodda, R.; Mandal, T.; Zhao, C.-G., *Tetrahedron Lett.*, (2008) **49**, 1899.
[90] Dong, X.-Q.; Fang, X.; Tao, H.-Y.; Zhou, X.; Wang, C.-J., *Chem. Commun. (Cambridge)*, (2012) **48**, 7238.
[91] Tan, B.; Candeias, N. R.; Barbas, C. F., III, *Nat. Chem.*, (2011) **3**, 473.
[92] Ding, D.; Zhao, C.-G.; Guo, Q.; Arman, H., *Tetrahedron*, (2010) **66**, 4423.
[93] Rueping, M.; Kuenkel, A.; Fröhlich, R., *Chem.–Eur. J.*, (2010) **16**, 4173.
[94] Tan, B.; Lu, Y.; Zeng, X.; Chua, P. J.; Zhong, G., *Org. Lett.*, (2010) **12**, 2682.
[95] Gao, Y.; Ren, Q.; Wang, L.; Wang, J., *Chem.–Eur. J.*, (2010) **16**, 13 068.
[96] Gao, Y.; Ren, Q.; Ang, S.-M.; Wang, J., *Org. Biomol. Chem.*, (2011) **9**, 3691.
[97] Wang, J.; Hu, Z.; Lou, C.; Liu, J.; Li, X.; Yan, M., *Tetrahedron*, (2011) **67**, 4578.
[98] Wang, X.; Yao, W.; Yao, Z.; Ma, C., *J. Org. Chem.*, (2012) **77**, 2959.
[99] Li, P.; Chai, Z.; Zhao, S.-L.; Yang, Y.-Q.; Wang, H.-F.; Zheng, C.-W.; Cai, Y.-P.; Zhao, G.; Zhu, S.-Z., *Chem. Commun. (Cambridge)*, (2009), 7369.
[100] Wang, H.-F.; Li, P.; Cui, H.-F.; Wang, X.-W.; Zhang, J.-K.; Liu, W.; Zhao, G., *Tetrahedron*, (2011) **67**, 1774.
[101] Rubush, D. M.; Morges, M. A.; Rose, B. J.; Thamm, D. H.; Rovis, T., *J. Am. Chem. Soc.*, (2012) **134**, 13 554.
[102] Waltz, K. M.; Gavenonis, J.; Walsh, P. J., *Angew. Chem.*, (2002) **114**, 3849; *Angew. Chem. Int. Ed.*, (2002) **41**, 3697.
[103] Jeon, S.-J.; Walsh, P. J., *J. Am. Chem. Soc.*, (2003) **125**, 9544.
[104] Kim, J. G.; Waltz, K. M.; Garcia, I. F.; Kwiatkowski, D.; Walsh, P. J., *J. Am. Chem. Soc.*, (2004) **126**, 12 580.
[105] Lurain, A. E.; Maestri, A.; Kelly, A. R.; Carroll, P. J.; Walsh, P. J., *J. Am. Chem. Soc.*, (2004) **126**, 13 608.
[106] Kelly, A. R.; Lurain, A. E.; Walsh, P. J., *J. Am. Chem. Soc.*, (2005) **127**, 14 668.
[107] Lurain, A. E.; Carroll, P. J.; Walsh, P. J., *J. Org. Chem.*, (2005) **70**, 1262.
[108] Salvi, L.; Jeon, S.-J.; Fisher, E. L.; Carroll, P. J.; Walsh, P. J., *J. Am. Chem. Soc.*, (2007) **129**, 16 119.
[109] Hussain, M. M.; Walsh, P. J., *Acc. Chem. Res.*, (2008) **41**, 883.
[110] Yoshino, T.; Morimoto, H.; Lu, G.; Matsunaga, S.; Shibasaki, M., *J. Am. Chem. Soc.*, (2009) **131**, 17 082.
[111] Wang, S.-X.; Wang, M.-X.; Wang, D.-X.; Zhu, J., *Angew. Chem.*, (2008) **120**, 394; *Angew. Chem. Int. Ed.*, (2008) **47**, 388.
[112] Yue, T.; Wang, M.-X.; Wang, D.-X.; Zhu, J., *Angew. Chem.*, (2008) **120**, 9596; *Angew. Chem. Int. Ed.*, (2008) **47**, 9454.

2.3.4 Additions to C=N Bonds and Nitriles

E. Kroon, T. Zarganes Tzitzikas, C. G. Neochoritis, and A. Dömling

General Introduction

Domino transformations allow the generation of several new bonds in a single synthetic operation starting from simple substrates, wherein the subsequent reactions result as a consequence of the functionality generated in a previous step. The quality of the domino reaction can be correlated to the number of chemical bonds formed, considering the complexity and diversity that can be achieved in the process, e.g. the bond-forming efficiency index defined by Tietze.[1] A multicomponent reaction is a process wherein three or more starting materials are combined into one compound in a single chemical operation; thus, it is a domino process. The rapid and facile access to biologically relevant compounds and the scaffold diversity of multicomponent reactions have been recognized by the synthetic community in industry and academia as a preferred method to both design and discover biologically active compounds.[2,3]

Ugi and Ugi-type multicomponent reactions, belonging to the most important transformations in the field, give access to biologically and pharmaceutically relevant compounds, with recent examples including boceprevir, retosiban, and mandipropamid.[4–6] Classical Ugi four-component,[7,8] Ugi four-component tetrazole,[9] Ugi lactam,[10] Ugi four-component hydantoin,[11] Ugi five-center four-component,[12] Ugi three-component,[13] and Ugi–Smiles[14] reactions are just a few examples that demonstrate the power of multicomponent reaction chemistry as a key set of domino transformations in organic synthesis. These kinds of reactions start with the formation of the corresponding imine and are followed by addition into the resulting C=N bond by the other reagents (e.g., isocyanide, acid), setting into motion additional events.

Very recently, a facile and convenient synthesis of indole derivatives based on a multicomponent reaction was published.[15] Employing novel isocyanide **1**, six different scaffolds (**2–7**) and 18 new compounds were described. In most of these reactions, the addition to the C=N bond is the initial reaction (Scheme 1).

for references see p 446

Scheme 1 The Power of Multicomponent Reaction Chemistry: One Isocyanide, Six Different Scaffolds, and Eighteen New Compounds[15]

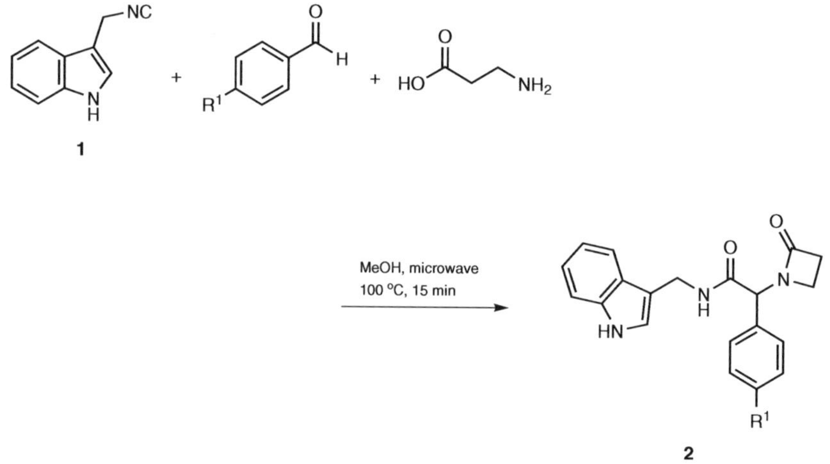

R^1	Yield (%)	Ref
Cl	59	[15]
H	45	[15]
Me	39	[15]

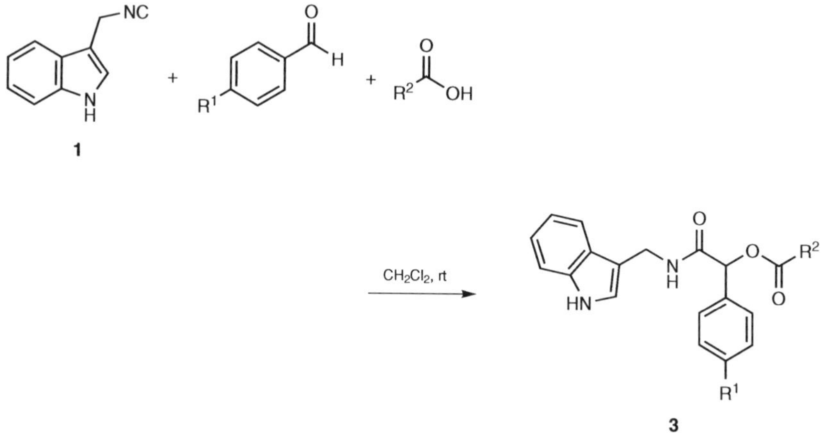

R^1	R^2	Yield (%)	Ref
Br	Me	67	[15]
H	Me	55	[15]
CN	2-ClC_6H_4	69	[15]

1

MeOH

4

R^1	R^2	Yield (%)	Ref
Br	Me	70	[15]
H	Me	55	[15]
CN	H	71	[15]

1

MeOH

5

R^1	R^2	Yield (%)	Ref
Br	Cl	85	[15]
H	H	82	[15]
CN	OMe	74	[15]

for references see p 446

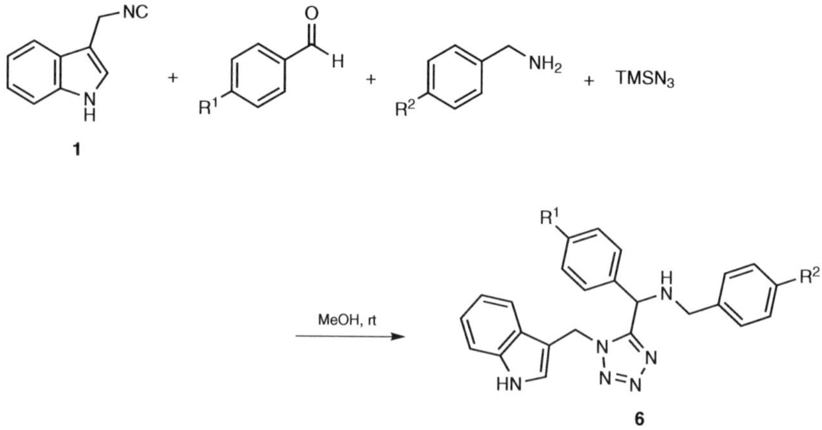

R^1	R^2	Yield (%)	Ref
Br	Cl	88	[15]
H	H	80	[15]
F	3-ClC_6H_4	71	[15]

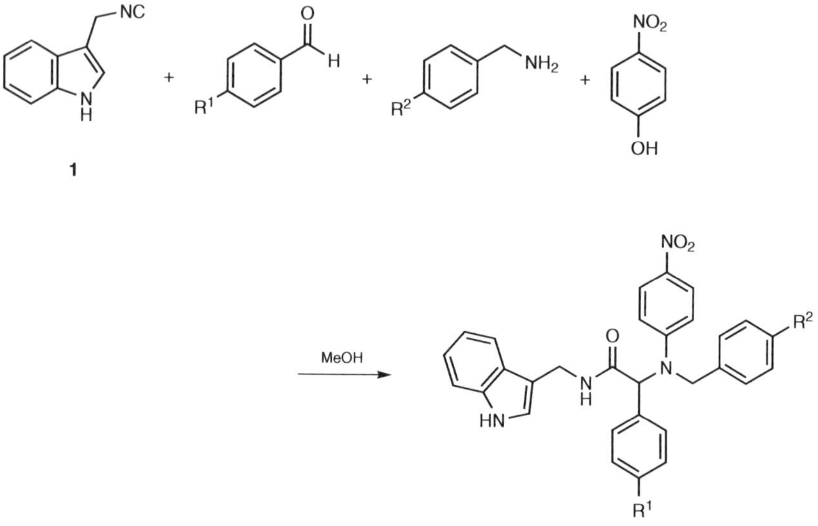

R^1	R^2	Yield (%)	Ref
Br	Cl	35	[15]
H	H	41	[15]
CN	OMe	45	[15]

2.3.4.1 Addition to C=N Bonds and the Pictet–Spengler Strategy

Chemical diversity and complexity of scaffolds are the keys for the design and/or screening-based discovery of useful materials. Multicomponent reactions are often highly compatible with a range of unprotected orthogonal functional groups; thus, even more "secondary" scaffolds can be obtained based on bifunctional starting materials or some various subsequent reactions of the initial products. This two-layered strategy has been extremely fruitful in the past, leading to a great array of scaffolds now routinely used in combinatorial and medicinal chemistry for drug discovery purposes.[3] One characteristic strategy is the combination of the venerable Ugi and Pictet–Spengler reactions, a process which results in polycyclic scaffolds similar to many classes of natural products (Scheme 2).

Scheme 2 Ugi and Pictet–Spengler Reaction Strategy

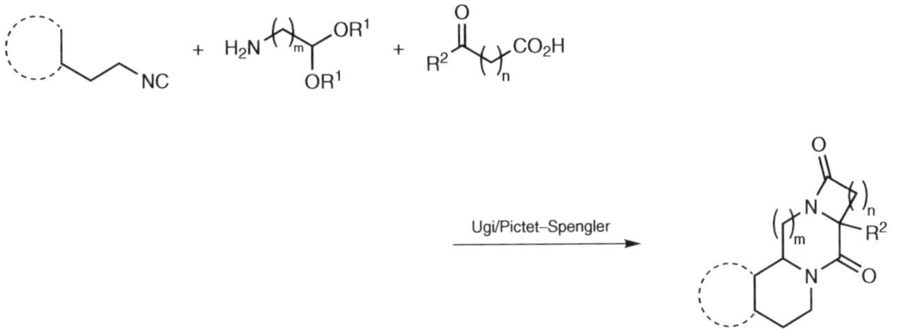

NC + H_2N (m) OR^1 OR^1 + R^2 O (n) CO_2H → Ugi/Pictet–Spengler → O N (m) (n) R^2 N O

In the first reported combination of the Ugi and Pictet–Spengler reactions, electron-rich 2-(1*H*-indol-3-yl)ethan-1-amine-derived isocyanide **8** reacts in the Ugi reaction with a diversity of bifunctional oxocarboxylic acid derivatives **10** and orthogonally protected aminoacetaldehyde **9**, followed by a Pictet–Spengler reaction. This process yields structurally intriguing polycyclic indole alkaloid type compounds **12** (Scheme 3). The yields for the Ugi products **11** vary from 50 to 80% (Ugi) and those for the Pictet–Spengler reactions are 48–90%.[16]

Scheme 3 Ugi and Pictet–Spengler Reactions Affording Polycyclic Indole Alkaloid Type Compounds[16]

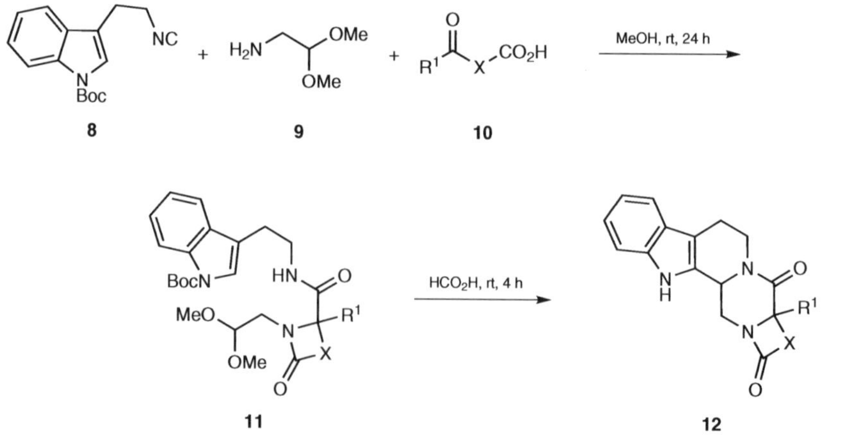

R^1	X	Yield (%) 11	Yield (%) 12	Ref
Me	$(CH_2)_2$	53	48	[16]
Me	$(CH_2)_3$	80	63	[16]
Me	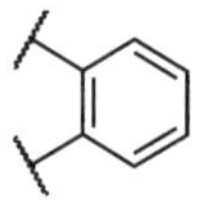	60	90	[16]
H	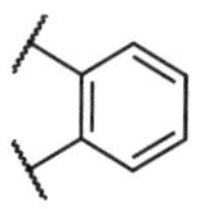	60	80	[16]

Furthermore, the reaction sequence has been extended to other isocyanides and the synthesis of a small focused library of polycyclic products **16** based on (2-phenylethyl)amine-derived isocyanides has been described (Scheme 4). Both syntheses naturally yield different scaffolds with different substitution patterns and likely different biological activities, though these results remain to be reported.

The first step is an Ugi three-component reaction of a (2-phenylethyl)amine-derived isocyanide **13** and aminoacetaldehyde dimethyl acetal (**9**) with a suitable bifunctional oxocarboxylic acid **14**. Both electron-neutral and electron-rich isocyanides can be employed, with the isocyanides being derived from the corresponding amines. The next step involves a Pictet–Spengler reaction of the dimethyl acetal protected Ugi intermediates **15** (Scheme 4). The conditions for this ring closure require formic acid or methanesulfonic acid and have been chosen according to previous optimizations of this reaction sequence, albeit using different isocyanide inputs. Using formic acid at room temperature for ring closure is good for the more-reactive dimethoxyphenyl Ugi products but is ineffective for the less-reactive monomethoxyphenyl or phenyl Ugi products; however, in these cases, anhydrous methanesulfonic acid at 70 °C with a longer reaction time is effective. Yields range from moderate to very good, and in some cases high diastereomeric ratios are obtained.[17]

Scheme 4 Ugi and Pictet–Spengler Reactions Affording Polycyclic Products Based on (2-Phenylethyl)amine-Derived Isocyanides[17]

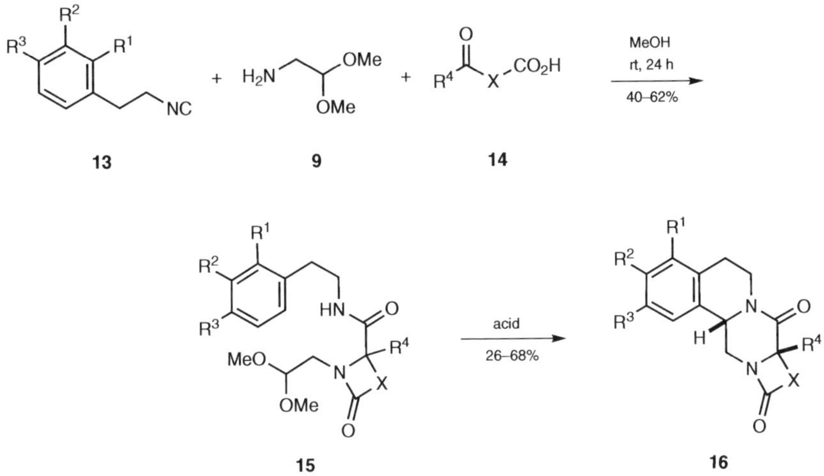

13 + 9 + 14 → (MeOH, rt, 24 h, 40–62%) 15 → (acid, 26–68%) 16

R^1	R^2	R^3	R^4	X	Conditions for Step 2	Yield (%) 15	Yield (%) 16	Ref
H	OMe	OMe	Me	$(CH_2)_2$	HCO_2H, rt, 4 h	62	62	[17]
H	OMe	OMe	Me	$(CH_2)_3$	HCO_2H, rt, 4 h	40	60	[17]
H	OMe	OMe	Me	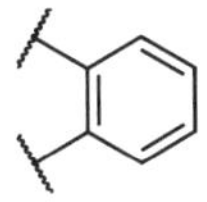	HCO_2H, rt, 4 h	55	26	[17]
OMe	H	H	Me	$(CH_2)_2$	MsOH, 70 °C, 24 h	60	68	[17]
H	H	H	Me	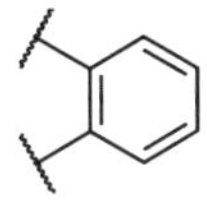	MsOH, 70 °C, 24 h	54	26	[17]

An efficient approach to access praziquantel (Scheme 5) has been described based on an Ugi four-component reaction followed by a Pictet–Spengler reaction.[17] Praziquantel, a tetrahydroisoquinoline derivative, is the only commercially available treatment of schistosomiasis, a high-volume neglected tropical disease affecting more than 200 million people worldwide, and is marketed by Bayer (Biltricide), Merck (Cysticide), and Shin Poong (Distocide), to name a few.

Scheme 5 Structure of Praziquantel[17]

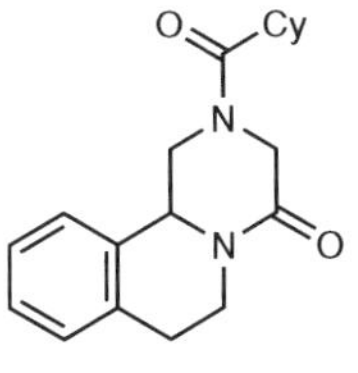

praziquantel

for references see p 446

This domino reaction approach comprises an Ugi four-component reaction as the key step, followed by a Pictet–Spengler ring closure. These events are carried out in a sequential one-pot, two-step procedure with an overall yield of 16–58%. Various isocyanides **17**, aldehydes **18**, and acids **19** have been used to define the scope and limitations of the reaction and thirty novel praziquantel derivatives **20** have been successfully synthesized (Scheme 6).[18] Novel tricyclic indole and isoquinoline derivatives have been synthesized based on a similar approach, thus extending the chemical space accessed even further.

Scheme 6 Ugi and Pictet–Spengler Reaction Affording Novel Praziquantel Derivatives[18]

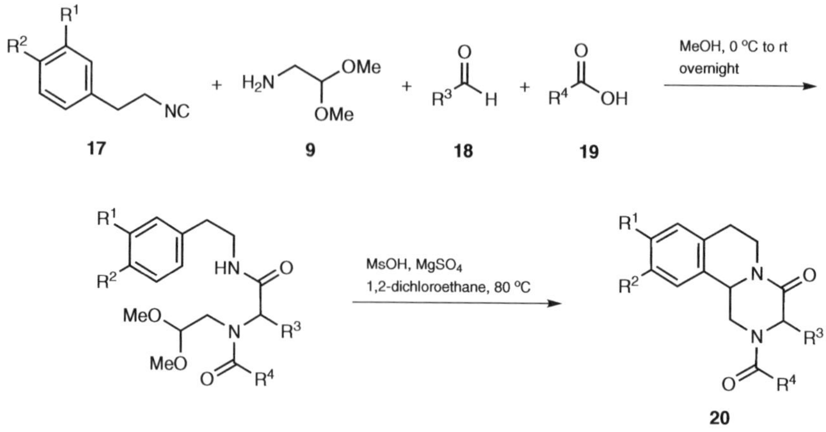

MeOH, 0 °C to rt overnight

MsOH, $MgSO_4$ 1,2-dichloroethane, 80 °C

R^1	R^2	R^3	R^4	Overall Yield (%) of **20**	Ref
OMe	OMe	H	Cy	34	[18]
H	H	Me	Cy	21	[18]
H	H	4-FC_6H_4	Cy	16	[18]
H	H	H	cyclopropyl	58	[18]
H	H	H	Ph	42	[18]

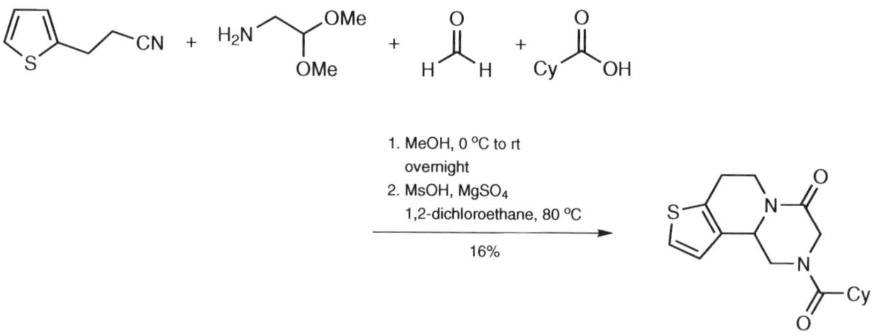

1. MeOH, 0 °C to rt overnight
2. MsOH, $MgSO_4$ 1,2-dichloroethane, 80 °C

16%

Recently, Khoury and co-workers developed an extension of the Ugi reaction of α-amino acids by introducing primary or secondary amines as additional reactants, thus rendering the Ugi five-center four-component reaction into a truly four-component reaction with four highly variable starting materials (see Section 2.3.4.2).[12] Based on that advance, chiral imino dicarboxamides can be formed by variation of an α-amino acid, an oxo component, an isocyanide, and a primary or secondary amine as previously unprecedented components for this Ugi variation. Critically, the reaction is compatible with many functional groups and different fragments.

A tryptophan amino acid derivative **23**, a ketone **22**, an isocyanide **21**, and aminoacetaldehyde dimethyl acetal (**9**) react in methanol/water (4:1) at room temperature to form the novel Ugi five-center four-component reaction products **24** (Scheme 7). Afterward, by using concentrated formic acid at room temperature, the corresponding Pictet–Spengler cyclization product **25** is formed. The reaction sequence is also conveniently performed in one pot without loss of yield by direct evaporation of the crude Ugi five-center four-component reaction mixture followed by addition of formic acid.[19]

Scheme 7 Ugi and Pictet–Spengler Reaction Affording Novel Tricyclic Indole Derivatives[19]

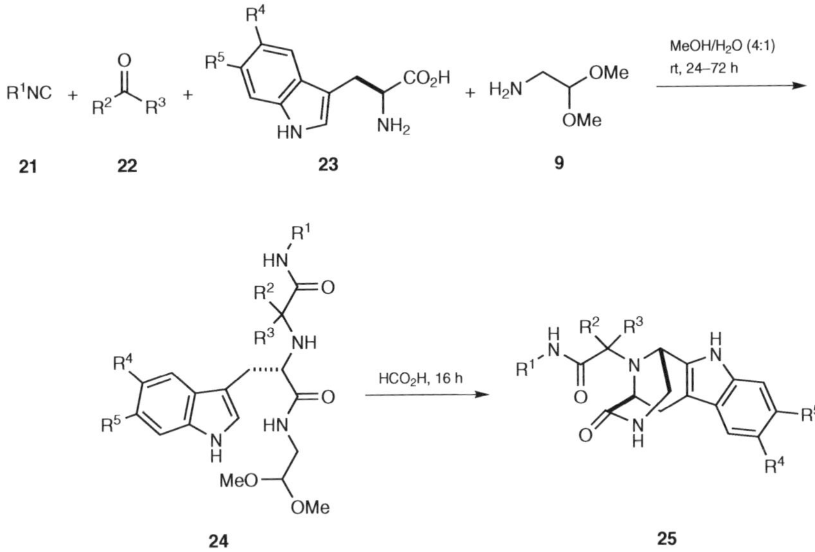

R^1	R^2	R^3	R^4	R^5	Overall Yield (%) of **25**	Ref
$(CH_2)_2Ph$	$(CH_2)_2NMe(CH_2)_2$		H	H	42	[19]
$(CH_2)_2Ph$	$(CH_2)_6$		H	H	52	[19]
$(CH_2)_2Ph$	$(CH_2)_5$		H	H	48	[19]
Bn	$(CH_2)_3$		OH	H	29	[19]
Bn	Ph	H	H	H	61	[19]
Cy	Me	Me	OMe	OMe	52	[19]

The advantage of the specific combination of Ugi and Pictet–Spengler reactions is the expedited and convergent access to polycyclic natural-product-like skeletons. The yields over two steps are acceptable to good. The stereochemistry of the amino acid is conserved and therefore stereochemically pure (using symmetrical ketones or formaldehyde) or diastereomeric (using aldehydes other than formaldehyde) compounds are obtained. Many variations in the oxo and isocyanide starting materials can lead to synthetic handles to further elaborate the primary structures.[19]

Pyrazino[1′,2′:1,2]pyrido[3,4-*b*]indolediones 12; **General Procedure:**[16]
Oxocarboxylic acid **10** (1 mmol) was dissolved in MeOH (1 mL) and isocyanide **8** (1 mmol) and aminoacetaldehyde dimethyl acetal (**9**; 1 mmol) were added. The soln was stirred at rt for 24 h and then concentrated. The residue was purified by column chromatography (silica gel) to give the corresponding Ugi product **11**.

for references see p 446

The Ugi product **11** (0.1 mmol) was dissolved in HCO_2H (0.5 mL) and the soln was stirred at rt for 4 h. Concentration and purification by preparative TLC or column chromatography gave the corresponding polycyclic product.

Pyrazino[2,1-*a*]isoquinolinediones 16; **General Procedure:**[17]
Oxocarboxylic acid **14** (1 mmol) was dissolved in MeOH (1 mL), and isocyanide **13** (1 mmol) and aminoacetaldehyde dimethyl acetal (**9**; 1 mmol) were added. The soln was stirred at rt for 24 h. The solvent was evaporated and the residue was purified by column chromatography (silica gel) to give the Ugi product **15**.

The Ugi intermediate **15** (0.1 mmol) was dissolved in HCO_2H (0.5 mL) and the soln was stirred at rt for 4 h. Excess HCO_2H was evaporated and the residue was purified by preparative TLC or column chromatography to give the corresponding polycyclic product.

Alternatively, the Ugi intermediate **15** was dissolved in MsOH (0.5 mL) and the soln was stirred at 70 °C for 24 h. Neutralization with 1 M aq Na_2CO_3, extraction with EtOAc, and purification by preparative TLC or column chromatography gave the corresponding polycyclic product.

Hexahydro-4*H*-pyrazino[2,1-*a*]isoquinolin-4-ones 20; **General Procedure:**[18]
A mixture of aldehyde **18** (1.2 mmol), aminoacetaldehyde dimethyl acetal (**9**; 1.0 mmol), and carboxylic acid **19** (1.0 mmol) in anhyd MeOH was treated with isocyanide **17** (1.0 mmol) at 0 °C. After the mixture had been stirred at rt overnight, it was concentrated under reduced pressure to give the Ugi four-component product, which was used directly in the next step without further purification. A mixture of the Ugi product (1.0 mmol) and anhyd $MgSO_4$ (2.0 mmol) in anhyd 1,2-dichloroethane was treated with MsOH (6.0 mmol) slowly under argon at rt. The mixture was then stirred at 80 °C until the reaction was complete, as indicated by TLC. The reaction was quenched with cold aq $NaHCO_3$ and the mixture was extracted with EtOAc. The combined organic layers were dried (Na_2SO_4) and concentrated under reduced pressure. The crude residue was purified by flash column chromatography.

(Hexahydro-1*H*-1,5-epiminoazocino[5,4-*b*]indol-12-yl)carboxamides 25;
General Procedure:[19]
A mixture of L-amino acid **23** (0.5 mmol), ketone **22** (0.5 mmol), isocyanide **21** (0.5 mmol), and aminoacetaldehyde dimethyl acetal (**9**; 0.5 mmol) in MeOH/H_2O (4:1; 0.1 M) was stirred for 24–72 h at rt. Solvents were evaporated under reduced pressure. The crude Ugi product **24** was dissolved in HCO_2H (2 mL) and the soln was stirred for another 16 h at rt. The mixture was diluted with CH_2Cl_2 and solid K_2CO_3 was added in portions to neutralize the mixture. Excess K_2CO_3 was filtered off by using Celite and solvents were evaporated. The crude product was purified by supercritical fluid chromatography, flash column chromatography, or preparative TLC.

2.3.4.2 Ugi Five-Center Four-Component Reaction Followed by Postcondensations

In 1996, Ugi and co-workers described the scope and limitations of a novel variant of the original Ugi reaction.[20] The reaction was termed a five-center four-component reaction because of the use of bireactive α-amino acids, oxo components, isocyanides, and an alcohol both as solvent and reactant. However, this reaction only comprises a multicomponent reaction where three components show great variability: the α-amino acid, the oxo component, and the isocyanide. The variability of the alcohol component is rather restricted to low-molecular-weight alcohols such as methanol and ethanol. This restriction is likely a result of the poor solubility of the amino acids in other alcohols as well as the reduced nucleophilicity of the alcohols in general terms. The key reaction intermediate is the six-membered α-adduct **26** of the α-amino acid, the oxo component, and the isocyanide. This α-adduct undergoes nucleophilic attack by the solvent to give the linear product **27** (Scheme 8).

Scheme 8 The Five-Center Four-Component Reaction

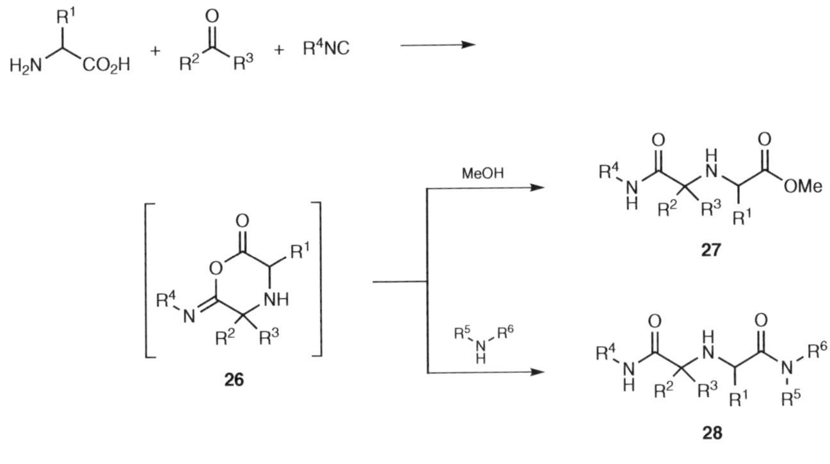

Recently, a novel, stereoselective Ugi-type reaction of four highly variable starting materials has been described.[12] In this report, the authors envisioned that nitrogen-based nucleophiles, such as primary or secondary amines, could potentially work by successfully competing with the alcohol solvent to attack the six-membered adduct **26**, thus leading to iminodicarboxamide derivatives **28** (Scheme 8). The optimal solvent for the reaction is a mixture of methanol/water (4:1), one that allows full solubilization of the starting materials. The reaction is performed at room temperature for 24–72 hours to circumvent the formation of unwanted side products that are formed when the reaction is carried out under microwave conditions instead.

The scope of the reaction has been investigated by using representative starting materials of each class (Scheme 9). Not surprisingly, the identity of the starting materials and their specific combinations plays a role in the overall yield and selectivity of the reaction. Virtually all the natural α-amino acids **29** and some non-natural ones can be used in the reaction; nevertheless, α-amino acids with reactive side chains (e.g., Ser, Glu, Asp, Lys) give lower yields with unprotected side chains and therefore are employed in their side-chain-protected form. As oxo components **30**, symmetrical ketones are used to obtain only one stereoisomer; however, both aldehydes and ketones work equally well in the reaction. Isocyanides **31** that can be used include (hetero)aromatic, aliphatic, and bulky examples, and (mostly primary) amines **32** include functionalized, heterocyclic, and (hetero)aromatic amines (Scheme 9). All reactions have been performed on a 0.5-mmol scale and the products **33** are separated into their component diastereomers by efficient and fast supercritical fluid carbon dioxide technology to afford 40–60% yield of the final products. The influence of the amine component was also investigated in relation to the integrity of the stereocenter, and it was found that primary amines lead to retention of stereochemistry, whereas secondary amines lead to partial racemization. The one-pot synthesis toward this scaffold constitutes a major advance and is superior to reported stepwise sequential or multicomponent reaction approaches.[12]

for references see p 446

Scheme 9 A "Truly" Five-Center Four-Component Reaction[12]

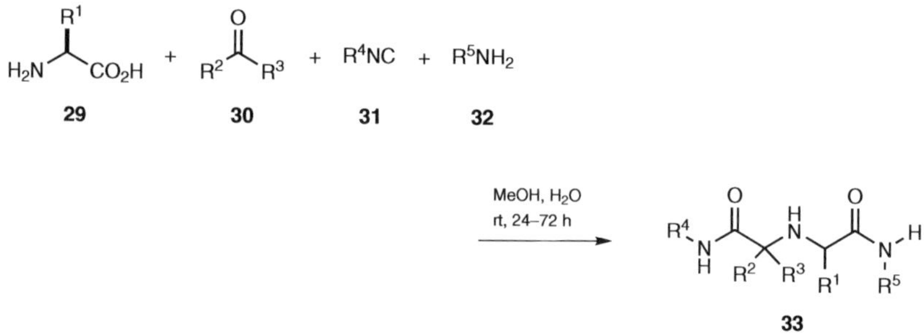

29 + 30 + R^4NC (31) + R^5NH_2 (32) → 33

MeOH, H_2O
rt, 24–72 h

R^1	R^2	R^3	R^4	R^5	Yield (%)	Ref
Me	$(CH_2)_2O(CH_2)_2$		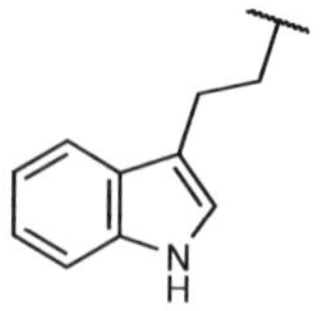	3,4-$Cl_2C_6H_3CH_2$	54	[12]
 Et	$(CH_2)_3$		Bn	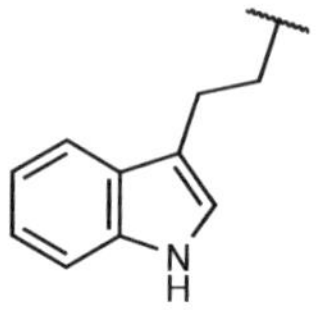	54	[12]
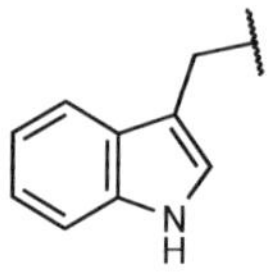	Me	Me	$CH_2P(O)(OEt)_2$	3,4-$Cl_2C_6H_3CH_2$	52	[12]
iPr	$(CH_2)_5$		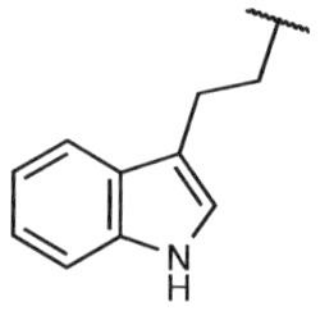	4-$FC_6H_4CH_2$	54	[12]
Bn	$(CH_2)_5$		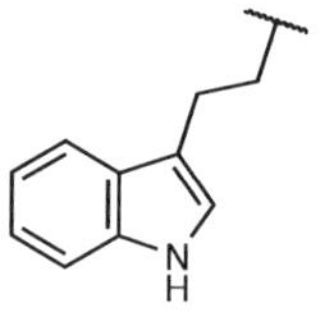	CH_2CO_2Me	63	[12]
$(CH_2)_4NHBoc$	$(CH_2)_5$		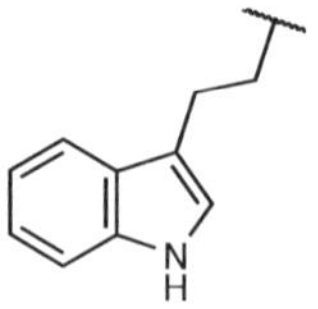	4-$MeOC_6H_4$	62	[12]

In further elaboration of the Ugi four-component reaction, the synthesis of four heterocyclic scaffolds based on the Ugi four-component reaction of α-amino acids, oxo components, isocyanides, and primary or secondary amines has been described by suitably functionalizing the starting materials of the reaction.[21]

Amongst the different strategies for the design of molecular complexity using multicomponent reaction chemistry,[22] post Ugi secondary cyclization has been a very fruitful strategy to accomplish novel scaffolds. Due to the well-known functional group compatibility of the Ugi starting materials, it was envisioned that the construction of heterocyclic scaffolds involving the unique and reactive secondary amine formed during the Ugi reaction variation was feasible. Several intramolecular lactamizations based on different Ugi scaffolds have been reported in the past;[23–26] however, the current synthesis differs from the rest because it yields a unique scaffold under much milder cyclization conditions. When methyl 2-formylbenzoates **37** are used as an oxo component in the Ugi four-component reaction along with equivalent amounts of an amino acid **34**, an isocyanide **35**, and a primary or secondary amine **36** with heating (85 °C, 24–72 h), the expected isoindolinone scaffold **38** is formed in moderate yield (Scheme 10).[21]

Scheme 10 Cyclization toward Isoindolones[21]

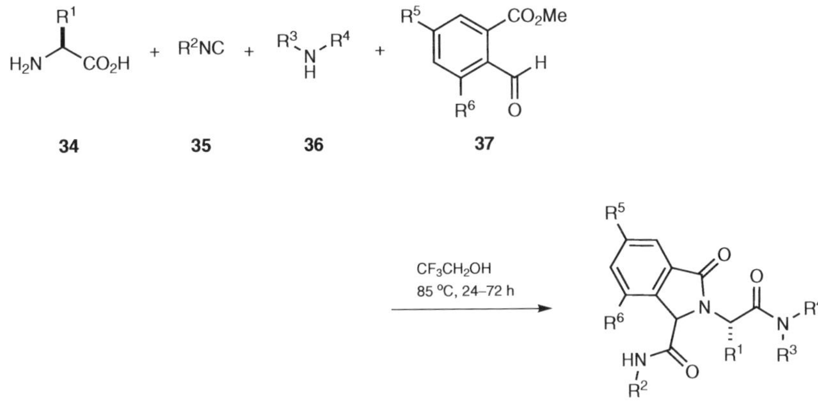

CF_3CH_2OH, 85 °C, 24–72 h

R^1	R^2	R^3	R^4	R^5	R^6	Yield (%)	Ref
iBu	$CHPh_2$	CH_2CHPh_2	H	H	H	35	[21]
Bn	$CHPh_2$	$(CH_2)_3Ph$	H	H	H	50	[21]
iBu	Bn	$(CH_2)_2O(CH_2)_2$		OMe	OMe	40	[21]

When a β-oxo ester **42** is employed as a bifunctional orthogonal building block along with an amino acid **39**, an isocyanide **40**, and a primary amine **41** under heating at 85 °C, complete formation of the Ugi product is observed. This material is transformed to the corresponding pyrrolidinedione scaffold **43** upon treatment with 3 equivalents of cesium carbonate (Scheme 11).[21] Several analogous reactions yield pyrrolidinediones in satisfactory yields with formation predominantly of the *cis*-stereoisomer, as confirmed by X-ray analysis. The discovery of this reaction is remarkable because few other orthogonal methods exist to access a pyrrolidinedione scaffold by isocyanide-based multicomponent reactions.[27]

Scheme 11 Cyclization toward Pyrrolidinediones[21]

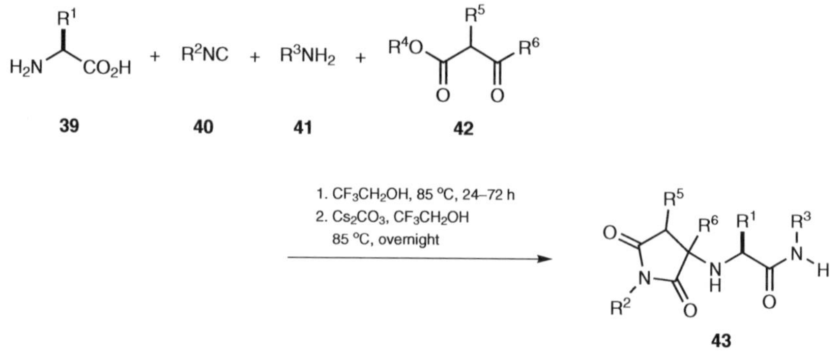

R^1	R^2	R^3	R^4	R^5	R^6	Yield (%)	Ref
H	Bn	Bu	Et	$(CH_2)_4$		40	[21]
Me	Cy	$(CH_2)_2Ph$	Et	$(CH_2)_4$		49	[21]
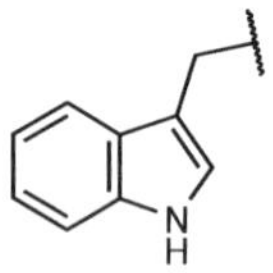 3-indolylmethyl	$(CH_2)_2Ph$	$(CH_2)_2OEt$	Et	$(CH_2)_4$		45	[21]
Bn	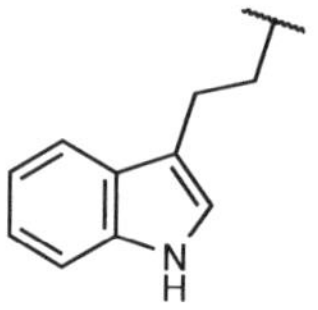 2-(3-indolyl)ethyl	iPr	Bn	H	Me	34	[21]

In a final example, the use of electron-rich aromatic α-amino acids, such as phenylalanine (**46**), shows applicability in a new Ugi variation to produce compounds **47** in good to high yields using aminoacetaldehyde dimethyl acetal (**9**), ketones **44**, and isocyanides **45** (Scheme 12).[21] In all cases, only the *trans*-diastereomer is observed, as confirmed by X-ray analysis of compounds **47**.

Scheme 12 Cyclization toward Bicyclic Tetrahydroimidazo[1,2-*a*]pyrazine-2,6(3*H*,5*H*)-diones[21]

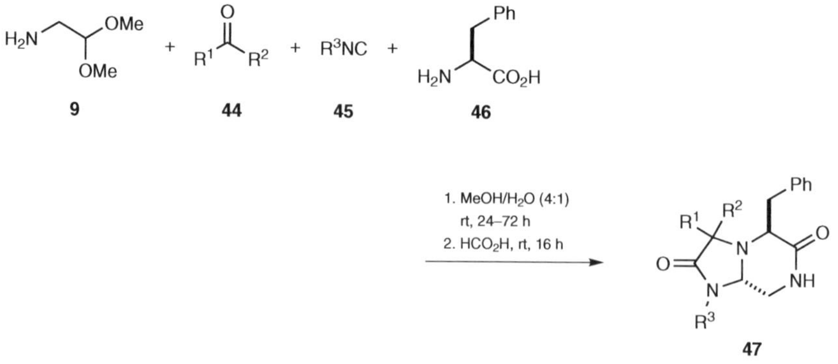

R^1	R^2	R^3	Yield (%)	Ref
Me	Me	Bn	63	[21]
$(CH_2)_5$		Bn	87	[21]
$(CH_2)_4$		$(CH_2)_2Ph$	40	[21]
$(CH_2)_5$		$(CH_2)_2Ph$	54	[21]

The Ugi four-component reaction, along with its postcondensation modifications, can also serve as an excellent way to access diverse 1,4-benzodiazepine scaffolds.[28] Such scaffolds are of particular interest in drug design due to a balanced ensemble of beneficial physicochemical properties which includes a semi-rigid and compact diazepine ring with spatial placements of several substituents, combined with a low number of rotatable bonds, hydrogen bond donors and acceptors, and intermediate lipophilicity. As an alternative to traditional multistep sequential syntheses of these materials, new routes have been designed employing one-pot multicomponent reactions to accelerate access to diverse 1,4-benzodiazepine scaffolds. Novel applications of [(*tert*-butoxycarbonyl)amino]-acetaldehyde in the synthesis of 1,4-benzodiazepines utilizing the Ugi/deprotection/cyclization strategy have been described.[29]

1,4-Benzodiazepin-5-ones **53** can be accessed using methyl 2-aminobenzoate (**49**) as a building block. This compound serves as an amine component for the Ugi four-component reaction together with an isocyanide **48**, [(*tert*-butoxycarbonyl)amino]acetaldehyde (**50**), and a carboxylic acid **51** to form the Ugi product **52**, which is cyclized to the 1,4-diazepine scaffold (Scheme 13).[28] The Ugi/deprotection/cyclization strategy allows access to 1,4-benzodiazepin-5-ones **53** with different substitutions derived from the isocyanide and carboxylic acid starting materials.

Scheme 13 Ugi Four-Component Reaction Route to 1,4-Benzodiazepin-5-ones[28]

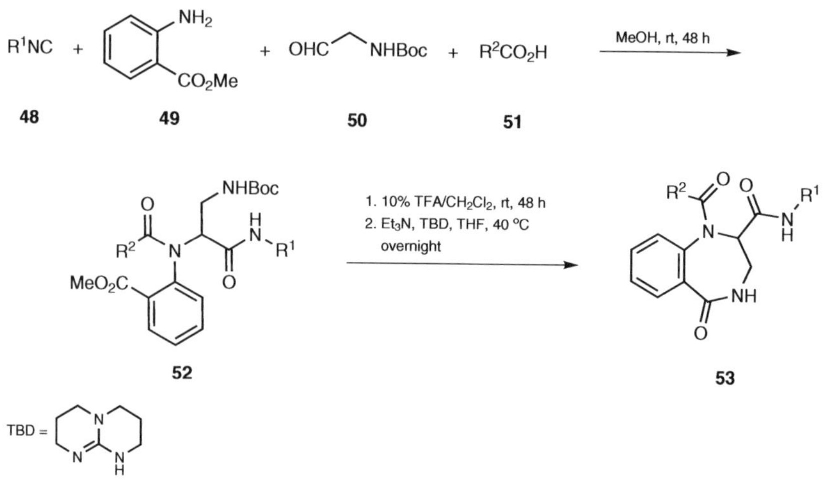

R^1NC + (2-aminobenzoate, NH_2, CO_2Me) + OHC–CH_2–NHBoc + R^2CO_2H → MeOH, rt, 48 h

48 49 50 51

52 → 1. 10% TFA/CH_2Cl_2, rt, 48 h; 2. Et_3N, TBD, THF, 40 °C overnight → 53

NHBoc, O, R^2, N, H, N, R^1, MeO_2C, O

R^2, O, O, R^1, N, N, H, NH, O

TBD = N, N, N, H

R^1	R^2	Overall Yield (%) of **53**	Ref
t-Bu	Me	41	[28]
t-Bu	Cy	28	[28]
Mes	Me	16	[28]
t-Bu	Pr	20	[28]
t-Bu	cyclopropyl	38	[28]
t-Bu	4-FC_6H_4	22	[28]

Because aminophenyl ketones show good reactivity in multicomponent reactions as an amine component,[30–32] they are employed in a new Ugi/deprotection/cyclization strategy for the rapid access to a second 1,4-benzodiazepine scaffold **58** (Scheme 14).[28] Initially, aminophenyl ketones **56** react with an isocyanide **54**, [(*tert*-butoxycarbonyl)amino]acetaldehyde (**50**), and a carboxylic acid **55** to form the Ugi product **57**. In the second step, the deprotected amino group is immediately cyclized with the ketone functionality to form 1,4-benzodiazepines **58** with four points of diversification.

Scheme 14 Ugi Four-Component Reaction Route to 1,4-Benzodiazepines[28]

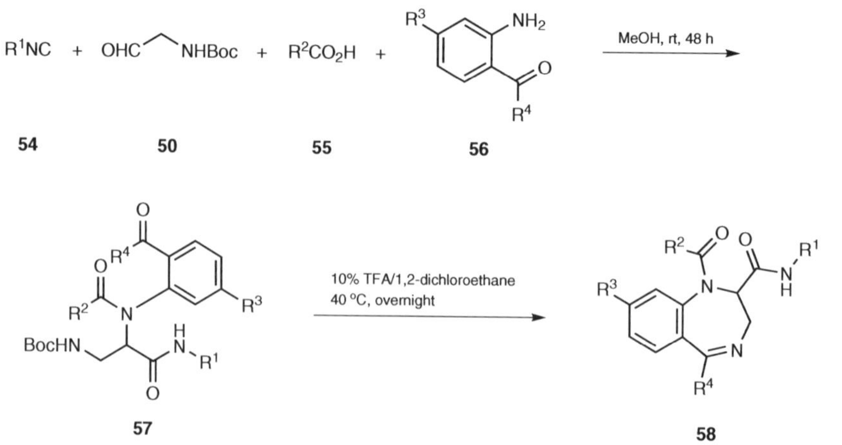

R^1NC + OHC–CH_2–NHBoc + R^2CO_2H + (R^3, NH_2, O, R^4) → MeOH, rt, 48 h

54 50 55 56

57 → 10% TFA/1,2-dichloroethane, 40 °C, overnight → 58

O, R^4, O, R^2, N, R^3, BocHN, H, N, R^1, O

R^2, O, O, R^1, R^3, N, N, H, N, R^4

R^1	R^2	R^3	R^4	Overall Yield (%)	Ref
t-Bu	Me	H	Ph	47	[28]
Cy	Me	H	Ph	36	[28]
Cy	Me	H	Me	43	[28]
Bn	Me	H	Ph	40	[28]
t-Bu	CH_2OH	Cl	Ph	22	[28]
t-Bu	Pr	Cl	Ph	24	[28]
t-Bu	cyclopropyl	Cl	Ph	25	[28]
Bn	Me	Cl	2-ClC_6H_4	13	[28]

The Ugi/deprotection/cyclization strategy is further applied for the synthesis of a third 1,4-benzodiazepine scaffold **62**. In this case, an aminophenyl ketone **60** reacts with an isocyanide **59**, [(*tert*-butoxycarbonyl)amino]acetaldehyde (**50**), and azidotrimethylsilane to obtain the Ugi tetrazole product **61**, followed by deprotection/cyclization forming substituted 1,4-benzodiazepines **62** with three points of diversification (Scheme 15).[28] The scaffolds **58** and **62** are unprecedented in the chemical literature, whereas scaffold **53** is accessed in a new and convenient way.

Scheme 15 Ugi Four-Component Reaction Route to 2-Tetrazole-Substituted 1,4-Benzodiazepines[28]

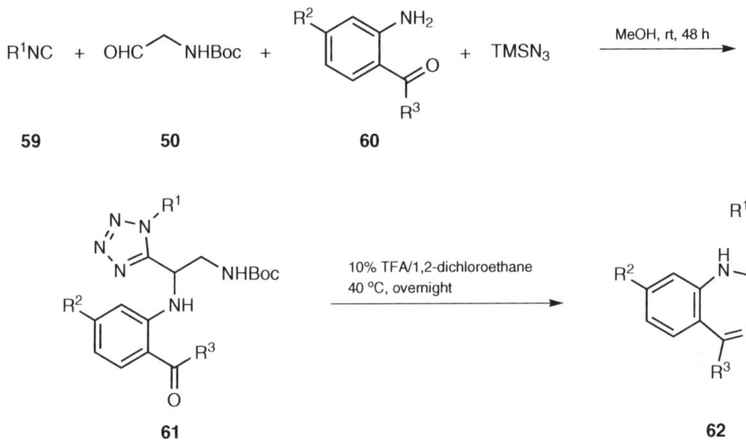

R^1	R^2	R^3	Overall Yield (%)	Ref
t-Bu	H	Ph	29	[28]
Cy	H	Ph	35	[28]
Cy	H	Me	49	[28]
t-Bu	Cl	Ph	32	[28]
Bn	Cl	2-FC_6H_4	12	[28]
Bn	H	Ph	26	[28]

Further development of an alternative approach toward the 1,4-benzodiazepine scaffold with an additional point of diversification led to the introduction of an "anchor" fragment to the diazepine ring.[28] A technology, referred to as AnchorQuery, performs an exact pharmacophore search of anchor-oriented virtual libraries of explicit conforma-

for references see p 446

tions. The anchor-oriented libraries are based upon key amino acid residues in, for example, protein–protein interactions where the amino acid residue (e.g., phenylalanine) is used as the anchor.[33] An *N-tert*-butoxycarbonyl-protected amino acid is an ideal building block to introduce anchor fragments, which can then be incorporated into drug-like compounds via multicomponent reactions.[34,35] Hence, the Ugi/deprotection/cyclization strategy is employed to assemble the orthogonal *N-tert*-butoxycarbonyl-protected amino acid building block for the synthesis of a 1,4-benzodiazepine scaffold. Aminophenyl ketones **65**, isocyanides **63**, *N*-(*tert*-butoxycarbonyl)glycine (**64**), and aldehydes **66** are employed to give the crude Ugi products **67**, which are not isolated but immediately treated with trifluoroacetic acid in 1,2-dichloroethane to produce 1,4-benzodiazepines **68** in a one-pot procedure (Scheme 16).[28] These 1,4-benzodiazepines with four points of diversification are isolated in reasonable to good yields.

Scheme 16 Ugi Four-Component Reaction Route to 1,4-Benzodiazepin-2-ones[28]

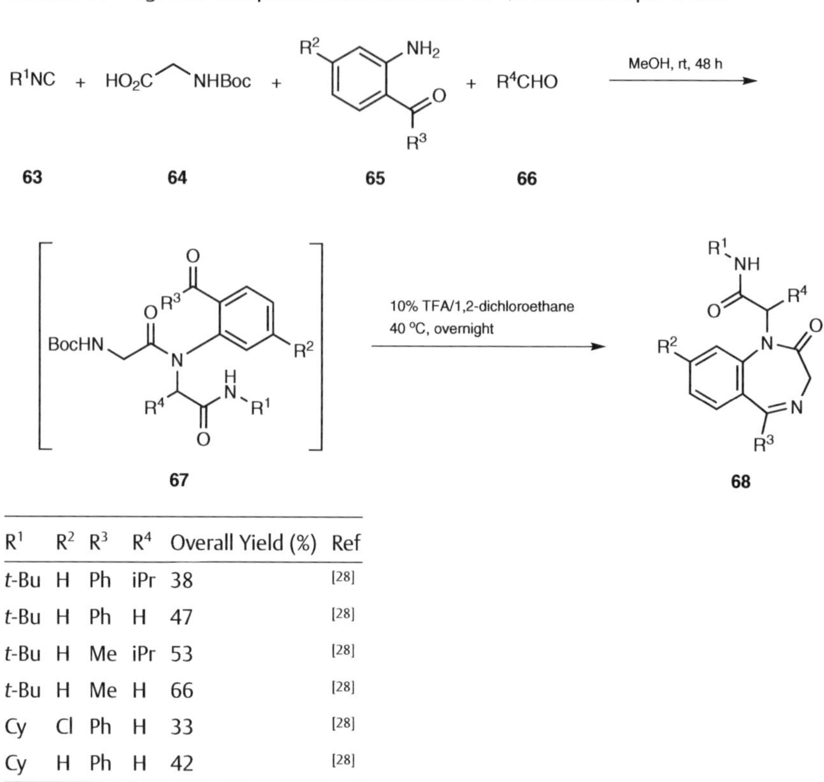

R^1	R^2	R^3	R^4	Overall Yield (%)	Ref
t-Bu	H	Ph	iPr	38	[28]
t-Bu	H	Ph	H	47	[28]
t-Bu	H	Me	iPr	53	[28]
t-Bu	H	Me	H	66	[28]
Cy	Cl	Ph	H	33	[28]
Cy	H	Ph	H	42	[28]

As a final example, the synthetic feasibility of "anchor"-based compound libraries using *N-tert*-butoxycarbonyl-protected amino acid derivatives have also been tested.[28] Phenylalanine, leucine, tryptophan, and tyrosine, amino acids which are abundant in the protein–protein interaction interface, were selected. As shown in Scheme 17, *N-tert*-butoxycarbonyl-protected amino acids **71** were subjected to the same protocol as for the synthesis of 1,4-benzodiazepines **68** with variable aminophenyl ketones **70** and isocyanides **69**. Compounds **72**, with three points of diversification, were isolated in 22–69% overall yield. These examples further demonstrate the overall ease of synthesis and increase in molecular complexity during the two-step, one-pot procedure, features which are remarkable for the efficient synthesis of 1,4-benzodiazepines containing a variety of "anchor" residues.

Scheme 17 Synthesis of Anchor-Directed 1,4-Benzodiazepines[28]

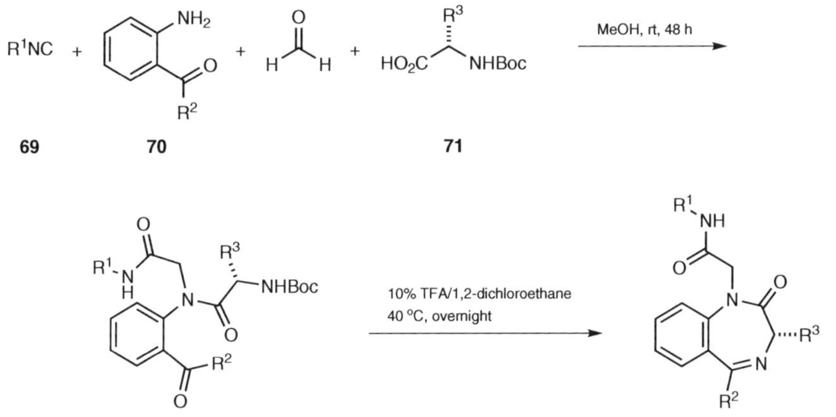

R^1	R^2	R^3	Yield (%)	Ref
Cy	Ph	Bn	50	[28]
Cy	Ph	iBu	69	[28]
t-Bu	Me	iBu	46	[28]
Cy	Ph	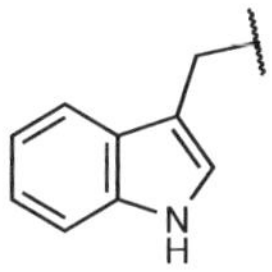	65	[28]
Cy	Ph	$4\text{-}HOC_6H_4CH_2$	60	[28]

Diethyl {[2-({1-[(3,4-Dichlorobenzyl)amino]-3-(1*H*-indol-3-yl)-1-oxopropan-2-yl}amino)-2-methylpropanamido]methyl}phosphonate [33, **R^1 = (1*H*-Indol-3-yl)methyl; $R^2 = R^3$ = Me; $R^4 = CH_2P(O)(OEt)_2$; $R^5 = 3,4\text{-}Cl_2C_6H_3CH_2$]; Typical Procedure:**[12]
A mixture of tryptophan [**29**, R^1 = (1*H*-indol-3-yl)methyl; 0.5 mmol], acetone (**30**, $R^2 = R^3$ = Me; 0.5 mmol), diethyl (isocyanomethyl)phosphonate [**31**, $R^4 = CH_2PO(OEt)_2$; 0.5 mmol], and 3,4-dichlorobenzylamine (**32**, $R^5 = 3,4\text{-}Cl_2C_6H_3CH_2$; 0.5 mmol), in $MeOH/H_2O$ (4:1; 0.1 M) was stirred for 24–72 h at rt. Solvents were evaporated under reduced pressure and the residue was dissolved in CH_2Cl_2. The unreacted amino acid was filtered off and the filtrate was concentrated to give the crude product, which was purified by supercritical fluid chromatography (MeOH); yield: 52%.

3-Oxoisoindoline-1-carboxamides 38; **General Procedure:**[21]
A mixture of an L-amino acid **34** (0.5 mmol), a methyl 2-formylbenzoate **37** (0.5 mmol), an isocyanide **35** (0.5 mmol), and an amine **36** (0.5 mmol) in 2,2,2-trifluoroethanol (5 mL) was stirred at 85 °C for 24–72 h. Supercritical fluid chromatography/MS showed both the Ugi product as well as the desired cyclized product **38**. 2,2,2-Trifluoroethanol was evaporated under reduced pressure and the residue was dissolved in CH_2Cl_2. The unreacted amino acid **34** was filtered off and the filtrate was concentrated. The residue was dissolved in EtOH (1 mL) and let sit in an oil bath at 60 °C for 24 h to allow for the remainder of the Ugi product to cyclize. The precipitate was collected by filtration.

for references see p 446

Pyrrolidinediones 43; **General Procedure:**[21]
A mixture of an L-amino acid **39** (0.5 mmol), a β-oxo ester **42** (0.5 mmol), an isocyanide **40** (0.5 mmol), and an amine **41** (0.5 mmol) in 2,2,2-trifluoroethanol (5 mL) was stirred at 85 °C for 24–72 h. Solvents were removed under N_2 to 0.5–1.0 mL and then Cs_2CO_3 was added to the mixture. The resultant mixture was heated at 85 °C overnight, then diluted with CH_2Cl_2 and H_2O, and extracted with CH_2Cl_2. The organic layer was dried ($MgSO_4$), filtered, and concentrated. The crude product was purified by supercritical fluid or flash chromatography.

(5*S*,8a*R*)-5-Benzyltetrahydroimidazo[1,2-*a*]pyrazine-2,6(3*H*,5*H*)-diones 47;
General Procedure:[21]
A mixture of L-phenylalanine (**46**; 0.5 mmol), a ketone **44** (0.5 mmol), an isocyanide **45** (0.5 mmol), and aminoacetaldehyde dimethyl acetal (**9**; 0.5 mmol) in MeOH/H_2O (4:1; 0.1 M) was stirred for 24–72 h at rt. Solvents were evaporated under reduced pressure and the residue was dissolved in CH_2Cl_2. Unreacted L-phenylalanine (**46**) was filtered off and the filtrate was concentrated. The residue was dissolved in formic acid (2 mL) and stirred at rt for 16 h. The mixture was quenched with aq $NaHCO_3$ and extracted with CH_2Cl_2. The extracts were dried (Na_2SO_4), filtered, and concentrated. The crude product was purified by supercritical fluid or flash chromatography.

2,3,4,5-Tetrahydro-1*H*-benzo[*e*][1,4]diazepine-2-carboxamides 53; **General Procedure:**[28]
A mixture of methyl 2-aminobenzoate (**49**; 0.2 mmol), [(*tert*-butoxycarbonyl)amino]acetaldehyde (**50**; 0.2 mmol), an isocyanide **48** (0.2 mmol), and a carboxylic acid **51** (0.2 mmol) in MeOH (0.5 mL) was stirred at rt for 2 d. The Ugi product **52** was isolated by chromatography (silica gel) and then treated with a 10% TFA soln in CH_2Cl_2 (0.5 mL), and the mixture was stirred at rt for 2 d. The solvent was evaporated and the residue was treated with Et_3N (100 μL) and 1,3,4,6,7,8-hexahydro-2*H*-pyrimido[1,2-*a*]pyrimidine (TBD; 10 mg) in THF (0.5 mL) and stirred overnight at 40 °C. The product was purified by chromatography (silica gel).

2,3-Dihydro-1*H*-benzo[*e*][1,4]diazepine-2-carboxamides 58; **General Procedure:**[28]
A mixture of a 2-aminoaryl ketone **56** (0.2 mmol), [(*tert*-butoxycarbonyl)amino]acetaldehyde (**50**; 0.2 mmol), an isocyanide **54** (0.2 mmol), and a carboxylic acid **55** (0.2 mmol) in MeOH (0.5 mL) was stirred at rt for 2 d. The Ugi product **57** was purified by chromatography (silica gel). A mixture of the isolated Ugi product **57** and a 10% TFA soln in 1,2-dichloroethane (0.5 mL) was stirred at 40 °C overnight. The residue was treated with Et_3N (100 μL) and the product was purified by chromatography (silica gel).

2-(1-*tert*-Butyl-1*H*-tetrazol-5-yl)-5-phenyl-2,3-dihydro-1*H*-benzo[*e*][1,4]diazepine
(62, **R^1 = *t*-Bu; R^2 = H; R^3 = Ph); Typical Procedure:**[28]
A mixture of 2-aminobenzophenone (**60**, R^2 = H; R^3 = Ph; 0.2 mmol), [(*tert*-butoxycarbonyl)-amino]acetaldehyde (**50**; 0.2 mmol), *tert*-butyl isocyanide (**59**, R^1 = *t*-Bu; 0.2 mmol), and $TMSN_3$ (0.3 mmol) in MeOH (0.5 mL) was stirred at rt for 2 d. The Ugi product **61** was purified by chromatography (silica gel). A mixture of the Ugi product **61** (26 mg) and a 10% TFA soln in 1,2-dichloroethane (0.5 mL) was stirred at 40 °C overnight. The residue was treated with Et_3N (100 μL), and the product was purified by chromatography (silica gel, petroleum ether/EtOAc 3:1); yield: 20 mg (29%).

2-(2-Oxo-2,3-dihydro-1*H*-benzo[*e*][1,4]diazepin-1-yl)acetamides 68 **($R^4 = H$); General Procedure:**[28]

CAUTION: *Formaldehyde is a probable human carcinogen, a severe eye, skin, and respiratory tract irritant, and a skin sensitizer.*

A mixture of a 2-aminoaryl ketone **65** (0.2 mmol), *N*-(*tert*-butoxycarbonyl)glycine (**64**; 0.2 mmol), an isocyanide **63** (0.2 mmol), and formaldehyde (**66**, $R^4 = H$; 0.2 mmol) in MeOH (0.5 mL) was stirred at rt for 2 d. After evaporation of the solvent, the residue of the Ugi product **67** was treated with a 10% soln of TFA in 1,2-dichloroethane (0.75 mL) and stirred at 40 °C overnight. Then it was treated with Et_3N (150 μL), and the product was purified by chromatography (silica gel).

(*S*)-2-(2-Oxo-5-phenyl-2,3-dihydro-1*H*-benzo[*e*][1,4]diazepin-1-yl)acetamides 72 **($R^2 = Ph$); General Procedure:**[28]

CAUTION: *Formaldehyde is a probable human carcinogen, a severe eye, skin, and respiratory tract irritant, and a skin sensitizer.*

A mixture of 2-aminobenzophenone (**70**, $R^2 = Ph$; 0.2 mmol), an *N*-*tert*-butoxycarbonyl-protected *S*-amino acid **71** (0.2 mmol), an isocyanide **69** (0.2 mmol), and formaldehyde (0.2 mmol) in MeOH (0.5 mL) was stirred at rt for 2 d. After evaporation of the solvent, the residue was treated with a 10% TFA soln in 1,2-dichloroethane (0.75 mL) and stirred at 40 °C overnight. Then it was treated with Et_3N (150 μL), and the product was purified by chromatography (silica gel).

2.3.4.3 Addition to Nitriles

Cyanoacetic acid derivatives are key starting materials in a plethora of multicomponent reactions yielding carbocycles and their heterocyclic analogues.[36] For instance, in the Gewald multicomponent reaction, the α-acidic character of the cyanoacetic acid derivatives **73** is exploited in a Knoevenagel condensation with α-acidic carbonyl compounds **74** (aldehydes, ketones, or 1,3-dicarbonyls) to form an acrylonitrile derivative **75**. After sulfanylation with elemental sulfur, the cyano group in **76** is available for intramolecular attack by the sulfur atom, leading to the final Gewald scaffold (e.g., **77**), which possesses an exocyclic amine moiety that can be utilized in subsequent chemistry (Scheme 18).[36–38]

Scheme 18 The Gewald Multicomponent Reaction[36–38]

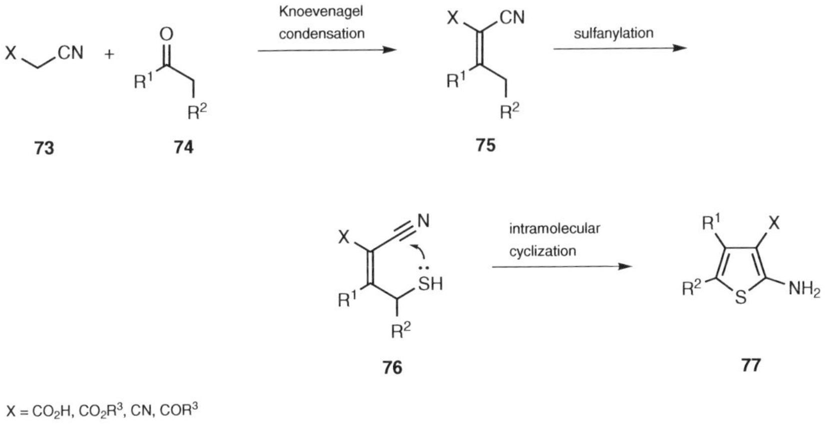

X = CO_2H, CO_2R^3, CN, COR^3

for references see p 446

Unfortunately, only a select few cyanoacetic derivatives, including cyanoacetic acid and esters, malononitrile, and cyanomethyl ketones, have been described in such multicomponent reactions, thereby limiting the scope of the reaction to the diversification of the activated carbonyl compounds. Recently, however, an inexpensive, mild, scalable, and simple parallel formation of cyanoacetamides **73** ($X = CONR^3_2$) as cyanoacetic acid derivatives in domino multicomponent reactions has been described.[39] With this new access to multigram quantities of a wide range of cyanoacetamides, the application in multicomponent reaction chemistry is described.

The Gewald multicomponent reaction scaffold, 2-aminothiophene, is an important and versatile building block in several drugs (e.g., olanzapine) and other biologically active compounds. Moreover, the Gewald multicomponent reaction itself constitutes an elegant, convenient, and effective reaction compared to traditional methods for the preparation of such compounds.[40] Wang and co-workers have described the use of their parallel cyanoacetamides methodology in the Gewald multicomponent reaction to broaden its scope. Reacting α-acidic carbonyl compounds **78**, cyanoacetamides **79**, and elemental sulfur in ethanol gives the corresponding 2-aminothiophenes **80** in yields ranging from 9 to 95% (Scheme 19).[41]

Acetaldehyde is, due to its low boiling point, not a suitable starting material. However, 1,4-dithiane-2,5-diol **81** [the dimer of 2-sulfanylacetaldehyde (**78**, $R^1 = H$; $R^2 = SH$)] is an effective and commercially available acetaldehyde substitute in the Gewald multicomponent reaction with cyanoacetamides **82**, and no elemental sulfur is required for formation of products **83**.[41,42] Other aldehydes show reasonable Gewald reactivity with a range of cyanoacetamides to give the thiophene products in good yields.

Scheme 19 Gewald Reaction Employing Cyanoacetamides[41]

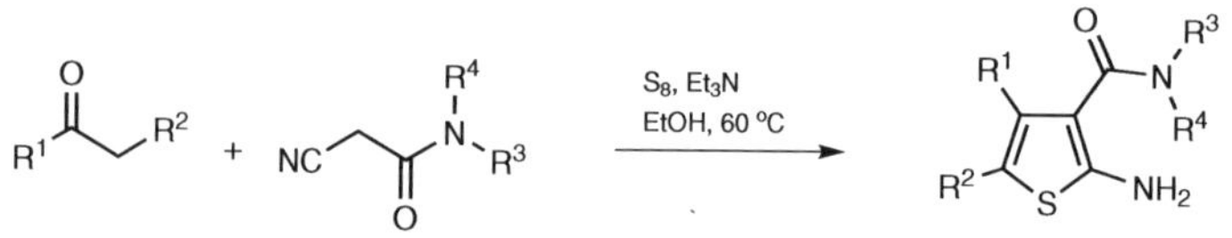

78 + 79 → (S_8, Et_3N, EtOH, 60 °C) → 80

R^1	R^2	R^3	R^4	Yield (%)	Ref
H	Ph	H	$CH_2CH{=}CH_2$	82	[41]
H	Ph	H	$CH_2C{\equiv}CH$	77	[41]
H	Ph	H	$CH(iPr)CO_2Me$	78	[41]
H	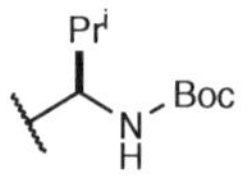 $CH(NHBoc)CH(CH_3)$-Pr^i	$(CH_2)_2O(CH_2)_2$		75	[41]
Ph	H	H	Bu	9	[41]
$(CH_2)_4$		H	$(CH_2)_2Ph$	20	[41]

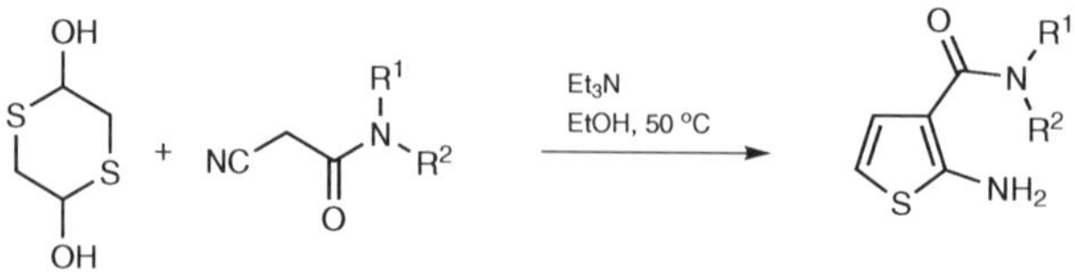

81 + 82 → (Et_3N, EtOH, 50 °C) → 83

R^1	R^2	Yield (%)	Ref
H	cyclopropyl	35	[41]
H	Bn	56	[41]
H	Bu	60	[41]
H	$CH_2CH(OMe)_2$	65	[41]
$(CH_2)_2NPh(CH_2)_2$		70	[41]

tert-Butoxycarbonyl-protected chiral β-amino aldehyde **84** reacts nicely with different cyanoacetamides in good to excellent yields (70–90%). Interestingly, when chiral aldehyde **84** and chiral cyanoacetamide **85** are used, only one diastereomer of 2-aminothiophene **86** is obtained without epimerization in the base-promoted Gewald multicomponent reaction. Nevertheless, when chiral valine methyl ester derived cyanoacetamide **87** reacts with the same aldehyde, the product **88** is generated as two diastereomers in a 2:1 ratio indicating strong epimerization at the valine isopropyl group (Scheme 20). Cyclohexanone and acetophenone give low yields, whereas other ketones do not react at all, presumably due to inferior reactivity compared to the aldehydes. Hence, the use of precondensed acrylonitrile derivatives is described in the literature for less reactive alkyl aryl ketones.[43,44]

Scheme 20 Stereochemical Outcome in the Gewald Reaction[41]

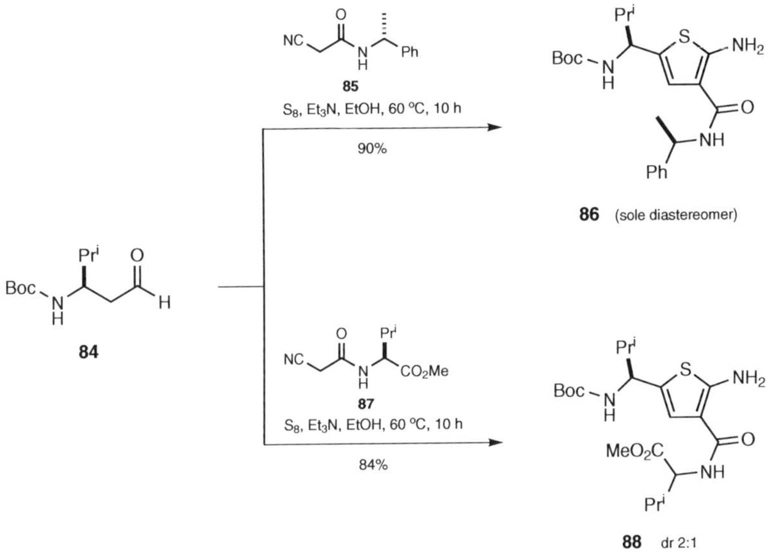

85, S_8, Et_3N, EtOH, 60 °C, 10 h; 90% → 86 (sole diastereomer)

87, S_8, Et_3N, EtOH, 60 °C, 10 h; 84% → 88 dr 2:1

Cyanoacetamides are useful reagents in the formation of 2-aminoindole-3-carboxamides (Scheme 21).[45] The 2-aminoindole fragment obtained is a key fragment in several biologically active compounds.[46,47] The reaction of 2-halonitrobenzenes or heterocyclic analogues **89** and cyanoacetamide **90** in one pot produces 2-aminoindole-3-carboxamides **91** in moderate to good yield (42–85%). Various functional groups, including alcohols, alkenes, and alkynes, are introduced into the 2-aminoindole scaffold without any protecting groups, allowing for direct follow-up chemistry to reach additional materials. In addition to 2-fluoronitrobenzene derivatives, some heterocyclic starting materials offer easy access to pyrrolopyridines, a result indicating that this one-pot procedure is quite general.

for references see p 446

Scheme 21 Synthesis of 2-Aminoindole-3-carboxamides Employing Cyanoacetamides[45]

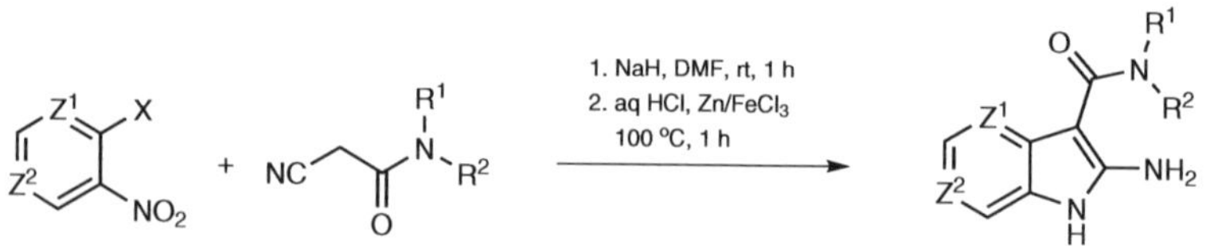

89 90 91

X	Z^1	Z^2	R^1	R^2	Yield (%)	Ref
F	CH	CH	H	Bu	85	[45]
F	CH	CH	H	$CH_2CH=CH_2$	77	[45]
F	CH	CH	H	$(CH_2)_2OH$	52	[45]
F	CH	CCF_3	H	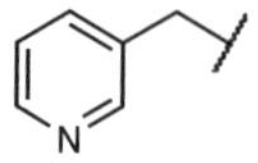	71	[45]
F	CH	CCl	$(CH_2)_2O(CH_2)_2$		55	[45]
F	N	CH	$(CH_2)_4$		45	[45]
Cl	CH	N	H	$CH_2C\equiv CH$	42	[45]

Classical methods for the synthesis of quinoline derivatives include the Combes, Conrad–Limpach, Doebner–Miller, and Gould–Jacobs syntheses to name a few.[48] The Friedländer reaction is another named reaction for the general preparation of quinolines,[49] and the Friedländer annulation is used in the synthesis of 2-aminoquinoline-3-carboxamides **94** (Scheme 22).[50]

Scheme 22 Synthesis of 2-Aminoquinoline-3-carboxamides Using Cyanoacetamides[50]

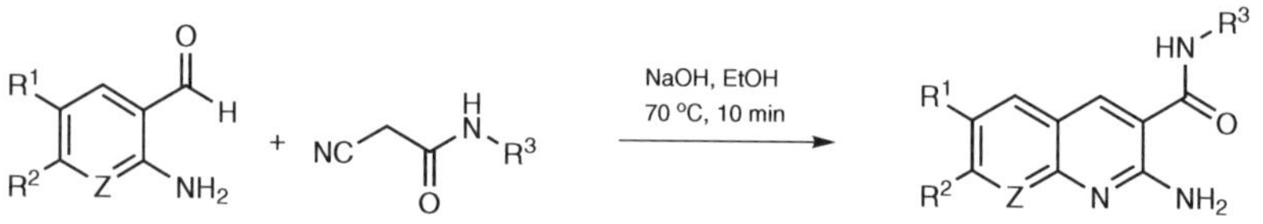

92 93 94

R^1	R^2	Z	R^3	Yield (%)	Ref
H	H	CH	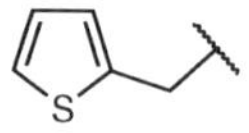	67	[50]
H	H	CH	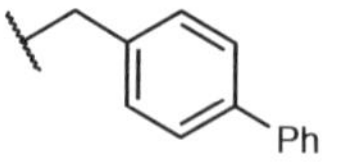	96	[50]
Cl	H	CH	$(CH_2)_2OH$	66	[50]
Cl	H	CH	$4-H_2NC_6H_4CH_2$	87	[50]
H	Cl	CH	$4-MeOC_6H_4(CH_2)_2$	75	[50]

R^1	R^2	Z	R^3	Yield (%)	Ref
H	CF_3	CH	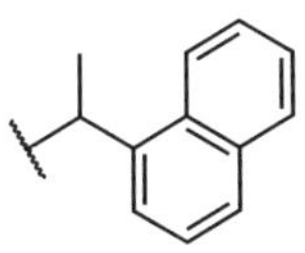	88	[50]
OCH_2O		CH	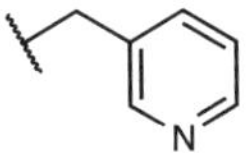	67	[50]
OMe	OMe	CH	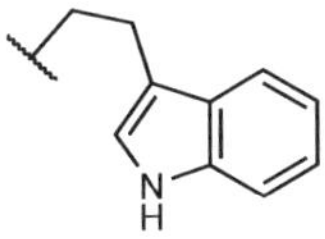	87	[50]
H	H	N	cyclopropyl	78	[50]

Heating a mixture of a 2-aminobenzaldehyde **92** and a cyanoacetamide **93** with sodium hydroxide in ethanol (70 °C, 10 min) yields the corresponding 2-aminoquinoline-3-carboxamides **94**, following a simple filtration step, in good to excellent yields (60–96%). For more expensive, or commercially limited 2-aminobenzaldehydes, the nitro analogue can be reduced (Fe/HCl) in situ.[51] Remarkably, the reaction of 2-cyano-*N*-(prop-2-ynyl)acetamide (**95**) with 2-amino-5-chlorobenzaldehyde (**98**) shows isomerization of the propargyl amide to the more thermodynamically stable prop-1-ynyl amide **99** under strongly basic conditions. However, with a mild base (such as *N*-methylpiperidine) the corresponding 2-aminoquinoline-3-carboxamide **97** can be obtained in 97% yield from 2-amino-3,6-dibromobenzaldehyde (**96**; Scheme 23).[50]

Scheme 23 Isomerization of 2-Cyano-*N*-(prop-2-ynyl)acetamide under Basic Conditions[50]

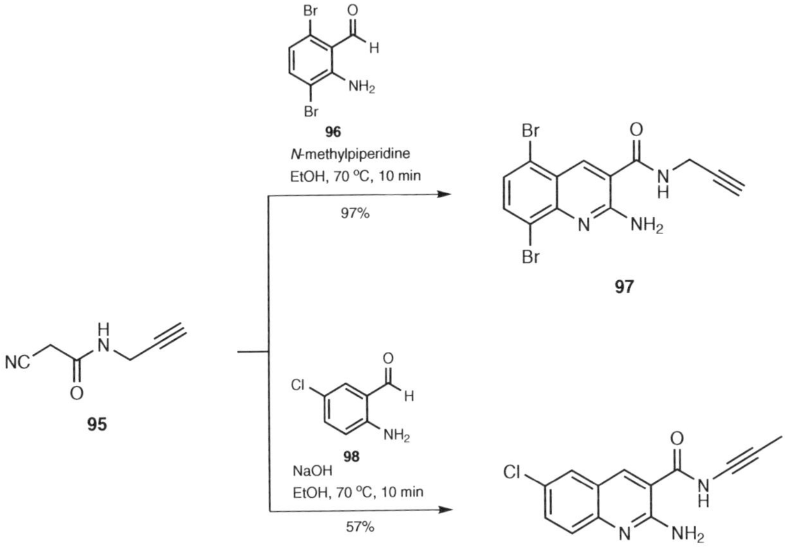

2-Aminoquinoline-3-carboxamides (e.g., **100**) can be further transformed into scaffold **103** by heating the carboxamide neat with 1,1-dimethoxy-*N*,*N*-dimethylmethanamine (**101**, dimethylformamide dimethyl acetal). The products are precipitated upon the addi-

for references see p 446

tion of ethanol, thus simplifying the workup procedure. With bulky substituents [R^3 = Cy or (1-naphthyl)ethyl], the expected product is not obtained; instead, the dimethylamino adduct **102** is isolated (Scheme 24).[50]

Scheme 24 Synthesis of Pyrimido[4,5-*b*]quinolin-4(3*H*)-ones from 2-Aminoquinoline-3-carboxamides[50]

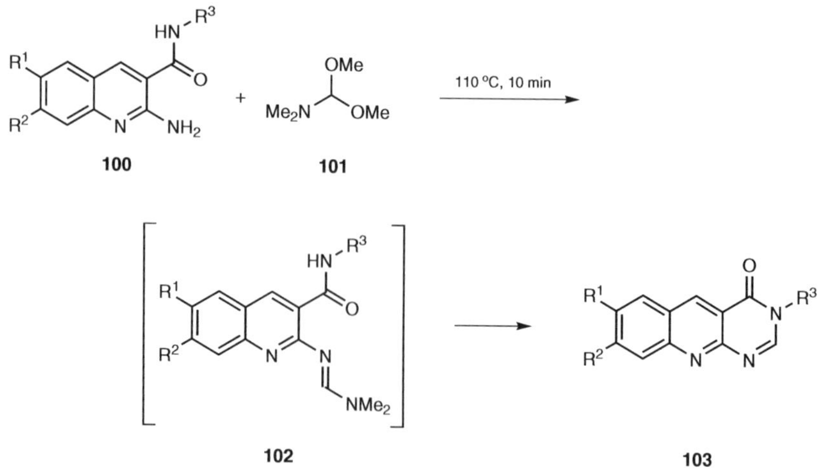

R^1	R^2	R^3	Yield (%)	Ref
H	H	4-$ClC_6H_4(CH_2)_2$	65	[50]
Cl	H	piperidino	82	[50]
Cl	H	$CH_2CH{=}CH_2$	56	[50]
H	Cl	4-$MeOC_6H_4(CH_2)_2$	78	[50]
OCH_2O		$(CH_2)_2Ph$	65	[50]

2-Aminothiophenes 80; **General Procedure:**[41]
A 20-mL vial equipped with a stirrer bar was charged with an aldehyde or a ketone **78** (5 mmol), a cyanoacetamide **79** (5 mmol), sulfur (5 mmol), and Et_3N (505 mg, 5 mmol) in EtOH (5 mL, 1.0 M soln). The mixture was heated in an oil bath at 60 °C for 10 h. Then, the mixture was cooled to rt and a batch of ice water (50 mL) was poured into it to yield a precipitate, which was collected by filtration and washed with cold EtOH.

2-Aminothiophenes 83; **General Procedure:**[41]
In a 20-mL glass vial was added a cyanoacetamide **82** (10 mmol), 1,4-dithiane-2,5-diol (**81**; 5 mmol), Et_3N (5 mmol), and EtOH (10 mL, 1.0 M soln). The mixture was heated at 50 °C for 12 h. Then, ice water (50 mL) was added and the mixture was extracted with CH_2Cl_2 (3 × 20 mL). The combined organic phase was dried (Na_2SO_4) and concentrated. The crude product was purified by column chromatography (silica gel).

2-Aminoindole-3-carboxamides 91; **General Procedure:**[45]
A cyanoacetamide **90** (2.0 mmol) in anhyd DMF (0.5 M) and NaH (60% dispersion in mineral oil; 2.2 mmol) were added to a 50-mL flask equipped with a stirrer bar. After 10 min, a 2-halonitrobenzene **89** (2.0 mmol) was added, and the mixture was stirred at rt for 1 h. The mixture became deep purple. 1.0 M aq HCl (4.0 mmol) was added, followed by $FeCl_3$ (6.0 mmol), and Zn dust (20 mmol). The mixture was heated to 100 °C for 1 h and then cooled before H_2O (20 mL) was added. The crude mixture was filtered to remove any

solid residue, and the filter was washed with EtOAc (25 mL). The layers were separated and the aqueous layer was extracted with EtOAc (2 × 20 mL). The combined organic phase was washed with sat. aq $NaHCO_3$ (10 mL) and brine (10 mL). The organic phase was dried (Na_2SO_4) and the solvent was removed. The crude product was purified by chromatography on a short column (silica gel).

2-Aminoquinoline-3-carboxamides 94; **General Procedure:**[50]
A 2-aminobenzaldehyde **92** (1.0 mmol), a cyanoacetamide **93** (1.0 mmol), and NaOH (0.2 mmol) in EtOH (2 mL) were added to a 20-mL vial, and the resulting mixture was heated with stirring in an oil bath at 70 °C for 10 min. The mixture was cooled to 0 °C. The precipitate was collected by filtration and washed with cold EtOH. The title compounds were obtained as solids.

Pyrimido[4,5-*b*]quinolin-4(3*H*)-ones 103; **General Procedure:**[50]
A mixture of 1,1-dimethoxy-*N*,*N*-dimethylmethanamine (**101**; 0.5 mL) and a 2-aminoquinoline-3-carboxamide **100** (0.2 mmol) was heated at 110 °C for 10 min. The mixture was cooled to rt and EtOH (1 mL) was added. The precipitate was collected by filtration and washed with EtOH (1 mL). The title compounds were obtained as solids.

for references see p 446

References

[1] Tietze, L. F., *Chem. Rev.*, (1996) **96**, 115.

[2] Dömling, A., *Chem. Rev.*, (2006) **106**, 17.

[3] Dömling, A.; Wang, W.; Wang, K., *Chem. Rev.*, (2012) **112**, 3083.

[4] Venkatraman, S.; Bogen, S. L.; Arasappan, A.; Bennett, F.; Chen, K.; Jao, E.; Liu, Y.-T.; Lovey, R.; Hendrata, S.; Huang, Y.; Pan, W.; Parekh, T.; Pinto, P.; Popov, V.; Pike, R.; Ruan, S.; Santhanam, B.; Vibulbhan, B.; Wu, W.; Yang, W.; Kong, J.; Liang, X.; Wong, J.; Liu, R.; Butkiewicz, N.; Chase, R.; Hart, A.; Agrawal, S.; Ingravallo, P.; Pichardo, J.; Kong, R.; Baroudy, B.; Malcolm, B.; Guo, Z.; Prongay, A.; Madison, V.; Broske, L.; Cui, X.; Cheng, K.-C.; Hsieh, Y.; Brisson, J.-M.; Prelusky, D.; Korfmacher, W.; White, R.; Bogdanowich-Knipp, S.; Pavlovsky, A.; Bradley, P.; Saksena, A. K.; Ganguly, A.; Piwinski, J.; Girijavallabhan, V.; Njoroge, F. G., *J. Med. Chem.*, (2006) **49**, 6074.

[5] Liddle, J.; Allen, M. J.; Borthwick, A. D.; Brooks, D. P.; Davies, D. E.; Edwards, R. M.; Exall, A. M.; Hamlett, C.; Irving, W. R.; Mason, A. M.; McCafferty, G. P.; Nerozzi, F.; Peace, S.; Philp, J.; Pollard, D.; Pullen, M. A.; Shabbir, S. S.; Sollis, S. L.; Westfall, T. D.; Woollard, P. M.; Wu, C.; Hickey, D. M. B., *Bioorg. Med. Chem. Lett.*, (2008) **18**, 90.

[6] Lamberth, C.; Jeanguenat, A.; Cederbaum, F.; De Mesmaeker, A.; Zeller, M.; Kempf, H.-J.; Zeun, R., *Bioorg. Med. Chem.*, (2008) **16**, 1531.

[7] Ugi, I.; Steinbrückner, C., *Angew. Chem.*, (1960) **72**, 267.

[8] Sheehan, S. M.; Masters, J. J.; Wiley, M. R.; Young, S. C.; Liebeschuetz, J. W.; Jones, S. D.; Murray, C. W.; Franciskovich, J. B.; Engel, D. B.; Weber, W. W., II; Marimuthu, J.; Kyle, J. A.; Smallwood, J. K.; Farmen, M. W.; Smith, G. F., *Bioorg. Med. Chem. Lett.*, (2003) **13**, 2255.

[9] Nixey, T.; Hulme, C., *Tetrahedron Lett.*, (2002) **43**, 6833.

[10] Ribelin, T. P.; Judd, A. S.; Akritopoulou-Zanze, I.; Henry, R. F.; Cross, J. L.; Whittern, D. N.; Djuric, S. W., *Org. Lett.*, (2007) **9**, 5119.

[11] Barrow, J. C.; Stauffer, S. R.; Rittle, K. E.; Ngo, P. L.; Yang, Z.; Selnick, H. G.; Graham, S. L.; Munshi, S.; McGaughey, G. B.; Holloway, M. K.; Simon, A. J.; Price, E. A.; Sankaranarayanan, S.; Colussi, D.; Tugusheva, K.; Lai, M.-T.; Espeseth, A. S.; Xu, M.; Huang, Q.; Wolfe, A.; Pietrak, B.; Zuck, P.; Levorse, D. A.; Hazuda, D.; Vacca, J. P., *J. Med. Chem.*, (2008) **51**, 6259.

[12] Khoury, K.; Sinha, M. K.; Nagashima, T.; Herdtweck, E.; Dömling, A., *Angew. Chem. Int. Ed.*, (2012) **51**, 10280.

[13] Chandra Pan, S.; List, B., *Angew. Chem. Int. Ed.*, (2008) **47**, 3622.

[14] El Kaïm, L.; Grimaud, L.; Oble, J., *Angew. Chem. Int. Ed.*, (2005) **44**, 7961.

[15] Neochoritis, C. G.; Dömling, A., *Org. Biomol. Chem.*, (2014) **12**, 1649.

[16] Wang, W.; Herdtweck, E.; Dömling, A., *Chem. Commun. (Cambridge)*, (2010) **46**, 770.

[17] Wang, W.; Ollio, S.; Herdtweck, E.; Dömling, A., *J. Org. Chem.*, (2011) **76**, 637.

[18] Liu, H.; William, S.; Herdtweck, E.; Botros, S.; Dömling, A., *Chem. Biol. Drug Des.*, (2012) **79**, 470.

[19] Sinha, M. K.; Khoury, K.; Herdtweck, E.; Dömling, A., *Chem.–Eur. J.*, (2013) **19**, 8048.

[20] Demharter, A.; Hörl, W.; Herdtweck, E.; Ugi, I., *Angew. Chem.*, (1996) **108**, 185; *Angew. Chem. Int. Ed. Engl.*, (1996) **35**, 173.

[21] Sinha, M. K.; Khoury, K.; Herdtweck, E.; Dömling, A., *Org. Biomol. Chem.*, (2013) **11**, 4792.

[22] Huang, Y.; Dömling, A., In *Isocyanide Chemistry*; Nenajdenko, V., Ed.; Wiley: New York, (2012); p 431.

[23] Marcos, C. F.; Marcaccini, S.; Menchi, G.; Pepino, R.; Torroba, T., *Tetrahedron Lett.*, (2008) **49**, 149.

[24] Zimmer, R.; Ziemer, A.; Gruner, M.; Brüdgam, I.; Hartl, H.; Reissig, H.-U., *Synthesis*, (2001), 1649.

[25] Gunawan, S.; Ayaz, M.; De Moliner, F.; Frett, B.; Kaiser, C.; Patrick, N.; Xu, Z.; Hulme, C., *Tetrahedron*, (2012) **68**, 5606.

[26] Gunawan, S.; Keck, K.; Laetsch, A.; Hulme, C., *Mol. Diversity*, (2012) **16**, 601.

[27] Bossio, R.; Marcos, C. F.; Marcaccini, S.; Pepino, R., *Synthesis*, (1997), 1389.

[28] Huang, Y.; Khoury, K.; Chanas, T.; Dömling, A., *Org. Lett.*, (2012) **14**, 5916.

[29] Keating, T. A.; Armstrong, R. W., *J. Am. Chem. Soc.*, (1996) **118**, 2574.

[30] Gordon, C. P.; Young, K. A.; Hizartzidis, L.; Deane, F. M.; McCluskey, A., *Org. Biomol. Chem.*, (2011) **9**, 1419.

[31] He, P.; Nie, Y.-B.; Wu, J.; Ding, M.-W., *Org. Biomol. Chem.*, (2011) **9**, 1429.

[32] Marcaccini, S.; Pepino, R.; Pozo, M. C.; Basurto, S.; García-Valverde, M.; Torroba, T., *Tetrahedron Lett.*, (2004) **45**, 3999.

[33] Koes, D.; Khoury, K.; Huang, Y.; Wang, W.; Bista, M.; Popowicz, G. M.; Wolf, S.; Holak, T. A.; Dömling, A.; Camacho, C. J., *PLoS One*, (2012) **7**, e32839.

[34] Hulme, C.; Cherrier, M. P., *Tetrahedron Lett.*, (1999) **40**, 5295.
[35] Lecinska, P.; Corres, N.; Moreno, D.; García-Valverde, M.; Marcaccini, S.; Torroba, T., *Tetrahedron*, (2010) **66**, 6783.
[36] Shestopalov, A. M.; Shestopalov, A. A.; Rodinovskaya, L. A., *Synthesis*, (2008), 1.
[37] Gewald, K., *Chem. Ber.*, (1965) **98**, 3571.
[38] Gewald, K.; Schinke, E.; Böttcher, H., *Chem. Ber.*, (1966) **99**, 94.
[39] Wang, K.; Nguyen, K.; Huang, Y.; Dömling, A., *J. Comb. Chem.*, (2009) **11**, 920.
[40] Huang, Y.; Dömling, A., *Mol. Diversity*, (2011) **15**, 3.
[41] Wang, K.; Kim, D.; Dömling, A., *J. Comb. Chem.*, (2010) **12**, 111.
[42] Puterová, Z.; Krutšíková, A.; Végh, A., *ARKIVOC*, (2010), i, 209.
[43] Mohareb, R. M.; Ho, J. Z.; Alfarouk, F. O., *J. Chin. Chem. Soc.*, (2007) **54**, 1053.
[44] Kim, M.-H.; Park, C.-H.; Chun, K.; Oh, B.-K.; Joe, B.-Y.; Choi, J.-H.; Kwon, H.-M.; Huh, S.-C.; Won, R.; Kim, K. H.; Kim, S.-M., WO 2007 102 679, (2007); *Chem. Abstr.*, (2007) **147**, 365 516.
[45] Wang, K.; Herdtweck, E.; Dömling, A., *ACS Comb. Sci.*, (2011) **13**, 140.
[46] Enomoto, H.; Kawashima, K.; Kudou, K.; Yamamoto, M.; Murai, M.; Inaba, T.; Ishizaka, N., WO 2008 087 933, (2008); *Chem. Abstr.*, (2008) **149**, 176 178.
[47] Eggenweller, H.-M.; Baumgarth, M.; Schelling, P.; Beier, N.; Christadler, M., DE 101 48 883, (2003); *Chem. Abstr.*, (2003) **138**, 304 293.
[48] Wolfe, J. P.; Pflum, D. A.; Curran, T. T.; Orahovats, P. A.; Moore, A. J.; Holsworth, D. D.; Tinsley, J. M.; Hudson, A., In *Name Reactions in Heterocyclic Chemistry*, Li, J. J., Ed.; Wiley: Hoboken, NJ, (2005); p 375.
[49] Cheng, C.-C.; Yan, S.-J., *Org. React. (N. Y.)*, (1982) **28**, 37.
[50] Wang, K.; Herdtweck, E.; Dömling, A., *ACS Comb. Sci.*, (2012) **14**, 316.
[51] Diedrich, C. L.; Haase, D.; Christoffers, J., *Synthesis*, (2008), 2199.

Keyword Index

In this keyword index, which should be used in conjunction with the Table of Contents, starting material entries are indicated in an *italic font*, product entries are identified by an arrow (→), and all other entries are given in a roman font.

A

→Absouline, key synthetic step, domino intramolecular diamination of 1-alk-4-enylureas, iodonium salt catalyzed 299

→ABT-341, one-pot synthesis from nitroalkenes and acetaldehyde, via domino sequence 372–375

→Abyssomycin C, key intermediate, by elimination reaction from dienyl silyl ether, Lewis acid catalyzed, then intramolecular Diels–Alder macrocyclization 9, 10

Acetaldehydes, (pentan-3-yloxy)-, enantioselective three-component reaction with nitroacrylates and vinylphosphonates, chiral diarylprolinol silyl ether catalyzed, via domino process, (–)-oseltamivir functionalized intermediate synthesis 366–369

Acetaldehydes, (pentan-3-yloxy)-, enantioselective three-component reaction with N*-(2-nitrovinyl)-acetamide and vinylphosphonates, chiral diarylprolinol silyl ether catalyzed, via domino process, (–)-oseltamivir functionalized intermediate synthesis* 371, 372

Acetaldehydes, phenyl-, enantioselective reaction with cyclohex-2-enone, using various chiral catalysts, via domino conjugate addition/aldol sequence, chiral bicyclo[2.2.2]octane synthesis 377–379

Acetals, dienyl, chiral, double domino intramolecular iodoetherification, using N*-iodosuccinimide, octahydrofuro[2,3-*e*][1,4]dioxocin synthesis* 314, 315

Acetone, dihydroxy-, dimer, enantioselective reaction with α,β-unsaturated aldehydes, chiral proline derivative catalyzed, via domino oxo-Michael/aldol/hemiacetalization sequence, chiral bicyclic hemiacetal synthesis 404

Acrylates, domino asymmetric reductive aldol reaction with acetophenone, chiral phosphinocopper/phosphinoferrocene complex catalyzed, chiral β-hydroxy ester synthesis 390, 391

Acrylates, domino asymmetric reductive aldol reaction with benzaldehyde, chiral rhodium complex catalyzed, anti*-aldol product synthesis* 390

ACT-280778, retrosynthetic analysis 379, 380

→Adamantane-2,4,9-triones, polysubstituted, chiral, from enantioselective alkylative dearomatization/annulation of clusiaphenone B, using enals/chiral *Cinchona* alkaloid-derived phase-transfer catalyst 261–263

→Adamantane-2,4,9-triones, polysubstituted, from reaction of clusiaphenone B with 3-formyl-2-methylbut-3-en-2-yl acetate/base, via tandem alkylative dearomatization/intramolecular Michael addition/aldol cyclization 256, 257

Alanine, phenyl-, Ugi four-component reaction with aminoacetaldehyde dimethyl acetal, isocyanides, and ketones, substituted imidazo[1,2-a]pyrazine-2,6-dione synthesis 432, 433

Alcohols, allylic – *see* Allylic alcohols

→Alcohols, bishomoallylic, chiral, from 1,3-dienes, via domino asymmetric reductive aldol reaction with benzaldehyde, nickel/chiral spirocyclic phosphoramidite catalyzed 391, 392

→Alcohols, dienyl, chiral, from 1,3-enynes, via domino asymmetric reductive aldol reaction with picolinaldehyde, rhodium/chiral bisphosphine catalyzed 391

→Alcohols, epoxy, allylic, chiral, from cyclic enones, via tandem asymmetric allylation/epoxidation sequence, using tetraallylstannane, titanium/chiral ligand catalyzed 411

→Alcohols, epoxy, allylic, chiral, from cyclohex-1-ene-1-carbaldehyde, by enantioselective reaction with divinylzinc reagents, titanium/chiral ligand catalyzed, via tandem asymmetric alkylation/epoxidation sequence 412, 413

→Alcohols, epoxy, chiral, from α,β-unsaturated aldehydes, via tandem asymmetric alkylation/epoxidation sequence, using diethylzinc, titanium/chiral diamine catalyzed 411, 412

Alcohols, methallyl – *see* Methallyl alcohols

→Alcohols, β-nitro, chiral, from indoles, via domino enantioselective Friedel–Crafts/Henry reaction with nitroalkenes and benzaldehyde, copper/chiral amine catalyzed 394

Alcohols, propargyl – *see* Propargyl alcohols

Aldehydes, alkenyl(chloroalkyl) substituted, imine formation by acid-catalyzed reaction with glycine, then cyclization/decarboxylation to give azomethine ylide, and intramolecular dipolar cycloaddition, (±)-aspidospermidine precursor synthesis 75

Aldehydes, β-amino, N-protected, chiral, Gewald reaction with chiral cyanoacetamides, chiral substituted 2-aminothiophene synthesis 441

Aldehydes, 2-azidobenzenesulfonamido substituted, reaction with Bestmann–Ohira reagent to give terminal alkyne, then intramolecular dipolar cycloaddition with azido group, tricyclic 1,2,3-triazole synthesis 82

Aldehydes, enantioselective reaction with propanal, proline catalyzed, via two-step domino aldol/aldol/hemiacetalization sequence, chiral cyclic hemiacetal synthesis 396
Aldehydes, α-oxy, enantioselective domino aldol/Mukaiyama aldol/hemiacetalization sequence with silyl enol ethers, proline catalyzed, chiral protected carbohydrate synthesis 395, 396
Aldehydes, α,β-unsaturated – *see also* Enals
Aldehydes, α,β-unsaturated, asymmetric functionalization, via organocatalytic domino reactions, iminium based 338, 339
Aldehydes, α,β-unsaturated, asymmetric functionalization, via organocatalytic domino reactions, via iminium–enamine catalysis 338
Aldehydes, α,β-unsaturated, enantioselective domino conjugate addition/amination, using thiols/azodicarboxylates, chiral silyl prolinol ether catalyzed, then reduction, chiral functionalized oxazolidin-2-one synthesis 342, 343
Aldehydes, α,β-unsaturated, enantioselective domino nucleophilic/electrophilic addition, using aromatic π-nucleophiles/hexachlorocyclohexa-2,4-dienone, chiral imidazolidinone catalyzed, via iminium–enamine catalysis, mechanism 340–342
Aldehydes, α,β-unsaturated, enantioselective multicomponent reaction with nitroalkenes and linear aldehydes, chiral diarylprolinol silyl ether catalyzed, via triple domino sequence, chiral functionalized cyclohexene-1-carbaldehyde synthesis 347, 348
Aldehydes, α,β-unsaturated, enantioselective reaction with 1,3-dinitroalkanes, chiral diarylprolinol silyl ether catalyzed, via domino process, chiral functionalized cyclohexanol synthesis 349
Aldehydes, α,β-unsaturated, enantioselective reaction with 1,3-dinitropropanes, chiral proline derivative catalyzed, via domino Michael/Henry sequence, chiral 2,4-dinitrocyclohexanol synthesis 404
Aldehydes, α,β-unsaturated, enantioselective [2+2] reaction with α-(hydroxymethyl)nitroalkenes, chiral proline derivative/substituted thiourea catalyzed, via domino Michael/Michael/hemiacetalization sequence, chiral 1-nitro-3-oxabicyclo[4.2.0]octan-4-ol synthesis 401, 402
Aldehydes, α,β-unsaturated, enantioselective three-component reaction with active methylene compounds, chiral diarylprolinol silyl ether catalyzed, via triple domino sequence, chiral functionalized cyclohexene-1-carbaldehyde synthesis 348
Aldehydes, α,β-unsaturated, enantioselective three-component reaction with benzofuranones and linear aldehydes, chiral diarylprolinol silyl ether catalyzed, via domino Michael/Michael/aldol/dehydration sequence, chiral spiro[benzofuranone-cyclohexene] synthesis 352
Aldehydes, α,β-unsaturated, enantioselective three-component reaction with oxindoles and linear aldehydes, chiral diarylprolinol silyl ether catalyzed, via triple domino cascade, chiral spiro[cyclohexene-oxindole] synthesis 352
Aldehydes, α,β-unsaturated, reaction with arynes, via tandem [2+2] cycloaddition/thermal electrocyclic ring opening/oxa-electrocyclization sequence, 1-benzopyran synthesis 153
Aldehydes, α,β-unsaturated, reaction with thiols, via enantioselective domino Michael/aldol sequence, chiral diarylprolinol silyl ether catalyzed, chiral functionalized tetrahydrothiophene synthesis 345
Aldehydes, α,β-unsaturated, reaction with β,γ-unsaturated thiols, chiral diarylprolinol silyl ether catalyzed, via enantioselective domino Michael/Michael sequence, chiral functionalized tetrahydrothiophene synthesis 346
Aldehydes, α,β-unsaturated, tandem asymmetric alkylation/epoxidation sequence, using diethylzinc, titanium/chiral diamine catalyzed, chiral epoxy alcohol synthesis 411, 412
Aldehydes, α,β,γ,δ-unsaturated, imine formation/electrocyclization/aromatization, 2,4-disubstituted pyridine synthesis 139, 140
→ Aldehydes, δ,ε-unsaturated, from diallylic ethers, via Wittig rearrangement/oxy-Cope rearrangement cascade, base catalyzed 168
→ *anti*-Aldol products, from acrylates, via domino asymmetric reductive aldol reaction with benzaldehyde, chiral rhodium complex catalyzed 390
→ *syn*-Aldol products, from acrylates, via domino asymmetric reductive aldol reaction with aldehydes, rhodium/chiral bisphosphine catalyzed 389, 390
→ Alkaloid natural products, from condensation of homoallylic amino alcohols and aldehydes, via aza-Cope rearrangement/Mannich reaction domino sequence 205, 206
→ Alkaloids, polycyclic indole-type, from three-component reaction of 3-(2-isocyanoethyl)indoles with oxocarboxylic acids and aminoacetaldehyde dimethyl acetal, via Ugi/Pictet–Spengler sequence 423, 424
Alkenes, bis(substituted), desymmetrization, via domino iodoetherification, using bis(2,4,6-collidine)iodonium perchlorate, asimicin core intermediate synthesis 315, 316
Alkenes, nitro-, one-pot conversion into ABT-341, via domino sequence 372–375
Alkenes, triepoxy-, carbonate functionalized, reaction with N-bromosuccinimide, via bromonium ion initiated ring-opening cascade, polycyclic bromooxepane synthesis 309, 310
Alkynes, domino asymmetric reductive aldol reaction with aldehydes, nickel/chiral phosphine catalyzed, chiral allylic alcohol synthesis 391

Alk-2-ynones, 3-(2-aminophenyl)-, dipolar cycloaddition with nitrile oxides, then intramolecular cyclization, isoxazolo[4,5-c]quinoline synthesis 65, 66
Allenes, 2-(1,3-dioxolan-2-yl)phenyl-, thermal 1,5-hydrogen shift/6π-electrocyclization/aromatization domino sequence, functionalized naphthalene synthesis 118, 119
Allenes, vinyl-, reaction with spirocyclic dienes, via Diels–Alder cycloaddition/ene reaction cascade, chloropupukeanolide D intermediate synthesis 184
→ Allylic alcohols, chiral, from alkynes, via domino asymmetric reductive aldol reaction with aldehydes, nickel/chiral phosphine catalyzed 391
Allylic alcohols, quinone monoepoxide containing, oxidation/electrocyclization/intramolecular Diels–Alder cascade, using Dess–Martin periodane, torreyanic acid ester synthesis 14, 15
Allylic alcohols, tertiary, base-promoted anionic oxy-Cope rearrangement/intramolecular S_N' displacement domino sequence, tetracyclic azulen-8-one synthesis 198, 199
Amides, N*-allenyl, thermal [1,3]-sigmatropic rearrangement/6π-electrocyclization cascade,* N*-cyclohexadienyl amide synthesis* 180, 181
→ Amides, *N*-cyclohexadienyl, from *N*-allenyl amides, via thermal [1,3]-sigmatropic rearrangement/6π-electrocyclization cascade 180, 181
→ Amides, *N*-trienyl, from *N*-allenyl amides, via thermal [1,3]-sigmatropic rearrangement/ [1,7]-sigmatropic rearrangement cascade 180, 181
→ Amidines, bromo-, from cyclohexenes, using *N*-bromosuccinimide/nitriles/4-toluenesulfonamide, via domino bromoamidination cascade 301, 302
Aminals, cyclohexa-1,4-dienyl, chiral, stereoselective reaction with N*-bromosuccinimide, via bromoamination/oxidation/elimination domino sequence, chiral imidazo[1,2-a]indole synthesis* 304, 305
Amines, dialkenyl-, intramolecular domino hydroamination/cyclization sequence, organolanthanide catalyzed, hexahydropyrrolizine synthesis 295, 296
Amines, homoallylic, condensation with aldehydes, iminium ion formation/azonia-Cope rearrangement/enamine nucleophilic trapping domino sequence, substituted 3-acylpyrrolidine synthesis 205
α-Amino acids, five-center Ugi four-component reaction with ketones, isocyanides, and amines, substituted iminodicarboxamide synthesis 428–430
α-Amino acids, N-protected, Ugi four-component reaction with aminophenyl ketones, isocyanides, and formaldehyde, then deprotection/cyclization, substituted 1,4-benzodiazepin-2-one synthesis 436, 437
→ β-Amino acids, pyrrolidine, polyhydroxylated, chiral, key synthetic step, stereoselective reaction of chiral unsaturated β-amino esters with iodine, via iodoamination/deprotection domino cascade 304
Amino alcohols, homoallylic, condensation with aldehydes, via aza-Cope rearrangement/Mannich reaction domino sequence, substituted 3-acylpyrrolidine synthesis 205
Amino alcohols, spirocyclic, vinylogous elimination reaction, 4-toluenesulfonic acid catalyzed, then intramolecular Diels–Alder reaction, pseudotabersonine synthesis 8, 9
β-Amino esters, unsaturated, chiral, stereoselective reaction with iodine, via iodoamination/deprotection domino cascade, chiral polysubstituted pyrrolidine synthesis 304
Ammonium salts, benzyltrimethyl-, base-catalyzed Sommelet–Hauser rearrangement, via [2,3]-/ [1,3]-sigmatropic rearrangements, N,N*-dimethyl-1-(2-tolyl)methylamine synthesis* 176, 177
Anilines, N*-alkynylidene-2-iodo-, dipolar cycloaddition of alkyne group with sodium azide, copper/ proline catalyzed, then N-arylation, triazolo-[1,5-a]quinoxaline synthesis* 83
Anilines, N*-allyl-*N*-benzyl-2-vinyl-, ring-closing metathesis, ruthenium carbene complex catalyzed, then oxidation to azomethine ylide, intermolecular dipolar cycloaddition with benzoquinone, and oxidation, tetracyclic isoindolo[2,1-a]quinoline synthesis* 77, 78
Anilines, geranylated, selenium-initiated polyene cyclization cascade, using benzeneselenenyl chloride, scandium trifluoromethanesulfonate catalyzed, 8-(phenylselanyl)octahydrophenanthridine synthesis 330, 331
Anilines, 2-iodo-, reaction with N*-tosylenynamines, palladium catalyzed, via heteroannulation/elimination/electrocyclization cascade, δ-carboline synthesis* 135
→ [8]Annulen-10-ones, tricyclic, from reaction of bicyclic cyclobutanones with cyclopent-1-enyllithium, then iodomethane, via carbonyl addition/anionic oxy-Cope rearrangement/enolate alkylation domino sequence 197
→ Anthraquinones, from 2,3-dibromonaphthalene-1,4-diones, via domino two-fold Heck/6π-electrocyclization sequence, using acrylates, palladium–phosphine ligand catalyzed 98, 99
→ Anthraquinones, functionalized, averufin precursor, from deprotonation of aryl lactones, using lithium tetramethylpiperidide, then benzyne precursor addition and Diels–Alder cyclization 23, 24

→Anthrones, key intermediates for dynemicin A, from deprotonation of 2-benzopyran-1,3-dione with lithium hexamethyldisilazanide, then Diels–Alder/retro-Diels–Alder cascade with multicyclic quinone imine 7, 8
→Aplysin-20, key intermediate, from polyenes, via bromonium-initiated cyclization cascade, using bromodiethylsulfonium bromopentachloroantimonate 327, 328
→Aquaticol, chiral, from asymmetric oxidation/dimerization of 3-(trimethylcyclopentane)-substituted lithium phenoxide, copper/sparteine catalyzed, via Diels–Alder reaction 3
→Aquaticol, from oxidative dimerization of 3-(trimethylcyclopentane)-substituted phenol, using stabilized 1-hydroxy-1,2-benziodoxol-3-one 1-oxide, via Diels–Alder reaction 2, 3
→Arenes, polyfunctional, from dienynones, via 1,3-tautomerization/6π-electrocyclization domino sequence, base catalyzed 116, 117
→(+)-Arteannuin M, bicyclic core, from 1,4-dien-3-ols, via thermal intramolecular [3,3]-sigmatropic rearrangement/ene reaction cascade, base catalyzed 185, 186
→Asimicin core, key intermediate, from desymmetrization of bis(substituted) alkenes, via domino iodoetherification, using bis(2,4,6-collidine)iodonium perchlorate 315, 316
→(±)-Aspidophytine, polycyclic intermediate, from intramolecular cyclization of an amide-substituted diazo ketone to give carbonyl ylide, rhodium catalyzed, then intramolecular cycloaddition with alkene 69, 70
→(±)-Aspidospermidine, precursor, by imine formation from reaction of alkenyl(chloroalkyl)-substituted aldehydes with glycine, acid catalyzed, then cyclization/decarboxylation to give azomethine ylide, and intramolecular dipolar cycloaddition 75
→Averufin, key intermediate, from deprotonation of aryl lactones, using lithium tetramethylpiperidide, then benzyne precursor addition and Diels–Alder cyclization 23, 24
→Avrainvillamide, key intermediate, from (propargyloxy)benzenes, via thermal propargyl Claisen rearrangement/[1,5]-hydrogen shift/6π-electrocyclization domino sequence 220, 221
→1-Azabicyclo[2.2.1]heptanes, from intramolecular asymmetric double domino hydroamination/cyclization of dienamines, chiral diaminobinaphthyl catalyzed 294, 295
→Azabicyclo[4.2.0]oct-5-enes, from 1,7-enyne benzoates, via 1,3-migration/intramolecular [2+2] cycloaddition cascade, cationic gold complex catalyzed 56
3-Aza-1,5-enynes, thermal aza-Claisen rearrangement/6π-electrocyclization domino reaction, 1,2-dihydropyridine synthesis 147
Azanorbornene, dienyl/indol-2-one functionalized, retrocycloaddition/Diels–Alder cycloaddition, Lewis acid catalyzed, spirocyclic pseudotabersonine intermediate synthesis 27
→Azatricycles, from benzynes, via enamide [2+2] cycloaddition/electrocyclic ring opening/intramolecular [4+2] cycloaddition cascade 50
→Azepanes, chiral, from stereoselective reaction of chiral 2-(arylalkenyl)aziridines with *N*-bromosuccinimide/4-nitrobenzenesulfonamide, via bromonium-initiated aminocyclization/ring expansion domino cascade 302, 303
Aziridines, 2-(arylalkenyl)-, chiral, stereoselective reaction with N*-bromosuccinimide/4-nitrobenzenesulfonamide, via bromonium-initiated aminocyclization/ring expansion domino cascade, chiral substituted azepane synthesis* 302, 303
→Aziridinomitosenes, precursors, from internal alkylation of alkynyl[(iodomethyl)aziridinyl]-substituted oxazoles, silver promoted, then cyanide ion addition/ring opening to give azomethine ylides, intramolecular dipolar cycloaddition, and loss of hydrogen cyanide 73, 74
→Azocin-2-ones, substituted, from domino [2+2] cycloaddition of 1,5-diarylpenta-1,4-dien-3-imines with a ketene/Cope rearrangement 51
→Azulen-8-ones, tetracyclic, from tertiary allylic alcohols, via base-promoted anionic oxy-Cope rearrangement/intramolecular S_N' displacement domino sequence 198, 199
→Azuleno[2,1-*b*]pyrans, from intermolecular [5+2] cycloaddition of pyranyl-substituted ynones with vinylcyclopropanes, rhodium complex catalyzed, then Nazarov cyclization 86

B

→Basiliolides, key intermediates, from reaction of 6-substituted pyran-2-ones with *N,O*-bis(trimethylsilyl)acetamide/triethylamine, via Ireland–Claisen rearrangement/Diels–Alder cycloaddition domino sequence 219, 220
Benzaldehydes, 2-alkynyl-, reaction with tosylhydrazine, silver catalyzed, then intramolecular cyclization to give azomethine imine, intermolecular dipolar cycloaddition with acrylate, and aromatization, pyrazolo[5,1-a]isoquinoline synthesis 80
Benzaldehydes, 2-allyl-, reaction with alkenyl sulfones, base catalyzed, via aldol condensation/alkene migration/8π-electrocyclization/6π-electrocyclization cascade, cyclobuta[a]naphthalene synthesis 124, 125
Benzaldehydes, 2-amino-, reaction with cyanoacetamides, via Friedländer annulation, 2-aminoquinoline-3-carboxamide synthesis 442, 443
Benzaldehydes, 2-(2-bromoethyl)-, hydrazine addition, then intramolecular cyclization to give azomethine imine, and intermolecular dipolar cycloaddition with fumarate, pyrazolo[5,1-a]isoquinoline synthesis 80, 81

Benzaldehydes, highly functionalized, photochemical 1,5-hydrogen abstraction, then intermolecular Diels–Alder reaction with acrylates, bicyclic hybocarpone intermediate synthesis 19, 20

→Benzaldehydes, 2-hydroxy-, substituted, from propargyl vinyl ethers, via thermal propargyl Claisen rearrangement/tautomerization/acylketene generation/6π-electrocyclization/aromatization domino sequence 222

Benzaldehydes, intramolecular cyclization with attached alkyne to give carbonyl ylide, platinum catalyzed, then intramolecular dipolar cycloaddition with terminal alkene and C—H insertion, pentacyclic furan synthesis 72

Benzaldehydes, 2-sulfanyl-, enantioselective reaction with α,β-unsaturated N-*acyloxazolidin-2-ones, chiral thiourea derivative catalyzed, via Michael/aldol sequence, chiral benzothiopyran synthesis* 406

→Benz[7]annulen-7-ones, 4a,8-epoxy-, from conversion of alkenyl-substituted diazo ketones into carbonyl ylides, rhodium catalyzed, then intramolecular cycloaddition 66, 67

Benzenes, allyl-, ene reaction with nitrosobenzene, copper catalyzed, oxidation to nitrone, and intermolecular cycloaddition with N-*alkylmaleimide, pyrrolo[3,4-*d*]isoxazole synthesis* 60

*Benzenes, 1-(azidomethyl)-2-iodo-, intermolecular coupling with terminal alkyne, palladium/copper catalyzed, then intramolecular dipolar cycloaddition, [1,2,3]triazolo[5,1-*a*]isoindole synthesis* 81, 82

Benzenes, (chloroalkyl)(2-oxocyclopentenyl) substituted, reaction with sodium azide, then intramolecular dipolar cycloaddition, nitrogen extrusion, and rearrangement, tetracyclic enamine synthesis 83

Benzenes, 2-halo-1-nitro-, reaction with cyanoacetamides, 2-aminoindole-3-carboxamide synthesis 441, 442

Benzenes, (2-isocyanoethyl)-, four-component reaction with aldehydes, carboxylic acids, and aminoacetaldehyde dimethyl acetal, via Ugi/Pictet–Spengler sequence, praziquantel derivative synthesis 425, 426

Benzenes, (2-isocyanoethyl)-, three-component reaction with oxocarboxylic acids and aminoacetaldehyde dimethyl acetal, via Ugi/Pictet–Spengler sequence, polycyclic compound synthesis 424, 425

Benzenes, (propargyloxy)-, thermal propargyl Claisen rearrangement/[1,5]-hydrogen shift/6π-electrocyclization domino sequence, 1-benzopyran synthesis 220, 221

Benzenes, 1,2,3-trivinyl-, reaction with propanamine, via intermolecular domino hydroamination/cyclization sequence, organolanthanide catalyzed, hexahydrocyclopenta[ij]isoquinoline synthesis 295, 296

Benzoates, 2-amino-, Ugi four-component reaction with N-*protected aminoacetaldehyde, isocyanides, and carboxylic acids, then deprotection/cyclization, substituted 1,4-benzodiazepin-5-one synthesis* 433, 434

Benzoates, 2-formyl-, Ugi four-component reaction with α-amino acids, isocyanides, and amines, substituted isoindolinone synthesis 431

→Benzo[*a*]carbazoles, from 2-aryl-3-(1-tosylalkyl)-indoles, via photoinduced homolytic cleavage/elimination/6π-electrocyclization/aromatization domino sequence 128, 129

→Benzo[*g*]chromene-5,10-diones, chiral, from enantioselective [3+3] reaction of β,γ-unsaturated α-oxo esters with 2-hydroxynaphtho-1,4-quinone, chiral thiourea derivative catalyzed, via domino Michael/hemiacetalization sequence 408, 409

Benzocyclobutenes, imino-, tandem thermal electrocyclic ring opening/6π-electrocyclization, 3-aryl-3,4-dihydroisoquinoline synthesis 140

Benzocyclobutenes, trisiloxy substituted, thermal reaction with cyclic enedione, then dehydration, rishirilide B intermediate synthesis, via intermolecular Diels–Alder cascade 13, 14

Benzocyclobutenes, vinylcyclopentanone substituted, thermal electrocyclization/intramolecular Diels–Alder cascade, estrone core synthesis 13

Benzocyclobutenones, reaction with lithiated diazo compounds, via 4π-electrocyclization/8π-electrocyclization cascade, 2,3-benzodiazepin-5-one synthesis 145, 146

Benzodecapentaenes, thermal 8π-electrocyclization/6π-electrocyclization cascade, [4.2.0]bicyclooctadiene synthesis 121

→1,4-Benzodiazepines, substituted, from Ugi four-component reaction of aminophenyl ketones with N-protected aminoacetaldehyde, isocyanides, and azidosilanes, then deprotection/cyclization 435

→1,4-Benzodiazepines, substituted, from Ugi four-component reaction of aminophenyl ketones with N-protected aminoacetaldehyde, isocyanides, and carboxylic acids, then deprotection/cyclization 434, 435

→1,4-Benzodiazepin-2-ones, substituted, from Ugi four-component reaction of aminophenyl ketones with N-protected glycine, isocyanides, and aldehydes, then deprotection/cyclization 435, 436

→1,4-Benzodiazepin-2-ones, substituted, from Ugi four-component reaction of N-protected α-amino acids with aminophenyl ketones, isocyanides, and formaldehyde, then deprotection/cyclization 436, 437

→Benzo[*d*][1,3]diazepin-5-ones, from intermolecular cycloaddition of isatoic anhydride with azomethine ylides, then ring opening, decarboxylation, and cyclization 78

→ 1,4-Benzodiazepin-5-ones, substituted, from Ugi four-component reaction of 2-aminobenzoates with N-protected aminoacetaldehyde, isocyanides, and carboxylic acids, then deprotection/cyclization 433, 434

→ 2,3-Benzodiazepin-5-ones, from reaction of benzocyclobutenones with lithiated diazo compounds, via 4π-electrocyclization/8π-electrocyclization cascade 145, 146

1,3-Benzodioxol-5-ol, 6-propenyl-, oxidative dimerization/intramolecular Diels–Alder reaction, palladium catalyzed, carpanone synthesis 21

Benzo[b]furans, 2,3-dibromo-, domino two-fold Heck/6π-electrocyclization sequence, using acrylates, palladium–phosphine ligand catalyzed, dibenzofuran synthesis 98, 99

→ Benzo[c]furans, from intermolecular [2+2+2] cycloaddition of oxadiynes with alkynes, via propargyl ene reaction/Diels–Alder cyclization cascade 162

Benzooctatetraenes, thermal 8π-electrocyclization/6π-electrocyclization cascade, [4.2.0]bicyclooctadiene synthesis 121

2-Benzopyran-1,3-diones, deprotonation with lithium hexamethyldisilazanide, then Diels–Alder/retro-Diels–Alder cascade with multicyclic quinone imine, dynemicin A anthrone intermediate synthesis 7, 8

→ 1-Benzopyrans, from 2-dienylphenols, via thermal [1,7]-sigmatropic hydrogen shift/6π-electrocyclization cascade 178, 179

→ 1-Benzopyrans, from (propargyloxy)benzenes, via thermal propargyl Claisen rearrangement/[1,5]-hydrogen shift/6π-electrocyclization domino sequence 220, 221

→ 1-Benzopyrans, from reaction of α,β-unsaturated aldehydes with arynes, via tandem [2+2] cycloaddition/thermal electrocyclic ring opening/oxa-electrocyclization sequence 153, 154

Benzo[b]thiophenes, 2,3-dibromo-, domino two-fold Heck/6π-electrocyclization sequence, using acrylates, palladium–phosphine ligand catalyzed, dibenzothiophene synthesis 98, 99

→ Benzothiopyrans, chiral, from enantioselective reaction of 2-sulfanylbenzaldehyde with α,β-unsaturated N-acyloxazolidin-2-ones, chiral thiourea derivative catalyzed, via Michael/aldol sequence 406

→ Benzo[c]xanthen-7-ones, 5-aryl-, from 3-styrylflavones, via photoinduced 6π-electrocyclization/oxidation 127, 128

Benzynes, enamide [2+2] cycloaddition/electrocyclic ring opening/intramolecular [4+2] cycloaddition cascade, azatricycle synthesis 50

→ Bicyclic adducts, chiral, from diazo diones, via domino conversion into carbonyl ylides/enantioselective intermolecular cycloaddition with acetylenedicarboxylates, chiral dirhodium complex catalyzed 389

→ Bicyclo[3.1.0] compounds, functionalized, from intramolecular [5+2] cycloaddition of alkynyl-tethered vinylic oxiranes, rhodium carbene complex catalyzed, then Claisen rearrangement 85

→ Bicyclo[5.3.0]decanes, from intermolecular [5+2] cycloaddition of pyranyl-substituted ynones with vinylcyclopropanes, rhodium complex catalyzed, then Nazarov cyclization 86

→ Bicyclo[4.3.1]dec-6-ene-2,10-diones, from α-vinyl-β-oxo esters, via alkenyl Grignard reagent addition/anionic oxy-Cope rearrangement/Dieckmann cyclization domino sequence 199, 200

→ Bicyclo[3.3.1]nonanes, polysubstituted, from reaction of acylphloroglucinols with α-acetoxymethyl acrylate/base, via tandem alkylative dearomatization/annulation 254–256

→ Bicyclo[3.3.1]nonanes, polysubstituted, from reaction of clusiaphenone B with α-acetoxymethyl acrylate/base, via tandem alkylative dearomatization/annulation 254–256

→ Bicyclo[3.3.1]nonanes, polysubstituted, from reaction of type A adamantane core polyprenylated acylphloroglucinols with organocerium reagents 257, 258

Bicyclo[2.2.2]octadienes, industrial synthesis via partial domino process 380–382

Bicyclo[2.2.2]octadienes, retrosynthetic analysis 379, 380

→ Bicyclo[4.2.0]octadienes, from benzodecapentaenes, via thermal 8π-electrocyclization/6π-electrocyclization cascade 121

→ Bicyclo[4.2.0]octadienes, from benzooctatetraenes, via thermal 8π-electrocyclization/6π-electrocyclization cascade 121

→ Bicyclo[2.2.2]octanes, chiral, from enantioselective reaction of phenylacetaldehyde with cyclohex-2-enone, using various chiral catalysts, via domino conjugate addition/aldol sequence 377–379

→ Bicyclo[3.2.1]oct-3-en-2-ones, 3-methyl-, 1,5-disubstituted, from tandem alkylative dearomatization/hydrogenation of 2,6-dimethylphenol, then reaction with ethylaluminum chloride, via tandem cationic 1,4-addition/two consecutive Wagner–Meerwein 1,2-rearrangements 252, 253

→ Bicyclo[4.2.0]octenones, from trienones, via thermal 4π-electrocyclization/retro-4π-electrocyclization/enolization/8π-electrocyclization/6π-electrocyclization cascade, base catalyzed 122, 123

→ Bicyclo[5.4.0]undecanes, from conversion of alkenyl-substituted diazo ketones into carbonyl ylides, rhodium catalyzed, then intramolecular cycloaddition 66, 67

→ Bispirooxindoles, chiral, from enantioselective reaction of 3-methyleneindol-2-ones with 3-(isothiocyanato)oxindoles, chiral thiourea catalyzed, via Michael/cyclization cascade sequence 362
→ Bispirooxindoles, chiral, from enantioselective reaction of 3-substituted oxindoles with methyleneindolinones, chiral thiourea derivative catalyzed, via Michael/aldol sequence 358, 359, 406, 407
→ Bis(trichloroacetamides), from double tandem [3,3]-sigmatropic rearrangement of allylic homoallylic bis(trichloroacetimidates) 207, 208
Bis(trichloroacetimidates), allylic homoallylic, double tandem [3,3]-sigmatropic rearrangement, bis(trichloroacetamide) synthesis 207, 208
Black cascade, thermal 8π-electrocyclization/6π-electrocyclization cascade of benzodecapentaenes, [4.2.0]bicyclooctadiene synthesis 121
→ Bolivianine, from tetracyclic lactone, via thermal intermolecular Diels–Alder cyclization with skipped triene, then intramolecular hetero-Diels–Alder cyclization 31
Boronates, amidoallyl-, domino hydroformylation/allylboration/hydroformylation sequence, using hydrogen/carbon monoxide, rhodium/bis(phosphite) catalyzed, perhydropyrano[3,2-b]pyridine synthesis 318, 319
Buta-1,3-dienes, 2,3-bis(phenylsulfonyl)-, conjugate addition with indole oximes to give nitrone, then intramolecular cycloaddition and reduction, (±)-yohimbenone precursor 59
Buta-1,3-dienes, N-*sulfonyl-, enantioselective reaction with enamines, chiral proline derivative catalyzed, via inverse electron demand aza-Diels–Alder domino sequence, chiral piperidin-2-ol synthesis* 399
But-2-enal, [2+2] cycloaddition with ketene, palladium complex catalyzed, then allylic rearrangement, δ-lactone synthesis 54, 55
But-3-enamides, domino hydroformylation/reductive amination sequence, using hydrogen/carbon monoxide, rhodium/bis(phosphite) catalyzed, hexahydro[1,3]oxazolo[3,2-a]pyridin-5-one synthesis 319, 320
But-2-ene, hydroaminomethylation, via isomerization to but-1-ene, rhodium/bisphosphine catalyzed, then conversion to pentanal using carbon monoxide, amination using piperidine, and hydrogenation, 1-pentylpiperidine synthesis 297, 298
But-2-enoates, 3-hydroxy-4-(2-oxoindolin-3-ylidene)-, enantioselective three-component reaction with α,β-unsaturated aldehydes, chiral diarylprolinol silyl ether catalyzed, via domino Michael/Michael/Michael/aldol sequence, spiro[indenone-oxindole] synthesis 355, 356
Butenolides, terminal aldehyde substituted, reaction with 1,2-bis(trimethylsiloxy)ethane, Lewis acid catalyzed, then intramolecular Diels–Alder cyclization/acetal formation, tetracyclic himgravine intermediate synthesis 41, 42
But-3-ynamide, N-*phenyl-*N-*(prop-2-enyl)-, base treatment, then thermal intramolecular Himbert Diels–Alder cyclization, tricyclic lactam synthesis* 22, 23

C

→ Camptothecin, tetracyclic intermediate, from intramolecular cyclization of diazoimide to give isomünchnone carbonyl ylide, rhodium catalyzed, then intramolecular cycloaddition with activated alkene 70, 71
Carbamates, furanyl-, cyclopentene substituted, thermal intramolecular Diels–Alder cyclization/fragmentation, fused tricyclic dendrobine intermediate synthesis 39
→ Carbazoles, from 2,3-dibromoindoles, via domino two-fold Heck/6π-electrocyclization sequence, using acrylates, palladium–phosphine ligand catalyzed 98, 99
Carbazoles, 2-hydroxy-, reaction with α,β-unsaturated aldehydes, via aldol reaction/oxa-electrocyclization cascade reaction, pyrano[3,2-b]carbazole synthesis 152, 153
→ Carbazoles, tetracyclic, from arylation of 2-(bromomethyl)indoles, using arenes, zinc catalyzed, then [1,5]-sigmatropic rearrangement/6π-electrocyclization/aromatization sequence 109, 110
Carbinols, bicyclic tertiary, reaction with base, then iodomethane, via anionic oxy-Cope rearrangement/enolate alkylation domino sequence, taxol intermediate synthesis 196, 197
→ Carbocycles, tetracyclic, from electron-withdrawing group substituted 2-bromoenediynes, via tricyclization cascade sequence, palladium catalyzed 95, 96
→ Carbocycles, tricyclic, from reaction of dipropargylic diols with benzenesulfenyl chloride, via Mislow–Evans rearrangement/6π-electrocyclization/[4+2] cycloaddition cascade 175
Carbodiimides, aryl, [1,5]-sigmatropic rearrangement/6π-electrocyclization cascade, substituted quinazoline synthesis 179, 180
→ Carbohydrates, protected, chiral, from β-oxyaldehydes, via enantioselective domino aldol/Mukaiyama aldol/hemiacetalization sequence with silyl enol ethers, proline catalyzed 395, 396
→ δ-Carbolines, from reaction of 2-iodoanilines with *N*-tosylenynamines, palladium catalyzed, via heteroannulation/elimination/electrocyclization cascade 135
Carbonyl compounds, activation via enamines or iminium ions 337, 338
Carbonyl compounds, diazo – *see also* Diazo ketones

Carbonyl compounds, α-diazo, copper-catalyzed conversion into carbonyl ylides, then cycloaddition with benzaldehydes, via domino reaction, tricyclic adduct synthesis 387
Carbonyl compounds, α-diazo, rhodium-catalyzed conversion into carbonyl ylides, then enantioselective 1,3-dipolar cycloaddition with aldehydes, chiral scandium complex catalyzed, via domino reaction, chiral tricyclic adduct synthesis 388
→Carnitine acetyltransferase inhibitor, from multicomponent reaction of 2-(chloromethyl)oxirane with 2-methylenepropane-1,3-diol and 4-nitrobenzenesulfonamide, via bromonium ion/oxonium ion/ring opening domino sequence, then ring closure 313, 314
→Carpanone, from 6-propenyl-1,3-benzodioxol-5-ol, via oxidative dimerization/intramolecular Diels–Alder reaction, palladium catalyzed 21
→(±)-Chelidonine, azatricyclic intermediate, from benzynes, via enamide [2+2] cycloaddition/electrocyclic ring opening/intramolecular [4+2] cycloaddition cascade 50
→Chlorothricolide, key intermediate, from heating protected hexaenoate with 5-methylene-1,3-dioxolan-4-one, via tandem intermolecular/intramolecular Diels–Alder cyclizations 31, 32
Cinnamaldehyde, enantioselective reaction with Nazarov reagents, chiral proline derivative catalyzed, via domino Michael/Morita–Baylis–Hillman sequence, chiral multifunctionalized cyclohexene synthesis 404, 405
→Clusianone framework, from cationic biomimetic cyclization of polysubstituted cyclohexa-2,5-dienones derived from dearomatization of acylphloroglucinols 285, 286
Clusiaphenone B, enantioselective alkylative dearomatization/annulation, using enals/chiral Cinchona *alkaloid-derived phase-transfer catalyst, chiral polysubstituted adamantane-2,4,9-trione synthesis* 260–263
Clusiaphenone B, intermolecular alkylative dearomatization, using prenyl bromide/base, phenyl lupone synthesis 244
Clusiaphenone B, reaction with α-acetoxymethyl acrylate/base, via tandem alkylative dearomatization/annulation, polysubstituted bicyclo[3.3.1]-nonane synthesis 254–256
Clusiaphenone B, reaction with α-acetoxymethyl acrylate/base, via tandem alkylative dearomatization/annulation, type B polyprenylated acylphloroglucinol synthesis 256, 257
Clusiaphenone B, reaction with 3-formyl-2-methylbut-3-en-2-yl acetate/base, via tandem alkylative dearomatization/intramolecular Michael addition/aldol cyclization, type B polyprenylated acylphloroglucinol adamantane core synthesis 257, 258
→(±)-Coerulescine, spiropyrrolenine intermediate, from tryptamine-based functionalized enaminone, via intramolecular [2+2]-photocycloaddition/retro-Mannich tandem sequence 48
Cohumulone, deoxy-, alkylative dearomatization, then acylation and mesylation/intramolecular cyclization, type A polyprenylated acylphloroglucinol synthesis 259
Cohumulone, deoxy-, intermolecular alkylative dearomatization, using γ-hydroxy alkenyl halides/base/phase-transfer catalyst, colupulone analogue synthesis 245, 246
→Colchicine, analogues, tetracyclic intermediate, from conversion of alkynyl-substituted diazo ketones into carbonyl ylides, rhodium catalyzed, then intramolecular cyclization/cycloaddition 68
→Colombiasin A, key intermediate, from sulfolene-masked 5-alkadienylnaphthalene-1,4-dione, via thermal and pressure unmasking/intramolecular Diels–Alder cascade 17
→Colupulone, analogues, from intermolecular alkylative dearomatization of deoxycohumulone, using γ-hydroxy alkenyl halides/base/phase-transfer catalyst 245, 246
→(−)-Coprinolone core, from trienones, via thermal enolization/8π-electrocyclization/6π-electrocyclization cascade 121, 122
→Coralydine, key intermediate, from iminobenzocyclobutenes, via tandem thermal electrocyclic ring opening/6π-electrocyclization 140, 141
→CP-263,114 core architecture, from oxidation/intramolecular Diels–Alder cascade sequence of 2-(alkenyloxy) substituted-phenols, using (diacetoxyiodo)benzene/alcohols 3, 4
p-*Cresol, intermolecular alkylative dearomatization, osmium complex mediated, via reaction with methyl vinyl ketone and decomplexation,* para-*disubstituted cyclohexa-2,5-dienone synthesis* 234, 235
→Cyclobuta[*a*]naphthalenes, from reaction of 2-allylbenzaldehydes with alkenyl sulfones, base catalyzed, via aldol condensation/alkene migration/8π-electrocyclization/6π-electrocyclization cascade 124, 125
→Cyclobutanes, pyridin-2-one fused, from *N*-vinyl-β-lactams, via copper-catalyzed [3,3]-sigmatropic rearrangement/6π-electrocyclization domino reaction 149, 150
→Cyclobutanes, tetracyclic γ-lactone-fused, from indole-substituted propargylic esters, via intramolecular [3,3]-sigmatropic rearrangement/[2+2] cycloaddition, gold complex catalyzed 53, 54
Cyclobutanones, bicyclic, reaction with cyclopent-1-enyllithium, then iodomethane, via carbonyl addition/anionic oxy-Cope rearrangement/enolate alkylation domino sequence, tricyclic [8]-annulen-10-one synthesis 197

→ Cyclobutenes, 3,4-dimethylene-, from reaction of dipropargylic diols with trichloromethanesulfenyl chloride, via [2,3]-/[2,3]-sigmatropic rearrangements/4π-electrocyclization cascade 174

Cyclohexa-1,3-diene-1-carboxylates, 5-amino-, domino bromoamidation sequence, using N-bromoacetamide, Lewis acid catalyzed, 3-(acetylamino)-5-amino-4-bromocyclohex-1-ene-1-carboxylate synthesis 300, 301

→ Cyclohexadienes, tricyclic, from dienynes within bridged bicyclic skeleton, via semireduction/6π-electrocyclization cascade 115

→ Cyclohexadienes, tricyclic, from propargyl alcohols attached to bridged bicyclic skeleton, via carbomagnesiation/6π-electrocyclization cascade, using vinylmagnesium chloride 115, 116

→ Cyclohexa-1,3-dienes, bicyclic, from [2+2+2] cycloaddition of 1,6-diynes with alkenes, ruthenium catalyzed, then 6π-electrocyclization 110, 111

→ Cyclohexa-1,3-dienes, tricyclic, from terminal diene-substituted 1,6-diynes, via ene–yne–yne ring-closing metathesis, ruthenium carbene complex catalyzed, then 6π-electrocyclic ring closure/sigmatropic 1,5-hydride shift 111

→ Cyclohexa-1,3-dienes, tricyclic, functionalized, from 1,6-disubstituted hexa-1,3,5-trienes, via domino 6π-electrocyclization/intermolecular Diels–Alder cycloaddition, using various dienophiles 126

→ Cyclohexa-2,4-dienones, polysubstituted, from intermolecular alkylative dearomatization of acylphloroglucinols, using unactivated electrophiles/base 248–250

Cyclohexa-2,5-dienone 5,6-monoepoxides, 2-alkenyl-3-(hydroxymethyl)-, oxidation/oxa-electrocyclization/oxidation cascade reaction, using oxoammonium salts, enzyme inhibitor EI-1941-2 synthesis 152

→ Cyclohexa-2,5-dienones, 4,4-diallyl-3,5-dimethoxy-, from intermolecular alkylative dearomatization of 3,5-dimethoxyphenol, via bisallylation, using allyl alcohol, palladium/phosphine/titanium catalyzed 236

→ Cyclohexa-2,5-dienones, *para*-disubstituted, from intermolecular alkylative dearomatization of *p*-cresol, osmium complex mediated, via reaction with methyl vinyl ketone and decomplexation 234, 235

Cyclohexa-2,5-dienones, 4-(hydroperoxy)-4-methyl-, desymmetrization, using aldehydes, chiral phosphoric acid/thiourea derivative catalyzed, via domino acetalization/oxa-Michael sequence, chiral bicyclic 1,2,4-trioxane synthesis 410

Cyclohexa-2,5-dienones, polysubstituted, from dearomatization of acylphloroglucinols, cationic biomimetic cyclization, clusianone framework synthesis 285, 286

Cyclohexane-1,2-diones, enantioselective reaction with cinnamaldehyde, chiral proline derivative catalyzed, via domino Michael/aldol sequence, functionalized bicyclic ketone synthesis 403

Cyclohexane-1,2-diones, enantioselective [3+2] reaction with nitroalkenes, chiral thiourea derivative catalyzed, via domino Michael/Henry sequence, chiral functionalized bicyclic ketone synthesis 407, 408

Cyclohexane-1,4-diones, enantioselective [3+2] reaction with nitroalkenes, chiral thiourea derivative catalyzed, via domino Michael/Henry sequence, chiral functionalized bicyclic ketone synthesis 408

→ Cyclohexanes, 1-bromo-2-(2-bromoethoxy)-, from reaction of cyclohexene with ethylene oxide/bromine, via bromonium ion/oxonium ion/ring opening domino sequence 306, 307

→ Cyclohexanes, multifunctionalized, chiral, from enantioselective reaction of pentane-1,5-dial with nitroalkenes, chiral proline derivative catalyzed, via domino Michael/intramolecular Henry sequence 399

→ Cyclohexane-1,3,5-triones, 2,2,4,4,6,6-hexaallyl-, from intermolecular alkylative dearomatization of phloroglucinol, via allylation, using allyl alcohol, palladium catalyzed, triethylborane activated 236, 237

→ Cyclohexanols, 2,4-dinitro-, chiral, from enantioselective reaction of α,β-unsaturated aldehydes with 1,3-dinitropropanes, chiral proline derivative catalyzed, via domino Michael/Henry sequence 404

→ Cyclohexanols, functionalized, chiral, from enantioselective reaction of α,β-unsaturated aldehydes with 1,3-dinitroalkanes, chiral diarylprolinol silyl ether catalyzed, via domino process 349

→ Cyclohexanols, 3-methylene-, from reaction of α,β-unsaturated allyloxy esters with lithium diisopropylamide, then Wittig rearrangement/oxy-Cope rearrangement/intramolecular ene reaction cascade 171

→ Cyclohexanones, epoxy-, chiral, from reaction of enals with γ-chloro-β-oxo esters, via enantioselective domino Michael/aldol sequence, chiral prolinol silyl ether catalyzed, then cyclization 344

→ Cyclohexanones, functionalized, chiral, from enantioselective reaction of β-oxo esters with α,β-unsaturated ketones, chiral imidazolidine catalyzed, via domino Michael/aldol sequence 402

→ Cyclohexanones, functionalized, chiral, from α,β-unsaturated ketones, via enantioselective Michael/aldol domino reaction with β-diketones, β-oxo esters, or β-oxo sulfones, chiral imidazolidine catalyzed 339

→Cyclohexanones, substituted, by radical abstraction from allylic sulfoxides, using tributyltin radical, then [2,3]-sigmatropic rearrangement/radical ring expansion cascade 177, 178

Cyclohexene-1-carbaldehydes, enantioselective reaction with divinylzinc reagents, titanium/chiral ligand catalyzed, via tandem asymmetric alkylation/epoxidation sequence, chiral allylic epoxy alcohol synthesis 411, 412

→Cyclohexene-1-carbaldehydes, functionalized, chiral, from enantioselective multicomponent reaction of α,β-unsaturated aldehydes with nitroalkenes and linear aldehydes, chiral diarylprolinol silyl ether catalyzed, via triple domino sequence 347, 348

→Cyclohexene-1-carbaldehydes, functionalized, chiral, from enantioselective three-component reaction of α,β-unsaturated aldehydes with active methylene compounds, chiral diarylprolinol silyl ether catalyzed, via triple domino sequence 348, 349

→Cyclohexene-1-carboxylates, 3-(acetylamino)-5-amino-4-bromo-, from domino bromoamidation of 5-aminocyclohexa-1,3-diene-1-carboxylates, using *N*-bromoacetamide, Lewis acid catalyzed 300, 301

Cyclohexenes, domino bromoamidination cascade, using N*-bromosuccinimide/nitriles/4-toluenesulfonamide, bromoamidine synthesis* 301, 302

→Cyclohexenes, multifunctionalized, chiral, from enantioselective reaction of cinnamaldehyde with Nazarov reagents, chiral proline derivative catalyzed, via domino Michael/Morita–Baylis–Hillman sequence 404, 405

Cyclohexenes, reaction with ethylene oxide/bromine, via bromonium ion/oxonium ion/ring opening domino sequence, 1-bromo-2-(2-bromoethoxy)cyclohexane synthesis 306, 307

→Cyclohex-2-enols, polysubstituted, from reaction of polysubstituted dihydropyrans with triethyl phosphite/base, via Claisen rearrangement/Mislow–Evans transposition domino sequence 215, 216

→Cyclooctatrienes, bicyclic, from 1,2,4,7-tetraenes, via thermal 1,5-hydrogen shift/8π-electrocyclization domino sequence 117, 118

→Cyclooctatrienes, tetracyclic, from in situ generated 1,3,5,8-tetraenes, via 1,7-hydrogen shift/8π-electrocyclization domino sequence 120

→Cyclooctatrienes, tetracyclic, from trienynes, via Lindlar semihydrogenation/6π-electrocyclization cascade 111, 112

Cyclopentadienes, alkenyl-, thermal [1,5]-hydrogen sigmatropic rearrangement/Diels–Alder cycloaddition cascade, isocedrene intermediate synthesis 190

→Cyclopenta[*ij*]isoquinolines, hexahydro-, from reaction of 1,2,3-trivinylbenzene with propanamine, via intermolecular domino hydroamination/cyclization sequence, organolanthanide catalyzed 295, 296

→Cyclopentanecarboxylates, 2-alkenyl-1-hydroxy-, from reaction of α,β-unsaturated allyloxy esters with lithium diisopropylamide, then Wittig rearrangement/oxy-Cope rearrangement/intramolecular ene reaction cascade 169, 170

→Cyclopenta[*c*]pyridines, from cycloisomerization of diynols, ruthenium catalyzed, then tandem imine formation/electrocyclization/dehydration process, using hydroxylamine 139

Cyclopent-4-ene-1,3-dione monoacetals, domino enantioselective Michael/aldol reaction with benzaldehydes and dialkylzinc reagents, copper/chiral phosphoramidite catalyzed, chiral β-diketone synthesis 393

Cyclophanes, highly functionalized, oxidation/transannular intramolecular Diels–Alder cascade, using iodosobenzene, longithorone A synthesis 25, 26

Cyclopropanes, methylene-, reaction with 4-toluenesulfonamide, via intermolecular domino ring opening/ring closing hydroamination sequence, gold catalyzed, substituted pyrrolidine synthesis 296, 297

→Cyclostreptin, pentacyclic intermediate, by selenoxide elimination from 19-membered macrocycle, via oxidation, then tandem elimination/double transannular Diels–Alder reaction 10, 11

D

→*proto*-Daphniphylline, 1,2-dihydro-, from Swern oxidation of squalene diol, then reaction with methylamine and cyclization/Diels–Alder reaction/hydride transfer cascade 6

→*Daphniphyllum* alkaloid core, tricyclic, by oxime formation from β-bromoalkyl-substituted hept-6-enal, using hydroxylamine, then intramolecular alkylation to give nitrone and intramolecular cycloaddition 58

→(±)-Demethoxyschelhammericine, precursor, by imine formation from dienyl(iodoalkyl)-substituted ketones, using (tributylstannyl)-methanamine, then cyclization/destannylation to give azomethine ylide, and intramolecular dipolar cycloaddition 74, 75

→Dendrobine, fused tricyclic intermediate, from cyclopentene-substituted furanylcarbamate, via thermal intramolecular Diels–Alder cyclization/fragmentation 39

Diamines, allylic, reaction with ketenes, Lewis acid catalyzed, via double Bellus–Claisen rearrangement, 1,7-diamino-4-methyleneheptane-1,7-dione synthesis 213, 214

Diazo ketones – *see also* Carbonyl compounds, diazo
Diazo ketones, alkenyl(amido) substituted, intramolecular cyclization with amide to give carbonyl ylide, rhodium catalyzed, then intramolecular cycloaddition with alkene, (±)-aspidophytine polycyclic intermediate synthesis 69, 70
Diazo ketones, alkenyl(amido) substituted, intramolecular cyclization with amide to give carbonyl ylide, rhodium catalyzed, then intramolecular cycloaddition with alkene, (±)-tacamonine polycyclic intermediate synthesis 69
Diazo ketones, alkenyl(imino) substituted, intramolecular cyclization to give azomethine ylide, rhodium catalyzed, then acid-catalyzed isomerization and intramolecular dipolar cycloaddition, didehydrostemofoline precursor synthesis 76
Diazo ketones, alkenyl substituted, conversion into carbonyl ylides, rhodium catalyzed, then intramolecular cyclization/cycloaddition, polygalolide tricyclic intermediate synthesis 67, 68
Diazo ketones, alkenyl substituted, conversion into carbonyl ylides, rhodium catalyzed, then intramolecular cycloaddition, bicyclo[5.4.0]undecane synthesis 66, 67
Diazo ketones, alkynyl substituted, conversion into carbonyl ylides, rhodium catalyzed, then intramolecular cyclization/cycloaddition, colchicine tetracyclic intermediate synthesis 68
Diazo ketones, intramolecular cyclization to give carbonyl ylides, rhodium catalyzed, then intermolecular dipolar cycloaddition with γ-alkylidenebutenolides, spirotricyclic furan-2-one synthesis 68
→Dibenzofurans, from 2,3-dibromobenzo[*b*]furans, via domino two-fold Heck/6π-electrocyclization sequence, using styrenes, palladium–phosphine ligand catalyzed 98, 99
→Dibenzothiophenes, from 2,3-dibromobenzo[*b*]thiophenes, via domino two-fold Heck/6π-electrocyclization sequence, using acrylates, palladium–phosphine ligand catalyzed 98, 99
→Dicarboxamides, imino-, substituted, from five-center Ugi four-component reaction of α-amino acids with ketones, isocyanides, and amines 428–430
Dienamines, intramolecular asymmetric double domino hydroamination/cyclization sequence, chiral diaminobinaphthyl catalyzed, 1-azabicyclo[2.2.1]heptane synthesis 294, 295
Dienediynediols, conversion into diimines by Mitsunobu reaction, then retro-ene/6π-electrocyclization/Diels–Alder cycloaddition cascade, octahydrophenanthrene synthesis 166
Dienes, nonconjugated, reaction with formaldehyde, via ene reaction/hetero-Diels–Alder cycloaddition cascade, pseudomonic acid A intermediate synthesis 183, 184
Dienes, phenol substituted, asymmetric cascade cyclization, chiral platinum–phosphine complex catalyzed, chiral tetrahydroxanthene synthesis 324
1,3-Dienes, 1-bromo-, Stille cross-coupling/8π-electrocyclization/6π-electrocyclization domino sequence, using 1-stannyl-1,3-dienes, palladium catalyzed, kingianin A intermediate synthesis 106, 107
1,3-Dienes, domino asymmetric reductive aldol reaction with aldehydes and silylboranes, nickel/chiral phosphoramidite catalyzed, chiral allylsilane synthesis 392
1,3-Dienes, domino asymmetric reductive aldol reaction with benzaldehyde, nickel/chiral spirocyclic phosphoramidite catalyzed, chiral bishomoallylic alcohol synthesis 391, 392
1,3-Dienes, 1-stannyl-, Stille cross-coupling/8π-electrocyclization/6π-electrocyclization domino sequence, using 1-iodo-1,3-dienes, palladium catalyzed, elysiapyrone synthesis 105
1,3-Dienes, 1-stannyl-, Stille cross-coupling/8π-electrocyclization/6π-electrocyclization domino sequence, using 1-iodo-1,3-dienes, palladium catalyzed, SNF4435 C/D synthesis 103, 104
1,4-Dien-3-ols, thermal intramolecular [3,3]-sigmatropic rearrangement/ene reaction cascade, base catalyzed, (+)-arteannuin M bicyclic core synthesis 185, 186
Dienones, α,β,γ,δ-unsaturated, enantioselective reaction with oxindoles, chiral Cinchona-based *alkaloid catalyzed, via vinylogous organocascade process, chiral spiro[cyclopentane-oxindole] synthesis* 353
2,6-Dienones, 10-azido-, microwave heating/dipolar cycloaddition, then nitrogen elimination and cyclization, substituted indole synthesis 83, 84
Dienynes, bromo-, tricyclization cascade sequence, palladium catalyzed, strained aromatic tetracyclic synthesis 97, 98
Dienynes, within bridged bicyclic skeleton, semireduction/6π-electrocyclization cascade, tricyclic cyclohexadiene synthesis 115
Dienynones, 1,3-tautomerization/6π-electrocyclization domino sequence, base catalyzed, polyfunctional arene synthesis 116, 117
→β-Diketones, chiral, from cyclopent-4-ene-1,3-dione monoacetals, via domino enantioselective Michael/aldol reaction with benzaldehydes and dialkylzinc reagents, copper/chiral phosphoramidite catalyzed 393
Diols, dipropargylic, reaction with benzenesulfenyl chloride, via Mislow–Evans rearrangement/6π-electrocyclization/[4+2] cycloaddition cascade, tricyclic carbocycle synthesis 174, 175
Diols, dipropargylic, reaction with trichloromethanesulfenyl chloride, via [2,3]-/[2,3]-sigmatropic rearrangements/4π-electrocyclization cascade, 3,4-dimethylenecyclobutene synthesis 174

Diones, diazo, domino conversion into carbonyl ylides/enantioselective intermolecular cycloaddition with acetylenedicarboxylates, chiral dirhodium complex catalyzed, chiral bicyclic adduct synthesis 389

→ *ent*-Dioxepandehydrothyrsiferol, key intermediate, from reaction of carbonate-functionalized triepoxyalkenes with *N*-bromosuccinimide, via bromonium ion initiated ring-opening cascade 309, 310

→ 1,2-Dioxine-3-carboxylates, from photoreaction of hexa-2,4-dienoates with singlet oxygen, via ene reaction/hetero-Diels–Alder cycloaddition cascade 164

Dioxinones, highly functionalized, thermal acyl ketene addition/intramolecular Diels–Alder double cyclization, hirsutellone B intermediate macrolactam synthesis 29, 30

→ Dispiro[oxindole-furan-pyrrolopyridines], chiral, from enantioselective three-component reaction of isatin with propenal and pyrrolo[2,3-*b*]pyridine-2,3-diones, chiral tertiary amine catalyzed, via Morita–Baylis–Hillman/bromination/[3+2] annulation domino sequence 362, 363

Disulfenates, propargylic, [2,3]-/[2,3]-/[2,3]-/[3,3]-sigmatropic rearrangements/[2+2] cycloaddition cascade, 4-alkylidene-6,7-dithiabicyclo[3.1.1]heptan-2-one 6-oxide synthesis 173, 174

→ Disulfides, bis(3-thienylmethyl), from dipropargyl disulfides, via thermal [2,3]-/[2,3]-/[3,3]-sigmatropic rearrangements/double Michael addition/oxidation cascade 172, 173

Disulfides, dipropargyl, thermal [2,3]-/[2,3]-/[3,3]-sigmatropic rearrangements/double Michael addition/aromatization cascade, thieno[3,4-c]-thiophene synthesis 172, 173

→ 6,7-Dithiabicyclo[3.1.1]heptan-2-one 6-oxides, 4-alkylidene-, from propargylic disulfenates, via [2,3]-/[2,3]-/[2,3]-/[3,3]-sigmatropic rearrangements/[2+2] cycloaddition cascade 173, 174

1,4-Dithiane-2,5-diol, Gewald reaction with cyanoacetamides, substituted 2-aminothiophene synthesis 440

Diyne nitriles, propargyl ene reaction/Diels–Alder cycloaddition cascade, tricyclic pyridine synthesis 163

1,6-Diynes, [2+2+2] cycloaddition with alkenes, ruthenium catalyzed, then 6π-electrocyclization, bicyclic cyclohexa-1,3-diene synthesis 110, 111

1,6-Diynes, terminal diene substituted, ene–yne–yne ring-closing metathesis, ruthenium carbene complex catalyzed, then 6π-electrocyclic ring closure/sigmatropic 1,5-hydride shift, tricyclic cyclohexa-1,3-diene synthesis 111

Diynols, cycloisomerization, ruthenium catalyzed, then tandem imine formation/electrocyclization/dehydration process, using hydroxylamine, pyridine-fused carbocycle synthesis 139

→ Doitunggarcinones, from oxidative radical cyclization of dearomatized acylphloroglucinols, using manganese acetate/copper acetate 264, 267

→ Dynemicin A, polycyclic intermediate, from Yamaguchi macrolactonization of bis(enyne) acid/alkenyl alcohol-substituted quinoline, using 2,4,6-trichlorobenzoyl chloride, then proximity-induced intramolecular Diels–Alder cyclization 29

→ Dynemicin A, key synthetic step, deprotonation of 2-benzopyran-1,3-dione with lithium hexamethyldisilazanide, then Diels–Alder/retro-Diels–Alder cascade with multicyclic quinone 7, 8

E

→ Elysiapyrones, from 1-stannyl-1,3-dienes, via Stille cross-coupling/8π-electrocyclization/6π-electrocyclization domino sequence, using 1-iodo-1,3-dienes, palladium catalyzed 104

Enals – *see also* Aldehydes, α,β-unsaturated

Enals, 2-alkyl-, reaction with β-keto esters, via Knoevenagel condensation/6π-electrocyclization domino reaction, substituted pyran-5-carboxylate synthesis 150, 151

Enals, reaction with γ-chloro-β-oxo esters, via enantioselective domino Michael/aldol sequence, chiral silyl prolinol ether catalyzed, then cyclization, chiral epoxycyclohexanone synthesis 344

Enamides, enantioselective reaction with α,β-unsaturated aldehydes, chiral proline derivative catalyzed, via domino Michael/hemiacetalization sequence, chiral cyclic hemiaminal synthesis 405

→ Enamines, tetracyclic, from (chloroalkyl)(2-oxocyclopentenyl)-substituted benzenes, via reaction with sodium azide, intramolecular dipolar cycloaddition, nitrogen extrusion, and rearrangement 83

Enaminones, functionalized, tryptamine based, intramolecular [2+2]-photocycloaddition/retro-Mannich tandem sequence, spiropyrrolenine synthesis 48

→ Endiandric acid esters, from pericyclic electrocyclizations/intramolecular Diels–Alder cascade of acyclic pentaenediynes, via Lindlar hydrogenation 15, 16

→ Endiandric ester A, from tetraenediynes, via twofold Lindlar hydrogenation/8π-electrocyclization cascade, then intramolecular Diels–Alder cycloaddition 112, 113

Ene reaction/Diels–Alder cycloaddition cascade, intermolecular, using cyclohexa-1,4-dienes and alkynes 160, 161

Enediynes, bromo-, electron-donating group substituted, tricyclization cascade sequence, palladium catalyzed, substituted phenanthrene synthesis 95, 96

Enediynes, bromo-, electron-withdrawing group substituted, tricyclization cascade sequence, palladium catalyzed, tetracyclic carbocycle synthesis 95, 96
Enediynes, bromo-, tricyclization cascade sequence, palladium catalyzed, strained aromatic pentacycle synthesis 97, 98
Enetriones, domino enantioselective Michael/aldol reaction with arylboronic acids, rhodium/chiral bisphosphine catalyzed, chiral bicyclic ketone synthesis 394
Enols, bicyclic, enantioselective reaction with α,β-unsaturated aldehydes, chiral proline derivative catalyzed, via domino Michael/hemiacetalization sequence, chiral tricyclic hemiacetal synthesis 405
Enol trifluoromethanesulfonates, reaction with diazoacetates, palladium catalyzed, via tandem cross coupling/electrocyclization, 3,4,5-trisubstituted pyrazole synthesis 136, 137
Enones, cyclic, tandem asymmetric allylation/epoxidation sequence, using tetraallylstannane, titanium/chiral ligand catalyzed, chiral allylic epoxy alcohol synthesis 411
1,7-Enyne benzoates, 1,3-migration/intramolecular [2+2] cycloaddition cascade, cationic gold complex catalyzed, cyclobutane-fused piperidine synthesis 56
Enynes, aza-, base-catalyzed propargyl–allenyl isomerization/aza-electrocyclization domino reaction, polyfunctionalized quinoline synthesis 148
1,3-Enynes, domino asymmetric reductive aldol reaction with picolinaldehyde, rhodium/chiral bisphosphine catalyzed, chiral dienyl alcohol synthesis 391
→ Enzyme inhibitor EI-1941-2, from 2-alkenyl-3-(hydroxymethyl)cyclohexa-2,5-dienone 5,6-monoepoxides, via oxidation/oxa-electrocyclization/oxidation cascade reaction, using oxoammonium salt 152
Epoxides – *see also* Oxiranes
→ Epoxides, chiral, from enantioselective reaction of γ-chloro-β-oxo esters with α,β-unsaturated aldehydes, chiral proline derivative catalyzed, via domino Michael/aldol/S_N2 sequence 403
Epoxides, dienyl, reaction with N-*bromosuccinimide/water, via bromonium ion/oxonium ion/bromonium ion/oxonium ion domino sequence, hexahydrolaureoxanyne synthesis* 308, 309
Epoxides, reaction with silyl cyanides to give isocyanides, then cycloaddition with nitroalkenes to yield nitronates, ring opening to nitrile oxides, and cycloaddition with acrylates, functionalized isoxazole synthesis 64, 65
Epoxides, trienyl, acyclic, epoxide-opening allylation/intramolecular Diels–Alder cascade, using diethylaluminum chloride, hirsutellone B core synthesis 12
Esters, allyloxy, α,β-unsaturated, reaction with lithium diisopropylamide, then Wittig rearrangement/oxy-Cope rearrangement/intramolecular ene reaction cascade, 2-alkenyl-1-hydroxycyclopentanecarboxylate synthesis 169, 170
Esters, γ-chloro-β-oxo, enantioselective reaction with α,β-unsaturated aldehydes, chiral proline derivative catalyzed, via domino Michael/aldol/S_N2 sequence, chiral epoxide synthesis 403
Esters, α-diazo-β-oxo, domino conversion into carbonyl ylides/enantioselective intramolecular cycloaddition, chiral dirhodium complex catalyzed, chiral tricyclic adduct synthesis 388
Esters, α-dienyl-β-oxo, alkenyl Grignard reagent addition/anionic oxy-Cope rearrangement/Dieckmann cyclization domino sequence, phomoidride B intermediate synthesis 200, 201
→ Esters, β-hydroxy, chiral, from acrylates, via domino asymmetric reductive aldol reaction with acetophenone, chiral phosphinocopper/phosphinoferrocene complex catalyzed 390, 391
Esters, α-oxo-, β,γ-unsaturated, enantioselective reaction with aldehydes, chiral enamine catalyzed, via domino oxy-Diels–Alder reaction sequence, chiral lactone synthesis 397, 398
Esters, α-oxo-, β,γ-unsaturated, enantioselective reaction with cyclohexanone, chiral proline derivative catalyzed, via domino Michael/intramolecular aldol reaction sequence, chiral bicyclic ketone synthesis 400
Esters, α-oxo, β,γ-unsaturated, enantioselective [3+3] reaction with 2-hydroxynaphtho-1,4-quinone, chiral thiourea derivative catalyzed, via domino Michael/hemiacetalization sequence, chiral benzo[g]chromene-5,10-dione synthesis 408, 409
Esters, α-oxo, β,γ-unsaturated, enantioselective [4+2] reaction with 4,4,4-trifluoroacetoacetates, chiral thiourea derivative catalyzed, via domino Michael/hemiacetalization sequence, chiral multifunctionalized 3,4-dihydropyran synthesis 409
Esters, β-oxo, enantioselective reaction with α,β-unsaturated ketones, chiral imidazolidine catalyzed, via domino Michael/aldol sequence, chiral functionalized cyclohexanone synthesis 402
Esters, β-oxo, Ugi four-component reaction with α-amino acids, isocyanides, and amines, substituted pyrrolidine-2,5-dione synthesis 431, 432
Esters, propargylic – *see* Propargylic esters
Esters, α,β-unsaturated, domino asymmetric reductive aldol reaction with aldehydes, rhodium/chiral bisphosphine catalyzed, syn-*aldol product synthesis* 389, 390
Esters, α-vinyl-β-oxo, alkenyl Grignard reagent addition/anionic oxy-Cope rearrangement/Dieckmann cyclization domino sequence, bicyclo[4.3.1]dec-6-ene-2,10-dione synthesis 199, 200

β-Estradiol, intermolecular alkylative dearomatization, osmium complex mediated, via reaction with methyl vinyl ketone and decomplexation, para*-substituted cyclohexadienone steroid synthesis* 232, 233

→Estrone core, from vinylcyclopentanone-substituted benzocyclobutene, via thermal electrocyclization/intramolecular Diels–Alder cascade 13

→Ethane-1,2-diamine, *N,N,N′,N′*-tetramesyl-1-phenyl-, chiral, from enantioselective domino intermolecular diamination of styrene, using N-(methylsulfonyl)methanesulfonamide, chiral iodine(III) reagent catalyzed 299, 300

Ethanones, 1-phenyl-2-sulfanyl-, enantioselective reaction with α,β-unsaturated aldehydes, chiral proline derivative catalyzed, via domino Michael/aldol sequence, chiral functionalized tetrahydrothiophene synthesis 402, 403

Ethene, tetracyano-, reaction with aniline-capped buta-1,3-diynes, via double [2+2]-cycloaddition/retro-electrocyclization domino sequence, substituted hexa-1,5-diene-1,1,6,6-tetracarbonitrile synthesis 130

Ethers – *see also* Oxime ethers, Silyl ethers

Ethers, α-acetoxy, reaction with Lewis acids, via ionization/oxonia-Cope rearrangement/oxocarbenium ion trapping domino sequence, polysubstituted tetrahydropyran-4-one synthesis 202

Ethers, allyl, thermal oxy-Cope rearrangement/Claisen rearrangement/ene reaction domino sequence, 1-methyleneoctahydronaphthalen-4a-ol synthesis 212

Ethers, allyl, thermal oxy-Cope rearrangement/ene reaction/Claisen rearrangement domino sequence, tricyclic lactol synthesis 211, 212

Ethers, allyl vinyl, thermal intramolecular [3,3]-sigmatropic rearrangement/ene reaction cascade, partial steroid ring skeleton synthesis 185

→Ethers, bromo, substituted, from multicomponent aminoalkoxylation of styrene with cyclic ethers/*N*-bromosuccinimide/4-nitrobenzenesulfonamide, via bromonium ion/oxonium ion/ring opening domino sequence 312, 313

Ethers, diallylic, Wittig rearrangement/oxy-Cope rearrangement cascade, base catalyzed, δ,ε-unsaturated aldehyde synthesis 168

Ethers, propargyl – *see also* Propargylic ethers

Ethers, propargyl vinyl, reaction with primary amines, via propargyl Claisen rearrangement/imine formation/tautomerization/6π-electrocyclization domino sequence, substituted 1,6-dihydropyridine synthesis 223, 224

Ethers, propargyl vinyl, thermal propargyl Claisen rearrangement/tautomerization/acylketene generation/6π-electrocyclization/aromatization domino sequence, substituted 2-hydroxybenzaldehyde synthesis 222, 223

Ethers, vinyl, reaction with methyl vinyl ketone, Lewis acid catalyzed, via Mukaiyama reaction/Michael addition/oxonia-Cope rearrangement/oxocarbenium ion trapping domino sequence, polysubstituted 3-acyltetrahydropyran synthesis 203, 204

F

→[4.6.4.6]Fenestradienes, from alkynyl-substituted cyclic vinyl bromides, via cyclocarbopalladation/Sonogashira-type coupling/alkynylation/8π-electrocyclization/6π-electrocyclization domino sequence, using enynes, palladium/copper/phosphine catalyzed 107, 108

→[4.6.4.6]Fenestrenes, from trienynes, via semihydrogenation/8π-electrocyclization/6π-electrocyclization/oxidation/1,4-addition domino sequence 113, 114

Fischer chromium carbene complex, alkynyl(alkoxy), [2+2]/[2+1]-cycloaddition cascade with 2,3-dihydrofuran, oxabicyclohexane-oxabicycloheptane adduct synthesis 50

Flavones, 3-styryl-, photoinduced 6π-electrocyclization/oxidation, 5-arylbenzo[c]xanthen-7-one synthesis 127, 128

→Fluoren-9-ones, from 2,3-dibromoindenones, via domino two-fold Heck/6π-electrocyclization sequence, using styrenes, palladium–phosphine ligand catalyzed 98, 100

→Fluoren-9-ones, substituted, from triynes, via thermal Diels–Alder cyclization/desaturation cascade, involving hexadehydro-Diels–Alder variant 36

→Forbesione, from 3,5,6-tris(allyloxy)1-hydroxyxanthen-9-one, via thermal aryl Claisen rearrangement/intramolecular Diels–Alder domino sequence 18, 217, 219

Friedländer annulation, reaction of 2-aminobenzaldehydes with cyanoacetamides, 2-aminoquinoline-3-carboxamide synthesis 442, 443

→Furan-2-ones, spirotricyclic, from intramolecular cyclization of diazo ketones to give carbonyl ylides, rhodium catalyzed, then intermolecular dipolar cycloaddition with γ-alkylidenebutenolides 68

→Furans, epoxytetrahydro-, from alcohol-functionalized oxetanes, via *N*-bromosuccinimide-initiated rearrangement domino sequence 311, 312

→Furans, pentacyclic, from intramolecular cyclization of benzaldehydes with attached alkyne to give carbonyl ylide, platinum catalyzed, then intramolecular dipolar cycloaddition with terminal alkene and C—H insertion 72

Furans, tetrahydro-, alkene functionalized, bromonium ion initiated ring-opening cascade, using bromodiethylsulfonium bromopentachloroantimonate, Laurencia *species core synthesis* 310, 311

→ Furans, tricyclic fused, from 1,3-dienyl-substituted propargylic ethers, via intramolecular [2+2] cycloaddition/[3,3]-sigmatropic rearrangement, base catalyzed 52

→ Furo[2,3-*e*][1,4]dioxocins, octahydro-, chiral, from double domino intramolecular iodoetherification of chiral diene acetals, using *N*-iodosuccinimide 314, 315

→ Furo[3,4-*f*]indolizines, 3a,8a-epoxy-, from intramolecular cyclization of diazoimides to give isomünchnone carbonyl ylides, rhodium catalyzed, then intramolecular cycloaddition with activated alkenes 70, 71

G

→ Garcibracteatones, from oxidative radical cyclization of dearomatized acylphloroglucinols, using manganese acetate/copper acetate 264, 266

→ *Garcinia* xanthones, caged, from 3,5,6-tris(allyloxy)-1-methoxyxanthen-9-ones, via thermal Claisen rearrangement/Diels–Alder cycloaddition domino sequence 18, 217, 218

→ Garsubellin A core, from intermolecular alkylative dearomatization of 3,5-dimethoxyphenols, via allylation, using allyl methyl carbonate, palladium/phosphine/titanium catalyzed, then deketalization/Michael addition/β-elimination/demethylation domino sequence 237, 238

Gewald multicomponent reaction, using cyanoacetic acid derivatives, α-acidic carbonyl compounds, and elemental sulfur, substituted 2-aminothiophene synthesis 441

→ Guanidines, bromo-, from cyclohexenes, using *N*-bromosuccinimide/*N*,*N*-dimethylcyanamide/4-toluenesulfonamide, via bromoamidination domino cascade 301, 302

H

→ Halenaquinone core, from Wessely oxidation/coupling of 2-methoxyphenols with dienyl alcohols, then intramolecular Diels–Alder reaction/Cope rearrangement 4

→ Haouamine A, from terminal alkyne-substituted 2-pyrone, via thermal Diels–Alder/retro-Diels–Alder cascade, then potassium carbonate treatment 33

→ Hemiacetals, bicyclic, chiral, from enantioselective reaction of dihydroxyacetone dimer with α,β-unsaturated aldehydes, chiral proline derivative catalyzed, via domino oxo-Michael/aldol/hemiacetalization sequence 404

→ Hemiacetals, cyclic, chiral, from enantioselective reaction of aldehydes with propanal, proline catalyzed, via two-step domino aldol/aldol/hemiacetalization sequence 395

→ Hemiacetals, tricyclic, chiral, from enantioselective reaction of bicyclic enols with α,β-unsaturated aldehydes, chiral proline derivative catalyzed, via domino Michael/hemiacetalization sequence 405

→ Hemiacetals, tricyclic, chiral, from enantioselective reaction of naphtho-1,4-quinone with aldehydes, chiral proline derivative catalyzed, via domino Michael/hemiacetalization sequence 400, 401

→ Hemiaminals, chiral, from enantioselective reaction of enamides with α,β-unsaturated aldehydes, chiral proline derivative catalyzed, via domino Michael/hemiacetalization sequence 405

→ Heptane-1,7-diones, 1,7-diamino-4-methylene-, from reaction of allylic diamines with ketenes, Lewis acid catalyzed, via double Bellus–Claisen rearrangement 213–215

Hept-6-enals, β-bromoalkyl substituted, oxime formation, using hydroxylamine, then intramolecular alkylation to give nitrone and intramolecular cycloaddition, tricyclic daphniphyllum *alkaloid core synthesis* 58

Hept-6-enals, α-chloroalkyl substituted, oxime formation, using hydroxylamine, then intramolecular alkylation to give nitrone and intramolecular cycloaddition, tricyclic isoxazolidine synthesis 58

Hexa-1,5-dienes, 3-hydroxy-, thermal oxy-Cope rearrangement/ene reaction domino sequence, 1-methyleneoctahydronaphthalen-4a-ol synthesis 209, 210

→ Hexa-1,5-diene-1,1,6,6-tetracarbonitriles, substituted, from reaction of tetracyanoethene with aniline-capped buta-1,3-diynes, via double [2+2]-cycloaddition/retro-electrocyclization domino sequence 130

Hexa-2,4-dienoates, photoreaction with singlet oxygen, via ene reaction/hetero-Diels–Alder cycloaddition cascade, 1,2-dioxine-3-carboxylate synthesis 164

Hexaenoates, protected, reaction with 5-methylene-1,3-dioxolan-4-one, via tandem intermolecular/intramolecular Diels–Alder cyclizations, chlorothricolide intermediate synthesis 31, 32

Hexa-1,3,5-trienes, 1,6-disubstituted, domino 6π-electrocyclization/intermolecular Diels–Alder cycloaddition, using various dienophiles, functionalized tricyclic cyclohexa-1,3-diene synthesis 126

→ Himandrine, tetracyclic intermediate, from aryl dienylsulfonamides, via oxidative amidation/intramolecular Diels–Alder cyclization/epimerization cascade, using (diacetoxyiodo)benzene then heat 24, 25

→ Himgravine, tetracyclic intermediate, from reaction of terminal aldehyde-substituted butenolide with 1,2-bis(trimethylsiloxy)ethane, Lewis acid catalyzed, then intramolecular Diels–Alder cyclization/acetal formation 41, 42

→ Hirsutellone B, macrolactam intermediate, from highly functionalized dioxinone, via thermal acyl ketene addition/intramolecular Diels–Alder double cyclization 29, 30
→ Hirsutellone B core, from acyclic trienyl epoxides, via epoxide-opening allylation/intramolecular Diels–Alder cascade, using diethylaluminum chloride 12
→ Histrionicotoxin alkaloids, tricyclic precursor, by oxime formation from 7-oxotrideca-2,11-dienedinitrile, using hydroxylamine, then intramolecular conjugate addition to give nitrone and intramolecular cycloaddition 58, 59
Hopene, from squalene via biogenetic conversion, hypothetical representation 322
→ Hybocarpone, bicyclic intermediate, by photochemical 1,5-hydrogen abstraction from highly functionalized benzaldehyde, then intermolecular Diels–Alder reaction with acrylate 19, 20
Hydrazines, 1-aryl-2-(2,2,2-trifluoroethylidene)-, asymmetric reaction with α,β-unsaturated aldehydes, chiral phosphonic acid catalyzed, via iminium ion formation/6π-electrocyclization, chiral 1,3,4-trisubstituted 1,4-dihydropyridazine synthesis 141–143
→ (±)-Hyperguinone B, from oxidative radical cyclization of dearomatized acylphloroglucinols, using (diacetoxyiodo)benzene 267, 268

I

→ Ialibinones, from oxidative radical cyclization of dearomatized acylphloroglucinols, using manganese acetate/copper acetate 263, 266, 267
→ Imidazo[1,2-*a*]indoles, chiral, from stereoselective reaction of cyclohexa-1,4-dienyl aminals with *N*-bromosuccinimide, via bromoamination/oxidation/elimination domino sequence 304, 305
→ Imidazoles, dihydro-, from cyclohexenes, using *N*-bromosuccinimide/nitriles/4-toluenesulfonamide, via bromoamidination/cyclization domino cascade 301, 302
→ Imidazo[1,2-*a*]pyrazine-2,6-diones, substituted, from Ugi four-component reaction of phenylalanine with aminoacetaldehyde dimethyl acetal, isocyanides, and ketones 432, 433
Imides, diazo-, intramolecular cyclization to give isomünchnone carbonyl ylide, rhodium catalyzed, then intramolecular cycloaddition with activated alkene, camptothecin tetracyclic intermediate synthesis 70, 71
Imides, diazo-, intramolecular cyclization to give isomünchnone carbonyl ylide, rhodium catalyzed, then intramolecular cycloaddition with unactivated alkene, (±)-lycopodine tricyclic intermediate synthesis 70
Imides, α-isothiocyanato, enantioselective reaction with aldehydes, chiral thiourea derivative catalyzed, via domino aldol/cyclization sequence, chiral 2-thioxooxazolidine synthesis 405, 406
Imines, alkynyl-N-aryl-, reaction with sulfonyl azides, copper catalyzed, via 1,3-dipolar cycloaddition/ketenimine formation/6π-electrocyclization/1,3-hydrogen shift domino sequence, 4-sulfonamidoquinoline synthesis 136
Imines, N-aryl-, α,β-unsaturated, reaction with internal alkynes, cobalt catalyzed, via C—H bond activation/migratory insertion/6π-electrocyclization cascade, polysubstituted dihydropyridine synthesis 134, 135
→ Imines, highly substituted, bicyclic, from *N*-allylynamides, via allyl transfer/intramolecular [2+2] cycloaddition cascade, palladium catalyzed 55, 56
Imines, α,β-unsaturated, reaction with alkynes, rhodium/P,N-ligand catalyzed, via C—H alkenylation/electrocyclization cascade, then aromatization, polysubstituted pyridine synthesis 132
→ Inden-7-ols, hexahydro-, from cyclization of propenal with alkylidenecyclopentanes, via intermolecular ene reaction/intramolecular ene reaction cascade annulation 164, 165
Indenones, 2,3-dibromo-, domino two-fold Heck/6π-electrocyclization sequence, using styrenes, palladium–phosphine ligand catalyzed, fluoren-9-one synthesis 98, 100
→ (−)-Indicol, key synthetic step, conversion of alkenyl-substituted diazo ketone into carbonyl ylide, rhodium catalyzed, then intramolecular cycloaddition 66, 67
→ Indole-3-carboxamides, 2-amino-, from reaction of 2-halo-1-nitrobenzenes with cyanoacetamides 441, 442
Indoles, 2-aryl-3-(1-tosylalkyl)-, photoinduced homolytic cleavage/elimination/6π-electrocyclization/aromatization domino sequence, benzo[a]carbazole synthesis 128, 129
Indoles, 2-(bromomethyl)-, arylation, using arenes, zinc catalyzed, then [1,5]-sigmatropic rearrangement/6π-electrocyclization/aromatization sequence, tetracyclic carbazole synthesis 109, 110
Indoles, 2,3-dibromo-, domino two-fold Heck/6π-electrocyclization sequence, using acrylates, palladium–phosphine ligand catalyzed, carbazole synthesis 98, 99
Indoles, domino enantioselective Friedel–Crafts/Henry reaction with nitroalkenes and benzaldehyde, copper/chiral amine catalyzed, chiral β-nitro alcohol synthesis 394
Indoles, enantioselective reaction with α,β-unsaturated ketones, chiral alkaloid catalyzed, via domino Michael/ketone aldol/dehydration sequence, chiral spiro[cyclohex-2-enone-oxindole] synthesis 359

Indoles, 3-(2-isocyanoethyl)-, three-component reaction with oxocarboxylic acids and aminoacetaldehyde dimethyl acetal, via Ugi/Pictet–Spengler sequence, polycyclic indole-type alkaloid synthesis 423, 424
Indoles, 3-(isocyanomethyl)-, multicomponent reaction with benzaldehydes and carboxylic or amino acids, N-*(indol-3-ylmethyl)amide synthesis* 419, 420
→ Indoles, 1-phenyl-2,3-dihydro-, from reaction of 2-chlorostyrene with aniline, via base-catalyzed domino hydroamination/aryne cyclization sequence 294
→ Indoles, substituted, from microwave heating/dipolar cycloaddition of 10-azido-2,6-dienones, then nitrogen elimination and cyclization 83, 84
→ Indoles, tricyclic derivatives, from four-component reaction of isocyanides with ketones, tryptophan amino acid derivatives, and aminoacetaldehyde dimethyl acetal, via Ugi/Pictet–Spengler sequence 427
Indolin-2-ones, 3-methylene-, enantioselective reaction with α-isothiocyanato imides, chiral thiourea catalyzed, via Michael/cyclization domino sequence, chiral spiro[oxindole-pyrrolidine] synthesis 363, 364
Indolin-2-ones, 3-methylene-, enantioselective reaction with 3-(isothiocyanato)oxindoles, chiral thiourea catalyzed, via Michael/cyclization cascade sequence, chiral bispirooxindole synthesis 362
Isatin, derivatives, enantioselective three-component reaction with malononitrile and 1,3-diketones, chiral cupreine catalyzed, via domino Knoevenagel/Michael/cyclization process, chiral spiro[pyran-oxindole] synthesis 354, 355
Isatin, enantioselective three-component reaction with propenal and pyrrolo[2,3-b]pyridine-2,3-diones, chiral tertiary amine catalyzed, via Morita–Baylis–Hillman/bromination/[3+2] annulation domino sequence, chiral dispiro[oxindole-furan-pyrrolopyridine] synthesis 362, 363
Isatoic anhydride, intermolecular cycloaddition with azomethine ylides, then ring opening, decarboxylation, and cyclization, benzo[d][1,3]-diazepin-5-one synthesis 78
→ Isocedrene, key intermediate, from alkenylcyclopentadienes, via thermal [1,5]-hydrogen sigmatropic rearrangement/Diels–Alder cycloaddition cascade 190
→ Isochrysohermidin, intermediate, from twofold hetero-Diels–Alder reaction of a 1,2,4,5-tetrazine-3,6-dicarboxylate with a substituted 1,3-diene, then double retro-hetero-Diels–Alder reaction 33, 34
Isocyanides, four-component reaction with ketones, tryptophan amino acid derivatives, and aminoacetaldehyde dimethyl acetal, via Ugi/Pictet–Spengler sequence, tricyclic indole derivative synthesis 427
→ Isoindolinones, substituted, from Ugi four-component reaction of 2-formylbenzoates with α-amino acids, isocyanides, and amines 431
→ Isoindolo[2,1-*a*]quinolines, from ring-closing metathesis of *N*-allyl-*N*-benzyl-2-vinylaniline, ruthenium carbene complex catalyzed, then oxidation to azomethine ylide, intermolecular dipolar cycloaddition with benzoquinone, and oxidation 77, 78
→ Isoquinolines, 3-aryl-3,4-dihydro-, from iminobenzocyclobutenes, via tandem thermal electrocyclic ring opening/6π-electrocyclization 140, 141
→ Isoxazoles, functionalized, from reaction of epoxides with silyl cyanides to give isocyanides, then cycloaddition with nitroalkenes to yield nitronates, ring opening to nitrile oxides, and cycloaddition with acrylates 64, 65
→ Isoxazolidines, bicyclic, from reaction of nitrosobenzene with 3-arylpropenes and maleimides, via nitroso-ene reaction/oxidation/[3+2] cycloaddition cascade, copper catalyzed 166, 167
→ Isoxazolidines, tricyclic, by oxime formation from α-chloroalkyl-substituted hept-6-enal, using hydroxylamine, then intramolecular alkylation to give nitrone/intramolecular cycloaddition 58
→ Isoxazolidines, tricyclic, from intramolecular hydroamination of alkenyl(phenylpent-4-enyl)-substituted oximes to give nitrones, then intramolecular cycloaddition 59, 60
→ Isoxazolo[4,5-*c*]quinolines, from dipolar cycloaddition of 3-(2-aminophenyl)alk-2-ynones with nitrile oxides, then intramolecular cyclization 65, 66

J

→ Jasmone, dihydro-, bicyclic β-lactone intermediate, from [2+2] cycloaddition of ynolates with δ-dicarbonyl compounds/Dieckmann condensation sequence 48, 49
→ (−)-Joubertinamine, key intermediate, from vinyl sulfides, via Claisen rearrangement, then Wittig reaction, oxidation, and [2,3]-sigmatropic rearrangement 188, 189

K

Ketene imines, aryl, [1,5]-sigmatropic rearrangement/6π-electrocyclization cascade, substituted quinoline synthesis 179
Ketones, aminophenyl, Ugi four-component reaction with N-protected aminoacetaldehyde, isocyanides, and azidosilanes, then deprotection/cyclization, substituted 1,4-benzodiazepine synthesis 435
Ketones, aminophenyl, Ugi four-component reaction with N-protected aminoacetaldehyde, isocyanides, and carboxylic acids, then deprotection/cyclization, substituted 1,4-benzodiazepine synthesis 434, 435

Ketones, aminophenyl, Ugi four-component reaction with N-protected glycine, isocyanides, and aldehydes, then deprotection/cyclization, substituted 1,4-benzodiazepin-2-one synthesis 435, 436
Ketones, aryl, photoinduced [1,5]-hydrogen sigmatropic rearrangement/intramolecular Diels–Alder cycloaddition cascade, steroid tetracyclic framework synthesis 190, 191
→Ketones, bicyclic, chiral, from enantioselective reaction of β,γ-unsaturated α-oxo esters with cyclohexanone, chiral proline derivative catalyzed, via domino Michael/intramolecular aldol reaction sequence 400
→Ketones, bicyclic, chiral, from enetriones, via domino enantioselective Michael/aldol reaction with arylboronic acids, rhodium/chiral bisphosphine catalyzed 393, 394
→Ketones, bicyclic, functionalized, chiral, from enantioselective [3+2] reaction of cyclohexane-1,2-dione with nitroalkenes, chiral thiourea derivative catalyzed, via domino Michael/Henry sequence 407, 408
→Ketones, bicyclic, functionalized, chiral, from enantioselective [3+2] reaction of cyclohexane-1,4-dione with nitroalkenes, chiral thiourea derivative catalyzed, via domino Michael/Henry sequence 407
→Ketones, bicyclic, functionalized, from enantioselective reaction of cyclohexane-1,2-dione with cinnamaldehyde, chiral proline derivative catalyzed, via domino Michael/aldol sequence 403
Ketones, diazo – *see* Diazo ketones
Ketones, dienyl(iodoalkyl) substituted, imine formation, using (tributylstannyl)methanamine, then cyclization/destannylation to give azomethine ylide, and intramolecular dipolar cycloaddition, (±)-demethoxyschelhammericine precursor synthesis 74, 75
→Ketones, saturated, from methallyl alcohol derivatives, via domino hydroformylation/Wittig alkenation/hydrogenation sequence, using Wittig ylides/hydrogen/carbon monoxide, rhodium catalyzed 317, 318
→Ketones, tricyclic, fused, from tandem alkylative dearomatization/hydrogenation of 2,6-dimethylphenol, then reaction with ethylaluminum chloride, via tandem cationic 1,4-addition/two consecutive Wagner–Meerwein 1,2-rearrangements 252, 253
Ketones, α,β-unsaturated, enantioselective Michael/aldol domino reaction with β-diketones, β-oxo esters, or β-oxo sulfones, chiral imidazolidine catalyzed, chiral functionalized cyclohexanone synthesis 339
Ketones, α,β-unsaturated, enantioselective reaction with oxindoles, chiral Cinchona*-based alkaloid catalyzed, via Michael/Michael domino sequence, chiral spiro[cyclohexanone-oxindole] synthesis* 351
Ketoxime O-pentafluorobenzoates, α,β-unsaturated, reaction with alkenylboronic acids, copper catalyzed, via C—N cross coupling/electrocyclization/aromatization cascade, polysubstituted pyridine synthesis 131
Ketoximes, α,β-unsaturated, reaction with alkynes, rhodium catalyzed, via C—H alkenylation/electrocyclization/aromatization cascade, polysubstituted pyridine synthesis 132–134
→Kingianin A, key intermediate, from 1-bromo-1,3-dienes, via Stille cross-coupling/8π-electrocyclization/6π-electrocyclization domino sequence, using 1-stannyl-1,3-dienes, palladium catalyzed 107

L

→Lactams, tricyclic, from base treatment of N-phenyl-N-(prop-2-enyl)but-3-ynamide, then thermal intramolecular Himbert Diels–Alder cyclization 22, 23
→Lactams, tricyclic, from reaction of an azide-substituted triene with methylaluminum dichloride, via Diels–Alder cyclization/Schmidt rearrangement cascade 37, 38
β-Lactams, N*-vinyl-, copper-catalyzed [3,3]-sigmatropic rearrangement/6π-electrocyclization domino reaction, pyridin-2-one-fused cyclobutane synthesis* 149, 150
→δ-Lactams, cyclobutane fused, from N-vinyl-β-lactams, via copper-catalyzed [3,3]-sigmatropic rearrangement/6π-electrocyclization domino reaction 149, 150
→Lactols, caged tetracyclic, from enal-substituted acetoxypyranones, via tandem intramolecular [5+2] cycloaddition/conjugate addition 86, 87
→Lactols, chiral, from propanal trimerization, via domino enantioselective aldol/aldol/hemiacetalization sequence, proline catalyzed 395
→Lactols, tricyclic, from allyl ethers, via thermal oxy-Cope rearrangement/ene reaction/Claisen rearrangement domino sequence 211
Lactones, aryl, deprotonation using lithium tetramethylpiperidide, then benzyne precursor addition and Diels–Alder cyclization, averufin intermediate synthesis 23, 24
→Lactones, chiral, from enantioselective reaction of β,γ-unsaturated α-oxo esters with aldehydes, chiral pyrrolidine catalyzed, via domino oxy-Diels–Alder reaction sequence 397
Lactones, tetracyclic, thermal intermolecular Diels–Alder cyclization with skipped triene, then intramolecular hetero-Diels–Alder cyclization, heptacyclic bolivianine synthesis 31
→β-Lactones, bicyclic, from [2+2] cycloaddition of ynolates with δ-dicarbonyl compounds/Dieckmann condensation sequence 48, 49
→δ-Lactones, from [2+2] cycloaddition of but-2-enal with ketene, palladium complex catalyzed, then allylic rearrangement 54, 55

→ Laureatin, from reaction of laurediol with sodium bromide/hydrogen peroxide, bromoperoxidase catalyzed, via domino bromoetherification/bromoetherification sequence 307, 308

Laurediol, reaction with sodium bromide/hydrogen peroxide, bromoperoxidase catalyzed, via domino bromoetherification/bromoetherification sequence, laureatin synthesis 307, 308

→ Laurefucin, key intermediate, from alkene-functionalized tetrahydrofurans, via bromonium ion initiated ring-opening cascade, using bromodiethylsulfonium bromopentachloroantimonate 310, 311

→ *Laurencia* species, core, from alkene-functionalized tetrahydrofurans, via bromonium ion initiated ring-opening cascade, using bromodiethylsulfonium bromopentachloroantimonate 310, 311

→ Laureoxanyne, hexahydro-, from reaction of dienyl epoxides with *N*-bromosuccinimide/water, via bromonium ion/oxonium ion/bromonium ion/oxonium ion domino sequence 308, 309

→ (*S*)-Levamisole, key synthetic step, enantioselective domino intermolecular diamination of styrene, using *N*-(methylsulfonyl)methanesulfonamide, chiral iodine(III) reagent catalyzed 299, 300

→ Longithorone A, from highly functionalized cyclophane, via oxidation/transannular intramolecular Diels–Alder cascade, using iodosobenzene 25, 26

→ Lupone, phenyl, from intermolecular alkylative dearomatization of clusiaphenone B, using prenyl bromide/base 244

→ (−)-Lyconadin C, precursor, from a tricyclic penta-2,4-dienoyl azide, via thermal tandem Curtius rearrangement/aza-electrocyclization domino reaction 148, 149

→ (±)-Lycopodine, tricyclic intermediate, from intramolecular cyclization of a diazoimide to give an isomünchnone carbonyl ylide, rhodium catalyzed, then intramolecular cycloaddition with an unactivated alkene 70

→ (−)-γ-Lycorane, key synthetic step, stereoselective reaction of chiral cyclohexa-1,4-dienyl aminals with *N*-bromosuccinimide, via bromoamination/oxidation/elimination domino sequence 304, 305

M

Macrocycles, 19-membered, selenoxide elimination via oxidation, then tandem elimination/double transannular Diels–Alder reaction, pentacyclic cyclostreptin intermediate synthesis 10, 11

→ Manzamine A, tricyclic intermediate, from Stille cross coupling of a highly functionalized vinyl bromide with a vinylstannane, palladium catalyzed, then intramolecular Diels–Alder reaction 20, 21

→ (−)-Mesembrine, key intermediate, from vinyl sulfides, via Claisen rearrangement, then Wittig reaction, oxidation, and [2,3]-sigmatropic rearrangement 188, 189

Methallyl alcohols, derivatives, domino hydroformylation/Wittig alkenation/hydrogenation sequence, using Wittig ylides/hydrogen/carbon monoxide, rhodium catalyzed, saturated ketone synthesis 317, 318

→ Methylamine, *N*,*N*-dimethyl-1-(2-tolyl)-, from Sommelet–Hauser rearrangement of benzyltrimethylammonium salts, base catalyzed, via [2,3]-/[1,3]-sigmatropic rearrangements 176, 177

Morita–Baylis–Hillman bromides, reaction with arylacetylenes, via tandem alkynylation/propargyl–allenyl isomerization/6π-electrocyclization sequence, copper catalyzed, substituted naphthalene synthesis 109

Morita–Baylis–Hillman cinnamaldehyde adducts, Wittig alkenation/6π-electrocyclization/oxidation sequence, ortho-*terphenyl synthesis* 123, 124

→ Morpholines, substituted, from multicomponent reaction of 2-(chloromethyl)oxirane with ethene and 4-nitrobenzenesulfonamide, via bromonium ion/oxonium ion/ring opening domino sequence 313, 314

→ (±)-Myrioxazine A, by oxime formation from α-chloroalkyl-substituted hept-6-enal, using hydroxylamine, then intramolecular alkylation to give nitrone, intramolecular cycloaddition, reductive cleavage, and acetal formation 58

N

Naphthalene-1-carboxylates – *see* 1-Naphthoates

Naphthalene-1,4-diones, 5-alkadienyl-, sulfolene masked, thermal and pressure unmasking/intramolecular Diels–Alder cascade, colombiasin A intermediate synthesis 17

Naphthalene-1,4-diones, 2,3-dibromo-, domino twofold Heck/6π-electrocyclization sequence, using acrylates, palladium–phosphine ligand catalyzed, anthraquinone synthesis 98, 99

→ Naphthalenes, functionalized, from [2-(1,3-dioxolan-2-yl)phenyl]allenes, via thermal 1,5-hydrogen shift/6π-electrocyclization/aromatization domino sequence 118, 119

→ Naphthalenes, substituted, from reaction of Morita–Baylis–Hillman bromides with arylacetylenes, via tandem alkynylation/propargyl–allenyl isomerization/6π-electrocyclization sequence, copper catalyzed 109

Naphthalen-1-ols – *see* 1-Naphthols

Naphthalen-2-ols – *see* 2-Naphthols

→ Naphthalen-4a-ols, 1-methyleneoctahydro-, from allyl ethers, via thermal oxy-Cope rearrangement/Claisen rearrangement/ene reaction domino sequence 212

→Naphthalen-4a-ols, 1-methyleneoctahydro-, from 3-hydroxyhexa-1,5-dienes, via thermal oxy-Cope rearrangement/ene reaction domino sequence 209, 210
→Naphthalen-1-ones, 2-methyl-2-(3-methylbut-2-enyl)-, chiral, from enantioselective alkylative dearomatization of 2-methylnaphthalen-1-ol, using chiral α-isosparteine/prenyl chloride 243
→Naphthalen-2-ones, 1-allyl-, chiral, from intermolecular asymmetric allylic dearomatization of 2-naphthols, using allyl carbonates, palladium/chiral bisphosphine catalyzed 268–275
→Naphthalen-2-ones, substituted, chiral, from enantioselective intermolecular alkylative dearomatization of 2-naphthols, using *meso*-aziridine electrophiles, magnesium/chiral ligand catalyzed 275–280
→1-Naphthoates, 1-allyl-2-oxo-1,2-dihydro-, from intermolecular allylic dearomatization of 1-(alkoxycarbonyl)-2-naphthols, using allyl carbonates, palladium catalyzed, lithium carbonate base 274
→1-Naphthoates, 1-allyl-2-oxo-1,2-dihydro-, from intermolecular allylic dearomatization of 2-hydroxy-1-naphthoates, using allyl carbonates, palladium catalyzed, optimized conditions 273
1-Naphthoates, 2-hydroxy-, intermolecular allylic dearomatization, using allyl carbonates, palladium catalyzed, 1-allyl-2-oxo-1,2-dihydro-1-naphthoate synthesis, optimized conditions 273
→Naphtho[2,3-*c*]furans, from alkynyl-substituted propargylic carbonates, via tandem Suzuki cross coupling/6π-electrocyclization/aromatization, using arylboronic acids, palladium catalyzed 94, 95
1-Naphthols, 2-methyl-, enantioselective alkylative dearomatization, using chiral α-isosparteine/prenyl chloride, chiral 2-methyl-2-(3-methylbut-2-enyl)naphthalen-1-one synthesis 243
→1-Naphthols, octahydro-, from cyclization of propenal with alkylidenecyclohexanes, via intermolecular ene reaction/intramolecular ene reaction cascade annulation 164, 165
2-Naphthols, 1-(alkoxycarbonyl)-, intermolecular allylic dearomatization, using allyl carbonates, palladium catalyzed, lithium carbonate base, 1-allyl-2-oxo-1,2-dihydro-1-naphthoate synthesis 273, 274
2-Naphthols, enantioselective intermolecular alkylative dearomatization, using meso-*aziridine electrophiles, magnesium/chiral ligand catalyzed, chiral substituted naphthalen-2-one synthesis* 275–280
2-Naphthols, intermolecular asymmetric allylic dearomatization, using allyl carbonates, palladium/chiral bisphosphine catalyzed, chiral 1-allylnaphthalen-2-one synthesis 268–275
Naphtho-1,4-quinones, enantioselective reaction with aldehydes, chiral proline derivative catalyzed, via domino Michael/hemiacetalization sequence, chiral tricyclic hemiacetal synthesis 400, 401
→Natural product scaffolds, from oxidative radical cyclization of dearomatized acylphloroglucinols, using manganese acetate/copper acetate 263–267
Nazarov reagents, enantioselective reaction with methyleneindolones, chiral urea catalyzed, via domino Michael/Michael sequence, chiral spiro[cyclohexene-oxindole] synthesis 357, 358
→Nimbidiol, key synthetic step, from polyenes via sulfenium-initiated cyclization cascade, using methyl phenyl sulfoxide/boron trifluoride 330
Nitriles, Blaise reaction with bromoacetates, zinc catalyzed, then chemoselective addition with 1,3-enynes and isomerization/6π-electrocyclization/aromatization cascade, polysubstituted pyridine synthesis 144, 145
→Nitrogen heterocycles, bicyclic, from *N*-allenyl amides, via thermal [1,3]-sigmatropic rearrangement/Diels–Alder cycloaddition cascade 181, 182
→Nitrogen heterocycles, bridged tricyclic, from *N*-allenyl amides, via [1,3]-sigmatropic rearrangement/6π-electrocyclization/Diels–Alder cycloaddition cascade 182, 183
→Nitrogen heterocycles, tetracyclic, from intramolecular cyclization of nitrones onto alkynes, gold catalyzed, then N—O cleavage, cyclization to give azomethine ylides, and intramolecular dipolar cycloaddition with alkenes 76, 77
→Nitrogen heterocycles, tetracyclic, from triazatriynes, via intramolecular propargyl ene reaction/Diels–Alder cycloaddition cascade 161
Nitrones, alkenyl(alkynyl) substituted, intramolecular cyclization onto alkyne, gold catalyzed, then N—O cleavage, cyclization to give azomethine ylide, and intramolecular dipolar cycloaddition with alkene, tetracyclic nitrogen heterocycle synthesis 76, 77
Nitrones, aryl oxazolidinyl, cycloaddition with disubstituted ketenes, then [3,3]-sigmatropic rearrangement, aromatization, ring opening, and ring closure, oxindole synthesis 62, 63
Nitrones, diaryl, cycloaddition with 2-diazobut-3-enoates, rhodium catalyzed, then pericyclic cascade, 1-oxa-4-azacyclopenta[cd]azulene synthesis 62
Nitrosobenzene, reaction with 3-arylpropenes and maleimides, via nitroso-ene reaction/oxidation/[3+2] cycloaddition cascade, copper catalyzed, bicyclic isoxazolidine synthesis 166, 167

→ Norsecurinine, key intermediate, from thermolysis of alkyne-substituted oxazoles, via annulative Diels–Alder/retro-Diels–Alder cascade 34, 35

O

Obtusallene II analogue, reaction with N-bromosuccinimide/acetic acid, tetramethylguanidine catalyzed, via bromonium ion/oxonium ion domino sequence, obtusallene VII framework synthesis 308

→ Ocellapyrones, from 1-stannyl-1,3-dienes, via Stille cross-coupling/8π-electrocyclization/6π-electrocyclization domino sequence, using 1-iodo-1,3-dienes, palladium catalyzed 105, 106

(−)-Oseltamivir, fully functionalized intermediates 365, 366

→ (−)-Oseltamivir, functionalized intermediate, from enantioselective three-component reaction of (pentan-3-yloxy)acetaldehyde with nitroacrylates and vinylphosphonates, chiral diarylprolinol silyl ether catalyzed, via domino process 366–369

→ (−)-Oseltamivir, functionalized intermediate, from enantioselective three-component reaction of (pentan-3-yloxy)acetaldehyde with *N*-(2-nitrovinyl)acetamide and vinylphosphonates, chiral diarylprolinol silyl ether catalyzed, via domino process 371, 372

→ (−)-Oseltamivir, key synthetic step, domino bromoamidation of 5-aminocyclohexa-1,3-diene-1-carboxylates, using *N*-bromoacetamide, Lewis acid catalyzed 300, 301

→ (−)-Oseltamivir, total synthesis from functionalized intermediate, via one-pot operation 372

→ (−)-Oseltamivir, total synthesis from functionalized intermediate, via three one-pot operations 366–370

Overman aza-Cope rearrangement/Mannich reaction domino sequence, condensation of homoallylic amino alcohols with aldehydes, substituted 3-acylpyrrolidine synthesis 205

→ 1-Oxa-4-azacyclopenta[*cd*]azulenes, from cycloaddition of diaryl nitrones with 2-diazobut-3-enoates, rhodium catalyzed, then pericyclic cascade 62

→ Oxabicyclohexane-oxabicycloheptane, adduct from [2+2]/[2+1]-cycloaddition cascade of an alkynyl(alkoxy) Fischer chromium carbene complex with 2,3-dihydrofuran 50

→ 9-Oxabicyclo[3.3.1]nonanes, from 9-oxabicyclo[6.1.0]non-4-enes, via domino transannular O-heterocyclization sequence, halonium ion initiated, using *N*-halosuccinimide/triethylamine tris(hydrofluoride) 306

9-Oxabicyclo[6.1.0]non-4-enes, domino transannular O-heterocyclization sequence, halonium ion initiated, using N-halosuccinimide/triethylamine tris(hydrofluoride), 9-oxabicyclo[3.3.1]nonane synthesis 306

→ 3-Oxabicyclo[4.2.0]octan-4-ols, 1-nitro-, chiral, from enantioselective [2+2] reaction of α,β-unsaturated aldehydes with α-(hydroxymethyl)nitroalkenes, chiral proline derivative/substituted thiourea catalyzed, via domino Michael/Michael/hemiacetalization sequence 401, 402

→ 8-Oxabicyclo[3.2.1]octene-6,7-dicarboxylates, chiral, from diazo diones, via domino conversion into carbonyl ylides/enantioselective intermolecular cycloaddition with acetylenedicarboxylates, chiral dirhodium complex catalyzed 389

Oxadiynes, intermolecular [2+2+2] cycloaddition with alkynes, via propargyl ene reaction/Diels–Alder cyclization cascade, benzo[c]furan synthesis 162

→ Oxataxane skeleton, from conversion of Wieland–Miescher ketone into a propargylic ether, then intramolecular [2+2] cycloaddition/[3,3]-sigmatropic rearrangement, base catalyzed 52, 53

Oxazoles, alkyne substituted, thermolysis, annulative Diels–Alder/retro-Diels–Alder cascade, norsecurinine intermediate synthesis 34, 35

Oxazoles, alkynyl[(iodomethyl)aziridinyl] substituted, internal alkylation, silver promoted, then cyanide ion addition/ring opening to give azomethine ylide, intramolecular dipolar cycloaddition, and loss of hydrogen cyanide, aziridinomitosene precursor synthesis 73, 74

→ Oxazolidines, 2-thioxo-, chiral, from enantioselective reaction of α-isothiocyanato imides with aldehydes, chiral thiourea derivative catalyzed, via domino aldol/cyclization sequence 405, 406

→ Oxazolidin-2-ones, functionalized, chiral, from α,β-unsaturated aldehydes, via enantioselective domino conjugate addition/amination, using thiols/azodicarboxylates, chiral silyl prolinol ether catalyzed, then reduction 342, 343

→ [1,3]Oxazolo[3,2-*a*]pyridin-5-ones, hexahydro-, from but-3-enamides, via domino hydroformylation/reductive amination sequence, using hydrogen/carbon monoxide, rhodium/bis(phosphite) catalyzed 319, 320

→ Oxepanes, bromo-, polycyclic, from reaction of carbonate-functionalized triepoxyalkenes with *N*-bromosuccinimide, via bromonium ion initiated ring-opening cascade 309, 310

Oxetanes, alcohol functionalized, N-bromosuccinimide-initiated rearrangement domino sequence, epoxytetrahydrofuran synthesis 311, 312

Oxime ethers, dibromo, reaction with Grignard reagents, via 1,5-diazatriene formation/6π-electrocyclization/elimination domino sequence, 2,4,6-trisubstituted pyrimidine synthesis 144

Oximes, alkenyl(phenylpent-4-enyl) substituted, intramolecular hydroamination to give nitrone, then intramolecular cycloaddition, tricyclic isoxazolidine synthesis 59, 60
Oximes, O-*propargylic,* α,β*-unsaturated, intramolecular [2,3]-sigmatropic rearrangement/6*π*-electrocyclization tandem reaction, copper catalyzed, polysubstituted pyridine* N-*oxide synthesis* 137, 138
Oxindoles, 3-bromo-, dehydrobromination, using cesium carbonate, then intermolecular hetero-Diels–Alder cycloaddition with substituted indole, and fragmentation, perophoramidine intermediate synthesis 38, 39
Oxindoles, enantioselective reaction with α,β*-unsaturated aldehydes, chiral diarylprolinol silyl ether catalyzed, via domino Michael/Michael/aldol sequence, chiral spiro[cyclohexene-oxindole] synthesis* 359, 360
Oxindoles, enantioselective reaction with α,β*-unsaturated ketones, chiral* Cinchona*-based alkaloid catalyzed, via Michael/Michael domino sequence, chiral spiro[cyclohexanone-oxindole] synthesis* 351
Oxindoles, enantioselective three-component reaction with nitroalkenes and α,β*-unsaturated aldehydes, chiral diarylprolinol silyl ether and chiral thiourea catalyzed, via Michael/Michael/aldol domino sequence, chiral spiro[cyclohexane-oxindole] synthesis* 364
Oxindoles, enantioselective three-component reaction with α,β*-unsaturated aldehydes and linear aldehydes, chiral diarylprolinol silyl ether catalyzed, via triple domino cascade, chiral spiro[cyclohexene-oxindole] synthesis* 352
→Oxindoles, from cycloaddition of aryl oxazolidinyl nitrones with disubstituted ketenes, then [3,3]-sigmatropic rearrangement, aromatization, ring opening, and ring closure 62, 63
Oxindoles, 3-methylene-, enantioselective reaction with bromo(nitro)methane, chiral thiourea-quinidine catalyzed, via domino sequence, chiral spiro[cyclopropane-oxindole] synthesis 360, 361
Oxindoles, 3-substituted, enantioselective reaction with methyleneindolinones, chiral thiourea derivative catalyzed, via domino Michael/aldol sequence, chiral multifunctionalized bispirooxindole synthesis 358, 359, 406, 407
Oxindoles, 3-substituted, enantioselective reaction with nitroalkenes, chiral Cinchona*-based alkaloid catalyzed, via Michael/Henry cascade sequence, chiral spiro[cyclopentane-oxindole] synthesis* 361, 362
Oxindoles, spirocyclic, alkaloids and anticancer agents containing 349, 350
Oxiranes – *see also* Epoxides
Oxiranes, 2-(chloromethyl)-, multicomponent reaction with ethene and 4-nitrobenzenesulfonamide, via bromonium ion/oxonium ion/ring opening domino sequence, substituted morpholine synthesis 313, 314
Oxiranes, vinylic, alkynyl tethered, intramolecular [5+2] cycloaddition, rhodium carbene complex catalyzed, then Claisen rearrangement, functionalized bicyclo[3.1.0] compound synthesis 85
→Oxygen heterocycles, tetracyclic, from intramolecular [5+2] cycloaddition of terminal alkene tethered γ-pyranones, then intermolecular [4+2] cycloaddition with 1,3-dienes 87

P

→Pentacycles, aromatic, strained, from bromoenediynes, via tricyclization cascade sequence, palladium catalyzed 97
Penta-1,4-dien-3-imines, 1,5-diaryl-, domino [2+2] cycloaddition with a ketene/Cope rearrangement, substituted azocin-2-one synthesis 51
Penta-2,4-dienoates, 2-(bromomethyl)-, Wittig alkenation/6π*-electrocyclization/oxidation sequence,* ortho-*terphenyl synthesis* 123, 124
Penta-2,4-dienoyl azides, tricyclic, thermal tandem Curtius rearrangement/aza-electrocyclization domino reaction, (–)-lyconadin C precursor synthesis 148, 149
Pentaenediyne, acyclic, pericyclic electrocyclizations/intramolecular Diels–Alder cascade, via Lindlar hydrogenation, endiandric acid ester synthesis 15, 16
→Pentaenes, from reaction of trienynols with benzenesulfenyl chloride, via [2,3]-/[2,3]-sigmatropic rearrangements/[1,5]-sigmatropic hydrogen shift cascade 175, 176
Pentane-1,5-dial, enantioselective reaction with nitroalkenes, chiral proline derivative catalyzed, via domino Michael/intramolecular Henry reaction sequence, chiral multifunctionalized cyclohexane synthesis 399
→Perophoramidine, key intermediate, from dehydrobromination of a 3-bromooxindole, using cesium carbonate, then intermolecular hetero-Diels–Alder cycloaddition with a substituted indole, and fragmentation 38, 39
→Perovskone, polycyclic intermediate, from Diels–Alder intermolecular cycloaddition of tricyclic quinone with doubly-skipped triene, Lewis acid catalyzed, then alkene isomerization/Prins cyclization/etherification cascade 40, 41
→(–)-PF-1018 core, from 1-iodo-1,3,7-trienes, via Stille cross-coupling/8π-electrocyclization/Diels–Alder cascade, using 1-stannyl-1,3-dienes, palladium/copper thiophene-2-carboxylate catalyzed 102, 103

→ Phenanthrenes, 2-iodooctahydro-, chiral, from polyprenoids, via asymmetric iodonium-initiated polyene cyclization cascade, using N-iodosuccinimide, chiral phosphoramidite mediated, then cyclization using chlorosulfuric acid 326, 327
→ Phenanthrenes, octahydro-, from dienediynediols, via conversion into diimines by Mitsunobu reaction, then retro-ene/6π-electrocyclization/Diels–Alder cycloaddition cascade 166
→ Phenanthrenes, 2-(phenylsulfanyl)octahydro-, from polyenes, via sulfenium-initiated cyclization cascade, using methyl phenyl sulfoxide/boron trifluoride 330
→ Phenanthrenes, substituted, from electron-donating group-substituted 2-bromoenediynes, via tricyclization cascade sequence, palladium catalyzed 95, 96
→ Phenanthridines, 8-(phenylselanyl)octahydro-, from geranylated anilines, via selenium-initiated polyene cyclization cascade, using benzeneselenenyl chloride, scandium trifluoromethanesulfonate catalyzed 330, 331
Phenolates, 2,6-dimethyl-, reaction with dienyl bromide, via tandem alkylative dearomatization/[4+2] cycloaddition, then hydrogenation, methyllithium addition, and acid-catalyzed rearrangement, seychellene synthesis 252
Phenolates, 3-(trimethylcyclopentane) substituted, lithium salt, asymmetric oxidation/dimerization, copper/sparteine catalyzed, chiral aquaticol synthesis, via Diels–Alder reaction 3
Phenols, 2-(alkenyloxy) substituted, oxidation/intramolecular Diels–Alder cascade sequence, using (diacetoxyiodo)benzene/alcohols, CP-263,114 core architecture synthesis 3, 4
Phenols, alkylative dearomatization, effect of substituents, electrophiles, and counterion 240, 241
Phenols, 2-dienyl-, thermal [1,7]-sigmatropic hydrogen shift/6π-electrocyclization cascade, 1-benzopyran synthesis 178, 179
Phenols, 3,5-dimethoxy-, intermolecular alkylative dearomatization, via allylation, using allyl methyl carbonate, palladium/phosphine/titanium catalyzed, then deketalization/Michael addition/β-elimination/demethylation domino sequence, garsubellin A core synthesis 237, 238
Phenols, 3,5-dimethoxy-, intermolecular alkylative dearomatization, via bisallylation, using allyl alcohol, palladium/phosphine/titanium catalyzed, 4,4-diallyl-3,5-dimethoxycyclohexa-2,5-dienone synthesis 236
Phenols, 2,6-dimethyl-, tandem alkylative dearomatization/hydrogenation, then reaction with ethylaluminum chloride, via tandem cationic 1,4-addition/two consecutive Wagner–Meerwein 1,2-rearrangements, 1,5-disubstituted 3-methylbicyclo[3.2.1]oct-3-en-2-one synthesis 252, 253
Phenols, 2-methoxy-, Wessely oxidation/coupling with dienyl alcohols, then intramolecular Diels–Alder reaction/Cope rearrangement, halenaquinone core synthesis 4
Phenols, 3-(trimethylcyclopentane) substituted, oxidative dimerization, using stabilized 1-hydroxy-1,2-benziodoxol-3-one 1-oxide, aquaticol synthesis, via Diels–Alder reaction 2, 3
Phenoxides – *see* Phenolates
Phloroglucinols, acyl-, dearomatized, oxidative radical cyclization, using manganese acetate/copper acetate, natural product scaffold synthesis 263–267
Phloroglucinols, acyl-, domino dearomative allylation using trifluoromethanesulfonate electrophile, effect of base counterion 284, 285
Phloroglucinols, acyl-, domino dearomative conjunctive allylic annulation, using protected 2-methylenepropane-1,3-diol, palladium catalyzed, substrate scope 283, 284
Phloroglucinols, acyl-, intermolecular alkylative dearomatization, using unactivated electrophiles/base, polysubstituted cyclohexa-2,4-dienone synthesis 248–250
Phloroglucinols, acyl-, methoxy substituted, domino dearomative conjunctive allylic annulation, using protected 2-methylenepropane-1,3-diol, palladium catalyzed, type A polyprenylated acylphloroglucinol analogue synthesis 281, 282
Phloroglucinols, acyl-, methoxy substituted, reaction with 3-formyl-2-methylbut-3-en-2-yl acetate/base, then acid-catalyzed 1,4-conjugate addition/demethylation/aldol cyclization domino sequence, type A polyprenylated acylphloroglucinol adamantane core synthesis 257, 258
Phloroglucinols, acyl-, polyprenylated, natural products, biosynthesis from dearomatized adduct 229–231
Phloroglucinols, acyl-, reaction with α-acetoxymethyl acrylate/base, via tandem alkylative dearomatization/annulation, polysubstituted bicyclo[3.3.1]nonane synthesis 255, 256
→ Phloroglucinols, acyl-, type A polyprenylated, adamantane core, from reaction of 4-methoxy-substituted acylphloroglucinols with 3-formyl-2-methylbut-3-en-2-yl acetate/base, then acid-catalyzed 1,4-conjugate addition/demethylation/aldol cyclization domino sequence 257, 258
→ Phloroglucinols, acyl-, type A polyprenylated, analogues, from domino dearomative conjunctive allylic annulation of methoxy-substituted acylphloroglucinols, using protected 2-methylenepropane-1,3-diol, palladium catalyzed 281, 282
→ Phloroglucinols, acyl-, type A polyprenylated, from alkylative dearomatization of deoxycohumulone, then acylation and mesylation/intramolecular cyclization 259

→Phloroglucinols, acyl-, type B polyprenylated, adamantane core, from reaction of clusiaphenone B with 3-formyl-2-methylbut-3-en-2-yl acetate/base, via tandem alkylative dearomatization/intramolecular Michael addition/aldol cyclization 256, 257

→Phloroglucinols, acyl-, type B polyprenylated, analogues, from domino dearomative conjunctive allylic annulation of acylphloroglucinols, using protected 2-methylenepropane-1,3-diol, palladium catalyzed 282, 283

→Phloroglucinols, acyl-, type B polyprenylated, from reaction of clusiaphenone B with α-acetoxymethyl acrylate/base, via tandem alkylative dearomatization/annulation 254, 255

Phloroglucinols, intermolecular alkylative dearomatization, via allylation, using allyl alcohol, palladium catalyzed, triethylborane activated, 2,2,4,4,6,6-hexaallylcyclohexane-1,3,5-trione synthesis 236, 237

→Phomoidride B, key intermediate, from α-dienyl-β-oxo esters, via alkenyl Grignard reagent addition/anionic oxy-Cope rearrangement/Dieckmann cyclization domino sequence 200, 201

Phosphonodithioacetates, reaction with α,β-unsaturated aldehydes, via Knoevenagel condensation/thia-electrocyclization domino reaction, 5-phosphono-substituted thiopyran synthesis 154, 155

Phosphoranes, imino-, reaction with α,β-unsaturated carbonyl compounds, via aza-Wittig/6π-electrocyclization tandem reaction, fused pyridine synthesis 143

→Pinnatoxin A, key intermediate, from vinylic sulfoxides, via Claisen rearrangement/[2,3]-sigmatropic rearrangement cascade 187, 188

Piperidines, 4-(bromomethylene)-3-methylene-, tetracyclic, tandem Suzuki cross coupling/6π-electrocyclization, using alkenylboronates, palladium catalyzed, reserpine pentacyclic core synthesis 93, 94

→Piperidines, cyclobutane fused, from 1,7-enyne benzoates, via 1,3-migration/intramolecular [2+2] cycloaddition cascade, cationic gold complex catalyzed 56

→Piperidines, highly substituted, from intramolecular cyclization of amine-substituted propargylic ester to give carbonyl ylide, gold catalyzed, then intramolecular dipolar cycloaddition with terminal alkene, and hydrolysis 72, 73

→Piperidines, 1-pentyl-, from hydroaminomethylation of but-2-ene, via isomerization to but-1-ene, rhodium/bisphosphine catalyzed, then conversion to pentanal using carbon monoxide, amination using piperidine, and hydrogenation 297, 298

→Piperidin-2-ols, chiral, from enantioselective reaction of N-sulfonylbuta-1,3-dienes with enamines, chiral proline derivative catalyzed, via inverse-electron-demand aza-Diels–Alder domino sequence 398, 399

→Polycyclic compounds, from three-component reaction of (2-isocyanoethyl)benzenes with oxocarboxylic acids and aminoacetaldehyde dimethyl acetal, via Ugi/Pictet–Spengler sequence 424, 425

→6,4,8,5-Polycyclic systems, from silylalkynyl-substituted cyclic vinyl bromides, via cyclocarbopalladation/Stille cross-coupling/8π-electrocyclization sequence 101

Polyenes, bromonium-initiated cyclization cascade, using bromodiethylsulfonium bromopentachloroantimonate, aplysin-20 intermediate synthesis 327, 328

Polyenes, halo-initiated cyclization, general challenges 325

Polyenes, reaction with bis(pyridine)iodonium tetrafluoroborate/tetrafluoroboric acid, via iodonium-initiated rearrangement/carbocyclization domino sequence, 2-iodohexahydroxanthene synthesis 320, 321

Polyenes, sulfenium-initiated cyclization cascade, using methyl phenyl sulfoxide/boron trifluoride, 2-(phenylsulfanyl)octahydrophenanthrene synthesis 330

→(+)-Polygalolides, tricyclic intermediate, from conversion of alkenyl-substituted diazo ketones into carbonyl ylides, rhodium catalyzed, then intramolecular cyclization/cycloaddition 67, 68

Polyprenoids, asymmetric iodonium-initiated polyene cyclization cascade, using N-iodosuccinimide, chiral phosphoramidite mediated, then cyclization using chlorosulfuric acid, chiral 2-iodooctahydrophenanthrene synthesis 326, 327

→Praziquantel, derivatives, from four-component reaction of (2-isocyanoethyl)benzenes with aldehydes, carboxylic acids, and aminoacetaldehyde dimethyl acetal, via Ugi/Pictet–Spengler sequence 425, 426

Propanal, enantioselective reaction with nitroalkenes and glyoxylates, chiral proline derivative catalyzed, via Michael/intermolecular Henry/hemiacetalization domino sequence, chiral functionalized tetrahydropyran-2-ol synthesis 399, 400

Propanal, trimerization, via domino enantioselective aldol/aldol/hemiacetalization sequence, proline catalyzed, chiral lactol synthesis 395

Propargyl alcohols, alkenyl/cyclobutenyl substituted, conversion into sulfenate ester, then [2,3]-sigmatropic rearrangement/intramolecular Diels–Alder cascade, sterpurene intermediate synthesis 18, 19, 186, 187

Propargyl alcohols, attached to bridged bicyclic skeleton, carbomagnesiation/6π-electrocyclization cascade, using vinylmagnesium chloride, tricyclic cyclohexadiene synthesis 115, 116
Propargyl ene reaction/Diels–Alder cycloaddition cascade, intramolecular, using hexa-1,6-diynes 161, 162
Propargylic carbonates, alkynyl substituted, tandem Suzuki cross coupling/6π-electrocyclization/aromatization, using arylboronic acids, palladium catalyzed, naphtho[2,3-c]furan synthesis 94, 95
Propargylic esters, amine substituted, intramolecular cyclization to give carbonyl ylide, gold catalyzed, then intramolecular dipolar cycloaddition with terminal alkene and hydrolysis, highly substituted piperidine synthesis 72, 73
Propargylic esters, indole substituted, intramolecular [3,3]-sigmatropic rearrangement/[2+2] cycloaddition, gold complex catalyzed, tetracyclic γ-lactone-fused cyclobutane synthesis 53, 54
Propargylic ethers, 1,3-dienyl substituted, intramolecular [2+2] cycloaddition/[3,3]-sigmatropic rearrangement, base catalyzed, tricyclic fused furan synthesis 52, 53
Propenal, cyclization with alkylidenecyclohexanes, via intermolecular ene reaction/intramolecular ene reaction cascade annulation, octahydronaphthalen-1-ol synthesis 164, 165
Propenal, cyclization with alkylidenecyclopentanes, via intermolecular ene reaction/intramolecular ene reaction cascade annulation, hexahydroinden-7-ol synthesis 164, 165
→ Pseudomonic acid A, key intermediate, from reaction of nonconjugated diene with formaldehyde, via ene reaction/hetero-Diels–Alder cycloaddition cascade 183, 184
→ Pseudotabersonine, key synthetic step, vinylogous elimination reaction of spirocyclic amino alcohol, 4-toluenesulfonic acid catalyzed, then intramolecular Diels–Alder reaction 8, 9
→ Pseudotabersonine, spirocyclic intermediate, from dienyl/indol-2-one-functionalized azanorbornene, via retrocycloaddition/Diels–Alder cycloaddition, Lewis acid catalyzed 27
→ Pupukeanolide D, chloro-, key intermediate, from reaction of vinylallenes with spirocyclic dienes, via Diels–Alder cycloaddition/ene reaction cascade 184, 185
→ Pyran-5-carboxylates, substituted, from reaction of 2-alkyl-2-enals with β-keto esters, via Knoevenagel condensation/6π-electrocyclization domino reaction 150, 151
→ Pyrano[3,2-*b*]carbzaoles, from reaction of 2-hydroxycarbazoles with α,β-unsaturated aldehydes, via aldol reaction/oxa-electrocyclization cascade reaction 152, 153
→ Pyran-2-ols, tetrahydro-, functionalized, chiral, from enantioselective reaction of propanal with nitroalkenes and glyoxylates, chiral proline derivative catalyzed, via Michael/intermolecular Henry/hemiacetalization domino sequence 399, 400
Pyran-2-ones, functionalized, tandem Ireland–Claisen/intramolecular Diels–Alder cascade, using N,O-bis(trimethylsilyl)acetamide/triethylamine, transtaganolide core synthesis 27, 28
Pyran-2-ones, 6-substituted, reaction with N,O-bis(trimethylsilyl)acetamide/triethylamine, via Ireland–Claisen rearrangement/Diels–Alder cycloaddition domino sequence, basiliolide/transtaganolide intermediate synthesis 219, 220
Pyran-2-ones, terminal alkyne substituted, thermal Diels–Alder/retro-Diels–Alder cascade, then potassium carbonate treatment, haouamine A synthesis 33
Pyran-3-ones, acetoxy-, enal substituted, tandem intramolecular [5+2] cycloaddition/conjugate addition, tetracyclic caged lactol synthesis 86, 87
Pyran-4-ones, terminal alkene tethered, intramolecular [5+2] cycloaddition, then intermolecular [4+2] cycloaddition with 1,3-dienes, tetracyclic oxygen heterocycle synthesis 87
Pyran-4-ones, terminal alkyne tethered, intramolecular [5+2] cycloaddition, then intermolecular [4+2] cycloaddition with 1,3-dienes, tetracyclic oxygen heterocycle synthesis 87, 88
→ Pyran-4-ones, tetrahydro-, polysubstituted, from reaction of α-acetoxy ethers with Lewis acids, via ionization/oxonia-Cope rearrangement/oxocarbenium ion trapping domino sequence 202
→ Pyrano[3,2-*b*]pyridines, perhydro-, from amidoallylboronates, via domino hydroformylation/allylboration/hydroformylation sequence, using hydrogen/carbon monoxide, rhodium/bis(phosphite) catalyzed 318, 319
→ Pyrano[3,2-*c*]pyridin-5-ones, from reaction of 4-hydroxypyran-5-carboxylates with enals, via Knoevenagel condensation/oxa-electrocyclization domino reaction 151, 152
→ Pyrans, 3-acyltetrahydro-, polysubstituted, from reaction of vinyl ethers with methyl vinyl ketone, Lewis acid catalyzed, via Mukaiyama reaction/Michael addition/oxonia-Cope rearrangement/oxocarbenium ion trapping domino sequence 203, 204
→ Pyrans, 3,4-dihydro-, multifunctionalized, chiral, from enantioselective [4+2] reaction of β,γ-unsaturated α-oxo esters with 4,4,4-trifluoroacetoacetates, chiral thiourea derivative catalyzed, via domino Michael/hemiacetalization sequence 409

Pyrans, 3,4-dihydro-, polysubstituted, reaction with triethyl phosphite/base, via Claisen rearrangement/Mislow–Evans transposition domino sequence, polysubstituted cyclohex-2-enol synthesis 215, 216
→ Pyrazoles, 3,4,5-trisubstituted, from reaction of enol trifluoromethanesulfonates with diazoacetates, palladium catalyzed, via tandem cross coupling/electrocyclization 136, 137
→ Pyrazolo[5,1-*a*]isoquinolines, from hydrazine addition to 2-(2-bromoethyl)benzaldehyde, then intramolecular cyclization to give azomethine imine, and intermolecular dipolar cycloaddition with fumarate 80, 81
→ Pyrazolo[5,1-*a*]isoquinolines, from reaction of 2-alkynylbenzaldehydes with tosylhydrazine, silver catalyzed, then intramolecular cyclization to give azomethine imine, intermolecular dipolar cycloaddition with acrylate, and aromatization 80
→ Pyridazines, 1,4-dihydro-, 1,3,4-trisubstituted, chiral, from asymmetric reaction of 1-aryl-2-(2,2,2-trifluoroethylidene)hydrazines with α,β-unsaturated aldehydes, chiral phosphonic acid catalyzed, via iminium ion formation/6π-electrocyclization 141–143
→ Pyridine *N*-oxides, polysubstituted, from α,β-unsaturated *O*-propargylic oximes, via intramolecular [2,3]-sigmatropic rearrangement/6π-electrocyclization tandem reaction, copper catalyzed 137, 138
→ Pyridines, 1,2-dihydro-, from 3-aza-1,5-enynes, via thermal aza-Claisen rearrangement/6π-electrocyclization domino reaction 147
→ Pyridines, 1,2-dihydro-, polysubstituted, from reaction of α,β-unsaturated *N*-arylimines with internal alkynes, cobalt catalyzed, via C—H bond activation/migratory insertion/6π-electrocyclization cascade 134, 135
→ Pyridines, 1,6-dihydro-, substituted, from reaction of propargyl vinyl ethers with primary amines, via propargyl Claisen rearrangement/imine formation/tautomerization/6π-electrocyclization/aromatization domino sequence 223, 224
→ Pyridines, 2,4-disubstituted, from α,β,γ,δ-unsaturated aldehydes, via imine formation/electrocyclization/aromatization 139, 140
→ Pyridines, fused, from reaction of iminophosphoranes with α,β-unsaturated carbonyl compounds, via aza-Wittig/6π-electrocyclization tandem reaction 143
→ Pyridines, polysubstituted, from Blaise reaction of nitriles with bromoacetates, zinc catalyzed, then chemoselective addition with 1,3-enynes and isomerization/6π-electrocyclization/aromatization cascade 144, 145
→ Pyridines, polysubstituted, from reaction of α,β-unsaturated imines with alkynes, rhodium/P,N-ligand catalyzed, via C—H alkenylation/electrocyclization cascade, then aromatization 131, 132
→ Pyridines, polysubstituted, from reaction of α,β-unsaturated ketoxime *O*-pentafluorobenzoates with alkenylboronic acids, copper catalyzed, via C—N cross coupling/electrocyclization/aromatization cascade 131
→ Pyridines, polysubstituted, from reaction of α,β-unsaturated ketoximes with alkynes, rhodium catalyzed, via C—H alkenylation/electrocyclization/aromatization cascade 132, 133
→ Pyridines, tricyclic, from diyne nitriles, via propargyl ene reaction/Diels–Alder cycloaddition cascade 163
Pyridin-2-ones, 4-hydroxy-, reaction with enals, via Knoevenagel condensation/oxa-electrocyclization domino reaction, pyrano[3,2-c]pyridin-5-one synthesis 151, 152
→ Pyrimidines, 2,4,6-trisubstituted, from reaction of dibromo oxime ethers with Grignard reagents, via 1,5-diazatriene formation/6π-electrocyclization/elimination domino sequence 144
→ Pyrimido[4,5-*b*]quinolin-4-ones, from reaction of 2-aminoquinoline-3-carboxamides with dimethylformamide dimethyl acetal 444, 445
→ Pyrroles, tetracyclic, from allyl-substituted tricyclic pyrroles, via N-oxidation, elimination, and oxidation to nitrone, using 3-chloroperoxybenzoic acid, then intramolecular cycloaddition 60, 61
Pyrroles, tricyclic, allyl substituted, N-oxidation, elimination, and oxidation to nitrone, using 3-chloroperoxybenzoic acid, then intramolecular cycloaddition, tetracyclic pyrrole synthesis 60, 61
→ Pyrrole-2-thiones, substituted, from reaction of ethynyl propargyl sulfoxides with amines, via Mislow–Evans reaction/[3,3]-sigmatropic rearrangement cascade 171, 172
→ Pyrrolidine-2,5-diones, substituted, from Ugi four-component reaction of β-oxo esters with α-amino acids, isocyanides, and amines 431, 432
→ Pyrrolidines, 3-acyl-, substituted, from condensation of homoallylic amines with aldehydes, via iminium ion formation/azonia-Cope rearrangement/enamine nucleophilic trapping domino sequence 205, 206
→ Pyrrolidines, 3-allyl-, chiral, from intramolecular asymmetric domino hydroamination/cyclization of dienamines, chiral diaminobinaphthyl catalyzed 294, 295
→ Pyrrolidines, polysubstituted, chiral, from stereoselective reaction of chiral unsaturated β-amino esters with iodine, via iodoamination/deprotection domino cascade 304

→ Pyrrolidines, substituted, from reaction of methylenecyclopropanes with 4-toluenesulfonamide, via intermolecular domino ring opening/ring closing hydroamination sequence, gold catalyzed 296, 297
→ Pyrrolizines, hexahydro-, from dialkenylamines, via intramolecular domino hydroamination/cyclization sequence, organolanthanide catalyzed 295, 296
→ Pyrrolo[1,2-*c*]imidazol-3-ones, hexahydro-, from domino intramolecular diamination of 1-alk-4-enylureas, iodonium salt catalyzed 299
→ Pyrrolo[3,4-*d*]isoxazoles, from ene reaction of allylbenzene with nitrosobenzene, copper catalyzed, oxidation to nitrone, and intermolecular cycloaddition with N-substituted alkylmaleimide 60

Q

→ Quinazolines, substituted, from carbodiimide imines, via [1,5]-sigmatropic rearrangement/6π-electrocyclization cascade 179, 180
→ Quinoline-3-carboxamides, 2-amino-, from reaction of 2-aminobenzaldehydes with cyanoacetamides, via Friedländer annulation 442, 443, 445
*Quinoline-3-carboxamides, 2-amino-, reaction with dimethylformamide dimethyl acetal, pyrimido[4,5-*b*]quinolin-4-one synthesis* 443–445
Quinolines, bis(enyne) acid/alkenyl alcohol substituted, Yamaguchi macrolactonization, using 2,4,6-trichlorobenzoyl chloride, then proximity-induced intramolecular Diels–Alder cyclization, polycyclic dynemicin A intermediate synthesis 29
→ Quinolines, from azaenynes, via base-catalyzed propargyl–allenyl isomerization/aza-electrocyclization domino reaction 148
→ Quinolines, substituted, from aryl ketene imines, via [1,5]-sigmatropic rearrangement/6π-electrocyclization cascade 179, 180
→ Quinolines, 4-sulfonamido-, from reaction of alkynyl *N*-aryl imines with sulfonyl azides, copper catalyzed, via 1,3-dipolar cycloaddition/ketenimine formation/6π-electrocyclization/1,3-hydrogen shift domino sequence 136
→ Quinoline-2-thiones, from reaction of 2-isocyanostyrenes with sulfur, selenium catalyzed, then 6π-electrocyclization 141
→ Quinolin-4-ones, from 2-(ketenimino)thiobenzoates, via thermal 1,5-migration/6π-electrocyclization domino reaction 146
Quinones, tricyclic, Diels–Alder intermolecular cycloaddition with doubly skipped triene, Lewis acid catalyzed, then alkene isomerization/Prins cyclization/etherification cascade, polycyclic perovskone intermediate synthesis 40, 41

R

→ Reboxetine, key intermediate, from multicomponent reaction of 2-(chloromethyl)oxirane with ethene and 4-nitrobenzenesulfonamide, via bromonium ion/oxonium ion/ring opening domino sequence, then ring closure 313, 314
→ Reserpine, pentacyclic core, from 4-(bromomethylene)-3-methylenepiperidines, via tandem Suzuki cross coupling/6π-electrocyclization, using alkenylboronates, palladium catalyzed 93, 94
→ Rishirilide B, key intermediate, from thermal reaction of trisiloxy-substituted benzocyclobutene with cyclic enedione, then dehydration, via intermolecular Diels–Alder cascade 13, 14
→ (+)-Rubrenolide, key intermediate, from double domino intramolecular iodoetherification of chiral diene acetals, using *N*-iodosuccinimide 314, 315

S

Salicylaldehyde, enantioselective reaction with 3-methylbut-2-enal, chiral proline derivative catalyzed, via domino aldol/Michael/hemiacetalization sequence, chiral tricyclic adduct synthesis 396, 397
→ Seychellene, from reaction of 2,6-dimethylphenolate with dienyl bromide, via tandem alkylative dearomatization/[4+2] cycloaddition, then hydrogenation, methyllithium addition, and acid-catalyzed rearrangement 252
→ Shimalactones, from 1-stannyl-1,3,6-trienes, via Stille cross-coupling/8π-electrocyclization/6π-electrocyclization domino sequence, using 1-iodo-1,3,5-trienes, palladium catalyzed 105, 106
→ Silanes, allyl-, chiral, from 1,3-dienes, via domino asymmetric reductive aldol reaction with aldehydes and silylboranes, nickel/chiral phosphoramidite catalyzed 392
Silyl ethers, dienyl, silanol elimination, Lewis acid catalyzed, then intramolecular Diels–Alder macrocyclization, abyssomycin C intermediate synthesis 9, 10
→ SNF4435 C/D, from 1-stannyl-1,3-dienes, via Stille cross-coupling/8π-electrocyclization/6π-electrocyclization domino sequence, using 1-iodo-1,3-dienes, palladium catalyzed 103–105
→ Snyderols, from cyclization of terpenes, vanadium bromoperoxidase catalyzed 326
→ Sorbic acid, from [2+2] cycloaddition of but-2-enal with ketene, palladium complex catalyzed, then allylic rearrangement, hydrolysis, and saponification 54, 55

→ Spiro[benzofuranone-cyclohexenes], chiral, from enantioselective three-component reaction of α,β-unsaturated aldehydes with benzofuranones and linear aldehydes, chiral diarylprolinol silyl ether catalyzed, via domino Michael/Michael/aldol/dehydration sequence 352
→ Spiro[cyclohexane-oxindoles], chiral, from enantioselective three-component reaction of oxindoles with nitroalkenes and α,β-unsaturated aldehydes, chiral diarylprolinol silyl ether and chiral thiourea catalyzed, via Michael/Michael/aldol domino sequence 364
→ Spiro[cyclohexanone-oxindoles], chiral, from enantioselective reaction of α,β-unsaturated ketones with oxindoles, chiral *Cinchona*-based alkaloid catalyzed, via Michael/Michael domino sequence 351
→ Spiro[cyclohexene-oxindoles], chiral, from enantioselective reaction of Nazarov reagents with methyleneindolinones, chiral urea catalyzed, via domino Michael/Michael sequence 357, 358
→ Spiro[cyclohexene-oxindoles], chiral, from enantioselective reaction of oxindoles with α,β-unsaturated aldehydes, chiral diarylprolinol silyl ether catalyzed, via domino Michael/Michael/aldol sequence 359, 360
→ Spiro[cyclohexene-oxindoles], chiral, from enantioselective reaction of α,β-unsaturated aldehydes with oxindoles and linear aldehydes, chiral diarylprolinol silyl ether catalyzed, via triple domino cascade 352
→ Spiro[cyclohexenone-oxindoles], chiral, from enantioselective reaction of indoles with α,β-unsaturated ketones, chiral alkaloid catalyzed, via domino Michael/ketone aldol/dehydration sequence 359
→ Spiro[cyclopentane-oxindoles], chiral, from enantioselective reaction of 3-substituted oxindoles with nitroalkenes, chiral *Cinchona*-based alkaloid catalyzed, via Michael/Henry cascade sequence 361, 362
→ Spiro[cyclopentane-oxindoles], chiral, from enantioselective reaction of α,β,γ,δ-unsaturated dienones with oxindoles, chiral *Cinchona*-based alkaloid catalyzed, via vinylogous organocascade process 353
→ Spiro[cyclopropane-oxindoles], chiral, from enantioselective reaction of 3-methyleneoxindoles with bromo(nitro)methane, chiral thiourea-quinidine catalyzed, via domino sequence 360, 361
→ Spiroindanes, phenylselanyl substituted, from trienols, via selenium-initiated polyene cyclization cascade, using benzeneselenenyl chloride/silver hexafluorophosphate 329, 330
→ Spiro[indenone-oxindoles], chiral, from enantioselective three-component reaction of 3-hydroxy-4-(2-oxoindolin-3-ylidene)but-2-enoates with α,β-unsaturated aldehydes, chiral diarylprolinol silyl ether catalyzed, via domino Michael/Michael/Michael/aldol sequence 355, 356
→ Spiro[oxindole-pyrrolidines], chiral, from enantioselective reaction of 3-methyleneindol-2-ones with α-isothiocyanato imides, chiral thiourea catalyzed, via Michael/cyclization domino sequence 363, 364
Spirooxindoles – *see* Oxindoles, spirocyclic
→ Spiro[pyran-oxindoles], chiral, from enantioselective three-component reaction of isatin derivatives with malononitrile and 1,3-diketones, chiral cupreine catalyzed, via domino Knoevenagel/Michael/cyclization process 354, 355
→ Spiropyrrolenine, from tryptamine-based functionalized enaminone, via intramolecular [2+2]-photocycloaddition/retro-Mannich tandem sequence 48
Squalene, biogenetic conversion into hopene, hypothetical representation 322
*Squalene diol, Swern oxidation, then reaction with methylamine and cyclization/Diels–Alder reaction/hydride transfer cascade, 1,2-dihydro-*proto-*daphniphylline synthesis* 6
→ Stemofoline, didehydro-, precursor, from intramolecular cyclization of alkenyl(imino)-substituted diazo ketones to give azomethine ylides, rhodium catalyzed, then acid-catalyzed isomerization and intramolecular dipolar cycloaddition 76
→ Stephacidin B, key intermediate, from (propargyloxy)benzenes, via thermal propargyl Claisen rearrangement/[1,5]-hydrogen shift/6π-electrocyclization domino sequence 221
→ Steroids, partial ring skeleton, from allyl vinyl ethers, via thermal intramolecular [3,3]-sigmatropic rearrangement/ene reaction cascade 185
→ Steroids, *para*-substituted cyclohexadienone, from intermolecular alkylative dearomatization of β-estradiol, osmium complex mediated, via reaction with methyl vinyl ketone and decomplexation 232, 233
→ Steroids, tetracyclic framework, from aryl ketones, via photoinduced [1,5]-hydrogen sigmatropic rearrangement/intramolecular Diels–Alder cycloaddition cascade 190, 191
→ (+)-Sterpurene, key tricyclic intermediate, from alkenyl-/cyclobutenyl-substituted propargyl alcohol, via conversion into sulfenate ester, then [2,3]-sigmatropic rearrangement/intramolecular Diels–Alder cascade 18, 19, 186, 187

→ Strychnine, key intermediate, from condensation of homoallylic amino alcohols and formaldehyde, via aza-Cope rearrangement/Mannich reaction domino sequence 205–207
Styrenes, 2-chloro-, reaction with aniline, via base-catalyzed domino hydroamination/aryne cyclization sequence, 1-phenyl-2,3-dihydroindole synthesis 294
Styrenes, enantioselective domino intermolecular diamination, using N-(methylsulfonyl)methanesulfonamide, chiral iodine(III) reagent catalyzed, chiral N,N,N′,N′-tetramesyl-1-phenylethane-1,2-diamine synthesis 299, 300
Styrenes, 2-isocyano-, reaction with sulfur, selenium catalyzed, then 6π-electrocyclization, quinoline-2-thione synthesis 141
Styrenes, multicomponent aminoalkoxylation with cyclic ethers/N-bromosuccinimide/4-nitrobenzenesulfonamide, via bromonium ion/oxonium ion/ring opening domino sequence, then ring closure, substituted bromo ether synthesis 312, 313
Sulfenates, propargylic, intramolecular [2,3]-sigmatropic rearrangement/Diels–Alder cycloaddition cascade, (+)-sterpurene tricyclic intermediate synthesis 18, 19, 186, 187
Sulfonamides, aryl dienyl-, oxidative amidation/intramolecular Diels–Alder cyclization/epimerization cascade, using (diacetoxyiodo)benzene then heat, tetracyclic himandrine intermediate synthesis 24, 25
Sulfoxides, allylic, radical abstraction, using tributyltin radical, then [2,3]-sigmatropic rearrangement/radical ring expansion cascade, substituted cyclohexanone synthesis 177, 178
Sulfoxides, ethynyl propargyl, reaction with amines, via Mislow–Evans reaction/[3,3]-sigmatropic rearrangement cascade, substituted pyrrole-2-thione synthesis 171, 172
Sulfoxides, vinylic, Claisen rearrangement/[2,3]-sigmatropic rearrangement cascade, pinnatoxin A intermediate synthesis 187, 188

T

→ (±)-Tacamonine, polycyclic intermediate, from intramolecular cyclization of amide-substituted diazo ketone to give carbonyl ylide, rhodium catalyzed, then intramolecular cycloaddition with alkene 69
Tamiflu – *see also* (−)-Oseltamivir
→ Tamiflu, key synthetic step, domino bromoamidation of 5-aminocyclohexa-1,3-diene-1-carboxylates, using *N*-bromoacetamide, Lewis acid catalyzed 300, 301
→ Taxol, key intermediate, from reaction of bicyclic tertiary carbinol with base, then iodomethane, via anionic oxy-Cope rearrangement/enolate alkylation domino sequence 196, 197
Terpenes, cyclization, vanadium bromoperoxidase catalyzed, snyderol synthesis 326
→ *ortho*-Terphenyls, from 2-(bromomethyl)penta-2,4-dienoates, via Wittig alkenation/6π-electrocyclization/oxidation sequence 123, 124
→ Tetracycles, aromatic, strained, from bromodienynes, via tricyclization cascade sequence, palladium catalyzed 97, 98
→ Tetradecomycin, key intermediate, from allyl ethers, via thermal oxy-Cope rearrangement/ene reaction/Claisen rearrangement domino sequence 211
Tetraenediynes, twofold Lindlar hydrogenation/8π-electrocyclization cascade, then intramolecular Diels–Alder cycloaddition, tetracyclic endiandric ester A synthesis 112, 113
1,2,4,7-Tetraenes, thermal 1,5-hydrogen shift/8π-electrocyclization domino sequence, bicyclic cyclooctatriene synthesis 117, 118
1,3,5,7-Tetraenes, acyclic, domino 6π-electrocyclization/intramolecular Diels–Alder cycloaddition, tricyclooct-3-ene synthesis 126, 127
1,3,5,8-Tetraenes, in situ generated, 1,7-hydrogen shift/8π-electrocyclization domino sequence, tetracyclic cyclooctatriene synthesis 120
Tetraynes, hexadehydro-Diels–Alder reaction, manganese dioxide promoted, then cyclization/Brook rearrangement, functionalized fused tricycle synthesis 42
1,2,4,5-Tetrazine-3,6-dicarboxylate, twofold hetero-Diels–Alder reaction with substituted 1,3-diene, then double retro-hetero-Diels–Alder reaction, isochrysohermidin intermediate synthesis 33, 34
→ Thieno[3,4-*c*]thiophenes, from dipropargyl disulfides, via thermal [2,3]-/[2,3]-/[3,3]-sigmatropic rearrangements/double Michael addition/aromatization cascade 172, 173
Thiobenzoates, 2-(ketenimino)-, thermal 1,5-migration/6π-electrocyclization domino reaction, quinolin-4-one synthesis 146
→ Thiophenes, 2-amino-, substituted, chiral, from Gewald reaction of chiral N-protected β-amino aldehydes with chiral cyanoacetamides 441
→ Thiophenes, 2-amino-, substituted, from Gewald multicomponent reaction, using cyanoacetic acid derivatives, α-acidic carbonyl compounds, and elemental sulfur 440
→ Thiophenes, 2-amino-, substituted, from Gewald reaction of 1,4-dithiane-2,5-diol with cyanoacetamides 440
→ Thiophenes, tetrahydro-, functionalized, chiral, from enantioselective reaction of thiols with α,β-unsaturated aldehydes, chiral diarylprolinol silyl ether derivative catalyzed, via domino Michael/aldol sequence 345, 402, 403
→ Thiophenes, tetrahydro-, functionalized, chiral, from reaction of α,β-unsaturated aldehydes with β,γ-unsaturated thiols, chiral diarylprolinol silyl ether catalyzed, via enantioselective domino Michael/Michael sequence 346

→Thiopyrans, 5-phosphono substituted, from reaction of phosphonodithioacetates with α,β-unsaturated aldehydes, via Knoevenagel condensation/thia-electrocyclization domino reaction 154, 155

→Torreyanic acid ester, from oxidation/electrocyclization/intramolecular Diels–Alder cascade of quinone monoepoxide-containing allylic alcohol, using Dess–Martin periodane 14, 15

→Transtaganolide core, key intermediates, from reaction of functionalized pyran-2-ones with *N,O*-bis(trimethylsilyl)acetamide/triethylamine, via tandem Ireland–Claisen rearrangement/intramolecular Diels–Alder cycloaddition domino sequence 27, 28, 219, 220

Triazatriynes, intramolecular propargyl ene reaction/Diels–Alder cycloaddition cascade, tetracyclic nitrogen heterocycle synthesis 161

→1,2,3-Triazoles, tricyclic, from reaction of 2-azidobenzenesulfonamido-substituted aldehydes with Bestmann–Ohira reagent to give terminal alkyne, then intramolecular dipolar cycloaddition with azido group 82

→[1,2,3]Triazolo[5,1-*a*]isoindoles, from intermolecular coupling of 1-(azidomethyl)-2-iodobenzene with terminal alkyne, palladium/copper catalyzed, then intramolecular dipolar cycloaddition 81, 82

→Triazolo[1,5-*a*]quinoxalines, from dipolar cycloaddition of alkyne group of 2-iodo-*N*-alkynylideneaniline with sodium azide, copper/proline catalyzed, then N-arylation 83

→Tricyclic adducts, chiral, from α-diazo carbonyl compounds, by rhodium-catalyzed conversion into carbonyl ylides, then enantioselective 1,3-dipolar cycloaddition with aldehydes, chiral scandium complex catalyzed, via domino reaction 388

→Tricyclic adducts, chiral, from α-diazo-β-oxo esters, via domino conversion into carbonyl ylides/enantioselective intramolecular cycloaddition, chiral dirhodium complex catalyzed 388

→Tricyclic adducts, chiral, from enantioselective reaction of salicylaldehyde with 3-methylbut-2-enal, chiral proline derivative catalyzed, via domino aldol/Michael/hemiacetalization sequence 396, 397

→Tricyclic adducts, from α-diazo carbonyl compounds, by copper-catalyzed conversion into carbonyl ylides, then cycloaddition with benzaldehydes, via domino reaction 387

→Tricyclic compounds, fused, functionalized, from hexadehydro-Diels–Alder reaction of tetraynes, manganese dioxide promoted, then cyclization/Brook rearrangement 42

→Tricyclooct-3-enes, from acyclic 1,3,5,7-tetraenes, via domino 6π-electrocyclization/intramolecular Diels–Alder cycloaddition 126, 127

Trideca-2,11-dienedinitrile, 7-oxo-, oxime formation, using hydroxylamine, then intramolecular conjugate addition to give nitrone and intramolecular cycloaddition, tricyclic histrionicotoxin alkaloid precursor synthesis 58, 59

Trienes, azide substituted, reaction with methylaluminum dichloride, Diels–Alder cyclization/Schmidt rearrangement cascade, tricyclic lactam synthesis 37, 38

1,3,7-Trienes, 1-iodo-, Stille cross-coupling/8π-electrocyclization/Diels–Alder cascade, using 1-stannyl-1,3-dienes, palladium/copper thiophene-2-carboxylate catalyzed, (–)-PF-1018 core synthesis 102, 103

Trienols, selenium-initiated polyene cyclization cascade, using benzeneselenenyl chloride/silver hexafluorophosphate, phenylselanyl-substituted spiroindane synthesis 329, 330

Trienones, thermal 4π-electrocyclization/retro-4π-electrocyclization/enolization/8π-electrocyclization/6π-electrocyclization cascade, base catalyzed, bicyclo[4.2.0]octenone synthesis 122, 123

Trienones, thermal enolization/8π-electrocyclization/6π-electrocyclization cascade, (–)-coprinolone core synthesis 121, 122

Trienynes, Lindlar semihydrogenation/6π-electrocyclization cascade, tetracyclic cyclooctatriene synthesis 111, 112

Trienynes, semihydrogenation/8π-electrocyclization/6π-electrocyclization/oxidation/1,4-addition domino sequence, pentacyclic [4.6.4.6]fenestrene synthesis 113, 114

Trienynols, reaction with benzenesulfenyl chloride, via [2,3]-/[2,3]-sigmatropic rearrangements/[1,5]-sigmatropic hydrogen shift cascade, pentaene synthesis 175, 176

→2,3,5-Trioxa-2a-azapentaleno[1,6-*ab*]naphthalenes, from domino process of sulfur ylide addition to nitroalkenes to give nitronates, then intramolecular dipolar cycloaddition 63, 64

→Trioxanes, bicyclic, chiral, from desymmetrization of 4-(hydroperoxy)-4-methylcyclohexa-2,5-dienones, using aldehydes, chiral phosphoric acid/thiourea derivative catalyzed, via domino acetalization/oxa-Michael sequence 410

Triynes, thermal Diels–Alder cyclization/desaturation cascade, involving hexadehydro-Diels–Alder variant, substituted fluoren-9-one synthesis 36

Tsuji–Trost mechanism, decarboxylative allylation of phloroglucinols, palladium catalyzed 238, 239

U

Ugi five-center four-component reaction, using α-amino acids, ketones, isocyanides, and amines, substituted iminodicarboxamide synthesis 428–430

Ureas, 1-alk-4-enyl-, domino intramolecular diamination, iodonium salt catalyzed, hexahydropyrrolo[1,2-c]imidazol-3-one synthesis 299

V

Vinyl bromides, cyclic, alkynyl substituted, cyclocarbopalladation/Sonogashira-type coupling/alkynylation/8π-electrocyclization/6π-electrocyclization domino sequence, using enynes, palladium/copper/phosphine catalyzed, [4.6.4.6]fenestradiene synthesis 107, 108

Vinyl bromides, cyclic, silylalkynyl substituted, cyclocarbopalladation/Stille cross coupling/8π-electrocyclization sequence, 6,4,8,5-polycyclic system synthesis 101, 102

Vinyl bromides, highly functionalized, Stille cross coupling with vinylstannane, palladium catalyzed, then intramolecular Diels–Alder reaction, tricyclic manzamine A intermediate synthesis 20, 21

Vinyl sulfides, [3,3]-sigmatropic rearrangement, then Wittig reaction, oxidation, and [2,3]-sigmatropic rearrangement, (–)-joubertinamine intermediate synthesis 188, 189

→Vitamin D derivative, tetracyclic cyclooctatriene, from trienynes, via Lindlar semihydrogenation/6π-electrocyclization cascade 111, 112

W

Wessely oxidation, dimerization of 3-(trimethylcyclopentane)-substituted phenols, using stabilized 1-hydroxy-1,2-benziodoxol-3-one 1-oxide, aquaticol synthesis, via Diels–Alder reaction 2, 3

Wieland–Miescher ketone, conversion into propargylic ether, then intramolecular [2+2] cycloaddition/[3,3]-sigmatropic rearrangement, base catalyzed, oxataxane skeleton synthesis 52, 53

X

→Xanthenes, 2-iodohexahydro-, from reaction of polyenes with bis(pyridine)iodonium tetrafluoroborate/tetrafluoroboric acid, via iodonium-initiated rearrangement/carbocyclization domino sequence 320, 321

→Xanthenes, tetrahydro-, chiral, from phenol-substituted dienes, via asymmetric cascade cyclization, chiral platinum–phosphine complex catalyzed 324

Xanthen-9-ones, 3,5,6-tris(allyloxy)-1-hydroxy-, thermal aryl Claisen rearrangement/intramolecular Diels–Alder domino sequence, forbesione synthesis 18, 217, 219

Xanthen-9-ones, 3,5,6-tris(allyloxy)-1-methoxy-, thermal Claisen rearrangement/Diels–Alder cycloaddition domino sequence, caged Garcinia *xanthone synthesis* 217, 218

Y

Ylides, sulfur, addition to nitroalkenes to give nitronates, then intramolecular dipolar cycloaddition, tetracyclic 2,3,5-trioxa-2a-azapentaleno[1,6-ab]-naphthalene synthesis 63, 64

Ynamides, N*-allyl-, allyl transfer/intramolecular [2+2] cycloaddition cascade, palladium catalyzed, bicyclic highly substituted imine synthesis* 55, 56

Ynolates, [2+2] cycloaddition with δ-dicarbonyl compounds/Dieckmann condensation sequence, bicyclic β-lactone synthesis 48, 49

Ynones, pyranyl substituted, intermolecular [5+2] cycloaddition with vinylcyclopropanes, rhodium complex catalyzed, then Nazarov cyclization, bicyclo[5.3.0]decane derivative synthesis 86

→(±)-Yohimbenone precursor, from conjugate addition of 2,3-bis(phenylsulfonyl)buta-1,3-diene with indole oxime to give a nitrone, then intramolecular cycloaddition and reduction 59

Author Index

In this index the page number for that part of the text citing the reference number is given first. The number of the reference in the reference section is given in a superscript font following this.

A

Abbiati, G. 65[48], 66[48]
Abdel-Magid, A. F. 376[87]
Abe, H. 77[86], 78[86], 79[86]
Abe, I. 322[138]
Abe, T. 389[8]
Abele, S. 379[101], 380[101], 380[105], 380[106], 380[107], 380[108], 380[109], 381[109], 381[112], 382[112], 382[113], 383[112]
Aben, R. W. M. 219[94]
Abrahams, Q. M. 62[41], 63[41]
Aburel, P. S. 339[20], 339[21], 402[69]
Acebey, L. 31[73]
Achari, B. 82[93]
Adam, F. 190[59], 191[59]
Adam, P. 229[31], 230[31], 231[31]
Adam, W. 160[9]
Adams, H. 58[28], 75[76], 79[76], 80[91], 81[91]
Adams, J. P. 58[31], 59[31]
Adlington, R. M. 121[55], 122[55], 229[9], 248[9], 250[9], 251[9], 263[9], 264[9], 267[9]
Adrio, L. A. 293[6]
Agami, C. 205[45], 206[45]
Agopcan, S. 59[33]
Agrawal, S. 419[4]
Aguilar, E. 50[10]
Aguilar, H. R. 136[80], 137[80]
Ahmad, R. 98[10], 99[10], 101[10]
Ahmed, M. 297[39], 297[40], 297[41], 298[39]
Ahrendt, K. A. 337[13], 395[37]
Airiau, E. 317[113], 319[128], 319[129], 320[128]
Aitken, L. 376[93]
Akrawi, O. A. 98[12], 100[12]
Akritopoulou-Zanze, I. 419[10]
Akssira, M. 217[78]
Alajarin, M. 118[51], 119[51], 120[51], 146[95], 146[96], 146[97], 147[95], 147[97], 179[43], 179[44], 179[45], 179[46], 180[43], 180[44], 180[46], 183[43]
Alba, A. N. 359[53], 360[53]
Albadri, H. 154[117], 155[117]
Albertshofer, K. 361[55], 362[55]
Albini, A. 128[65], 129[65]
Alder, K. 1[1], 27[1]
Alemán, J. 400[57], 401[57], 404[75], 405[75]
Alexakis, A. 401[68]
Alfarouk, F. O. 441[43]
Ali, A. 20[55], 21[55], 22[55]
Allen, J. 83[100]
Allen, J. G. 13[38], 14[38]
Allen, M. J. 419[5]
Almássy, A. 370[69], 371[69]
Alnajjar, M. 196[9]
Alvarez-Pérez, M. 320[132]
Alvernhe, G. 306[76]
Amiri, S. 113[42], 114[42]
Amorelli, B. 390[18]
An, J. 63[45], 63[46], 64[45]
An, X.-L. 63[45], 64[45]
Anada, M. 68[56], 389[7], 389[8], 389[9], 389[10]
Ananikov, V. P. 282[81]
Anderson, E. A. 102[15]
Anderson, R. 70[63], 71[63]
Andreatta, J. R. 293[10]
Andrews, D. R. 197[20], 198[20]
Andrews, G. C. 168[26], 215[72]
Ang, S.-M. 408[96], 409[96]
Angle, S. R. 205[42], 206[42]
Annunziata, R. 376[81]
Antoniotti, S. 293[7]
Antony, A. 293[9]
Aoki, S. 38[82]
Aoyagi, S. 171[33], 172[33], 294[21]
Appella, D. H. 389[13]
Appendino, G. 27[66], 217[79]
Arai, T. 394[32]
Araki, T. 137[82], 138[82]
Arasappan, A. 419[4]
Aratake, S. 399[52]
Arcadi, A. 65[48], 66[48]
Arepally, A. 160[13], 161[13]
Arezki, N. 376[93]
Arigela, R. K. 82[95]
Arigoni, D. 229[31], 230[31], 231[31], 322[142]
Arini, L. G. 58[32], 59[32]
Arisawa, M. 77[86], 78[86], 79[86]
Arman, H. 407[92], 408[92]
Armstrong, A. 205[46], 206[46], 378[96], 378[97], 378[98], 379[96], 379[97], 379[98]
Armstrong, P. 58[30]
Armstrong, R. W. 433[29]
Arnold, L. A. 393[27], 393[28]
Arns, S. 209[57], 211[57], 212[64]
Arrastia, I. 54[19]
Asensio, G. 320[131]
Ashek, L. 201[26]
Aso, K. 52[12], 53[12]
Ates, A. 295[26]
Atodiresei, I. 141[87], 142[87], 143[87]
Aubé, J. 37[80], 38[80], 40[80]
Aviyente, V. 59[33]
Ayaz, M. 431[25]
Aye, M. 307[80], 308[80]

B

Babiash, E. S. C. 50[8], 51[8]
Babinski, D. J. 136[80], 137[80]
Babjak, M. 70[63], 71[63]
Bacher, A. 229[31], 230[31], 231[31]
Backes, B. J. 372[72], 373[72]
Bader, F. E. 1[4]
Badía, D. 404[73]
Badillo, J. J. 349[31]
Badine, D. M. 371[71]
Bai, J.-F. 359[52]
Baier, H. 190[59], 191[59]
Bailey, S. 196[16], 196[17], 197[17], 198[17]
Baire, B. 36[79], 42[87], 43[87]
Baldino, C. M. 33[77], 34[77], 35[77]
Baldwin, J. E. 41[86], 42[86], 121[55], 122[55], 229[9], 248[9], 250[9], 251[9], 263[9], 264[9], 267[9]
Ballaron, S. J. 372[72], 373[72]
Ballero, M. 27[66], 217[79]
Ballini, R. 128[65], 129[65]
Ball-Jones, N. R. 349[31]
Bandaranayake, W. M. 15[42]
Bandini, M. 293[12]
Bandyopadhyay, D. 57[24]
Banerji, A. 57[24]
Banfield, J. E. 15[42]
Baran, P. S. 33[76], 35[76]
Barange, D. K. 83[99]
Barbarow, J. E. 105[23]
Barbas, C. F., III 351[41], 354[46], 358[51], 359[51], 361[55], 362[55], 395[36], 395[38], 399[54], 407[91]
Barcan, G. A. 93[1], 93[2], 94[1], 94[2]
Barkley, J. V. 168[30]
Barluenga, J. 299[49], 320[131], 320[132], 320[133], 321[133]
Baroudy, B. 419[4]
Barriault, L. 185[54], 186[54], 209[56], 209[57], 209[58], 209[59], 210[56], 211[57], 211[60], 211[61], 211[62], 212[60], 212[63], 212[64], 212[65], 213[60], 213[63]
Barrow, J. C. 419[11]
Bartels, K. 18[52], 19[52]
Bartlett, E. S. 229[17]
Bartoli, G. 350[39], 351[39], 351[40], 351[42], 352[39], 360[54], 361[54], 365[54], 401[59]
Basabe, P. 151[108]
Basak, A. 229[23]
Basurto, S. 434[32]
Basutto, J. 78[87], 79[87]
Batova, A. 217[77]
Bats, J. W. 403[72], 404[72]
Baudoin, O. 140[85], 141[85]
Baumgarth, M. 441[47]
Bautista, D. 146[95], 147[95]
Baxendale, I. R. 164[19]
Beaudry, C. M. 103[22], 104[22], 108[22]
Beck, A. K. 371[71]

Beck, S. 31[73]
Becker, J. J. 324[153]
Becker, M. W. 300[59]
Bee, C. 390[16]
Beeler, A. B. 229[7], 260[7], 261[7], 262[7], 263[7]
Behenna, D. C. 239[54], 239[55]
Beier, N. 441[47]
Bella, M. 337[9], 376[9], 377[95], 378[95], 379[100], 381[95]
Beller, M. 293[1], 294[16], 294[17], 294[22], 297[38], 297[39], 297[40], 297[41], 297[43], 298[39], 317[118], 317[120]
Bellus, D. 213[66], 213[67]
Belot, S. 401[68]
Bencivenni, G. 350[39], 351[39], 352[39], 360[54], 361[54], 365[54], 401[59]
Bender, C. F. 293[5]
Bennett, F. 419[4]
Bennett, J. N. 1[14]
Bennett, W. D. 376[92]
Beno, D. W. A. 372[72], 373[72]
Benson, C. L. 115[44]
Bergman, R. G. 131[71], 131[72], 131[73], 132[71], 132[76], 133[76], 134[76]
Bergmeier, S. C. 295[28]
Berkessel, A. 337[2], 376[2], 378[2]
Berrigan, P. J. 240[62], 244[62], 250[62]
Berson, J. A. 196[10]
Bertelsen, S. 338[16], 339[16], 344[16], 345[16], 403[71], 404[75], 405[75]
Besnard, C. 401[68]
Betancort, J. M. 354[46]
Bettray, W. 141[87], 142[87], 143[87]
Bharate, S. B. 229[25], 230[25]
Bhargava, G. 51[11], 52[11]
Bhattacharya, D. 83[101]
Bhonoah, Y. 205[46], 206[46]
Bickel, H. 1[4]
Bickley, J. F. 168[31]
Bigger, S. 196[9]
Billington, R. A. 27[66], 217[79]
Bischofberger, N. 300[58], 365[61]
Bista, M. 436[33]
Black, D. St. C. 15[42]
Blackmond, D. G. 378[96], 378[97], 378[98], 379[96], 379[97], 379[98]
Blechert, S. 380[111], 381[111]
Bleeker, N. 294[19]
Bloch, K. 322[134]
Bloch, W. M. 229[10], 248[10], 250[10], 263[10], 264[10]
Blond, G. 97[7], 100[7], 101[14], 107[27], 107[28], 107[29], 108[27], 113[41], 113[42], 113[43], 114[41], 114[42], 116[41]
Blum, C. A. 34[78], 35[78]
Board, J. 164[18]
Bocknack, B. M. 393[31], 394[31]
Bogdanowich-Knipp, S. 419[4]
Bogen, S. L. 419[4]
Boger, D. L. 33[77], 34[77], 35[77]
Bohlmann, F. 190[58]
Bolla, M. L. 203[28], 204[28]
Bolleddula, J. 295[31]
Bolze, P. 404[75], 405[75]
Bonderoff, S. A. 70[61]
Bondzic, B. P. 370[67]
Bonillo, B. 118[51], 119[51], 120[51], 146[97], 147[97], 179[43], 179[44], 179[45], 179[46], 180[43], 180[44], 180[46], 183[43]
Bonini, C. 143[89]
Bonjoch, J. 376[79]
Bonney, K. J. 306[75], 308[85], 309[85], 310[85]
Börner, A. 317[114]
Borths, C. J. 337[13], 395[37]
Borthwick, A. D. 419[5]
Bossio, R. 431[27]
Botros, S. 426[18], 428[18]
Böttcher, H. 439[38]
Boudhar, A. 107[27], 107[28], 107[29], 108[27]
Boudon, C. 130[66], 130[67], 130[68]
Bour, C. 97[7], 100[7], 101[14]
Bourdet, D. L. 295[31]
Bowler, J. M. 229[23]
Boyce, J. H. 281[80], 282[80], 283[80], 284[80], 284[82], 285[82], 286[82], 287[80], 288[82]
Braddock, D. C. 306[75], 308[82], 308[83], 308[84], 308[85], 309[85], 310[85], 329[180], 329[181], 329[182]
Bradley, P. 419[4]
Bradshaw, B. 376[79]
Bradshaw, J. S. 234[49], 248[49], 250[49]
Brahma, K. 82[93]
Brajeul, S. 234[48], 248[48]
Brandau, S. 338[17], 339[17], 345[17], 346[17], 402[70], 403[70]
Bräse, S. 95[4], 396[42], 397[42]
Brassil, P. J. 295[31]
Brath, H. 370[69], 371[69]
Braverman, S. 172[34], 173[34], 173[35], 174[35], 174[36], 178[34]
Bravo, F. 309[88]
Breindl, C. 294[16], 294[17], 294[22]
Breit, B. 317[111], 317[115], 317[116], 317[122], 317[123], 317[124], 318[122], 318[123], 318[126]
Breiten, B. 130[67]
Brennan, M. K. 350[36]
Brennführer, A. 317[118]
Breugst, M. 248[67], 250[67], 272[67]
Brieger, G. 1[14]
Brisbois, R. G. 222[103]
Brisson, J.-M. 419[4]
Britton, J. E. 40[84], 41[84], 43[84]
Broach, V. 160[11]
Brodbeck, S. 379[101], 380[101]
Brodney, M. A. 70[62]
Bröhmer, M. C. 396[42], 397[42]
Bronger, R. P. J. 297[41]
Brönnimann, R. 382[113]
Brooks, D. P. 419[5]
Broske, L. 419[4]
Brown, T. L. 234[43], 240[43]
Brückner, D. 318[127], 319[127]
Brucks, A. P. 310[89], 310[90], 311[89], 311[90], 327[178], 328[178], 329[178]
Brüdgam, I. 431[24]
Brummond, K. M. 54[16]
Brunner, G. 229[13], 252[13], 253[13], 254[13]
Buchwald, S. L. 389[13]
Burés, J. 378[96], 379[96]
Burger, E. C. 239[56]
Burgstahler, A. W. 1[3], 322[141]
Burk, R. M. 205[39], 206[39]
Burnley, J. 107[26]
Burns, C. J. 388[3]
Burns, N. Z. 33[76], 35[76]
Burrell, A. J. M. 57[23], 58[25], 58[26], 58[27], 61[27], 75[75], 75[76], 75[78], 75[79], 79[76]
Butkiewicz, N. 419[4]
Butler, A. 326[173]

C

Caballero, M. J. A. 59[35]
Cabrera, S. 348[29], 349[29], 400[57], 401[57], 404[75], 405[75]
Cai, Y.-P. 410[99]
Camacho, C. J. 436[33]
Campbell, A. N. 324[152]
Campos, P. J. 320[131]
Campos-Gómez, E. 299[49]
Candeias, N. R. 358[51], 359[51], 407[91]
Canu, F. 168[28]
Cao, B. 297[42]
Cao, C.-L. 400[56]
Cao, J. 135[78]
Cao, Y. 277[79], 363[58], 364[58], 405[84], 405[86]
Cárdenas, J. 229[33], 230[33], 231[33]
Cardona, F. 298[45]
Carlone, A. 348[29], 349[29], 351[40]
Carlsmith, L. A. 23[60]
Carlson, R. G. 1[12]
Carrillo, L. 401[60], 402[60], 404[73]
Carroll, P. J. 411[105], 411[107], 411[108], 412[105]
Carroll, W. A. 8[30], 9[30], 27[30]
Carruthers, W. 1[10]
Carson, C. A. 29[72], 30[72]
Carter-Franklin, J. N. 326[173]
Casas, J. 396[40]
Cases, M. 205[45], 206[45]
Cassani, C. 352[43], 353[43]
Castedo, L. 87[109], 87[110], 88[110], 110[32], 110[33], 111[33]
Castet, F. 2[24], 3[24], 4[24]
Castillo, B. F., II 311[94], 312[94]
Castro, A. 175[38], 176[38]
Cauble, D. F. 393[30], 394[30]
Cavaleiro, J. A. S. 127[64], 128[64]
Cederbaum, F. 419[6]
Çelebi-Ölçüm, N. 62[43], 63[43]
Chai, Z. 410[99]

Chakraborty, S. 389[12]
Chambers, C. S. 82[97]
Chan, B. 327[176]
Chan, C.-K. 124[60]
Chan, P. W. H. 56[21]
Chanas, T. 433[28], 434[28], 435[28], 436[28], 437[28], 438[28], 439[28]
Chandra Pan, S. 419[13]
Chandrasekhar, S. 401[61]
Chang, M. 297[42]
Chang, M.-Y. 124[59], 124[60], 125[59]
Chantarasriwong, O. 217[77]
Chapman, H. H. 300[59]
Chapman, O. L. 21[56], 22[56]
Charpenay, M. 107[27], 107[28], 107[29], 108[27]
Chase, R. 419[4]
Chattopadhyay, P. 83[101]
Chaumontet, M. 140[85], 141[85]
Chavasiri, W. 217[77]
Chávez, P. 299[51], 299[54]
Chemin, C. 317[113]
Chen, C. 199[23], 200[23], 200[25], 201[23], 201[25]
Chen, C.-H. 80[89]
Chen, D. Y.-K. 229[29], 230[29], 232[29]
Chen, F. 312[99]
Chen, H. 60[39], 61[39], 166[25], 167[25], 215[70]
Chen, H.-J. 20[55], 21[55], 22[55]
Chen, J. 116[46], 117[46], 301[65], 301[66], 302[65], 302[66], 312[96], 313[96]
Chen, J. S. 199[24], 200[24]
Chen, K. 419[4]
Chen, L. 362[57], 363[57]
Chen, M. S. 300[58], 365[61]
Chen, Q. 72[70]
Chen, S. 65[49]
Chen, W.-B. 354[47]
Chen, X. 83[98], 84[98]
Chen, Y.-C. 355[49], 356[49], 398[48], 398[49], 398[50], 398[51], 401[58]
Chen, Y.-L. 124[59], 125[59]
Cheng, B. 70[61]
Cheng, C.-C. 442[49]
Cheng, C.-H. 132[74], 132[75], 133[74], 133[75]
Cheng, G. 136[79]
Cheng, J. 94[3], 95[3], 100[3]
Cheng, K.-C. 419[4]
Cheng, X. 148[101], 149[101]
Cheng, X.-M. 43[88]
Cheng, Y. A. 312[99]
Cherkinsky, M. 172[34], 173[34], 174[36], 178[34]
Cherrier, M. P. 436[34]
Chesworth, R. 41[86], 42[86]
Cheung, L. L. W. 149[102], 150[102]
Chiarucci, M. 293[12]
Chida, N. 207[49], 207[50], 207[51], 208[51]
Chiu, P. 66[54], 67[54], 71[54]
Chng, S. 312[96], 313[96]
Choi, J.-H. 441[44]
Chorlton, A. P. 209[52]
Chou, T. 215[70], 215[71]
Chou, T. C. 7[29], 8[29]
Chougnet, A. 396[41]
Chouraqui, G. 68[58]
Chowdari, N. S. 395[38]
Chowdhury, C. 82[93], 217[84], 217[85]
Choy, A. L. 74[74], 75[74]
Choy, W. 14[39]
Christadler, M. 441[47]
Christensen, S. B. 23[62], 24[62]
Christoffers, J. 443[51]
Chua, P. J. 408[94]
Chuang, H.-Y. 61[40]
Chuard, R. 177[41], 178[41]
Chun, K. 441[44]
Chun, S. 382[113]
Chun, Y. S. 144[92], 144[93], 145[92], 150[92]
Churyukanov, V. V. 295[25]
Chuzel, O. 390[17], 391[17], 393[17]
Cini, E. 319[129]
Ciochina, R. 229[24], 230[24]
Ciufolini, M. A. 24[63], 25[63], 26[63]
Claisen, L. 18[47], 195[2], 234[44], 240[44]
Clardy, J. 14[40]
Clardy, J. C. 21[56], 22[56]
Clark, J. S. 388[3]
Clarke, M. L. 159[4], 160[4]
Claver, C. 317[121]
Clay, M. D. 115[45], 116[45]
Clement, J. A. 109[31], 110[31]
Clément, R. 212[65]
Cobb, A. A. 376[93]
Cochrane, N. A. 324[153], 324[154]
Cockfield, D. M. 397[45], 398[45]
Colapret, J. A. 197[20], 198[20]
Colby, D. A. 131[71], 132[71]
Coldham, I. 57[23], 58[25], 58[26], 58[27], 58[28], 61[27], 73[72], 75[75], 75[76], 75[77], 75[78], 75[79], 79[76], 80[91], 81[91]
Collum, D. B. 248[69], 250[69]
Colombano, G. 27[66], 217[79]
Colonna, S. 376[81]
Colussi, D. 419[11]
Combrink, K. D. 196[14], 196[15]
Combrisson, S. 19[53]
Commeiras, L. 68[58]
Companyo, X. 359[53], 360[53]
Cope, A. C. 195[3]
Córdova, A. 395[38], 396[40], 400[55]
Cordova, R. 183[51], 184[51]
Corey, E. J. 1[19], 43[88], 229[16], 259[16], 300[57], 300[63], 301[57], 322[140]
Corres, N. 436[35]
Cossio, F. P. 54[19]
Cotos, L. 221[99], 222[101], 223[101]
Couladouros, E. A. 229[4], 245[4], 246[4], 247[4], 248[4], 250[4], 251[4], 259[4]
Courtney, A. K. 20[55], 21[55], 22[55]
Couty, F. 205[45], 206[45]
Craig, D. 1[17]
Crawford, R. J. 240[58]
Cross, J. L. 419[10]
Crouse, D. M. 178[42], 179[42]
Cuesta, J. A. 320[132]
Cuesta-Rubio, O. 229[33], 230[33], 231[33]
Cui, H.-F. 410[100]
Cui, X. 136[79], 419[4]
Cui, Y. 135[78]
Cun, L.-F. 354[47]
Curran, D. P. 1[18]
Curran, T. T. 442[48]
Curtin, D. Y. 234[42], 234[43], 240[42], 240[43], 240[57], 240[58]
Cusella, P. P. 377[95], 378[95], 381[95]

D

Dachs, A. 161[14], 162[14]
Dahmen, A. 127[63]
Dai, M. 229[14], 229[15]
Dakanali, M. 229[4], 229[27], 230[27], 245[4], 246[4], 247[4], 248[4], 250[4], 251[4], 259[4]
Dalgard, J. E. 202[27], 203[27]
Dalko, P. I. 337[1], 337[6], 376[1], 376[6]
Dalley, N. K. 234[49], 248[49], 250[49]
Dalrymple, D. L. 178[42], 179[42]
Danheiser, R. L. 161[15], 162[15], 163[15], 163[16], 164[15], 167[15], 222[103]
Danishefsky, S. J. 7[29], 8[29], 13[38], 14[38], 229[1], 229[2], 229[14], 229[15], 229[26], 230[26], 237[1], 237[2], 238[2], 239[1], 239[2], 349[34]
da Rocha, D. R. 151[105]
Das, A. 80[89], 141[87], 142[87], 143[87]
da Silva, F. de C. 151[105]
da Silva, I. M. C. B. 151[105]
Davies, D. E. 419[5]
Davies, H. M. L. 195[8]
Davies, J. E. 58[31], 59[31]
Davies, S. G. 304[69], 306[78], 306[79], 307[78], 307[79]
Davis, P. 376[88]
Davis, S. G. 23[62], 24[62]
Davison, E. C. 58[31], 59[31]
Deane, F. M. 434[30]
Defauw, J. 160[13], 161[13]
de Figueiredo, R. M. 298[44]
De Jong, S. 298[46]
de Kiff, A. 164[20]
de Koning, H. 204[29]
Dekorver, K. A. 55[20], 56[20]
Delacroix, O. 293[3]
de Lera, A. R. 175[38], 175[39], 176[38], 176[39]
Dell'Isola, A. 376[93]
Delpech, B. 234[47], 234[48], 244[47], 248[47], 248[48], 250[47]
del Río, I. 317[121]
Demadis, K. D. 229[4], 245[4], 246[4], 247[4], 248[4], 250[4], 251[4], 259[4]

de Meijere, A. 95[4], 95[5], 95[6], 96[5], 100[5], 126[61], 127[61]
DeMent, K. 295[31]
De Mesmaeker, A. 419[6]
Demharter, A. 428[20]
de Miguel, I. 83[103], 83[104], 84[104]
De Moliner, F. 431[25]
Deng, L. 405[81]
Deng, W.-Y. 72[69]
Deng, Y. 135[78]
Denissova, I. 211[60], 211[62], 212[60], 213[60]
Denmark, S. E. 331[188], 331[189]
Deon, D. H. 185[54], 186[54], 209[59]
Deschamp, J. 390[17], 391[17], 393[17]
DeSimone, R. W. 34[78], 35[78]
Deutsch, E. A. 164[21], 164[22], 165[22]
Devan, B. 209[54]
Dhaenens, M. 341[22]
Dhara, S. 139[84], 140[84]
Dhayalan, V. 109[31], 110[31]
Diederich, F. 130[66], 130[67], 130[68]
Diedrich, C. L. 443[51]
Diels, O. 1[1], 27[1]
Díez, D. 151[108]
Dimitroff, M. 39[83], 40[83]
Ding, D. 407[92], 408[92]
Ding, M.-W. 434[31]
Ding, Q. 59[37], 232[36]
Dixon, D. J. 397[45], 398[45], 401[67]
Djuric, S. W. 419[10]
Dodda, R. 406[89]
Dolling, U.-H. 376[88], 376[89]
Domingo, L. R. 87[111]
Dömling, A. 419[2], 419[3], 419[12], 419[15], 420[15], 421[15], 422[15], 423[3], 423[16], 424[16], 424[17], 425[17], 426[12], 426[18], 427[16], 427[19], 428[17], 428[18], 428[19], 429[12], 430[12], 431[21], 431[22], 432[21], 433[21], 433[28], 434[28], 435[28], 436[28], 436[33], 437[12], 437[21], 437[28], 438[21], 438[28], 439[28], 440[39], 440[40], 440[41], 441[41], 441[45], 442[45], 442[50], 443[50], 444[41], 444[45], 444[50], 445[50]
Dondoni, A. 337[8], 376[8]
Dong, S. 3[25], 5[25]
Dong, V. M. 213[68], 213[69], 214[69], 215[69]
Dong, X.-Q. 406[90]
Doye, S. 293[4]
Doyle, M. P. 62[41], 63[41], 388[5]
Drahl, C. 9[31], 10[31], 11[31]
Driscoll, J. P. 295[31]
D'Souza, A. M. 78[87], 79[87]
Du, B. 31[74], 32[74]
Du, W. 398[51]
Duan, H.-F. 392[25]
Dübon, P. 317[125]
Dudley, G. B. 219[92]
Duffey, M. O. 389[14], 390[14]
Duñach, E. 293[7]
Durmis, J. 370[69], 371[69]
Dürner, G. 190[59], 191[59]
Duttwyler, S. 131[73]
Duvall, J. R. 111[34]
Dziedzic, P. 400[55]

E

Eckert-Maksić, M. 121[53]
Edayadulla, N. 151[106]
Eder, U. 337[12], 376[12]
Edstrom, E. D. 330[184]
Edwards, R. M. 419[5]
Eggenweller, H.-M. 441[47]
Eichberger, M. 294[16]
Eisenreich, W. 229[31], 230[31], 231[31]
El Kaïm, L. 419[14]
Ellman, J. A. 131[71], 131[72], 131[73], 132[71], 132[76], 133[76], 134[76]
Elmore, S. W. 196[14], 196[15]
Elsner, P. 338[19], 339[19], 349[19], 404[74]
Enders, D. 195[1], 196[1], 347[27], 348[27], 348[28], 349[27], 350[27], 376[94], 401[64]
Engel, D. B. 419[8]
Engel, M. R. 21[56], 22[56]
England, D. B. 69[59], 71[59]
Engler, T. A. 87[112]
Engqvist, M. 396[40]
Enomoto, H. 441[46]
Ent, H. 204[29]
Er, J. C. 300[64]
Eriksson, L. 400[55]
Eschenmoser, A. 1[5], 322[142]
Escudero-Adán, E. C. 299[51], 352[43], 353[43]
Espeseth, A. S. 419[11]
Evans, D. A. 10[34], 168[26], 196[11], 215[72]
Evans, P. A. 195[7]
Exall, A. M. 419[5]

F

Fagnoni, M. 128[65], 129[65]
Fallis, A. G. 1[16], 115[45], 116[45]
Fan, C.-A. 83[102]
Fan, R. 232[36]
Fañanás, F. J. 320[132]
Fang, B. 72[71], 73[71]
Fang, C. 76[80], 76[81]
Fang, X. 406[90]
Farjas, J. 161[14], 162[14]
Farmen, M. W. 419[8]
Farwick, A. 317[125], 319[130]
Faulkner, D. J. 325[160]
Fédou, N. M. 64[47], 65[47]
Feltenberger, J. B. 50[8], 50[9], 51[8], 180[47], 180[48], 181[47], 182[48]
Feng, J.-J. 85[106]
Feringa, B. L. 393[27], 393[28]
Fernández, S. 111[35], 112[35]
Ferreira, E. M. 389[13]
Ferreira, S. B. 151[105]
Ferreira, V. F. 151[105]
Fettinger, J. C. 330[187], 331[187]
Fevig, J. M. 205[42], 206[42]
Fielenbach, D. 344[25]
Fierz, H. 380[107]
Fisher, E. L. 411[108]
Fleischer, I. 317[120]
Fleming, I. 1[9], 295[27]
Flick, A. C. 59[35]
Flynn, B. L. 120[52]
Foley, W. J. 229[25], 230[25]
Foote, K. M. 329[182]
Forbes, D. C. 388[5]
Forbes, I. T. 58[31], 59[31]
Fotiadou, A. D. 151[109], 152[109], 154[109]
Fow, M. E. 58[31], 59[31]
Fox, D. N. A. 388[3]
Fox, J. M. 311[92]
Foxman, B. M. 405[81]
France, M. B. 159[4], 160[4]
France, S. 66[52]
Franciskovich, J. B. 419[8]
Frank, D. 126[61], 127[61]
Frank, G. 190[59], 191[59]
Franke, R. 297[43]
Frankowski, K. J. 37[80], 38[80], 40[80]
Frantz, D. E. 136[80], 137[80]
Franz, A. K. 349[31]
Franzén, J. 338[15], 342[15], 343[15], 344[25]
Fraser, R. R. 234[42], 234[43], 240[42], 240[43], 240[57]
Fráter, G. 229[34], 234[46], 240[46], 242[46], 243[46], 246[46], 252[34], 253[34], 262[46]
Freerks, R. L. 205[35], 206[35]
Freshwater, D. 196[9]
Frett, B. 431[25]
Fretwell, M. 388[3]
Frey, A. J. 1[4]
Friedman, L. J. 1[3]
Fröhlich, R. 407[93], 408[93]
Frontana-Uribe, B. A. 229[33], 230[33], 231[33]
Frontier, A. J. 116[48]
Fryer, R. M. 372[72], 373[72]
Fu, D. 405[86]
Fu, W. 148[100]
Fu, X. 25[64]
Fuchs, J. R. 38[81], 39[81], 40[81]
Fujii, Y. 77[86], 78[86], 79[86]
Fujioka, H. 304[73], 305[73], 314[101], 314[102], 314[103], 314[104], 314[105], 314[106], 314[107], 314[108], 314[109], 315[108]
Fujita, S. 141[86]
Fujita, T. 314[105]
Fujiwara, S. 325[166], 325[168]
Fukamachi, S. 141[86]
Fukasaka, M. 236[52], 237[52]
Fukaya, C. 205[34], 206[34]
Fukuda, H. 77[86], 78[86], 79[86]
Fukuzawa, A. 307[80], 307[81], 308[80], 308[81]
Fukuzawa, S. 215[70]

Funel, J.-A. 379[101], 380[101], 380[105], 380[106], 380[107], 380[108], 380[109], 381[109]
Funicello, M. 143[88], 143[89]
Funk, R. L. 13[37], 14[37], 38[81], 39[81], 40[81]
Furukawa, T. 405[80]

G
Gagné, M. R. 324[149], 324[150], 324[151], 324[152], 324[153], 324[154], 324[155], 324[156], 324[157]
Gagnepain, J. 2[24], 3[24], 4[24]
Galliford, C. V. 349[32]
Galzerano, P. 401[59]
Ganai, S. 82[96]
Ganguly, A. 419[4]
Gao, O. 304[72]
Gao, X. 300[63]
Gao, Y. 217[88], 217[89], 217[90], 408[95], 408[96], 409[95], 409[96]
Garcia, I. F. 411[104]
García-García, P. 50[10], 373[77]
García-Granda, S. 320[132]
García-Tellado, F. 221[99], 222[101], 223[101], 223[104], 224[104]
García-Valverde, M. 434[32], 436[35]
Gaspar, B. 102[16], 103[16]
Gasperi, T. 337[9], 376[9], 377[95], 378[95], 381[95]
Gato, K. 38[82]
Gaumont, A. C. 293[3]
Gaunt, M. J. 229[21]
Gavenonis, J. 411[102]
Geier, M. J. 324[155], 324[157]
Geller, T. 376[82], 376[83], 376[84], 376[85]
Genazzani, A. A. 27[66], 217[79]
George, J. H. 229[9], 229[10], 229[11], 248[9], 248[10], 248[11], 250[9], 250[10], 250[11], 251[9], 251[11], 263[9], 263[10], 263[11], 264[9], 264[10], 264[11], 267[9]
Gerlach, A. 376[82]
Gescheidt, G. 130[67]
Gewald, K. 439[37], 439[38]
Ghorai, A. 83[101]
Giannichi, B. 350[39], 351[39], 352[39]
Gibbs, R. A. 18[51], 18[52], 19[51], 19[52], 186[55], 187[55]
Giguere, R. J. 160[13], 161[13]
Gimenez, A. 31[73]
Gipson, J. D. 393[30], 394[30]
Girard, N. 317[113], 319[128], 319[129], 320[128]
Giraud, A. 177[41], 178[41]
Girijavallabhan, V. 419[4]
Gisselbrecht, J.-P. 130[66], 130[67], 130[68]
Godenschwager, P. F. 248[69], 250[69]
Goeke, A. 229[13], 234[46], 240[46], 242[46], 243[46], 246[46], 252[13], 252[74], 253[13], 253[74], 254[13], 262[46]
Goldberg, I. 174[36]
Golden, J. E. 37[80], 38[80], 40[80]
Goldman, B. E. 164[23], 165[23]
Golob, A. M. 196[11]
Gong, L.-Z. 357[50], 358[50], 365[50]
Gong, Y. 83[98], 84[98], 401[63]
González, I. 161[14], 162[14]
Gonzalez, J. 232[41]
González, J. M. 299[49], 320[131], 320[133], 321[133]
González, Y. 299[55], 300[55]
González-Rodríguez, C. 110[33], 111[33]
Goodell, J. R. 86[108], 87[108]
Gopalan, A. S. 323[148]
Gordon, C. P. 434[30]
Gordon, J. R. 27[69], 28[69], 219[98]
Goti, A. 298[45]
Gotoh, H. 369[65], 405[78]
Gottlieb, H. E. 173[35], 174[35]
Goulet, N. 211[62]
Grabowski, E. J. J. 376[87], 376[88], 376[89]
Graham, S. L. 419[11]
Gray, D. 19[54], 20[54]
Greene, A. E. 70[63], 71[63]
Greenwood, J. M. 325[162]
Greeves, N. 168[30], 168[31]
Grenning, A. J. 281[80], 282[80], 283[80], 284[80], 287[80]
Grieco, P. A. 8[30], 9[30], 27[30]
Griesbeck, A. G. 164[20]
Griesser, M. 130[67]
Grigg, R. 58[29], 58[30]
Grillet, F. 70[63], 71[63]
Grimaud, L. 419[14]
Grimmett, M. R. 295[30]
Grisé, C. M. 212[65]
Grisé-Bard, C. M. 209[58]
Gröger, H. 337[2], 376[2], 378[2]
Grondal, C. 347[27], 348[27], 349[27], 350[27], 376[94]
Grošelj, U. 371[71]
Gross, M. 130[66], 130[68]
Gross, T. 72[68], 297[39], 298[39]
Grossman, R. B. 229[24], 230[24]
Grossmann, A. 195[1], 196[1]
Gruner, K. K. 153[114], 153[115]
Gruner, M. 431[24]
Gschwind, R. M. 378[99], 379[99]
Gu, P. 83[102]
Gu, Z. 117[49], 117[50], 118[49], 118[50], 119[49]
Guan, H. 389[12]
Guerra, F. M. 217[78]
Guerrand, H. D. S. 58[27], 61[27], 75[79], 80[91], 81[91]
Guggisberg, Y. 379[101], 380[101]
Guillena, G. 344[24]
Guixer, J. 376[81]
Gülak, S. 297[43]
Gulea, M. 154[117], 155[117]
Gunawan, S. 431[25], 431[26]
Gunnoe, T. B. 293[10]
Guo, C. 405[81]
Guo, L.-N. 82[94]
Guo, Q. 217[88], 217[89], 217[90], 407[92], 408[92]
Guo, S. 77[85]
Guo, Z. 77[85], 419[4]
Gusevskaya, E. V. 317[114]
Gutierrez, A. C. 139[83], 144[83]
Gwaltney, S. L., II 373[74]

H
Haase, D. 443[51]
Haberfield, P. 240[63]
Hagiwara, H. 311[91]
Haino, T. 215[70]
Hajos, Z. G. 337[10], 337[11], 376[10], 376[11]
Hajzer, V. 370[69], 371[69]
Hakuba, H. 166[24]
Halder, R. 373[77]
Halland, N. 339[20], 339[21], 402[69]
Hama, N. 207[49], 207[51], 208[51]
Hamada, Y. 229[19], 229[20]
Hamaker, C. G. 86[108], 87[108]
Hamamura, K. 248[68], 250[68]
Hamlett, C. 419[5]
Hampel, F. 294[23], 295[23]
Hamza, A. 405[82]
Han, B. 398[48], 398[49], 398[50], 398[51], 401[58]
Han, F. 275[78], 276[78], 277[78], 277[79], 278[78], 279[78], 280[78], 281[78]
Han, J. W. 72[66]
Han, S. B. 390[16]
Han, X. 293[5]
Hanari, T. 68[56], 389[10]
Handa, S. 395[35]
Hannedouche, J. 293[11], 390[17], 391[17], 393[17]
Hardcastle, K. I. 309[88]
Hardy, E. M. 195[3]
Harman, W. D. 232[38], 232[39], 232[40], 232[41], 233[38], 234[39], 234[40], 235[40]
Harring, S. R. 330[185], 330[186]
Harrison, B. A. 9[31], 10[31], 11[31]
Harrison-Marchand, A. 154[117], 155[117]
Hart, A. 419[4]
Hartl, H. 431[24]
Harvey, T. C. 122[56], 123[56]
Hashimoto, S. 66[53], 67[55], 68[55], 68[56], 71[55], 389[7], 389[8], 389[9], 389[10], 389[11]
Hassan, A. 390[16]
Hassner, A. 295[29]
Hatae, N. 52[13]
Hatano, S. 131[70]
Hattori, T. 54[17], 54[18], 55[17]
Haufe, G. 306[74], 306[76], 306[77]
Hauser, C. R. 176[40], 177[40]
Hayakawa, K. 52[12], 53[12]
Hayashi, R. 50[8], 51[8], 111[35], 112[35], 180[47], 180[49], 181[47], 183[49]

Hayashi, T. 369[65], 380[102], 380[103], 380[104], 382[104], 393[29]
Hayashi, Y. 152[112], 323[143], 323[145], 323[146], 323[147], 365[62], 365[63], 365[64], 366[62], 366[63], 366[64], 367[62], 369[62], 369[63], 369[64], 369[65], 370[63], 370[64], 370[67], 371[64], 371[71], 372[64], 373[75], 373[76], 374[75], 375[75], 376[78], 399[52], 399[53], 400[53], 405[78]
Hazelard, D. 399[52]
Hazell, R. G. 338[19], 339[19], 349[19], 404[74]
Hazuda, D. 419[11]
He, M. 304[72]
He, P. 434[31]
He, Y. 72[70]
He, Z.-Q. 398[49], 398[50], 401[58]
Heaney, F. 58[30]
Heathcock, C. H. 6[28], 7[28]
Hehre, W. J. 1[11]
Hein, J. E. 378[96], 378[97], 379[96], 379[97]
Heindel, N. D. 19[53]
Helmchen, G. 317[125], 319[130]
Hemming, K. 82[97]
Hendrata, S. 419[4]
Henn, L. 22[57]
Henry, R. F. 419[10]
Herdtweck, E. 419[12], 423[16], 424[16], 424[17], 425[17], 426[12], 426[18], 427[16], 427[19], 428[17], 428[18], 428[19], 428[20], 429[12], 430[12], 431[21], 432[21], 433[21], 437[12], 437[21], 438[21], 441[45], 442[45], 442[50], 443[50], 444[45], 444[50], 445[50]
Hermans, C. 295[26]
Hermitage, S. A. 329[180]
Hernandez-Juan, F. A. 397[45], 398[45]
Hernández-Torres, G. 399[54]
Herradón, B. 83[103], 83[104], 84[104]
Herzig, J. 250[72], 250[73]
Herzon, S. B. 221[100]
Hesp, K. D. 293[8]
Hesse, M. D. 229[9], 248[9], 250[9], 251[9], 263[9], 264[9], 267[9]
Hesse, R. 153[115]
Hessler, E. J. 325[163], 325[165]
Hickey, D. M. B. 419[5]
Hickey, E. R. 196[15]
Hiemstra, H. 201[26]
Hiersemann, M. 159[3], 169[32], 170[32], 171[32], 195[7]
Higashino, C. 207[49]
Hii, K. K. 293[6]
Hill, R. K. 1[8]
Himbert, G. 22[57]
Hipple, W. G. 232[39], 234[39]
Hirabaru, T. 150[103], 151[103], 154[103]
Hirai, G. 184[52]
Hiramatsu, A. 314[105]
Hirano, K. 314[103], 314[106]
Hirano, T. 325[166]
Hirata, K. 29[71]
Hirose, H. 304[73], 305[73], 314[103], 314[108], 314[109], 315[108]
Hizartzidis, L. 434[30]
Ho, J. Z. 441[43]
Ho, T.-L. 248[65], 250[65]
Hoashi, Y. 405[80]
Höck, S. 380[106]
Hodgson, D. M. 388[6]
Hoffmann, R. W. 318[127], 319[127]
Holak, T. A. 436[33]
Holloway, M. K. 419[11]
Holmes, A. B. 58[31], 59[31], 59[33], 78[87], 79[87]
Holsworth, D. D. 442[48]
Hong, B.-C. 401[65]
Hong, C. S. 72[64], 72[65], 72[66], 72[67], 73[64]
Hong, S. 300[57], 301[57]
Hong, X. 304[72]
Honga, L. 349[33]
Honma, M. 373[75], 374[75], 375[75]
Horeau, A. 341[22]
Hörl, W. 428[20]
Hoshi, T. 311[91]
Hosoya, K. 248[68], 250[68]
Hossain, M. B. 25[64]
Hotta, D. 54[18]
Houk, K. N. 22[58], 59[33], 62[43], 63[43], 93[1], 93[2], 94[1], 94[2], 378[96], 379[96]
Hövelmann, C. H. 299[49], 299[51], 299[54]
Howell, A. R. 311[94], 312[94]
Hoye, T. R. 36[79], 42[87], 43[87], 325[158]
Hsieh, Y. 419[4]
Hsung, R. P. 50[8], 50[9], 51[8], 55[20], 56[20], 151[104], 180[47], 180[48], 180[49], 181[47], 182[48], 183[49]
Hu, J. 82[94]
Hu, X. 304[72]
Hu, Y.-Y. 82[94]
Hu, Z. 135[78], 409[97]
Hua, Q.-L. 63[45], 64[45]
Huang, H. 395[33]
Huang, K. 297[42]
Huang, L. 77[85]
Huang, Q. 419[11]
Huang, Q.-C. 359[52]
Huang, W.-S. 61[40], 391[21]
Huang, X. 153[116]
Huang, Y. 60[39], 61[39], 166[25], 167[25], 338[14], 340[14], 342[14], 419[4], 431[22], 433[28], 434[28], 435[28], 436[28], 436[33], 437[28], 438[28], 439[28], 440[39], 440[40]
Huang, Y.-D. 362[56]
Huber, H. 127[63]
Hudson, A. 442[48]
Hufton, R. 73[72]
Hughes, D. L. 58[32], 59[32], 376[89]
Hughes, M. 378[96], 378[98], 379[96], 379[98]
Huh, S.-C. 441[44]
Huisgen, R. 127[63], 294[18], 294[19], 294[20]
Hulme, C. 419[9], 431[25], 431[26], 436[34]
Hulot, C. 107[28], 113[41], 113[42], 113[43], 114[41], 114[42], 116[41]
Hultzsch, K. C. 293[2], 294[23], 295[23]
Humphrey, J. M. 20[55], 21[55], 22[55]
Humphreys, P. G. 204[30]
Hung, N. T. 98[9], 99[9]
Huo, X. 72[71], 73[71]
Huple, D. B. 80[89]
Hussain, M. 98[8], 98[9], 98[10], 98[11], 98[12], 99[8], 99[9], 99[10], 99[11], 100[12], 101[10]
Hussain, M. M. 412[109], 413[109]
Hutchins, R. 160[11]
Hut'ka, M. 370[69], 371[69]
Hüttl, M. R. M. 347[27], 348[27], 348[28], 349[27], 350[27]

I

Ibata, T. 387[2]
Ibrahem, I. 396[40]
Ichiki, M. 207[50]
Ichinose, I. 325[159], 325[164], 325[171], 326[171]
Iglesias, B. 175[39], 176[39]
Ihle, D. C. 48[1], 48[2]
Iida, H. 294[21], 390[16]
Ikeda, M. 236[51]
Ilardi, E. A. 188[57], 189[57]
Imagawa, H. 323[144]
Imai, H. 152[112]
Imashiro, R. 399[54]
Inaba, T. 441[46]
Inanaga, J. 29[71]
Inauen, R. 380[109], 381[109], 381[112], 382[112], 383[112]
Ingravallo, P. 419[4]
Inokuchi, T. 150[103], 151[103], 154[103]
Inoue, K. 388[4]
Inoue, M. 3[26], 4[26], 5[26]
Inoue, S. 388[4]
Iqbal, J. 98[8], 99[8]
Irving, W. R. 419[5]
Isaacman, M. J. 188[57], 189[57]
Ischay, M. A. 131[72]
Ishige, Y. 229[19], 229[20]
Ishihara, J. 307[81], 308[81]
Ishihara, K. 326[174], 327[174], 327[175]
Ishii, K. 325[159]
Ishikawa, H. 365[62], 365[63], 365[64], 366[62], 366[63], 366[64], 367[62], 369[62], 369[63], 369[64], 370[63], 370[64], 370[67], 371[64], 372[64], 373[75], 373[76], 374[75], 375[75], 399[53], 400[53], 405[78]

Ishizaka, N. 441[46]
Ito, M. 77[86], 78[86], 79[86]
Ito, Y. 54[18], 77[86], 78[86], 79[86]
Itoh, T. 373[76]
Iwakiri, K. 184[52]
Iwasawa, N. 77[82]
Iyengar, R. 87[112]
Izatt, R. M. 234[49], 248[49], 250[49]

J

Jackson, J. J. 229[23]
Jackstell, R. 297[39], 297[40], 297[41], 297[43], 298[39], 317[120]
Jacobi, P. A. 34[78], 35[78]
Jacobsen, M. F. 121[55], 122[55]
Jagadeesh, B. 401[61]
Jamison, T. F. 309[86], 309[87], 310[86], 310[87], 391[21], 391[22]
Jana, S. 75[77]
Janardhanam, S. 209[53], 209[54]
Jao, E. 419[4]
Jarowski, P. D. 130[66], 130[67]
Jaunet, A. 331[188], 331[189]
Jawalekar, A. 229[21]
Jayakumar, S. 143[90]
Jeanguenat, A. 419[6]
Jeanty, M. 376[94]
Jeganmohan, M. 132[74], 133[74]
Jeon, S.-J. 411[103], 411[108]
Jia, M.-Q. 229[22]
Jia, Z.-J. 355[49], 356[49]
Jiang, H. 338[19], 339[19], 349[19], 404[74]
Jiang, K. 355[49], 356[49], 398[49], 398[51]
Jiang, W. 338[18], 346[18], 406[87], 406[88], 407[87]
Jiang, X. 363[58], 364[58], 405[84], 405[86]
Jiang, Z. 217[89]
Jiménez-Pinto, J. 317[114]
Jin, Z. 304[70], 304[71]
Jing, P. 72[71], 73[71]
Jiricek, J. 380[111], 381[111]
Joe, B.-Y. 441[44]
Johansson, M. 27[67], 219[93], 219[94]
Johnson, B. A. 59[33]
Johnson, R. P. 152[111], 164[17]
Johnson, W. S. 322[135], 322[136]
Johnston, J. N. 322[137]
Johnstone, C. 388[6]
Jokisz, P. 78[87], 79[87]
Jones, M., Jr. 196[10]
Jones, S. D. 419[8]
Jorge, Z. D. 217[78]
Jørgensen, K. A. 338[15], 338[16], 338[17], 338[19], 339[16], 339[17], 339[19], 339[20], 339[21], 342[15], 343[15], 344[16], 344[23], 344[25], 345[16], 345[17], 346[17], 348[29], 349[19], 349[29], 397[43], 399[43], 400[57], 401[57], 402[69], 402[70], 403[70], 403[71], 404[74], 404[75], 405[75]
Jourdain, P. 295[26]
Judd, A. S. 372[72], 373[72], 419[10]
Juhl, K. 397[43], 399[43]
Julia, S. 376[80], 376[81]
Jullian, V. 31[73]
Jung, E. J. 151[107]

K

Kabalka, G. W. 160[11]
Kabuto, C. 171[33], 172[33]
Kahn, S. D. 1[11]
Kaiser, C. 431[25]
Kaji, A. 248[68], 250[68]
Kakehi, A. 388[4]
Kakeya, H. 152[112]
Kakimoto, M. 204[31], 205[31], 205[33], 206[31], 206[33], 207[33]
Kakiya, H. 144[91]
Kamer, P. C. J. 297[41]
Kamo, O. 215[71]
Kamoshida, A. 325[164]
Kampf, J. W. 74[74], 75[74]
Kan, S. B. J. 102[15]
Kanai, M. 390[19], 390[20]
Kanazawa, A. 70[63], 71[63]
Kanematsu, K. 52[12], 52[13], 52[14], 53[12]
Kanematsu, M. 229[19]
Kang, Y.-B. 400[56]
Kani, M. 300[62]
Kanoh, N. 307[81], 308[81]
Kantor, S. W. 176[40], 177[40]
Kataeva, O. 153[114], 153[115]
Kataoka, O. 389[7]
Kath, J. C. 6[28], 7[28]
Kato, H. 77[86], 78[86], 79[86]
Kato, T. 325[159], 325[164], 325[166], 325[167], 325[168], 325[171], 326[171]
Katoh, K. 175[37]
Katritzky, A. R. 295[28], 295[29], 295[30]
Katsuki, T. 29[71]
Katsumura, S. 131[70]
Kavala, V. 83[99]
Kawafuchi, H. 150[103], 151[103], 154[103]
Kawashima, K. 441[46]
Kaynak, B. 396[40]
Ke, Z. 312[98], 313[98]
Keating, T. A. 433[29]
Keck, K. 431[26]
Keller, K. 13[36]
Kelly, A. R. 411[105], 411[106], 412[105], 412[106], 413[106]
Kelly, D. E. 300[59]
Kelsh, L. P. 232[38], 233[38]
Kemp, J. E. G. 295[27]
Kempf, H.-J. 419[6]
Kempf-Grote, A. J. 372[72], 373[72]
Kent, K. M. 300[59]
Kerr, D. J. 120[52]
Keshipeddy, S. 311[94], 312[94]
Khan, A. 98[12], 100[12]
Kharkevich, D. A. 295[25]
Khoury, K. 419[12], 426[12], 427[19], 428[19], 429[12], 430[12], 431[21], 432[21], 433[21], 433[28], 434[28], 435[28], 436[28], 436[33], 437[12], 437[21], 437[28], 438[21], 438[28], 439[28]
Kibayashi, C. 294[21]
Kierstead, R. W. 1[4]
Kikuchi, F. 67[55], 68[55], 71[55]
Kikuchi, H. 311[93]
Kim, C. U. 300[58], 365[61]
Kim, D. 440[41], 441[41], 444[41]
Kim, J. G. 411[104]
Kim, J. H. 72[66], 144[92], 144[93], 145[92], 150[92]
Kim, J. I. 72[64], 73[64]
Kim, J. N. 109[30], 123[57], 123[58], 124[57], 125[57], 125[58]
Kim, K. 103[19], 104[19]
Kim, K. H. 109[30], 123[57], 123[58], 124[57], 125[57], 125[58], 441[44]
Kim, M.-H. 441[44]
Kim, S. H. 109[30], 123[57], 123[58], 124[57], 125[57], 125[58], 217[85]
Kim, S.-M. 441[44]
Kimura, M. 236[52], 237[52]
Kimura, Y. 168[29]
Kirsch, J. 299[51]
Kiss, G. 317[117]
Kita, Y. 304[73], 305[73], 314[101], 314[102], 314[103], 314[104], 314[105], 314[106], 314[107], 314[108], 314[109], 315[108]
Kitagaki, S. 166[24], 175[37], 389[7]
Kitagawa, H. 314[101], 314[102], 314[104]
Kitagawa, I. 38[82]
Kitahara, Y. 325[164], 325[171], 326[171]
Kitawaki, T. 161[15], 162[15], 163[15], 164[15], 167[15]
Kivala, M. 130[68]
Kjaersgard, A. 344[25]
Klapars, A. 73[73], 74[73], 79[73]
Klauber, E. G. 405[83], 406[83]
Klein, H. 297[39], 298[39]
Kleinke, A. S. 152[110]
Kleinpeter, E. 306[77]
Klose, P. 371[71]
Knepper, I. 98[10], 99[10], 101[10]
Knight, S. D. 205[40], 205[41], 205[42], 206[40], 206[41], 206[42], 207[40], 207[41]
Knölker, H.-J. 153[114], 153[115]
Ko, Y. O. 144[92], 145[92], 150[92]
Kobayashi, A. 392[26]
Kobayashi, K. 141[86]
Kobayashi, M. 38[82]
Kobayashi, S. 184[52]
Kobayashi, T. 131[70]
Kociolek, M. G. 164[17]
Koes, D. 436[33]
Koh, J. H. 324[149], 324[150]
Kohn, H. 73[73], 74[73], 79[73]

Kohno, Y. 229[19]
Komanduri, V. 391[23]
Kong, J. 419[4]
Kong, R. 419[4]
Kong, W. 277[79]
Kong, Y. 135[78]
König, H. 294[18], 294[19], 294[20]
König, M. 68[57]
Konishi, H. 141[86]
Kopach, M. E. 232[38], 232[39], 232[40], 232[41], 233[38], 234[39], 234[40], 235[40]
Korfmacher, W. 419[4]
Kornblum, N. 240[60], 240[61], 240[62], 240[63], 241[61], 244[62], 250[62]
Koshimizu, M. 184[52]
Koshino, H. 365[64], 366[64], 369[64], 370[64], 371[64], 372[64]
Kotame, P. 401[65]
Kotoku, N. 314[101], 314[104], 314[105], 314[106], 314[107]
Kowalczyk, J. J. 222[103]
Kozytska, M. V. 219[92]
Kranich, R. 17[45]
Kranz, D. P. 68[57]
Krause, J. 397[44], 398[44]
Krause, J. A. 389[12]
Krause, N. 401[68]
Kremers, F. 234[44], 240[44]
Krenske, E. H. 59[33]
Krische, M. J. 390[16], 391[23], 393[30], 393[31], 394[30], 394[31]
Krishnamurthi, J. 389[11]
Kropf, J. E. 74[74], 75[74]
Krossing, I. 371[71]
Krüger, C. M. 376[82], 376[83], 376[84], 376[85]
Krutšíková, A. 440[42]
Kudo, Y. 59[38], 60[38], 137[82], 138[82]
Kudou, K. 441[46]
Kuenkel, A. 403[72], 404[72], 407[93], 408[93]
Kumagai, T. 325[159]
Kumar, B. 82[95]
Kumar, E. V. K. S. 174[36]
Kumar, N. S. 109[31], 110[31]
Kumar, V. 143[90]
Kumar, V. P. 153[114]
Kumazawa, S. 325[171], 326[171]
Kundu, B. 82[95]
Kuntamukkula, M. 196[9]
Kuo, C.-W. 83[99]
Kurdyumov, A. V. 151[104]
Kurosaki, Y. 68[56], 389[10]
Kurosawa, E. 311[93]
Kurosu, M. 38[82]
Kurth, M. J. 325[158]
Kusama, H. 77[82]
Kwiatkowski, D. 411[104]
Kwok, L. 329[180]
Kwon, E. 137[82], 138[82]
Kwon, H.-M. 441[44]
Kwon, O. 93[1], 93[2], 94[1], 94[2]
Kyle, J. A. 419[8]

L

Laboragine, V. 143[88]
Ladépéche, A. 373[77]
Laetsch, A. 431[26]
Lai, G. 135[78]
Lai, M.-T. 419[11]
Lalic, G. 199[24], 200[24]
Lam, H. C. 229[10], 248[10], 250[10], 263[10], 264[10]
Lam, J. K. 22[58], 22[59], 23[59]
Lam, S. K. 66[54], 67[54], 71[54]
Lam, Y. 62[43], 63[43]
Lam, Y.-H. 378[96], 379[96]
Lamberth, C. 419[6]
Landa, A. 338[16], 339[16], 344[16], 345[16], 403[71]
Langer, N. 130[67]
Langer, P. 98[8], 98[9], 98[10], 98[11], 98[12], 99[8], 99[9], 99[10], 99[11], 100[12], 101[10]
Larsen, C. H. 338[14], 340[14], 342[14]
Larsson, R. 27[67], 219[93], 219[94]
Latika, A. 370[69], 371[69]
Lauher, J. W. 103[19], 104[19]
Laurent, A. 306[76]
Lauterbach-Keil, G. 2[22]
Laver, W. G. 300[58], 365[61]
Lawrence, A. L. 121[55], 122[55]
Layton, M. E. 25[65], 26[65], 199[23], 200[23], 200[25], 201[23], 201[25]
Lecinska, P. 436[35]
Lee, C.-S. 219[91]
Lee, I. Y. 74[74], 75[74]
Lee, J. C. 14[40]
Lee, J.-E. 77[83], 79[83]
Lee, J. H. 72[64], 72[65], 73[64], 144[92], 145[92], 150[92]
Lee, R. W. K. 18[52], 19[52]
Lee, S.-g. 144[92], 144[93], 145[92], 150[92]
Lee, S. J. 72[64], 73[64]
Lee, S. M. 72[65], 72[67]
Lee, W.-M. 168[30], 168[31]
Lee, Y. R. 151[106], 151[107], 153[113]
Lei, Y. 37[80], 38[80], 40[80]
Leimgruber, W. 1[5]
Lennartz, C. 130[67]
le Noble, W. J. 240[62], 244[62], 250[62]
Leon, R. 229[21]
Lerner, R. A. 395[36]
Lesuisse, D. 168[28]
Leung, G. Y. C. 300[64]
Levorse, D. A. 419[11]
Lew, W. 300[58], 300[59], 365[61]
Lewis, C. P. 23[62], 24[62]
Ley, S. V. 164[19]
Li, C. 15[41], 16[41], 29[41], 152[110], 152[111]
Li, D. 275[78], 276[78], 277[78], 277[79], 278[78], 279[78], 280[78], 281[78]
Li, F. 63[45], 64[45]
Li, G. Y. 296[35]
Li, H. 70[61], 217[84], 217[85], 299[50], 338[18], 346[18], 405[79], 405[81], 406[87], 406[88], 407[87]
Li, J. 18[49], 217[80], 218[80]
Li, J. J. 442[48]
Li, J.-L. 398[48], 398[49], 398[51]
Li, L. 405[83], 405[85], 406[83]
Li, L.-H. 248[66], 250[66]
Li, N. 217[88]
Li, P. 410[99], 410[100]
Li, R. 398[49], 398[50]
Li, S. 83[98], 84[98], 297[42]
Li, V.-S. 73[73], 74[73], 79[73]
Li, X. 147[98], 217[88], 217[89], 217[90], 372[72], 373[72], 409[97]
Li, Y. 40[84], 41[84], 43[84], 48[1], 295[32], 295[33], 296[32], 296[33], 298[33], 364[59], 365[59], 401[63]
Li, Y.-B. 370[70], 371[70]
Li, Y.-J. 61[40]
Lian, B. W. 295[28]
Lian, X. 103[17]
Lian, Y. 195[8]
Liang, H. 24[63], 25[63], 26[63]
Liang, X. 419[4]
Liang, Y.-M. 72[69], 82[94]
Liao, C.-C. 2[23]
Liao, J.-H. 401[65]
Liao, Y. 20[55], 21[55], 22[55]
Liddle, J. 419[5]
Liebeschuetz, J. W. 419[8]
Liebeskind, L. S. 131[69], 138[69]
Lim, C. H. 123[57], 124[57], 125[57]
Lim, J. W. 109[30]
Lim, Y.-H. 103[18], 103[20], 104[18], 104[20]
Lin, C. 217[88]
Lin, H. 349[34]
Lin, K.-C. 196[19]
Ling, B. L. 62[43], 63[43]
Link, J. C. 23[62], 24[62]
Lipshutz, B. H. 390[18]
Lishchynskyi, A. 299[55], 300[55]
List, B. 337[4], 337[5], 354[45], 373[77], 376[4], 376[5], 395[36], 419[13]
Litjens, R. 379[101], 380[101]
Little, R. D. 326[173]
Liu, B. 31[74], 32[74], 39[83], 40[83]
Liu, C. 293[5]
Liu, D. 294[13]
Liu, D. R. 322[140]
Liu, F. 80[90]
Liu, G. 62[42], 297[42]
Liu, H. 62[42], 300[58], 365[61], 426[18], 428[18]
Liu, H.-M. 350[35], 350[37], 350[38]
Liu, J. 409[97]
Liu, J.-T. 397[47], 398[47]
Liu, K. 396[41]
Liu, L. 148[99], 363[58], 364[58], 405[84]
Liu, L.-P. 296[37], 297[37]
Liu, Q. 297[43], 317[120]
Liu, R. 365[60], 419[4]
Liu, R.-S. 80[89]
Liu, S. 131[69], 138[69]

Liu, T.-Y. 398[49]
Liu, W. 410[100]
Liu, W.-B. 229[18]
Liu, X. 217[90], 219[91], 405[81]
Liu, X.-Y. 72[69]
Liu, Y.-L. 362[57], 363[57]
Liu, Y.-T. 419[4]
Liu, Y.-Z. 401[67]
Livinghouse, T. 330[184], 330[185], 330[186]
Lloveras, V. 161[14], 162[14]
Lobkovsky, E. 14[40], 15[41], 16[41], 29[41]
Lohse, A. G. 180[47], 181[47]
Lolkema, L. D. M. 201[26]
Longenecker, K. 372[72], 373[72]
Lonzi, G. 317[113]
Lopez, B. O. 160[13], 161[13]
López, M. 217[78]
López, S. 175[38], 176[38]
Lou, C. 409[97]
Louie, M. S. 300[59]
Lovchik, M. A. 234[46], 240[46], 242[46], 243[46], 246[46], 262[46]
Lovey, R. 419[4]
Lu, C. 131[73]
Lu, C.-D. 215[75]
Lu, D. 401[63]
Lu, G. 370[70], 371[70], 413[110]
Lu, L.-Q. 63[45], 63[46], 64[45]
Lu, Y. 116[47], 117[47], 408[94]
Lubben, T. H. 372[72], 373[72]
Lucas, I. 295[26]
Luo, Y.-C. 401[67]
Luo, Z. 364[59], 365[59]
Luque-Corredera, C. 376[79]
Lurain, A. E. 411[105], 411[106], 411[107], 412[105], 412[106], 413[106]
Lurie, A. P. 240[61], 241[61]
Lutz, R. P. 195[5]

M

Ma, C. 398[48], 410[98]
Ma, D. 294[15], 369[66], 370[68], 371[68], 397[46], 398[46], 401[62]
Ma, S. 103[17], 117[49], 117[50], 118[49], 118[50], 119[49]
Ma, Y. 277[79]
Ma, Z.-X. 50[9], 180[49], 183[49]
Macaev, F. 355[48]
McCafferty, G. P. 419[5]
McCluskey, A. 434[30]
McDonald, F. E. 295[34], 296[34], 309[88]
McDonald, I. M. 3[26], 4[26], 5[26]
McErlean, C. S. P. 327[176]
McGaughey, G. B. 419[11]
McGee, L. R. 300[59]
McGrath, N. A. 229[17]
McKeown, B. A. 293[10]
McLachlan, S. P. 168[31]
MacMillan, D. W. C. 213[68], 213[69], 214[69], 215[69], 337[13], 338[14], 340[14], 342[14], 395[37], 395[39], 396[39]
Madar, D. J. 372[72], 373[72]
Maddaluno, J. 154[117], 155[117]
Madhavachary, R. 401[66]
Madison, V. 419[4]
Maerten, E. 338[17], 339[17], 345[17], 346[17], 400[57], 401[57], 402[70], 403[70]
Maestri, A. 411[105], 412[105]
Magerlein, W. 17[45]
Magull, J. 126[61], 127[61]
Mahajan, M. P. 51[11], 52[11], 143[90]
Maiti, N. C. 83[101]
Majetich, G. 40[84], 40[85], 41[84], 41[85], 43[84], 160[13], 161[13]
Majumdar, K. C. 81[92], 82[96], 159[2]
Makabe, M. 171[33], 172[33]
Malcolm, B. 419[4]
Malherbe, R. 213[66], 213[67]
Malik, I. 98[8], 99[8]
Mallikarjun, K. 401[61]
Mandadapu, A. K. 82[95]
Mandal, T. 397[44], 398[44], 406[89]
Mann, A. 317[113], 319[128], 319[129], 320[128]
Mann, E. 83[103], 83[104], 84[104]
Marazano, C. 234[47], 234[48], 244[47], 248[47], 248[48], 250[47]
Marcaccini, S. 431[23], 431[27], 434[32], 436[35]
Marcos, C. F. 431[23], 431[27]
Marcos, I. S. 151[108]
Margetić, D. 121[53], 121[54]
Margulies, H. 240[59], 241[59]
Marigo, M. 338[15], 338[16], 339[16], 342[15], 343[15], 344[16], 344[23], 344[25], 345[16], 348[29], 349[29], 351[40], 403[71]
Marimuthu, J. 419[8]
Marinelli, F. 65[48], 66[48]
Marinić, Ž. 121[53], 121[54]
Marin-Luna, M. 118[51], 119[51], 120[51], 179[45]
Marino, J. P., Jr. 70[62]
Markandu, J. 58[29]
Marklew, J. S. 329[181], 329[182]
Markó, I. E. 295[26]
Marks, T. J. 295[32], 295[33], 295[34], 296[32], 296[33], 296[34], 296[35], 298[33]
Marquis, R. W., Jr. 205[42], 206[42]
Marti, R. 380[106], 382[113]
Martin, C. L. 205[43], 205[44], 206[43], 206[44]
Martin, E. 299[51]
Martin, J. G. 1[8]
Martin, N. G. 58[26], 58[28], 75[75], 75[77], 75[78], 75[79]
Martin, R. M. 132[76], 133[76], 134[76]
Martin, S. F. 20[55], 21[55], 22[55], 76[80], 76[81]
Martínez, C. 298[47]
Martínez, I. 311[94], 312[94]
Martínez, P. H. 294[23], 295[23]
Martínez-Belmonte, M. 299[51]
Masamune, S. 14[39]
Masana, J. 376[80], 376[81]
Mascareñas, J. L. 87[109], 87[110], 88[110]
Mascarenhas, C. 324[149]
Mashkovsky, M. D. 295[25]
Mason, A. M. 419[5]
Massanet, G. M. 217[78]
Massi, A. 337[8], 376[8]
Masson, S. 154[117], 155[117]
Masters, J. J. 419[8]
Masui, R. 405[78]
Matsuda, I. 390[15]
Matsumoto, K. 49[7]
Matsumoto, T. 77[86], 78[86], 79[86]
Matsumura, S. 254[75]
Matsunaga, S. 395[34], 413[110]
Matsunami, K. 38[82]
Matsuno, K. 389[7]
Matsuya, Y. 145[94], 146[94]
Mayer, P. 102[16], 103[16]
Mayr, H. 248[67], 250[67], 272[67]
Mazzanti, A. 350[39], 351[39], 352[39], 359[53], 360[53], 360[54], 361[54], 365[54], 401[59]
Meier, G. P. 205[33], 206[33], 207[33]
Mejía-Oneto, J. M. 69[60], 70[60]
Melchiorre, P. 350[39], 351[39], 351[40], 351[42], 352[39], 352[43], 353[43], 353[44], 356[44], 401[59]
Mellouki, F. 217[78]
Menchi, G. 431[23]
Mendel, D. B. 300[58], 365[61]
Mendelson, L. T. 205[36], 206[36]
Méndez-Abt, G. 221[99], 222[101], 223[101], 223[104], 224[104]
Meridor, D. 172[34], 173[34], 178[34]
Merino, E. 405[76], 405[77]
Mertl, D. 229[13], 252[13], 253[13], 254[13]
Metz, P. 72[68]
Meyer, A. G. 78[87], 79[87]
Meyer, F. E. 95[5], 96[5], 100[5]
Mielgo, A. 344[26]
Mihara, H. 395[34]
Mika, A. M. 372[72], 373[72]
Mikami, K. 159[5], 159[6], 159[7], 168[6], 168[7], 185[53]
Mikhkina, E. E. 295[24]
Milelli, A. 338[19], 339[19], 349[19], 404[74]
Militzer, H.-C. 376[82], 376[83], 376[84], 376[85]
Millan, D. S. 308[83]
Miller, A. K. 105[23], 105[24]
Miller, B. 240[59], 241[59]
Miller, B. G. 229[23]
Miller, K. M. 391[21], 391[22]
Miller, R. F. 222[103]
Minnaard, A. J. 393[27], 393[28]
Minter, D. E. 48[3]
Mitasev, B. 229[5], 229[8], 244[5], 244[8], 245[5], 247[5], 247[8], 248[8], 251[8], 257[5], 258[5], 263[8], 264[8], 284[8]
Mitchell, T. A. 86[108], 87[108]

Miura, M. 236[51]
Miwa, S. 207[50]
Miwa, Y. 254[75]
Miyajima, Y. 184[52]
Miyano, S. 54[17], 54[18], 55[17]
Miyashita, Y. 77[82]
Mochizuki, M. 325[166]
Moessner, C. 381[112], 382[112], 383[112]
Moga, I. 310[90], 311[90]
Mohammed, H. H. 98[12], 100[12]
Mohanakrishnan, A. K. 109[31], 110[31]
Mohareb, R. M. 441[43]
Mohr, J. T. 239[54]
Moisan, L. 337[1], 376[1]
Molander, G. A. 296[36]
Molčanov, K. 121[54]
Molinari, H. 376[81]
Momose, T. 207[49]
Montagnon, T. 1[20], 112[36], 113[36], 159[1]
Montgomery, F. J. 196[16], 196[18]
Moore, A. J. 442[48]
Moore, J. T. 330[187], 331[187]
Mootoo, D. R. 315[110], 316[110]
Morales, C. A. 25[65], 26[65]
Moran, A. 378[98], 379[98]
Moreno, D. 436[35]
Moreno-Dorado, F. J. 217[78]
Morges, M. A. 410[101]
Mori, M. 391[24]
Morimoto, H. 413[110]
Moritani, Y. 389[13]
Morken, J. P. 389[14], 390[14]
Moro, R. F. 151[108]
Morris, G. A. 209[52]
Morton, A. D. 311[94], 312[94]
Moscona, A. 300[60], 300[61]
Moses, J. E. 107[26]
Mosny, K. K. 7[29], 8[29]
Mota, K. 151[105]
Moulis, C. 31[73]
Moyano, A. 359[53], 360[53]
Mueller, B. 323[148]
Mukai, C. 166[24], 175[37]
Mukaida, M. 152[112]
Mukaiyama, T. 365[64], 366[64], 369[64], 370[64], 371[64], 372[64]
Mulhern, M. 372[72], 373[72]
Mullen, C. A. 324[151], 324[152]
Muller, C. D. 113[43]
Müller, T. E. 293[1]
Muñiz, K. 298[47], 298[48], 299[49], 299[51], 299[54], 299[55], 300[55]
Munshi, S. 419[11]
Murai, A. 307[80], 307[81], 308[80], 308[81]
Murai, K. 304[73], 305[73], 314[106], 314[108], 314[109], 315[108]
Murai, M. 441[46]
Murakami, K. 219[97], 220[97]
Murray, C. W. 419[8]
Myers, A. G. 221[100]

N

Na, Y. 73[73], 74[73], 79[73]
Naasz, R. 393[27], 393[28]
Nagashima, T. 419[12], 426[12], 429[12], 430[12], 437[12]
Nagatomi, Y. 314[101], 314[102], 314[104], 314[105], 314[106], 314[107]
Nagatsu, A. 215[70]
Naidu, B. N. 73[73], 74[73], 79[73]
Nakahara, K. 314[103], 314[106], 314[108], 315[108]
Nakai, T. 159[6], 159[7], 168[6], 168[7], 168[29], 185[53]
Nakai, Y. 325[159]
Nakajima, M. 389[8]
Nakamura, I. 59[38], 60[38], 137[81], 137[82], 138[82]
Nakamura, M. 307[80], 308[80]
Nakamura, S. 67[55], 68[55], 71[55], 389[8], 389[9]
Nakano, Y. 229[11], 248[11], 250[11], 251[11], 263[11], 264[11]
Nakatsuji, H. 327[175]
Nakayama, Y. 207[51], 208[51]
Nambu, H. 66[53], 68[56], 389[8], 389[9], 389[10], 389[11]
Namen, A. M. 160[13], 161[13]
Nandi, R. K. 159[2]
Narkunan, K. 248[66], 250[66]
Natale, N. R. 160[11]
Navarro, G. 105[25]
Nay, B. 229[27], 230[27]
Neiwert, W. A. 309[88]
Nelson, H. M. 27[68], 27[69], 28[69], 219[95], 219[96], 219[97], 219[98], 220[97]
Nemoto, H. 145[94], 146[94]
Nemoto, T. 229[19], 229[20]
Nenajdenko, V. 431[22]
Neochoritis, C. G. 419[15], 420[15], 421[15], 422[15]
Nerio, T. 254[75]
Nerozzi, F. 419[5]
Neudörfl, J.-M. 68[57]
Neumann, H. 317[118], 317[119]
Ng, S.-S. 309[86], 309[87], 310[86], 310[87]
Ngo, P. L. 419[11]
Nguyen, H. 324[154], 324[156]
Nguyen, K. 440[39]
Nguyen, N. 115[45], 116[45]
Nicholson, R. L. 304[69]
Nicolaou, K. C. 1[20], 12[35], 15[43], 15[44], 16[44], 17[44], 17[45], 18[49], 19[54], 20[54], 29[35], 112[36], 112[37], 112[38], 112[39], 112[40], 113[36], 113[37], 113[38], 113[39], 113[40], 159[1], 217[80], 217[81], 217[82], 218[80]
Nie, Y.-B. 434[31]
Nieger, M. 299[49], 396[42], 397[42]
Nishide, H. 323[143], 323[146]
Nishiguchi, T. 329[183], 330[183]
Nishimata, T. 239[54]
Nishiyama, H. 390[15]
Nishizawa, M. 323[143], 323[144], 323[145], 323[146], 323[147]
Niu, D. 36[79], 42[87], 43[87]
Niu, Y.-N. 82[94]
Nixey, T. 419[9]
Njardarson, J. T. 3[26], 4[26], 5[26], 229[17], 229[28], 230[28]
Njoroge, F. G. 419[4]
Nomura, M. 236[51]
Northrup, A. B. 395[39], 396[39]
Nosal, D. G. 298[46]
Nuhant, P. 234[47], 244[47], 248[47], 250[47]

O

Oakes, G. H. 168[31]
Oble, J. 419[14]
O'Donnell, M. J. 376[90], 376[91], 376[92]
Ofial, A. R. 248[67], 250[67], 272[67]
Ogasawara, M. 393[29]
Oh, B.-K. 441[44]
Oh, C. H. 72[64], 72[65], 72[66], 72[67], 73[64]
Ohba, Y. 304[73], 305[73], 314[105], 314[106], 314[108], 314[109], 315[108]
Ohdachi, K. 175[37]
Ohkata, K. 254[75]
Ohkubo, M. 373[76]
Ohsawa, N. 145[94], 146[94]
Ohshima, T. 380[110], 381[110]
Oi, S. 54[17], 55[17]
Oisaki, K. 390[19], 390[20]
Oishi, H. 207[51], 208[51]
Okamoto, K. 380[103], 380[104], 382[104]
Okamura, W. H. 18[51], 18[52], 19[51], 19[52], 111[35], 112[35], 186[55], 187[55]
Okano, K. 161[15], 162[15], 163[15], 164[15], 167[15]
Okano, T. 399[52]
Okazaki, M. E. 205[33], 206[33], 207[33]
Oki, T. 314[103], 314[106]
Okino, T. 405[80]
Ollio, S. 424[17], 425[17], 428[17]
Olsher, U. 234[49], 248[49], 250[49]
Onitsch, C. 130[67]
Oohara, T. 66[53], 389[11]
Oppolzer, W. 1[9], 1[13], 13[36], 160[12]
Orahovats, P. A. 442[48]
Oram, N. 58[25], 58[26], 58[27], 61[27], 75[75], 75[76], 75[78], 75[79], 79[76]
Orenes, R.-A. 179[44], 179[45], 180[44]
Orita, H. 365[63], 366[63], 369[63], 370[63]
Oritani, T. 325[170]
Ortin, M.-M. 146[95], 146[96], 146[97], 147[95], 147[97], 179[43], 179[44], 179[46], 180[43], 180[44], 180[46], 183[43]
Osada, H. 152[112]
Osbourn, J. M. 54[16]
Oshima, K. 144[91]

Otomaru, Y. 380[103], 380[104], 382[104]
Overgaard, J. 400[57], 401[57]
Overman, L. E. 1[11], 204[30], 204[31], 205[31], 205[32], 205[33], 205[34], 205[35], 205[36], 205[37], 205[38], 205[39], 205[40], 205[41], 205[42], 205[43], 205[44], 206[31], 206[33], 206[34], 206[35], 206[36], 206[37], 206[38], 206[39], 206[40], 206[41], 206[42], 206[43], 206[44], 207[33], 207[40], 207[41], 207[47], 207[48]

P

Pack, S. K. 296[36]
Padwa, A. 39[83], 40[83], 57[22], 59[34], 59[35], 59[36], 61[36], 66[50], 66[51], 69[59], 69[60], 70[60], 70[61], 70[62], 71[59], 387[1]
Pairaudeau, G. 205[40], 205[41], 206[40], 206[41], 207[40], 207[41]
Pal, U. 83[101]
Palmieri, A. 128[65], 129[65]
Palomo, C. 344[26]
Pan, W. 419[4]
Pandit, R. P. 153[113]
Pandolfo, R. 143[88]
Papai, I. 405[82]
Paquette, L. A. 196[12], 196[13], 196[14], 196[15], 196[16], 196[17], 196[18], 197[17], 197[20], 198[12], 198[17], 198[20], 198[21], 198[22], 199[21], 199[22]
Parekh, T. 419[4]
Parella, T. 161[14], 162[14]
Park, B. H. 151[107]
Park, C.-H. 441[44]
Park, P. K. 229[15]
Park, S. 123[58], 125[58]
Parker, J. S. 41[86], 42[86]
Parker, K. A. 103[18], 103[19], 103[20], 103[21], 104[18], 104[19], 104[20], 104[21]
Parrain, J.-L. 68[58]
Parrish, D. R. 337[10], 337[11], 376[10], 376[11]
Parrish, J. D. 326[173]
Parsons, P. J. 64[47], 65[47], 95[5], 95[6], 96[5], 100[5], 164[18]
Parthasarathy, K. 132[74], 132[75], 133[74], 133[75]
Patel, A. 93[1], 93[2], 94[1], 94[2]
Patel, N. 82[97]
Patonay, T. 127[64], 128[64]
Patrick, N. 431[25]
Patterson, B. 203[28], 204[28]
Pau, C. F. 1[11]
Paull, D. H. 76[80], 76[81]
Pavankumarreddy, G. 401[61]
Pavlovsky, A. 419[4]
Pavošević, F. 121[53]
Peace, S. 419[5]
Pearson, R. G. 248[64], 250[64]
Pearson, W. H. 57[22], 74[74], 75[74], 295[28]
Pechenick, T. 173[35], 174[35]
Pei, Q.-L. 354[47]
Pei, Z. 372[72], 373[72]
Pelc, M. J. 187[56], 188[56], 215[73], 215[74], 216[73]
Peluso, J. 113[43]
Peña, J. 151[108]
Peng, L. 359[52]
Peng, W. 150[103], 151[103], 154[103]
Pepino, R. 431[23], 431[27], 434[32]
Pepper, H. P. 229[10], 229[11], 248[10], 248[11], 250[10], 250[11], 251[11], 263[10], 263[11], 264[10], 264[11]
Pérez-Anes, A. 50[10]
Pérez-Fuertes, Y. 308[83]
Pesaro, M. 1[5]
Pesciaioli, F. 350[39], 351[39], 352[39], 360[54], 361[54], 365[54]
Petasis, N. A. 15[43], 15[44], 16[44], 17[44], 112[37], 112[38], 112[39], 112[40], 113[37], 113[38], 113[39], 113[40]
Peters, D. 323[148]
Peters, R. J. 322[139]
Peterson, C. S. 388[5]
Petricci, E. 317[113]
Petrini, M. 128[65], 129[65]
Pfau, M. 19[53]
Pflum, D. A. 442[48]
Pham, H. V. 22[58]
Phillips, G. B. 183[50], 183[51], 184[50], 184[51]
Phillips, R. 40[84], 41[84], 43[84]
Philp, J. 419[5]
Phun, L. H. 66[52]
Piccardi, R. 140[85], 141[85]
Pichardo, J. 419[4]
Pietrak, B. 419[11]
Piettre, S. 6[28], 7[28]
Pike, R. 419[4]
Pilgram, C. D. 75[75]
Pinto, D. C. G. A. 127[64], 128[64]
Pinto, P. 419[4]
Piovesana, S. 379[100]
Piwinski, J. 419[4]
Pla-Quintana, A. 161[14], 162[14]
Pogrebnoi, V. 355[48]
Pollard, D. 419[5]
Polywka, M. E. C. 306[78], 306[79], 307[78], 307[79]
Popov, V. 419[4]
Popowicz, G. M. 436[33]
Porco, J. A., Jr. 3[25], 5[25], 15[41], 16[41], 29[41], 152[110], 152[111], 229[3], 229[5], 229[6], 229[7], 229[8], 232[35], 244[3], 244[5], 244[8], 245[5], 247[3], 247[5], 247[8], 248[8], 251[8], 254[3], 255[3], 256[3], 257[3], 257[5], 257[6], 258[5], 258[6], 259[3], 260[3], 260[7], 261[7], 262[7], 263[7], 263[8], 264[8], 281[80], 282[80], 283[80], 284[8], 284[80], 284[82], 285[82], 286[82], 287[80], 288[82]
Pordea, A. 317[112]
Postich, M. J. 300[59]
Poulain-Martini, S. 293[7]
Poulin, J. 209[58]
Poupon, E. 229[27], 230[27]
Pouwer, R. 329[180]
Pouwer, R. H. 229[29], 230[29], 232[29], 308[83]
Pozo, M. C. 434[32]
Prasad, M. S. 401[66]
Prein, M. 160[9]
Prelusky, D. 419[4]
Press, N. J. 58[31], 59[31]
Prestwich, G. D. 325[172]
Preusser, L. C. 372[72], 373[72]
Price, E. A. 419[11]
Price, P. D. 304[69]
Prieto, R. 323[148]
Prisbe, E. J. 300[59]
Prokop, A. 68[57]
Prongay, A. 419[4]
Prosperini, S. 27[66], 217[79]
Protti, S. 128[65], 129[65]
Pu, S. 62[42]
Pujadas, A. J. 217[78]
Pulkkinen, J. 339[21]
Pullen, M. A. 419[5]
Purdie, M. 168[31]
Puterová, Z. 440[42]

Q

Qi, J. 229[3], 229[7], 244[3], 247[3], 254[3], 255[3], 256[3], 257[3], 259[3], 260[3], 260[7], 261[7], 262[7], 263[7]
Qi, P.-P. 350[35], 350[37], 350[38]
Qian, B. 72[69]
Qian, W. 83[100]
Qin, A. Y. 294[14]
Qin, Y.-c. 188[57], 189[57]
Qu, C. 304[72]
Quideau, S. 2[24], 3[24], 4[24]
Quigley, D. 196[9]
Quillinan, A. J. 18[48], 217[83]
Quinet, C. 295[26]
Quinkert, G. 190[59], 191[59]

R

Raabe, G. 347[27], 348[27], 348[28], 349[27], 350[27], 401[64]
Rabasso, N. 152[110]
Rabjohn, N. 160[10]
Ragan, J. A. 6[28], 7[28], 376[87]
Rai, K. M. L. 295[29]
Raikar, S. B. 234[47], 244[47], 248[47], 250[47]
Raithby, P. R. 58[31], 59[31]
Rajagopalan, K. 209[53], 209[54], 209[55]
Rajzmann, M. 68[58]
Ramachary, D. B. 395[38], 401[66]
Ramirez, M. A. 222[101], 223[101]
Ramón, D. J. 344[24]
Ramos, D. E. 160[13], 161[13]
Rao, K. V. 401[61]
Rao, T. S. 293[9]
Rao, W. 56[21]
Raulins, N. R. 195[4]

Ray, D. 72[66]
Ray, J. K. 139[84], 140[84]
Ray, K. 81[92]
Reagan, J. 198[21], 199[21]
Reber, K. P. 29[72], 30[72], 222[102]
Recsei, C. 327[176]
Reddy, B. V. S. 293[9]
Redert, T. 229[21]
Redmond, J. M. 329[180]
Reed, L. A., III 14[39]
Rees, C. W. 295[28], 295[29], 295[30]
Rehák, J. 370[69], 371[69]
Rehbein, J. 159[3]
Reich, H. J. 248[70], 250[70]
Rein, T. 20[55], 21[55], 22[55]
Reinhart, G. A. 372[72], 373[72]
Reissig, H.-U. 431[24]
Ren, H. 80[90]
Ren, J. 65[49]
Ren, Q. 408[95], 408[96], 409[95], 409[96]
Renaud, P. 177[41], 178[41]
Reyes, E. 338[19], 339[19], 349[19], 401[60], 402[60], 404[73], 404[74]
Rheingold, A. L. 131[73]
Rhoads, S. J. 195[4]
Riant, O. 390[17], 391[17], 393[17]
Ribelin, T. P. 419[10]
Richard, J.-A. 229[29], 229[32], 230[29], 230[32], 231[32], 232[29]
Richmond, E. 62[43], 63[43]
Rickards, R. W. 127[62]
Riermeier, T. H. 294[16], 294[17]
Righi, P. 360[54], 361[54], 365[54]
Rios, R. 349[30], 359[53], 360[53]
Rist, G. 213[67]
Ritson, D. J. 107[26]
Ritter, S. 68[57]
Rittle, K. E. 419[11]
Riu, A. 154[117], 155[117]
Röben, C. 299[55], 300[55]
Roberts, J. D. 23[60]
Roberts, P. M. 304[69]
Robinson, J. M. 161[15], 162[15], 163[15], 164[15], 167[15]
Rocas, J. 376[81]
Rocha, D. H. A. 127[64], 128[64]
Roche, S. P. 232[35]
Rodier, F. 68[58]
Rodinovskaya, L. A. 439[36]
Rodrigo, R. G. A. 4[27], 5[27]
Rodríguez, F. 320[132]
Rodríguez, J. R. 87[109], 87[110], 88[110]
Rogers, R. D. 196[14], 196[15]
Roglans, A. 161[14], 162[14]
Rohde, J. M. 205[43], 205[44], 206[43], 206[44]
Rohloff, J. C. 300[59]
Rong, Z.-R. 229[18]
Rose, B. J. 410[101]
Rossi, E. 65[48], 66[48]
Roth, F. 234[44], 240[44]
Roughley, S. D. 58[31], 59[31]
Roura, P. 161[14], 162[14]
Roush, W. R. 1[18], 31[75], 32[75]
Rovis, T. 410[101]
Rowe, J. E., Jr. 19[53]
Ruan, S. 419[4]
Ruan, Z. 315[110], 316[110]
Rubal, J. J. 217[78]
Rubín, S. G. 110[32], 110[33], 111[33]
Rubio, E. 320[133], 321[133]
Rubtsov, M. V. 295[24]
Rubush, D. M. 410[101]
Rueping, M. 141[87], 142[87], 143[87], 403[72], 404[72], 405[76], 405[77], 407[93], 408[93]
Ruggeri, R. B. 6[28], 7[28]
Rumbo, A. 87[109]
Runsink, J. 348[28]
Russell, A. T. 41[86], 42[86]
Ryan, E. F. 376[89]
Ryan, J. H. 78[87], 79[87]
Rychnovsky, S. D. 202[27], 203[27], 203[28], 204[28]
Ryu, J.-S. 295[34], 296[34], 296[35]

S

Saá, C. 110[32], 110[33], 111[33]
Saaby, S. 164[19]
Sabot, C. 70[63], 71[63]
Saeki, H. 29[71]
Saito, N. 391[24], 392[26]
Saito, R. 207[50]
Sakai, K. 152[112]
Sakai, T. 161[15], 162[15], 163[15], 163[16], 164[15], 167[15]
Sakakura, A. 326[174], 327[174], 327[175]
Saksena, A. K. 419[4]
Sakthivel, K. 354[46]
Salem, B. 97[7], 100[7], 101[13], 101[14], 103[13]
Salman, G. A. 98[11], 99[11]
Salvadori, J. 317[113], 319[128], 319[129], 320[128]
Salvi, L. 411[108]
Samanta, S. 397[44], 398[44]
Sanchez-Andrada, P. 118[51], 119[51], 120[51], 146[95], 146[96], 146[97], 147[95], 147[97], 179[43], 179[44], 179[45], 179[46], 180[43], 180[44], 180[46], 183[43]
Sankaranarayanan, S. 419[11]
Santhanam, B. 419[4]
Saouf, A. 217[78]
Sarlah, D. 12[35], 29[35]
Sasmal, P. K. 217[81]
Satake, A. 236[50]
Sato, H. 207[49]
Sato, M. 325[169]
Sato, T. 207[50], 207[51], 208[51]
Sato, Y. 49[4], 49[5], 49[6], 49[7], 391[24], 392[26]
Satoh, T. 236[51]
Sattler, W. 310[89], 311[89]
Sauer, E. L. O. 212[63], 213[63]
Sauer, G. 337[12], 376[12]
Sauer, J. 1[6], 1[7]
Sauvain, M. 31[73]
Sawama, Y. 314[106], 314[107]
Sawamura, Y. 327[175]
Sawano, S. 399[53], 400[53]
Sayo, N. 168[29]
Sbircea, D.-T. 308[84]
Scarpino Schietroma, D. M. 377[95], 378[95], 379[100], 381[95]
Schall, A. 328[179]
Scheeren, H. W. 219[94]
Scheibe, C. M. 87[112]
Scheidt, K. A. 349[32]
Scheinmann, F. 18[48], 217[83]
Schelling, P. 441[47]
Schinke, E. 439[38]
Schmalz, H.-G. 68[57]
Schmidt, A. W. 153[115]
Schmidt, G. 380[105], 380[106], 380[107]
Schmidt, M. B. 378[99], 379[99]
Schmidt, T. 239[53]
Schmidt, Y. 22[58], 22[59], 23[59]
Schmitz, F. J. 25[64]
Schoenewaldt, E. F. 376[89]
Schoenfelder, A. 319[128], 320[128]
Schreiber, J. 1[5]
Schreiber, S. L. 29[70], 111[34], 198[21], 199[21]
Schreiner, P. 113[42], 114[42]
Schroeder, F. C. 111[34]
Schubert, G. 405[82]
Schudel, P. 1[5]
Schulte, T. 338[15], 342[15], 343[15]
Schultze, L. M. 300[59]
Schulz, E. 293[11]
Schulz, G. E. 322[140]
Schwaninger, M. 380[107]
Schwartz, M. A. 325[165]
Schwartz, U. 190[59], 191[59]
Schwede, W. 190[58]
Schweizer, E. E 178[42], 179[42]
Schweizer, S. 95[5], 95[6], 96[5], 100[5]
Schweizer, W. B. 130[66], 130[67], 371[71]
Scialpi, R. 143[89]
Sciotti, R. J. 31[75], 32[75]
Scriven, E. F. V. 295[28], 295[29], 295[30]
Seayad, A. 297[39], 298[39]
Seayad, A. M. 297[40]
Seayad, J. 337[4], 376[4]
Šebesta, R. 370[69], 371[69]
Seebach, D. 371[71]
Seiche, W. 317[116]
Seidel, D. 405[83], 405[85], 406[83]
Seidel, T. 379[101], 380[101]
Seiler, P. 130[68]
Seki, M. 52[13]
Sekiya, R. 207[51], 208[51]
Selnick, H. G. 419[11]
Seltzer, R. 240[60], 240[63]
Semeyn, C. 201[26]
Severin, R. 293[4]
Shabbir, S. S. 419[5]

Shair, M. D. 7[29], 8[29], 25[65], 26[65], 199[23], 199[24], 200[23], 200[24], 200[25], 201[23], 201[25]
Sham, H. L. 372[72], 373[72]
Shan, L.-H. 350[38]
Shanahan, C. S. 76[80], 76[81]
Shanahan, S. E. 205[46], 206[46]
Shanmugam, P. 209[55]
Sharif, M. 98[11], 99[11]
Sharma, P. 107[26]
Sharma, S. K. 82[95]
Shaw, J. T. 330[187], 331[187]
She, X. 72[70], 72[71], 73[71]
Sheehan, S. M. 70[62], 199[24], 200[24], 200[25], 201[25], 419[8]
Shelly, S. A. 188[57], 189[57]
Shen, F. 363[58], 364[58], 405[84], 405[86]
Shepeli, F. 355[48]
Shepherd, N. E. 395[34]
Sheppard, R. N. 308[83]
Shestopalov, A. A. 439[36]
Shestopalov, A. M. 439[36]
Shi, D. 395[33]
Shi, J. 364[59], 365[59]
Shi, M. 62[44], 296[37], 297[37], 299[52], 299[53]
Shi, X.-J. 350[37], 350[38]
Shi, Y.-J. 198[22], 199[22]
Shibasaki, M. 300[62], 380[110], 381[110], 390[19], 390[20], 395[34], 413[110]
Shibata, Y. 399[53], 400[53]
Shieh, H.-M. 325[172]
Shim, J. 205[37], 205[38], 206[37], 206[38]
Shimada, K. 171[33], 172[33]
Shimada, N. 66[53], 68[56], 389[8], 389[9], 389[10], 389[11]
Shimada, Y. 307[81], 308[81]
Shimizu, I. 236[50]
Shimizu, M. 159[5]
Shin, H. 144[93]
Shin, S. 77[83], 79[83]
Shindo, M. 49[4], 49[5], 49[6], 49[7]
Shing, T. K. M. 248[66], 250[66]
Shinokubo, H. 144[91]
Shintani, R. 380[103], 380[104], 382[104], 389[13]
Shiomi, T. 390[15]
Shiro, M. 52[12], 52[14], 53[12], 389[10]
Shishido, K. 49[4], 49[5], 49[6], 49[7]
Shoji, M. 152[112], 369[65]
Shu, X.-Z. 82[94]
Shuto, S. 77[86], 78[86], 79[86]
Sidana, J. 229[25], 230[25]
Siegel, D. R. 229[1], 229[2], 229[26], 230[26], 237[1], 237[2], 238[2], 239[1], 239[2]
Silva, A. M. S. 127[64], 128[64]
Simanis, J. A. 86[108], 87[108]
Simmons, F. 196[9]
Simmons, H. E., Jr. 23[60]
Simon, A. J. 419[11]
Simpkins, N. S. 229[12], 229[30], 230[30], 250[12], 263[12], 264[12]
Sinclair, A. 58[32], 59[32]
Šindler-Kulyk, M. 121[53]
Singh, I. P. 229[25], 230[25]
Singh, P. 51[11], 52[11]
Singha, R. 139[84], 140[84]
Sinha, M. K. 419[12], 426[12], 427[19], 428[19], 429[12], 430[12], 431[21], 432[21], 433[21], 437[12], 437[21], 438[21]
Sinwel, F. 2[22]
Sirois, L. E. 86[107]
Sittihan, S. 229[17]
Skeean, R. W. 329[183], 330[183]
Škorić, I. 121[53], 121[54]
Skropeta, D. 127[62]
Slaughter, L. M. 395[35]
Slobodov, I. 18[50], 217[86], 217[87], 219[87], 220[87]
Smallwood, J. K. 419[8]
Smith, A. D. 62[43], 63[43], 304[69]
Smith, C. J. 58[31], 59[31]
Smith, G. F. 419[8]
Snider, B. B. 122[56], 123[56], 164[21], 164[22], 164[23], 165[22], 165[23], 183[50], 183[51], 184[50], 184[51]
Snieckus, V. 160[12]
Snyder, S. A. 1[20], 112[36], 113[36], 159[1], 310[89], 310[90], 311[89], 311[90], 327[177], 327[178], 328[177], 328[178], 328[179], 329[178]
Sodeoka, M. 184[52]
Sofiyev, V. 105[25]
Sokol, J. G. 324[153]
Solà, M. 161[14], 162[14]
Solanki, S. 308[83]
Soldi, C. 330[187], 331[187]
Sollis, S. L. 419[5]
Sondheimer, F. 1[2]
Song, H. 294[14]
Song, M.-P. 350[39], 351[39], 352[39]
Song, W.-Z. 55[20], 56[20]
Soos, T. 405[82]
Sorensen, E. J. 9[31], 10[31], 10[32], 10[33], 11[31], 11[32], 11[33], 29[72], 30[72], 222[102]
Souto, J. A. 299[55], 300[55]
Souza, F. E. S. 4[27], 5[27]
Spagnolo, P. 143[88], 143[89]
Spangenberg, T. 319[128], 320[128]
Spangler, C. W. 159[8], 178[8], 179[8]
Speckamp, W. N. 201[26], 204[29]
Spencer, W. T. 116[48]
Spiccia, N. 78[87], 79[87]
Spiegel, D. A. 3[26], 4[26], 5[26], 111[34]
Spielvogel, D. 381[112], 382[112], 383[112]
Sprecher, M. 172[34], 173[34], 174[36], 178[34]
Springer, J. P. 21[56], 22[56]
Stafford, J. A. 373[74]
Stark, H. 190[59], 191[59]
Starr, J. T. 10[34]
Stashko, M. A. 372[72], 373[72]
Staudinger, H. 17[46]
Stauffer, S. R. 419[11]
Stearman, C. J. 59[36], 61[36]
Stein, A. R. 250[71]
Steinbrückner, C. 419[7]
Steinmeyer, A. 190[58]
Stemmler, R. T. 86[107]
Sterner, O. 27[66], 27[67], 217[79], 219[93], 219[94]
Stevens, R. C. 300[58], 365[61]
Steward, K. D. 372[72], 373[72]
Still, R. 136[80], 137[80]
Stivala, C. E. 215[76]
Stocking, E. M. 1[21]
Stockman, R. A. 58[32], 59[32]
Stoessel, F. 380[107]
Stoltz, B. M. 23[61], 27[68], 27[69], 28[69], 219[95], 219[96], 219[97], 219[98], 220[97], 239[54], 239[55]
Stork, G. 1[3], 322[141]
Stork, K. C. 232[38], 233[38]
Storni, A. 325[165]
Stradiotto, M. 293[8]
Streuff, J. 299[49], 299[51], 299[54]
Strobel, G. A. 14[40]
Struijk, M. 379[101], 380[101]
Stryker, J. M. 84[105]
Stupple, P. A. 388[6]
Su, Z. 196[16]
Suárez-Sobrino, Á. L. 50[10]
Sucman, N. 355[48]
Suffert, J. 97[7], 100[7], 101[13], 101[14], 103[13], 107[27], 107[28], 107[29], 108[27], 113[41], 113[42], 113[43], 114[41], 114[42], 116[41]
Suga, H. 388[4]
Sugano, Y. 67[55], 68[55], 71[55]
Sugihara, T. 323[144]
Sugimoto, M. 311[91]
Sugiono, E. 405[76], 405[77]
Sun, H. 217[89], 217[90]
Sun, Q. 277[79]
Sun, X.-L. 400[56]
Sünnemann, H. W. 126[61], 127[61]
Surendra, K. 229[16], 259[16]
Surendrakumar, S. 58[29], 58[30]
Sureshbabu, R. B. R. 109[31], 110[31]
Susanti, D. 56[21]
Sutherland, H. S. 4[27], 5[27]
Sutherland, J. K. 209[52], 325[162]
Suzuki, K. 184[52]
Suzuki, M. 311[93]
Suzuki, T. 184[52], 311[91], 311[93], 365[62], 365[63], 366[62], 366[63], 367[62], 369[62], 369[63], 370[63]
Suzuki, Y. 54[17], 54[18], 55[17]
Swaminathan, S. 300[58], 365[61]
Sworin, M. 196[19], 205[39], 206[39]
Sydorenko, N. 151[104]
Szeto, P. 58[32], 59[32]

T

Taber, D. F. 1[15]
Tada, Y. 236[50]
Taddei, M. 317[113], 319[128], 319[129], 320[128]
Tadross, P. M. 23[61]
Tai, C. Y. 300[58], 365[61]
Takagi, R. 254[75]
Takahashi, K. 185[53], 323[144]
Takao, H. 323[144]
Takase, M. K. 131[72]
Takasugi, Y. 307[80], 307[81], 308[80], 308[81]
Takaya, J. 77[82]
Takeda, K. 66[53], 325[168], 389[10]
Takemoto, Y. 405[80]
Takenaka, H. 323[143], 323[145]
Takikawa, Y. 171[33], 172[33]
Takita, R. 380[110], 381[110]
Talavera, G. 401[60], 402[60], 404[73]
Tamaru, Y. 236[52], 237[52]
Tamura, M. 307[80], 308[80]
Tan, B. 358[51], 359[51], 361[55], 362[55], 407[91], 408[94]
Tan, C. K. 301[65], 302[65], 312[95], 313[95], 316[95]
Tanaka, A. 325[169], 325[170]
Tang, J. 296[37], 297[37]
Tang, L. 405[81]
Tang, Y. 50[8], 51[8], 400[56]
Tanikaga, R. 248[68], 250[68]
Tanimoto, H. 207[50]
Tanuwidjaja, J. 309[86], 309[87], 310[86], 310[87]
Tao, H.-Y. 406[90]
Tao, L. 217[89]
Tato, F. 403[72], 404[72]
Taub, D. 1[2]
Taunton, J. 29[70]
Tay, D. W. 300[64]
Taylor, J. G. 293[6]
Taylor, M. T. 311[92]
Taylor, S. T. 389[14], 390[14]
Tejedor, D. 221[99], 222[101], 223[101], 223[104], 224[104]
Teleha, C. A. 198[21], 199[21]
Terada, M. 59[38], 60[38], 137[81], 137[82], 138[82]
Terada, Y. 325[161]
Termath, A. O. 68[57]
Tessier, P. E. 115[45], 116[45]
Thamm, D. H. 410[101]
Thayumanavan, R. 354[46]
Theodorakis, E. A. 18[50], 217[77], 217[84], 217[85], 217[86], 217[87], 219[87], 220[87], 229[27], 230[27]
Thomas, A. J. F. 329[181]
Thomas, S. E. 306[78], 306[79], 307[78], 307[79]
Thompson, M. J. 58[31], 59[31]
Thornton-Pett, M. 58[29]
Tian, X. 40[84], 41[84], 43[84], 352[43], 353[43], 353[44], 356[44]
Tian, Z.-Q. 362[56]
Tietze, E. 234[44], 240[44]
Tietze, L. F. 419[1]
Tillack, A. 294[17]
Tilley, S. D. 29[72], 30[72], 222[102]
Tinsley, J. M. 442[48]
Tisdale, E. J. 18[50], 217[84], 217[85], 217[86], 217[87], 219[87], 220[87]
Toguem, S.-M. T. 98[8], 98[10], 99[8], 99[10], 101[10]
Toita, A. 67[55], 68[55], 71[55]
Tokan, W. M. 95[5], 95[6], 96[5], 100[5]
Tokunaga, N. 380[102], 380[103]
Toma, Š. 370[69], 371[69]
Tong, X. 94[3], 95[3], 100[3]
Topgi, R. S. 234[45], 241[45]
Torrado, A. 175[38], 175[39], 176[38], 176[39]
Torre, A. 325[162]
Torroba, T. 431[23], 434[32], 436[35]
Townsend, C. A. 23[62], 24[62]
Toyoda, J. 387[2]
Trauner, D. 102[16], 103[16], 103[22], 104[22], 105[23], 105[24], 105[25], 108[22]
Trauthwein, H. 294[16]
Treitler, D. S. 310[89], 310[90], 311[89], 311[90], 327[177], 327[178], 328[177], 328[178], 328[179], 329[178]
Trevillyan, J. M. 372[72], 373[72]
Tric, B. 168[28]
Trincado, M. 320[133], 321[133]
Trost, B. M. 1[9], 139[83], 144[83], 239[53], 295[27], 300[56], 350[36]
Tschirret-Guth, R. A. 326[173]
Tsoi, I. T. 300[64]
Tsuchikawa, H. 131[70]
Tsuchiya, Y. 390[15]
Tsukano, C. 229[1], 229[26], 230[26], 237[1], 239[1]
Tsutsui, H. 389[8], 389[9]
Tu, Y.-C. 83[99]
Tu, Y.-Q. 83[102]
Tugusheva, K. 419[11]
Tulip, S. J. 229[11], 248[11], 250[11], 251[11], 263[11], 264[11]
Tung, T. D. 98[10], 99[10], 101[10]
Tunge, J. A. 239[56]

U

Uchimaru, T. 365[63], 366[63], 369[63], 370[63], 371[71]
Udodong, U. E. S. 34[78], 35[78]
Uehara, H. 399[54]
Uemura, D. 215[70], 215[71]
Uenishi, J. 15[43], 112[37], 112[39], 113[37], 113[39]
Uesugi, O. 54[17], 55[17]
Ueyama, K. 380[102], 380[103]
Ugi, I. 419[7], 428[20]
Ukai, A. 326[174], 327[174]
Ullah, F. 400[55]
Umeda, C. 389[7]
Umemiya, S. 376[78]
Underwood, B. S. 309[87], 310[87]
Unger, J. B. 390[18]
Uyehara, T. 325[166], 325[167], 325[168]

V

Vacca, J. P. 419[11]
Valdivia, C. 27[66], 217[79]
van der Helm, D. 25[64]
Vanderwal, C. D. 10[32], 10[33], 11[32], 11[33], 22[58], 22[59], 23[59]
van Leeuwen, P. W. N. M. 297[41], 317[121]
van Tamelen, E. E. 1[3], 325[163], 325[165]
Varela, J. A. 110[32], 110[33], 111[33]
Vasbinder, M. M. 388[5]
Vassilikogiannakis, G. 1[20], 17[45], 159[1]
Vaughan, C. W. 23[60]
Vazdar, M. 121[53]
Vecchione, M. K. 405[85]
Vedejs, E. 73[73], 74[73], 79[73]
Vega, J. 376[80]
Végh, A. 440[42]
Velado, M. 83[104], 84[104]
Velez-Castro, H. 229[33], 230[33], 231[33]
Venkatraman, S. 419[4]
Vibulbhan, B. 419[4]
Vicario, J. L. 401[60], 402[60], 404[73]
Vidal, A. 118[51], 119[51], 120[51], 146[95], 146[96], 146[97], 147[95], 147[97], 179[43], 179[44], 179[45], 179[46], 180[43], 180[44], 180[46], 183[43]
Vidal-Gancedo, J. 161[14], 162[14]
Vidali, V. P. 229[4], 245[4], 246[4], 247[4], 248[4], 250[4], 251[4], 259[4]
Vidović, D. 126[61], 127[61]
Vigneron, J. P. 341[22]
Villinger, A. 98[8], 98[10], 98[11], 98[12], 99[8], 99[10], 99[11], 100[12], 101[10]
Virgil, S. C. 27[69], 28[69], 219[97], 219[98], 220[97]
Visca, V. 377[95], 378[95], 381[95]
Viseux, E. M. E. 64[47], 65[47]
Vogt, K. A. 401[68]
Volla, C. M. R. 141[87], 142[87], 143[87]
Vollhardt, K. P. C. 13[37], 14[37]
Volz, N. 396[42], 397[42]
von Essen, R. 126[61], 127[61]
Vong, B. G. 217[84], 217[85]
von Golden, T. W. 372[72], 373[72]
von Zezschwitz, P. 95[4]
Vosburg, D. A. 10[32], 10[33], 11[32], 11[33]
Vuk, D. 121[54]

W

Wabnitz, T. C. 344[25]
Wakabayashi, A. 323[144]
Wakamatsu, S. 314[106]
Walji, A. M. 338[14], 340[14], 342[14]
Walsh, P. J. 411[102], 411[103], 411[104], 411[105], 411[106], 411[107], 411[108], 412[105], 412[106], 412[109], 413[106], 413[109]

Walter, D. S. 164[18]
Walton, M. C. 55[20], 56[20], 180[47], 181[47]
Waltz, K. M. 411[102], 411[104]
Wan, B. 147[98]
Wang, C. 401[64]
Wang, C.-H. 362[57], 363[57]
Wang, C.-J. 406[90]
Wang, D. 147[98]
Wang, D.-X. 413[111], 413[112]
Wang, D.-Z. 72[69]
Wang, F. 94[3], 95[3], 100[3]
Wang, G. 135[78]
Wang, H. 83[100]
Wang, H.-F. 410[99], 410[100]
Wang, J. 168[27], 338[18], 346[18], 397[46], 398[46], 405[79], 406[87], 407[87], 408[95], 408[96], 409[95], 409[96], 409[97]
Wang, K. 419[3], 423[3], 440[39], 440[41], 441[41], 441[45], 442[45], 442[50], 443[50], 444[41], 444[45], 444[50], 445[50]
Wang, L. 275[78], 276[78], 277[78], 277[79], 278[78], 279[78], 280[78], 281[78], 408[95], 409[95]
Wang, L.-C. 393[31], 394[31]
Wang, L.-L. 359[52]
Wang, L.-X. 359[52], 392[25]
Wang, M.-X. 413[111], 413[112]
Wang, R. 275[78], 276[78], 277[78], 277[79], 278[78], 279[78], 280[78], 281[78], 349[33], 363[58], 364[58], 405[84], 405[86]
Wang, R.-B. 370[70], 371[70]
Wang, S.-X. 413[111]
Wang, T. 63[46]
Wang, T.-L. 314[106]
Wang, T.-Z. 196[18]
Wang, W. 77[84], 338[18], 346[18], 405[79], 406[87], 406[88], 407[87], 419[3], 423[3], 423[16], 424[16], 424[17], 425[17], 427[16], 428[17], 436[33]
Wang, X. 62[41], 63[41], 72[70], 217[88], 217[90], 362[57], 363[57], 410[98]
Wang, X.-C. 82[94]
Wang, X.-J. 397[47], 398[47]
Wang, X.-N. 55[20], 56[20]
Wang, X.-W. 410[100]
Wang, Y. 217[90], 370[68], 371[68], 401[62], 401[67], 405[81], 405[84]
Wang, Y.-M. 362[56]
Wang, Z. 59[37], 60[39], 61[39], 62[42], 65[49], 103[21], 104[21], 148[100], 166[25], 167[25], 229[15]
Ward, T. R. 317[112]
Wardrop, D. J. 298[46]
Warner, D. L. 73[73], 74[73], 79[73]
Warnock, W. J. 58[29], 58[30]
Warrington, J. M. 209[56], 210[56], 211[61]
Wartmann, M. 217[82]
Watanabe, N. 389[7]
Waters, A. J. 164[18]
Waters, S. P. 148[101], 149[101]
Watson, L. 58[26], 58[28], 75[75], 75[77], 75[78], 75[79]
Weber, A. E. 373[73]
Weber, W.-D. 190[59], 191[59]
Weber, W. W., II 419[8]
Webster, R. 102[16], 103[16]
Wegner, H. A. 121[55], 122[55]
Wei, C.-P. 2[23]
Wei, H. 401[67]
Wei, Q. 357[50], 358[50], 365[50]
Wei, Y. 62[44]
Weiler, S. 10[33], 11[33]
Weingarten, M. D. 387[1]
Weller, M. D. 229[12], 250[12], 263[12], 264[12]
Weller, T. 380[109], 381[109]
Welmaker, G. S. 204[30]
Wender, P. A. 86[107]
Wendt, B. 348[28]
Wendt, K. U. 322[140]
Wenig, J. 370[70], 371[70]
Wessely, F. 2[22]
West, F. G. 115[44]
Westfall, T. D. 419[5]
White, A. J. P. 306[75], 308[83], 329[180], 329[182]
White, J. D. 48[1], 48[2], 329[183], 330[183]
White, J. M. 78[87], 79[87]
White, L. E. 75[76], 79[76]
White, R. 419[4]
Whitlock, G. A. 388[3]
Whittern, D. N. 419[10]
Whittle, A. J. 64[47], 65[47]
Widenhoefer, R. A. 293[5], 299[50]
Wiechert, R. 337[12], 376[12]
Wiley, M. R. 419[8]
Wilhelm, M. 240[58]
William, S. 426[18], 428[18]
Williams, G. M. 58[31], 59[31]
Williams, M. A. 300[58], 365[61]
Williams, R. M. 1[21]
Willis, A. C. 120[52]
Willoughby, P. H. 36[79], 42[87], 43[87]
Wilson, M. 59[36], 61[36]
Wilson, M. S. 59[34]
Wilson, S. R. 195[6]
Winslow, C. D. 48[3]
Woggon, W.-D. 396[41]
Wolf, S. 436[33]
Wolfe, A. 419[11]
Wolfe, J. P. 442[48]
Wolinsky, L. E. 325[160]
Won, R. 441[44]
Wong, J. 419[4]
Wong, L. S.-M. 78[87], 79[87]
Wong, Y.-L. 20[55], 21[55], 22[55]
Wood, J. L. 3[26], 4[26], 5[26], 29[70]
Woodall, E. L. 86[108], 87[108]
Woods, B. P. 36[79], 42[87], 43[87]
Woodward, R. B. 1[2], 1[4], 322[134]
Woollard, P. M. 419[5]
Worboys, P. 295[31]
Wu, C. 419[5]
Wu, F. 147[98], 405[81]
Wu, H. 362[56]
Wu, J. 59[37], 80[88], 80[90], 81[88], 434[31]
Wu, L. 153[116], 297[43], 317[120], 355[49], 356[49], 398[50]
Wu, L.-Y. 72[69], 350[39], 351[39], 352[39]
Wu, M.-H. 124[59], 124[60], 125[59]
Wu, Q.-F. 229[18]
Wu, S. 376[92]
Wu, T. R. 12[35], 29[35]
Wu, W. 219[91], 297[42], 299[52], 299[53], 419[4]
Wu, X.-F. 317[119]
Wu, Y. 83[98], 84[98]
Wu, Y.-L. 130[66], 130[67]
Wu, Z.-J. 354[47]

X

Xia, C. 395[33]
Xiang, L. 294[13]
Xiao, Q. 229[23]
Xiao, W.-J. 63[45], 63[46], 64[45]
Xiao, Y.-C. 401[58]
Xie, H. 83[98], 84[98], 338[18], 346[18], 405[79], 406[87], 406[88], 407[87]
Xie, Y. 395[33]
Xie, Y.-X. 72[69]
Xin, X. 147[98]
Xing, Y. 116[46], 116[47], 117[46], 117[47]
Xiong, G. 77[85]
Xu, F. 148[100]
Xu, H. 217[81], 217[82]
Xu, J. 239[53]
Xu, M. 419[11]
Xu, P.-F. 401[67]
Xu, S. 148[99], 299[53]
Xu, X. 405[80]
Xu, X.-Y. 359[52]
Xu, Y. 135[78], 380[110], 381[110], 395[34]
Xu, Z. 431[25]

Y

Yadav, J. S. 293[9]
Yagi, K. 144[91]
Yakhontov, L. N. 295[24], 295[25]
Yamada, H. 323[147]
Yamaguchi, M. 29[71]
Yamaguchi, Y. 325[167]
Yamakawa, T. 134[77], 135[77], 138[77]
Yamamoto, A. 236[50]
Yamamoto, M. 441[46]
Yamamura, S. 325[161]
Yamashita, K. 325[169]
Yamazaki, M. 207[51], 208[51]
Yan, M. 409[97]
Yan, S.-J. 442[49]
Yan, Z.-Y. 72[69]

Yang, D. 275[78], 276[78], 277[78], 277[79], 278[78], 279[78], 280[78], 281[78]
Yang, D. T. C. 160[11]
Yang, H.-B. 62[44]
Yang, J. 72[71], 73[71]
Yang, L. 31[74], 32[74]
Yang, Q.-Q. 63[46]
Yang, W. 419[4]
Yang, X. 80[88], 81[88], 401[64]
Yang, Y. 217[90], 364[59], 365[59], 392[25]
Yang, Y.-Q. 410[99]
Yang, Z. 72[70], 419[11]
Yao, C.-F. 83[99]
Yao, J. 116[46], 116[47], 117[46], 117[47]
Yao, W. 410[98]
Yao, Z. 410[98]
Yap, G. P. A. 209[56], 210[56]
Yaqoob, M. 306[75]
Yasui, Y. 399[53], 400[53]
Ye, K.-Y. 229[18]
Ye, S. 80[88], 80[90], 81[88]
Ye, Y. 232[36]
Yeh, S.-M. 61[40]
Yeo, S.-K. 52[13], 52[14]
Yeom, H.-S. 77[83], 79[83]
Yeung, Y.-Y. 300[57], 300[63], 300[64], 301[57], 301[65], 301[66], 302[65], 302[66], 302[67], 302[68], 303[67], 303[68], 312[95], 312[96], 312[97], 312[98], 312[99], 313[95], 313[96], 313[97], 313[98], 313[100], 314[100], 316[95]
Yi, H. J. 72[65]
Yin, X. 355[49], 356[49]
Ylijoki, K. E. O. 84[105]
Yoder, R. A. 322[137]
Yokoyama, N. 394[32]
Yong, H. 372[72], 373[72]
Yoon, T. P. 213[68]
Yoon, T. Y. 7[29], 8[29]
Yoshida, K. 380[102], 393[29]
Yoshida, M. 229[19], 229[20]
Yoshikai, N. 134[77], 135[77], 138[77]
Yoshino, T. 413[110]
You, Q. 217[88], 217[89], 217[90]
You, S.-L. 229[18], 229[22], 232[37], 268[76], 269[76], 270[76], 271[76], 271[77], 272[76], 273[77], 274[77], 275[77], 280[76], 280[77]
Young, K. A. 434[30]
Young, S. C. 419[8]
Yu, B. 72[70], 350[35], 350[37], 350[38]
Yu, C. 77[84]
Yu, D.-F. 401[67]
Yu, D.-Q. 350[35], 350[37], 350[38]
Yu, F. 397[46], 398[46]
Yu, J. 148[100]
Yu, R. H. 300[59]
Yu, S. 297[42], 369[66], 370[68], 371[68]
Yu, W. Z. 312[99]
Yu, X. 405[79]
Yu, Z. 217[88], 350[35]
Yuan, C. 31[74], 32[74]
Yuan, W.-C. 354[47]
Yudin, A. K. 149[102], 150[102]
Yue, T. 413[112]

Z

Zahn, S. K. 317[122], 317[123], 317[124], 318[122], 318[123], 318[126]
Zakarian, A. 187[56], 188[56], 188[57], 189[57], 215[73], 215[74], 215[75], 215[76], 216[73], 229[23]
Zalesskiy, S. S. 282[81]
Zapf, C. W. 9[31], 10[31], 11[31]
Zaragozá, R. J. 87[111]
Zavalij, P. Y. 62[41], 63[41]
Zea, A. 359[53], 360[53]
Zeh, J. 195[7]
Zeisel, S. 250[72], 250[73]
Zeitler, K. 378[99], 379[99]
Zeller, M. 419[6]
Zeng, S. 217[90]
Zeng, X. 408[94]
Zeng, X.-P. 362[57], 363[57]
Zeng, Y. 37[80], 38[80], 40[80]
Zeun, R. 419[6]
Zhan, W. 12[35], 29[35]
Zhang, D. 137[81], 137[82], 138[82], 294[14]
Zhang, F. 363[58], 364[58]
Zhang, G. 299[52], 299[53], 405[86]
Zhang, H. 77[85]
Zhang, H.-J. 83[102]
Zhang, J. 85[106], 299[52], 299[53], 365[60]
Zhang, J.-J. 63[45], 64[45]
Zhang, J.-K. 410[100]
Zhang, L. 53[15], 54[15], 300[58], 300[59], 365[61]
Zhang, L.-L. 362[56]
Zhang, Q. 229[5], 229[6], 229[7], 244[5], 245[5], 247[5], 257[5], 257[6], 258[5], 258[6], 260[7], 261[7], 262[7], 263[7]
Zhang, Q.-W. 83[102]
Zhang, S. 77[84], 217[90]
Zhang, S.-J. 398[48]
Zhang, T. 153[116], 300[56]
Zhang, W. 232[37]
Zhang, X. 103[17], 217[88], 217[89], 217[90], 297[42], 299[52], 299[53], 397[46], 398[46]
Zhang, X.-M. 354[47]
Zhang, Y. 40[84], 40[85], 41[84], 41[85], 43[84], 60[39], 61[39], 77[84], 166[25], 167[25], 168[27]
Zhang, Z. 94[3], 95[3], 100[3]
Zhao, C. 72[71], 73[71]
Zhao, C.-G. 397[44], 398[44], 406[89], 407[92], 408[92]
Zhao, D. 275[78], 276[78], 277[78], 277[79], 278[78], 279[78], 280[78], 281[78], 390[19], 390[20]
Zhao, G. 294[13], 400[55], 410[99], 410[100]
Zhao, J. 77[85]
Zhao, L. 217[90]
Zhao, S.-L. 410[99]
Zhao, X.-L. 362[57], 363[57]
Zhao, Y. 397[47], 398[47]
Zhao, Y.-L. 362[57], 363[57]
Zhao, Y.-M. 83[102]
Zhao, Z. 229[20]
Zheng, C.-W. 410[99]
Zheng, H. 72[70], 72[71], 73[71]
Zheng, J. 72[70]
Zheng, S. 215[70]
Zhong, G. 408[94]
Zhou, B. 364[59], 365[59]
Zhou, C.-Y. 392[25]
Zhou, H. 116[46], 116[47], 117[46], 117[47], 148[99], 395[33]
Zhou, J. 301[65], 301[66], 302[65], 302[66], 302[67], 302[68], 303[67], 303[68], 312[95], 312[97], 313[95], 313[97], 313[100], 314[100], 316[95], 362[57], 363[57]
Zhou, L. 297[42], 301[65], 301[66], 302[65], 302[66], 312[95], 312[96], 312[97], 313[95], 313[96], 313[97], 316[95]
Zhou, Q.-L. 392[25]
Zhou, X. 219[91], 406[90]
Zhu, C. 103[17]
Zhu, F. 362[57], 363[57]
Zhu, J. 3[25], 5[25], 83[98], 84[98], 413[111], 413[112]
Zhu, M. 148[100]
Zhu, S. 369[66], 370[68], 371[68], 401[62]
Zhu, S.-F. 392[25]
Zhu, S.-Z. 410[99]
Zhuo, C.-X. 229[18], 232[37], 268[76], 269[76], 270[76], 271[76], 271[77], 272[76], 273[77], 274[77], 275[77], 280[76], 280[77]
Zi, W. 294[15]
Ziemer, A. 431[24]
Zimmer, R. 431[24]
Zinad, D. S. 98[11], 99[11]
Zinker, B. A. 372[72], 373[72]
Zipkin, R. E. 15[43], 15[44], 16[44], 17[44], 112[37], 112[38], 112[39], 112[40], 113[37], 113[38], 113[39], 113[40]
Zografos, A. L. 151[109], 152[109], 154[109]
Zu, L. 338[18], 346[18], 405[79], 406[87], 406[88], 407[87]
Zuck, P. 419[11]
Zuo, Z. 294[15]
Zveaghintseva, M. 355[48]

Abbreviations

Chemical

Name Used in Text	Abbreviation Used in Tables and on Arrow in Schemes	Abbreviation Used in Experimental Procedures
(*R*)-1-amino-2-(methoxymethyl)pyrrolidine	RAMP	RAMP
(*S*)-1-amino-2-(methoxymethyl)pyrrolidine	SAMP	SAMP
ammonium cerium(IV) nitrate	CAN	CAN
2,2′-azobisisobutyronitrile	AIBN	AIBN
barbituric acid	BBA	BBA
benzyltriethylammonium bromide	TEBAB	TEBAB
benzyltriethylammonium chloride	TEBAC	TEBAC
N,*O*-bis(trimethylsilyl)acetamide	BSA	BSA
9-borabicyclo[3.3.1]nonane	9-BBNH	9-BBNH
borane–methyl sulfide complex	BMS	BMS
N-bromosuccinimide	NBS	NBS
tert-butyldimethylsilyl chloride	TBDMSCl	TBDMSCl
tert-butyl peroxybenzoate	TBPB	*tert*-butyl peroxybenzoate
10-camphorsulfonic acid	CSA	CSA
chlorosulfonyl isocyanate	CSI	chlorosulfonyl isocyanate
3-chloroperoxybenzoic acid	MCPBA	MCPBA
N-chlorosuccinimide	NCS	NCS
chlorotrimethylsilane	TMSCl	TMSCl
1,4-diazabicyclo[2.2.2]octane	DABCO	DABCO
1,5-diazabicyclo[4.3.0]non-5-ene	DBN	DBN
1,8-diazabicyclo[5.4.0]undec-7-ene	DBU	DBU
dibenzoyl peroxide	DBPO	dibenzoyl peroxide
dibenzylideneacetone	dba	dba
di-*tert*-butyl azodicarboxylate	DBAD	di-*tert*-butyl azo-dicarboxylate
di-*tert*-butyl peroxide	DTBP	DTBP
2,3-dichloro-5,6-dicyanobenzo-1,4-quinone	DDQ	DDQ
dichloromethyl methyl ether	DCME	DCME
dicyclohexylcarbodiimide	DCC	DCC
N,*N*-diethylaminosulfur trifluoride	DAST	DAST
diethyl azodicarboxylate	DEAD	DEAD
diethyl tartrate	DET	DET
2,2′-dihydroxy-1,1′-binaphthyllithium aluminum hydride	BINAL-H	BINAL-H
diisobutylaluminum hydride	DIBAL-H	DIBAL-H
diisopropyl tartrate	DIPT	DIPT

Chemical (cont.)

Name Used in Text	Abbreviation Used in Tables and on Arrow in Schemes	Abbreviation Used in Experimental Procedures
1,2-dimethoxyethane	DME	DME
dimethylacetamide	DMA	DMA
dimethyl acetylenedicarboxylate	DMAD	DMAD
2-(dimethylamino)ethanol	$Me_2N(CH_2)_2OH$	2-(dimethylamino)ethanol
4-(dimethylamino)pyridine	DMAP	DMAP
dimethylformamide	DMF	DMF
dimethyl sulfide	DMS	DMS
dimethyl sulfoxide	DMSO	DMSO
1,3-dimethyl-3,4,5,6-tetrahydro-pyrimidin-2(1*H*)-one	DMPU	DMPU
ethyl diazoacetate	EDA	EDA
ethylenediaminetetraacetic acid	edta	edta
hexamethylphosphoric triamide	HMPA	HMPA
hexamethylphosphorous triamide	HMPT	HMPT
iodomethane	MeI	MeI
N-iodosuccinimide	NIS	NIS
lithium diisopropylamide	LDA	LDA
lithium hexamethyldisilazanide	LiHMDS	LiHMDS
lithium isopropylcyclohexylamide	LICA	LICA
lithium 2,2,6,6-tetramethylpiperidide	LTMP	LTMP
lutidine	lut	lut
methylaluminum bis(2,6-di-*tert*-butyl-4-methyl-phenoxide)	MAD	MAD
methyl ethyl ketone	MEK	methyl ethyl ketone
methylmaleimide	NMM	NMM
4-methylmorpholine *N*-oxide	NMO	NMO
1-methylpyrrolidin-2-one	NMP	NMP
methyl vinyl ketone	MVK	methyl vinyl ketone
petroleum ether	PE[a]	petroleum ether
N-phenylmaleimide	NPM	NPM
polyphosphoric acid	PPA	PPA
polyphosphate ester	PPE	polyphosphate ester
potassium hexamethyldisilazanide	KHMDS	KHMDS
pyridine	pyridine[b]	pyridine
pyridinium chlorochromate	PCC	PCC
pyridinium dichromate	PDC	PDC
pyridinium 4-toluenesulfonate	PPTS	PPTS
sodium bis(2-methoxyethoxy)aluminum hydride	Red-Al	Red-Al
tetrabutylammonium bromide	TBAB	TBAB

[a] Used to save space; abbreviation must be defined in a footnote.
[b] py used on arrow in schemes.

Chemical (cont.)

Name Used in Text	Abbreviation Used in Tables and on Arrow in Schemes	Abbreviation Used in Experimental Procedures
tetrabutylammonium chloride	TBACl	TBACl
tetrabutylammonium fluoride	TBAF	TBAF
tetrabutylammonium iodide	TBAI	TBAI
tetracyanoethene	TCNE	tetracyanoethene
tetrahydrofuran	THF	THF
tetrahydropyran	THP	THP
2,2,6,6-tetramethylpiperidine	TMP	TMP
trimethylamine *N*-oxide	TMANO	trimethylamine *N*-oxide
N,*N*,*N*′,*N*′-tetramethylethylenediamine	TMEDA	TMEDA
tosylmethyl isocyanide	TosMIC	TosMIC
trifluoroacetic acid	TFA	TFA
trifluoroacetic anhydride	TFAA	TFAA
trimethylsilyl cyanide	TMSCN	TMSCN

Ligands

acetylacetonato	acac
2,2′-bipyridyl	bipy
1,2-bis(dimethylphosphino)ethane	DMPE
2,3-bis(diphenylphosphino)bicyclo[2.2.1]hept-5-ene	NORPHOS
2,2′-bis(diphenylphosphino)-1,1′-binaphthyl	BINAP
1,2-bis(diphenylphosphino)ethane	dppe (not diphos)
1,1′-bis(diphenylphosphino)ferrocene	dppf
bis(diphenylphosphino)methane	dppm
1,3-bis(diphenylphosphino)propane	dppp
1,4-bis(diphenylphosphino)butane	dppb
2,3-bis(diphenylphosphino)butane	Chiraphos
bis(salicylidene)ethylenediamine	salen
cyclooctadiene	cod
cyclooctatetraene	cot
cyclooctatriene	cte
η^5-cyclopentadienyl	Cp
dibenzylideneacetone	dba
6,6-dimethylcyclohexadienyl	dmch
2,4-dimethylpentadienyl	dmpd
ethylenediaminetetraacetic acid	edta
isopinocampheyl	Ipc
2,3-*O*-isopropylidene-2,3-dihydroxy-1,4-bis(diphenylphosphino)butane	Diop
norbornadiene (bicyclo[2.2.1]hepta-2,5-diene)	nbd
η^5-pentamethylcyclopentadienyl	Cp*

Radicals

acetyl	Ac
aryl	Ar
benzotriazol-1-yl	Bt
benzoyl	Bz
benzyl	Bn
benzyloxycarbonyl	Cbz
benzyloxymethyl	BOM
9-borabicyclo[3.3.1]nonyl	9-BBN
tert-butoxycarbonyl	Boc
butyl	Bu
sec-butyl	*s*-Bu
tert-butyl	*t*-Bu
tert-butyldimethylsilyl	TBDMS
tert-butyldiphenylsilyl	TBDPS
cyclohexyl	Cy
3,4-dimethoxybenzyl	DMB
ethyl	Et
ferrocenyl	Fc
9-fluorenylmethoxycarbonyl	Fmoc
isobutyl	iBu
mesityl	Mes
mesyl	Ms
4-methoxybenzyl	PMB
(2-methoxyethoxy)methyl	MEM
methoxymethyl	MOM
methyl	Me
4-nitrobenzyl	PNB
phenyl	Ph
phthaloyl	Phth
phthalimido	NPhth
propyl	Pr
isopropyl	iPr
tetrahydropyranyl	THP
tolyl	Tol
tosyl	Ts
triethylsilyl	TES
triflyl, trifluoromethanesulfonyl	Tf
triisopropylsilyl	TIPS
trimethylsilyl	TMS
2-(trimethylsilyl)ethoxymethyl	SEM
trityl [triphenylmethyl]	Tr

General

absolute	abs
anhydrous	anhyd
aqueous	aq
boiling point	bp
catalyst	no abbreviation
catalytic	cat.
chemical shift	δ
circular dichroism	CD
column chromatography	no abbreviation
concentrated	concd
configuration (in tables)	Config
coupling constant	*J*
day	d
density	*d*
decomposed	dec
degrees Celsius	°C
diastereomeric ratio	dr
dilute	dil
electron-donating group	EDG
electron-withdrawing group	EWG
electrophile	E^+
enantiomeric excess	ee
enantiomeric ratio	er
equation	eq
equivalent(s)	equiv
flash-vacuum pyrolysis	FVP
gas chromatography	GC
gas chromatography-mass spectrometry	GC/MS
gas–liquid chromatography	GLC
gram	g
highest occupied molecular orbital	HOMO
high-performance liquid chromatography	HPLC
hour(s)	h
infrared	IR
in situ	in situ
in vacuo	in vacuo
lethal dosage, e.g. to 50% of animals tested	LD_{50}
liquid	liq
liter	L
lowest unoccupied molecular orbital	LUMO
mass spectrometry	MS
medium-pressure liquid chromatography	MPLC
melting point	mp
milliliter	mL
millimole(s)	mmol
millimoles per liter	mM
minute(s)	min
mole(s)	mol
nuclear magnetic resonance	NMR
nucleophile	Nu^-
optical purity	op
phase-transfer catalysis	PTC
proton NMR	^1H NMR

General (cont.)

quantitative	quant
reference (in tables)	Ref
retention factor (for TLC)	R_f
retention time (chromatography)	t_R
room temperature	rt
saturated	sat.
solution	soln
temperature (in tables)	Temp (°C)
thin layer chromatography	TLC
ultraviolet	UV
volume (literature)	Vol.
via	via
vide infra	*vide infra*
vide supra	*vide supra*
yield (in tables)	Yield (%)

List of All Volumes

Science of Synthesis, Houben–Weyl Methods of Molecular Transformations

Category 1: Organometallics

1 Compounds with Transition Metal—Carbon π-Bonds and Compounds of Groups 10–8 (Ni, Pd, Pt, Co, Rh, Ir, Fe, Ru, Os)

2 Compounds of Groups 7–3 (Mn···, Cr···, V···, Ti···, Sc···, La···, Ac···)

3 Compounds of Groups 12 and 11 (Zn, Cd, Hg, Cu, Ag, Au)

4 Compounds of Group 15 (As, Sb, Bi) and Silicon Compounds

5 Compounds of Group 14 (Ge, Sn, Pb)

6 Boron Compounds

7 Compounds of Groups 13 and 2 (Al, Ga, In, Tl, Be ··· Ba)

8a Compounds of Group 1 (Li ··· Cs)

8b Compounds of Group 1 (Li ··· Cs)

Category 2: Hetarenes and Related Ring Systems

9 Fully Unsaturated Small-Ring Heterocycles and Monocyclic Five-Membered Hetarenes with One Heteroatom

10 Fused Five-Membered Hetarenes with One Heteroatom

11 Five-Membered Hetarenes with One Chalcogen and One Additional Heteroatom

12 Five-Membered Hetarenes with Two Nitrogen or Phosphorus Atoms

13 Five-Membered Hetarenes with Three or More Heteroatoms

14 Six-Membered Hetarenes with One Chalcogen

15 Six-Membered Hetarenes with One Nitrogen or Phosphorus Atom

16 Six-Membered Hetarenes with Two Identical Heteroatoms

17 Six-Membered Hetarenes with Two Unlike or More than Two Heteroatoms and Fully Unsaturated Larger-Ring Heterocycles

Category 3: Compounds with Four and Three Carbon—Heteroatom Bonds

18 Four Carbon—Heteroatom Bonds: X—C≡X, X=C=X, X_2C=X, CX_4

19 Three Carbon—Heteroatom Bonds: Nitriles, Isocyanides, and Derivatives

20a Three Carbon—Heteroatom Bonds: Acid Halides; Carboxylic Acids and Acid Salts

20b Three Carbon—Heteroatom Bonds: Esters and Lactones; Peroxy Acids and R(CO)OX Compounds; R(CO)X, X = S, Se, Te

21 Three Carbon—Heteroatom Bonds: Amides and Derivatives; Peptides; Lactams

22 Three Carbon—Heteroatom Bonds: Thio-, Seleno-, and Tellurocarboxylic Acids and Derivatives; Imidic Acids and Derivatives; Ortho Acid Derivatives

23 Three Carbon—Heteroatom Bonds: Ketenes and Derivatives

24 Three Carbon—Heteroatom Bonds: Ketene Acetals and Yne—X Compounds

Category 4: Compounds with Two Carbon—Heteroatom Bonds

25 Aldehydes

26 Ketones

27 Heteroatom Analogues of Aldehydes and Ketones

28 Quinones and Heteroatom Analogues

29 Acetals: Hal/X and O/O, S, Se, Te

30 Acetals: O/N, S/S, S/N, and N/N and Higher Heteroatom Analogues

31a Arene—X (X = Hal, O, S, Se, Te)

31b Arene—X (X = N, P)

32 X—Ene—X (X = F, Cl, Br, I, O, S, Se, Te, N, P), Ene—Hal, and Ene—O Compounds

33 Ene—X Compounds (X = S, Se, Te, N, P)

Category 5: Compounds with One Saturated Carbon—Heteroatom Bond

34 Fluorine

35 Chlorine, Bromine, and Iodine

36 Alcohols

37 Ethers

38 Peroxides

39 Sulfur, Selenium, and Tellurium

40a Amines and Ammonium Salts

40b Amine *N*-Oxides, Haloamines, Hydroxylamines and Sulfur Analogues, and Hydrazines

41 Nitro, Nitroso, Azo, Azoxy, and Diazonium Compounds, Azides, Triazenes, and Tetrazenes

42 Organophosphorus Compounds (incl. RO—P and RN—P)

Category 6: Compounds with All-Carbon Functions

43 Polyynes, Arynes, Enynes, and Alkynes

44 Cumulenes and Allenes

45a Monocyclic Arenes, Quasiarenes, and Annulenes

45b Aromatic Ring Assemblies, Polycyclic Aromatic Hydrocarbons, and Conjugated Polyenes

46 1,3-Dienes

47a Alkenes

47b Alkenes

48 Alkanes